8th International Space Safety Conference 2016

Safety First, Safety for All

Melbourne, Florida, USA
18 - 20 May 2016

ISBN: 978-1-5108-7530-2

Printed from e-media with permission by:

Curran Associates, Inc.
57 Morehouse Lane
Red Hook, NY 12571

Some format issues inherent in the e-media version may also appear in this print version.

Printed by Curran Associates, Inc. (2018)

International Association for the Advancement of Space Safety (IAASS)
Kapteynstraat 1
2201BB Noordwijk
The Netherlands

Phone: +31(0)712020023

spacesafety@iaass.org

Additional copies of this publication are available from:

Curran Associates, Inc.
57 Morehouse Lane
Red Hook, NY 12571 USA
Phone: 845-758-0400
Fax: 845-758-2633
Email: curran@proceedings.com
Web: www.proceedings.com

PROCEEDINGS OF THE 8TH INTERNATIONAL SPACE SAFETY CONFERENCE

SAFETY FIRST SAFETY FOR ALL

MELBOURNE - FLORIDA (USA) - 18-20 MAY 2016

U.S. AIR FORCE

NASA

cnes
CENTRE NATIONAL D'ETUDES SPATIALES

AIRBUS DEFENCE & SPACE

JAXA Japan Aerospace Exploration Agency

LOCKHEED MARTIN

ISSF INTERNATIONAL SPACE SAFETY FOUNDATION

FLORIDA INSTITUTE OF TECHNOLOGY 1958

American Astronautical Society

8TH INTERNATIONAL SPACE SAFETY CONFERENCE

SAFETY FIRST
SAFETY FOR ALL

Conference Organizing Committee

Chairs:

I. Rongier (F)
R. DeLoach (USA)
M. Glissman (USA)
N. Takeuchi (JP)
M. Ciancone (USA)

Members:

W. Ailor (USA)
K. Amsden (USA)
C. Botts (USA)
G. Boy (USA)
E. Browne (USA)
C. Cazaux (F)
D. Cone (USA)
P. Contoveros (USA)
T. Erikson (USA)
H. Ernst (D)
C. Felix (NL)
D. Finkleman (USA)
W. Frazier (USA)
T. Fukatsu (JP)
G. Gafka (USA)
T. Heimann (USA)
R. Jakhu (CA)
J. Jeevarajan (USA)
M. Jones (USA)
B. Kanki (USA)
S. Kaul (IN)
R. Kelley (USA)
M. Kezirian (USA)
P. Kirkpatrick (USA)
J. Kreimer (D)
B. Lazare (F)
C. Leveau (F)
T. Lips (D)
E. Mango (USA)
W. McArthur (USA)
D. McKnight (USA)
A. Menzel (D)
C. Moura (BR)
M. Nogami (JP)
P. Omaly (F)
N. Packham (USA)
S. Pearson (USA)
J. Pelton (USA)
T. Pfitzer (USA)
A. Quinn (UK)
L. Ren (CH)
K. Schubert (USA)
T. Sgobba (NL)
G. Sinnema (NL)
J. Souders (USA)
A. Stampfel (USA)
S. Takada (JP)
G. Trinchero (F)
S. Weikert (D)
P. Wilde (USA)
U. Wirth (D)

Organizing Committee:

G. Elfering (NL)
T. Pfitzer (USA)
P. Kirkpatrick (USA)
J. Rudolph (USA)

Publication: Proceedings of the 8th IAASS Conference "Safety First, Safety for All"
Melbourne, USA 18-20 May 2016
Cover: Orion Exploration Flight Test (201412040006HQ) - Credit: (NASA/Bill Ingalls)

Table of Contents

Acknowledgements

The editors would like to thank all authors of papers presented at the 8th IAASS Conference. The quality of their work made the conference a successful and outstanding event. The editors would like also to thank all keynote speakers and sponsors.

Special thanks to the organising committee of the 8th IAASS Conference and in particular to Carmen Felix. She was a key contributor to the smooth running and success of the conference, and her patient work has been essential for the timely issuing of these proceedings. Many thanks to Gonnie Elfering, Kristhian Mason, Paul Kirkpatrick, Megan Kane, Camilo Reyes, and FIT Staff in special to Kimberly Schaefer, for their valuable and continuous work in preparing and running the conference.

Preface

The International Association for the Advancement of Space Safety (IAASS) is a non-profit organisation dedicated to furthering international cooperation and scientific advancement in the fields of space systems safety and sustainability. The IAASS is a member of the International Astronautical Federation (IAF) and Permanent Observer at the United Nations Committee on the Peaceful Uses of Outer Space (UN COPUOS). The association exists to help shape and advance an international culture of space safety (technical, organisational, and socio-political), which could contribute to make space missions, vehicles, stations, extra-terrestrial habitats, equipment, and payload safer for the general public, ground personnel, crews, and flight participants. The association also pursues the safeguarding of the on-orbit, atmospheric, and ground environment during space systems and associated ground infrastructure operations.

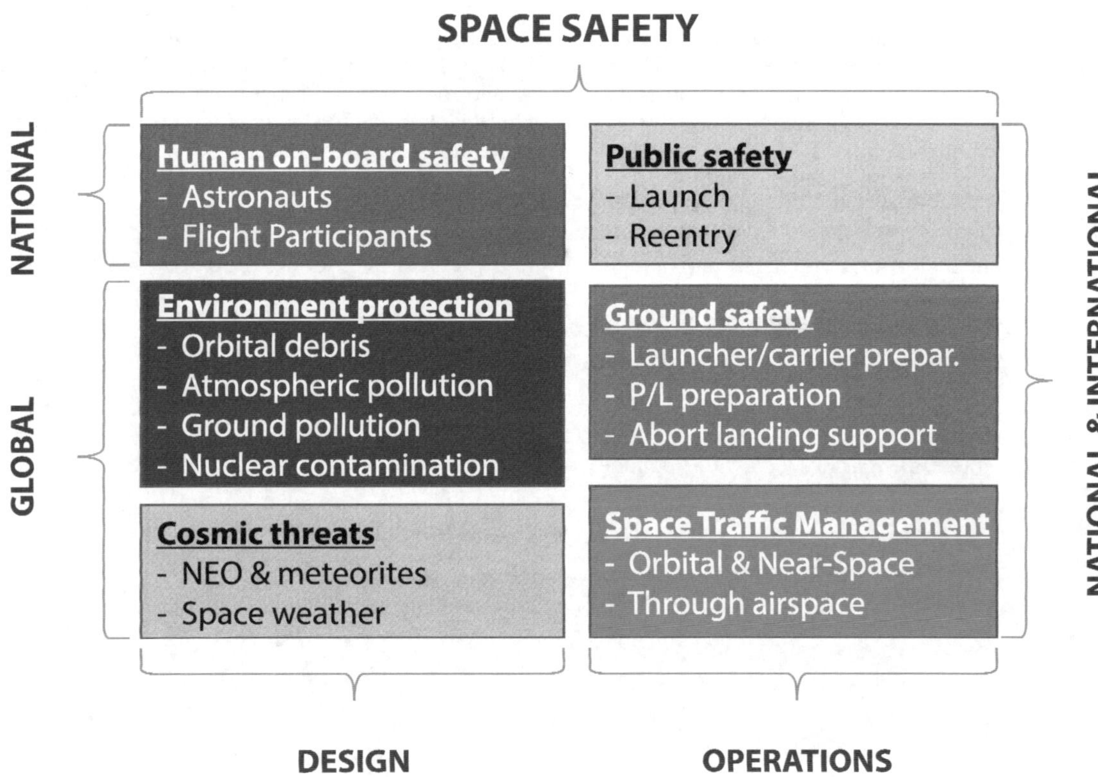

Conference Purpose and Goals

The 8th IAASS Conference "Safety First, Safety For All" is an invitation to reflect and exchange information on a number of topics in space safety and sustainability of national and international interest. The conference is also a forum to promote mutual understanding, trust, and the widest possible international cooperation in such matters. The once exclusive "club" of nations with autonomous sub-orbital and orbital space access capabilities is becoming crowded with fresh and ambitious new entrants. New commercial spaceports are starting operations and others are being built.

In the manned spaceflight arena a commercial market is becoming a tangible reality with sub-orbital spaceflights and government use of commercial services for cargo and crew transportation to orbit. Besides the national ambitions in space, the international cooperation both civil and commercial is also gaining momentum. In the meantime robotic space exploration will accelerate and with it the need to internationally better regulate the usage of nuclear power sources. Space bound systems and aviation traffic will share more and more a crowded airspace, while aviation will increasingly rely on space-based safety-critical services. Finally, most nations own nowadays space assets, mainly satellites of various kinds and purpose, which are under the constant threat of collision with other spacecraft and with the ever-increasing number of space debris. Awareness is increasing internationally (as solemnly declared since decades in space treaties) that space is a mankind asset and that we all have the duty of caring for it. Without proactive and courageous international initiatives to organise space, we risk to negate access and use of space to future generations.

The 8th IAASS Conference will in addition dedicate a set of specialised sessions to three topics which need to get better attention in space programmes: Space Debris Remediation, Space Situational Awareness and International Space Traffic Control, and Commercial Human Spaceflight Safety. Active removal from orbit of "dead" satellites and spent launcher upper stages, and international coordination of space traffic are mandatory for safe orbital operations, and for the safety of the public on ground. Instead the safety of commercial human spaceflight is a critical element for the expansion, and perhaps even continuation, of human spaceflight. The success of commercial spaceflight will be determined by a substantially higher level of safety and not solely by cost reductions.

These special sessions will provide participants insight, innovative strategies, and open discussion into these important aspects of space safety.

PROCEEDINGS OF THE 8TH INTERNATIONAL SPACE SAFETY CONFERENCE

PLENARY SESSION

NASA ADMINISTRATOR CHARLES BOLDEN OPENING REMARKS

May 19, 2016

Thank you, Mr. [Mark] Glissman for your very kind introduction. It's a pleasure to be here. I also want to thank IAASS President Isabelle Rongier for guiding this organization in all the good work it does.

You know, since I've become NASA Administrator, there has been an incredible expansion of our capabilities, and our vision. When I started, commercial space was a concrete program, but its future was still a foggy notion. I don't think anyone would have predicted 7 years ago that today we'd not only have two fully functional commercial resupply partners, but that they would already be critical components in our ISS supply chain.

It's also important to note that both of them have recovered from mishaps, which is the true test of their mettle and ability to learn from mistakes, correct them and get better. Nor would we have guessed that a third commercial cargo partner would have been added for the second phase of the program to resupply an International Space Station that we've now committed to through at least 2024.

In commercial crew, it also was not a foregone conclusion that we would be successful, but now The Boeing Company and SpaceX are well into the development of their crew transportation systems, we've let contracts for those services, astronauts are training for the first test flights and next year, NEXT year, is our targeted first launch of each of these vehicles. *Dragon* and *Starliner* are the new names that will go down in history along with the Space Shuttle – and *Orion*! – as the carriers of humanity's aspirations.

Wow! And commercial space is just one piece of our comprehensive and collaborative journey to Mars.

Many people may not think of commercial space vehicles and capability as part of the journey to Mars, but it most definitely is. Without this, NASA couldn't begin to look at the farther horizons with the Space Launch System rocket and the *Orion* spacecraft. Without this, we couldn't have seeded a whole new segment of the economy, giving industry the chance to use its own hard work and innovation to create jobs and new markets for their capabilities and growing the ways the next generation could become excited and inspired by the many facets of exploration. From business to science to engineering and design, there's a reflection of your passions in our modern space program.

All that said, today we're focused on a very necessary topic.

The launches and the waving astronauts on TV; the incredible homecoming after a one-year mission; the prospect of humans finally traveling to Mars – none of that can happen without a strong safety presence.

I've said from day one that our commercial space transportation systems will fly only when they meet NASA's stringent safety requirements. Some fault NASA for being slow in its development programs, but a lot of that is because of our early and often focus on the safety factor. While commercial companies have moved rapidly to develop their systems, it has also been with careful insight by NASA – insuring they meet important safety milestones and conduct testing and more testing.

We've progressed at the rate at which funding allowed. We aren't going to rush the development of our commercial cargo and crew vehicles no matter how much people want it.

We really need our own U.S. crew transport capability because we should not be totally dependent on our Russian partners and we should not have to continue paying an increasingly high rate for seats aboard the Soyuz. But I want to be very clear, once again, we won't fly until we're convinced the new commercial transportation systems meet our safety requirements.

Our teams are working hard to ensure the safety standards for our Commercial Crew Program are consistent with our Exploration Program's safety standards and where there are necessary differences we understand and acknowledge those differences (e.g. Deep space long duration vs. LEO short duration safety criteria).

We have 50 years of history launching humans to space, and there's an incredible knowledge base in there. Believe me, our commercial partners are making use of it. We've always been about lessons learned – both good and bad.

In our history, every mishap, whether involving crew or not, has always made us stronger, more resilient and with one more dataset to help us be safer in the future.

I've lost friends and colleagues in space accidents. I don't want anyone else to have to go through that. The bottom line is that space travel is extremely difficult and risky. It always has been and always will be and there's no way to make it risk free any more than getting into your car and driving down the highway is ever going to be 100% safe. But that doesn't mean that we don't try our hardest to continually minimize risk; that we don't check and re-check and that we don't encourage everyone to speak up when they don't think it's safe to proceed. I've worked hard in my tenure to develop a culture of safety at NASA where dissenting voices are heard and considered. It's not a catch all, but it's another check and balance and in our business, that's essential.

So, our partners have begun tests of their launch abort systems. The rockets on which their spacecraft will fly also are being tested. The list of things to overcome to make commercial crew a reality is extensive, as you can imagine, but I am confident in our partners' abilities and their commitment to safety is just as strong as ours is.

Because the market in low Earth orbit is going to be deeper and broader than just NASA, there's going to be an increasing need for cargo, and yes, for human transportation in the coming years. So all of us are working to get it right here on the ground floor. NASA's horizon mission may be Mars, but that doesn't mean at all that we're taking our eyes off of LEO and the work of our commercial partners.

As I mentioned, the work of our commercial partners is essential to getting to Mars.

Our stepping stone approach involves getting the most out of the International Space Station (ISS), where we're learning to live and work in space for the long term, making breakthroughs in human health and demonstrating technologies for traveling farther. From there, we'll shake out some of the cutting edge new technologies such as solar electric propulsion in the proving ground of the regions around our moon (cis-lunar) and out in deep space. This will phase will involve missions using *Orion* and the SLS to cis-lunar space and to an asteroid and then we get to the Earth independent phase when we finally make the decision to move our missions out to the Martian region and to its surface. That's going to be a huge international undertaking where everything we need we'll either have to take, make enroute or send on ahead. From Mars, Earth will no longer be a matter of hours or days away and many challenges will have to be solved by our technology and our human ingenuity millions of miles and months away from home base.

It's daunting, but exciting! The journey to Mars is the biggest thing humanity has ever undertaken and it's thrilling to think that somebody in my granddaughters' generation, maybe even one of them, will be the first humans at Mars.

So that stepping stone approach builds on our capabilities and expands the uses of each milestone as we go and our culture of safety must always be present in every phase of this journey.

I mentioned our international partners and it will be critically important that they're equally as concerned about safety as we are. Our astronauts in space are our most precious resource up there. They represent our brightest, most talented individuals and they're selected in a highly competitive process. At NASA, for instance, we received 18,300 applications for our last call for astronauts this past February.

We'll make the selection of the Class of 2017 next year, but the previous Class of 2013 astronauts numbered 8, selected from some 6,800 applications, so you get an idea of the level to which these folks are scrutinized for their skills, talents, passion and dedication. By the way, Class 2013 was the most diverse group we have ever selected for anything with 4 women and 4 men.

Your organization awarded us the IAASS Safety-by-Design-Award in 2013 for the International Space Station as one of the safest and most successful space programs in history and we're very proud of that. Because it took a lot of work and consensus building and it acknowledged that whatever we do, if nothing else, we operate with a strong safety consciousness out of an abundance of caution.

There have been a lot of space-related movies lately, from *Gravity* to *The Martian* that depict astronauts in danger.

But beyond the science fiction -- and we often work with producers to help make their stories as accurate as possible – the common thread is that there were multiple systems in all these scenarios for astronauts to ensure their safety, even if sometimes they had to jury-rig them. That is a fact!

At the end of the day, our men and women in space are going to have to rely on their wits and training, but we can give them the benefit of the very best technology, the most thorough testing and the strongest chances for safe missions.

A lot of experts at this gathering are going to give you much more detailed insight on the broadest picture of safety in space, from orbital debris mitigation, to getting into and returning from space and to living on other planetary bodies. What I want to emphasize, though, is that I believe our future is very bright. We will continue to push the envelope because we have to.

More and more partners, both in industry and with other nations are going to require new ways of thinking and our journey to Mars is the biggest thing humanity has undertaken. The hard won knowledge of the past is going to have to be applied to new systems and new missions that were not even conceived when first we began this long journey to space.

It's a journey that really has no end, so we will continue to be teachable. To move forward in a sustainable manner even as political winds come and go and budgets sustain or change, what is inalterable is our desire to explore. The sacrifices of the men and women who have come before will never be forgotten and the practical, day to day challenges of making space as safe as possible, will never be overlooked. The passion and commitment that you in IAASS bring for establishing and maintaining a strong safety mindset and culture of safety must remain strong. At NASA and among our industry and international partners, we must follow the example you set in each and every one of our programs whether they involve humans or robotic systems.

It will be essential that we all serve as safety officers in our individual work sites and our leadership must be committed to promoting safety as every level of effort. This we must and will do – I promise you that.

Thanks so much for allowing me to join your forum this morning and God Speed!

Thank you.

PROCEEDINGS OF THE 8TH INTERNATIONAL SPACE SAFETY CONFERENCE

SESSION 01 LAUNCH SAFETY RISK

SOURCE DATA IMPACTS ON EPISTEMIC UNCERTAINTY FOR LAUNCH VEHICLE FAULT TREE MODELS

Mohammad Al Hassan, Steven D. Novack, Robert W. Ring

Bastion Technologies Incorporated, 17625 El Camino Real #330, TX 77058, USA
Emails: mohammad.i.alhassan@nasa.gov, steven.d.novack@nasa.gov, robert.w.ring@nasa.gov

ABSTRACT

Launch vehicle systems are designed and developed using both heritage and new hardware. Design modifications to the heritage hardware to fit new functional system requirements can impact the applicability of heritage reliability data. Risk estimates for newly designed systems must be developed from generic data sources such as commercially available reliability databases using reliability prediction methodologies, such as those addressed in MIL-HDBK-217F. Failure estimates must be converted from the generic environment to the specific operating environment of the system in which it is used. In addition, some qualification of applicability for the data source to the current system should be made. Characterizing data applicability under these circumstances is crucial to developing model estimations that support confident decisions on design changes and trade studies. This paper will demonstrate a data-source applicability classification method for suggesting epistemic component uncertainty to a target vehicle based on the source and operating environment of the originating data. The source applicability is determined using heuristic guidelines while translation of operating environments is accomplished by applying statistical methods to MIL-HDK-217F tables.

The paper will provide one example for assigning environmental factors uncertainty when translating between operating environments for the microelectronic part-type components. The heuristic guidelines will be followed by uncertainty-importance routines to assess the need for more applicable data to reduce model uncertainty.

INTRODUCTION

Today's space launch vehicles are typically evolved from a combination of both heritage hardware and new technology. This developmental approach is often driven by cost, schedule, and reliability considerations. The Probabilistic Risk Assessment (PRA) team is developing PRA models for use in risk-informed decision making. PRA is a methodology for quantifying the risk of high-consequence events such as loss of crew and loss of mission. The process involves developing fault-tree logic models based on the current design and then quantifying the basic events in the model. Basic events in the model represent failure events, which can be functional (critical component failures), phenomenological (structural, fire/explosion, etc.), or human caused. This paper focuses on developing estimates for functional failure of components. Component failure rates are derived from a wide variety of data sources such as demonstrated reliability data for heritage hardware, reliability predictions developed by the prime contractors, and component failure databases, such as RIAC EPRD/NPRD, and NUCLARR. An important consideration in PRA modeling is an explicit treatment of uncertainty. Reliability prediction methodologies typically do not address uncertainty. Therefore, when using prediction data sources, methods need to be developed for consistently characterizing the uncertainty of component reliability predictions. Uncertainties in a PRA can be aleatory (random variation) or epistemic (lack of knowledge). Often epistemic uncertainties are the dominant contributors. Therefore, characterizing epistemic uncertainty is crucial to risk-informed decision making to support design trade studies and flight readiness decisions.

The team developed a two-part approach for quantifying epistemic uncertainty of component basic events in the PRA model. The first part reviews the data sources used in the component reliability prediction, evaluates the applicability of the data sources used in the prime contractor's estimate, and assesses the uncertainty based on a heuristic approach. The second part of the approach accounts for epistemic uncertainty associated with translating failure rate estimates from the data-source environment to the launch vehicle's operating environment.

By reducing the fault tree logic to cutsets through Boolean logic reduction, basic event uncertainties are propagated to the top event using Monte Carlo simulation methods. It then becomes important to identify those basic events that are important contributors to the uncertainty of the top event. Once identified, additional effort can focus on ways to reduce their uncertainty, such as by identifying additional data sources, reviewing and analyzing test data, and recommending additional testing. Therefore, the two step process is followed with a process to identify basic-event contributions to the uncertainty of the top event in the logic model through the use of uncertainty-importance routines.

BACKGROUND ON UNCERTAINTY

Failure rates of components cannot be measured directly, and components used in space applications are highly reliable. Hence, system-specific failure data is rare. Consequently, estimates developed from generic sources must be used extensively, but are uncertain due to lack of knowledge and applicability. Typically in PRA applications, components are proof tested for flight and qualified for operation within their service life. Under these constraints, the exponential failure model is used, which has a constant hazard function equal to the mean failure rate (λ). The exponential distribution is a single parameter (λ) model. The uncertainty of the failure-rate parameter is represented as a Lognormal probability density function (p.d.f.). Unique to the Lognormal distribution is a measure of dispersion called the Error Factor (EF). The EF is defined in terms of the 5th, 50th (median), and 95th percentiles of the probability distribution. Specifically, the EF is equal to the 95th divided by the 50th (median). The Lognormal failure rate uncertainty p.d.f. is illustrated in Figure 1.

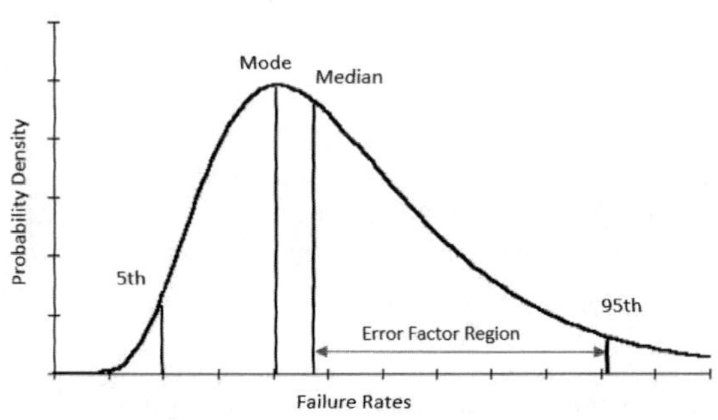

Figure1. Lognormal Probability Density Function (PDF)

As noted above, uncertainty has two sources typically inherent in every system, aleatory and epistemic. Aleatory uncertainty is due to random variation, which is an inherent characteristic of the system and as such cannot be reduced except through physical changes to the system, such as quality improvement. Epistemic uncertainty can be reduced through acquisition of additional knowledge, such as better data sources, additional testing, flight experience, etc. Epistemic uncertainties stem from the modeling context, such as component reliability data (failure rates), model assumptions, and model completeness (e.g., missing scope or scenarios). This paper focuses on epistemic uncertainty associated with component reliability data. This context was selected based on the fact that model completeness and model assumptions are specific to their launch vehicle design whereas component reliability is more general to any launch vehicle design and is extensible to other systems as well.

1. APPROACH TO ASSESS PARAMETER EPISTEMIC UNCERTAINTY

The approach described below aims to consistently assess epistemic uncertainty across launch vehicle subsystems (e.g., booster, core stage, upper stage, engine, thrust vector control, avionics) by providing heuristic guidelines for assessing uncertainty based on the data-source applicability and operating environment. The discussion of this approach will be divided into two parts, Data-Source Applicability and Data-Source Operating Environment.

1.1 Source-Data Applicability

It is important to note that the guidelines in this part of the approach are tailored to new launch vehicle systems or subsystems that lack historical flight data. Failure rates in this case often come from generic sources, such as reliability databases, and are usually provided as point estimates (mean or median). This section uses the point estimate and the applicability guidelines to estimate the parameters of the Lognormal failure rate uncertainty distribution for use in basic events in the PRA logic model. Bayesian reliability requires a prior distribution to represent degree of belief about the value of a component failure rate before system specific data become available from testing or operations. Hence, the uncertainty distributions developed with this method are prior distributions that will be updated using Bayes theorem when system specific information becomes available through testing or flight experience.

This section describes the heuristic uncertainty classification method for assessing uncertainty due to data-source applicability. Applicability refers to the relevance of the source-data, used in developing the point estimate of the component's failure rate, to the specific launch vehicle system being modeled. Table 1. *Data Source Applicability & Error Factor Assignment*, lists the typical reliability data sources for new components. It is used to assess data-source uncertainty and assign an error factor, which along with the provided mean or median, completely specifies the basic event distribution in the PRA fault tree model. Notice that the error factors increase from the most applicable source (Category A) to the least applicable source (Category E) as one would expect.

Table 1. Data Source Applicability & Error Factor Assignment

Note: This table is intended to be used for point estimates that lack distribution data. Do not use if the distribution is already known

Source	Category	Source Descrption	Source Application	Source Application Error Factor	Adjusted Environment
Legacy Hardware	A	Other Launch Vehicle Data (Most Applicable)	Same component	3	Increases the Error Factor
			Like component	4	
	B	Aerospace Data	Same component	5	
			Like component	6	
	C	Other Industry Data	Same component	6	
			Like component	7	
New Hardware	D	MIL-HDBK-217F Methods	Same component	8	
			Like component	9	
	E	Non-expert Engineering Judgment (Least Applicable)	Documented Process	10	
			Undocumented Process	15	

Table 2 below provides an example of how to apply the guidelines based on Table 1. Consider a simple launch vehicle subsystem comprised of four components operating in the Airborne Uninhabited Fighter (AUF) environment. The failure rate and the assessed Error Factor for each component is listed in Table 2 based on Table 1 uncertainty guidelines.

Table 2. Example for Assigning Failure Rates EFs of a LV System

Component	Data Source	Mean (Point Estimtae)	Error Factor
1	Engineering Judgement (Documentent Process)	3.00E-06	10
2	MIL-KBK-217F Piece Part Method	6.01E-06	8
3	Aerospace Historical Data (Same Component)	1.00E-06	5
4	Engineering Judgement (Undocumentent Process)	3.50E-07	15

1.2 Environmental Factors & Uncertainty

Reliability data for a particular component operating in a specific environment, such as Missile Launch (ML) or AUF, may not be available from the desired operating environment to the extent necessary to develop an adequate prior distribution. MIL-HDBK-217F provides an environmental factor conversion method which allows for converting the failure rate from one environment to another. These conversion factors are presumably mean values based on data, but are also uncertain. The purpose of this section of the approach is to estimate this source of epistemic uncertainty and propagate it to the failure rate prediction. The process followed begins with a derivation of the equation for the environmental conversion factor, identifies the variables in this equation, generates an uncertainty distribution for each variable, and finally propagates uncertainty to the resulting failure rate through the environmental equation. To implement this process, it was necessary to derive the equation for the environmental factor from the general failure rate, Equation [1], for microelectronics as given in MIL-HBK-217F:

$$\lambda_P = (C_1\, \pi_T + C_2\, \pi_E)\, \pi_Q\, \pi_L \qquad [1]$$

Where

λ_P is the component failure rate in million hours

C_1 is the circuit complexity

C_2 is the packaging complexity

π_E is the environmental factor

π_T is the component joint temperature factor

π_Q is the component quality factor

π_L is the learning factor (assumed 1 by the handbook)

Solving for π_E , the equation becomes

$$\pi_E = \frac{\left(\frac{\lambda_P}{\pi_Q}\right) - C_1 \pi_T}{C_2} \qquad [2]$$

The challenge with the MIL-HDBK-217F tables was that values for λ_P, C_1, C_2, π_Q, and π_T were provided as mean estimates only. Information about the standard deviations of the variables was necessary in order to develop uncertainty distributions for each of the variables in the equation. The next subsections explain how this was accomplished.

1.2.1 Estimating C_1, C_2, π_Q, and π_T Uncertainty Distributions

A literature research was conducted using the references cited in the MIL-HDBK-217F but yielded no insight into the standard deviations of the variables C_1, C_2, π_T and π_Q. This forced the team to make engineering assumptions about the distribution of the means to estimate uncertainty. There is no basis for assuming skewed distributions, and physical parameters tend towards Normality [Reference 3, Page 144], therefore, it was deemed appropriate to assumed normality.

The coefficient of variation (CV), which is defined as the ratio of the standard deviation to the mean and is often expressed as a percentage, was used to estimate a reasonable relationship between the mean and the standard deviation

$$\left(CV = \frac{\sigma}{\mu}\right)$$

A value of 20% was assumed. Standard deviation was then calculated by multiplying the assumed CV by the provided mean. Microcircuits Example 1, Section 5.13, Page 5-20 of the MIL-HBK-217F was used to illustrate the approach to generating uncertainty distributions for the variables in the equation for π_E. See Table 2 for distribution results.

Table 2. Uncertainty Distribution for C_1, C_2, π_Q, and π_T

Variable	Mean	SD(σ)
C1	2.00E-02	4.00E-03
C2	1.10E-02	2.20E-03
π_T	2.90E-01	5.80E-02
π_Q P.V.	80.00	16.00

1.2.2 Estimating The Environmental Factor (π_E) Failure Rate (λ_P) Uncertainty Distribution

Reference [3] *Revision of Environmental Factors for MIL-HDBK-217B for Microelectronics* provides revised data for estimation of the environmental factors for the microelectronics part type. The report summarizes the data analysis methodology in Section 5.0 and reports the data on Tables 5.5-1 and 5.5-2. The report used demonstrated failure rate data to determine λ_P for five environments with ample historical data, namely Ground Benign (GB), Space Flight (SF), Ground Fixed (GF), Ground Fixed Non-Operating (GF-Non), and Naval Submarine (NSB). The average of the constant failure rate and the adjusted failure rate was assumed in this analysis to be Normally distributed and the standard deviation was calculated based on this assumption. Equation [2] was then modeled to a statistical software scripted in R language called Programmable Uncertainty Parameter Propagation into Equation Software (PUPPIES) to use a Monte Carlo routine to solve for the environmental factor using a random seed and 20000 simulations. The results of the Monte Carlo simulations for the five environments utilizing PUPPIES are depicted in Figure 2. The distribution results for the five environments do not appear to be normally distributed, since the median and mean are not identical. It is evident from the figure that the distributions are skewed and it was assumed that the data fit a Lognormal distribution.

GB environment was used in Reference [2] as the reference environment for π_E. Since the results shown in Figure 2 appear to fit lognormal distribution, the error factor (a measure of uncertainty for lognormal distribution) for the GB π_E equation was calculated to be 3.00 using the formula (95th/median).

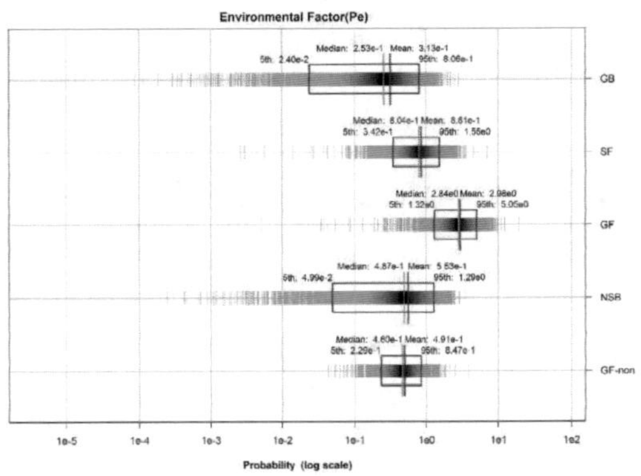

Figure 2. Environmental Factor Epistemic Uncertainty for Five Environments

2. PROCESS TO REDUCE UNCERTAINTY

The purpose of this section is to demonstrate the process the team used to reduce model's epistemic uncertainty by focusing on data-source applicability of the key contributors to the uncertainty. This process conforms to the flow chart in Figure 3.

The collected failure rate data for each component basic event of the model was compared to the different categories found in Table 1 and was assigned a Lognormal distribution by selecting the appropriate EF. After solving the models fault tree, an uncertainty analysis was conducted using Monte Carlo simulations. This step created an uncertainty distribution for the entire model (as opposed to the uncertainty distribution for a single basic event). In cases where the Monte Carlo analyses yielded a high model uncertainty, uncertainty-importance analyses routines were used to identify the basic events that drive the uncertainty bounds. This step assessed the degree of need for more applicable data to reduce uncertainty by showing where additional resources need to be placed to the PRA model. Finally, the iterative loop should end when the model uncertainty results are satisfactory.

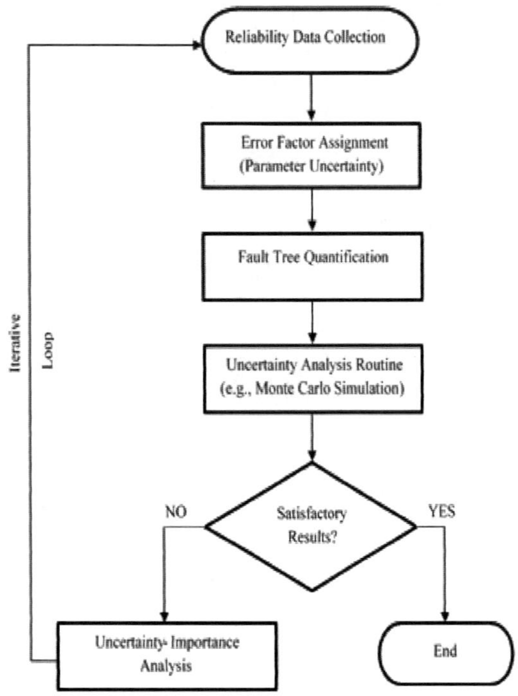

Figure 3. Process Flow Chart for Reducing Epistemic Uncertainty

2.1 Case Study

The purpose of this section is to provide an example for applying the uncertainty-importance process shown in Figure 4. Consider the simple LV system example given in Section 1.1. Table 2 in the same section shows components failure rates. Assume the fault tree logic is implemented using a PRA software.

Applying Monte Carlo simulation to the fault tree yields a median estimate of 1.59E-07 and 95th percentile of 1.63E-06. A quantification of the model error factor (95th/median) equates to 10.25. This is considered a high model uncertainty. According to the flow chart in Figure 3, uncertainty-importance analyses identifies Component1 to be responsible for driving this high uncertainty. A more advanced data search is conducted and finds a failure rate from a historical aerospace data of 3.00E-06. Based on the new source applicability, this new failure rate is assigned an error factor of 5. Another trial of Monte Carlo simulation, with the same number of samples and seed, is simulated and the model error factor is now reduced to 5.15 (2.15E-06/4.23E-07). This is considered satisfactory uncertainty and the process of the flow chart ends here.

3. CONCLUSION

Parameter epistemic uncertainty applied in PRA represents the lack of knowledge of the component failure rate used in the logic model. As evident in Figure 1, the wider the distribution the larger the uncertainty. The heuristic guidelines developed for use in launch vehicle design risks and discussed in this paper provide a standard approach for better traceability of the epistemic uncertainty associated with the environmental factors and parameter failure rate data source. Once parameter uncertainty is categorized the proposed process flow chart in Figure 3 can be followed to prioritize the need to collect additional parameter data in order to reduce uncertainty if needed.

The uncertainty about the environmental factor conversion, Equation2, was statistically estimated with an error factor of about 3. There were assumptions made about the variables uncertainty distributions due to the fact that very little data was provided about the mean values and nature of their distributions. Possible future work includes reaching out to the authors of MIL-HBK-217F to confirm the uncertainties about the variables mean estimates that were not supplied in the handbook. This part of the approach only assessed the microelectronics part type, and future work will assess other part types pertinent to launch vehicles control systems to ultimately develop an environmental conversion uncertainty matrix for use in PRA.

4. REFERENCES

1. MIL-HDBK-217F, Military Handbook, Reliability Prediction of Electronic Equipment, December 1991

2. Ronald E. Walpole, Raymond H. Myers, Sharon L. Myers, & Keying Ye (2002). Probability & Statistics for Engineers & Scientists, Seventh Edition, New Jersey, Prentice Hall

3. RADC-TR-80-299, AD A091837, Revision of Environmental Factors for MIL-HDBK-217B, Martin Marietta Corporation, September 1980

PITFALLS AND PRECAUTIONS WHEN USING PREDICTED FAILURE DATA FOR QUANTITATIVE ANALYSIS OF SAFETY RISK FOR HUMAN RATED LAUNCH VEHICLES

Glen S. Hatfield[(1)], Frank Hark[(2)], James Stott, Ph.D.[(3)]

(1) Bastion Technologies Incorporated, 17625 El Camino Real #330, TX 77058, USA,
Emails: glen.s.hatfield@nasa.gov, frank.hark@nasa.gov
(2) NASA Marshall Space Flight Center, Huntsville, AL 35812, USA, Email: james.e.stott@nasa.gov

ABSTRACT

Launch vehicle reliability analysis is largely dependent upon using predicted failure rates from data sources such as MIL-HDBK-217F. Reliability prediction methodologies based on component data do not take into account risks attributable to manufacturing, assembly, and process controls. These sources often dominate component level reliability or risk of failure probability.

While consequences of failure is often understood in assessing risk, using predicted values in a risk model to estimate the probability of occurrence will likely underestimate the risk. Managers and decision makers often use the probability of occurrence in determining whether to accept the risk or require a design modification. Due to the absence of system level test and operational data inherent in aerospace applications, the actual risk threshold for acceptance may not be appropriately characterized for decision making purposes. This paper will establish a method and approach to identify the pitfalls and precautions of accepting risk based solely upon predicted failure data. This approach will provide a set of guidelines that may be useful to arrive at a more realistic quantification of risk prior to acceptance by a program.

INTRODUCTION

In today's environment, cost, schedule, and safety risk are paramount in the aerospace industry. Many companies and government agencies are allowing predicted reliability data to be used to determine if a component, assembly, system, and ultimately the vehicle, meet safety requirements. Typically, Safety engineers use subjective methods or they may use quantitative methods to assess the likelihood of a specific failure scenarios occurrence. Quantifying the risks would be the first choice in assessing risk; however, some programs allow the use of predicted failure data to quantify the risk accomplished by the modeling of failure scenarios.

Failure data typically come from MIL-HDBK-217F, which is outdated and has been discontinued by the Department of Defense (DoD). The values are generally overly optimistic, as can be seen by comparison with demonstrated data. These values are adjusted for the specific failure mode distributions that causes a specific failure event. These values often misrepresent the actual risk when managers are considering accepting or rejecting the risk. In either case, the cost of redesign or loss of the asset may represent a significant failure for the program. This paper will examine the differences in using predicted vs. demonstrated data methods. One method uses epistemic error factors, and the other uses calculated error factors per the formulation found in SAPHIRE version 8 [2]. Examining these differences will identify the pitfalls and demonstrate the need for precautions.

In addition, testing which is used to validate the product is also being reduced to realize even further cost and schedule savings, often at the expense of reliability and safety. Decreased electrical, electronic, and electro- mechanical (EEE) parts grades have also been allowed in certain instances. In other areas, it has been proposed to reduce thermal cycling, burn in, and thermal vacuum testing to provide even further cost and schedule reduction. These measures may actually increase the risk of failure by not knowing if the reliability has been met as well as allowing infant mortality failures to occur while on the launch pad. Unidentified latent failures that are undetected due to reduced testing may manifest during flight. All of these issues may very well increase the overall risk to a point that the true risk is unknown, especially when predicted values are used in modeling the risk, and uncertainty surrounding the model is now in question.

Precautions should be made aware when using predicted data to assess safety risks due to pitfalls of unidentified or incorrectly identified failure data and associated uncertainty. Predicted data does not take into account manufacturing, assembly, and operational processes. When quantifying risk, the actual risk reported and used in the decision making process may misrepresent both the outcome and associated uncertainty.

The predicted and demonstrated data models utilize epistemic and aleatory error factors3 (EF), which represent the uncertainty in each model.. A lognormal distribution is assumed for both models and the resulting mean risk and system uncertainties are then compared.

1. ESTIMATING FAILURE RATES

1.1 MIL-HDBK-217F

This approach quantifies failure rates at the component level, which is modified, based on a component's, environment, temperature range, and quality. When used in similar environments, the differences between system applications may be significant. Correct application by the user is a limitation of the prediction method.

The example model below demonstrates the effect of predicted reliability data.

The Model basis: MIL-HDBK-217F [5] stress method, environment is Airborne Uninhabited Fighter (AUF). Quality factors are those of the highest level for the specific part analyzed. Temperatures, where applicable, are baselined at 130 deg. C, stress loads, where applicable, are between 90 and 100%. Devices are procured with normal manufacturer's screening consisting of temperature cycling, constant acceleration, electrical testing, seal test, and external visual inspection. It is assumed that the component manufacturer also performs all screens and tests to the applicable MIL-PRF [6] or equivalent MIL-STD-883 screening method.

Component types modeled are CMOS digital gate arrays, low frequency diodes, bipolar transistors, resistors and capacitors.

The applicable adjustment factors for the prediction model components may be found in MIL-HDBK-217F [5]
The predicted failure rates and MTTF for each component is shown in Table 1

Table 1. MIL-HDBK-217F Failure Rates

Component	Failure Rate	MTTF
Digital Gate Array	2.47e-8	40.485,830
Diode (LF)	4.25e-7	2,350,012
Transistor	3.53e-7	2,829,562
Resistors (RCR)	1.42e-7	7,058,767
Capacitors (CCR)	1.37e-8	72,878,861

A representative system fault tree was built using SAPHIRE 8 for predicted values as is seen in Figure 2.

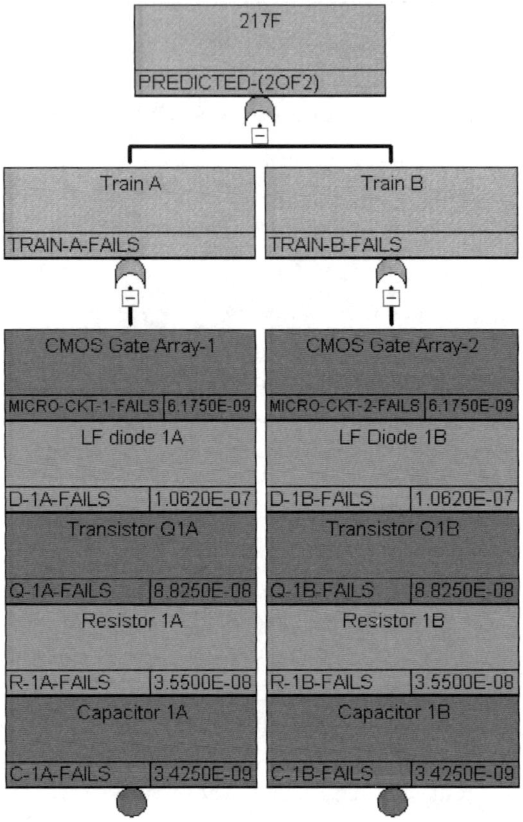

Figure 2 Predicted Value Fault Tree

An epistemic error factor value of 8 was selected for all components, which were taken from the Data Source Classification Application [1], as seen in Table 2.

Table 2. Error Factor Data Source Classification Approach[1]

Data Source Classification Approach			
Source	**Source Description**	**Source Application**	**Error Factor**
New Hardware	MIL-HDBK-217F	Same Component	8
		Like Component	9

1.2 Demonstrated Failure Rates

This approach quantifies failure rates by using same components taken from the Quanterion Automated Databook using EPRD-2014, NPRD-2016, and where applicable FMD-2016 [3]. The same components used in the predicted model are selected where several years of failure data was found. Adjustments for varying environments were adjusted to the AUF environment by using MIL-HDBK-338 [4]. The mean values and standard deviations were calculated. The Error Factors were calculated using Eq.1. [2].

$$EF = e^{1.645}\sqrt{\ln(1 + (\sigma_{ln}/\mu_{ln})^2)} \qquad (1)$$

SAPHIRE 8 Error Factor Calculation

The demonstrated failure rates and MTTF for each component are shown in Table 3.

Table 3. Demonstrated Failure Rates

Component	Failure Rate	MTTF
Digital Gate Array	2.93e-6	341,491
Diode (LF)	2.49e-6	400,267
Transistor	9.38e-7	1,065,719
Resistors (RCR)	2.77e-6	360,577
Capacitors (CCR)	1.06e-6	934,798

A representative system fault tree was built using SAPHIRE 8 for demonstrated values is seen in Figure 4

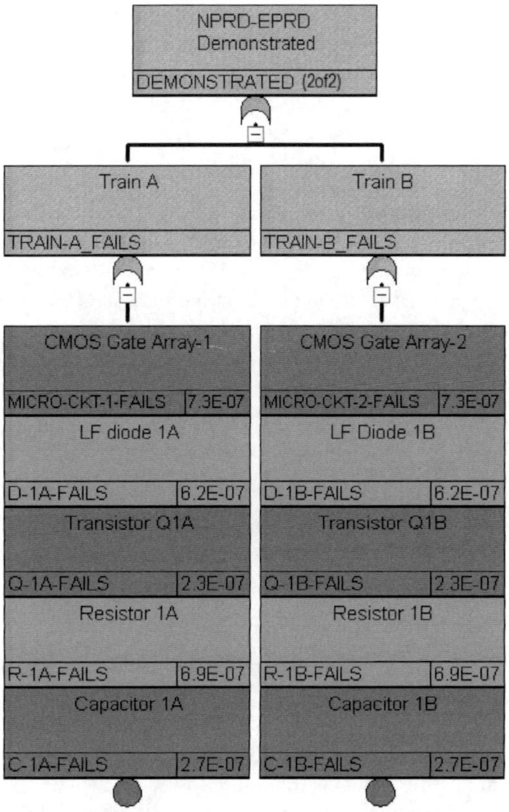

Figure 4 Demonstrated Value Fault Tree

Demonstrated lognormal uncertainty (EF) was calculated using Eq. 1. The values in Table 4, were applied to their respective components.

Table 4. Demonstrated Error Factors

Component	Error factor
Digital Gate Array	4.61
Diode (LF)	4.1
Transistor	4.09
Resistors (RCR)	4.67
Capacitors (CCR)	3.91

1.3 Model Parameters

Both models were constructed in the same fashion. Model logic was based on two parallel strings requiring a two of two output logic to achieve functional output. Each strings components are in serial with a single failure causing a loss of function for that string. The mission time is one quarter of an hour. All components were correlated to each individual type. Fifty thousand Monte Carlo trials were ran with identical seed values.

1.4 System Uncertainty Calculations

The system level uncertainty for each model type was then computed based upon the method in Figure 5 [1]. Quantitatively, the error factor represents the spread of the lognormal distribution about the median. The Error Factor is represented as the 95th divided by the median [1]. System uncertainty for each model type is based on this method.

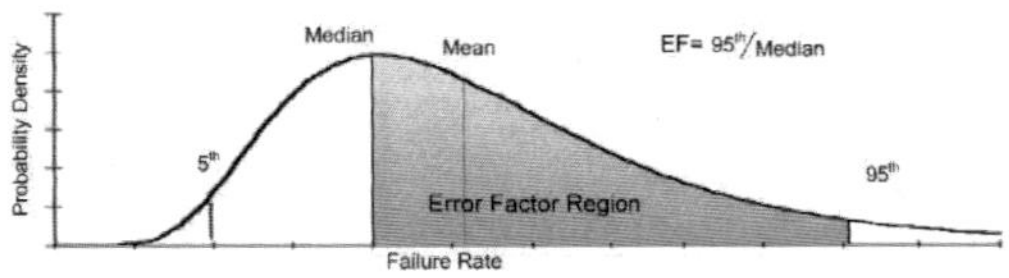

Figure 5.
Lognormal Probability Density Function

2. RESULTS

The two models were then solved in SAPHIRE 8 by performing Monte Carlo trials, the results of which were used to calculate the system level uncertainty. The end results were then compared with the results as shown in Table 5.

Table 5. Predicted vs. Demonstrated results

Model Type	Risk (1/N)	Uncertainty
Predicted	1/2,111,041	4.08
Demonstrated	1/196,657	2.3

3. CONCLUSION

As one can see, the results in Table 5 are vastly different. The risk value is 11 times less for the predicted model.
If used for a quantitative assessment when considering the risk in a decision scenario, the risk may be misrepresented. In the case of uncertainty, the dispersion of the risk is two times greater in the predicted model.
This indicates that demonstrated data lowers uncertainty and predicted uncertainty range may necessitate further effort to collect additional data.
This difference gives rise to identify the pitfalls and precautions when developing quantitative models, which are used to assess and accept risk.

3.1 Pitfalls

The pitfalls one may experience are:

- Predicted failure data are overly optimistic.
- Predicted failure data does not take into account manufacturing, assembly and quality process controls, which are primary failure drivers.
- Predicted failure data may mislead managers into accepting a level of risk that is not commensurate with the actual risk.
- Predicted data should not be used to assess system reliability against reliability requirements.

3.2 Precautions

Precautions when using predicted failure data.

- If a concerted effort to obtain realistic data is not done, the resulting risk model may not be valid.
- Failure databases are difficult to locate and may be more difficult to obtain permission to access.
- Equal quality components must be used in developing risk models.
- Evaluate and adjust environments, if necessary.
- The source of data must be documented for traceability.

Obtaining valid data to perform risk modeling requires an understanding of the nature of the problems in obtaining and analyzing data to be used in modeling a system or in performing a system analysis to determine if the risk of a particular failure scenario is worth accepting, or if a redesign or system modification is warranted.

4. SUMMARY

System design is constrained by Safety, Reliability, and Quality requirements as well as design standards. The purpose being to assure safety, which is generally related to quality of product, design, and testing.
These bounds are where the risks lie, and must be fully recognized, understood, minimized, and eventually accepted or redesigned to a level, which then meets acceptable risk.
As programs and projects proceed, risks are accepted at these bounds. The bounds over time become eroded by the acceptance of risk, and at some time, the aggregate risks may well exceed these bounds. This may result in a falsely perceived level of confidence, and allow a project to proceed to a state of potential disaster, which may result in a loss of life and physical property.

5. REFERENCES

1. Mohammad AL Hassan, Steven Novack, Rob Ring, *Data Applicability of Heritage and New Hardware for Launch Vehicle Reliability models*, Huntsville Society of Reliability Engineers RAM VIII Training Summit, November 3-4, 2015

2. SAPHIRE 8.0.9, Idaho National Laboratory Battelle Energy Alliance Idaho Falls, ID 83415
Quanterion Automated Databook, Version 4.2.1.67, Quanterion Solutions Incorporated, Copyright 2013-2015

3. Electronic Reliability Design handbook (MIL-HDBK-338B), October 1, 1998

4. Military Handbook, Reliability Prediction of Electronic Equipment, (MIL-HDBK-217 Rev. F Notice 2)

5. Military Performance Specifications, (MIL-PRF-xxxxx),

CHARACTERIZING EPISTEMIC UNCERTAINTY FOR LAUNCH VEHICLE DESIGNS

Steven D. Novack[(1)], Jim Rogers[(2)], Mohammad Al Hassan[(1)], Frank Hark[(1)]

[(1)] Bastion Technologies Incorporated, 17625 El Camino Real #330, TX 77058, USA,
Emails: steven.d.novack@nasa.gov, mohammad.i.alhassan@nasa.gov, frank.hark@nasa.gov
[(2)] NASA Marshall Space Flight Center, Huntsville, AL 35812, USA, Email: jim.rogers@nasa.gov

ABSTRACT

NASA Probabilistic Risk Assessment (PRA) has the task of estimating the aleatory (randomness) and epistemic (lack of knowledge) uncertainty of launch vehicle loss of mission and crew risk, and communicating the results. Launch vehicles are complex engineered systems designed with sophisticated subsystems that are built to work together to accomplish mission success. Some of these systems or subsystems are in the form of heritage equipment, while some have never been previously launched. For these cases, characterizing the epistemic uncertainty is of foremost importance, and it is anticipated that the epistemic uncertainty of a modified launch vehicle design versus a design of well understood heritage equipment would be greater. For reasons that will be discussed, standard uncertainty propagation methods using Monte Carlo simulation produce counter intuitive results, and significantly underestimate epistemic uncertainty for launch vehicle models. Furthermore, standard PRA methods, such as Uncertainty-Importance analyses used to identify components that are significant contributors to uncertainty, are rendered obsolete, since sensitivity to uncertainty changes are not reflected in propagation of uncertainty using Monte Carlo methods.

This paper provides a basis of the uncertainty underestimation for complex systems and especially, due to nuances of launch vehicle logic, for launch vehicles. It then suggests several alternative methods for estimating uncertainty and provides examples of estimation results. Lastly, the paper describes how to implement an Uncertainty-Importance analysis using one alternative approach, describes the results, and suggests ways to reduce epistemic uncertainty by focusing on additional data or testing of selected components.

1. INTRODUCTION

Uncertainty is an important aspect of any PRA to reflect the level of confidence in the models and data used in the analysis, and to provide information used for engineering and management decisions. In launch vehicles, risks are typically considered high [1], compared to most industry applications. Uncertainty, especially the upper bounds, can aid in anticipated launch commit decisions for NASA.

As technology advances and new system designs populated with first launch hardware are developed for more ambitious space applications, estimating uncertainty of these designs becomes paramount to mission success. Additionally, the ability to characterize the contribution of components to the overall uncertainty provides a means for supporting testing plans and system development.

To perform a proper uncertainty analysis for new designs, the development of epistemic uncertainty (uncertainty due to lack of knowledge) represents a difficult task, both in assigning failure probability distribution bounds to new components, and in propagating this uncertainty through the PRA logic model.

One would believe that standard Monte Carlo simulation provides an appropriate methodology to propagate uncertainty through subsequent logic models. However, upon further investigation, this method appears to be highly questionable, and becomes more of a concern in logic models with structures like those that reflect launch vehicle designs.

2. UNCERTAINTY PROGAGATION

Monte Carlo simulation provides a process to propagate uncertainty through a reduced Boolean equation created by PRA software to solve logic trees, such as built in the Systems Analysis Programs for Hands-on Integrated Reliability Evaluations (SAPHIRE) tool [2]. Probability distributions, representing the uncertainty of component failure rates, can be randomly sampled using Monte Carlo simulation and combined to estimate an uncertainty distribution for the top event in the PRA fault-tree model.

Uncertainty propagation is affected by the logic of the tree and in general, AND gates tend to preserve or increase the bounds of two numerically similar combined basic events, while OR gates reduce this uncertainty. Sensitivity studies show the top event in a fault tree with many OR gates is insensitive to changes in the epistemic uncertainty of the tree's basic events. Although this phenomenon was known, we did not fully realize the resulting implications on PRA until fairly recently, which prompted this paper for peer discussion.

Several simple tests were performed to understand the nature and extent of uncertainty changes due to propagating in this manner. It was understood that when performing Monte Carlo routines on a Boolean expression, the mean results of the PRA mode top event could be determined by dominating basic events. Therefore in these test cases the basic event failure probabilities are all set to the same magnitude. Moderate variances in the magnitude of basic events, such as found in typical complex systems, will not significantly affect the results of these test cases. This also applies to uncertainty, where dominance by any basic event or small set of basic events will result in the uncertainty also being dominated by those same basic events.

To facilitate understanding of the nature and extent of this reduction, several sample sets were run in SAPHIRE. Each sample set was selected to minimize the effect of other variables in the solution of the top event and emphasize the changes in uncertainty. Initial test cases focused on determining the number of similar events with specific logic potentially causing this decrease in uncertainty.

To set up the test cases, two small logic trees of ten similar events were developed. Each independent basic event was given the same probability distribution. One logic tree contained all OR gates and the other contained all AND gates. All basic-event probabilities for these test cases used a lognormal distribution with a mean of 1.0E-5 and an Error Factor (EF) and were uncorrelated (The EF for the lognormal distribution is defined as the 95th percentile divided by the median.) The EF was varied for each test case as was the number of basic events in the fault tree logic. For each case, 99999 random SAPHIRE trials were run using a random seed, and in all of the cases the top event converged. In each case, the uncertainty distribution of the top event was a good fit to a lognormal distribution that we parameterized using moment matching. Tables 1 and 2 below show the test cases and top event uncertainty results for the OR gate tree and the AND gate tree, respectively.

Table 1: EF Results from OR Logic Uncertainty Test Cases

	Error Factor				
Number of Basic Events	5	10	15	20	100
2	3.4	6.1	8.6	11.1	44.5
5	2.3	3.8	5.1	6.3	21.3
10	1.9	2.8	3.7	4.5	13.8
20	1.6	2.2	2.8	3.3	9.5

As can be seen from Table 1, the EF significantly decreases as the number of similar basic events increases under the top event. Even as the EF is increased to 100, a very large and potentially unreasonable EF, the reduction in uncertainty of uncorrelated events added together in the Boolean expression reduces significantly and is essentially diminished compared to the basic event uncertainty after ten such events. Test runs were also conducted at a higher mean value (5.0E-5) showing similar results.

Table 2: EF Results from AND Logic Uncertainty Test Cases

	Error Factor				
Number of Basic Events	5	10	15	20	100
2	9.6	26.5	47.1	70.9	700.3
3	17.5	NA	NA	NA	NA
4	NA	NA	NA	NA	NA
20	NA	NA	NA	NA	NA

Table 2 shows some of the AND logic test cases. As the number of basic events increases, the ability to characterize the uncertainty becomes more difficult, since the result of combining the basic events using Monte Carlo becomes unbounded.

3. CORRELATION AND PRACTICAL APPLICATIONS

Correlation effects were as anticipated in the test cases. By correlating some of these events using 100 percent positive correlation, the effect was essentially to reduce the number of distributions, therefore; the number of basic events used by SAPHIRE to determine the uncertainty of the top event; Table 1 and Table 2 applies accordingly. In our launch vehicle design PRA models, the degree of correlation between similar component types (e.g., different cable and connector pairs) had some minor effects in increasing uncertainty. However, the increase was not enough to show sensitivity to increased epistemic uncertainty at the component level.

Our belief is that due to the launch vehicle environments, for example in ascent, most components at least at the system level are partially correlated in some manner. Partially correlating all of these components would result in greater uncertainty in the logic model. The possibility of adding partial correlation was investigated, considering the complexity and number of component basic events affected. Even at a system level, this approach was both beyond the standard current tool set available and introduced additional model uncertainty in ascertaining the appropriate partial correlation factors. In addition, even if we developed alternative tools and addressed the added model uncertainty in assigning partial correlation factors, the method of applying partial correlation and creating an N-by-N basic event matrix would be daunting to develop and apply. Therefore, alternative methods were sought to capture epistemic uncertainty. These methods are introduced below.

4. LAUNCH VEHICLES AND COMPLEX SYSTEMS WITH MULTIPLE SAFETY BARRIERS DIFFERENCES

One pertinent question arises concerning uncertainty bounds when comparing Monte Carlo simulations used in solving complex systems employing multiple safety systems and launch vehicle PRAs. Namely, why is this uncertainty reduction effect not as apparent in these PRAs and is it an issue?

The answer probably lies in the nature of functionality and safety design in these complex systems. If multiple safety barriers can be called on to mitigate accident transients, then this redundancy is at the system level and results in AND logic (safety redundancy) at the top of the fault tree logic. Launch vehicle design, due to weight reduction, optimized aerodynamics/volume, and cost considerations, usually features single fault tolerance (redundancy) at the subsystem level. Loss of the avionics system, for example, will result in loss of the launch vehicle. In other words, launch vehicle redundancy is built-in typically at the subsystem level rather than the system or element level. This will be reflected in redundancy, or AND gates at a lower level in the logic tree, while the upper portion of the tree will have multiple OR gate logic, depicting that any loss at the system or element level results in loss of the vehicle. After the discussion in Section 2, one can see where the differences in the tree logic structure between complex systems with multiple safety barriers and launch vehicles result in differences in epistemic uncertainty propagation. Furthermore, dividing the logic trees into phases to ascertain functional and environmental differences, as is experienced with launch vehicle missions, can only perpetuate the uncertainty reduction problem.

Although these complex systems typically preserve or increase some uncertainty due to the logic structure of these designs using Monte Carlo simulation, it was suspected that the uncertainty and uncertainty-importance results are as sensitive as needed to be for analytical purposes. In any case, the reduction in epistemic uncertainty for systems that have redundancy at a lower level using Monte Carlo uncertainty propagation techniques in PRA, may be occurring to some extent. PRA analysts should be aware of this potential. It is very apparent that this reduction is occurring in PRA models of launch vehicle designs.

5. IMPLICATIONS TO LAUNCH VEHICLE PRA UNCERTAINTY AND RELATED ANALYSES

One of the main implications of this reduced uncertainty effect is the reduced ability to determine which basic events, and thus components, in the model represent a significant contributor to uncertainty due to lack of knowledge about that component. This is an artifact of the propagation of the logic tree for launch vehicles and basic events become insensitive to changes in the uncertainty. Often in cases where new equipment is used and demonstrated historical data is unavailable, PRA component reliability information may reflect like equipment or the use of a component in a different environment by assigning an increase in the component epistemic uncertainty. This increase would reflect use of the component in a different environment, a different purpose, or a modified function. The magnitude of the increased uncertainty should provide insight into how much the "like" component differs from the design component and how much lack of knowledge exists related to that piece of the design.

Uncertainty-Importance analyses can often identify and prioritize components, especially in a design setting, where reduced component uncertainty could substantially reduce the system-level uncertainty and, therefore; improve the credibility bounds of the model. This can be accomplished by further investigation and using more applicable failure rate sources for the component, or invoking focused component testing. It can be time consuming and expensive to research every component in a complex system to reduce the level of uncertainty. Minimizing the number of components that require further investigation could prove to be a time and cost saving method, which would avoid performing this for every component.

6. ALTERNATIVE APPROACHES TO ESTIMATE EPISTEMIC UNCERTAINTY

Alternative approaches [3] were explored to estimate and maintain uncertainty and express sensitivity to changes in component or components groups' epistemic uncertainty.

6.1 Partial Correlation Based On Environmental Factors

One such method explored was to apply a partial correlation factor across physical or interactive systems. The concept is based on environmental factors coupling failure rates of unlike components together either due to a similar location or similar environmental stress factors. Thus partially correlating failure rates would have the effect of increasing uncertainty by not treating the events as completely independent.. A launch vehicle will often have co-located systems due to limited space and volume requirements. For example, the engine compartment may include the engine, portions of the main propulsion system, and some avionics electronics hardware. All these systems may have some correlated failures associated with being co-located in a launch vehicle.

The challenge with this approach is assigning a generic partial correlation factor and then implementing the correlation. Implementing a generic correlation factor in a custom code to build matrices for a reduced Boolean expression was explored. This was a more straight forward approach then applying multiple partial correlation factors. However, it was found that there was a lack of defensible engineering basis to establish generic correlation factors for specific systems at this time. Although this approach was not finalized, it still remains a potential option for improving accuracy of uncertainty estimates. The true test of this approach will be determined when modifications to the assigned epistemic uncertainty are propagated through the logic model and exhibit sensitivity to these changes.

6.2 Interval Approach

Another approach that we discovered when investigating epistemic uncertainty sensitivity was the Interval Analysis technique [4]. Interval Analysis claims to characterize epistemic uncertainty, as distinguished from variability or randomness that is propagated through Monte Carlo routines. Although we do not necessarily adhere to this aspect of the concept, we have found the Interval Analysis method in our case to be useful in characterizing uncertainty and in showing responsiveness to assigned changes in epistemic uncertainty of our logic models.

Unfortunately, standard methods of uncertainty propagation using Monte Carlo techniques do not provide this sensitivity for typical launch vehicle design PRA models, due to the inherent logic structure that represents launch vehicle risks. While this is not a new phenomenon, it becomes very apparent when attempting to apply Interval Analysis uses a reduced Boolean expression to solve a logic model for the top event. It does this by calculating the uncertainty ranges as an upper and lower bound (5th and 95th) for each basic event, and then using these bounding values to estimate an upper and lower bound for the top event [5].

Results from an example complex launch vehicle system showed an uncertainty EF (The EF is specific to the lognormal distribution and is only used as an approximation of the uncertainty) of about 1.5. Furthermore, applying a large EF to all basic events in this system, such as an EF of 10, showed minimal effects on the propagated uncertainty (i.e., EF~2). This effect was apparent in all major systems models of the launch vehicle.

Epistemic uncertainty for basic events ranged from EFs of 4 to 15 with many basic events having an uncertainty of around 6. Using the Interval Analysis technique, the uncertainty results showed an uncertainty value of 5.1. Uncertainty-Importance measurements [6] were run to identify dominant basic event contributors to uncertainty. These were grouped according to similarity to each other (e.g., Electrical Power, Navigation, and Flight Computer Software) and the EF modified for sensitivity to epistemic uncertainty.

The Interval Analysis approach responded to changes in the initial epistemic uncertainty in a reasonable manner. As one important group of basic events were increased from an EF of 5 to 10, the EF of the Interval approach went from 5.1 to 7.5. When this same group's EF was decreased to 3, the resulting EF was 3.5. Less important groups, as determined by the Uncertainty-Importance routines, had much smaller effects on the uncertainty as anticipated.

7. SUMMARY AND CONCLUSIONS

Uncertainty Analysis plays an important role in any PRA by describing the confidence around risk estimates. Much of the uncertainty in the design of launch vehicles is due to epistemic uncertainty or "lack of knowledge" either due to new equipment or technology, or from use of heritage equipment in a new environment. In either case, the ability to estimate uncertainty at the basic event level and establish a method for targeting this uncertainty requires an approach that is sensitive to changes in uncertainty levels.

Alternative options were suggested to support implementing sensitivity strategies. We found partial correlation between unlike components could potentially solve this issue and result in an increase in uncertainty while maintaining the standard Monte Carlo uncertainty propagation methods. However, after further investigation, this option was hindered due to the lack of engineering data and derivation to estimate partial correlation and the lack of tools to accomplish this task.

The Interval Analysis option was also introduced as an alternative that may aid in epistemic uncertainty approximation and support sensitivity studies on the uncertainty to help focus component failure rate data investigation, testing, and collection. This analysis methodology was shown to have reasonable uncertainty results on a test system PRA model, and be sensitive to modifications of the epistemic uncertainty at the basic event level.

It is recognized the preferred uncertainty propagation approach is to implement partial correlation and use Monte Carlo routines. However, implementation of this approach will likely require integration of physics-based risk models within PRA and advancement of PRA techniques to accommodate these models. Although further testing is warranted, the Interval Analysis option is a good alternative to provide uncertainty bounds for launch vehicle design PRA models and appropriate for uncertainty sensitivity studies.

8. REFERENCES

1. Keith Edward; The New Rocket Science; page 89; Lulu.com; July 2010.

2. Wood, S.T., Smith, C.L., Kvarfordt, K.J, Beck, S.T., NUREG/CR-6952, Volume 1, Systems Analysis Programs for Hands-on Integrated Reliability Evaluations (SAPHIRE), US Nuclear Regulatory Commission, September, 2008.

3. Helton, J.C., Johnson, J.D.,Oberkampf, W.L., Sorlie, C. B., A sampling-based computational strategy for the representation of epistemic uncertainty in model predictions with evidence theory; Computational Methods Applied Mechanical Engineering 196 (2007) 3980-3998.

4. Ferson Scott, Ginzburg Lev R.; Different Methods are needed to propagate ignorance and variability; Reliability Engineering and System Safety 54 (1996) 133-144; 1996 Elsevier Science Limited.

5. R. Flage, Terje Aven, Piero Baraldi, Enrico Zio; An Imprecision importance measure for uncertainty representation interpreted as lower and upper probabilities, with special emphasis on possibility theory; HAL Id hal-00765296; Dec 14, 2012.

6. Van der Borst, M, Schoonakker, H.;An Overview of PSA Importance Measures; Reliability Engineering and System Safety 72 (2001) 241-245; Elsevier Science Limited

PROCEEDINGS OF THE 8TH INTERNATIONAL SPACE SAFETY CONFERENCE

SESSION 02 LESSONS LEARNED

ROLE OF GROUND SAFETY REVIEWS IN ISS PAYLOAD MISSION SUCCESS

Ravi Margasahayam [(1)] **and Theodore Meade** [(2)]

(1) National Aeronautics and Space Administration (NASA), Co-Chair GSRP, John F. Kennedy Space Center mail code SA-B, Kennedy Space Center, FL 32899, Ravi.N.Margasahayam@nasa.gov
(2) National Aeronautics and Space Administration (NASA), Safety Engineer, John F. Kennedy Space Center mail code SA-F, Kennedy Space Center, FL 32899, Theodore.B.Meade@nasa.gov

ABSTRACT

The International Space Station (ISS) represents mankind's ambition to live and work in Space using a habitable artificial satellite in low Earth orbit. The Safety and Mission Assurance (S&MA) organization has been in the forefront of this goal by providing measured ground safety reviews, thereby significantly reducing the hazards during processing and integration. The role of ground safety reviews that have helped in maintaining process safety and ensuring payload mission success on the ISS are highlighted [1-5]. Many safety requirements have been derived from mishaps, accidents, and close calls to ensure proper safety protocol is followed. It is recommended that, lessons learned from avoiding unknown knowns (hazards known to others but not communicated to us), known unknowns (the data needed to evaluate hazards) and unknown unknowns (unknown hazards that have not yet been identified or hidden hazards) in systems engineering, test, and integration must be documented to ensure continued quest towards payload mission success [6-8]. S&MA's motto is "Mission success starts with safety; safety begins with engineering excellence."

1. INTRODUCTION

In 2003, the Defense Science Board observed, "Space is unforgiving; thousands of good decisions can be undone by a single by a single engineering flaw or workmanship error, and these flaws and errors can result in catastrophe [12]."

NASA astronauts have been living and working on the International Space Station (ISS) *(Figure 1)* for over 15 years now. Over 1700+ experiments from over 60 countries around the world have been flown to the ISS, a world-class microgravity and space environment research laboratory. The quality of life of the astronauts conducting these experiments truly depends on how safely we performed ground integration, testing, and operations on Earth. The ISS payload mission success is in some ways attributable to the role of flight and ground safety reviews to ensure all hazards are identified and mitigated.

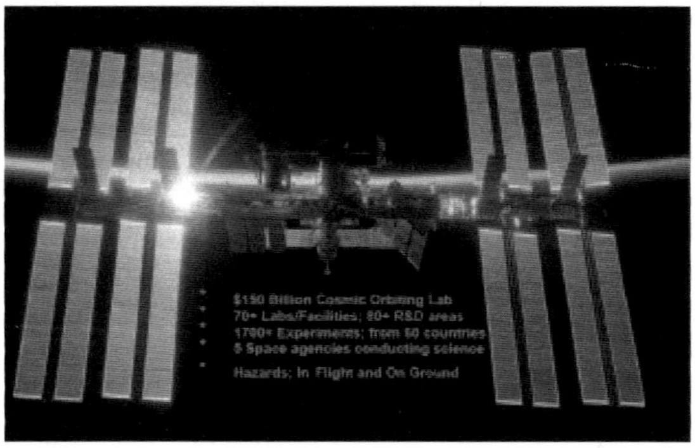

Figure 1. ISS – Assembly Complete

The NASA Kennedy Space Center served as the premier launch site during the Space Shuttle program and was responsible for ensuring the ground operations safety of the all of the elements of the Space Shuttle ground systems as well as payloads bound for the ISS in the Shuttle payload compartment. Space launch vehicles have millions of unique components that all must work in unison to ensure launch success. Successfully sending payloads to the ISS on a rocket system involves overcoming a multitude of hazards associated with the sound and vibration issues at lift-off and overcoming gravity and the effects of vacuum in space.

Ensuring the safety of ISS payloads is the realm of the KSC Ground Safety Review Panel (GSRP). This is where systematic use of systems engineering and integration tools and years of design engineering and safety review experience combine during the crucial development and mitigation of hazards that could affect any and all aspects of the payload.

2. PAYLOAD GROUND PROCESSING ISSUES

In the past, all large payloads (e.g., Node 1, MPLM) were launched into low Earth orbit in the Shuttle payload compartment. Once in space, they were integrated and assembled onto the ISS. With the retirement of the Space Shuttle in 2011, the KSC ground safety function has evolved from ground safety reviews of payloads that are transported to the ISS on a single rocket to launching on a diverse fleet of space vehicles and cargo modules.

There has been paradigm shift in how the payloads reach the ISS. As shown in *Figure 2*, this includes a variety of rockets and launch sites, different types of payloads (pressurized and unpressurized), payloads with varied ground processing needs, catering to myriad of experiments and support requirements, and last but not least, customers from all over the world who have had less stringent training while operating in KSC SSPF facilities when compared to the highly-trained workforce with comprehensive experience working at the SSPF. These complications are compounded by the fact that many payload owners (PO) who process ISS payloads/experiments work in close proximity with and their contractor's internal researchers, other NASA-hosted ISS partners, and commercial customers that could include both ISS or non-ISS partners from around the world.

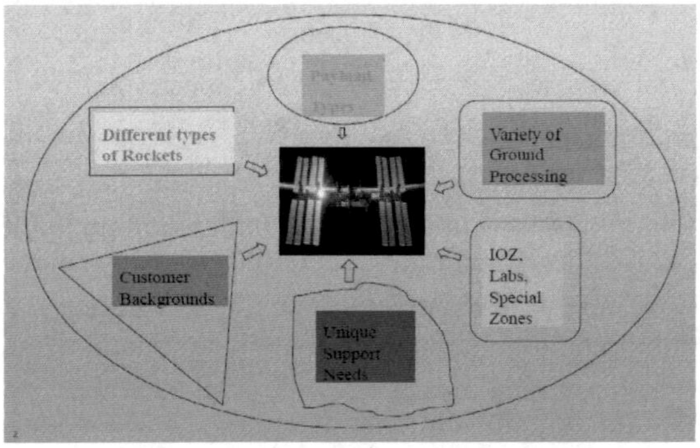

Figure 2. One Destination: Diverse Issues

Integrating payloads into the cargo modules of the Space X Dragon or the Cygnus modules of Orbital ATK rockets is also a unique activity, especially when late load experiments have to be incorporated in the last hours prior to launch. These space vehicles are part of an international fleet that delivers propellant, supplies, and science payloads to the ISS but carry much less cargo to the ISS than the Shuttle. Use of this fleet requires many more launches requiring more integration and ground processing activities. Additional complexities include placement of payloads both internally (inside the ISS – typically experiments) and externally (exposed to the environment of vacuum and space radiation).

For the ISS payloads, off-line labs are provided to perform post-delivery checkout of their payload. Payloads then go "on-line" and a series of tests are performed to ensure interface compatibility as well as ground control of the payload through the Payloads Operations Center. The ISS program provides a Payload Test and Checkout System to perform end-to-end testing. An EXPRESS Functional Checkout Unit and the on-orbit EXPRESS Rack is also provided. Special processing of payloads requiring unique conditioning can also be supported at the SSPF facilities.

Since the completion of the Space Shuttle program, KSC is shaping evolving into a multi-user spaceport, and facilities such as SSPF *(Figure 3)* are part of this transition. The procedural requirements for safety depend on whether the facilities are used exclusively by NASA or by a combination of NASA, contractors, and tenants. The High-bay and some offline labs where the PO is working with other KSC partners have aptly been termed as joint-use facilities. POs operating in exclusive-use facilities require a different set of safety requirements. NASA's biggest challenge is determining how to protect NASA personnel and property without dictating how other non-NASA organizations protect their own employees. Additional challenges exist when NASA completely turns over a government-owned facility over to a non-NASA entity or to non-NASA entities working in a NASA facility titled "Industrial Operations Zone (IOZ)" [5].

Figure 3. International Docking Adapter Last Minute Hardware Modifications

3. GROUND SAFETY REVIEW PROCESS

Engineers face two primary obstacles: uncertainties and people [9]. Uncertainties challenge the engineer since the design must be deterministic, but the world in which we operate is uncertain, random, and chaotic at best. People challenge the engineer since they are unpredictable and uncontrollable. They do not fit the equations and analytical models that an engineer uses in design. Many failure-causing hazards are attributable to uncertainties in the process and the associated activities of people in processing.

Understanding the hazards and risks they pose is vital to the GSRP's role in ground safety review process. Many safety requirements have been derived from mishaps, accidents, and close calls to ensure proper safety protocol is followed. The main purpose of the ground safety review process is to evaluate the ground safety data packages provided by the payload owner that identify hazards, evaluate their severity, and reduce the hazards to acceptable levels.

GSRP has resorted to a systems safety engineering approach to evaluate actual and probable hazards in the ground process. This has led to much efficiency in avoiding operational hazards and thereby ensuring payload safety. Important lessons learned from avoiding known unknowns and unknown unknowns in systems safety engineering, test, and integration activities have helped us in maintaining overall process safety and assuring mission success for hundreds of payloads bound for the ISS [2].

The GSRP follows a simplified approach to the ground safety review process to ensure the safety of all ISS bound payloads integrated at KSC SSPF. The requirements stem from two specific documents: SSP 30599 that defines the Safety Review Process for the ISS payloads, and KNPR 8715.3, Chapter 20, that defines the ground safety protocol to be followed. The review process ensures payload success by foreseeing potential failures. A continuous practice of failure analysis is essential for the pursuit of engineering excellence and thereby safety.

The five W's of GSRP payload processing provide clarity on who, what, where, when, and why the safety requirements exist and to whom they apply [1]. Using this approach, there are two basic questions: "What are the payload owners bringing to KSC?" and "What are they doing at KSC (i.e., are operations are hazardous)?"

Based on the answers for above questions, the GSRP determines whether it is safe for the payload owners to process at the KSC SSPF. The simple Ground Safety Checklist (GSCL) *(Figure 4)* facilitates documentation requirements including hazard identification and mitigation.

The primary objectives of the safety review process are to:

- Identify potential hazards applicable to a payload (including its flight support hardware and ground support equipment).
- Provide evaluation of all hazards for means of eliminating, reducing, and controlling, the risk to ground facilities, equipment, and personnel.
- Ensure that the hazard controls and verifications of those controls are adequate and in compliance with safety requirements.

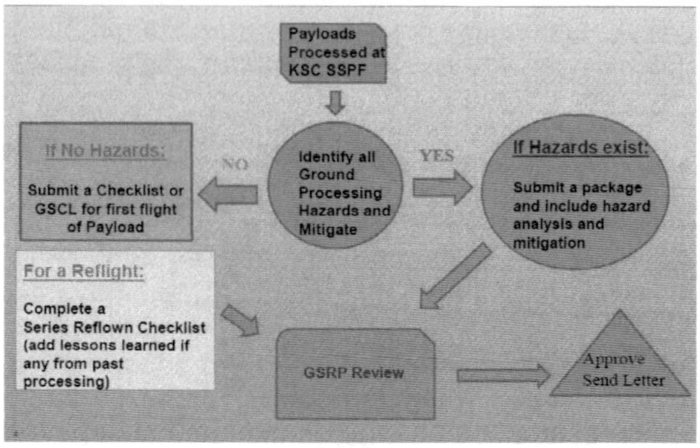

Figure 4. Current Ground Safety Review Process

Simply put, a hazard analysis (HA) is used to determine the significance of hazards so that safety design measures can be established to eliminate or mitigate them. A hazard is defined as a "Condition, event, or circumstance that could lead to or contribute to an unplanned or undesirable event." In cases where a risk cannot be eliminated, a certain degree of risk must be accepted via flight readiness review panels [7].

System safety engineering aims for the timely identification of hazards and initiation of actions necessary to eliminate or control hazards or associated risks. These include selecting designs to lower risk by reducing the probability/severity of hazards, incorporating safety devices (interlocks, barriers, and guards), providing warning and clears, and developing procedures and training. During design assurance, system safety assists with the assessment of HA studies, the assessment of critical hazards mitigation, and the review of required safety documentation. During integration, test, and ground operations, system safety reviews the GSCL and hazardous testing processes, handling and transportation procedures, and physically witnessing critical lift operations.

Typical hazards that can be expected during ground processing of payloads include: flight hardware hazards in the ground environment, ground support hazards, chemicals, biomedical systems and materials, electrical, ground pressure systems, flight pressure systems (e.g., COPV, CPV), explosive commodities, radiation, battery, lifting devices, mechanical and electromechanical devices, propellants, cryogenic fluids, oxygen, GSE and software safety [2]. Less widely known incidents have involved flammables, toxic materials, high voltages, cables, batteries, asphyxiates, COPVs, plastics, films and adhesives (PFAs), ionizing and non-ionizing radiation sources [12]. Integrated hazards which occur across multiple items such flight hardware to ground support equipment (GSE), or GSE-to-GSE, or GSE-to-facilities must also be considered. The idea is to ensure the safety of all payloads, GSE, and personnel working in close proximity or labs.

For example, Mechanical Ground Support Equipment (MGSE) may generate many generic hazard sources including thermodynamic and fluidic, electrical and electromagnetic, radiation, chemical, mechanical, and noise. The idea is to certify that at least three barriers exist in order to mitigate the hazard [e.g., physical (design related, safety margins), administrative barriers (qualification tests, monitoring operations, and training of personnel), and lastly hazard reduction measures (elimination methods, safety verification, and status verification)] would provide a triple redundancy to any failures [7].

4. BENEFITS OF GROUND SAFETY REVIEW IN MISSION SUCCESS

A manned rocket mission is considered a success when no loss of crew (LOC), loss of vehicle (LOV), or loss of science mission (LOM) occurs. It is imperative that successful missions occur on regular and frequent basis. The role of the GSRP is to ensure to the greatest extent possible the PO has a safe and successful campaign through KSC [1,2].

KSC SSPF facilities provide a variety of test capabilities for the payload that will eventually be integrated into cargo modules of rockets from Space X and Orbital ATK on commercial resupply missions to the ISS. Many payload customers derive a variety of benefits from ground safety reviews prior to ground processing their hardware, improving their chances of success. The GSRP's role is to conduct a measured safety review based on individual PO's complexity of integration and checkout activities. Safety reviews covers many facets including ground handling, interface tests, complex COPV fill operations, and Payload Rack Checkout Unit tests (PRCU) and Express Logistics Carrier tests (ELC).

As Henry Petroski said in "To Engineer is Human: The Role of Failure in Successful Design," "To understand what engineering is and what engineers do, is to understand how failures can happen and how they contribute more than successes to advance technology." This exemplifies importance of ground safety reviews in ground processing of payloads.

It is the sole responsibility of a payload safety engineer to ensure that he eliminates the "fear of the unknown" in a science mission or a payload. Many types of processing unknowns (known unknown, unknown known, unknown unknown) can contribute to towards mission failure. Fear of the unknown leads to risk due to complex interaction of systems leading to hazards caused by human error and miscommunication as the most probable cause. Some major payloads that have benefitted immensely from ground safety reviews include:

- SAGE III – Identified Oxygen deficiency issues during GN2 purges
- HDEV – Performed functional test before flight
- IDA – Assessed the integrity of Peripheral Docking Target Frequencies
- OPALS – PRCU test to check interface to ISS, ELC to check interface with HOSC
- Fruit Fly Experiment – FTR, post-flight analyses, ground control experiments *(Figure 5)*
- Orbital ATK OA-4 – PVS certification, Laser usage, and helium tank fill with COPV

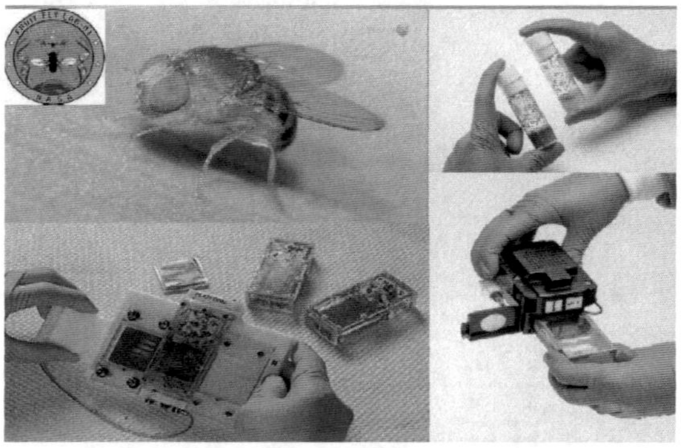

Figure 5. Fruit Flies: LT Effects of Spaceflight

A recent report by the NASA ELV Payload Safety Working Group identified various incidents that contributed to failures during ground processing of payloads. Out of 95 recorded incidents since 1985, lifting and handing contributed to over 50% of the incidents, followed by testing related incidents of almost 26%. Ground safety reviews diligently try to assimilate and evaluate many facets of ground handling and lifting-related issues to prevent a bad day from turning into a catastrophe such as the NOAA N-prime satellite mishap and Orbiting Solar Observatory disasters. The evaluation provides a second set of eyes to review safety of engineered systems that have been "hidden" or "missed" in prior reviews. Some key benefits derived from ground safety reviews include:

- Improves consistency of how payload safety is advocated
- Clarity of hazards and their interaction with other systems
- Improves communication between all parties involved
- Tries to eliminate human errors by understanding the big picture
- Systematic hazard elimination and reduction during ground processing
- Option to use safety devices if the design could not reduce risk to an acceptable level
- Warning devices to alert personnel and detect hazardous conditions
- Enhance procedure reviews and personnel training to eliminate human errors

5. LESSONS LEARNED FROM GROUND SAFETY REVIEWS

"Engineers today, like Galileo three and a half centuries ago, are not superhuman. They make mistakes in their assumptions, in their calculations, in their conclusions. That they make mistake is forgivable; that they catch them is imperative." "To Engineer is Human: The Role of Failure in Successful Design,*" by Henry Petroski* [9].

The key areas of lessons learned from payload ground safety reviews, though not encompassing, is highlighted below:

- Most failures occur due to human errors and miscommunication.
- Despite barriers, failures occur when errors are aligned (Swiss cheese model).
- Understand unknown unknowns, known unknowns, unknown knowns in the system *(Figure 6)*.
- Systems engineering must harmonize various disciplines – get everyone involved.
- Understand the physics of the problem prior to addressing solutions.
- Ensure that the design has been wetted in the reviews – allow no changes thereafter.
- If you see something wrong, say something – don't be a nobody – speak up.

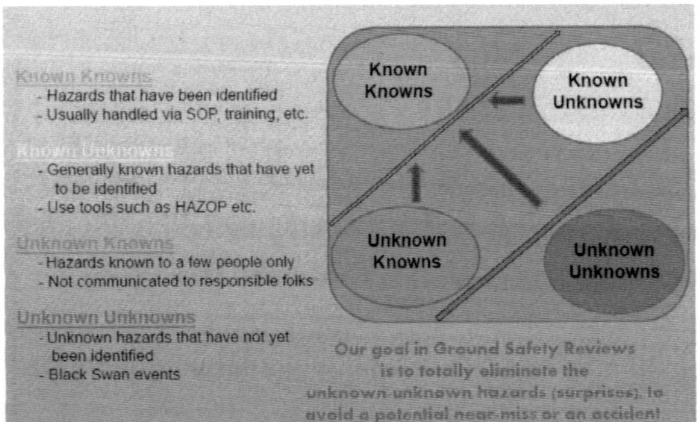

Figure 6. Ground Processing Risks: Hazards

6. CONCLUDING REMARKS

Hazard analysis and risk assessment are not a one-time "do" documents. We often need to revisit or revise those documents when there is a near-miss, when a system has been modified, when maintenance on the system has been altered, when duty cycles change on the system, and when the operating environment changes and there can be catastrophic effects. A hazard analysis and risk assessment should be a living document that changes as the system changes.

Emphasis on Safety Management Systems and Safety Culture have been well documented in literature. Mission success starts with safety, and safety begins with engineering excellence. Safety mindset has to be instilled into the organizational culture to ensure against the risk of accidents. Safety organizations should practice continuous risk management and avoid being a silent safety program.

Lessons learned from examining the underlying causes of Space Launch Failures [10,11,13] include failures due to design problems, failures due to human errors, and failures due to component malfunction. If failures occur in the wake of prolonged success, it encourages lower margins of safety. The following recommendations can help mitigate failures, especially during manufacturing, testing, and integration processes [8] as noted below:

- Extensive quality control procedures should be implemented.
- Beware of state-of-the-art components with space heritage.
- Educate design, manufacturing, and integration personnel.
- Test all subsystems according to their operating environment.
- Minimize the number of parts used in the payloads if at all possible.
- Mistakes should be treated as valuable educational materials.
- Provide a feedback mechanism from operation experience in HA.

Even though we design space missions, spacecraft systems, and their payloads to operate in the presence of multiple failures, these systems still fail in a spectacular manner due to human error, miscommunication, or both. Failure is a common occurrence and future mission success depends on our reaction to failures and ability to find the root cause *(Figure 7)*. Hazard analysis is instrumental in identifying, classifying, and managing risk and plays a pivotal role in foreseeing and preventing critical system failures. Failure has played an important role in engineering, has added value to design, and has affected our ability to select margins of safety for complex systems. Failure can also be an engine of innovation. Engineering is a human endeavor and thereby subject to error. "Every single calculation that an engineer makes is a failure calculation. Successful engineering is all about understanding how things break or fail" [9].

Figure 7: Expect the Unexpected: Unknown Unknowns

Despite all these efforts, failure of complex systems requires us to understand how systems fail. It is highly recommended that tomorrow's engineers should study:

- Systems engineering process and integration aspects.
- Understand relation between hazards, failures, and risk.
- Mission failure attributable to environmental factors.
- Failures that can be accountable due to design errors.
- Eliminate sources of human error and miscommunication.
- Lessons learned from past systems engineering failures.

7. REFERENCES

1. Dollberg, J.C., Kirkpatrick, P.D., (2002), NASA's Payload Safety Review Process at Kennedy Space Center, ESA-NASA-Spaceflight Safety Conference, ESTEC, Netherlands.

2. Kirkpatrick, P., Dollberg, J., Trinchero, J., (2009) Safety During Payload Ground Processing, Chapter IV, NASA Internal Document.

3. Wahlberg, J., (2010), Kennedy Space Center Launch and Landing Support: ISS and Spacecraft Processing, NASA Internal Document.

4. Rarick, D., Wahlberg, J., (2011), Space-X Overview for CRS Payload Integration at Kennedy Space Center.

5. Kirkpatrick, P., McDaniel, L., Smith, M. (2011) Shaping NASA's Kennedy Space Center Safety for the Future. NASA Internal Document.

6. Brekke, M., Hartman, D., (2002), Payload Integration and Safety Review Process, ESA-NASA-Spaceflight Safety Conference, ESTEC, Netherlands.

7. Gonzalez, J., Bastias, C., Barberis. E., and Delbianco, D., (2015), Mechanical Ground Support Equipment's (MGSE) for ARSAT-1 Hazard Analysis, Space Safety is no Accident, Springer Verlag.

8. Takada, S., Sasaki.S., Miki.M., Iwata, Y., and Takahashi, S., (2015), Effective S&MA Activities Based on the Experiences during JEM Operation, Space Safety in No Accident, Springer Verlag.

9. Petroski, H., (1992) To Engineer Is Human, The Role of Failure in Successful Design.

10. Jayaram, S., and Swartwout, M., (2012), AC2012-3201: A dual undergraduate/graduate course in Space Mission Failures.

11. Arif, G.K., Bukley, A., (2015), Examining the underlying causes of space launch failures, Space Safety is no accident, Springer Verlag.

12. Tosney, W.F., and Cheng, P.G., (2015) Space Safety is No Accident, How the Aerospace Corporation promotes Space Safety, Springer Verlag.

13. Baskcomb, S. Ouedghiri, D., (2015) Safety in Numbers? Lessons Learned from Aviation Safety Assessment Techniques, Springer Verlag.

ICARUS LAUGH. RISK AND SPACE EXPLORATION

Jacques Arnould

Centre national d'études spatiales – 2 place Maurice Quentin 75001 Paris (France) – jacques.arnould@cnes.fr

ABSTRACT

The risk perception depends on historical and cultural factors; it is the case for risk perception linked to space activities. As a result, risk management and safety for future space programs (planetary exploration, space resources exploitation, tourism, etc.) cannot only be based on the past experience of space, in particular the period of Apollo missions or the beginning of the American space shuttle program. It must take account of new stakeholders (NewSpace), of a growing diversity of historical and cultural realities. Risk management and safety should also integrate a real ethical question in order to define the goals of space activities and not merely their destinations, also to clarify the meaning of its exploration efforts for humanity.

1. INTRODUCTION

It is not even a question of regretting the Moon landing era. The risks undertaken by the astronauts still seem inordinate, especially given the undesirable socio-political context of the Cold War. The years went by, the iron curtain was lifted and the global geopolitical situation changed. The fierce battle between the Soviet Union and the United States subsided into a more rational balance creating competition and cooperation between the dozen space powers.

As a result, the risk perception linked to space exploration is not so easy to evaluate and to manage. It requests to elaborate a valuable philosophy of risk and to precise the value of space exploration.

2. A PHILOSOPHY OF RISK

Is risk specific to humans? Does the simple fact of existing not lie at the root of most risks faced by all living beings? There are natural risks against which scientific progress can only ever offer partial or temporary protection, risks for which only destiny, biological chance, fatality or God may be summoned. However, technological, industrial and military accidents and professional misconduct belong to the cultural sphere and were created by man due to his inability to master all the elements of the systems he has designed, in his haste to apply on a large scale solutions or products that have not been tried and tested, through his failure, recklessness, violence or unreasonableness. Effectively, if we look at it this way and consider the role played and the place occupied by technology, the human being is an animal singularly doomed to risk.

We have a very strange relationship today with the notion of risk. On the one hand, we demand that public and government institutions, at least those in western societies, manage and increasingly control risks to which individuals and groups are constantly and inevitably exposed. On the other hand, we are ready to choose attitudes, to undertake high-risk activities - in these I include the taking of drugs and the practice of extreme sports. This strange relation may become paradoxical or a source of conflict when an individual who consciously engages in a high-risk activity expects society to help them out of danger. Many debates are taking place around Europe concerning the practice of mountain sports and the financing of rescue teams which these extreme sportsmen and women may need to call upon. This sensitive question is linked to the relationship between the private sphere and the public sphere; a relationship that cannot be fixed once and for all; a relationship that legislation can only partially fix. This is particularly true when it comes to landing on new planets, acquiring new knowledge and exploring.

The risk, this "fear of evil" referred by scholars and philosophers in the 17th century who introduced the notion, is proportional to the seriousness of a danger and to its possibility of occurrence. Any decision made in this respect is based both on the objective probability of occurrence and the subjective inappropriateness of surrendering to it. Inevitably associated with the design of a world entirely controlled by the Gods or by destiny, associated with the modern introduction of statistical calculations, the notion of risk has been or has been seen to be ignored by engineers for a long time. The zero-risk cult, from the absence of failure long ago clamped down on, on behalf of the single objective to be achieved, the single success to be planned. Thinking of failure seemed, to say the least, inconvenient, a waste of time and energy, a weakening of the ideology of progress and success. After release of the *Apollo XIII* film, the words commonly, and probably falsely, attributed to Eugene F. Kranz, the famous director of flights at NASA, "Failure is not an option", became the slogan for a generation of engineers.

Times have changed. The people responsible today for developing and operating space launch vehicles and vessels now recognise that whilst a certain level of risk is inevitable, risks can be managed and reduced. They must abandon simplistic and determinist explanations of phenomena and the ideal of unlimited rational knowledge, to focus on an understanding and an apprehension of reality under the mode of complexity and the tangle of factors, most often using statistical calculations. It is not a question of losing the voluntarist character that is true of the profession of engineer. It is important to consider the increasing complexity of systems implemented and of sciences and techniques in general. It is also important to make sense of growth, at the interface, the intricacy of equipment, or organisational matters and of humans, in a way that the supposed skills and responsibilities seem to disappear. This is a very strange alchemy as now an object, a technology, an operation and a treatment are often declared certain, especially when their associated risks are known and deemed acceptable by their users! This is why we fear aircraft accidents more than we fear road accidents. Is it enough to invoke the enlightened or informed consent to ward off fear of evil? Not necessarily, as knowledge can lead to other fears. A dilemma without an apparent solution or outcome or a dilemma that is associated with the single truly responsible situation, a dilemma based on absolute conviction and the double stubborn refusal of fatality and the scandalous "After me, the deluge!" This courage and this sense of price to be paid and of dignity to be defended are necessary, when, tomorrow, astronauts could encounter new dangers, new risks and new sacrifices. Will they be ready to accept them? Will we be ready?

3. EXPLORATION IS NOT AN OPTION

Exploration is unique to humans. While the human species shares a sense of curiosity with many living creatures, it alone has the ability to imagine – to project oneself beyond the immediate boundaries of space and time. Thus, the notion could be suggested that humans have never set out to explore worlds that they have not first imagined or seen in their dreams. And thus, through exploration, humans have never stopped examining their identity, their origins and their destiny. There are different types of answers to these ancestral questions: those which are the focus of the sciences – whether cosmology, particle physics, biology, anthropology or psychology, and those that belong to our own individual philosophical and spiritual quests.

Those who worked on establishing a legal framework for space in the 1960s seized on this unique aspect of exploration. The first Article of the Outer Space Treaty, drawn up in 1967, states: "Article I: § 1. The exploration and use of outer space, including the Moon and other celestial bodies, shall be carried out for the benefit and in the interests of all countries, irrespective of their degree of economic or scientific development, and shall be the province of all mankind." Thus, before giving space a status of its own (this status will be discussed below), the treaty first focused on the status of the human activities. What does "province" mean? During the era of royalty in France, the term province (in French, *apanage*) referred to the share of the royal kingdom granted to the younger sons of the royal family in compensation for their exclusion from the throne. Since then, the term has become more generalised, meaning property, inheritance; it retains a notion of elitism. The use of the notion of province, to refer to space, is interesting. On the one hand, it offers mankind a rightful position: not that of domination (mankind does not rule the universe), or that of submission (no less deserving of its share), but more that of an heir benefiting from the work, power and potential troubles of its predecessors, who would like to be able to provide its descendants with the same possibility. On the other hand, still referring to space law, what mankind claims as its inheritance is not actually a territory, but a mission: that of exploring, exploiting and using a portion of the universe, for its own interests and that of future generations.

4. HUMAN DIGNITY

Space exploration activities have no equivalent, which means that they have to be evaluated in other terms than price. What I suggest is to apply the notion of dignity to them, as introduced by Emmanuel Kant in the second section of *Fundamental Principles of the Metaphysic of Morals*: "In the kingdom of ends everything has either value or dignity. Whatever has a value can be replaced by something else which is equivalent; whatever, on the other hand, is above all value, and therefore admits of no equivalent, has a dignity." To increase human knowledge of the universe, about life and its origins, to face worlds of which we were previously unaware, to risk challenging ideas, theories and certainties: what would have become of our species if, since its birth, it had not stayed true to its natural curiosity and penchant for exploration? Space exploration is a noble activity, a province of all mankind which, even more than major financial investments, has cost sweat, tears, and sometimes even human lives. But is this not a way of acknowledging its worth beyond its financial cost, I mean: its dignity?

5. PLEA FOR A NEW HORIZON

Exploration is not optional for our species. Exploring is part of our condition: it feeds our curiosity, nourishes our imagination, and is even necessary to our survival by providing other resources for our needs, other horizons for our creativity, other fields for us to study, and other possibilities for the existence of others. Exploring is one of the roots and foundations of our possible future. Yet, exploring is not an imposed obligation, an inevitable fate or a predestined certainty: as a human being, one is free to say no to one's desires, impulses and passions for exploration, free to weigh the pros and cons, to set a time limit for oneself and to revise one's goals. Moreover, despite advances in science and technology, our human horizon has not yet broken free from the bounds of Earth. Human exploration has not yet "broken free" of Earth's horizon. However, with humans continuing to face these two infinities, the silence of which filled Blaise Pascal with dread, a breakthrough could finally end up happening – triggered by advances in astronautics or human genetics. Is humanity prepared to go the distance? Does it really want this? Is the constant thirst for new discoveries, sensational experiences and emotional thrills that drives some of its members enough to push humanity towards an adventure such as this? Is the motivation high enough to accept the dangers? Taking this step would require no less than for all forms of enthusiasm and all minds, all knowledge and all technologies, all means and all purposes to come together and be combined: Tsiolkovsky, Goddard, Esnault-Pelterie, and all scientists and engineers, Latécoère, Musk and all bold entrepreneurs, Wright, Lindbergh, Mermoz, Armstrong and all human explorers, Cyrano de Bergerac, Verne, Clark and all masters of the imaginary. If we are to take humanity's exploratory tradition to this new level, moving far beyond the confines of African lands, as in times past, and truly breaking free of Earth's horizon… if we are to make this happen, we clearly need to forget and eliminate any distinction between public and private that would be a handicap or obstacle rather than an additional asset. On the contrary, we need to align and ally the individual and collective efforts which, since the dawn of humanity, have allowed humans to explore or realise the unknown lands they have imagined.

I believe it is essential to defend and encourage exploration within our human societies: exploration is made for man. It demands and allows us to determine the meaning of sacred that is not a taboo or a protection but instead a driving force, a projection. It requires and allows us to dream of new horizons and to expand our internal and external desires. Does exploration not honour this? Does it not ensure the dual act of shrinking the universe to fit human dimensions and of expanding human dimensions to fit those of the universe? Thus, exploration is an essential element of the humanism of space that I champion with all my heart.

THE DEMISE OF THE SAFETY ENGINEER – OVERCOMING THE POTENTIAL IMPACT TO SPACE SAFETY

James A. Runnells

HX5, LLC – JSC Engineering and Technical Services Contract (JETS)
2224 Bay Area Blvd, Houston, TX 77058 Email: james.runnells@jacobs.com

ABSTRACT

As flight hardware development projects are challenged to reduce costs and development time, the risks associated with safety of human space flight increase due to a loss of trained safety professionals on the development team. Historically, flight hardware development projects engage a dedicated Safety Engineer who is trained in proven safety analysis techniques, understands and prudently applies safety requirements for the project, and documents how a project is controlling all potential hazards through design features or other acceptable mitigation methods. The current budget climate now appears to be compelling projects to reduce Engineering staff or otherwise condense the resources used to develop flight hardware. Thus, Project teams may not include a trained Safety Engineer, and this may be increasing the risk of failures in human space flight programs. New methods for ensuring safety in a compressed development environment must be developed to minimize the risk to manned space flight.

1. THE CHALLENGE OF SPACE SAFETY FOR PAYLOAD AND EXPERIMENT PROJECTS

The current economic climate, combined with efforts to encourage commercial space ventures, is compelling space exploration projects to condense resources used to design and develop space flight hardware. This shift, along with a goal to increase the amount of science and engineering activity on space research platforms such as the International Space Station (ISS) and commercial space launches has resulted in efforts to reduce project time-to- flight and development costs for spaceflight hardware and software.

One popular topic of interest for development cost and schedule reduction is that the current human space safety requirements are complex and overly conservative, making it difficult and expensive for projects to fly payloads (or experiments). This area of interest by leaders in the space industry makes the safety process and requirements an obvious target for change. Current activities to partially address payload/experiment project cost/schedule challenges include clarifying, reducing or combining safety requirements resulting in a more cost-effective safety process.

However, reducing or combining safety requirements does not inherently make a program safer, or even as safe as before. In fact, there does not appear to be clear evidence the safety requirements as they exist today unduly burden a Project due to the intent of the requirements. Indeed, what is evident from recent safety reviews is that the most significant impact to Projects today is from a lack of knowledge and early application and implementation of the existing safety requirements. Current safety requirements have been developed over decades based upon experience which often included failures, some catastrophic. Changing the current safety requirements or the safety process could translate to an increase in the risks associated with human space flight. Addressing project safety requirements deficits, or the misapplication of the safety requirements by projects, by softening, minimizing, or simplifying the current safety requirements will most likely lead to an increase in risk to the overall project and the space vehicle it operates on. More importantly, the lack of trained, specialized, safety professionals on space hardware and software Project development teams can allow the potential risk increase to go unnoticed by the project and the program, especially for less complex projects that most believe to pose little to no safety risk. One must recognize a risk exists for it to be addressed, and Flight Safety Engineer's experience and training make them uniquely qualified to recognize missed requirements and added risk.

An area of cost and schedule reduction recently observed on less complex flight hardware development projects is the elimination of the Safety Engineer. Historically, flight hardware and software development projects engage at Project inception a dedicated Flight or System Safety Engineer who is trained in proven safety analysis techniques and safety requirements. Effective Flight Safety Engineers understand and prudently apply safety requirements early in the development process, and provide early input to the Project team on design issues and solutions for potential hazards. They also document early and effectively how the system design and operating procedures control potential hazards through design features or other acceptable risk mitigation methods. The Safety Engineer also documents adequate verifications for the documented hazard controls, ensuring the controls function as expected. In order to provide adequate safety deliverables required by the program under diminishing budgets and development timelines, trained System Safety Engineers are being excluded and replaced on the Project Team by Project Managers or Engineers in an effort to reduce costs.

The safety deliverables, such as Safety Data Packages or Hazard Analyses and Reports, are used by the Project and the Program to assess the level of safety of a payload or experiment and the risks associated with failures on-orbit. Using untrained or inadequately trained personnel to perform the function of a Flight Safety Engineer can lead to an increase in risk to the project and the program.

The efforts to clarify, reduce, or combine space safety requirements are intensifying within current manned space programs. However, from recent safety review experience it is apparent that Project teams that do not include a trained Safety Engineer often experience project costs overruns, development schedule extensions, and even a potential increase in the number of failures in human space flight programs due to missed or overlooked safety requirements. This is evident in the amount of rework (design and/or safety documentation) that has resulted when safety reviews uncover design issues or flaws that can impact the relative safe use of a payload or experiment. This impact to projects can be traced to the lack of trained safety professionals on the project team, and inexperienced design/project Engineers attempting to replicate the results usually achieved by trained safety Engineers. To offset the reduction or elimination of trained safety professionals from Project Teams, new methods for ensuring that safety is effectively implemented early in compressed development environments must be established in order to prevent an increase in risk to manned space flight.

The initial cost and schedule reduction from excluding or minimizing System Safety Engineer resources is quickly offset by the substantial cost and schedule impact resulting from rework of safety documentation, and from hardware/software design updates resulting from misapplied or missed safety requirements. Casual observation and informal data collection from Safety Data Package (SDP) reviews, Hazard Report (HR) reviews, and Safety Review Meetings suggests that an increasing number of Projects are either unaware or unfamiliar with the applicable safety requirements, such as SSP 51700 [1], SSP 50021 [2], and the safety process requirements are captured in SSP 30599 [3] (and related safety documents for the ISS referenced therein).

The impact to the Project, as well as the Program, appears to result in an increased design and development time, and an extension of the safety review process (i.e., recurring meetings) to ensure compliance to relevant safety requirements. This often translates to additional safety reviews such as delta reviews, working groups, or substantial documentation updates that require repeating technical reviews of the documentation to ensure updates address previously documented discrepancies. These delta reviews or substantial documentation updates (rework) add cost to the Project Team in the form of safety document rework (e.g. SDPs, HRs, etc.), and possibly redesign or design changes. The rework, update, or design changes also negatively impact independent reviewer/assessment resources (i.e., Program costs) that help guide the Project and advise the Program. Therefore, the rework or updates results in an expansion in cost and/or development time instead of the desired reduction originally intended for basic or intermediate projects.

2. A COST-EFFECTIVE SOLUTION SPACE PROGRAMS CAN PROVIDE TO PAYLOAD/EXPERIMENT PROJECTS

One approach to minimizing the impact of extensive rework or updates to ineffective project safety deliverables that capture project design or operational changes (typically resulting from missed safety requirements) is to combine and reduce safety requirements, and combine phased safety reviews for experiments or payloads. This typically results in diminished data deliverable content (i.e., SDPs, supporting technical documentation and diagrams, etc.) from the project. The reduction of data and safety review meetings results in limited input from safety experts at combined phased safety reviews. The interaction with safety experts is critical to independent assessments and risk reduction or elimination for the project and program.

The subsequent reduction in technical data and safety information provided to safety experts for an independent assessment increases the need for reworked (or updated) HRs and supporting documentation, and eventually results in additional safety review meetings to assess and approve the rework/corrections. This consolidation of reviews and diminishment of data complicates and, quite frankly, limits independent reviews of the payload or experiment safety.

An alternative to changing, reducing, or minimizing safety requirements is to provide an application that captures the knowledge of trained Flight Safety Engineers and replicates the safety assessment process through a well written and user friendly web-based application accessible to all payloads and experiments. An application as described above would compile and centrally locate safety requirements, making them more accessible and easily searchable. The application could also guide a design team through the safety assessment process with a methodical, intuitive mapping of an effective safety assessment process for use by the project team. This approach could also convey clarity for a project team that has limited safety resources by adding rationale that supports and simplifies the requirements application in the design and development of space related projects.

Obviously, having a Flight/Systems Safety expert on the Project Team that has knowledge of the applicable requirements is critical to the success of the project, especially on more complex projects. However, acknowledging the fact that safety requirements are widely scattered and sometimes nebulous should lead one to agree that a new approach is warranted, especially when Flight Safety Engineering resources are reduced or excluded from a project team. Addressing the ease of access to requirements, guiding project teams in the application and focus of the safety assessment process, and providing clear guidance on when or how to apply the requirements can result in a decisive improvement that can reduce or eliminate the impact to project teams and the program from safety assessment rework, or even missed safety requirements.

The approach advocated herein is to capture the knowledge of a Flight Safety Engineer and automate the assessment techniques to ensure an effective and economical project safety assessment. Knowing where to find the appropriate safety requirements, much less having the ability to include all relevant safety requirements, in the overall project requirements list at project inception can be a significant challenge. Providing a logical web-based application with a comprehensive list of all safety requirements in a single database (or library) that provides clear direction on the applicable requirements sorted by subsystem applicability or other traceable and relevant selection criteria, will both guide and educate projects teams through an effective safety assessment process. Providing a specialized database of safety requirements that leverages the captured knowledge of trained Flight Safety Engineers can assist Projects in selecting and interpreting the appropriate safety requirements. Requirements that exist in a single repository, are easily searchable, and provide clear application objectives and implementation recommendations, could provide significant impact early in a project. The earlier a project understands and applies the safety requirements in the development timeline, the more unlikely it is for any significant impact (cost and schedule) from missed or misapplied safety requirements. The database could also provide Program contacts for each subsystem, or a specific requirements section, that would allow Projects to seek guidance on unresolved or unsettled safety requirements, including their applicability or implementation.

The mechanics of such a database can be a web- based graphical user interface (GUI) that leads the user step-by-step through the requirements selection process for ease of access and ubiquitous use by Projects. Having a web-based, or cloud-based, application also allows for simple maintenance or updates to the requirements database as requirements change or evolve. Collocating all flight safety requirements in one database and automating search and applicability functions by sub-system, hazard category, or functional area (i.e., by engineering discipline such as electrical, structural, etc.), will allow a Project to rapidly assess the requirements applicable to their device. This would ensure early and insightful application of the safety requirements, and help projects determine cost benefits of one type of design solution versus alternative solutions based upon the applicable requirements, application rational, or lessons learned from previous users or Program direction (all critical areas that should be captured in requirements and verification rationale statements in the database).

A simple example of a web-based Flight Safety Requirements application could be to ask if the payload/experiment is stowed and operated internally using Intravehicular Activity (IVA) or externally to the space vehicle through ExtraVehicular Robotic (EVR) or ExtraVehicular.

Activity (EVA) operations. The application would also ask what sub-systems are applicable (structures, mechanical, electrical, etc.), and ask system questions about the use of Radio Frequency [RF], pressure systems, mechanisms, etc.. For example, a sub-system could be electrical IVA. If the device uses electrical power IVA, then the next question could be: what is the source of power for the payload/experiment electrical system, and does the electrical system contain any voltages above 32V or any current capability above 65A. Dependent upon the answers, the database would select the appropriate safety requirements (i.e., wire sizing and circuit protection requirements, bonding/grounding safety requirements, power connector mate/demate requirements, battery safety requirements (including if batteries are used as a primary or secondary power source for charging), capacitor requirements, corona requirements (if applicable for high voltages), etc.).

This guided search using a web-based application will result in a clear list of applicable safety requirements for the project team to use during the early design stages. Including the requirements and verification rationale can also provide insight into the typical design application and any implementation challenges for certain design solutions. As the project evolves through the preliminary and critical design reviews (PDR/CDR), the requirements application can be revisited to ensure the evolution of the design has not resulted in different or new safety requirements. If so, this can be assessed by the project team for a cost effective solution that can be presented, assessed and approved during the safety review process. The impact of developing and providing a comprehensive Space Safety Analysis and Requirements application is reduced project costs and limited impact to design and development schedules resulting in a measureable improvement for project teams and for space programs.

The requirements database allows early inclusion of requirements in a project. A process now used in some space programs provides generic hazard reports to guide a project team. However, generic hazard reports, with generic causes, controls, and verifications, are cumbersome, challenging to apply (too specific), and are used later in the design process (typically not at concept/ inception) when phased safety reviews are combined. Generic hazard reports, causes, and controls also do not tend to support innovative approaches to safety requirements implementation in new designs, systems, or vehicles. Conversely, a requirements database can provide additional functionality, such as automated safety assessment processes and recommendations for conducting a thorough safety assessment (to industry recognized safety analysis practices), while still recommending a list of generic controls commonly used to address safety requirements. The database could even suggest typical verification methods for generic controls, along with rationale supporting each verification method as an effective solution for the recommended hazard control. Future versions of a safety requirements database could also automate production of SDPs and generate Hazard Reports (HRs) in program HR applications. These project safety deliverables could be imported to a Program database (such as the IHS) using information submitted by the project to generate design specific project HRs with project selected safety controls and verification solutions.

The generation of SDPs by a safety analysis and assessment application eliminates the cut and paste approach now used by many project teams who use previously generated SDPs as templates. It allows a fresh approach for small and intermediate projects lacking training in safety analysis and assessment techniques, and also moderates the need for a project-based Flight Safety Engineer to generate the safety assessment and resulting SDP. The lack of a safety analysis and assessment process for many projects is as much an impact as a lack of knowledge of the applicable safety requirements.

The initial function of a safety analysis application is to leverage the requirements database. The application would generate a list of top-level program safety requirements, as well as a list of specific safety requirements and potential design solutions dependent upon the applicable project sub- systems combined with the program/vehicle services used by the project. This is basically one of the principal sections of a SDP, and it allows reviewing Safety Engineers to perform a thorough evaluation of the safety requirements applied by the project. Other significant sections in a SDP are system and subsystem descriptions, as well as the safety analysis section. The system and sub-system sections could be generated by the web-based application with a series of targeted questions that steer the project through the generation of applicable descriptions, systematically created by a smart application plug-in maintained in the safety analysis application (and updated as space related system/sub-system knowledge evolves).

This same approach can also be used to generate the Safety Analysis/Assessment section for the SDP as well as initial hazard reports (HRs) that should result from the preliminary safety assessment. The data collected by the online application during the project system and subsystem description is also used in the safety analysis and safety requirements assessment, which can then assist the project in generating the system safety assessment conducted through the database. The safety analysis application can be used to help the project team select suitable hazards, causes, controls and verifications, possibly from generic program HRs. Using generic HRs created by a Program at this point is suitable, as the project safety analysis process is following a generally accepted space safety assessment. Instead of a project bypassing an acceptable safety analysis process by going directly to generic hazard reports to generate project HRs, a safety analysis application would lead a project through an acceptable safety analysis process that results in potential hazard identification prior to selecting appropriate hazard controls. This would greatly reduce the risk of missed hazards, causes, or controls. This would also help confirm applicable safety requirements are captured by the project early in the design and development process of the payload or experiment.

This automated HR generation process can provide rationale for causes and control(s), and list appropriate verifications that the project can select from. The generic controls can then be refined for specific application to the project hardware/software, resulting in control(s) and verifications appropriate to their specific design and the hazard severity level from the application provided assessment. The online system can even be designed to import the project data generated from the safety assessment application to a program hazard system such as the ISS Hazard System (IHS) to populate Program hazard reports.

3. THE EVOLUTION OF SAFETY ASSESSMENTS

It is clear that projects today are challenged with performing suitable space flight safety assessments. Budget and schedule challenges have resulted in the demise of safety engineers on smaller projects such as payloads and experiments, a lack of analysis and a lack of assessment training, and knowledge of industry practices and process. Projects need solutions, such as web-based applications that can provide the safety analysis expertise and capture the space safety requirements knowledge of a trained Flight Safety Engineer.

Another feature of a web-based flight safety requirements database and automated safety analysis application is its relevance to popular use of mobile and cloud based computing applications. The web-based requirements database and safety analysis application would be more germane in today's omnipresent internet search and cloud computing engineering research resources. By providing an application that leverages the computer and research skills of today's engineers, it is more likely it will be included in a project's toolbox of resources.

4. THE DEMISE OF THE SAFETY ENGINEER SHOULD NOT RESULT IN REDUCED SAFETY FOR SPACE FLIGHT

The loss of Flight Safety Engineers on manned space flight payload or experiment projects doesn't have to result in an increase in risk of a failure that results in a hazard. This paper does not advocate a project not employ trained Flight Safety Engineers. Indeed, this paper advocates for risk reduction impact by using trained Safety Engineers who follow industry accepted safety analysis processes, and suggests it is invaluable to flight safety. It's difficult to measure the odds of a failure that is counteracted, or never occurs, due to the expert implementation of a rigorous safety assessment and the resulting application of appropriate safety requirements. However, with a meticulously defined safety analysis application and a safety requirements database that captures the knowledge and expertise of recognized Flight Safety Engineers, small to medium sized projects that are economically challenged can still ensure a reasonable safety assessment and hazard analysis that minimizes the risk increase to the Program.

5. THE DEMISE OF THE SAFETY ENGINEER – OVERCOMING THE IMPACT TO SPACE SAFETY

Complex space systems and payloads require a clear, methodical approach to assessing and ensuring safe human space flight. Historically, a System Safety Engineer is involved from project kick-off to at least the end of the payload or experiment operations. Most complex experiments/payloads do include a Systems Safety Engineer on their team, at least for some safety activities, if not managing all aspects of safety for the Project. However, many small to medium projects (payloads and experiments) are moving toward eliminating trained Flight Safety Engineers as an integral teammate (or already have done so). This often leads to project impact from safety assessments becoming only back-end loaded, or later in the design development process.

This often results in dropped safety requirements, inadequate hazard controls for hazardous functions or potential hazards, and a significant impact to the cost and schedule of a project when a design is questioned or the safety assessment repeated and associated documentation has to be updated. Offering a small to medium project an alternative that replicates the impact of a trained Flight Safety Engineer significantly minimizes the potential for missed hazards andsafety requirements. The cost of the development, testing, and maintenance of a Flight Safety Analysis application and safety requirements database is minimal when spread amongst the hundreds of safety assessments performed annually by payloads and experiments over the life of a Program.

Eliminating or reducing the use of Flight Safety Engineers will add risk to human space flight programs. Replicating the safety analysis process for smaller payload or experimentprojects that may not employ trained Flight Safety Engineers or have sufficient safety analysis training will facilitate identification of hazards and potential controls for the project to incorporate into the design. Providing an easily searchable Flight Safety Requirements database will also minimize the potential for missed safety requirements while also providing risk reduction for the Program. Providing such an application and database translates into a reduction of safety analysis and safety documentation rework, and provides Projects and Programs with tangible results for minimizing project costs and schedules.

6. REFERENCES

1. SSP 51700, Payload Safety Policy and Requirements for the International Space Station, Feb. 9, 2016.

2. SSP 50021, Safety Requirements Document, International Space Station, Nov. 11, 2013.

3. SSP 30599, Safety Review Process, International Space Station, Feb. 1, 2015.

PROCEEDINGS OF THE 8TH INTERNATIONAL SPACE SAFETY CONFERENCE

SESSION 03
COMMERCIAL SPACEFLIGHT - I

THE EVOLUTION OF THE NASA COMMERCIAL CREW PROGRAM MISSION ASSURANCE PROCESS

Amy Canfield

NASA, Kennedy Space Center, Florida, 32899, USA, amy.c.canfield@nasa.gov

ABSTRACT

In 2010, the National Aeronautics and Space Administration (NASA) established the Commercial Crew Program (CCP) in order to provide human access to the International Space Station and low Earth orbit via the commercial (non-governmental) sector. A particular challenge to NASA has been how to determine that the Commercial Provider's transportation system complies with programmatic safety requirements. The process used in this determination is the Safety Technical Review Board which reviews and approves provider submitted hazard reports. One significant product of the review is a set of hazard control verifications. In past NASA programs, 100% of these safety critical verifications were typically confirmed by NASA. The traditional Safety and Mission Assurance (S&MA) model does not support the nature of the CCP. To that end, NASA S&MA is implementing a Risk Based Assurance process to determine which hazard control verifications require NASA authentication. Additionally, a Shared Assurance Model is also being developed to efficiently use the available resources to execute the verifications.

1. HISTORY OF THE COMMERICAL CREW PROGRAM

The National Aeronautics and Space Administration (NASA) established the Commercial Crew Program (CCP) in March 2010 to facilitate the development of a United States commercial crew space transportation capability with the goal of achieving safe, reliable, and cost effective access to and from low Earth orbit and the International Space Station (ISS). The CCP has and continues to fulfill this objective through multiple procurements and acquisitions in which commercial companies develop elements, subsystems, and systems in support of an integrated commercial crew transportation system [1].

CCP is currently administrating and executing two Commercial Crew Transportation Capability (CCtCap) contracts awarded to The Boeing Company and Space Exploration Technologies Corporation or SpaceX. These contracts will grant NASA certification to each of the Commercial Provider's Crew Transportation System (CTS) when compliance to NASA requirements is demonstrated. It also grants each of the Commercial Provider post certification missions to bring NASA crew to the International Space Station (ISS). Independent from and supporting the CCP are the three NASA Technical Authorities (TA); Engineering, Health and Human Performance, and Safety and Mission Assurance. Each of the TAs are responsible for agency level requirements and are responsible for ensuring NASA Programs comply with the applicable requirements. In addition to the TAs, the ISS Program independently verifies the CCP Commercial Providers comply with the visiting vehicle requirements for the ISS.

The CCP has provided NASA the opportunity for a transformation in how to assure requirement compliance. The transformation started with the first phase, Space Act Agreements, which allowed NASA to provide feedback on each of the Commercial Providers' CTS designs while NASA was refining the CTS requirements. This included obtaining feedback from the Commercial Providers on the CTS requirements during development. Then, once the CTS requirements were baselined, the second phase of contracts allotted NASA the ability to provide feedback on the compliance to the NASA CTS requirements. The next section on the history of mission assurance continues to describe how NASA transformed its methods of compliance.

2. DEVELOPMENT OF PROGRAMMATIC MISSION ASSURANCE REQUIREMENTS HISTORY

NASA's mission assurance program developed throughout the human spaceflight programs, learning from past experiences. After the Apollo 1 accident, separate safety and reliability offices within NASA were formed and Congress formed an independent safety organization called the Aerospace Safety Advisory Panel (ASAP) tasked to provide an independent review of policies and procedures that contribute to risk in the areas of operations, management, and systems [2]. With the Rogers Commission's recommendations from the Challenger accident, NASA created the Office of Safety, Reliability, and Quality Assurance responsible for safety-related policy; however, it was not until the Columbia Accident Investigation Board's recommendation that it became an independent technical authority [2].

In the development of the firm-fixed price CCtCap contract, the CCP was unwavering in discovering a new way of assuring contract compliance for the development work and services provided. The inspection clause, a deviation to the FAR Inspection of Services, allows inspection of both the services and the research and development work performed. In addition, this new direction also included a limited amount of data requirements deliverable (DRD) documents that necessitated the delivery to and approval by NASA. Therefore, to allow access to the data used in performance of the contract and to continue to cultivate the partnerships with the Commercial Providers, a special Government Insight clause was developed with a complementary DRD, *Insight Implementation Plan*. This Government Insight clause allows the Government to assure compliance to requirements through insight while the DRD allows the Commercial Provider to prescribe how NASA would have access to the data used in support of the CCtCap contract.

In addition, NASA contracts are required by the Federal Acquisition Regulation (FAR) Part 46, *Quality Assurance*, and NASA FAR Supplement (NFS) Part 1846, *Quality Assurance*, to ensure that the contractor conforms to the contract requirements. NASA Procedural Requirements (NPR) document 8735.2, *Management of Government Quality Assurance Functions for NASA Contracts*, was developed to implement quality assurance functions defined by these regulations. This document, depending on the criticality and complexity of the acquisition items, defines how NASA will determine its quality assurance functions, including product assurance actions (PAAs). These PAAs, or Government inspections, are selected based on risk factors that include criticality, complexity, maturity, supplier past performance, and personnel safety considerations. Also, PAAs were placed at the last opportunity for inspection in the assembly and integration of the space transportation system. These Government inspections are not a substitute for and do not relieve the Commercial Provider of its responsibility to perform quality inspections; these Government inspections are to ensure the final product is as promised by the Commercial Provider. Programs are required to assign a PAA for every product, processing, and/or performance attribute where a noncompliance could result in loss of human life; these are considered safety-critical [3].

During the Space Shuttle Program, NASA had PAAs, also called Government Mandatory Inspection Points (GMIPs), based on the failure modes and effects analyses and critical items lists for each of the shuttle system elements. Post-Challenger, there were approximately 44,000 GMIPs per flow at Kennedy Space Center [4], which were reduced to approximately 10,000 GMIPs by the end of the Space Shuttle Program. The performance of these GMIPs required the contractor to stop, inform, and wait for NASA quality assurance personnel to witness or verify the compliance to the requirements, which increased the ground processing timeline of the Space Shuttle elements. With the Space Shuttle GMIPs, the quality assurance program was performed mostly through direct oversight of the contractor.

With the move to the commercial environment, the CCP had to find a way of ensuring compliance without performing 100% inspection into the Commercial Providers' organizations, while still complying with NPR 8735.2 requirements. In 2013, the NPR was revised by the Office of Safety and Mission Assurance, permitting programs new ways of determining PAAs, including allowing an exemption of safety-critical PAAs based on either statistical process controls or a formally documented risk analysis [3]. The second exemption, a formally documented risk analysis, is how the CCP determined to proceed. The exemption allows the CCP to use a documented technical risk analysis based on many factors including hazard analysis controls/mitigations to determine and assign PAAs. The Government Insight clause is how NASA prescribed the use of a Risk Based Analysis (RBA) for determining the PAAs. The following section describes the CCP RBA Process.

3. RISK BASED ANALYSIS (RBA) PROCESS OVERVIEW

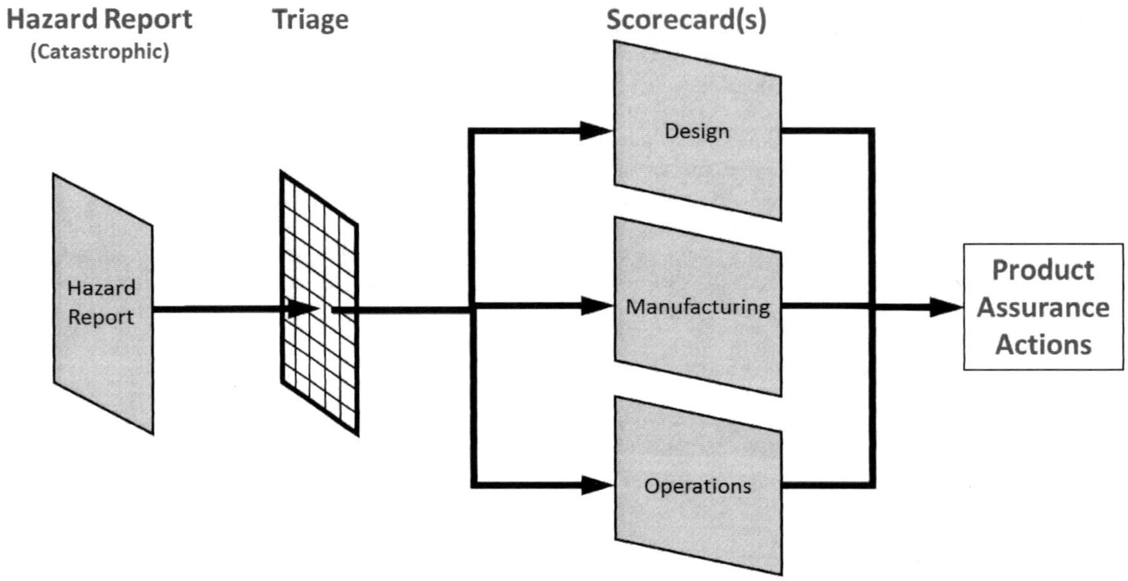

Figure 1. Commercial Crew Program Risk Based Analysis Process

The Safety and Mission Assurance Office supporting the CCP developed the CCP RBA process in *Figure 1* by using Safety Technical Review Board (STRB) approved catastrophic hazard reports. The STRB process is described in the previously mentioned paper, "The Evolution of the NASA Commercial Crew Program (CCP) Safety Process [1]." It was determined that catastrophic hazard reports would be beneficial in the process of identifying safety-critical attributes as they address hazards that, if not mitigated, results in loss of human life. The RBA team includes Engineering, program representatives, safety, and quality personnel, most of whom supported the STRB process. The RBA team reviews each catastrophic cause/control/verification set within the hazard report to determine the risk posed if it is not properly implemented. After the RBA team determines the risk of each cause/control/verification set, than based on the risk posed, it is determined whether a PAA(s) is assigned. Unlike the Space Shuttle Product Assurance Actions, the Commercial Provider does not have to wait on the Government to perform its inspection; however, the Commercial Provider is required to provide the Government with enough notice that the activity affected by the PAA is going to take place within a period of time. The RBA process consists of three phases: Triage, Scorecard(s), and Product Assurance. However, the most important part of the RBA process is that it is an iterative process. Once a risk profile is determined, it is re-evaluated after a time to determine if the risk has been abated, stayed constant, or has increased. If the risk has been abated, the PAA will be retired. If the risk is constant or increased, the PAA(s) may be changed. Each of these phases is defined further in the following sections.

3.1 Triage Phase

Upon receipt of the approved catastrophic hazard report, it goes through a Triage. This is a quick sorting of the control/verification sets associated with each catastrophic cause within the Commercial Provider's hazard report. The first two criteria the RBA team takes into account for each cause is an associated Program Risk and an associated risk identified by the Probabilistic Safety Assessment (PSA). The PSA informs CCP of potential risks to loss of crew and loss of mission. These two criteria and evaluation at the cause level allows the RBA team to understand where the CCP has identified potential risks. Next, each control/verification set is evaluated against four criteria and the likelihood of (or risk to) it not being properly implemented. The RBA team gives each of these four criteria a quick "yes" or "no." These four criteria are:

Complexity: whether the design and/or process described within the control/verification set is multifaceted, intricate, complicated, and/or difficult to perform;

Maturity in Aerospace Industry: whether the design and/or process is new or relatively new to industry;

Past Performance: whether the Commercial Provider or sub-tier supplier lacks experience in successfully implementing the control and/or verification; and

Subject Matter Expert and/or Quality Engineer's opinion: is there a high risk to correctly implementing the control/verification set.

If the RBA team answers "no" to all six criteria, then the risk is determined to be low and no PAA is assigned. If there is a single "yes" to any of the six criteria, the RBA team may provide rationale for not sending the cause/control/verification set into the Scorecard(s) phase. However, if the RBA team answers "yes" to more than one of the six criteria, the cause/control/verification set is brought forward to the Scorecard(s) Phase.

3.2 Scorecard(S) Phase

The Scorecard(s) phase is used to further evaluate the risk of the cause/control/verification set against criteria and to calculate a total risk score. A cause/control/verification set can be sent to one or more scorecards, depending on what is described within the control/verification set. The scorecards are: Design, Manufacturing, and Operations.
Design: evaluates the control/verification set with a focus on design, including processes and tools. The criteria include configuration management, design maturity, design complexity, Commercial Provider's past performance in controlling the design.

Manufacturing: evaluates the control/verification set with a focus on manufacturing, including fabrication, assembly, and the associated processes. The criteria include: maturity, complexity, personnel competency, degree of difficulty in the implementation of the verification, and Commercial Provider's past performance or similar experience in manufacturing, processing, materials, and tooling.

Operations: evaluates the control/verification set with a focus on operations, including integration, final assembly, and the associated processes. The criteria include: maturity, complexity, personnel competency, degree of difficulty in the implementation of the verification, and Commercial Provider's past performance or similar experience in operations or processing, materials, and tooling.

The RBA team determines the risk associated with each of the scorecard's criteria and a total risk score is calculated. For the above described criteria, the RBA team defines the risk based on a scale of one to three; with one being low and three being high. In addition, each scorecard includes the hazard cause likelihood from the hazard report and the overall subject matter's and/or quality engineer's opinion of the likelihood of the control/verification set being properly implemented as criteria. For the latter criteria, the RBA team defines the risk based on a scale of one to five; with one being low and five being high. The likelihood is as stated in the hazard report. The total calculated risk score will be between zero and 100. If the risk score is 60 or above, the cause/control/verification set is identified for Government surveillance. If the risk score is below 60, the RBA team may still determine to assign Government surveillance; if not, rationale is provided to support no Government surveillance. The RBA team may also provide to the Quality Engineering team what activity or action the Government surveillance might resemble for the Product Assurance Phase.

3.3 Product Assurance Phase

After the RBA is completed on the hazard report and determined that Government surveillance is necessary for assuring implementation, the Quality Engineering team considers the risk score results and determines the appropriate PAA for each of the cause/control/verification sets identified. These PAAs become the mandatory Government surveillances on the safety-critical attributes of the Commercial Provider's CTS. A cause/control/verification set that is identified for Government surveillance will be assigned a single or multiple PAA(s) that are developed from the descriptions within the hazard report. Quality Engineering may choose from four types of PAAs:

Record Review - A review and verification that recorded data properly evidences conformance to contract requirements (e.g., invoked drawings, specifications). Recorded data, including contractually required data deliverables, may document work performance, product attributes, product configuration, product performance, or quality assurance actions performed by each Commercial Provider. This also includes an assessment of document(s) to verify the planned work conforms to contract requirements [3];

Process Witness - A physical observation of each Commercial Provider's work processes or demonstrations (including tests) to ensure compliance with documented procedure(s) and contract requirements.

This includes processes related to manufacturing, fabrication, assembly, integration, repair, maintenance, refurbishment, test, and inspection [3];

Product Examination - A physical inspection, measurement, or test to ensure product conformity to prescribed technical and contract requirements. This method may also include Engineering's independent Verification and Validation of an analysis, model, simulation, or test results [3]; and

Process Audit - Provides assurance of general process health, confirmed by assessing factors such as, process yield, nonconformance(s), and/or repeatability; meaning that the process is clearly-defined and shows consistent adherence to requirements or the manufacturing of the product is stable (e.g., low nonconformance rate). Product/Process audits ultimately confirm that the process is capable of achieving the safety critical attribute(s). Process audits should also confirm that process drift has not occurred since baseline, that process changes are being adequately identified and assessed, and, where applicable, that control plans are adequately defined and implemented [5].

In the development of the PAAs, Quality Engineering assigns the most suitable function for performing the PAA and a CCP System Office the responsibility to ensure its execution. Quality Engineering will also assign a frequency for the performance of the PAAs and also indicates when a re-assessment of the risk profile determined by the cause/control/verification set is performed. The draft PAAs are then sent to the CCP System Office for the assignment of a point of contact (POC). Based on the description of the PAA the POC may be from the Program, Engineering, or Safety and Mission Assurance, allowing for shared assurance. This allows the most suitable personnel be assigned to execute the PAA.

As previously stated, the iterative nature of the CCP RBA process is important. The S&MA Office supporting CCP is responsible for ensuring that the RBA process continues to identify those areas requiring PAAs by understanding the risk profile of the executed PAAs and determining any risk profile changes from other sources, including nonconformance trends and audit findings. The CCP is still in the process of approving Phase II hazard reports, but there has been a large reduction in Government surveillance on safety-critical attributes through this documented risk based analysis.

In the long term, the S&MA Office is investigating a way to allow the RBA process to become a part of the STRB process, eliminating the duplication of meetings for the same resources to discuss the risk profiles of hazard reports. This combining of the STRB and RBA process will allow for a more streamlined approach where the triage and scorecards are combined, enabling a documented risk score for all control/verification sets that includes a weighting for Program risks and Probabilistic Safety Assessment impacts.

4. SHARED ASSURANCE MODEL

The key to making the PAAs for CCP even more powerful is a shared assurance model. The Shared Assurance model is where the most suitable person is allocated based on the skill and expertise which minimizes or eliminates organizational overlap and redundancy [1]. In Shared Assurance, S&MA relies on other organizations to provide some of the assurance functions that were traditionally performed by S&MA.

Traditionally, NASA accepted and owned the hardware/software. The Commercial Provider was responsible for ensuring compliance to NASA levied requirements of the hardware/software; however, NASA programs and TAs had a substantial role, not only in certification, but also in the assurance of flight safety [1].

In the CCP business model, NASA is purchasing a service. The Commercial Providers retain ownership of the hardware/software and are required to certify their CTS to NASA CTS requirements. As NASA is purchasing a service, its traditional role is reduced; however there still exists a responsibility to assure crew safety. As described in the previous sections, the PAAs developed from the risks identified from catastrophic hazard reports during the CCP RBA will be executed by the most suitable personnel; for example, a design-type PAA, like a stress analysis, engineering would be the most suitable. In addition, the CCP is sharing the resources and data from the ISS Program and the Launch Services Program. The key aspects of Shared Assurance are using the most suitable persons to perform the PAA, allowing the elimination of overlap among the NASA Program and TAs without losing the ability to assure the safety of NASA's crew.

5. SUMMARY

The Risk Based Assurance process that enables the CCP to use a technical analysis of risk to determine which hazard cause/control/verification sets require NASA surveillance has greatly reduced the number of mandatory Government surveillance points. Initial runs of the

RBA process confirm that this is indeed the case. The RBA process in conjunction with shared assurance, is allowing the CCP to efficiently use available resources to execute the mandatory Government verifications and provide sufficient surveillance to ensure each of the Commercial Providers are providing the Government a safe crew transportation system to the ISS.

6. REFERENCES

1 Kirkpatrick, P., and N. Vassberg. *The Evolution of the NASA Commercial Crew Program (CCP) Safety Process*. Proc. of 7th IAASS Conference "Space Safety Is No Accident", Friedrichshafen, Germany: October 2014. Print

2 Columbia Accident Investigation Board. *Columbia Accident Investigation Board Report*. By Harold W. Gehman. Vol. I. Arlington, VA: Columbia Accident Investigation Board, August 2003. Print. 178 to 193.

3 United States. National Aeronautics and Space Administration. Office of Safety and Mission Assurance. *NASA Procedural Requirements (NPR) 8735.2B, Management of Government Quality Assurance Functions for NASA Contracts. NODIS Library*. NASA Safety and Mission Assurance, 12 August 2013. Web. 8 April 2016.

4 United States. Government Mandatory Inspection Point Independent Assessment Team. *Government Mandatory Inspection Point (GMIP): Independent Assessment Final Report*. Washington, D.C.: National Aeronautics and Space Administration, January 2004. Print.

5 United States. National Aeronautics and Space Administration. Commercial Crew Program. *Commercial Crew Program Surveillance Plan*. Kennedy Space Center: CCP Program Control and Integration, December 2015. Print.

SAFETY CONSIDERATIONS REGARDING COMMERCIAL HUMAN SPACE FLIGHT TRAINING

Nicholas Mercury Carlstrom(1), Michael Long(2), Gavin James(3)

(1)Embry-Riddle Aeronautical University, 600 S. Clyde Morris Blvd. Daytona Beach, Florida, 32114 USA, Email: CARLSTRN@my.erau.edu
(2) Embry-Riddle Aeronautical University, 600 S. Clyde Morris Blvd. Daytona Beach, Florida, 32114 USA, Email: LONGM13@my.erau.edu
(3) Embry-Riddle Aeronautical University, 600 S. Clyde Morris Blvd. Daytona Beach, Florida, 32114 USA, Email: JAMESG5@my.erau.edu

ABSTRACT

Project PoSSUM provides human space flight education and training services focused on upper atmospheric science. The program's curriculum includes classroom lecture in atmospheric science principles, in-situ observational measurement techniques, and aerospace physiology. Aerobatic flight, hypobaric chamber testing, pressure suit operations, and space flight simulation are the primary training activities.

Commercial human space flight training regulations and standards were used to assess Project PoSSUM's efficacy and to provide recommendations for program growth. Selected candidates will be considered space flight participants under Federal law when carried within a launch vehicle. An honest and clear exchange of information pertaining to the liability of each PoSSUM candidate acting as a space flight participant is a necessity. Project PoSSUM should submit an application to the Federal Aviation Administration's Office of Commercial Space Transportation for a Safety Approval determination as a training service provider.

1. INTRODUCTION

Project PoSSUM (Polar Suborbital Science in the Upper Mesosphere), an international 501(C)(3) non-profit organization, offers the first manned suborbital space flight education and training program focused on performing upper atmospheric science. The purpose of this program is to prepare candidates for the opportunity to fly to space aboard a suborbital launch vehicle (Fig. 1). The goal of the space flight is to observe and sample high-altitude cloud structures, and study the relationship between polar mesospheric cloud formation and global climate change.

The selected launch site for PoSSUM missions is the Eielson Air Force Base, located 26 miles southeast of Fairbanks. An XCOR Aerospace Lynx Mark II is the proposed launch vehicle. The Lynx Mark II will have a single stage propulsion system, with a fixed-wing lifting-body structural design.

Figure 1: Screenshot of simulated XCOR Lynx Mark II suborbital launch vehicle (Figure courtesy of ERAU)

Embry-Riddle Aeronautical University hosts two versions of the program: the Scientist-Astronaut Qualification Program for experienced participants holding a bachelor's degree in a science, technology, engineering, or mathematics discipline; and PoSSUM Academy for high school and college students. All Project PoSSUM candidates are required to have a third-class airman medical certificate (14 CFR 67 Subpart D). [1]

The inaugural Project PoSSUM Scientist-Astronaut Qualification Program, titled Class 1501, was held at Embry-Riddle Aeronautical University in February 2015. Eight participants from around the world gathered to receive hands-on, fully immersive suborbital space flight education over a five-day period. In October 2015, thirty participants from six countries returned to Embry-Riddle Aeronautical University for Class 1502 and 1503, as well as the inaugural PoSSUM Academy. In April 2016, twelve participants from six countries returned to Embry-Riddle Aeronautical University for Class 1601, and three students participated in PoSSUM Academy.

The curricula for each program include classroom lectures in atmospheric science principles and in-situ PMC observational techniques by Dr. Jason Reimuller. Suborbital space flight simulation exercises (Fig. 2) are provided at Embry-Riddle Aeronautical University, Daytona Beach campus.

Figure 2: Photo of Project PoSSUM participant practicing payload activation sequence (Figure courtesy of ERAU)

High-G force training is provided in type certified (non-experimental) aerobatic aircraft by the Patty Wagstaff Aerobatic School at the Daytona Beach International Airport. Dr. Paul Buza of the Southern AeroMedical Institute (SAMI) in Melbourne, FL provides hypoxia awareness training and simulation using the SAMI hypobaric chamber. Pressure suit donning, doffing, and operational-use training is provided by Final Frontier Design with a functional intravehicular activity pressure suit at Embry-Riddle.

Project PoSSUM candidates have also completed additional microgravity research and testing of the Final Frontier Design pressure suit in parabolic flight (Fig. 3). Each microgravity flight included 18 to 24 parabolas in a modified Falcon 20. Each parabola has an approximate 30 to 40 second microgravity phase.

Figure 3: Photo of Project PoSSUM participant during microgravity flight (Figure courtesy of Project PoSSUM)

2. POLAR MESOSPHERIC CLOUD SCIENCE

PoSSUM candidates are trained to use the Polar Mesospheric Cloud Imaging and Tomographic Experiment payload package (Fig. 4). The payload package received funding from a NASA Flight Opportunities grant in March 2012. [2]

Figure 4: PoSSUM Cloud Imaging and Tomographic Experiment payload package (Figure courtesy of ERAU)

Polar Mesospheric Clouds (PMCs) are the highest occurring clouds in Earth's atmosphere located at an altitude of approximately 80 km (50 miles). Seasonal fluctuations effect PMC altitudes and latitudes. The PMC season is May-September in the northern hemisphere, and December-April in the southern hemisphere. PMCs were first observed in 1885, both satellite and ground observations over recent decades confirm PMC's have steadily increased in brightness, frequency and extension to lower latitudes. [3]

NASA launched the Aeronomy of Ice in the Mesosphere (AIM) satellite on 25 April, 2007. The AIM satellite monitored PMCs for six years using the Cloud Imaging and Particle Size (CIPS) instrument. The mesosphere environment is extremely cold (-125 C) and dry. PMCs consist of tiny ice particles and formation requires the presence of water molecules along with a dust or debris molecule to bind to. It is believed that debris from space drifts down into the mesosphere and combines with water vapor from upwelling summertime winds to form PMCs. Increased polar mesospheric cloud observations may be indicative of global climate change as greenhouse gases cause an increase in Earth's surface temperature, and a cooling effect in the upper atmosphere which supports ice particle formation. [4]

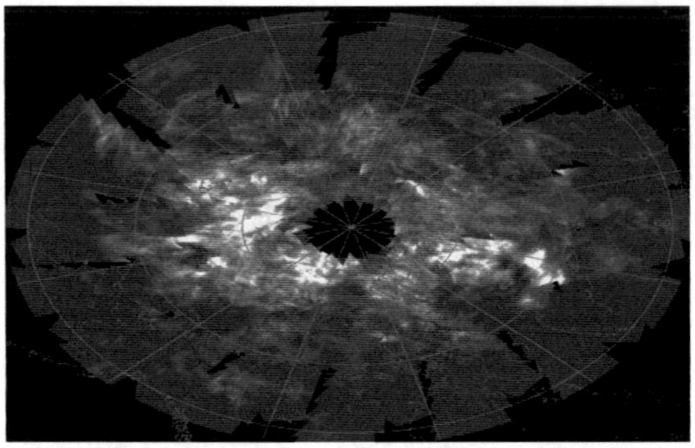

Figure 5: Image of PMCs over the North Pole taken from AIM satellite with CIPS instrument, 14 July, 2009 (Figure courtesy of NASA)

Commercial suborbital launch vehicles are ideal for the PoSSUM mission. The XCOR Lynx Mark II is projected to reach altitudes of 100 km during nominal flight. PMC formation occurs at approximately 80 km, which is too high to reach with aircraft and too low for orbiting satellites. The suborbital launch vehicle can provide access to this difficult-to-reach portion of Earth's atmosphere and enable in-situ analysis of PMCs.

3. SAFETY CONSIDERATIONS REGARDING COMMERCIAL SPACE FLIGHT TRAINING LAW AND REGULATIONS

Lawmakers recognize the importance of suborbital space flight science and mission operations. There have been cooperative efforts to support the use of suborbital launch vehicles for carrying out specific scientific investigations and educational activities as stated in Section 717 of the NASA Authorization Act of 2015. Currently, NASA is barred from proceeding with requests for proposals or contracting with launch service providers for a NASA-sponsored space flight participant until certain liability issues are addressed. [5] Commercial providers have begun responding to market demand for space flight training opportunities by offering various training services. Companies like NASTAR Center, Waypoint 2 Space, and Black Sky Training are among these providers.

The U.S. Commercial Space Launch Competiveness Act of 2015 addressed certain liability issues by extending the current U.S. risk-sharing structure through 2025. In addition, space flight participants (SFPs) were included as parties to the launch waiver of claims and given indemnification coverage by the Federal government. [6] The launch waiver of claims functions as a hold harmless agreement between the first-party (seller) and second-party (buyer) to the licensed launch operation – including contractors, subcontractors, customers, and any service providers. Each party assumes responsibility for any loss, damages, or personal injury.

SFPs still face legal and financial exposure from third-party (uninvolved public) claims. The U.S. Commercial Space Launch Competiveness Act of 2015 extended indemnification coverage to SFPs to limit the amount of financial exposure. The launch vehicle operator may elect to require a SFP to carry third-party liability coverage during the flight.

PoSSUM candidates must have a firm understanding of the risks associated with space flight. PoSSUM candidates will be required to provide written consent to the launch vehicle operator prior to making an agreement to fly.

Title 14 of the Code of Federal Regulations Part 460 – Human Space Flight Requirements Section 460.45 discusses the launch vehicle operator's duty to inform a SFP of the risks associated with space flight for the vehicle type. Currently, the launch vehicle operator (licensed under Title 14 of the Code of Federal Regulations Part 431) selected for the PoSSUM mission is legally required to satisfy the requirements found in Section 460.45.

The Federal Aviation Administration's Office of Commercial Space Transportation (FAA-AST) has authority to issue a Safety Approval determination (Title 14 of the Code of Federal Regulations Part 414 – Safety Approval) which allows for third-party personnel to assist the SFP in providing informed consent to participate as a safety critical service. [7]

Project PoSSUM should incorporate an informed consent and risk-sharing module into the program's curriculum, and submit an application to the FAA-AST requesting a Safety Approval determination. Specific safety-related training services provided in the PoSSUM curriculum would need to meet the definition of a safety element. [8] Project PoSSUM would be required to describe specific performance-based criteria in the application which demonstrates the adequacy of the safety element contained in application.

4. SAFETY CONSIDERATIONS REGARDING COMMERCIAL SPACE FLIGHT TRAINING POLICIES AND STANDARDS

The International Association for the Advancement of Space Safety (IAASS) Suborbital Safety Technical Committee (SS TC) Manual published guidelines in May 2014 for worldwide suborbital operations. [9] The SS TC Manual recommends medical and training guidelines for space flight participants who meet or exceed existing Federal regulations. PoSSUM curriculum currently follows multiple IAASS SS TC Manual recommendations. Physiological G-force training in an Extra 300 and Super Decathlon aerobatic aircraft is provided by Patty Wagstaff Aerobatic School. PoSSUM candidates were given detailed description of the flight profile. Anti-G straining maneuver (AGSM) breathing techniques were part of the pre-flight tutorial to help overcome in-flight physiological stresses including high-G force and rapid force onsets. Final Frontier Design provided anti-G pants to be worn by PoSSUM candidates for limiting blood pooling in the lower body.

Certain flight maneuvers generate forces, which may cause insufficient blood flow to the brain and induce a condition referred to as stagnant hypoxia. Hypoxia symptoms are related to arterial oxygen saturation levels. Normal arterial saturation percentage is in the high 90 percent range. Loss of peripheral vision and color vision (greyout), increased heart rate, discoloring of skin (cyanosis), and drowsiness are common symptoms which begin at a saturation level in the high 80 percent range. [10] Hypoxia awareness training is provided in a hypobaric chamber at the Southern AeroMedical Institute (SAMI) in Melbourne, Florida. PoSSUM candidates learn to recognize their unique hypoxia symptoms to mitigate consequences associated with nominal and off-nominal occurrences.

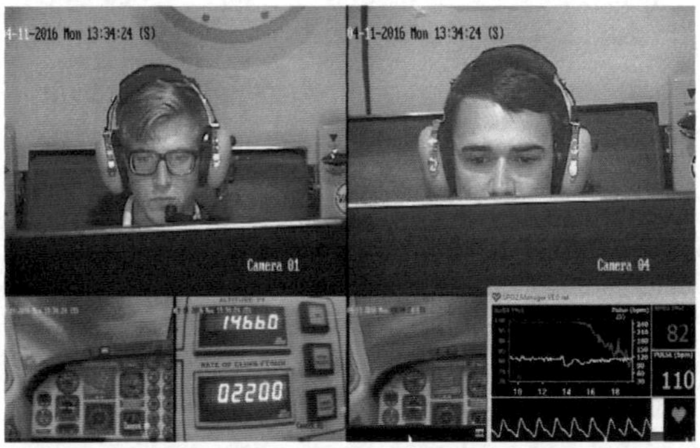

Figure 6: Screenshot photo of instructor station computer monitor during hypoxia awareness training (Figure courtesy of SAMI)

The suborbital space flight simulator at Embry-Riddle is used for ground school training; basic information regarding the space environment, flight profile, and human performance limitations are key learning objectives.

PoSSUM candidates experience comprehensive training using a functional pressure suit qualified for intravehicular activity. Participants learn donning and doffing procedures with the suit, suit pressurization, perform safety checks, and how to conduct basic operations while wearing the suit. The pressure suit provided by Final Frontier Design has an operating pressure of +5 pounds per square inch pressure differential (PSID), under which all procedures were operating. [11]

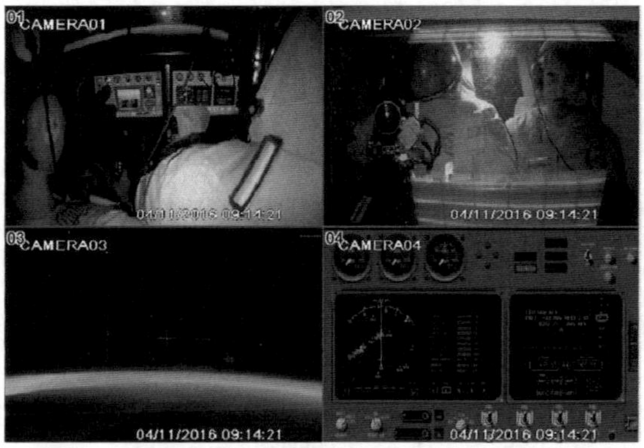

Figure 7: Screenshot photo of instructor station computer during ground school training in the space flight simulator (Figure courtesy of ERAU)

5. PRELIMINARY HAZARD ANALYSIS

A sample Preliminary Hazard Analysis (PHA) was performed for the re-entry phase of flight in a Lynx Mark II launch vehicle. The IAASS SS TC Manual recommends SFPs be provided wearable biomedical monitoring equipment during flight. Biomedical monitoring equipment was included in the PHA but a specific device has not been identified. Project PoSSUM curriculum should incorporate the use of flightworthy biomedical monitoring equipment in the program's curriculum to allow for improved situational safety awareness.

This phase of flight was selected for the PHA because the SFPs will experience a high amount of physiological stress as the vehicle transitions from a steep parabolic re-entry trajectory angle to a shallow aerodynamic glide attitude orientation. The in-flight vehicle energy management technique known as the 'pull-up' maneuver is the catalyst for the physiological stress.

In the sample PHA scenario, the PoSSUM candidate fails to respond to verbal instructions from the pilot following a successful completion of the pull-up maneuver. In addition, the biomedical monitoring equipment provides anomalous data to mission managers back on the ground. Guidelines for conducting a PHA were found in Section 3.1 Safety Criteria of the IAASS SS TC Manual, as well as NASA-RP-1358 *System Engineering "Toolbox" for Design-Oriented Engineers*. [12]

Table 1. Sample Preliminary Hazard Analysis

Brief Descriptive Title (Portion of System/Sub-system/Operational Phases covered by this analysis): Environmental Control & Life Support System/Biomedical Monitoring Equipment/Reentry Phase									
Probability Interval: 1 years	**Date:** 12 April 2016	**Hazard Target***	**Risk Before**			**Description of Counter Measures** Identify countermeasures by appropriate code letter(s): D = Design Alteration, E = Engineered Safety S = Safety Device, W = Warning Device P = Procedures/Training	**Risk After**		
System Number: ECLSS-0.410.16	**Analysis:** ✓ Initial o Revision o Addition		**Severity**	**Probability**	**Risk Code**		**Severity**	**Probability**	**Risk Code**
Hazard No. / Description									
A-410.001 — Excessive G-force duration and intensity during re-entry measurements recorded — physical injury; equipment damage; loss of life; threat to public safety.		P	I	C	3	Anti-gravity straining maneuver training for increased high positive g-load tolerance in vertical axis to prevent blood pooling in lower body (P).	III	C	5
		E	I	C	3	Require use of anti-G pants to assist in resisting blood flow to the lower body (S).	III	C	5
A-410.001a – Communication equipment failure – occupants unable to communicate verbally.		E	III	D	5	Crew resource management training in intravehicular non-verbal communication (P). Cross-training in checklist procedures for task recognition without verbal communication (P).	IV	D	6
A-410.001b – G-LOC induced myoclonic seizures – vehicle occupants may experience tremors following a loss of conscious.		P	III	D	5	Provide training in responding to and mitigating negative human-system interactions and erroneous flight control inputs (P).	IV	D	6
A-410.001c – Emesis within pressure suit or cockpit – contaminates cause task performance decrements at critical phase of flight.		P	II	C	4	Equipment for removing uncontained fluids from within the vehicle cockpit and pressure suit (S).	IV	D	7
Prepared by/Date:		*Target Codes: P = Personnel, T = Downtime, E = Equipment, R = Product, V = Environment				**Approved by/Date:**			

Tab 1. contains a PHA for the reentry phase including a description of the portion of the system, subsystem, and operation covered in the analysis; a declaration of the hazard probabilities which are based on the IAASS SS TC probability classification values; the system number; the date of analysis, the hazard description and identification number; hazard targets including personnel, equipment, downtime, product environment; risk assessment before countermeasures are considered including IAASS SS TC severity level definitions; risk priority code; description of countermeasure; risk assessment after countermeasures including updated severity levels, probability values, and risk priority code.

The probability or likelihood values for each hazard occurrence are provided in (Fig. 8). The severity/consequence level for each hazard is categorized as catastrophic, critical, moderate, or negligible.

A catastrophic event includes death or serious injury, equipment loss greater than $1M, down time greater than 4 months, or long-term (greater than 5 years) environmental effects. A critical event includes severe injury or temporary disability, equipment loss between $250K-1M, down time between 2 weeks – 4 months, or medium-term (1-5 year) environmental effects.

A moderate event includes minor injury or occupational illness, equipment loss between $1K – 250K, down time between 1 day – 2 weeks, or short-term environmental damage (less than 1 year) environmental effects. A negligible event includes some system stress, minor injury, equipment loss less than $1K, down time less than 1 day, or minor environmental effects readily repaired.

Likelihood of Occurrence	Quantitative Description	Qualitative Description
Frequent (A)	$X > 10^{-2}$	Continuously experienced
Probable (B)	$10^{-2} > X > 10^{-3}$	Will occur frequently
Occasional (C)	$10^{-3} > X > 10^{-4}$	Will occur sometime
Remote (D)	$10^{-4} > X > 10^{-5}$	May occur sometime
Extremely Remote (E)	$10^{-5} > X > 10^{-6}$	Unlikely, but reasonably expected
Extremely Improbable (F)	$X < 10^{-6}$	May not occur at all

Figure 8: Table of hazard probability classifications (Figure courtesy of IAASS)

6. SUMMARY

There are several advantages to performing a PHA for the Project PoSSUM mission. The PHA provides a structured methodology for recording and communicating safety hazards and corresponding risks, while identifying weak points in the overall mission. Ideally, preliminary hazard analysis results will be performed early in the mission development lifecycle to allow for design, engineering, and safety device mitigations rather than procedural and training changes.

Incorporating safety analysis tools into the PoSSUM curriculum helps mission managers allocate resources and prioritize training activities to control risk. These tools will also allow PoSSUM candidates to gain a better understanding of the risks involved with the mission. Sharing safety analysis results between Project PoSSUM and the launch vehicle operator allows for changes to the mission rules, procedures, emergency plans, checklists, and vehicle equipment to be included in the mission's Safety Directive. For example, the PoSSUM Flight Surgeon can verify each candidate's physiological performance during training activities and communicate those results to the launch vehicle operator prior to flight.

Analytical methods for recording and communicating safety hazards and corresponding risks, like preliminary hazard analysis, can be used as applicant-developed criteria in the Safety Approval determination application. PoSSUM candidates, as space flight participants, will benefit from the additional safety knowledge and be better prepared to give their informed consent before making a contractual agreement to fly.

7. ACKNOWLEDGEMENTS

A special thanks to Dr. Jason Reimuller, Executive Director of Project PoSSUM, for his transparency and willingness to support this study. This research would not have been possible without support from Embry-Riddle Aeronautical University faculty, staff, and students.

8. REFERENCES

1. Title 14: Aeronautics and Space, 2 Federal Aviation Administration § Chapter I, Subchapter D, Part 67, Subpart D—Third-Class Airman Medical Certificate (2016).

2. Reimuller, Jason. "Polar Mesospheric Cloud Imaging and Tomography Experiment." NASA Flight Opportunities. March 2012. Accessed April 18, 2016. https://flightopportunities.nasa.gov/technologies/46/.

3. Reimuller, Jason. "About Noctilucent Clouds." Project PoSSUM. 2016. Accessed April 18, 2016. http://projectpossum.org/research/noctilucent-clouds/.

4. Phillips, Tony. "AIM Mission Overview." National Aeronautics and Space Administration. July 30, 2015. Accessed April 18, 2016. http://www.nasa.gov/mission_pages/aim/overview/index.html.

5. National Aeronautics and Space Administration Authorization Act of 2015, H.R. H.R.810, 114th Cong., Representative Steven M. Palazzo (2015).

6. U.S. Congress. House. Science, Space, and Technology. *U.S. Commercial Space Launch Competitiveness Act, Senate Report 114-88, July 22, 2015, 114-1*. By Kevin McCarthy. 114th Cong. H. Doc. Deliberations on H.R.2262. 2015.

7. Title 14: Aeronautics and Space, 4 Federal Aviation Administration § Chapter III, Subchapter C, Part 460—Human Space Flight Requirements (2016).

8. Title 14: Aeronautics and Space, 4 Federal Aviation Administration § Chapter III, Subchapter C, Part 414, Subpart A—General (2016).

9. Rongier, Isabelle et al. *International Association for the Advancement of Space Safety*. Publication. May 2014. Accessed April 18, 2016. https://www.faa.gov/about/office_org/headquarters_offices/ast/advisory_committee/meeting_news/media/2014/may/15_IAASSSuborbitalSafetyGuidelinesManual_Dec2013_Master.pdf.

10. "Hypoxia and Oxygenation." *Alaska Air Medical Escort Training Manual*, 4th (February 2006): 71-82. Accessed April 18, 2016. http://dhss.alaska.gov/dph/Emergency/Documents/ems/assets/AirMedCourse/EMS-F_Chapter4.pdf.

11. "Space Suits for IVA." Final Frontier Design. 2015. Accessed April 18, 2016. http://www.finalfrontierdesign.com/space-suits-for-iva/.

12. United States. National Aeronautics and Space Administration. *System Engineering "Toolbox" for Design-Oriented Engineers*. By Benjamin E. Goldberg. MSFC, Ala.: National Aeronautics and Space Administration, Marshall Space Flight Center, 1994.

THE RIGHT STUFF 'v' THE RIGHT (SAFE) THING

Dr Andy Quinn MSc PhD MRAeS CEng (1), Dr Ivan Sikora MSc PhD MRAeS FHEA (2)

(1) Saturn Safety Management Systems Ltd, Bath, BA2 9BR, UK
(2) City University, London, EC1V 0HB, UK

ABSTRACT

This is not the 1950's where test pilots needed the 'right stuff' and certainly not the beginning of aviation where the Wright brother's early designs needed pilots with more than the right stuff. In those formative years of aviation and jet development, designers and pilots did not have the same design understanding and knowledge that we have today. In addition, they did not have the same understanding and knowledge of Systems Safety engineering and Human Factors expertise that we have today. Manned suborbital flights of today should be undertaken in vehicles that have been designed effectively with appropriately derived safety requirements including fault-tolerance, safe life and design-for-minimum risk approaches – and all to an acceptable level of safety.
Therefore, although initial suborbital pilots will originate from flight test schools and still possess similar traits to their earlier test pilot brethren, they should be protected by the *right (safe) thing* by design and analysis rather than rely on the *right stuff* due to ineffective design and operating procedures. The paper presents a review of the SpaceShip2 accident as a case study to highlight the right (safe) things that should be considered in the design, analysis and operations for suborbital operators. The authors of this paper contend that suborbital piloted vehicles should be designed with the knowledge and understanding and lessons learned from those early X-plane flights, lessons learned from general space safety, lessons learned from pilot Human Factors/ Crew Resource Management training and by understanding that safety management and safety engineering are essential disciplines that should be integrated with the design team from the concept phase.

1. INTRODUCTION

During the early days of airliners, military fast jet and rocket development, aircraft were plagued by technical issues that resulted in incidents and accidents. In the case of US fast jet and rocket development on the X-15 project, this meant relying on pilots with the 'right stuff'. Consequently, throughout the project, this meant reactive fixes to the reliability and design issues i.e. 'fly, fix, fly'. As technology and complexity improved over time, causal factors identified in incidents and accidents are more plagued by people and organisations. To counter this, technical designers are now being supported by other engineering disciplines including human factors and systems safety engineering:

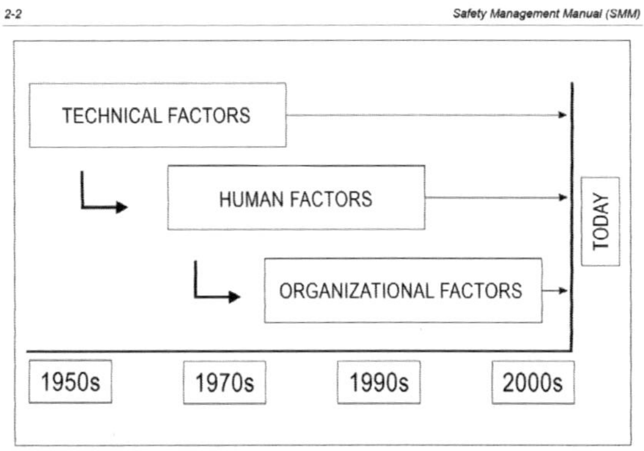

Figure 1: Evolution of Safety Thinking [1]

2. REVIEW OF ACCIDENT CREW CASUAL FACTORS

Based on National Transportation Safety Board (NTSB) sources of significant crew causal factors in 93 major hull losses from 1977-1984, Lautman and Gallimore [2] found that pilot deviation and inadequate cross-check were the main causal factors:

- **33% - Pilot deviated from SOP's**
- **26% - Inadequate cross check by second crew member**
- 9% - Crew's not trained for correct response in emergency situations
- 6% - Pilot did not recognise need for go-around
- 4% - Pilot Incapacitation
- 4% - Inadequate piloting skills
- 3% - Improper procedure during go-around
- 3% - Crew errors during training flight
- 3% - Pilot not conditioned to respond promptly to Ground Proximity Warning
- 3% - Inexperienced on ac type

The findings are backed up in a more recent Boeing study as presented by NTSB senior expert, Sumwalt [3]:

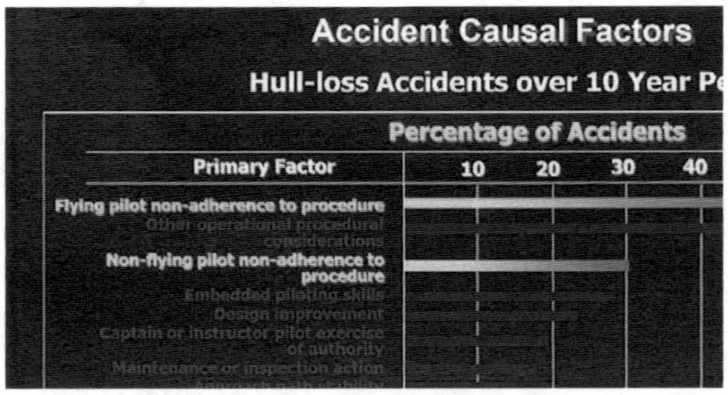

Figure 2: Accident Causal Factors – Extract from Sumwalt Presentation

3. REVIEW OF SPACESHIP 2 ACCIDENT

During Powered Flight number four (PF04) SpaceShip2 (SS2) suffered an in-flight break- up shortly after rocket motor ignition. This was due to uncommanded deployment of the feathering device at Mach 0.8 and under the aerodynamic loads tore the vehicle apart. The NTSB conclusions [4] had no doubt that the initiating (causal) factor was the co-pilot, as non-handling pilot, incorrectly unlocking the feathering device by operating the arming lever early (at Mach 0.8 instead of Mach 1.4). The NTSB concludes that the co-pilot was experiencing **high workload** as a result of recalling tasks from memory while performing under **time pressure** and with **vibration and loads** that he had not recently experienced, which increased the opportunity for errors.

The NTSB notes that Scaled did not recognize and mitigate the possibility that a test pilot could unlock the feather early.

The accident co-pilot was the co-pilot of SS2 on **seven glide flights** occurring between October 10, 2010, and August 28, 2014. He had been the co-pilot of SS2 for PF01 (**one powered flight**) on April 2013.

The co-pilot, age 39, held an airline transport pilot (ATP) certificate with a rating for airplane multi-engine land and commercial pilot privileges for airplane single-engine land and sea and glider. The co-pilot held a second-class medical certificate, dated May 22, 2014, with the limitation that he must wear corrective lenses. The cockpit image recording showed that the co-pilot was wearing glasses during the accident flight.

The NTSB concluded that the pilot and co- pilot were properly certificated and qualified.

3.1 Analysis versus NTSB known (aviation) crew casual factors

The NTSB research into crew casual factors in aircraft accidents (para. 2 above) found that the top two factors were (i) *pilot deviated from SOPs and (ii) Inadequate cross check by second crew member*. These causal factors appear to be the same for the SS2 accident.

The Scaled Composites' SS2 co-pilot came from the aviation domain, and he would have been aware of the need for effective Crew Resource Management (CRM). CRM is defined [5] as;

'a set of training procedures for use in environments ***where human error can have devastating effects****. Used primarily for improving air safety, CRM focuses on interpersonal communication, leadership, and decision making in the cockpit'*

As the definition implies, CRM is not related to the technical knowledge and flying skills of pilots but rather with the interpersonal skills and cognitive skills i.e. the ability to communicate, maintaining situational awareness and for solving problems and making effective decisions.

Designs should not rely on pilots having the 'right stuff' all of the time. Indeed, the Federal Aviation Administration Office of Commercial Space Transportation (FAA-AST) §460.15 [6] detail the following human factors requirements:

An operator must take the precautions necessary to account for human factors that can affect a crew's ability to performsafety-critical roles, including in the following safety critical areas—

- *(a) Design and layout of displays and controls;*
- *(b) Mission planning, which includes analysing tasks and allocating functions between humans and equipment;*
- *(c) Restraint or stowage of all individuals and objects in a vehicle; and*
- *(d) Vehicle operation, so that the vehicle will be operated in a manner that flight crew can withstand any physical stress factors, such as acceleration, vibration, and noise.*

Note item (d) above in relation to the NTSB findings detailed in bold further above in para. 3.

Additionally, in terms of being able to withstand any physical stress factors, FAA- AST launch licensing requirements detail the crew requirements, including pilot experience and medical requirements [6].

However, the IAASS [7] and Aerospace Medical Association (AsMA) [8] recommend that suborbital spaceplane pilots have a first- class medical certificate (not a second-class, as recommended by the FAA-AST) and be selected from military fast jet programs/astronauts i.e. pilots whom have experienced high g-forces and can cope with the environment; thereafter when mature suborbital operations are established, this could be relaxed to allow other (airline) pilots to be employed.

4. SAFETY BY DESIGN

Suborbital spaceplanes (or Reusable Launch Vehicles using the US terminology) are designed and operated using a mix of Aviation & Space attributes. So what design standards should be followed? A mixture of both or are forerunners designing on the edge of innovation and designing their way using their own practices, whilst being cognisant of eventual FAA-AST Launch Licensing requirements? This is stated as a question because the FAA-AST are NOT certifying these vehicles.

Space safety standards dictate a Design for Minimum Risk (DMR) philosophy. This includes deriving Fault Tolerance, Safe-Life and Fail Safe criteria (also for aviation). The FAA-AST DMR philosophy within Advisory Circular AC 437.55-1 [9] details the following Safety Precedence Sequence:

- Eliminate hazards (by design or operation)
- Incorporate safety devices
- Provide warning devices
- Develop and implement procedures and training.

Notice that in space the key term is 'hazard'. Hazards are analysed and then, as part of quantitative analysis (see section 5) a cumulative assessment is made in order to determine whether the Target Level of Safety has been met i.e. top-down analysis. In aviation, although the term hazard is recognised, the focus is on lower level failure conditions associated with failure modes i.e. in order to meet safety objectives such as $1x10^{-9}$ per flying hour for catastrophic events i.e. bottom-up analysis per AC1309 [10].

As part of the FAA-AST requirements for a launch license permit (CFR §437.55) the safety design analysis must consider the following:

(1) Identify and describe hazards, including but not limited to each of those that result from—

(i) Component, subsystem, or system failures or faults;
(ii) Software errors;
(iii) Environmental conditions;
(iv) ***Human errors****;*
(v) Design inadequacies; or
(vi) Procedural deficiencies.

The NTSB report [4] found that the designers (Scaled Composites) had not undertaken human error analysis (as the cause) in relation to decision errors and skill-based errors. The FAA-AST inspectors noted this and for PF04 provided a waiver against §437.55 (for not completing human error or software fault as a causal factor). We all want the nascent industry to succeed and we want innovative designs (we don't want to 'stifle' the industry) BUT we want safety driving success – not waivers for incomplete analysis.

4.1 Key safety requirements

By understanding space and aviation requirements the authorities and/or designer could identify key safety requirements that should be achieved in the design. For instance, a space requirement is that any Inadvertent Failure Modes that could result in a catastrophic outcome should have **3 INHIBITS**. This means 3 separate and independent inhibits which could be hardware (physical switches/guards etc.), software (latches) or combination thereof. The IAASS Space Safety Manual [11] is based on the NASA and (European) ECSS standards and rationalised/consolidated into one manual; in relation to 'Functions Resulting in Catastrophic Hazards' the following requirement is stated:

A system function whose inadvertent operation could result in a catastrophic hazard shall be controlled by a minimum of three independent inhibits, whenever the hazard potential exists. One of these inhibits shall preclude operation by a radio frequency (RF) command or the RF link shall be encrypted. In addition, the ground return for the function circuit must be interrupted by one of the independent inhibits. At least two of the three required inhibits shall be monitored.

Unfortunately, the Commercial Space Transportation Advisory Committee (COMSTAC), whom advise the FAA-AST, gave the IAASS Suborbital Safety Guidance Manual the 'Cold-Shoulder' [12]; COMSTAC comprise US space players including XCOR and Virgin Galactic.

Since the SS2 accident and following NTSB findings, Virgin Galactic have undertaken additional systems safety analysis and have implemented modifications to SS2 [13] including:

1. DESIGN: Feather locking pin, controlled by the vehicle flight computer, which prohibits pilots from unlocking the tail section early
   a. Pilots will have a mechanical override if the locking pin fails
2. PROCEDURAL: VG now deciding to keep the feather locked until after the rocket engine has shut down
   a. Pilots will have three- to five minutes to trouble shoot in case of problems before the feather would be needed for re-entry

5. SAFETY TOOLS & TECHNIQUES

There are a number of standard system safety analysis techniques such as Fault Tree Analysis (FTA), Event Tree Analysis (ETA), Failure Modes Effects & Criticality Analysis (FMECA) to name a few [14].

The FAA-AST hazard analysis AC [9] also refers to the System Safety Process AC 431.35-2A [1515], which includes an exemplar Safety Programme Plan (SPP); this AC dictates these standard techniques including the FMECA. This analysis would have identified an inadvertent failure mode, resulting in catastrophic outcome, and hence demanded further mitigation. The FAA-AST provide the following 3-pronged approach *(Figure 3)* and additional detailed methodology *(Figure 4)* within AC 431.35- 2A:

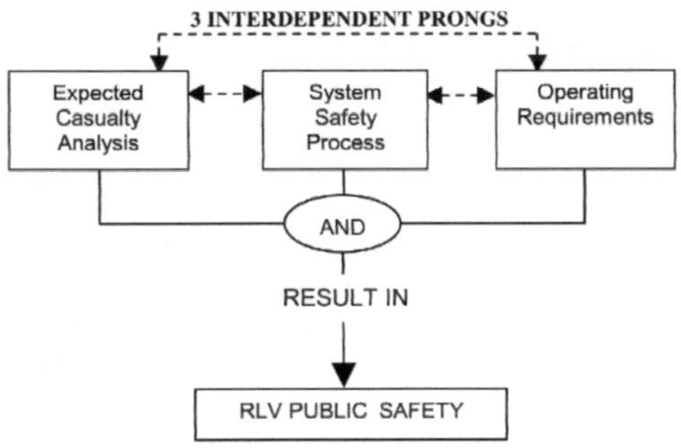

Figure 3: FAA-AST 3-Pronged Safety Approach

The 3-pronged approach includes:

- Expected Casualty Analysis; this relates to analysing the risks to the public in the event of a vehicle break-up, explosion or malfunction turn for instance
- Operating Requirements; this relates to deriving operating procedures and limitations to minimise risk to the public
- System Safety Analysis; this provides detailed guidance for vehicle design organisations in order to derive safety requirements that need to be considered as part of the design (as detailed in the following figure)

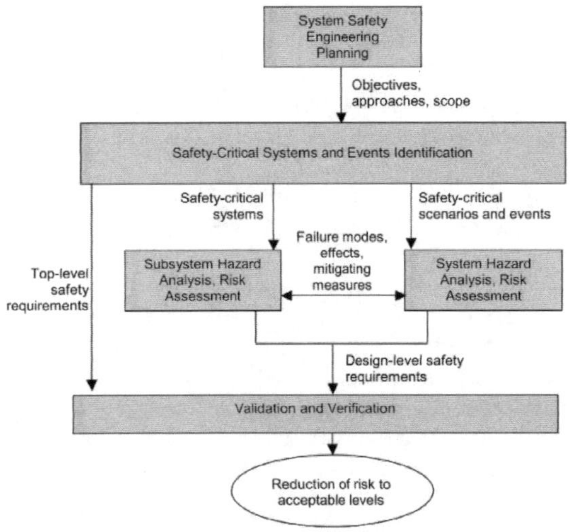

Figure 4: FAA-AST System Safety Engineering Guide

Note in the middle of the figure above, the requirement to undertake a FMECA.

Another safety technique is the Operating and Support Hazard Analysis (OSHA) [16] - here the aim is to analyse the pilot (and maintainer) procedures and the safety engineers then review these to determine whether any hazards are affected by the procedure and also to identify where the human can skip steps or do the steps incorrectly. The FAA OSHA guidelines state that:

This is performed by the Contractor primarily to identify and evaluate hazards associated with the interactions ***between humans and equipment/systems***

Back to the SS2 accident, it was clear that the co-pilot did the procedure step too early as indicated by the red circle below [17] *(Figure 5)*:

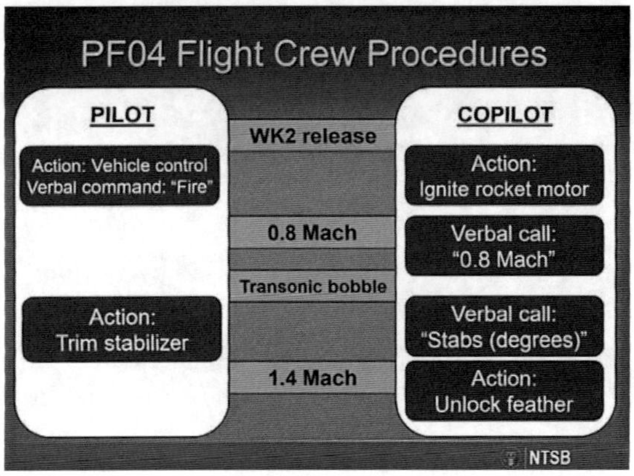

Figure 5: SS2 PF04 Flight Crew Procedures

Note the 'verbal call' at 0.8 Mach that the co- pilot had to make whilst in this busy stressful phase, as well as the trim stabilizer call, followed by the actual unlocking of the feather. The NTSB report notes that the co- pilot had memorised these tasks (from cockpit video footage) rather than read these critical steps from the procedures card (which was on his knee-pad) – which arguably could then have been 'checked' by the pilot.

6. SYSTEM SAFETY & HUMAN FACTORS ENGINEERING DISCIPLINES

This paper has focused on the system safety analysis that is performed (normally) in accordance with standard safety tools and techniques that would address human-machine interactions (both as causes and procedural controls). By undertaking diverse safety analysis, appropriate coverage is provided in order to identify all hazards and provide derived safety requirements (see figure 4) that will then be verified and validated during the development.

Human Factors experts also provide valuable analysis as part of a programme. Here the focus is on human factors integration (or human factors and ergonomics [HFE]) and this includes analysing the cockpit displays for instance i.e. anthropometry, cognitive psychology, display layouts including colours for cautions and warnings etc. and analysing test procedures.

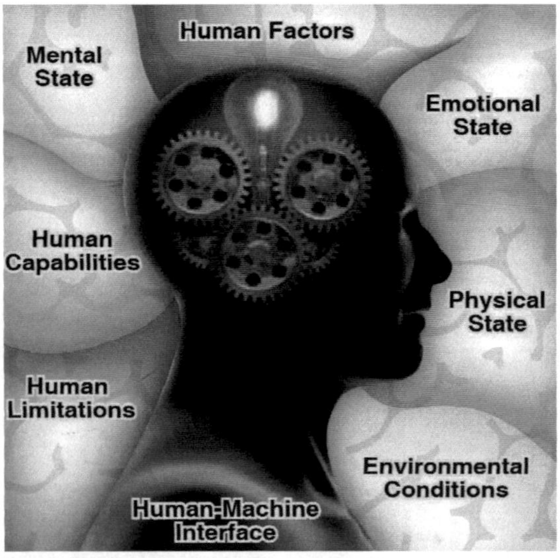

Figure 6: Human Factors that affect performance [18]

In relation to the SS2 PF04 the NTSB Human Performance Presentation noted the following Stressors contributing to the accident:

- *Memorization of tasks - Flight test data card not referenced*
- *Time pressure - Complete tasks within 26 seconds - Abort at 1.8 Mach if feather not unlocked*
- *Operational environment - No recent experience with SS2 vibration and load*

The Stressors noted above relate to 'human capabilities' in *Figure 6* in that the co-pilot memorized his tasks which needed to be completed by Mach 1.8 and arguably his 'mental state' may have been affected by the 'operational environmental' in the rocket phase of flight (during the transonic bubble) with vibration and noise.

The NTSB report [19] provides further details including details of the cockpit layout and instrumentation i.e. human-machine interface (or ergonomics) aspects. The NTSB found no major contributory factors.

On projects, a problem can exist when the separate disciplines do not work together effectively i.e. design engineers, safety engineers, human factors engineers, software engineers etc. Previous IAASS conferences have had presentations from NASA in their goal for continuous improvement stating that design and safety engineers are now working more closely; and then in a separate presentation stating that designers and HF engineers are working better; well how about all disciplines working together? And how about using safety and human factors specialists with appropriate qualifications, experience and training (from the beginning)?

7. FOCUSING ON THE RIGHT (SAFE) THING

The SS2 accident has highlighted that although initial suborbital pilots *should* originate from flight test schools and still possess similar traits to their earlier (1950s) test pilot brethren, they should be protected by the *right (safe) thing* by design and analysis rather than rely on the *right stuff* due to ineffective design and operating procedures. This paragraph details some aspects of the *right (safe) thing*:

Organisational factors. The co-pilot of SS2 PF04 did not mean to make a critical mistake and arguably contributory factors lay at the organisation level. This relates to the pilot procedures (which should be a safety net to prevent errors) and training; here the procedures were practiced in the simulator, however the simulator cannot realistically replicate the transonic phase with the vibration and noise etc. Professor Nancy Leveson's Systems-Theoretic Accident Model and Processes (STAMP) [20] takes a holistic organisational approach and, *instead of* *defining safety management in terms of preventing component failure events, it is defined as a continuous control task to impose the constraints (control actions) necessary to limit systems behaviour to safe changes and adaptations.* So here we can learn that particular focus should be spent on analysing the procedures (controls) more effectively to prevent errors – if this means adding design steps to prevent errors, all the better. The simple answer is to always have a check- response (feedback loop, per STAMP) for procedural steps during critical stages of the flight (or indeed design the system such that minimal pilot actions/verbal calls are required).

Human Factors. The suborbital flight profile involves rocket-powered flight at Mach 3, with g-forces, vibration, noise and mix of atmospheric and space loads. This is extreme interaction of man, the machine and the media (environment) is depicted well within the military-based Operational Risk Management as detailed in the FAA System Safety Handbook [21]:

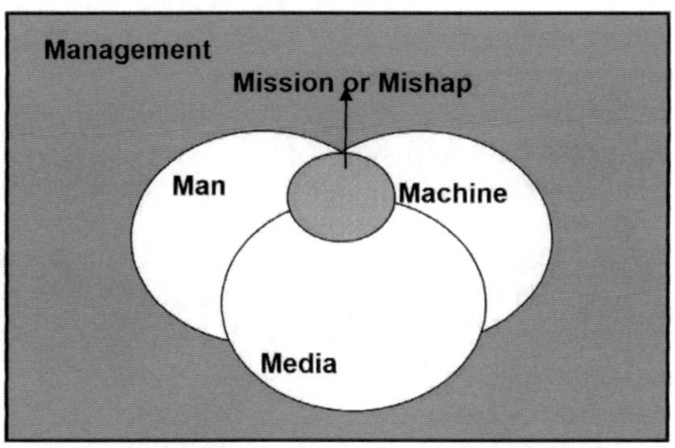

Figure 7: Human Factors 5-M model

The management can have a major influence on whether a suborbital flight results in a successful mission or a mishap (accident). It is the management (organisation) who make the decisions to fly at the Flight Readiness Reviews and who decide on the skillset of the pilots, the safety engineers and the human factors engineers.
A HF specialist should be employed to analyse the HFE aspects, particularly for suborbital flight. This can be based on existing military and civil knowledge base but adapted for the unique suborbital profile.

Safety by Design. It is of no use to have a 'safety officer' employed at some point in the programme to do some 'hazard analysis' to satisfy the FAA-AST requirements. System Safety Engineers should be employed from the concept phase so that they can follow best practice by undertaking detailed and diverse safety analysis. By doing this, derived safety requirements will then be able to influence the design i.e. these will then require design decisions as to acceptance or rejection (with rationale). Such requirements would include levels of fault tolerance, fail-safe and safe life design, and design and development assurance levels. Section 4 further above details the (space) safety precedence sequence using the Design for Minimum Risk approach. The Aerospace Recommended Practices detail the 'typical' development lifecycle from Concept phase to Design Validation & Verification (including testing); this then also is carried through to operations. It is imperative to have a SQEP safety engineer from the start who understands the diverse safety techniques and tools.

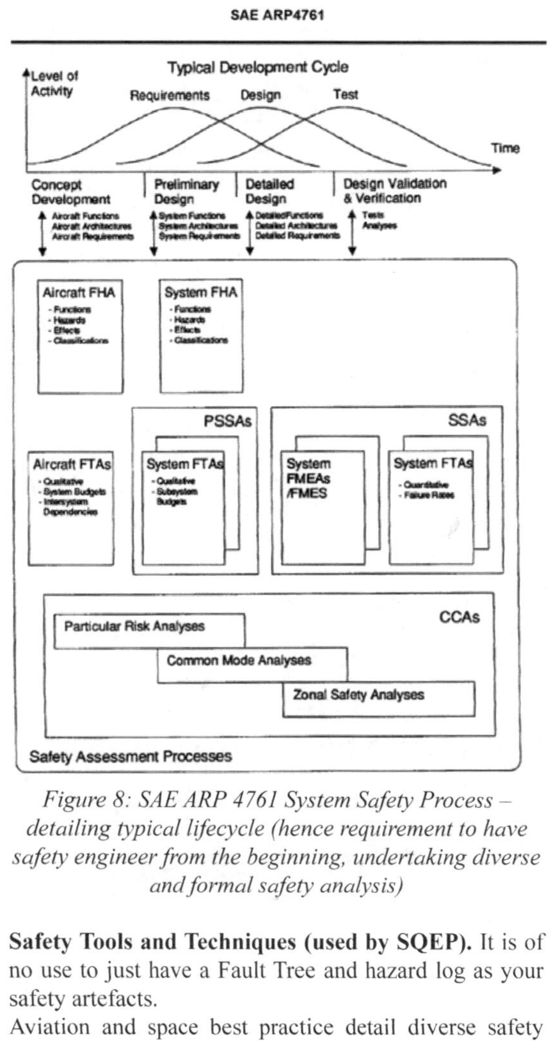

Figure 8: SAE ARP 4761 System Safety Process – detailing typical lifecycle (hence requirement to have safety engineer from the beginning, undertaking diverse and formal safety analysis)

Safety Tools and Techniques (used by SQEP). It is of no use to just have a Fault Tree and hazard log as your safety artefacts.
Aviation and space best practice detail diverse safety techniques (and tools) in order to identify hazards (as well as functional failure modes). Indeed, the FAA-AST have reasonable guidance, as detailed in section 5 above. The FAA-AST specifically state that human error analysis (as a cause) should be carried out.
You only know about diverse safety techniques if you have learned about them, have been trained to use them

and have used them in appropriate context. Also that your work has been independently checked by experts and/or authorities i.e. you have been involved in programmes using airworthiness/spaceworthiness standards (meaning that you know about standards and regulations).
So it is vitally important that suborbital operators/designers employ SQEP safety engineers.

8. CONCLUSIONS

This paper reviewed the SS2 accident in order to highlight some of the issues associated with suborbital flights. These flights involve a rocket-power, phase, g-forces and extreme environments and so pilots must be provided with the design and the tools (procedures) to cope with the exacting profile. The pilots must also be suitably fit (medically) and be provided with the 'realistic' training. Why? – so that they **do not** have to rely on the 'right stuff' i.e. flying on the seat of their pants and dealing with problems on their own. By undertaking effective systems safety analysis (per best practice/guidance) then the vehicle will be designed with consideration for safety (from derived safety requirements). The safety (and human factors) analysis will have included human error **as a cause** (as well as considering the human as a control i.e. pilot recovery to a malfunction). The safety analysis would also have covered inadvertent operation/function of a system (by fault or human incorrect selection) and, with a catastrophic outcome, would have derived that 3 Inhibits are required.

So suborbital operators/designers should ensure that the pilots are protected by the *right (safe) thing* by design and analysis rather than rely on the *right stuff.*

9. REFERENCES

1. Safety Management Manual (SMM), (Doc 9859). International Civil Aviation Organisation, Second Edition, Montreal, 2009

2. Lautman L and Gallimore P L, (1987), Control of the crew-caused accident: Results of a 12-operator survey, Boeing Airliner, April- June 1-6

3. Sumwalt R L, (20/9/2012), Obtaining Better Compliance with Standard Operating Procedures (SOPs), NTSB Presentation

4. NTSB/AAR-15/02 PB2015-10545, Aerospace Accident Report, In-Flight Breakup During Test Flight Scaled Composites SpaceShipTwo, N339SS Near Koehn Dry Lake, California October 31, 2014

5. https://en.wikipedia.org/wiki/Crew_resource_management

6. FAA Code of Federal Regulations, Title 14, Chapter III, Subchapter C, Part 460, Human Spaceflight Requirements

7. IAASS, Suborbital Safety Working Group, Safety Design & Operation of Suborbital Vehicles – Guidance, Issue 2. October 2015

8. Aerospace Medical Association, Commercial Spaceflight Working Group, Position Paper, Aviation, Space, and Environmental Medicine x Vol. 82, No. 4, April 2011

9. FAA-AST, Advisory Circular 437.55-1, Hazard Analyses for the Launch or Re-entry of a Reusable Suborbital Rocket Under an Experimental Permit, April 2007

10. FAA, Advisory Circular 25.1309-1A - System Design and Analysis, 1988

11. IAASS-ISSB-S-1700-Rev-B, Space Safety Standard, Commercial Human-Rated System, Requirement 201.2.2

12. http://spacenews.com/40615international-suborbital-safety-proposal-gets-cold-shoulder-in-us by Jeff Foust — May 19, 2014

13. http://news.discovery.com/space/private-spaceflight/new-virgin-galactic-spaceshiptwo-to-debut-160219.htm

14. ARP 4761 - Guidelines & Methods for Conducting SSA", Available from: www.sae.org/technical/standards/ARP4761

15. FAA-AST, Advisory Circular AC 431.35- 2A, Reusable Launch and Re-entry Vehicle System Safety Process, July 2005

16. FAA, FAA-DI-SAFT-105, Operating & Support Hazard Analysis

17. NTSB, Human and Organisational Issues, Human Performance Presentation, Supporting [4]

18. FAA, Aviation Maintenance Technician – Handbook, Chapter 14, Figure 14-1

19. NTSB, Human Factors and Organizational Issues, Human Performance Presentation

20. Leveson, N: A New Accident Model for Engineering Safer Systems, Massachusetts Institute of Technology paper

21. FAA System Safety Handbook, Chapter 15: Operational Risk Management December 30, 2000

10. ACRONYMS/ABBREVIATIONS

Abbreviation	Meaning
AAR	Air Accident Report
AC	Advisory Circular
AsMA	Aerospace Medical Association
ATP	Airline Test Pilot
CRM	Crew Resource Management
DMR	Design for Minimum Risk
FAA-AST	Federal Aviation Administration – Office for Commercial Space Transportation
FMECA	Failure Modes Effects & Criticality Analysis
FTA	Fault Tree Analysis
HF	Human Factors
HFE	Human Factors and Ergonomics
NASA	National Aerospace & Space Administration
NTSB	National Transportation Safety Board
OSHA	Operating & Support Hazard Analysis
PRA	Probabilistic Risk Analysis
SMM	Safety Management Manual
SMS	Safety Management System
SOP	Standard Operating Procedure
SPP	Safety Program Plan
SQEP	Suitably Qualified Experienced Personnel
SS2	SpaceShip2
STAMP	Systems-Theoretic Accident Model and Processes
VG	Virgin Galactic

PROCEEDINGS OF THE 8TH INTERNATIONAL SPACE SAFETY CONFERENCE

SESSION 05
SPACE DEBRIS - I

A SYSTEMATIC STUDY OF LASER ABLATION FOR SPACE DEBRIS MITIGATION

R. Battiston[1], W.J. Burger*[2], A. Cafagna[3], C. Manea[4], and B. Spataro[5]

[1] *University of Trento and TIFPA, via Sommarive 14, 38123 Trento Italy*
[2] *FBK, Via Sommarive 18, and TIFPA, via Sommarive 14, 38123 Trento, Italy*
[3] *University of Trento and TIFPA, via Sommarive 14, 38123 Trento Italy*
[4] *TIFPA, via Sommarive 14, 38123 Trento, Italy*
[5] *National Laboratory of Frascati, Via E. Fermi 40, 00044 Frascati, Italy*

ABSTRACT

The increase in the number of artificial objects in orbit around the Earth represents a serious threat to the future utilization of space. The scope and nature of the problem require an international effort. The Trento Institute of Fundamental Physics and Applications (TIFPA) participates in a study of laser ablation for space applications, propulsion and space debris mitigation, in the context of the New Reflections program initiated by the Italian Institute of Nuclear Physics (INFN). An evaluation of the performance of laser ablation for debris removal in Low Earth Orbit (LEO), for different scenarios of ground and space-based lasers is presented. The ultimate aim is to provide a performance evaluation of the technology over the wide range of debris parameters, for both space-based and ground systems.

Key words: laser ablation, space debris mitigation, simulation analysis.

1. INTRODUCTION

The likelihood that the increase of the number of objects launched in space results in an artificial debris belt around the Earth was pointed out by D.J. Kessler and B.G. Cour-Palais in 1978 [1]. The authors developed a model to predict the evolution of the debris population created by satellite collisions based on the available inventory of artificial objects in orbit around the Earth (size, speed, or-bit). The process of mutual collisions is similar to the mechanism responsible for the creation of asteroids from larger planet-like bodies. While the time scale of the latter is of the order of billions of years, the much smaller volume occupied by the satellites results in a significantly shorter time scale.

The *Kessler Syndrome*, the cascade in the number of collisions rendering the exploitation of space unfeasible for future generations, has led to a large consensus among national space agencies on the importance of the problem [2].

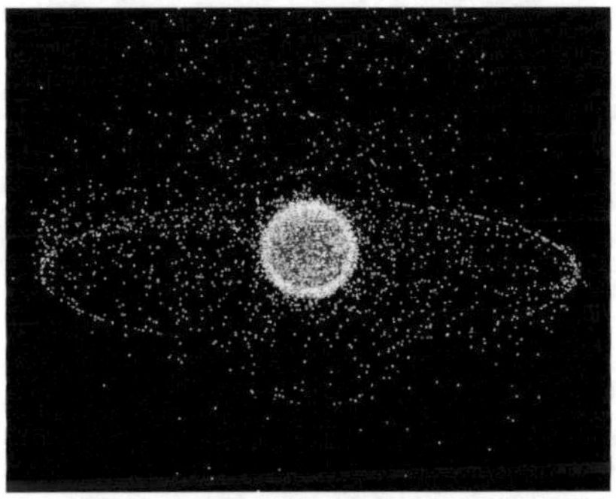

Figure 1. The principal space debris populations include a ring of objects in GEO and a cloud of objects in LEO [3].

The number of tracked objects, which exceeds 19,000. form two distinct populations: a ring in geostationary orbit (GEO) and a cloud in LEO (Fig.1). The information is used by NASA to defined collision avoidance processes for all human space flight missions and for maneuverable robotic satellites in LEO, and within 200 km of GEO [4].

The satellites and debris in orbit around the Earth are sub-ject to the frictional force of the atmosphere and the pres-sure due to solar activity. A priori the objects in LEO will eventually deorbit due to these natural effects. One of ap-proaches in active debris removal is to search for technological solutions which would shorten significantly the orbital lifetime of the debris.

2. ORBITAL MECHANICS

The impulse, applied in the direction opposite (parallel) to the orbital velocity, required to lower (raise) a ob-ject in a circular orbit is given by the Hohmann transfer. The momentum change is defined by the difference be-tween the original orbital velocity, and the velocity at the perigee (apogee) of the elliptical transfer orbit tangent to the circle of the lower (higher) altitude orbit (Appendix, Fig. 23). The expression for the momentum change required to displace a mass m in an circular orbit of radius R_1 to an orbit radius $R_2 < R_1$ is

$$m\Delta v = MG \cdot \left(\sqrt{\frac{1}{R_1}} - \sqrt{2\frac{R_2/R_1}{R_1 + R_2}} \right) \tag{1}$$

where M is the mass of the Earth and G is the gravitational constant.

The impulse required to lower the altitude of the satellite R_1 to R_2 equal to the Earth's radius[1] R_E , and R_E+150 km are shown in Fig. 2. The latter represents an altitude where the effect of atmospheric drag will deorbit the mass. The ratio of the impulse with and without atmospheric drag, i.e. the ratio of the impulses required to lower the mass to an altitude of 150 km and to the Earth's surface, are shown in Fig3.

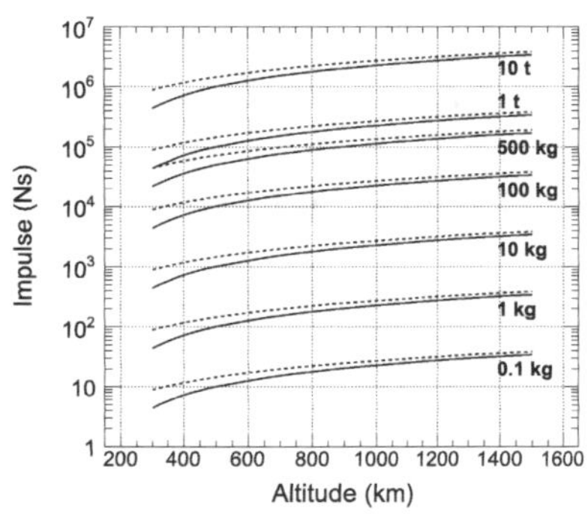

Figure 2. The impulse required to lower a given mass to a final orbital altitude of 150 km (solid lines) and to the Earth's surface (dashed lines).

Effect of the Impulse Direction

The Geant4 [5] application PLANETOCOSMICS was used to quantify the effect of the direction of the applied impulse. PLANETOCOSMICS performs a detailed simulation of the propagation and interaction of elementary particles in the Earth's magnetic field and atmosphere. The Earth's gravitational field was added to the program. Figures 4 and 5 show the orbits of a 500 kg point mass which is placed at an altitude of 600 km, with the velocities of 10.0 and 7.6 km/s.

The change of the perigee altitude produced by an impulse applied in the direction opposite to the orbital velocity, and in the directions along and opposite to the vector drawn from the Earth's center to the debris position at

[1] Mean equatorial radius 6378 km

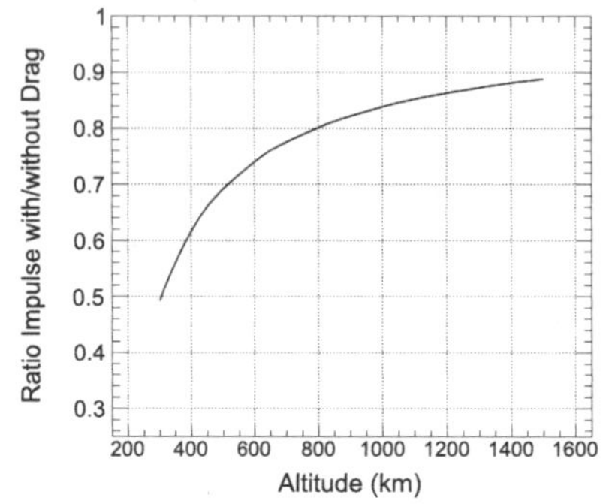

Figure 3. The ratio of the impulse required to deorbit the debris mass with and without the presence of the atmosphere.

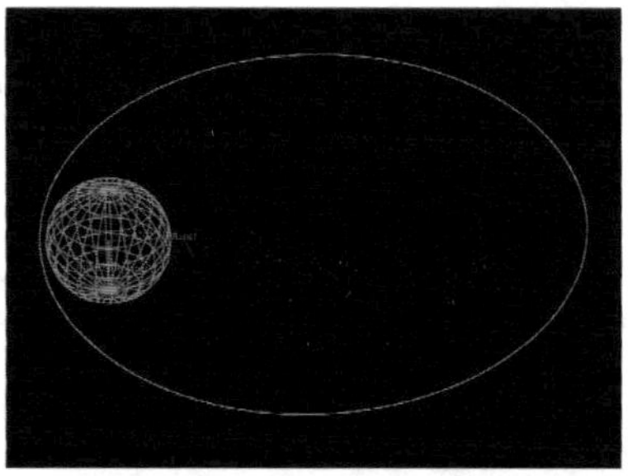

Figure 4. The elliptical orbit in the Geant4 simulation of the 500 kg mass with a velocity of 10.0 km/s. The trajectory corresponds to 13 h, the period of an elliptical orbit with a semi-major axis equal to ~ 4 Earth radii.

Figure 5. The circular orbit at an altitude of 600 km of a 500 kg mass with a velocity of 7.6 km/s. The trajectory corresponds 97 min, the period of the circular orbit.

the 600 km altitude in Fig. 5, are shown in Fig. 6. The simulated time is 97 min, corresponding to the period of the initial circular orbit of the 500 kg mass. The impulse is applied 16.7 min after the start of the simulation. The effect of atmospheric drag is absent.

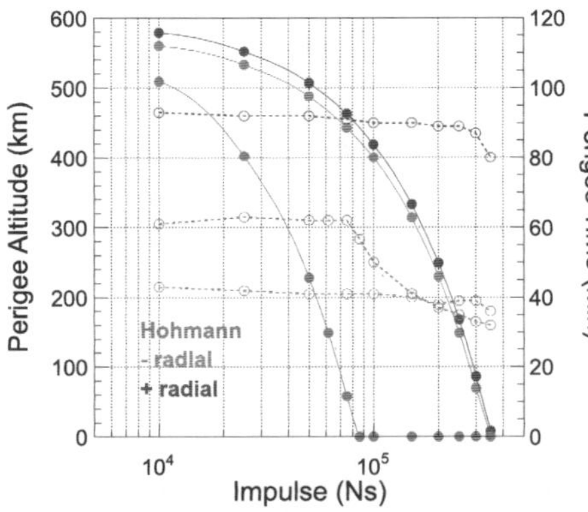

Figure 6. The change in the perigee altitude (•) for three different impulse directions, opposite to the orbital velocity of the 500 kg mass (Hohmann), towards the center of the Earth (negative radial) and the local zenith (positive radial), are indicated on the vertical axis on the left, the time required to attain the perigee (◦) on the vertical axis on the right.

The radially directed impulses result in an elliptical orbit with a perigee altitude lower than the initial 600 km altitude. A factor four larger radial impulse is required to produce a reduction in the perigee altitude comparable to the Hohmann transfer. The altitude reduction produced by the impulse in the negative radial direction is 5-10%larger than for the positive radial direction.

The perigee time reported in Fig. 6 reflects the relative positions of the perigee and apogee of the elliptical orbits produced by the different impulse directions. The perigee altitude is attained ~ 25 min after the negative radial impulse is applied (Fig. 7); the perigee is displaced to the opposite side of the orbit with the positive radial impulse (Fig. 8). The Hohmann transfer results in a perigee position, and corresponding perigee time, lying between the positions (Fig. 9) and times (Fig. 6) of the two radially directed impulses.

The time required to reach the perigee altitude varies monotonically with the applied impulse below the value required to lower the 500 kg mass to the Earth's surface, $8.6 \cdot 10^4$ Ns for the Hohmann transfer, and $3.5 \cdot 10^5$ Ns for the impulses applied in the radial directions. Above these values, the descent to the ground varies more rapidly with the impulse.

The simulation results in Fig. 6 are consistent with the impulse values of the Hohmann transfers (Eq. 1) of the 500 kg mass from 600 km to 150 km, $6.35 \cdot 10^4$ Ns, and to the Earth's surface, $8.6 \cdot 10^4$ Ns.

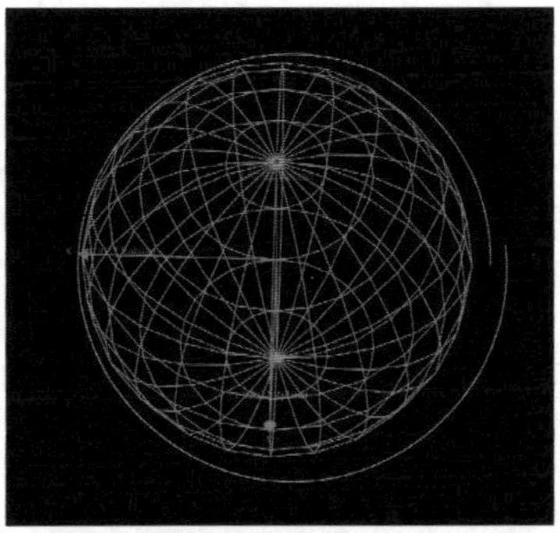

Figure 7. The modified trajectory of the 500 kg mass (Fig. 5) after the radial impulse of $3.0 \cdot 10^5$ *Ns in the direction of the Earth's center. The mass orbits in the clockwise direction.*

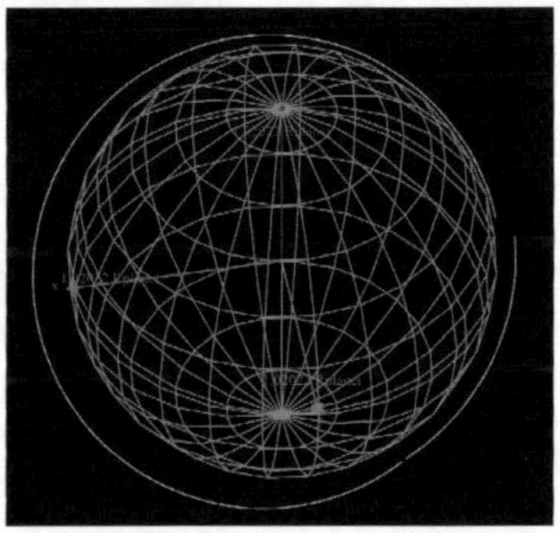

Figure 8. The modified trajectory of the 500 kg mass (Fig. 5) after the radial impulse of $3.0 \cdot 10^5$ *Ns in the zenith direction. The mass orbits in the clockwise direction.*

Figure 9. The modified trajectory of the 500 kg mass in a circular orbit at 600 km (Fig. 5) after the Hohmann transfer of $7.5 \cdot 10^4$ *Ns in the direction opposite to its orbital velocity. The mass orbits in the counter-clockwise direction.*

Effect of Atmospheric Drag

The frictional force acting on a satellite due to the presence of the atmosphere is given by the expression,

$$\mathbf{F} = -\frac{1}{2}\rho C_D A v^2 \hat{\mathbf{v}} \tag{2}$$

where ρ is the atmospheric density and C_D is the drag coefficient. A and v are the satellite's cross-sectional area and velocity. The force acts in the direction opposite to satellite's orbital velocity.

The effect of the atmosphere on the orbit is introduced in the Geant4 simulation by adding the continuous energy loss process described by Eq. 2. A surface area of 0.325 m^2, corresponding to a solid aluminum cube, and $C_D = 0.5$ are used in the simulation. The parameterizations for the troposphere, and the lower and upper stratospheres densities of Ref. [6] are used. The atmospheric density is shown Fig. 10.

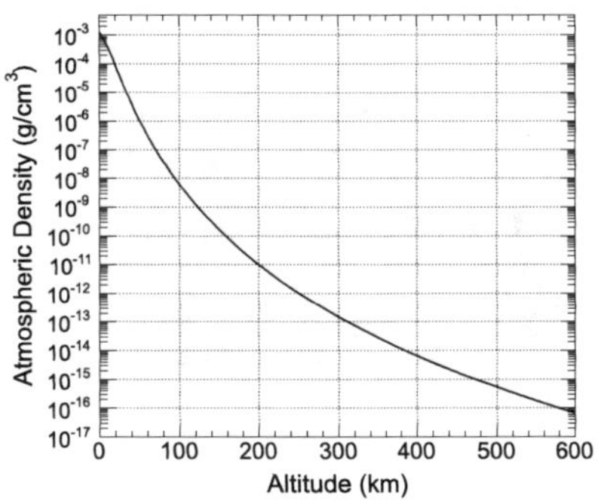

Figure 10. Atmospheric density from Ref. [6].

The effect of the atmospheric drag on the debris orbit for the three impulse directions is shown in Fig. 11. A Hofmann transfer of 4.0 · 10^4 Ns results in a deorbit of the 500 kg mass in 5 y. The minimal values required to deorbit the mass are 1.75 · 10^5 and 2.0 · 10^5 Ns, respectively for the negative and positive radial impulses; the corresponding deorbit times are 6.2 and 2.8 y. The threshold values for the negative and positive radial impulses are factors 4.3 and 5.0 larger than for the Hohmann transfer.

The perigee altitudes, 325-400 km for the Hohmann transfer and 285-508 km for the radial impulses, represent the altitudes attained after 6.34 y, the simulation time limit.

The Hohmann transfer applied to lower the perigee altitude is kinematically the optimal configuration. In practice, the impulse of a ground laser will always have a component directed away from the Earth. A space-based laser may engage debris targets at relatively lower and higher altitudes. The results for the radial direction transfers represent the performance lower limits.

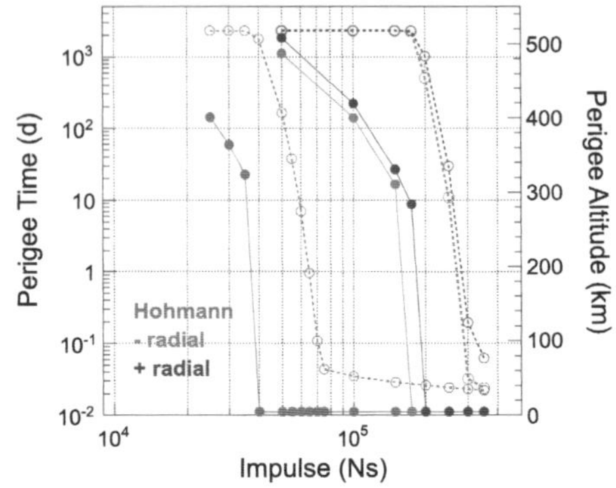

Figure 11. The perigee altitude (•) and time (◦) for the three impulse directions with the presence of the atmosphere.

The modified trajectory of the 500 kg mass after a Hohmann transfer of 7.5 · 10^4 Ns in the direction opposite to its orbital velocity is shown in Fig. 12. The applied impulse results in a series of elliptical orbits with progressively lower perigee altitudes. When the altitude attains the height of the atmosphere each successive passage through the atmosphere produces a further reduction of the orbital velocity. The 500 kg mass falls to the Earth in 7 d after the initial Hohmann transfer.

The corresponding scenarios for negative and positive radial impulses of 2.5·10^5 Ns are illustrated in Figs. 13 and 14. The starting position of the debris mass is located in the equatorial plane at -600 km x (red).

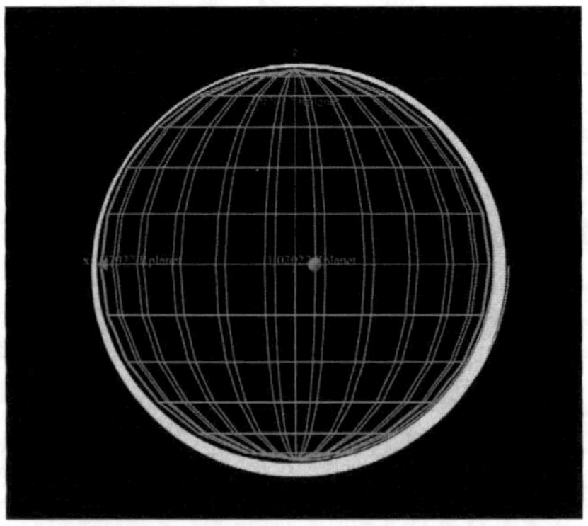

Figure 12. The 500 kg mass in a circular orbit at 600 km (Fig.5) returns to Earth 7 d after a Hohmann transfer of 6.0 · 10^4 Ns and multiple passes in the atmosphere. The mass orbits in the clockwise direction.

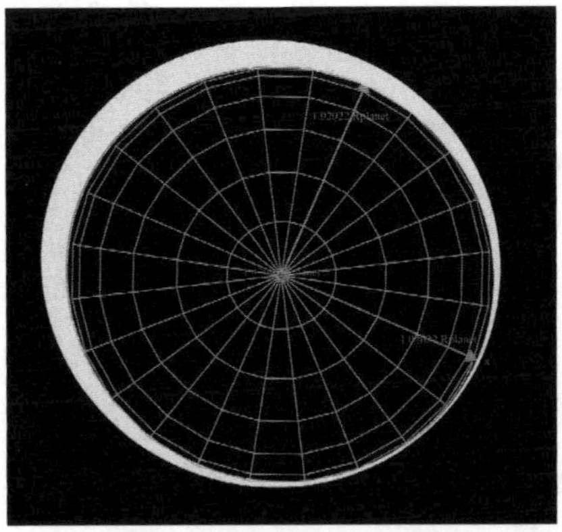

Figure 13. The 500 kg mass in a circular orbit at 600 km (Fig.5) returns to Earth 11 d after a negative radial impulse of 2.5 · 10⁵ Ns and multiple passes in the atmosphere. The mass orbits in the counter-clockwise direction.

Figure 14. The 500 kg mass in a circular orbit at 600 km (Fig.5) returns to Earth 30 d after a positive radial impulse of 2.5 · 10⁵ Ns and multiple passes in the atmo-sphere. The mass orbits in the clockwise direction.

3. DEBRIS MITIGATION WITH LASER ABLATION

The CLEANSPACE project has developed a space debris removal scheme based on a ground network of lasers [7]. The characteristics and important aspects affecting the performance of the system are reported here.

The relevant parameters for the CLEANSPACE ground laser are the energy per pulse ($E_o > 10$ kJ), the repeti-tion rate (> 10 Hz), the pulse length (1-50 ns), and the beam quality M2 (< 2.5). The expected power intensity at the surface of the debris is 0.1 – 0.8 GW/cm^2.

The impulse delivered to the debris mass is given by the expression,

$$m\Delta v = \left(\frac{E_o T_{tel} T_{atm}}{\pi\phi^2(R)}\right) S C_m \quad (3)$$

where $\phi^2(R)$ is the size of the beam at the range R of the target debris. S and m are the surface area and mass of the target. T_{tel} and T_{atm} are the transmission efficiency of the laser emitting telescope and the atmosphere. C_m is the coupling coefficient (N/W) for the conversion of the incident laser pulse power to kinetic energy.

CLEANSPACE has measured the coupling coefficient for aluminum, ~ $2 \cdot 10^{-5}$ N/W, for the power density range between 0.5 and 0.8 GW/cm^2, at $\lambda = 1.06\mu$m. The vaporization threshold P_o for aluminum is ~ 0.2 GW/cm^2, below threshold $m\Delta v = 0$.

The Rozenberg model is used to estimate the laser beam transmission through the atmosphere

$$T_m = (1 - A_{atm})^X \quad (4)$$

with

$$A_{atm} = 0.18 \text{ at } 1.06\,\mu\text{m}$$

$$X = \frac{1}{cos\psi_Z + 0.025 \cdot e^{(-11 \cdot cos\psi_Z)}}$$

where ψ_Z is the elevation angle. Figure 15 shows the variation in the atmospheric transmission with the elevation angle for a 1.06μm wavelength laser beam.

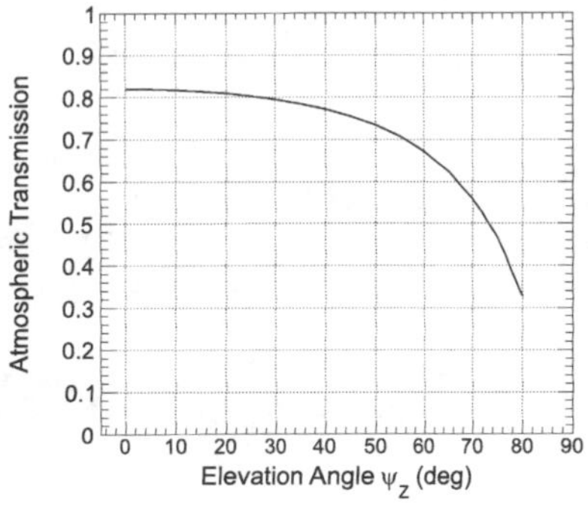

Figure 15. The variation of the transmission efficiency of the atmosphere for a 1.06μm wavelength laser beam with the elevation angle, defined with respect to the zenith [7].

Equation 3 and the CLEANSPACE value for C_m in aluminum are used to define the required laser performance corresponding to the simulation results presented in the next section.

4. LASER DEPLOYMENT SCENARIOS

The Simulation

A simulation developed in Matlab© is used to study the relative performance of space-based and ground laser systems. Four laser deployment scenarios are evaluated: a dedicated satellite (SAT), a laser system installed on the International Space Station (ISS), and two ground sys-tems located in the equatorial (GEQ) and polar (GPO) regions.

The debris is represented as a point mass M; all forces act on the center-of-mass. The surface-area-to-mass ratio, used for the computation of the frictional drag force, was chosen to yield a constant ballistic coefficient (C_D equal to 1-2)

$$\beta^* = C_D \cdot A/M = 0.1 . \quad (5)$$

The laser beam delivers a 1 N force acting along the line-of-sight between the laser and debris positions. The beam is switched on and off in order to limit the time of operation to periods with advantageous orientations of the beam with respect to the debris velocity vector.

The laser is switched on when the debris approaches the ISS or satellite, and the line-of-sight between the space platform and debris is not obscured by the Earth, defined as a spherical volume with radius $R = R_E + 200$ km. For the ground systems, the laser is turned on when the object approaches the laser position, and the angle of el-evation (Fig. 15) of the line-of-sight ψ_Z is less than 45°.

The Jacchia-Bowmann 2008 model [8] is used to calculate the air density at high altitudes. The model takes into account density fluctuations due to diurnal and seasonal thermal variations. The estimated accuracy of the computed densities is better than 4% based on comparisons with satellite accelerometer data. The epoch January 1, 2010, h.00:00:00.0 is chosen as the starting point in the simulation, in order to use known solar values for the calculation of the atmospheric density.

The debris mass orbital elements (Appendix, Fig. 24) at epoch are listed in Table 1, which corresponds to an initial position at the apogee of the orbit (800 km). The chosen inclination and altitude is representative of the major part of the debris population in LEO.

Table 1. The orbital elements of the debris mass at epoch: semi-major axis a, eccentricity e, inclination i, right as-cension of the ascending node Ω, the argument of the perigee ω and the true anomaly θ.

a	e	i	Ω	ω	θ
7170 km	0.001	100°	0°	180°	180°

The orbital elements of the satellite and ISS at epoch are listed in Table 2. The starting point of the satellite is the perigee at an altitude of 500 km. The satellite altitude and inclination angle may be chosen to optimize the number of debris objects in its line-of-sight.

Table 2. The starting point orbital elements for the dedicated satellite and ISS scenarios.

	a	e	i	Ω	ω	θ
SAT	6885 km	0.001	80°	180°	90°	0°
ISS	6780 km	0.0016	52°	128°	99°	313°

The difference in altitude of the satellite and the debris mass (Table 1) results in a relative rotation between their orbits due to the J_2 effect of the perturbed gravitational field. The rate of change in time (degree per mean so-lar day) of Ω, the Right Ascension of the Ascending Node (RAAN) is given by the expression [9]

$$\frac{d\Omega}{dt} = -\frac{9.9358}{(1-e^2)^2}\left(\frac{R_E}{R_E+\bar{h}}\right)^{3.5}\cos(i) \quad (6)$$

where $\bar{h}$ is the mean distance from the Earth.

For the two nearly circular orbits, with mean altitudes of 500 and 800 km, the difference in $\bar{h}$ results in an increase of ΔΩ with time. The initially co-planar orbits will become again co-planar in ~ 980 days. All debris objects orbiting at the same altitude, and nearby polar inclinations, will be seen by the satellite in this time interval.

The time interval required to return to coplanarity for the higher altitude (1000-1500 km), near-polar orbit debris objects is shorter. Therefore, the dedicated satellite would be able to target a significant fraction of the near-polar orbit debris in less than 3 y. A laser system installed on the ISS will observe the near-polar orbit debris due to the lower inclination angle of the space station.

The position of the equatorial ground laser station at epoch is on the x-axis of the Earth Centered Inertial (ECI) reference system, which is drawn between Earth's center to the Sun, at a distance of 6378 km from the Earth's center. The polar region laser station is at a distance of 6357 km from the Earth's center, placed to have the de-bris mass (i = 100°) at its zenith at epoch. The orbital direction of the debris mass (i > 90°) is opposite to the rotation of the Earth, and to the orbital directions of the satellite and ISS (Table 2).

Results

The results for the four scenarios are presented in Table 3 for a 500 kg mass, $A = 25$ m^2 in Eq. 5. The SAT (ISS) lasers provide the impulse required to lower the 500 kg mass at 800 km, to the altitude of the atmo-sphere at 200 km, in 3 weeks (3 months).

The altitude decrease is significantly slower for the two ground lasers. At the end of the 2 month simulated time, the debris perigee altitude decreases by 19 and 77 km. The total impulse delivered by the polar site laser is a factor 5.5 higher.

Table 3. The total simulated time t_{tot}, total time laser on t_{laser}, impulse delivered to the 500 kg mass and the final altitude at time t_{tot}. The simulated time was limited to 60 d for the two ground laser scenarios.

	t_{tot}	t_{laser}	Impulse	Alt. Final
	d	h	Ns	km
SAT	23.1	41.5	$1.5 \cdot 10^5$	200
ISS	92.9	66.0	$2.4 \cdot 10^5$	200
GEQ	60.0	2.3	$8.3 \cdot 10^3$	781
GPO	60.0	13.0	$4.7 \cdot 10^4$	723

The results may be compared to the impulse required by a Hohmann transfer to lower a 500 kg mass from 800 km to 150 km, $8.22 \cdot 10^4$ Ns (Fig. 2). The impulses required for the space-based lasers to lower the 500 kg debris mass to the 200 km perigee altitude are factors 1.8 and 2.9 larger. The fractions of time on target, t_{laser}/t_{tot} are 0.074 (SAT) and 0.029 (ISS).

The evolution in time of the semi-major axis length of the elliptical orbit of the 500 kg mass is shown in Fig. 16 for the two space-based lasers. A 0.5 s step size is used in the simulation to displace the orbiting objects in the gravitational field, and apply the effects due to atmospheric drag and the impulse delivered by the lasers.

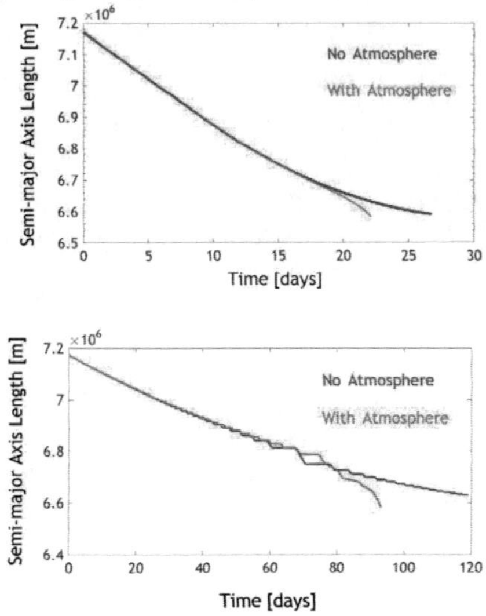

Figure 16. Variation with time of the semi-major axis length of the 500 kg mass for the space-based lasers, SAT (top) and ISS (bottom). Results are shown with and without the effect of atmospheric drag.

The structure observed in the ISS distribution reflects the variation in momentum transfer due to the relative positions of the debris mass and laser. The orbit of the dedi cated SAT is better adapted to the chosen debris orbit. In this case, a smooth and rapid descent is observed.

The corresponding time distributions for the two ground lasers are shown in Fig. 17. The effect of the atmosphere is negligible.

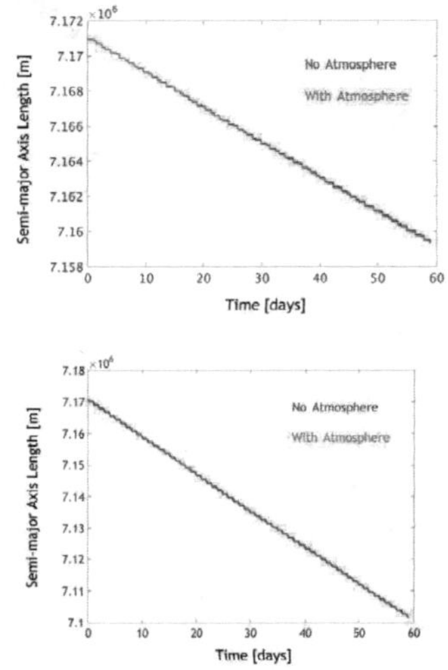

Figure 17. Variation with time of the semi-major axis length of the 500 kg mass for the equatorial laser (top) and the laser system at the polar site (bottom). No effect due to atmospheric drag is observed.

A cross-section of 2 m² is quoted for a large debris object in Ref. [7]. Large objects refer to the 2200 objects in LEO with a diameter ≥ 1 m and a mass of ~ 1 t. The small object population consists of 190 k objects with a diameter ≥ 1 cm.

The ground laser system of the CLEANSPACE project is designed for the smaller objects, 10 kg/m². A mass of 10 kg, A = 1.0 m² in Eq. 5 and the orbital elements of Table 1, was used in the simulation. The corresponding results for the four scenarios are presented in Table 4.

Table 4. The total simulated time t_{tot}, total time laser on t_{laser}, impulse delivered to 10 kg mass and the final altitude at time t_{tot}. The simulated time was limited to 60 d for equatorial site laser.

	t_{tot}	t_{laser}	Impulse	Alt. Final
	d	h	Ns	km
SAT	0.43	0.83	$3.0 \cdot 10^3$	200
ISS	1.16	1.32	$4.8 \cdot 10^3$	200
GEQ	60.0	1.51	$5.4 \cdot 10^3$	409
GPO	4.15	1.00	$3.5 \cdot 10^3$	200

The impulse required for the space-based lasers to lower the 10 kg mass to an altitude of 200 km are a factor 50 smaller than the corresponding values for the 500 kg mass, in agreement with the ratio of the corresponding Hohmann transfers in Fig. 2.

The results in Table 4 are comparable to those in Table 3 for the equatorial site laser. The lighter mass debris attains an altitude of 409 km after 60 d. The 10 kg mass descends to 200 km after ~ 4 d for the polar site laser.

The semi-major axis length of the elliptical orbits of the 10 kg mass are shown in Figs. 18 and 19 for the space-based and ground lasers respectively. No effect due to the presence of the atmosphere is observed for the 10 kg mass.

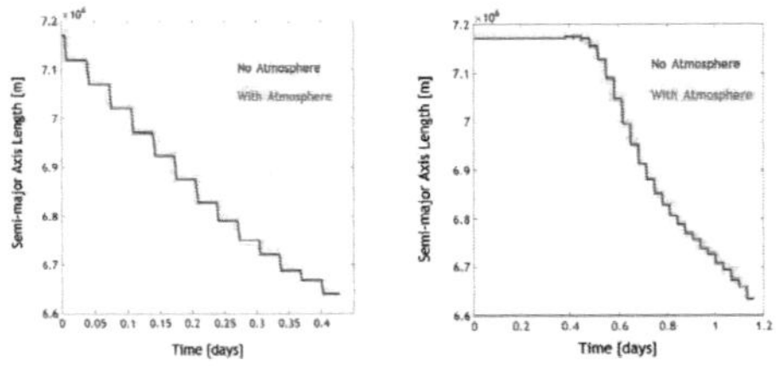

Figure 18. Variation with time of the semi-major axis length of the 10 kg mass for the space-based lasers, SAT (left) and ISS (right).

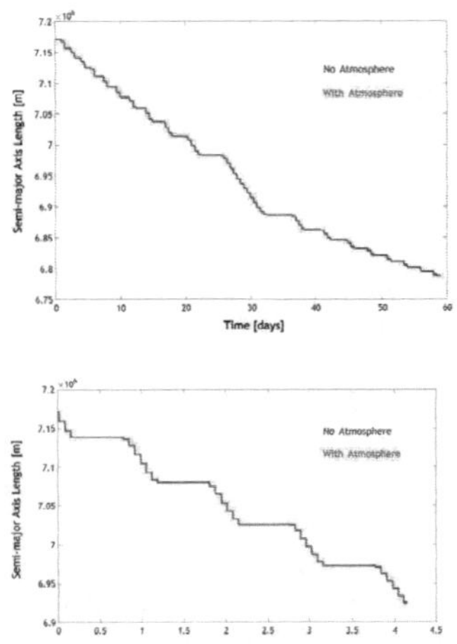

Figure 19. Variation with time of the semi-major axis length of the 10 kg mass for the equatorial laser (top) and the laser system at the polar site (bottom)

The polar orbit of the SAT, optimal with respect to the orbital elements of the debris object, results in a continuous decrease of the debris altitude (Fig. 18). The first impulse delivered to the 10 kg mass by the laser on the ISS occurs 10 h after the simulation starts.

In Fig. 19, the polar site ground laser delivers an impulse to the debris mass, which passes through the zenith of the polar site, at regular intervals during the four days. The effectiveness of the polar site is sensitive to the inclination angle of the orbit. A variation of a few degrees from the zenith is sufficient to remove the debris object from the line-of-site. In general, an equatorial site will deliver an impulse over a greater fraction of the orbit, as illustrated by the constant descent of the 10 kg mass in the top panel of Fig. 19. The fractions of time on tar-get, t_{laser}/t_{tot} in Table 4, are 0.081 (SAT), 0.047 (ISS), $4.4 \cdot 10^{-4}$ (GEQ) and $9.7 \cdot 10^{-3}$ (GPO).

Required Laser Performance

The total impulse quoted in Table 3 represents the sum of the 1 Ns impulse steps delivered by the laser in the simulated time t_{tot}. The required laser energy E_o can be estimated from Eq. 3 with $m\Delta v = 1$ Ns. For the space-based lasers ($T_{atm} = 1$),

$$E_o = \left(\frac{1\,\text{Ns}}{T_{tel} \cdot C_m}\right)\left(\frac{S}{\pi\phi^2(R)}\right). \qquad (7)$$

The second term in Eq. 7 is the ratio of areas of the debris surface and laser beam spot at the target. With the debris surface of 25 m^2 for the 500 kg mass, a 2.2 m diameter circular beam spot and a telescope transmission efficiency of 80%, E_o is 400 kJ and the average power 400 kW.

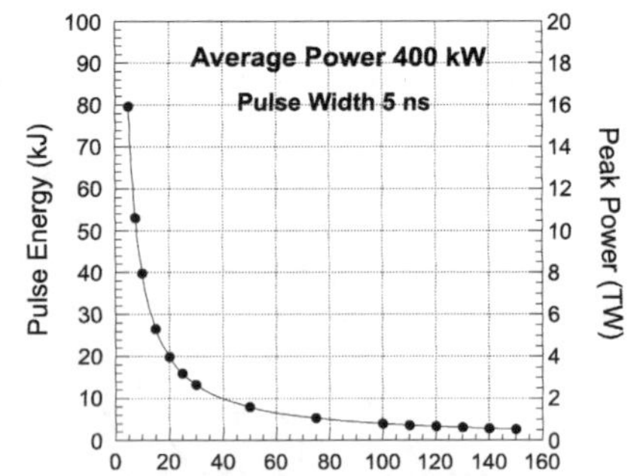

Figure 20. The variation with repetition rate of the pulse energy and peak power of the 400 kW space-based laser with a pulse width of 5 ns.

Figure 20 shows the variation of the pulse energy and peak power with repetition rate of a pulsed 400 kW laser, with a 5 ns pulse width. The laser peak power is plotted against the peak power density at the target in Fig. 21. The variation of the same parameters with the pulse width for a 5 Hz repetition rate is shown in Fig. 22.

The minimum power density required to produce a momentum transfer by the vaporization of the material on the surface of the debris (Al) is 0.2 GW/cm^2 [7]. The laser configurations with peak power densities between 0.2 and 1 GW/cm^2 are indicated in red in Figs. 22 and 23.

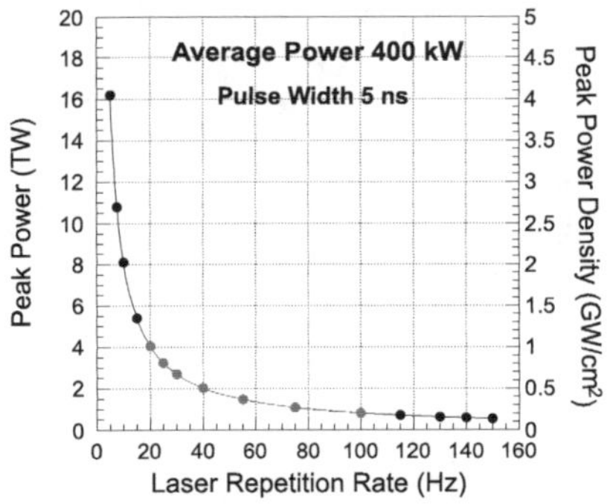

Figure 21. The variation of the peak power and the peak power density at the target with repetition rate, for a 5 ns pulse width and an average power of 400 kW. Peak power densities between 0.2 and 1 GW/cm^2 are indicated in red.

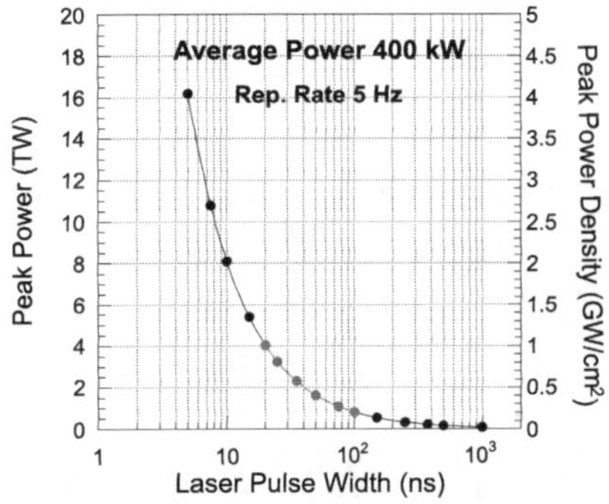

Figure 22. The variation of the peak power and the peak power density at the target with pulse width, for a 5 Hz repetition rate and an average power of 400 kW. Peak power densities between 0.2 and 1 GW/cm^2 are indicated in red.

The pulsed, 400 kW laser can deliver the required total impulse compatible with the performance presented in Table 3 for the two space-based lasers. Two possible laser configurations, which provide the necessary power density at the target are listed in Table 5.

The second term in Eq. 6 plays a fundamental role in the design of the laser system. The 2.2 m diameter spot size at the target was chosen to match the 25 m^2 surface area of the debris in the simulation study. The spot size approaches the dimensions of a laser designed for a military application.

The spot size could be redimensioned for the smaller mass debris. For example, with a 0.5 m diameter beam spot on a 1 m2 surface, the geometric factor in Eq. 7 is reduced by 20%.

Table 5. Pulsed laser configurations compatible with the performance of the space-based lasers in the Matlab© simulation for a 500 kg debris mass (Table 3).

P_{avg}	E_{pulse}	P_{peak}	width	rate	$P_{density}$
kW	kJ	TW	ns	Hz	$\frac{GW}{cm^2}$
400	16	3.2	5	25	0.85
400	80	3.2	25	5	0.85

Based on the results of the simulation for the 10 kg mass, a factor 50 reduction of the laser impulse would produce a performance similar to the results reported in Table 3 for the 500 kg mass. With a factor 30 reduction, the computed E_o is 10.5 kJ and the average power 10.5 kW.

Two configurations of the pulsed 10.5 kW laser, designed for the removal of small debris objects (10 kg/m^2) are listed in Table 6. The operation range is limited by the required peak power density at the debris surface.

Table 6. 10.5 kW space-based pulsed laser configurations for the removal of a small debris objects (10 kg/m^2).

P_{avg}	E_{pulse}	P_{peak}	width	rate	$P_{density}$
kW	kJ	TW	ns	Hz	$\frac{GW}{cm^2}$
10.5	0.84	0.15	5	12.5	0.85
10.5	2.1	0.15	12.5	5	0.85

5. CONCLUSION

The Hohmann transfer may be used to provide an estimate of the required laser performance for debris mitigation. The Geant4 simulation was used to compare the perigee altitude of the Hohmann transfer with the altitude change produced by radial impulses directed along, and opposite to the vector drawn from the Earth's center to the debris position, the latter considered as limits for effective operation. A factor four larger radial impulse is required to produce an altitude change comparable to the Hohmann transfer.

The Matlab© simulation was used to study the performance of space-based and ground lasers for 500 and 10 kg debris masses in a representative LEO orbit. A 1 Ns impulse transmitted to the 500 kg mass in a nearly circular orbit at 800 km results in an elliptical orbit. The perigee attains the altitude of the atmosphere (200 km) after 3 weeks for a laser on a satellite in polar orbit, and 3 months for a laser installed on the ISS. The corresponding altitude change for ground lasers at the polar and equato-rial sites was negligible at the end of the 60 d simulated time.

In the case of the 10 kg mass, the 200 km altitude was attained within 30 h for the space-based lasers, in 4 d for the polar site ground laser. After 60 d, the perigee altitude is 410 km for the ground laser at the equator.

The simulation results were used to define the general characteristics of a pulsed laser with the required performance using the coupling coefficient for aluminum of Ref. [7]. The parameters of the pulsed, space-based lasers for the 500 and 10 kg debris masses are presented in Tables 5 and 6.

The two simulations employed in the present study represent an initial framework. Matlab© provides a convenient environment for program development and analysis. The results presented are limited by the computational time since the program was run interactively. Geant4 is simulation program designed for elementary particle interactions. The program is often employed in batch mode on multi node processing farms adapted to the large-scale job processing for long computational time studies. The physics relevant to the present application can be implemented in the program.

The STELA [10] program of the French Space Agency (CNES) is another candidate for long computa-tional time calculations. STELA was developed to sup-port the French Space Operations Act which defines cri-teria for the end-of-life scenarios of satellites in order to reduce the debris population. The criteria are established on the time scales of 25 and 100 y, which require simula-tion times capable to ensure compliance with the imposed limits.

In view of the ultimate aim, a comprehensive study of the performance over a wide range of debris parameters, additional elements are required in the simulations, e.g. the perturbed gravitational field, details of the laser operation [7] and the momentum transfer between beam and target [11]. The three simulation programs, likely a combination, are considered for the future work.

6. REFERENCES

1. *Collision Frequency of Artificial Satellites: The Creation of a Debris Belt,* Journal of Geophysics Research, Vol 83, No. A6, 1979, 2637-2646.

2. Inter-Agency Space Debris Coordination Committee: www.iadc-online.org

3. NASA Orbital Debris Program Office: orbitaldebris.jsc.gov/index.html

4. *The Threat of Orbital Debris and Protecting NASA Space Assets from Satellite Collisions* images.spaceref.com/news/2009/ODMediaBriefing28Apr09-1.pdf

5. J. Allison et al., IEEE Transactions on Nuclear Science, Vol. 53, No. 1, 2006, 270-278

6. NASA Earth Atmosphere Model, http://www.grc.nasa.gov/WWW/k-12/airplane/atmos.html

7. *Changes of Space Debris Orbits after LDR Operation* E. Wnuk, J. Golebiewska, C. Jacquelard, and H. Haag CLEANSPACE project of EC Seventh Framework Programme (FP7)

8. *Jacchia-Lineberry Upper Atmosphere Density Model*, NASA Lyndon B. Johnson Space Center, 1982 *A NEW Empirical Thermospheric Density Model JB2008 Using New Solar and Geomagnetic Indices*, B. R. Bowman et al., AIAA 2008

9. *Orbital Mechanics*, V. A. Chobotov, AIAA 2002

10. https://logiciels.cnes.fr/content/stela

11. *Pulsed laser interactions with space debris: Target shape effect*, D. A. Liedahl *et al.* Advances in Space Research 52, 2013, 895-915

7. APPENDIX: ORBITAL MECHANICS

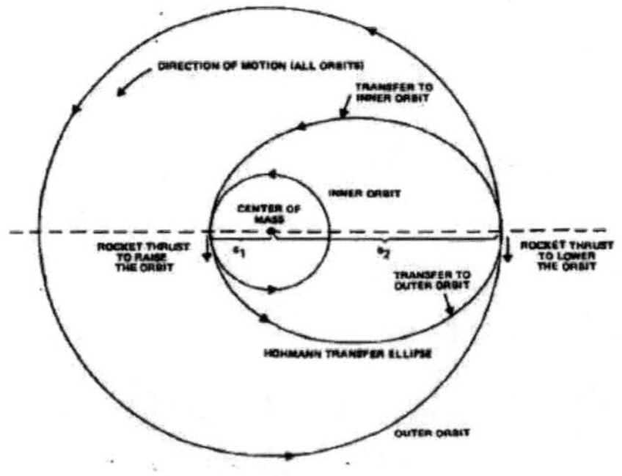

Figure 23. The Hohmann transfer [9].

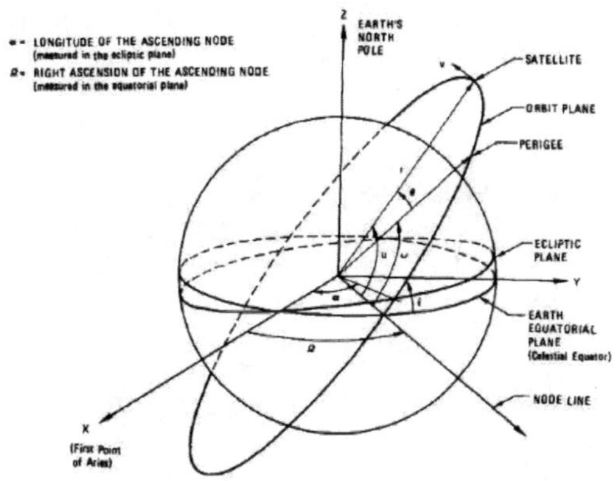

Figure 24. The orbital elements inclination i, right ascension of ascending node Ω , the argument of the perigee ω and the true anomaly θ [9].

LONG TERM ORBIT PROPAGATION: LESSONS LEARNT AND ON GOING ACTIVITES AT CNES

Vincent Morand[(1)], Clemence Le Fèvre[(2)], Hubert Fraysse[(2)], Juan Carlos Dolado Perez[(1)]

[(1)] CNES DCT/SB/SL, Centre Spatial de Toulouse, 18 Avenue Edouard Belin, 31400 Toulouse FRANCE, vincent.morand@cnes.fr
[(2)] CNES DCT/SB/MS, Centre Spatial de Toulouse, 18 Avenue Edouard Belin, 31400 Toulouse FRANCE

ABSTRACT

Space debris mitigation is one objective of the French Space Operations Act (FSOA), which came into force in 2010. At the end of their mission, space objects are to be placed on orbits that will minimize future hazards to space objects orbiting in the same region, which is in line with IADC (Inter-Agency Space Debris Coordination Committee) recommendations.

In order to address the compliance of disposal orbits with the law technical requirements, CNES draws up Good Practices as well as a dedicated tool, STELA (Semi-Analytical Tool for End of Life Analysis [1]).

Long term orbit propagation technics are required and have been studied over the last years. While the adequate dynamical model necessary to capture the main orbital dynamics is now well known, some uncertainties remain that sometimes makes the orbit evolution poorly know or unpredictable. The satellite attitude, for example, is not controlled anymore. Most of the time a random tumbling attitude is assumed or an equilibrium attitude is supposed. Taking such hypotheses allows neglecting the coupling between the orbit and the attitude, but they are not always verified. For Low Earth Orbits (LEO) or Geostationary Transfer Orbits (GTO), atmospheric density is of key importance to predict the long term orbit evolution. As a consequence, solar activity hypotheses can drastically change the orbit lifetime prediction and, therefore, the compliance with respect to the regulation. Finally, orbital resonances conditions exist in all orbital regimes (LEO, GTO, MEO…) and make the result of a propagation very sensitive to initial values; thus requiring a statistical approach [2] [3].

The paper will describe the on-going CNES activities related to these topics, towards a better understanding on the long term evolution of satellites' orbits after their end of mission.

1. FRAMEWORK

Space debris mitigation is considered today as an important contributor for the advancement in the field of space systems safety. With the growing number of debris into space, of various sizes and origins, the collision risk between one of them and an operational spacecraft becomes more and more likely, as illustrated in 2009 with the loss of the operational Iridium-33 satellite after the collision with the inactive satellite Kosmos-2251. The removal of non-operational objects from populated regions is one important answer to mitigate collisions risks, which is addressed from a legal point of view through the French Space Operations Act. In line with IADC, this Act requires that the disposal orbits of space objects (satellites or launcher pieces) shall be chosen to minimize future hazards to other objects orbiting in the more populated regions.

Two protected regions have been defined by IADC (Fig. 1): region A for Low Earth Orbits (altitude < 2000 km); region B for Geostationary Earth Orbits (GEO – 200km < altitude < GEO + 200km; absolute value of inclination < 15deg).

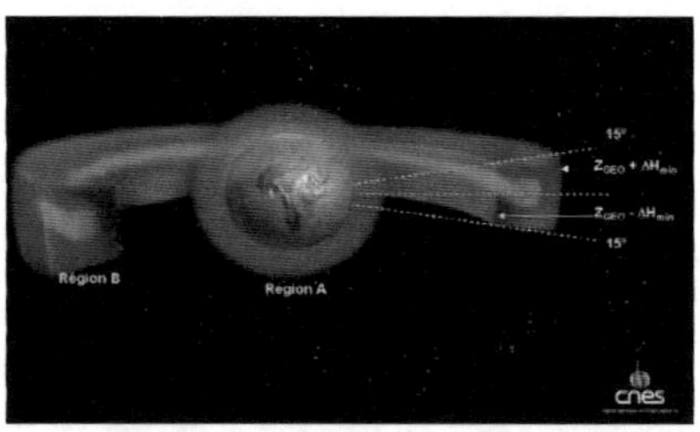

Figure 1. IADC protected regions

For space objects placed in orbits crossing the LEO protected region, the French Space Operations Act states that a direct and controlled atmospheric re-entry is the baseline. If the impossibility of meeting this requirement can be duly proven, as well as for orbits not crossing the LEO region, an uncontrolled re-entry, no later than 25 years after the end of mission, or a stable disposal orbit outside the protected areas can be chosen. Consequently some criteria were defined to verify that the chosen disposal orbits ensure the protection of these zones within a time period defined by the FSOA, for several orbit types (LEO, GEO, and GTO). The verification of these criteria requires long term orbit propagation to evaluate the evolution of the orbital elements over long time scales (up to more than 100 years). The Good Practices define the minimum dynamical model required to compute the orbital evolution with sufficient accuracy, and detail key computation hypotheses such as drag and reflecting areas, drag coefficient, reflectivity coefficient, solar activity, atmospheric density model and so on. They also recommend a methodology adapted to each orbit type (LEO, GEO, GTO) to assess the criteria of the French Space Act[2] [4]. Despite the large amount of work already done, some on-going activities focus on improving methods, tools and good practices to master space object disposal, thus ensuring "better" safety for all space missions.

2. DYNAMICAL MODELS AND TOOLS

Long term orbit propagation requires finding a compromise between the precision of the result and the computational time. As an example, a full numerical propagation considering a complete dynamical model takes about 4h versus few minutes for a semi-analytical propagation. As a consequence, semi-analytical or analytical technics are used to propagate orbit over long time scales. STELA is the CNES reference tool for long term orbit propagation. It can be freely downloaded on the CNES website [5]. Tab.1 gives the dynamical model implemented in the last STELA version:

Perturbation	STELA dynamical model
Earth's gravity field	Complete 15x15 model (Tesseral terms taken into account only when tesseral resonances are detected)
Solar and Lunar perturbation	Yes
Solid tides	Yes (Sun and Moon contribution)
Atmospheric drag	Yes (MSIS00 or US 76 atmospheric model)
Solar radiation pressure (SRP)	Yes, cylindrical earth shadow

Table 1. STELA dynamic model

The setting of this dynamical model, for example the degree and order of development of the Earth's gravity field, can be tuned as a function of the orbit type to gain computational time. The good practices give the minimum dynamical model that should be considered. Other softwares are available to perform long term orbit propagation: DAS v2.0.2 (NASA) and DRAMA v2.0.1 (ESA) have been compared to STELA on several test cases covering different orbit configurations. Fig 2. shows STELA and DRAMA orbit ephemeris for a LEO satellite (DAS only gives the lifetime). The differences that can be seen on the amplitudes of oscillations between DRAMA and STELA may be explained by a difference in the definition of the mean elements: the semi-analytical theory used are not the same and the propagation has been initialized using STELA mean elements.

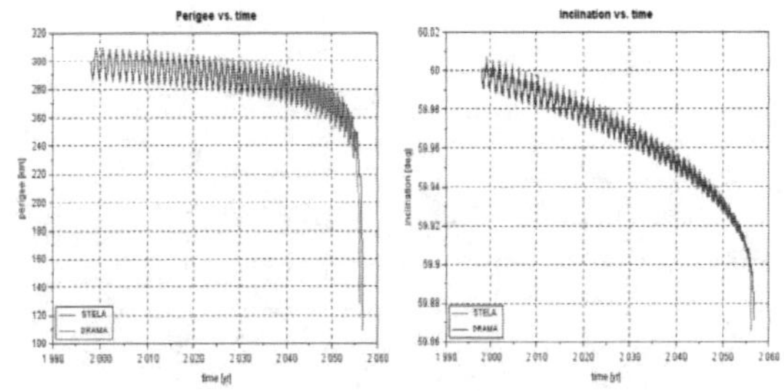

Figure 2. DRAMA and STELA propagation

On this test case, computed with a constant solar activity, STELA, DRAMA and DAS lifetime are 58y, 58.6y and 61.6y respectively, which show a good agreement between the three softwares. Higher gaps were observed on resonant orbits (see §5) or for GTOs where no statistical approach are implemented in DRAMA nor DAS making the comparison tedious.

Further comparisons with the DSST semi-analytical method implemented in OREKIT or with the King-Hele analytical model used by ESA are interesting and should be performed in near future. We also plan to keep improving STELA, for example by including a recursive formulation to be able to consider the Earth's potential development to higher order.

3. SPACECRAFT ATTITUDE

Spacecraft attitude is needed to compute the cross sectional area A that appears in the drag and SRP equations (Eq.1 and Eq2).

$$\vec{F}_{drag} = -\frac{1}{2}\rho A C_d V' \overrightarrow{V'} \qquad (1)$$

With V' the spacecraft velocity with respect to the atmosphere, Cd the drag coefficient and ρ the atmospheric density that will be discussed in §4.

$$\vec{F}_{SRP} = A C_r C_0 \left(\frac{d_0}{d}\right) \vec{u} \qquad (2)$$

With Cr the reflectivity coefficient, C0 and d0 are constant, d the sun-spacecraft distance and u the sun- spacecraft direction.

However, we are dealing in this case with objects that are not controlled anymore: their attitude is consequently a priori unknown. The righteous way would be to perform a 12 degree of freedom (DOF) propagation (Position, Velocity, Attitude and Spin) and to use the attitude to compute the cross sectional area. Ongoing activities at Cnes shows that 12 DOF propagation over long time scale are challenging due to computation time (semi-analytical approaches do not seem to be applicable in this case) and extreme sensitivity to initial conditions that are poorly known in practice [6]. As a consequence, a common hypothesis is to de-couple the attitude and orbit evolution, in other word to perform only 6-DOF propagation for the orbit.

The spacecraft attitude is assumed to be either randomly distributed (random tumbling motion) or stabilized (gravity gradient stabilization for example) which allows to consider a constant cross sectional area computed as a mean value. The validity of these hypotheses can be discussed, and several studies are ongoing at CNES. As, an example, the LEO report [7] allowed us to compare estimated Area to mass Ratio from TLE and theoretical Area to mass Ratio under tumbling motion hypothesis. We computed the area to mass ratio by the analysis of the temporal evolution of semi-major axis and eccentricity, derived from TLE, for French launcher elements for which we know very well the physical properties. We used STELA cross sectional area tool to compute the area to mass ratio under tumbling assumption. Fig. 3 depicts the relative error of estimation for all French launcher elements, as well as launcher elements with perigees lower than 500 km.

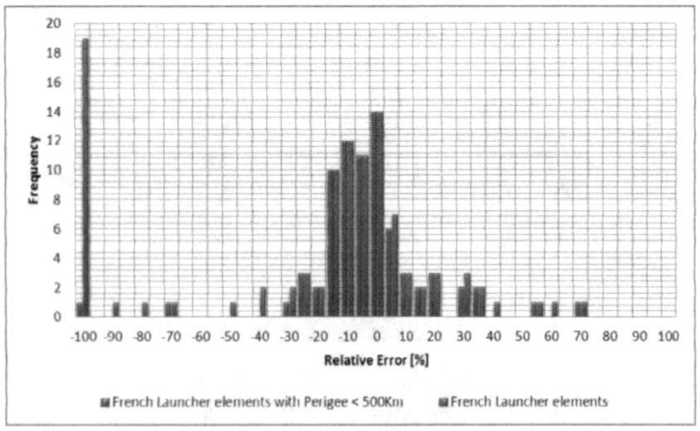

Figure 3. Relative error of S/m ratio estimation for about 100 French launcher elements

Two major conclusions were drawn from the data depicted in Fig 3:

1. The higher the perigee is, the harder it is to estimate the drag force, and therefore the higher is the relative error. The red bar at minus 100% relative error is the consequence of this lack of observability. The estimation error, also containing the atmospheric model error, is in most cases in the ±30% interval for perigee altitudes lower than 500 km;
2. For objects with perigee altitude lower than 500 km, the estimation error is centered in 0%, so there is no systematic over- or underestimation of the physical parameters.

These important results showed that for a large number of launcher elements the tumbling assumption seems adequate to perform long term orbit propagations.
We performed the same analysis for two CNES satellites: Fig 4. shows the evolution over time of the estimated area to mass ratio for DEMETER satellite while Fig.5 shows the same evolution for PARASOL satellite. Both are orbiting LEO at around 650 km altitude.

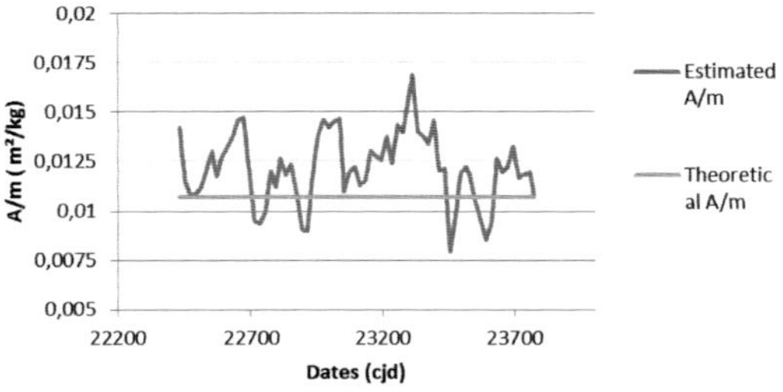

Figure 4. Demeter area to mass ratio evolution

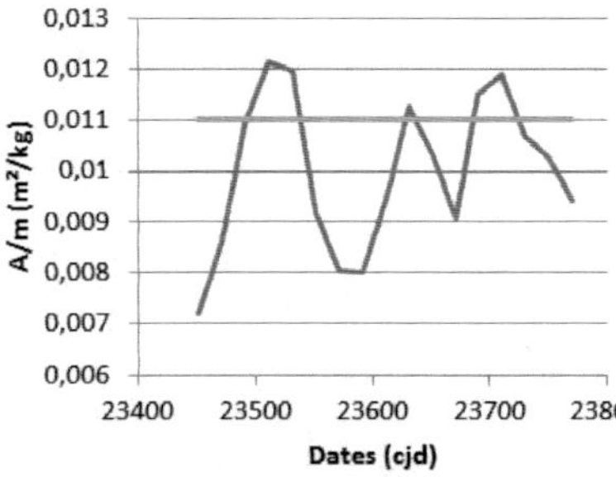

Figure 5. Demeter area to mass ratio evolution

We see some evolutions of the area to mass ratio (that also includes errors on atmospheric density or drag coefficient) that shows that even if the tumbling assumption is not senseless, a better precision would be obtained if the spacecraft attitude could be taken into account. Other examples for which the tumbling assumption seems less representative are given in [6] While no efficient 12-DOF propagation is expected in near future, ongoing activities focus on being able to identify or predict possible stable attitude and estimate the impact of attitude and orbit coupling.

4. DRAG PERTURBATION AND SOLAR ACTIVITY

For LEO and GTO the drag perturbation is a major contributor to the orbit evolution and consequently the lifetime. Several difficulties occur when trying to estimate the drag perturbation: the spacecraft attitude is unknown over long time scale (see §3), the drag coefficient of complex shapes in free molecular flow is poorly known, the atmospheric models such as MSIS00 are empirical and requires as input some solar activity proxies that are not predictable with a good accuracy over long time scales. As a consequence the drag perturbation, and consequently the lifetime, suffers of a huge amount of uncertainty: it is illusionary to think that it will be possible to know the actual re-entry date of a spacecraft in 25 years. Nevertheless, trying to do best is useful especially because we are dealing not with one but with a high number of objects. Improving our methods and making sure that the errors are centered makes the prediction of the evolution of the whole space debris population more accurate.

Solar activity is the main contributor of the error. The solar flux F10.7 is used as proxy for most of the atmospheric model, but its prediction is uncertain, especially after the end of the current 11-years solar cycle. The beginning, strength and duration of the next solar cycles (at least 2 are required to perform a 25 year propagation) are unknown and there is no consensus even between sunphysic experts. For long term orbit propagation, several solar activity hypotheses are commonly taken into account, with great impact on the orbit evolution, as illustrated on Fig. 6.

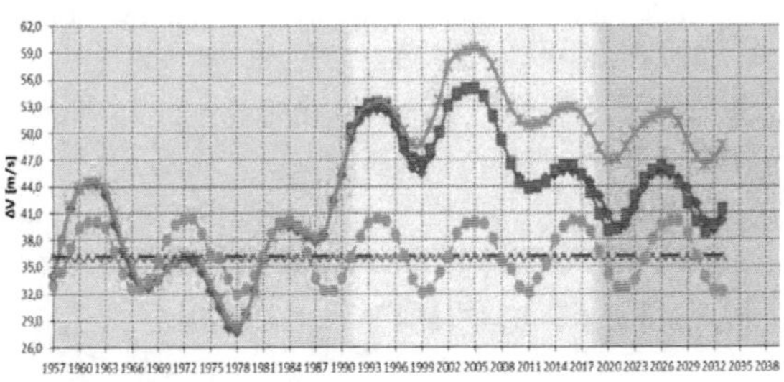

Figure 6. Cost of a 25y disposal for a LEO satellite as a function of date and solar activity hypothesis

Fig. 6 shows the cost of the maneuver (DV in m/s) a typical Earth-observation satellite in LEO should perform to be compliant with the 25y rule, as a function of date of disposal and for several solar activity hypotheses. Dark blue curve uses STELA default solar activity file, which is the same as DAS (red curve): it contains the solar activity measured from the 50's and then the NOAA prediction for future cycles. One can see that the cost of the disposal is not constant and can even vary by a factor 2 as a function of date. This is tedious for dimensioning studies as well as in a certification process since the amount of propellant needed to be compliant is a function of the date of disposal, which can change due to launch delay or mission prolongation, and also function of the solar activity prediction that will change in the future (updated several times a year). The sky blue curve uses the DRAMA solar activity prediction, with also measured value from the past and then an equivalent to NASA-Marshall prediction considering sunspot measurement since 18th century [8]. Orange curve is the so-called random draw approach presented in ISO standards [9]: this method is based on a random draw at each day within the past measurements, keeping an 11-years duration of solar cycles. Finally, the purple curve is the constant equivalent solar activity approach used in the FSOA [2] whose purpose is to allow the computation of a lifetime that does not depend on the date of disposal (that is why the cost of disposal is constant) and with the property that the real re-entry duration should be higher (resp. lower) than this value in 50% of the case (resp. 50%). It is worth noting that the results taking into account only past measurements (green area in Fig. 5 or random draw approach) oscillates around the constant equivalent result, which was expected.

Note that in this case the results coming from analysis based on past data lead to a lower cost of the disposal maneuver than the results using recent predictions. It can be explained by the fact that the current solar cycle is weak in term of values, thus reducing the predicted drag and increasing the cost of disposal. It might change in the future if the next solar cycle appears to be a strong one.

Without breakthrough in our understanding of solar physics, it is not expected that solar activity predictions with sufficient accuracy will be available over long time scale. Therefore, the presented hypotheses will still be used. From CNES point of view, the constant equivalent solar activity approach has been proven very useful both for engineers and for certifiers. To keep improving this method we will include in its computation the measured value from the actual solar cycle once it is over (around 2020) and it could also be interesting to include solar sunspot database measured from the 18th century.

5. ORBITAL RESONANCES

5.1. Theoretical explanation

Orbital resonances are often detected when a strong sensitivity of the orbit evolution with respect to initial conditions are observed. The "physical" explanation of a resonance comes most of the time from commensurability between two periods: one from the orbit and one from a perturbation in the force model, 3rd body for example. Without commensurability the combination of those two periods create a periodic signal whose effect on the orbit evolution is also periodic. But in case of commensurability the signal can become constant, secular, or still periodic but with a very long period which translates on the orbit evolution. The most noticeable effect concerns the evolution of the orbit eccentricity since it impacts the perigee altitude and then the drag perturbation (or the crossing of a protected region). The mean effect of the third body perturbation on the eccentricity is given in Eq. 3:

$$\left.\frac{d\bar{e}}{dt}\right)_{3rd} = -\frac{15}{4}\frac{e\sqrt{1-e^2}}{n}\frac{\mu}{d^3}\sum A_k sin(\Phi_k) \qquad (3)$$

With Ak function of the orbit and 3rd body inclination and Φk an angle function of the argument of perigee and RAAN of the orbit and 3rd body. The detailed expressions are given in [11]. A direct computation of the primitive of Eq. 3 gives, considering everything constant but Φk:

$$e(t) = \frac{15}{4}\frac{e\sqrt{1-e^2}}{n}\frac{\mu}{d^3}\sum\frac{A_k cos(\Phi_k)}{\dot{\Phi}_k} \qquad (4)$$

As can be seen in Eq. 4, problem occurs when the angle Φk becomes almost constant since its derivatives comes close to zero and we have a division by a small number: this is what we called commensurability or resonances

5.2. Examples for all orbit type

With the third body perturbation given in Eq. 3 we can find conditions for resonances for all orbit type. The most known one is for GTO and has been extensively studied [4]. In this case the resonant angle Φk can be seen as the local time of perigee, in other word the angle between the perigee direction and the sun direction. As some point during the evolution of a GTO at low inclination, the J2 drift of the orbit makes the perigee rotates at the same speed as the apparent movement of the Sun around the Earth. As a consequence, the local time of perigee becomes constant and we have an orbital resonance whose impact depends on the local time of perigee itself. Fig. 7 illustrates this effect, with very different orbit evolution after the resonance when the entry condition is slightly changed.

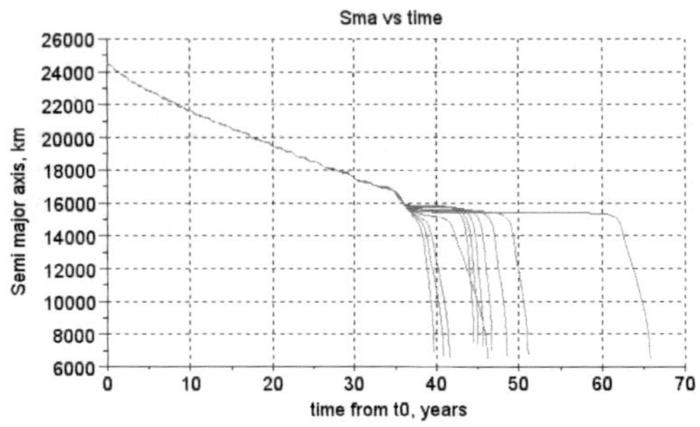

Figure 7: GTO evolution: impact of orbital resonances

GTO+, that is to say GTO with higher perigee altitude, are also subject to resonances, as illustrated with this very long term apogee altitude evolution due to a moon resonance (for more detail see [4])

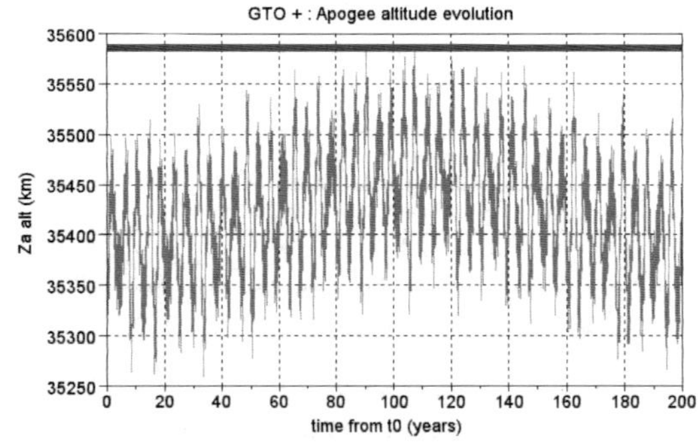

Figure 8. Apogee evolution of a GTO+ resonant orbit

One can see from Fig. 8 that we can have periodic evolutions with periods higher than 100 years, which is the time limitation commonly used to check the criteria of crossing of the protected region. It should make us consider the possibility to rethink this time limit in case of orbital resonances.

More surprisingly, even *a priori* "stable" orbit such as MEO are subject to strong resonances. J. Daquin finished a thesis sponsored by CNES, THALES and IMCCE to study such effects [10]. Galileo orbits for example are subject to a 3rd body resonance with Φk angle being 2ω+Ω (Argument of perigee and RAAN respectively). The very long term evolution of such orbits can be impressive, as can be seen in Fig 9. that shows the eccentricity evolutions of several MEOS with different values for the initial eccentricity vector.

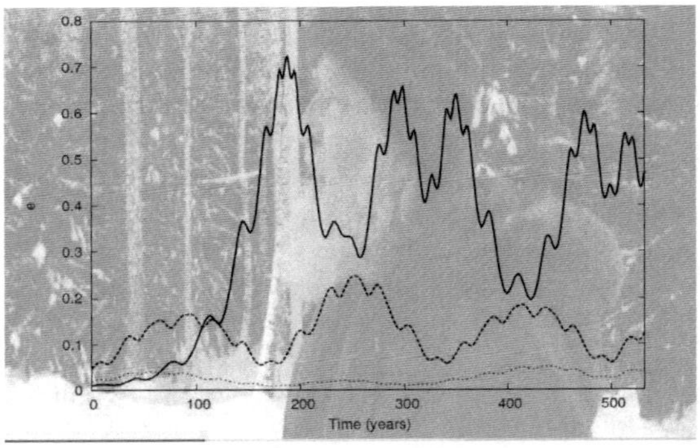

Figure 9. Eccentricity evolutions of MEOs

The picture in Fig. 9 Background shows Boris Chirikov whose work made significant advancement in resonances overlapping and Hamitolnian chaos.

It is worth noting that at this time no protected region or particular criteria seems to exist for MEOs, in the good practices attached to the FSOA nor in common international guidelines.

GEO [11] and LEO [12] are also facing the issue of resonances with possibly great impact on their long term evolution.

5.3. How to deal with resonances

First of all, one should know that even if the physical phenomenon can be understood and the underlying equation can be expressed in most of the case, a detailed analysis of all possible resonances seems unreachable due to the complexity of all combinations, configurations and couplings that may lead to such commensurabilities. As a consequence, theoretical analysis should not be sufficient to ensure that a resonance will or will not happen: simulations are required. During these simulations the appropriate force model has to be considered: if you decided to neglect the moon contribution because it is small most of the time, you can miss some commensurability with the moon with great impact on the orbit evolution. As a start, taking into account the most complete force model you have is recommended for a resonance analysis.

Chaos indicators, such as FLI or MEGNO [13], can be implemented in order to warn the user if a strong sensitivity to initial conditions is detected. Those indicators monitor the evolution of a small initial dispersion, using for example the state transition matrix. If a small dispersion starts at some point to grow fast, then an orbital resonance is to be suspected. Once a resonance has been detected and identified, a statistical method seems required since a worst case approach is hard to adopt and the result of only one propagation is irrelevant. It has been described in detailed for GTO in [4] and the same strategy could be adopted for other orbit types.

6. SUM UP

Extensive work has been perform since 2010 to establish some good practices, shared between CNES and French operators, to help both applicants and certifiers to follow the technical regulations attached to the French Space Operations Act. An efficient tool, STELA, has been released to be able to perform long term orbit propagation, thus helping to mitigate the debris growth in the LEO and GEO protected region. However, a lot of activities and studies are ongoing, to keep improving both tools and good practices.

Long term orbit propagation requires finding a compromise between precision and computation time. We still try to improve the STELA force model while keeping the short computing time that makes its use easy and convenient in early analysis phases. Implementing a recursive formulation for the Earth potential perturbation is a solid lead.

The long term evolution of the attitude of an uncontrolled spacecraft is hard to take into account and therefore attitude and orbit are in fact considered as de-coupled. While the tumbling assumption seems to remain valid most of the time, we believe there is room for some improvement in trying to be able to identify or predict possible stable attitude and estimate the impact of attitude and orbit coupling.

Drag estimation is of key importance for LEO and GTO evolution. The unpredictability of solar activity over such long time scale remains a challenge, so engineers have to find a way to deal with it in their analysis. The constant equivalent solar activity approach recommended by the good practices could be updated in the future to take into account recent solar activity measurement.

Cautious should be brought with respect to orbital resonances since they can have great impact on the orbit evolution. Using a complete enough force model is mandatory. Chaos indicators could be implemented in STELA software to warn the user if strong sensitivity to initial conditions is detected. In this case a statistical approach should be considered since the result of one propagation becomes irrelevant.

Also, while protected region has been defined for LEO and GEO, no common guidelines deal with MEO. The increasing number of satellite orbiting this region as well as the complex orbit evolution that can occur due to orbital resonances should motivate some studies toward the establishment of such guidelines.

7. REFERENCES

1. Inter-Agency Space Debris Co-ordination Committee, IADC Space Debris Mitigation Guidelines, IADC-02-01, 2007.

2. C. Le Fèvre, Compliance of disposal orbits with the French Space Operations Act: the good practices and the STELA Tool, ActaAstronaut.94(Issue1)(2014)234–245

3. J.C. Dolado, C. Pardini, L. Anselmo, Review of Uncertainty sources affecting the long term predictions of space debris evolutionary models, Acta Astronautica 113 (2015) 51-65.

4. Fraysse H. & al (2013). Statistical methods to address the compliance of GTO with the French Space Operations Act. IAC-13,A6.4,3x19616.

5. https://logiciels.cnes.fr/content/stela

6. Le Fèvre C. & al (2015). Integration of coupled orbit and attitude dynamics and impact on orbital evolution of space debris. AAS 15-335

7. Morand V. & al (2013). Mitigation Rules compliance in Low Earth orbit. IAASS 13

8. NASA Technical Memorandum 4759 http://sail.msfc.nasa.gov/

9. ISO 27852:2011 Space systems — Estimation of orbit lifetime

10. Daquin J. Essais sur le mouvement des corps artificiels résonants. Paris Observatory

11. Morand V. & al (2014). End of life disposal of satellites in the GEO region, the issue of high inclinations. ISSFD conference 2014

12. Lamy, A. et al, 2012. Resonance Effects on Lifetime of Low Earth Orbit Satellites. ISSFD conference 2012

13. Cincotta, 2002. Phase space structure of multi-dimensional systems by means of the mean exponential growth factor of nearby orbits

PROCEEDINGS OF THE 8TH INTERNATIONAL SPACE SAFETY CONFERENCE

SESSION 06
REGULATIONS AND STANDARDS - I

2016 UNITED STATES SAFETY REGULATIONS STATUS UPDATE

Diane Howard, J.D., LL.M., DCL[(1)]

[(1)] *Embry-Riddle Aeronautical University, 600 S. Clyde Morris Blvd., Daytona Beach FL U.S.A.*
Email: diane.howard@erau.edu

ABSTRACT

Safety remains a primary driver of regulation for space activities internationally. Consensus on appropriate safety metrics, calculated for whom, is elusive. The new U.S. law, the Commercial Space Launch Competitiveness Act of 2015 (CLSCA); the recent Federal Aviation Administration (FAA) Recommended Practices for Human Space Flight Occupants; and the National Transportation Safety Board (NTSB) report on last year's SpaceShip Two accident all contain implications for the space safety community. The immediate paper parses the U.S. law, the FAA guidance, as well as the NTSB recommendations and contrasts these with the spirit and intent of the International Association for the Advancement of Space Safety Suborbital Safety Guidelines Manual. Events are presented in chronological order.

1. INTRODUCTION

It has been inferred that the space treaties create a legal obligation for States to adopt and implement national space legislation, including safety standards and procedures.[1] As with other international treaties, these treaties are subject to various national treaty application systems.[2]

Safety is a policy driver in the majority of national space laws. Examination of the domestic space laws of nineteen different jurisdictions for common elements and policy drivers reveals the significance of safety as both an element of domestic space law and as a policy driver underpinning those laws.[3] Analysis showed that safety was the element most prevalently found in domestic space law (seventy-one percent of the jurisdictions included).[4] More importantly, the value on safety is of vital importance in most systems; safety was included as a value in seventy-five percent of the space laws and policies examined, second only to the value placed upon international obligations and cooperation.[5] However, safety can be assessed using different methodology and standards in different jurisdictions, an issue to be explored infra.

2. U.S. REGULATORY LANDSCAPE

The safety component of a U.S. license or permit fulfills the international obligations found in Articles VI, VII, VIII, and IX of the Outer Space Treaty.[6] Safety assessments are necessary to obtain either a spaceport, launch, or launch operator license, or an experimental permit.[7] The U.S. Congress, through the Commercial Space Launch Act of 1984 ("CSLA"), directs the Office of Commercial Space Transportation ("AST") to regulate private sector launches, reentries, and related services (this includes spaceport operations) "only to the extent necessary ... to ensure compliance with the international obligations of the United States and to protect the public health and safety, safety of property, and national security and foreign policy interests of the United States."[8] Safety assessment is part of the U.S. authorizing process.[9]
The requirements to receive a license include meeting the AST's parameters of acceptable or tolerable risk.[10]

2.1 CSLA & CSLAA & The Learning Period

Congress imposed a dual mandate upon the FAA AST in the Commercial Space Launch Act that has continued in all of that law's progeny.[11] The agency must protect third-party safety and property while it also encourages, facilitates and promotes U.S. commercial space transportation.[12] To this end, when enacting the CSLAA, Congress restricted the FAA's ability to promulgate regulations dealing with the health and safety of crew and spaceflight participants. The intention was to provide a "learning period" to newcomers to the space transportation industry without overly burdening the development of innovative technologies with premature regulation.[13] In this way, both prongs of the mandate could be realized.

The learning period was not without caveats. The CSLAA allowed the FAA to enact regulations prior to the expiration of the period if a flight resulted in a fatality or serious injury or in an incident that likely could.

Originally, this period was to last eight years. The thought was that this would give the industry time to begin operations and to accrue a bank of data from which tailored safety regulations could flow.[14] However, the timeline for operations was longer than expected. Despite the press and public's hunger for commercial operations to commence, many operators are still in experimental permit phase. Hence, Congress extended the period several times. The newest piece of U.S. space legislation, discussed below, enlarges this time frame.

2.2 FAA Recommended Practices for Human Space Flight Occupant Safety

The FAA first presented its Recommended Practices for Human Space Flight Occupant Safety in August 2014. Announcing the documents availability at a COMSTAC meeting in September, AST Associate Administrator Dr. George Nield explained that these were a starting point in the long and arduous rule-making process and urged industry to enter the discourse by compiling consensus industry standards pertinent to occupant safety.[15]

Best practices and standards are used with increasing frequency, particularly in technological sectors, such as aerospace, where more codified law is either not the goal or believed to be premature. [16] The reliance on standards and best practices in the absence of more positive law has been envisaged as symptomatic of emerging norms in either nascent or fast-changing areas of regulation.[17]

Presently, there are a number of space safety standards developing initiatives in various stages of work. The most recent COMSTAC Working Group was formed in 2015 and deals with Standards, and is tasked to identify and analyze existing industry standards and to develop and prioritize a list.[18] In 2013, the Commercial Spaceflight Federation, a consortium of U.S. member organizations working together for the common goal of democratized commercial human spaceflight, had approved the first U.S. industry standard drafted by its Technical Standards Committee.[19] However, since that time, standards development has been slow going, at best. Industry reluctance to part with closely held, proprietary information appears to be a hurdle.

2.3 Scaled Composites, the NTSB & the Learning Period

October 2014 ended somberly for two innovative spaceflight companies – Orbital Sciences and Scaled Composites. Interestingly, both suffered anomalies a mere two months after the AST presented its Recommended Practices for Human Space Flight Occupant Safety.

The Orbital crash, while unfortunate, did not result in fatality or serious bodily injury.[20] Nor did the company's investigation reveal conditions that likely would have. The triggering events that could set in motion FAA regulatory action were not present.

But the Virgin Galactic anomaly was another story entirely. The co-pilot died. The other pilot survived but was grievously injured. For the first time, the NTSB took the lead on the investigation,[21] finally exercising jurisdiction as per a Memorandum of Understanding between the FAA, the NTSB, and the U.S. Air Force that had been in place for ten years.[22] The commercial spaceflight industry was rocked. Two tragedies struck within days of one another. The sadness was heartfelt.

Once emotion began to settle, the questions began. How would the FAA respond? What, precisely, did the language of the CSLAA require? How would this impact the industry as a whole?

The actual text of the law limits safety regulations to "restricting or prohibiting design features or operating practices that have resulted in a serious or fatal injury to crew or space flight participants during a licensed or permitted commercial human space flight."[23] This means that any resulting regulation would have to be narrowly tailored to address only those design features and operating practices that were directly implicated in causation of the death and/or injury.

The NTSB concluded its investigation in late July 2015, finding probable cause in the company's (Scaled Composites) "failure to consider and protect against the possibility that a single human error could result in a catastrophic hazard" to the vehicle.[24] While the NTSB made eight recommendations to the FAA and two to the commercial spaceflight industry via the Commercial Spaceflight Federation, these deals primarily with human factors and the timing of pre-permit and pre-license consultations.[25] The ramifications for the industry have not been broad or onerous.

2.4 U.S. Commercial Space Launch Competitiveness Act (CLSCA)[26]

Enter the new law. To summarize, the new U.S. law consists of four sections, Titles I-IV. It is a sort of omnibus bill in that it covers several diverse issues but they are all related to commercial space activities in the U.S.. The entire law contains twelve different reporting requirements or "asks"[27] and references "safety" nineteen times.

Many of the safety allusions are an affirmation of safety as a key priority to be considered in the streamlining of the existing regulatory framework and the required reports dealing topics ranging from the appropriate entity to authorize on-orbit space activities to space traffic management and situational awareness. In fact, the law reaffirms existing policy that directs the Secretary of Transportation to promote and facilitate private sector space activities while protecting third party public health and safety, and safety of property along with national security and foreign policy interests of the United States.

However, the CSLCA goes further. First, it extends the learning period, described supra, until September 30, 2023. Next, it echoed the FAA invitation for participation in developing standards by directing the Secretary of Transportation to continue working with industry to develop consensus standards on a voluntary basis. Two branches of government, legislative and executive, are now aligned, urging industry to proactively engage in the rulemaking process.

Further, the new law directs the Secretary to provide reports from 2016 until 2021 on how these standards are evolving (an ongoing ask). These reports feed into other reports regarding industry metrics denoting readiness of the space sector and the Department of Transportation to transition to a safety framework including regulations covering space flight participants, crew, and the new government astronaut classification – first and second parties. The Secretary is expressly directed to work with the Commercial Space Transportation Advisory Committee (COMSTAC) or its successor in furtherance of standards development. COMSTAC is a group advising the FAA "on all matters relating to U.S. commercial space transportation industry activities," with support from the Office of Commercial Space Transportation.[28] In other words, there is a degree of accountability build into the law. Stonewalling will not be productive for the U.S. commercial spaceflight industry.

3. IAASS GUIDELINES FOR THE SAFE REGULATION, DESIGN AND OPERATION OF SUBORBITAL VEHICLES

How do the current U.S. regulatory directives fit into the bigger global picture, particularly in terms of interoperability? In 2011, the International Association for the Advancement of Space Safety (IAASS) formed a Technical Committee ("TC") to tackle the safety issues relevant to the suborbital industry.[29] The TC's membership included individuals representing regulatory authorities, space agencies, private industry, academia, and IAASS members with expertise. Its goal was to identify best practices and standards for technical issues pertaining to safety and operations.[30] In other words, industry would work with academics and regulators to develop some consensus standards. Although further back in time, this sounds quite a bit like the call Dr. Nield extended in 2014 and the requirement Congress codified in 2015.

The TC drafted guidelines and presented these with a report in Germany in October 2014.[31] The guidelines recommend safety assessment for all parties, not only the general uninvolved public. This position diverges from the current U.S. regulatory requirement. However, the risk assessments now utilized by the FAA are not completely incompatible with the methodology suggested by the IAASS (and a number of European experts). Both could be used simultaneously. One includes first and second parties, while the other deals only with third parties. Further divergence occurs as to which metrics are considered acceptable. A more conservative metric is used in the IAASS Suborbital Guidelines, similar to that used in the aviation industry. Despite these variables, the risk assessments are stated in the same terms (numerical), apples to apples, and do not represent an insurmountable hurdle.

4. CONCLUSION

Sometimes great ideas are lost in translation. One party can view an action as very obviously for the greater good and another will see it quite differently. While this diversity of viewpoint and independent opinion can be the antidote to groupthink,[32] it can also frustrate efforts toward coalition. This appears to be the case regarding the impasse between U.S. operators and their European counterparts regarding safety standards. However, impasse does not necessarily equate with failure; it can simply mean delay. Furthermore, the Secretary of Transportation and the FAA are now directed to consult with U.S. industry to move forward with consensus standards. Some of the legwork has commenced. Pragmatism suggests that U.S. industry could benefit from at least thinking about some of this preliminary work lest regulators move forward without their input.

All space stakeholders, whether they function in industry, academia, in any governmental capacity, or as regulators, prioritize safety in space activities. It is a shared value. While the details remain to be resolved, an acknowledgement of agreement on the priority of these issues is a necessary first step. The IAASS effort to create guidelines from applicant-developed criteria was simply an early initiative to begin the process of creating industry consensus standards. The U.S. operators, while invited to participate, chose not to. Perhaps this was a misunderstanding. Perhaps it was a timing issue. The fact remains that all the current initiatives described in the immediate article are voluntary, non-binding, efforts to identify existing protocols that can make space activities safer for human life. Consensus takes time. But, the ultimate pay-off is the satisfaction that all participants in building that consensus have in knowing that they contributed to a greater good.

5. REFERENCES

1. Jakhu & Nyampong, *supra* note 17, at 5.

2. John H. Jackson, *Status of Treaties in Domestic Legal Systems: A Policy Analysis*, 86 AM. J. INT 'L L. 310 (1992). The implementing statute may use different language and clarify or elaborate upon that found in the treaty; the domestic law is the transformatory act.

3. Diane Howard, The Emergence of an Effective National and International Spaceport Regime of Law, (Nov. 2014) (unpublished doctoral dissertation, McGill University) (on file with McGill Library, McGill University). The countries included in the study were Australia, Austria, Belgium, Brazil, Canada, China, France, Indonesia, Japan, Kazakhstan, the Netherlands, Nigeria, Norway, Russia, South Korea, Sweden, Ukraine, United Kingdom, and the United States.

4. *Id.* at 54.

5. *Id.* at 82.

6. 14 C.F.R. §§ 420 *et seq.*; *see also* J. Randall Repcheck, *Safety and Promotion in the Federal Aviation Administration—Enabling Safe and Successful Commercial Space Transportation*, in Proceedings Fourth IAASS Conference (May 2010) [hereinafter Repcheck I].

7. 51 U.S.C. § 50901(a)(7) (2012).

8. *Id.*

9. J. Randall Repcheck, FAA's *Implementation of the Commercial Space Launch Amendments Act of 2004 – the Experimental Permit*, in Proceedings Of The First IAASS Conference, Space Safety, A New Beginning, (Dec. 2005) [hereinafter, Repcheck II]. Although the paper describes a different aspect of the CSLAA of 2004, the experimental permit, than that with which we are here concerned, the FAA's role in quantifying risk levels is the same.

10. *Id.* The paper describes the experimental permit rather than license but the FAA's role in quantifying risk levels is the same.

11.. Commercial Space Launch Act, PUB. LAW 98-575 §3 (Oct. 30, 1984).

12. 51 U.S.C. §§ 50903 (b), (c) (2015).

13. While many refer to this part of the CSLAA as a moratorium on safety regulation, that statement is not accurate. Jeff Foust, *Industry, FAA at Odds over Extension of 'Learning Period' for Commercial Spaceflight Safety Regulations*, Space Politics (Feb. 6, 2014) *available at* http://www.spacepolitics.com/2014/02/06/industry-faa-at-odds-over-extension-of-learning-period-for-commercial-spaceflight-safety-regulations/.

14. Actually, Dr. Nield, Associate Administrator for Commercial Space Transportation, FAA, testified in 2014 that he believed the present to be the time to begin thinking about safety regulations for crew and participants as we now have 50+ years of NASA experience from which to draw information. Statement of Dr. George C. Nield, Associate Administrator for Commercial Space Transportation of the Federal Aviation Administration, before the House Committee on Science, Space, and Technology, Subcommittee on Space, on necessary updates to the Commercial Space Launch Act, Feb. 4, 2014, *available at* http://testimony.ost.dot.gov/test/nield1.pdf.

15. Prior to agency rule-making action, each proposed requirement must be justified, subject to economic analysis, and put before the public for comment. Exec. Order No. 12866, 58 Fed. Reg. 51735 (Oct. 4, 1993).

16. Daniel P. Murray & Andre Weil, *The FAA's Approach to Quality Assurance in the Flight Safety Analysis of Launch and Reentry Vehicles*, in Proceedings Fourth IAASS Conference, Making Safety Matter (May 2010).

17. Martha Finnemore & Kathryn Sikkink, *International Norm Dynamics and Political Change*, 52-4 INT'L ORG. 887, 895 (1998).

18. *Office of Commercial Space Transportation, COMSTAC Membership*, FAA.GOV, (Mar. 25, 2015, 4:51:26 PM), *available at* http://www.faa.gov/about/office_org/headquarters_offices/ast/advisory_committee/membership/.

19. Annual Report 2013, Commercial Spaceflight Federation at 4. This first standard deals with propellant handling; apparently three more are in process.

20. *Orbital-led Team to Investigate Antares Rocket Explosion*, Daily Press (Nov. 5, 2014), http://www.dailypress.com/news/science/dp-nws-orbital-accident-board-20141105-story.html.

21. *NTSB Investigates Virgin Galactic Test Flight Crash*, NTSB.Gov, *available at* http://www.ntsb.gov/investigations/Pages/2014_Virgin_Galactic.aspx (last visited Feb. 14, 2015). The NTSB is a completely independent investigative body that provides recommendations, not judgments of fault.

22. Memorandum of Understanding Between the National Transportation Safety Board, Department of the Air Force, and Federal Aviation Administration Regarding Space Launch Accidents, (2004), *available at* https://www.faa.gov/about/office_org/headquarters_offices/ast/media/mou_space_launch_accidents.pdf (last visited Feb. 14, 2015).

23. 51 U.S.C. ntsb.gov/)(2)(C) (2012).

24. *Aerospace Accident Report: In-Flight Breakup During Test Flight, Scaled Composites SpaceShipTwo, N339SS Near Koehn Dry Lake, California October 31, 2014*, (Adopted July 28, 2015) *available at*: http://www.ntsb.gov/investigations/AccidentReports/Reports/AAR1502.pdf (last visited September 6, 2015), pgs. 69-71.

25. *Id.*

26. Public Law 114-90-Nov. 25, 2015

27. The first Title addresses the methodology that the FAA has used to calculate maximum probable loss since 1988 when it was first introduced. This value is used to determine financial responsibility requirements for a launch or reentry license or a permit. This is the first ask for a report and is directed to the Secretary of Transportation. Title I also deals with indemnification for space flight participants (no ask), allowance of activity by an operator under a permit and a license concurrently (no ask), streamlining licensing and permitting processes and launch and reentry licenses (a few asks of the Secretary of Transportation again), determination of federal jurisdiction for claims by third parties and space flight participants, inclusion of space flight participants in the cross-waiver scheme used by contractors and subcontractors along the launch supply chain (no ask), appropriate authority within the USG over on-orbit activities (like servicing satellites or retrieval or extraction of resources or SSA) (an ask of a report by the Office of Science and Technology Policy working with other departments in the Executive Branch), an orbital space traffic management study by an independent organization contracted by NASA and other Departments (DOD, Commerce, etc.) (a big ask), a Space Situational Awareness study (another ask) by the Secretary of Transportation. The ISS is extended to at least 2024. Indemnification is contemplated for all government property, not just federal (another report on this), a report is required about the use of support vehicles. There are some housekeeping items like changing wording in statutes to reflect that the Shuttle is gone and the new SLS is with us, and changing the name of an agency to the Office of Space Commerce (Title III). The Department of Commerce is required to report on Remote sensing license applications (Title II). Title IV deals with resource exploration and utilization. It contains definitions, acknowledges US international obligations, very clearly describes the rights it believes a commercial entity is entitled to with regard to extracted resources and also very clearly states that these rights in no way infer a claim of sovereignty or ownership over a celestial body. This title also directs the President to conduct a study to determine what authorities would be necessary to meet US international obligations under the Outer Space Treaty Article VI and how to allocate those responsibilities among federal agencies.

28. COMSTAC Charter, U.S. Dep't. Trans. Order 1110.124H (Nov. 16, 2012). COMSTAC itself is comprised of twenty-five representatives from industry and academia, serving for a two-year renewable term. 29

29. Andy Quinn et al., New Suborbital Safety Technical Committee for the IAASS, in Proceedings Fifth IAASS Conference, A Safer Space For A Safer World, (Oct. 17-19, 2011) (ESA SP- 699, Jan. 2012). The author serves as the legal lead for the TC.

30. Ibid.

31. Guidelines for the Safe Regulation, Design and Operation of Suborbital Vehicles, IAASS Suborbital Safety Technical Committee Manual (May 2014), update presented at the Seventh IAASS Conference, Space Safety Is No Accident, (Oct. 2014, Germany).

32. Sharon Allen, The Death of Groupthink, Bloomberg Business (Feb. 5, 2008), available at http://www.bloomberg.com/news/articles/2008-02-05/the-death-of-groupthinkbusinessweek-business-news-stock-market-and-financial-advice.

REGULATION OF SAFETY OF SPACE MINING AND ITS IMPLICATIONS FOR SPACE SAFETY

Ram S. Jakhu [(1)] and Joseph N. Pelton [(2)]

[(1)] *Associate Professor, Institute of Air and Space Law, McGill University, Montreal, Canada.*
[(2)] *Executive Board, IAASS & former Dean International Space University.*

INTRODUCTION

The practical uses of satellites have grown over the past half-century. Such use was demonstrated with the deployment of telecommunications satellites, which now include broadcasting satellites, mobile communications satellites, search and rescue, and data relay satellites. Shortly after the first communications satellites were deployed and proved to be commercially viable other applications quickly followed. Next came, remote sensing satellites, weather satellites, and then navigation and timing satellites. Nor has this chain of space development come close to ending.

We may soon have operational systems for robotic repair and refueling of satellites, solar power satellites, as well as increasingly sophisticated satellites for various types of defense and security operations. But something entirely new is also now on the horizon for space applications and instead of operation in Earth orbits these systems may seek to operate in deep space. The next major commercial space application may redefine the future of space activities to include major off world activities. This new activity would constitute a serious attempt to reclaim natural resources from space, or quite simply, to engage in space mining. Some of those engaged in this activity also envision processing of materials in space and even space manufacturing. Such deep space operations may not only involve a new range of deep space technologies, smart robotics, autonomous operations, independently operated power supplies and space systems, but also completely new challenges in terms of space safety systems as well as new types of standards and approaches to space safety for deep space operations, where immediate Earth-based response systems may simply not be available.

There are at least four "space mining" companies that are seeking to implement space mining explorations and develop the deep space tools needed to undertake such future operations. This suggests that space mining is no longer an issue for the longer-term future, but a real for today's regulators. Deep Space Industries (DSI), Moon Express, Planetary Resources Inc., and Shackleton Energy Corporation - each with a different set of strategic plans - are primarily focused on developing the key technologies and developing target locations for their future operations. Yet each of these entities - all with a U.S. base of operations - also recognize that there are regulatory issues and safety standards questions that also must be addressed. This is why they actively supported new U.S. legislation that has sought to define the process whereby space mining might be undertaken and the development of a new U.S. law that would define how space mining might be conducted. This initiative at the national regulatory level to create new national law to cover future space-based mining operations as well as possible space-based processing and manufacturing have now started to come forward as one possible means to cope with the "regulatory gap" that has now started to become apparent. However, national space mining operations must also be conducted in accordance with the provisions of the currently applicable UN space treaties.

1. ADEQUACY OF INTERNATIONAL TREATIES ON SPACE FOR NEEDS RELATED TO SPACE MINING?

In addition to the regulatory system created by the Outer Space Treaty,[1] four other associated international treaties,[2] provide good governance foundation, but would also prove inadequate, to providing for an effective international safety regulatory environment and "rules of the road" that are fully responsive to the needs related to space mining operations. Particularly from the perspective of responsibility for safety in space mining, there are two issues that need to be addressed.

Firstly, who is entitled to exploit space resources and secondly what and by whom space safety standards and procedures should be adopted and implemented?

Under international space law, all States are free to explore and use outer space, including the moon and other celestial bodies, without discrimination of any kind, on a basis of equality and in accordance with international law.[3] Such freedom covers exploitation of space resources but does not extend to non-governmental entities (private corporations), which must carry out their space activities only pursuant to "authorization and continuing supervision" by their respective States. Therefore, States decide who, within their jurisdictions, should be allowed to carry out space mining operations and under what terms and conditions. Some of the standard conditions that are generally imposed are the requirements for compliance by the licensees with: (a) the international obligations of the respective States and their national security or foreign policy interests and (b) the safety standards as determined by the concerned States. Safety standards are imposed to protect public health and safety, and safety of property, or to implement international obligations. Moreover, imposition of safety standards is generally accompanied with the requirement of procurement of liability insurance to cover claims against government property damage and third party death, bodily injury, or property damage. Internationally, it is the launching State[4] that is liable for damage caused by space object of its private licensees, irrespective of the fact that the damage occurred in air space or in outer space, on the moon and other celestial bodies, including asteroids.[5]

States may undertake space mining operations on their own, allow their private companies to carry out such activities or participate in the space mining operations of international organization(s). However, in latter case the responsibility for adoption of and compliance with safety standards would be borne both by the international organization(s) and by the States participating in such organization(s).[6] Similarly, if two or more States jointly undertake space mining activities,[7] each State will be expected to adopt and impose space mining safety standards and procedures. Moreover, in order to avoid accidents, damage to property, death of and injury to humans, Earth and space environment damage, it would be imperative to adopt and implement complementary national and international space mining safety standards and procedures. Currently, there are no international safety standards and procedures for space mining.

2. THE U.S. SPACE RESOURCE EXPLORATION AND UTILIZATION ACT OF 2015 (SREU).

On 25th November 2015, the U.S. President Barack Obama signed the "Space Resource Exploration and Utilization Act of 2015" which is Title IV of the U.S. Commercial Space Launch Competitiveness Act (Space Act of 2015).[8] The Space Act of 2015 contains four formal parts designed to: (i) promote commercial space transportation and provide increased indemnification for space flight participants; (ii) deal with commercial remote sensing statutory particularly the updating of licensing of private remote sensing space systems; (iii) cover issues such as reporting requirements to Congress, the creation of a new Office of Space Commerce, and guidance to four key Departments of the US Government in the Space Arena namely Dept. of Defense, Dept. of Transportation, NASA, and the Department of Commerce; and (iv) address the legislative issues related to space mining (i.e. SREU Act).

The overall broad purpose of the new legislation is essentially to promote new space commercial transportation systems as well as "promote the right of United States citizens to engage in commercial exploration for and commercial recovery of space resources free from harmful interference, in accordance with the international obligations of the United States and subject to authorization and continuing supervision by the Federal Government."[9]

This national legislation, which is the first of its kind in the world, expresses the right of the U.S. government to "entitle an American company to possess, own, transport, use, and sell the space resources obtained through its commercial undertakings".[10] This legislative initiative has attracted unprecedented appreciation, primarily from the American private space sector, but has also aroused concerns[11] about the potential violation of the U.S. international obligations as well as possible safety risks such operations might pose.

It should be noted that the above-mentioned right of the private companies shall be granted only "in accordance with applicable law, including the international obligations of the United States."[12] This implies that there should be national law in the U.S. under which licenses for space mining can be issued. At present, no such law exists, though it has recently been reported that Rep. Jim Bridenstine (R-OK), is drafting a legislation "regarding property rights to resources mined from asteroids by U.S. companies."[13] Secondly, there is no clear official statement on the U.S. international obligations with respect to space mining. Thus one cannot determine precisely whether or not the SREU Act is in violation of the international obligations of the U.S. Such proclamation may be spelled out in a Report which the U.S. President is required and expected to submit to the U.S. Congress within 180 days of the adoption of the SREU Act; i.e. from 25 November 2015.[14] Though the U.S. is obliged to interpret and comply with its treaty obligations in good faith,[15] merely having a national law which is inconsistent with such obligation is not a violation of its international obligations. According to Ian Brownlie, "a breach [of an international obligation] arises only when the State concerned fails to observe its obligations on a specific occasion."[16] Therefore, if and when properly licensed U.S. private companies are engaged in space mining, which is determined to be contrary to the provisions of applicable international treaties, only then the U.S. would be considered in breach of its international obligations.

Article II of the Outer Space Treaty declares that "Outer space, including the moon and other celestial bodies, is not subject to national appropriation by claim of sovereignty, by means of use or occupation, or by any other means." This prohibition is considered to be strict and comprehensive covering appropriation by States, public and private companies and individuals. It is well known that the U.S. has been of the opinion that in accordance with this Article no State or individual or private company can validly claim sovereignty over or ownership of outer space and any celestial body, including asteroids. However, with respect to the ownership of natural resources of celestial bodies, there has been two different views internationally. The U.S. appears to be of the opinion that the legal status of natural resources of celestial bodies is different from that of their surfaces, thus natural resources can be appropriated. This is reflected, at least impliedly, by the provisions of the newly adopted SREU Act of 2015. On the other hand, several States and legal scholars have been expressing that Article II prohibits both the surfaces and the natural resources of all celestial bodies, including the moon and asteroids. The latter opinion appears to be more logical when one reads the provisions of the Outer Space Treaty by using established rules of treaty interpretations.[17] In this regard, Lyall and Larsen express more respective and well-established view:

> It is argued that the terms of Art. II of the [Outer Space Treat] prohibit only claims by states, therefore leaving open the possibility of a claim by an individual, and that such a claim creates a title which can then be passed on to a purchaser. We disagree. A valid right of property to immoveable estate can exist only with a legal system established by a state and in relation to property over which the state has sovereignty. Since state claims to sovereignty in space cannot exist, neither can title to immoveable property on celestial bodies in space.[18]

Moreover, the intention of the Outer Space Treaty is undoubtedly to expand the benefits of all forms of space exploration and utilization. Therefore, when Article II is interpreted in the light of other provisions of the Treaty, it becomes clear that prohibition of appropriation is necessary to allow the conduct of "exploration and use of outer space, including the moon and other celestial bodies for the benefit and in the interests of all countries, irrespective of their degree of economic or scientific development."[19] The Treaty establishes 'balance of interests' of all States. At the time of ratifying the Outer Space Treaty, the U.S. did not deny the requirement of sharing of space benefits, though it reserved the right to how to share these benefits.[20] Briefly, inclusive interests of States should prevail over the exclusive interests of some States and their respective private companies.

However, it can be said that currently there exists some ambiguity on the question of ownership of the resources extracted during space mining. The future of smooth and controversy-free conduct of space mining operations will depend upon a satisfactory answer to this question: What are potentially positive global benefits space mining that might be derived and equitably shared through future regulation?

One option to address the issue of uniform international regulatory system for space mining could be that all States should ratify by the Moon Agreement, which provides for phased development of global governance system as exploration and exploitation progress, allows expressly some appropriation of extracted natural resources, attempts to prevent celestial bodies from becoming an area of international conflict, expands the provisions for exploration and exploitation of celestial bodies exclusively for peaceful purposes, establishes the rule for help during emergency for humanitarian purposes, envisions equitable sharing of benefits derived from space natural resources, etc. The second option may be, though not exclusive of the first option, to create an international public-private consortium modelled on the original INTELSAT. For this purpose, the U.S. may adopt a national law, similar to the COMSAT Act of 1963, inviting other States to join such organisation. The U.S. private companies possessing significant financial and technological resources would be able to play dominant role in the development of the required technologies, organizational structure and business operations, safety standards and procedures, and to reap substantial benefits from space mining. It is believed that this approach will help eliminating, or at least minimizing international political tensions that are arising due to the adoption of the SREU Act of 2015, which is perceived to be a unilateral attempt to appropriate the celestial bodies.

The SREU Act of 2015 introduces an important distinction between 'asteroid resource' and 'space resource'. While the term 'asteroid resource' means a space resource found on or within a single asteroid, the term 'space resource' on the other hand means an abiotic resource in situ in outer space, including water and minerals.[21] Their status is not fully spelled out in the Act. It should be noted that under the international space treaties there is no clear definition of "celestial bodies" and the Moon is undoubtedly a celestial body. The question is whether asteroids are celestial bodies, and if yes, what size, weight and nature of asteroids or meteorite will be used to determine their legal status. Tons of "star dust" falls to Earth every week and there are billions of asteroids, bolides, meteorites, micrometeorites. No one knows at what level something becomes a "celestial body". In other words, the definitions used in the SREU Act of 2015 need to be assessed with their compatibility with those in the international space treaties, particularly the Outer Space Treaty.

Irrespective of the fact that billions (or even trillions) of dollars can be earned in profits and the right of States and their private companies to own space natural resources can be guaranteed, the success of space mining operations will heavily depend upon the formulation, adoption and implementation of safety standards and procedures for initiating and pursuing such operations. For this purpose, the most important questions need to be addressed include: What would the major safety risks in space mining, who and how should safety standards and procedures be formulated and implemented? What would be the legal consequences for non-compliance with such standards and procedures?

3. TECHNICAL CHALLENGES TO SPACE MINING AND SAFETY STANDARDS

The essential elements for space mining operations include robotic systems that can carry out the mining operations, a sustaining power source that can allow the robotic systems to operate, the sensing ability to locate valuable ores, water or minerals on the Moon or asteroids for possible recovery, transportation systems to and from the sites for the mining operations and to extract and process space resources without damaging the space and earth environment. The first stage to launching a credible effort to engage in space mining is to design, build and deploy surveyor probes that can go out at very low cost to identify likely sources for space mining and to create low cost transportation systems to support this effort. It is possible that in the future 3D printing systems equipped with the right "smart" software, robotic capabilities and raw materials could implement the infrastructure for an entirely functional space colony or other sophisticated tasks. For the nearer term space mining activities, advanced survey and prospecting, low cost commercial transportation systems, robotic mining systems, and self-sustaining power systems represent the true core technologies.

Clearly space mining to be successful must not only take on a wide range of technical challenges and find new solutions but along the way work with safety experts and standards organizations (both national and international) to ensure that both automated systems and human crewed transportation systems and habitats are provided at the highest level of safety. There are, unfortunately many different types of systems, each with their own unique technical, operational and safety challenges and safety. Some of these systems include:

(a) Continuity and sustainability of remote electrical power systems,
(b) Low cost and reliable deep space transportation systems,
(c) Reliable and precision low thrust maneuverability jets,
(d) Remote robotically-controlled drilling and excavation equipment that can be operated with reliably in deep space,
(e) Informational technologies, networking, and telecommunications systems that are "smart", efficient, and constantly sustainable, and
(f) Virtually all of these systems will also require a high degree of artificial intelligence, particularly if these systems are to operate remotely without human astronauts and at sufficiently great distances that tele-robotic systems are infeasible.

What is urgently needed is to adopt safety standards and procedures for extracting, processing, storing and using space natural resources on the celestial bodies, including asteroids, for orderly, safe and sustainable development of such resources. The nature of the technical challenges significantly increase if there are astronauts and remote human technical experts involved in the space mining operations. The additional requirements would include such features and life support systems for not only oxygen and water supply, but the even greater challenge of food and other expendables that humans expect in the course of their daily lives. The ability to create off world growing of nourishment and material processing to create sophisticated products such as television sets, furniture, and other items for a remote habitat is clearly some years away. Thus it is likely that the first decades of space mining and material processing would be based on automated systems. In time, however, there could develop capabilities for remote space colonies that can sustain human life.

4. CONCLUSIONS

With well-funded projects and full determination private companies can be expected to develop necessary technologies and infrastructures required for space mining. However, the success of their operations will depend upon appropriate and timely determination of solutions to the following issues:

(a) What are potentially positive global benefits space mining that might be derived and equitably shared through future regulation?

(b) What would the major safety risks in space mining, who and how should safety standards and procedures be formulated and implemented?

(c) What would be the legal consequences for non-compliance with such standards and procedures?

The adoption of national laws, like the U.S. SREU Act of 2015, are the first important steps for initiating governance system as they provide national regulatory basis for licensing process and continuous supervision of, and for imposition of safety standards and procedures upon, the space mining activities of private companies. However, the challenge will be to implement these laws in such a way that would not breach international obligations of the concerned States. Since space mining operations will be taking place in international environment of outer space or on celestial bodies (including asteroids), it would seem logical that safety standards and procedures be international and comprehensive in nature and scope so that all interests of countries (space faring and non-space faring) remain protected. It would also be prudent for these standards and procedures to use common metrics and interfaces so that rescue or repair operations could be more easily achieved. This is perhaps an areas where the IAASS could play a constructive future role.

6. REFERENCES

1. *The Treaty on Principles Governing the Activities of States in the Exploration and Use of Outer Space, Including the Moon and Other Celestial Bodies*, 27 January 1967, 610 UNTS 205, 18 UST 2410, TIAS No 6347, 6 ILM 386 (entered into force on 10 October 1967) [*Outer Space Treaty*] (as of 4 April 2016, there are 104 ratifications and 25 signatures).

2. *The Agreement on the Rescue of Astronauts, the Return of Astronauts and the Return of Objects Launched into Outer Space* [*Rescue and Return Agreement*], adopted by the General Assembly in its resolution 2345 (XXII), opened for signature on 22 April 1968, entered into force on 3 December 1968, there are 94 ratifications, 24 signatures, and 2 acceptance of rights and obligations (as of 4 April 2016); the *Convention on International Liability for Damage Caused by Space Objects*, 29 March 1972, 961 UNTS 187, 24 UST 2389, 10 ILM 965 (1971) (entered into force 1 September 1972) [*Liability Convention*] (as of 4 April 2016, there are 92 ratifications, 21 signatures, and 3 acceptances of rights and obligations); the *Convention on Registration of Objects Launched into Outer Space* [*Registration Convention*], adopted by the General Assembly in its resolution 3235 (XXIX), opened for signature on 14 January 1975, entered into force on 15 September 1976, there are 62 ratifications, 4 signatures, and 3 acceptances of rights and obligations (as of 4 April 2016); the *Agreement governing the Activities of States on the Moon and Other Celestial Bodies*, 5 December 1979, 1363 UNTS 3 (entered into force 11 July 1984) [*Moon Agreement*] (as of 4 April 2016, there are 16 ratifications and 4 signatures).

3. Outer Space Treaty, article I, para 2.

4. I.e. the State that launches or procures the launching of an object into outer space, including the moon and other celestial bodies, and each State from whose territory or facility an object is launched. The State whose private company launches or procures a launch will be considered a launching State.

5. Outer Space Treaty, article VII; Liability Convention articles II and III.

6. Outer Space Treaty, article VI.

7. It is recently reported that the "Luxembourg Government will work with Deep Space Industries [from the U.S.]...to develop the thechnology needed to mine asteroids and build a supply chain of valuable resources in space." See "Prospector-X™: An International Mission to Test Technologies for Asteroid Mining," online Deep Space Industries at: https://deepspaceindustries.com/prospector-x-an-international-mission-to-test-technologies-for-asteroid-mining/

8. 51 USC § 513 (2015).

9. 51 USC § 51302 (3).

10. 51 USC § 51303.

11. Marcia S. Smith, "Bridenstine Drafting Legislation to Implement CSLCA Asteroid Mining Provision," 7 May 2016, Spacepolicyonline at: http://www.spacepolicyonline.com/news/bridenstine-drafting-legislation-to-implement-cslca-asteroid-mining-provision

12. 51 USC § 51303.

13. Marcia S. Smith, "Bridenstine Drafting Legislation to Implement CSLCA Asteroid Mining Provision," 7 May 2016, Spacepolicyonline at: http://www.spacepolicyonline.com/news/bridenstine-drafting-legislation-to-implement-cslca-asteroid-mining-provision

14. 51 USC § 51302 (b) stipulates that: "Not later than 180 days after the date of enactment of this section, the President shall submit to Congress a report on commercial exploration for and commercial recovery of space resources by United States citizens that specifies (1) the authorities necessary to meet the international obligations of the United States, including authorization and continuing supervision by the Federal Government; and (2) recommendations for the allocation of responsibilities among Federal agencies for the activities described in paragraph (1)."

15. *Vienna Convention on the Law of Treaties*, 23 May 1969, 1155 UNTS 31, articles 26 and 31.

16. Ian Brownlie, *Principles of Public International Law*, Oxford: Oxford University Press, (2003), p. 35.

17. *Vienna Convention on the Law of Treaties*, 23 May 1969, 1155 UNTS 31, articles 31 and 32.

18. Francis Lyall and Paul B. Larsen, *Space Law: A Treatise*, (2009) at 184-185.

19. Outer Space treaty, article I, para 1.

20. When expressing its advise and consent to ratification of the Outer Space Treaty by the U.S, the U.S. Senate attached some understandings of the provisions of the Treaty, one of which related to Article I, which provided that: "It is the understanding of the Committee on Foreign Relations that nothing in article I, paragraph 1 of the treaty diminishes or alters the right of the United States to determine how it shares the benefits and results of its space activities." U.S., Senate Committee On Foreign Relations, 90th Cong, 1st Session, *Treaty on Outer Space* (S Exec Rep No 8) (18 April 1967), p.4.

21. 51 USC § 51301.

COMMERCIAL HUMAN SPACEFLIGHT: WHAT REGULATION?

Tommaso Sgobba[(1)], Michael Kezirian[(2)]

[(1)] *International Association for the Advancement of Space Safety, The Netherlands*
Email: tommaso.sgobba@iaass.org
[(2)] *International Space Safety Foundation, USA*
Email: michael.kezirian@spacesafetyfoundation.org

1. INTRODUCTION

The ever increasing number of international actors involved in civil, commercial and military launch, re-entry and on-orbit operations; the envisaged expansion of human access to space for tourism, commerce and point-to-point hypersonic transportation; and the placement on-orbit of global utilities, raises the central question of how to ensure the safety of public, flight participants, crew, ground personnel, and assets on orbit, and the protection of the environment. There is growing awareness that a harmonised framework of international policies and rules is needed. First they must be established, and then continuously monitored for proper implementation, by an accountable international body having the power to enforce compliance. There is, in other words, growing awareness that international governance of space operations is needed. Of all space operations? including exploration and scientific missions in deep space?

We have to separate environmental issues, which are in principle a global concern, like space debris or planetary contamination, from hazards that are of international nature, like launch, re-entry or space traffic, and from hazards that are convenient to address internationally to help commerce. We must also pragmatically consider that currently and for the foreseeable future there is little commercial (and military) interest beyond geostationary orbits (36,000 km), but growing interest for the so-called near-space region, defined as the region above the controlled airspace (i.e.18 km) up to 160-200 km of altitude. Therefore, the international governance discussed in this document is limited to commercial space systems taking place anywhere between 18 km above sea and 36,000 km.

It should be noted that the safety of a space system depends on its design, construction, and operations. Design and constructions are very much "in the hands" of industry, but operations safety cannot be treated neither "unilaterally" nor outside the wider scope of all space operations, manned, unmanned, commercial and non-commercial that take place in the region. Operational procedures have to take into account all possible interactions, multiple parties and sometimes conflicting interests, a variety of operating systems and environment, and also potential impacts on foreign countries. Thus, while in principle space systems safety design and construction rules could be left to industry to decide by means of industrial standards, with compliance certified by a third-part that is not necessarily a government organization, the definition of rules for space operations and the control of their implementation is a responsibility that always belongs to government regulatory bodies, and which needs to be internationally agreed and coordinated.

2. COMMERCIAL SPACEFLIGHT SYSTEMS

For the past half century humans have flown to space on suborbital or orbital space systems developed by governments. Within the next couple of years privately owned and operated systems should begin suborbital and orbital operations with human on-board. In the coming decades trans-atmospheric point-to-point transport may also become a reality. The missions of suborbital, orbital and trans-atmospheric space vehicles are aimed essentially to achieve three different goals: high altitudes (space tourism), prolonged periods of microgravity (research), and cross-range capabilities (fast mode of transportation) respectively.

2.1 Suborbital Spaceflight

A suborbital flight is defined as a flight up to a very high altitude which does not involve sending the vehicle into orbit (i.e. speed below 11.2 km/s). A 'sub-orbital trajectory' is defined as : *"The intentional flight path of a launch vehicle, re-entry vehicle, or any portion thereof, whose vacuum instantaneous impact point does not leave the surface of the Earth."* (US Code of Federal Regulations, 14 CFR 401.5 – Definitions). A typical suborbital trajectory is narrow parabolic trajectory (launch almost vertical).

Unmanned suborbital spaceflight has been common since the dawn of space-age by using sounding rockets for various science and technology research. Suborbital spaceflight is a matter of limits on speed not on altitude. A suborbital rocket can reach 700 km or even more, well above the orbit of the international space station, but without achieving orbital speed.

In the last decade suborbital spaceflight enjoyed new popularity because of the rise of commercial human spaceflight, leisure rather than true transportation, aimed around 100 km of altitude. An experience, sometimes referred to as *experiential spaceflight*, comprising view of Earth from the edge of space and few minutes of microgravity. Commercial suborbital spaceflight is opening the new frontier of the near-space region above Earth that's too high for airplanes but too low for satellites. The exploitation of the near-space region will gain importance in future by adding new commercial applications like stratospheric balloons and high altitude drones for space tourism and Internet services.

2.2 Orbital Spaceflight and Microgravity

Any spacecraft in low Earth orbit must reach speeds of about 28,000 kilometers per hour to remain in orbit. The exact speed depends on the actual spacecraft orbital altitude, which for human spaceflight ranges from about 300 kilometers to 500 kilometers depending on the mission.

It is important to understand the link between orbital spaceflight and microgravity. The pull of gravity gets weaker the farther apart two objects are, but this is not why things float on an orbital spaceflight. For example, the force of gravity at the altitude of the international space station is actually rather high: 89% of sea level. So why astronauts on-board experience almost zero gravity (microgravity)? Astronauts float in space experiencing what we call microgravity because they are in "free fall". In fact, a spacecraft in orbit moves at a speed such that the curve of its fall matches the curve of Earth. That speed is on average the 28,000 km/h mentioned above.

In the 1950s, aviation scientists discovered that they could simulate microgravity by flying in parabolic arcs. When a plane flies upward at an angle of 45 degrees, its passengers experience first hypergravity (commonly about 2 G's) as the force of the climb combines with the pull of Earth's gravity. Then to create a weightless environment (microgravity), the airplane flies in long parabolic arc, first climbing, then entering a powered dive. During the arc, the propulsion and steering of the aircraft are controlled such that the drag (air resistance) on the plane is cancelled out, leaving the plane to behave as it would if it were free-falling in a vacuum. A similar mechanism works during a suborbital spaceflight.

Therefore microgravity is not linked to being in space, but to being inside a spacecraft on "free fall" on orbit, or inside an airplane or suborbital vehicle on a parabolic trajectory, or inside a (space tourism) capsule released from a stratospheric balloon, before opening of the parachute.

2.3 Trans-atmospheric Spaceflight

Point-to-point hypersonic spaceflight is often presented as the next step of suborbital spaceflight, and even called suborbital point-to-point flight, thus somehow stretching too far the definition of suborbital spaceflight.

As a matter of fact, orbital winged vehicles have already the capability to be used on a trans-atmospheric point-to-point flight path. Instead current suborbital vehicles do not have such capability. When a sub-orbital space vehicle of current design reaches its maximum altitude at the vertex of the parabola the horizontal speed is almost zero. It may be possible to adapt a current sub-orbital design to cover few hundred kilometers, but for true intercontinental or "hemispheric" point-to-point space travel a new class of vehicle is needed: a hypersonic trans-atmospheric spaceplane, or in other words a crewed winged version of an ICBM. Compared with suborbital vehicles, it would be much more complex and technologically advanced, and orders of magnitude more expensive to develop and operate. The German space agency DLR is currently developing at conceptual level a trans-atmospheric hypersonic vehicle, the Spaceliner, as part of a research program funded by the European Union.

As mentioned, orbital winged vehicles have point-to-point transportation capability. The Space Shuttle was designed to satisfy a cross-range capability requirement, levied by the US military/reconnaissance community, of 1,100 nautical miles. Until its cancellation in 1963, the Boeing Dyna-Soar project of the USAF, a small lifting body vehicle to be launched on top of a Titan rocket, had a cross-range capability upon its re-entry in the atmosphere of 1,400 nautical miles. The current USAF X-37, and private Dream Chaser for cargo transportation to ISS, could have comparable point-to-point transportation capabilities.

3. REGULATING COMMERCIAL HUMAN SPACEFLIGHT

The commercial human spaceflight vehicles currently in development, and in particular sub-orbital vehicles, will be operated locally in the country of development but also in foreign countries. Furthermore, a substantial portion of each flight will be spent in the airspace. For such reason there is an interest of the manufacturers and of the national authorities that uniform legislations are enacted worldwide. Because the leading developments are in US, the FAA is actively promoting their regulatory approach as a model. In the next paragraphs we will discuss certain peculiarity of the US regulatory framework, and possible international regulatory set-up taking as model what has been done in civil aviation and maritime commerce.

3.1 US Legislation and Regulatory Framework

On December 23, 2004, President Bush signed into law the Commercial Space Launch Amendments Act of 2004 (CSLAA). The CSLAA aimed to promote the development of the emerging commercial space flight industry, and made the Department of Transportation and the Federal Aviation Administration (DOT/FAA) responsible for regulating commercial human space flight under 49 U.S.C. Subtitle IX, Chapter 701 (Chapter 701). The CSLAA forbids FAA to levy any regulation for the safety of the human on-board for an initial learning period with deadline in December 23, 2012 or until an accident happens, thus limiting the regulatory role of FAA to aspects related to public safety during launch and re-entry of commercial human spaceflight vehicles. The deadline was later postponed a first time to October 1, 2015, and then to 2020.

The CSLAA required that before receiving compensation from a space flight participant or making an agreement to fly a space flight participant, an space vehicle operator informs the space flight participant in writing that the US Government has not certified the vehicle as safe for carrying crew or space flight participants [49 U.S.C. § 70105(b)(5)(B)]. In addition, the space vehicle operator should have a placard displayed in the vehicle in full view of all space flight participants to warn that the launch vehicle does not meet aircraft certification standards.

The FAA guidelines for *Commercial Suborbital Reusable Launch Vehicle Operations with Space Flight Participants* issued in February 2005, provide details about the overall process of written informed consent. It requires that an operator should describe to each space flight participant the safety record of all launch or re-entry vehicles that have carried one or more persons on board, including both US government and private sector vehicles. The safety record should not be limited to only the vehicles of a particular operator. An operator should also describe the safety record of its vehicle to each space flight participant. The operator's safety record should include vehicle ground-test and flight-test information. This information should describe all safety-related anomalies or failures that occurred and corrective actions taken to resolve the anomalies or failures.

The FAA guidelines consider that the development of commercial launch vehicles to carry space flight participants is in the embryonic or early stages. Consequently, newly developed launch vehicles will not have the extensive flight-test history or operational experience that exists for commercial airplanes. Because of the lack of flight-test and operational experience, the risks of the RLV operator's particular launch vehicle and of vehicles like it, including both government and private sector vehicles, should be disclosed. The US House Committee on Science Report, H. Rep. 108-429, clarifies that Congress intended that all government and private sector vehicles to be included in this description. Because most human space flight to date has taken place under government auspices, the government safety record provides the most data. The operator should provide a record of all vehicles that have carried a person because they are the most relevant to what the operators propose. Regardless of whether humans travelled to space on board a vehicle destined for a suborbital or orbital mission, those persons travelled on vehicles based on technology as new then as what may be developed at current time. It was therefore as risky. Likewise, because it was intended for a human on board, greater care was likely to have been taken in its design and construction. The same should be expected for commercial human space flight. Furthermore the operator is required to provide space flight participants an opportunity to ask questions orally to acquire a better understanding of the hazards and risks of the mission.

Both sub-orbital and orbital vehicles in development in US are subjected to the current legislation, but there is a substantial difference for what pertains the safety of the human on-board because of the involvement of NASA as customer for the orbital transportation service to the International Space Station.

NASA's Commercial Crew Program (CCP) was formed to facilitate the development of a US commercial crew space transportation capability with the goal of achieving safe, reliable and cost-effective access to and from the International Space Station and low-Earth orbit. In the Commercial Crew Program, NASA performs safety certification activities without being a regulatory body. The International Space Station IGA (Inter-Governmental Agreement) and related MoUs assigned to NASA the responsibility to manage the ISS Program and the safety of US and international crews in all mission phases including transportation.
The companies involved in the CCP program are free to design the transportation system they think is best. They

are encouraged to apply their most efficient and effective manufacturing and business operating techniques throughout the process. They will own and operate their spacecraft and infrastructure. However the companies must meet (or exceed) NASA's pre-determined set of requirements for the safety of the human on-board. NASA engineers have no oversight role but the necessary insight into a company's development process while the agency's technical expertise and resources are accessible to a company. During safety reviews NASA will identify issues and make recommendations, but NASA cannot impose design or operational solutions.

In the initial phase of the Crew Commercial Program NASA made an inventory of existing technical standards and recommended them either as reference baseline (meet or exceed) or as good practices, stating that *"In the course of over forty years of human space flight, NASA has developed a working knowledge and body of standards that seek to guide both the design and the evaluation of safe designs for space systems"*.

Those standards could be considered also for non-NASA sub-orbital and orbital commercial projects. They can be easily adapted and re-used because they are either performance oriented or deal with generic sub-systems and technologies (e.g. pressurized systems, structural design, batteries, materials, etc.)

3.2 ICAO as Regulatory Model

The International Civil Aviation Organization (ICAO) was formed in 1944, with the signing of the Chicago Convention, which is also the Constitution of ICAO. ICAO is a specialized body of the United Nations (UN), with a mandate to "ensure the safe, efficient and orderly evolution of international civil aviation."

ICAO is composed of three sections, i.e. (a) the Assembly, made up of all contracting States, (b) the Council, composed of thirty-six States, and (c) the Secretariat. Both the Council and the Secretariat contain subordinate bodies; the Air Navigation Commission, Air Transport Committee, Committee on Joint Support of Air Navigation Services, and Finance Committee are part of the Council, while the Secretariat is divided into five bureaus - the Air Navigation Bureau, the Air Transport Bureau, the Technical Co-operation Bureau, the Legal Bureau, and the Bureau of Administration and Services. The Council is the permanent governing body, with the power to adopt and amend standards and recommended practices (SARPs), which are ultimately incorporated into the Chicago Convention as annexes.

SARPs, which cover all technical and operational aspects of civil aviation, are the basis of maintaining uniformity in the regulation and safety of civil aviation. A 'standard' is "any specification for physical characteristics, configuration, material, performance, personnel or procedure, *the uniform application of which is recognized as necessary for the safety or regularity of international air navigation* and to which Contracting States will conform in accordance with the Convention," while a 'recommendation' is "any specification for physical characteristics, configuration, material, performance, personnel or procedure, *the uniform application of which is recognized as desirable in the interest of safety, regularity or efficiency of international air navigation*, and to which Contracting States *will endeavor to conform* in accordance with the Convention."

The SARPs in the annexes to the convention contain only essential regulatory material; detailed technical specifications are spelled out in appendices to the annexes, or in separate manuals. For example, e very technical and detailed Procedures for Air Navigation Services, which are often elaborations of basic principles found in the corresponding SARPs, are listed separately. Also separate from the SARPs, which have global applicability, are the Regional Supplementary Procedures, which apply only in the designated regions.

Technical SARPs, the primary source of ICAO 'regulations,' are generally formulated in the Air Navigation Commission (ANC). The ANC consists of 15 individuals who are nominated by member States and appointed by the Council; the ANC members act in their personal expert capacity, not as representatives of the States which nominate them. After the ANC formulates the SARPs, it passes them on to the Council for review and adoption. Two-thirds of the Council members must be in favor of a proposed new or amended SARP in order for it to be adopted, though even after such approval and adoption member States may register their disapproval. If a majority of States does not register their disapproval by the effective date, the SARPs will enter into effect. In practice, however, standards are not binding on all States because any State is allowed to file a difference to any SARP.

ICAO does not have enforcement power; States have the responsibility for implementing SARPs. However, in order to further aviation safety, in 1999 ICAO established the Universal Safety Audit Program, which subjects all contracting States to "regular, mandatory, systematic and harmonized safety audits," which are carried out by ICAO.

3.3 Private-Government Partnership as Regulatory Model

In the second half of the 18th century, marine insurers, based at Lloyd's coffee house in London, developed a system for the independent inspection of the hull and equipment of ships presented to them for insurance cover. In 1760 a Committee was formed for this express purpose, the earliest existing result of their initiative being Lloyd's Register Book for the years 1764-65-66. The condition of each ship was "classified" on an annual basis. Hull condition was classified A, E, I, O or U, according to the excellence of its construction and its perceived continuing soundness (or otherwise).

Lloyd's Register of British and Foreign Shipping' was reconstituted as a self-standing 'classification society' in 1834. A number of classification societies were established worldwide in the 19th century (Bureau Veritas, Registro Italiano Navale, American Bureau of Shipping, Det Norske Veritas, Germanischer Lloyd and Nippon Kaiji Kyokai), and in the 20th century (Russian Maritime Register of Shipping, China Classification Society, Korean Register and Indian Register of Shipping).

In 1948 a new international organization was established to deal with maritime safety and environmental issues, the Inter-Governmental Maritime Consultative Organization (IMCO) later renamed International Maritime Organization (IMO).

The Inter-Governmental Maritime Consultative Organization (IMCO) was formed to fulfill a desire to bring the regulation of the safety of shipping into an international framework, for which the creation of the United Nations provided an opportunity. Hitherto such international conventions had been initiated piecemeal, notably the Safety of Life at Sea Convention (SOLAS), first adopted in 1914 following the Titanic disaster. IMCO's first task was to update that Convention; the resulting 1960 Convention was subsequently recast and updated in 1974 and it is that Convention that has been subsequently modified and updated to adapt to changes in safety requirements and technology.

In 1968 the major classification societies established the International Association of Classification Societies (IACS), which provides coordinated technical support and guidance to the International Maritime Organization (IMO). In particular the IACS provides and develops unified interpretations of the international statutory regulations developed by the member states of the IMO. Once adopted, these interpretations are applied by each IACS member society, when certifying compliance with the statutory regulations on behalf of authorizing flag States. The link between the international maritime regulations, developed by the IMO and the classification rule requirements for a ship's hull structure and essential engineering systems is codified in the SOLAS convention.

The Rules published by Classification Societies, together with the requirements set down in the various International Conventions of the International Maritime Organisation (IMO) and the marine legislation of the flag states, form a comprehensive and coherent set of standards for design, construction and maintenance in operation of ships. Basically the Classification Societies act as a Recognized Organization carrying out statutory surveys & certification as delegated by maritime administrations. In particular:

- Developing technical standards (rules) for design and construction of ships
- Approving designs against their standards
- Conducting technical surveillance during construction
- Performing in-service inspection and periodic survey during operation
- Reviewing and approving in-service modifications
- Performance of safety research and development programs

Classification societies maintain a leading role in all matters related to technical standards and certification activities, while national and international governmental organizations concentrate primarily on operational and environmental matters.

3.4 Self-Regulation Model

Space systems are highly innovative and it is simply impossible to prescribe detailed design safety solutions for something that never existed before. For such reason space agencies issue safety and technical requirements that are generally goal/performance oriented. The way they are applied, implemented, and verified goes under the name of safety-case regime as explained above.

Because of the broad, generic nature of top-level requirements, the design solutions need to be validated by the developer through an analytical process using techniques like Hazard Analysis, Fault-Three Analysis, etc. The results will be documented in a safety-case report that typically includes: a) the summary description of the system, and operational environment; b) identified hazards and their severity; c) performance requirements considered applicable to each hazard; d) possible causes of each hazards; e) description of how causes are controlled (i.e. eliminated or mitigated); e) description of relevant verification plans, procedures and methods for each control.

As previously mentioned, intrinsic in the concept of standard is that whenever it is made applicable, compliance must be monitored and enforced, otherwise requirements become simply a set of guidelines. Monitoring and enforcement can be done by any party to which such authority is assigned, non-necessarily by a government regulatory body. In any case, the organization must have the following three key prerequisites: authority, competence, and independence (from the specific project or program).

NASA, a government agency but not a regulatory body, is fulfilling such crew safety authority role for the current development of commercial orbital vehicles that will ferry astronauts to the ISS. The competence of NASA's multi-disciplinary safety review panels, and of the specialist teams and labs that support them is well known, but this is a rather unique circumstance, that has no match in traditional regulatory organizations. This means that an obvious substitute for NASA's technical skills does not currently exists for non-NASA commercial human spaceflight orbital and sub-orbital projects, although badly needed. If tomorrow FAA would be allowed to regulate crew and flight participant safety of commercial spaceflight, the problem would become apparent. Industry, collectively, has the means to solve it.

An alternative to government regulations are self-regulations. They are essentially motivated by the need to promote an 'acceptable' level of safety as business case.

Take the example of Formula 1 car racing. In the first three decades of the Formula 1 World Championship, inaugurated in 1950, a racing driver's life expectancy could often be measured in fewer than two seasons. It was accepted that total risk was something that went with the badge [8]. The turning point was the Imola Grand Prix of 1994 with the deaths of Roland Ratzenberger and Ayrton Senna (in direct TV) that forced the car racing industry to look seriously at safety or risk to be banned forever (and lose television rights in the process). In the days after the Imola crashes the FIA (Fédération Internationale de l'Automobile) established the safety Advisory Expert Group to identify innovative technologies to improve car and circuit safety, and to mandate implementation and certification testing. Few years later the advisory group became, the FIA Safety Institute. Nowadays Formula 1 car racing is a very safe multi-billion dollars business of sponsorships and global television rights. Entertainment for families that can be enjoyed without risking shocking sights.

Another example comes from the oil industry. The report of the Presidential Commission that investigated the Deepwater Horizon disaster in the Gulf of Mexico in April 2010 (11 workers killed plus an oil spill that caused an environmental catastrophe), made, among other, the recommendation that *"the gas and oil industry must move towards developing a notion of safety as a collective responsibility. Industry should establish a "Safety Institute"...this would be an-industry created, self-policing entity aimed at developing, adopting, and enforcing standards of excellence to ensure continuous improvement in safety and operational integrity offshore"*.

For commercial human spaceflight to flourish and expand, industry could cooperatively establish a coherent set of safety and technical standards, taking as starting reference the experience accumulated in more than 50 years of government programs.

Collectively, industry has all the necessary intellectual and organizational resources for the task, but sound business strategy always demands to advance space safety. For that end the standardization activities would greatly benefit from a systematic cooperation with universities and research centres. The overall cooperation, within industry and with academy, could take the form of a Space Safety Institute. The institute would network industry experts and university researchers, and perform ad-hoc studies, develop educational programs, run standardization efforts, and provide skills in support of independent safety certification of commercial space systems.

3.5 Organizing Commercial Human Spaceflight

Organizing commercial human spaceflight in the 21th century requires realism, pragmatism, and political will.

The key elements to consider are as follows:

- Commercial human spaceflight business survival and expansion will depend very much on companies individual and collective safety record;
- It is in the best interest of industry, and of all stakeholders, to publish the safety policies and technical best-practices they apply during design, manufacturing and operations of commercial space vehicles;
- For several years to come there will be no government regulation for the safety of those on-board in the country (US) that is spearheading commercial human spaceflight;
- Aviation projected growth will not tolerate economic losses due to proliferation of reserved airspace for commercial launch and re-entry operations. Air and space launch/re-entry traffic needs to be integrated;
- Suborbital spaceflight will not be confined to countries where vehicles are developed but will take place and expand internationally;

- Plans to allow landing at a number of civil airports worldwide, in case of launch abort or emergency of orbital commercial winged vehicle, requires international regulations;
- There is no appetite for creating a new international organization

A regulatory framework based on consideration of the above element could be structured as follows:

I. The ICAO should perform safety oversight of commercial space launch and reentry/return operations that make use of the international airspace;

II. United Nations COPUOS should promote the establishment of an international regulatory framework for space traffic management and space environment protection;

III. Commercial human spaceflight industry should establish a Space Safety Institute as self-policing entity aimed at developing, adopting, and enforcing safety and technical standards for the human onboard.

4. REFERENCES

1. R. Jakhu, T. Sgobba, P. Dempsey, "The Need for an Integrated Regulatory Regime for Aviation and Space - ICAO for Space?", Springer, 2011

2. IMO and The Environment", IMO, 2011 http://www.imo.org/en/OurWork/Environment/Documents/IMO%20and%20the%20Environment%202011.pdf

3. "Crew Transportation Technical Standards and Design Evaluation Criteria", NASA CCT-STD-1140

4. "Safety Review Process, International Space Station", NASA SSP 30599, Rev. F

5. Space Safety Standard Commercial Human-Rated System, IAASS-ISSB-S-1700-Rev-B http://iaass.space-safety.org/publications/standards/

6. "The Science of Formula 1 Design - Expert Analysis of the Anatomy of the Modern Grand Prix Car", David Tremayne, Haynes Publishing, 2009

7. Deep Water: The Gulf Oil Disaster and the Future of Offshore Drilling, Report to the President, January 2011

FILLING THE REGULATORY VOID FOR LAUNCH AND REENTRY SAFETY RESULTING FROM THE COMMERCIALIZATION OF SPACE OPERATIONS

Dr. Ruth Stilwell

*Adjunct Faculty, Norwich University College of Graduate and Continuing Studies,
158 Harmon Drive, Northfield, VT 05663 USA,
Email: rstilwel@norwich.edu*

ABSTRACT

The commercial use of space is not a new concept. Satellites have provided commercial services since the 1960s. However, the launch and recovery of space vehicles was dominated by State operators or under State Contract until the end of the last century. While commercial launches continue to increase, the international regulatory frameworks related to launch and reentry are fundamentally unchanged and are not structured to support a competitive commercial launch industry.

This paper examines the regulatory gaps in the existing frameworks, particularly in the international domain. While there remain unresolved questions with regard to ICAO as it applies to suborbital flight, the ability of ICAO to develop standards and recommended practices for the portion of the operation that takes place in civil airspace is clear. These standards are necessary to ensure the safety of civil airspace for both traditional airspace users and launch operators, and the interaction of spaceports with civil airspace.

1. DEFINING THE REGULATORY GAP

Commercial applications for space technology are not new, there have been commercial satellites in operation for decades and the provision of commercial services utilizing satellite technology is pervasive in our society. The United States opened launch activities to private operators through the Commercial Space Launch Act of 1984, but the operational treatment of private launches remained identical to that of State sponsored launches.1 The US has continued to advance the legal frameworks to allow expanded commercialization of space launch activities over the last three decades.2

Similarly, the European Space Agency supported the development of Arianespace to provide commercial launch services for the European market.3 However no new international treaties have been negotiated or advanced by the United Nations Committee on the Peaceful Uses of Outer Space in the commercial space era, creating new gaps between domestic policy and international frameworks.

The new commercial space industry is expanding rapidly into launch operations as private companies continue to develop not only new rocket technologies, but also new concepts of launch operations. These include horizontal take off, launch from aircraft or balloon, and launch from mobile and fixed sea platforms. Notably, the recent SpaceX Falcon 9 vertical landing illustrated a fundamentally new concept in reentry operations.4 The regulatory challenge is twofold, technology and policy. These distinct categories should be viewed and addressed independent from one another. This approach is necessary to ensure the policy issues are not disregarded in the regulatory process, which tends to have a natural focus on the technical side of safety regulation. Recent experience in the regulatory development for unmanned aircraft illustrates this difficulty, as rapid developments in technology have driven the debate, leaving many critical policy questions, particularly with regard to the right to regulate at very low altitudes, unanswered.

1.1 Policy

In the policy dimension, the transition from government operation to a regulated competitive industry presents a set of challenges, but is not without precedent. In the aviation industry, national airlines were privatized, air navigation service providers (ANSP) became corporatized and airports around the world have gone from State to private operation. In making the transition, each State needed to evaluate whether the appropriate regulatory regimes were in place. Both safety and economic considerations were addressed. As we examine the case of many ANSPs in the late 1990's, we see that the separation of the regulators from service providers either placed new burdens on existing regulators or required the creation of entirely new regulatory authorities. In addition, it became apparent that international civil aviation standards that had long been in place for airline operators did not necessarily exist for ANSPs and had to be created. ICAO continues to identify and fill these gaps. While we can look to these examples to help identify regulatory gaps, there are significant differences between the privatization of elements of a national transportation infrastructure and the development of a competitive commercial industry in a field once populated only by public operators.

In the commercial space question, we are not looking at the transition from public to private ownership, but rather the introduction of new commercial operators, which will likely co-exist with State operators. In addition, commercial space operators are unlikely to comprise a critical part of a State's national transportation infrastructure, alleviating the need for certain economic regulation that may have been considered necessary in the privatization of elements in traditional aviation. However, with regard to safety regulation, not only is there a need for States to consider the development of regulatory authorities, or assign the regulatory function for commercial space operators to existing regulatory authorities, but the international component is also of critical importance.

The policy questions regarding the operation of commercial space activities are not limited to the commercial space industry itself. Both the launch and recovery phases of commercial space operations occur in airspace that is shared with civil aviation. As these phases either disrupt, or pose as safety hazard to civil aviation, it is necessary to ensure that regulations include the civil aviation components. While there may be continued debate as to the need to distinguish between aviation and space, the authority to regulate, and other jurisdictional questions, there is no question that international civil aviation is affected by the activities of commercial space operators in civil airspace. At a minimum, regulations to prevent collision between civil aviation and commercial space operations are warranted.

In examining the international policy and regulatory issues, it is necessary to determine whether current definitions are appropriate. Launch operations themselves may not necessarily constitute international transport. It is traditionally accepted that air transport involves the transportation of people or goods between two points on the earth and that international air service involves the transit through the airspace or territory of more than one State.

For commercial space operations, the launch and recovery may occur within a single State and the transport of goods, i.e. satellites, is from the earth to space rather than between two points on the earth. However, new launch sites can create a circumstance where the trajectory of the launch may cross an international border. 5 The issue becomes clouded when the path of the trajectory crosses over the territory of another State at an altitude above that which is considered airspace, but below that, which is regarded as outer space. In addition, launch operations from sea launch platforms located on the high seas are necessarily international operations as it is established that the high seas and the overlying airspace is de facto international. Thus, it is subject to long recognized international principles in aviation including the principle that regulation is necessary.

The discussion of the International Commission for Air Navigation (ICAN) in 1929, found: "Any freedom that is to be exercised in the interests of all entitled to enjoy it must be regulated. Hence, the law of the high seas contains certain rules, most of them already recognized in positive international law, which are designed, not to limit or restrict the freedom of the high seas, but to safeguard its exercise in the interests of the entire international community. " 6

The concept that flight over the high seas is free, does not imply that it is without restriction, rather that the responsibility and authority to make rules regarding flight over the high seas, does not lie with an individual State but rests with International Civil Aviation Organization. For airspace over land or territorial water, States have the authority to vary from ICAO standards and recommended practices. This is not the case for airspace over the high seas. As referenced in Annex 2 to the ICAO Convention, Flight over the high seas, "It should be noted that the Council resolved, in adopting Annex 2 in April 1948 and Amendment 1 to the said Annex in November 1951, that the Annex constitutes Rules relating to the flight and manoeuvre of aircraft within the meaning of Article 12 of the Convention. Over the high seas, therefore, these rules apply without exception."7

The Annex 2 provisions are of critical consideration with regard to launch from a sea platform. As a launch necessarily requires the restriction of flight through a volume of airspace, ICAO must provide provisions that would allow for the closure of such airspace to civil aviation for a specified period to ensure safety. The Annex 2 definitions of both restricted area and prohibited area, the airspace categories routinely used to protect launch areas, specify that it applies to the airspace over land areas or territorial waters of a State. As a consequence of these definitions, there are no provisions for an air navigation service provider to create a restricted or prohibited area in high seas airspace. The only available option is the use of a danger area, defined in Annex 2 as "An airspace of defined dimensions within which activities dangerous to the flight of aircraft may exist at specified times." This designation does not preclude flight through that area and as a result does not provide the same level of protection for either aircraft or launch vehicles as is provided by the use of restricted or prohibited areas in sovereign airspace.

From a policy standpoint, a danger area is not the high seas equivalent of a restricted or prohibited area. In the ICAO Air Traffic Services Planning Manual (ICAO doc 9426) paragraph 3.3.2.2 states, "According to their definitions, a danger area implies the least degree of restriction, while the prohibited area constitutes its most stringent form. It should also be noted, however, that this definition is applied only with respect to airspace which is situated over the territory of a State. In areas where no sovereign rights are exercised (e.g. over the high seas) only danger areas may be established by that body responsible for the activities causing their establishment."

While the danger area provides the least degree of restriction, it is solely designed for safety purposes, and as such does not provide a means by which the competing demands for the civil use of the airspace are managed. In general, danger areas are used when unusual activities occur that pose a hazard to flight, it is not designed for routine civil operations. These areas are generally used to protect civil aircraft from military activity or other activities outside air traffic control where it is unlikely that a civil authority would be able to take steps to prevent or suspend the activity for the purpose of allowing civil aircraft to safety transit the area. Using the US example, there is a strong civil military cooperation, wherein the scheduling and release of airspace, including restricted areas can be adjusted to accommodate demands of civil users, particularly in the case of weather conditions that may pose a hazard to flight. The European concept of Flexible Use of Airspace, is to consider airspace as neither civil nor military and is allocated or segregated on a temporary and real time basis. Airspace allocation is increasingly dynamic on a global basis. In order to have this flexibility, there must be an underlying policy basis that allows for the establishment of protocols and agreements with the airspace users. This basis is not available in the construct of danger areas. While the use of danger areas is appropriate for falling debris or re-entry where planning and maneuver may not be possible, the same conditions do not exist for launch activity. Airspace planning should consider launch requirements as part of an airspace-planning model.

The ICAO Separation and Airspace Safety Panel considered several papers on the development of safety standards, separation standards, and the designation of aircraft hazard areas specifically related to space launch and recovery activities. In 2015, the United States member specifically requested that the Panel consider the creation of a new category of special use airspace designed for space activities. Without an airspace category that can be applied in high seas airspace, there is a significant regulatory gap that must be filled to permit the maximum safety benefit from a sea platform launch. The ability to create a launch facility far from populated land areas decreases the risk to life and property and while the technological barriers have been overcome, the policy barrier remains. Unlike other aviation provisions, where individual States can consider modifications to allow for specific operations, the provisions for operation over the high seas must be addressed by ICAO. Current provisions are limited to the coordination of activities and promulgation of information by the ATS authority.8 This language provides a de facto priority over civil aviation operators that may constitute an unreasonable restriction to the freedom of over flight of the high seas.

1.2 Technology

In developing proper regulatory frameworks to address the safety regulation, the rapid development of commercial space technology presents unusual, but not unprecedented challenges. The move in ICAO to shift from prescriptive to performance-based standards allows for the development of early requirements that are sufficiently flexible to accommodate technological developments. This approach allows for alternative means of compliance when compared to prescriptive standards that may hinder technological development. In the case of protected airspace for launch operations, the models used, generally adapt special use airspace designed for military aircraft operations and utilize it for a launch operation. While this may be adequate to provide a safety buffer, the approach may not take advantage of developments in space technology and could result in greater disruption to civil aviation than necessary. Developing new airspace models that not only consider the specific risk models of the intended launch vehicle, but also take advantage of modern risk modeling tools, provide an opportunity to reduce the operational disruption needed to accommodate commercial space launches. This is an important step for the commercial space industry. If commercial space launches are too disruptive to other users, their access to airspace may be limited, restricting growth in the industry.

The development of specific technical standards for commercial space launch operations has additional difficulties due to the diversity in technology and operational type. Unlike aviation, where a new standard is designed to apply to hundreds of aircraft with thousands of operations a year, commercial space operations may require standards for a launch vehicle type with only a few in existence and just a handful of operations. In addition, the amount of data available from testing may be considerably less than the aviation community is accustomed to. Consider the example of developing wake turbulence separation standards. Rather than a standard limiting the amount of wake an aircraft design can produce, the standard was to determine how far another aircraft had to remain from the wake generating aircraft to ensure safety of flight. This is the same concept that will be applied in creating a standard for protected airspace around a launch. That is, how far does an aircraft need to remain from a launch to ensure safety of flight?

For the development of wake turbulence separation standards, research has been ongoing since the late 1960's using data from airport sensors, aircraft manufacturers, and governments contributing to the process 9 This provides large amounts of data to support the development of wake turbulence separation standards. For example, by 2013 there were over 100,000 wake measurements per year. While wake turbulence separation is a standard to protect the safety of other aircraft (specifically, those following), if that standard is larger than necessary, it reduces system capacity and places an undue burden on other system users. In order to reduce the separation requirements, extensive testing and modeling is necessary to make a proper safety case. Similarly, the air traffic control separation required from a space launch or recovery operation is primarily to protect the safety of uninvolved aircraft operating in the vicinity. However, the amount of available data from the commercial space manufacturing community may not be comparable to that used in developing other aviation standards.

Designing a protected airspace volume tailored to meet the needs of, or more precisely to mitigate the risk imposed by, a specific launch or recovery operation requires sufficient data about the operation itself. The rapid technological innovation in this field will make it difficult, if not impossible for many government regulators to develop the models necessary. It will require significant and active input from industry to create timely airspace models. This is not to say that process standards cannot be developed that will facilitate the development of these models. The handful of States with launch experience can offer expertise to identify the elements that need to be considered and evaluated in building models for protected launch and recovery airspace.

The intersection of technology and policy occurs when launch specific modeling becomes sufficiently robust to create detailed boundaries for each launch type. The common models for restricted or prohibited airspace involve static boundaries and the airspace is active or inactive, but the boundaries themselves do not change. These boundaries are published in airspace charts that are known and available through well-established means. In a scenario where dynamic airspace modeling could be used, it will be necessary to determine the means by which airspace users will be notified of the airspace boundaries and how that information will be made available to flight planners and pilots. The publication requirements for danger areas can provide a foundation, but may need to be expanded as commercial space launches increase in frequency.

For organization structural purposes, the US example can be instructive. While there are specific legal restrictions on promulgating regulations on some elements of commercial space activity to allow the development of the industry, the Federal Aviation Administration has the oversight responsibility over launch and re-entry operations in order to protect the public. Its authority extends to protect aircraft from the threat of potential launch or re-entry vehicle failure. In developing aircraft hazard areas using a variety of modeling tools, the FAA determines which airspace or airways will be closed to accommodate the launch. The FAA has extensive experience using Temporary Flight Restrictions (TFR) for various purposes. Air traffic control is able to support the enforcement of these dynamic areas with the ability to depict areas defined by geographic points on a radar map. For procedural airspace, airways that traverse an aircraft hazard area are closed to traffic for the affected period. However, even with the extensive experience and modeling tools, the disruption in civil traffic may be greater than necessary to prevent transgression into the aircraft hazard areas.

The reliance by the FAA on intervention by air traffic control to ensure that aircraft do not enter a launch hazard area may not be a suitable model in all conditions. In areas with limited communications coverage, or those susceptible to communications disruption, it is important for the detailed information on the hazard area to be known in the cockpit. In addition to competition for airspace under planned conditions, there is increased competition for airspace when weather deviations are necessary. While the launch airspace may be free of convective activity, the surrounding airspace may contain a weather hazard, requiring civil aircraft to deviate. Considering that both are a hazard to safety of flight, how launch airspace will be managed in these circumstances must be included in the airspace planning process.

2. THE CASE FOR INTERNATIONAL STANDARDS

As the number of commercial spaceports and private launch facilities increase, there is a corresponding need for globally harmonized safety standards. There has been considerable debate as to whether the ICAO Chicago Convention allows for the regulation of space activities. While there remain some unresolved questions with regard to ICAO, particularly as it applies to suborbital flight, the ability of ICAO to develop standards and recommended practices for the portion of the operation that takes place in civil airspace is clear. Article 37 of the Chicago Convention grants ICAO the ability to develop standards and recommended practices for "all matters in which such uniformity will facilitate and improve air navigation." As the transit of civil airspace by spacecraft and other vehicles affect the safety, regularity, and efficiency of air navigation, the role of ICAO in developing these standards falls clearly within its existing mandate. In addition, considering that the work currently proposed in ICAO would address the requirements for aircraft to avoid hazard areas created by commercial space operations, the standards will apply more directly to civil aviation than to the space operators in the near term. However, these standards cannot be developed without substantial inputs from the space industry. As an integrated policy approach evolves, the standards should seek to balance the interests of the competing industries of aviation and space operators as airspace users, rather than treating space operators solely as a hazard.

2.1 New Space Faring States

Commercial development of space launch operations will allow for States that do not currently have launch capacity to participate in the space launch economy. We are seeing new States enter the space launch community now and we can expect that to expand rapidly in the near future. Like other technical developments, there is value in building structures where new entrants can benefit from the lessons learned from States that have been engaged in launch activities for decades. This premise is engrained in the ICAO processes, specifically in the development of guidance material to facilitate the implementation of new technologies and procedures. ICAO, working with the UN Office of Outer Space Affairs, can utilize this model to facilitate the safe expansion of launch capacity around the world.

2.2 The Transit User

If instead of instead of treating commercial space activities as a separate and distinct industry, the aviation community and ICAO in particular, regard it as a transit user of civil airspace, the regulatory processes become more clear. The requirement to determine either a functional or vertical demarcation between space activity and aviation activity is nullified. Rather, the regulatory questions align with the airspace and would apply to all actors within it. That is not to say that the type of operation does not warrant special considerations, but rather that there those considerations would fall into the myriad of different types and considerations required for comprehensive airspace design and planning. This approach is consistent with the current US federal model as well as the emerging European model for functional use of airspace.

By adopting the concept of transit user to apply to those operations that transit the airspace but do not operate in it, the regulatory framework is adaptable to accommodate new technologies and operational concepts as they emerge. To draw a parallel with traditional aviation, the regulatory frameworks accommodate a variety of operational types as diverse as helicopters and supersonic jets. It is by virtue of the fact that they all operate in civil airspace and not the technical capabilities of the aircraft that dictate the regulatory jurisdiction. If commercial space operators are regarded as civil airspace users the jurisdictional issues become equally clear. This multimodal approach allows for the linkage of airport and airspace policy in developing a comprehensive commercial space policy. In addition, by considering the transit user as a category, the mission type does not create a new regulatory or jurisdictional question. The methods by which airspace is allocated, hazard areas are determined and aviation operators are notified can be harmonized whether the mission purpose is launching satellites, commercial cargo, commercial crew, commercial space stations, satellite servicing, suborbital space tourism, research or other applications. Like aviation, questions of priority are determined by policy and operational limitations. Operational priority is a well-established principle in aviation, and policies can be adopted to consider the specific needs of different transit users.

3. INDUSTRY DEMANDS

Existing global launch capacity is not adequate to meet the projected growth in the commercial space sector. New launch facilities are planned and more will be developed. In order for a new launch site to provide optimal capacity, it is important to mitigate the amount of disruption a launch causes to other civil airspace users. Considering the location of launch facilities in the context of an overall airspace plan allows decision makers to optimize the investment decisions. Failure to do so can result in operational restrictions to new launch facilities, creating a need for additional facilities, delaying growth in the industry. To draw a comparison with aviation, an airport cannot take full advantage of its runway capacity if it has inadequate parking for aircraft once they have landed. Spaceports with inadequate access to airspace would be similarly limited.

In addition to the need for adequate launch capacity, internationally harmonized standards are necessary to support industry growth and technological development. In a presentation to ICAO, the FAA summed up the issue clearly, "International standards for launch and reentry operations will ensure clarity and fairness for all stakeholders."10

4. CURRENT ICAO ACTIVITIES

In June of 2014, ICAO initiated steps to form a Space Learning Group, which may lead to the establishment of a formal ICAO work program on space activities. The first joint symposium between ICAO and UN OOSA, the ICAO/ UNOOSA AeroSPACE Symposium was held in Montreal, Canada in March 2015 and the second held in Abu Dhabi, UAE in March 2016, with the theme, Emerging Space Activities and Civil Aviation – Challenges and Opportunities. These joint ICAO/ UN OOSA symposia serve to begin a process to build a relationship between the space and aviation communities.

Looking beyond the ICAO work that is specifically organized for space activities, within the ICAO Panel structure questions are emerging regarding specific applications. These include the potential creation of a new airspace category to separate launch and recovery operations, the need for separation standards between aircraft and launch vehicles or protected airspace and civil aircraft, and the introduction of risk modeling tools specific to space operations. The ICAO Separation and Airspace Safety Panel has been in consideration of these topics for the last few years.

4.1 Industry Engagement

The ICAO model is one of State and industry engagement to ensure that expert groups have access to complete and up to date information on the topics under discussion. As the ICAO structure is specific with regard to industry participation, the Panel members are comprised of those appointed by invited States and recognized international organizations. Currently there are no organizations representing the interest of commercial space operators on the List of International Organizations That May Be Invited to Attend Suitable ICAO Meetings.11 Without a recognized International Organization, the industry engagement in the development of standards is limited. ICAO has been actively working to bring in space industry expertise through various outreach efforts, including informal groups and the symposia. However, the formal work of the agency will be dominated by the inputs from States and recognized International Organizations.

The ICAO Assembly to be held in 2016 will set the formal work programme and budget for the coming triennium. Whether and how much of ICAO resources will be dedicated to these activities will be determined. As there are no formally recognized International Organizations representing the operational needs of transit users in civil airspace, the commercial space launch industry will have to rely on the States with an interest in advancing this work on the ICAO work programme and the International Organization representing the aerospace manufacturing community.

5. CONCLUSION

Commercial space launch operators are rapidly evolving technologies that will place specific demands on civil airspace, both domestic and international. As these operations increase, the regulatory gaps in both policy and technical regulation must be closed to support a growing industry that provides high value services to the global population. In order to develop standards to fill these regulatory gaps, it is critical to ensure that industry expertise is available to regulators and international bodies where the standards will be developed. Adopting the concept of airspace transit user can facilitate this process.

6. NOTES

1. The Act allowed for private operation of expendable launch vehicles, the operations occurred at federal launch facilities utilizing special use airspace designated for State operations.

2. Several Acts, including the Commercial Space Launch act of 1984, the Launch Services Purchase Act of 1990, the Commercial Space Act, the Commercial Space Launch Amendments Act of 2004, and the SPACE Act of 2015, sought to expand the commercial space industry in the US.

3. Company Profile, Arianespace Services and Solutions. 5 February 2016. Accessed March 2016 [online: www.arianespace.com]

4. Davis, Jason, (2015) SpaceX Falcon 9 Returns to Flight, Sticks Landing at Cape Canaveral. The Planetary Society. Accessed March 2016 [Online: www.planetary.org]

5. For example, Space X has broken ground on a private spaceport in Boca Chica Beach, within a few miles of the USA-Mexico Border.

6. The Law of the Air and the Draft Articles Concerning the Law of the Sea Adopted by the International

Law Commission at Its Eighth Session, United Nations Conference on the Law of the Sea. Geneva, Switzerland. 24 February to 27 April, 1958

7. Annex 2 to the Convention on International Civil Aviation: Rules of the Air (tenth edition) International Civil Aviation Organization, July 2005.

8. ICAO Annex 11, paragraph 2.18, Coordination of activities potentially hazardous to civil aviation.

9. Wake Turbulence Separation Standards for Aircraft. Volpe: The National Transportation Systems Center. Cambridge, MA Accessed March 2016 [online: www.volpe.dot.gov] !

10. Launch and Reentry Operation International Standards. Federal Aviation Administration Presentation to ICAO, 9 November 2015.!

11. As of 7 March 2014.

PROCEEDINGS OF THE 8TH INTERNATIONAL SPACE SAFETY CONFERENCE

SESSION 07 LAUNCH GROUND OPERATIONS SAFETY

DEALING WITH A SUBTLE DANGER DURING LAUNCH PREPARATION OPERATIONS: OXYGEN-DEFICIENT ATMOSPHERE LEADING TO ANOXIA

Miguel Morere [(1)]

[(1)] CNES, Centre Spatial Guyanais, BP n° 726, 97387 Kourou Cedex, French Guiana, Email: miguel.morere@cnes.fr

ABSTRACT

On launch bases, several toxic, explosive, flammable or combustive materials are used. The hazard arises from the presence of these materials on site. There are also life-threatening situations created by the lack of a vital matter: oxygen. The danger due to oxygen-deficient atmosphere is subtler than any other, because the lack of oxygen cannot be detected directly by humans and the danger of death appears without any warning. Indeed, the detection of that kind of atmosphere can only be made by detection devices. That article aims at describing a method to identify and classify operations leading possibly to oxygen-deficient atmosphere and at defining the related risk mitigation measures.

1. INTRODUCTION

Space launch preparation operations implement several toxic, explosive, flammable or combustive materials: solid propellant, hydrazine, nitrogen tetroxide, liquid hydrogen, hydrogen peroxide, etc. The hazard arises from the presence of these materials and from the operations in which they are involved. Once these materials are on site, personnel are aware of the related dangers.

There are also life-threatening situations created by the lack of a vital matter: oxygen. The hazard due to oxygen-deficient atmosphere is subtler than any other. In man-rated flights, oxygen deficiency is a key danger, mainly due to presence of astronauts before and during the flight. Nevertheless, on launch bases with only unmanned flight, that hazard is perhaps the less known.

Oxygen-reduced atmosphere may be the result of inert gas leakage or voluntary inerting operation. Inert gases, like nitrogen or helium, are present in very high volume on launcher stages and related support facilities, at high or low pressure, during the whole lifecycle of the launcher, from arrival at launch base to liftoff. The danger induced by too much nitrogen is well-known since the beginning of time: the French word "azote" or the Russian word "азот" for nitrogen comes from old Greek and means literally "which deprives life".

Oxygen-deficient atmosphere has quick effects on human beings and the first consequences of hypoxia are very fast to appear. From an oxygen concentration of 19%, worker's safety cannot be ensured anymore and first effects on workers are noted. As oxygen concentration decreases, more serious effects to humans occur and threat of death becomes imminent.

In case of inert gas leakage, the oxygen concentration drop is in the most situations very fast or even instantaneous. Accordingly, workers' response time to evacuate the oxygen-deficient zone is too short. Efficient protective measures have to be considered.

Space systems design is not and may not be focused on that risk. On other industrial plants, such as nitrogen production plant, that risk can be reduced by installing facilities in open space. Such a typical measure cannot be applied to space facilities, being subject to other constraints.

First of all, anoxia risk must be taken into account at every step of the risk analysis, from system design to system operations. That risk has to be assessed in hazardous operations analysis (such as ground-flight hardware connection) but also for non-hazardous operations and static phases (like launcher stage long-term storage). Leakage sources must be reduced and avoided by design as much as possible. Operations must be analyzed to reduce leakage risk and to decrease the consequences of leakage on workers. According to the identified oxygen-reduced risk zones, some protective measures such as gas detector, air breathing mask disposal, ventilation and firefighter's prepositioning must be set up.

At Guiana Space Centre (CSG), that risk is widely taken into account but there is always room for improvement, so as to ensure safety for all.

After explaining the phenomenon of anoxia and the related effects on human beings, that paper aims at:

- Describing a method to identify and classify operations leading possibly to oxygen-deficient atmosphere;
- Explaining how the remaining risks are minimized, using the safety basics: prevention, protection and rescue;

- Describing the safety devices suitable to detect oxygen-reduced atmosphere and to react in case of low oxygen rate;
- Showing examples implementing that method at CSG.

2. PHENOMENON OF ANOXIA

Why do humans need oxygen? According to [1] and [2], several bacteria and most cells can work without oxygen. Life began and remained during more than a billion of year without oxygen. Most cells can live with a very low amount of energy in anaerobic mode, generating lactate instead of carbon dioxide.

The anaerobic mode allows the cell to survive but produce almost 16 times less adenosine triphosphate (ATP), which is the main energy source for the majority of cellular functions. To work in full capability most cells need oxygen, but can survive without oxygen.

That kind of cells does not include neurons. Humans' neurons need oxygen to work in full-capacity mode and are not able to survive in anaerobic mode during a long time. Some animals can survive during few hours without oxygen. The reason why human neurons cannot survive is partly unknown. The human brain suffers the lack of oxygen in a whole but some neurons suffer more than others. Some neurons can survive only during a few minutes, some can survive during hours.

The brain's working conditions affect a lot its weakness. The brain activity needs a huge amount of energy. The production of energy in anaerobic mode is very low and not sufficient to maintain the brain activity. Without oxygen, a protection mechanism put the neurons in sleep mode. That safe mode interrupts brain activity and causes the loss of consciousness.

The lack of oxygen is detected very fast (in a few second). That fast detection allows to decrease very quickly the overall energy consumption and to maintain the main cellular functions, which are essential for life. Unfortunately, the brain safe mode cannot be held during a very long time. That timeframe depends on the kind of neuron and goes from a few seconds to a few minutes.

The sensitivity to a low oxygen concentration varies from a person to another. It can be explained by several adaptation mechanisms, like the loss of consciousness. Other slower mechanisms can act in two or three weeks. That adaptation phenomenon allows the alpinists to survive to great heights. The oxygen concentration at great heights is the same as at sea level. The main difference is that the air density is lower at great heights, so the oxygen input is lower than at sea level.

No one could survive being transported directly from sea level to Mount Everest. Indeed, exposing the brain to less oxygen starts an adaptation mechanism and increases the brain resistance to low-oxygen atmosphere. In this case, the irreversible damage appears later.

Air contains by volume around 21% of oxygen, 78% of nitrogen and other trace components (mainly argon and carbon dioxide). Alarm threshold of oxygen deficiency detection devices is commonly set at 19%.

From 17% of oxygen, air becomes more difficult to breathe normally and the human body starts adaptation phase. The person becomes less vivacious and has a reduced physical efficiency but no obvious sign can make the person aware of the low-oxygen atmosphere. There is no prior warning to anoxia.

Between 13% and 10% of oxygen, life-threatening situation occurs. A total loss of consciousness is very likely to happen. Below 10% of oxygen, the person suffocates and death appears in a few minutes.

Tab 1. summarizes the main effects of oxygen deficiency on human beings:

Table 1. Effects on human beings

Oxygen concentration	Effects on human beings
21%	Standard air, no effect
19%	Alarm threshold of oxygen deficiency detection devices at CSG
17% - 13%	Hard breathing, reduced awareness, disorientation, reduced physical efficiency, but no obvious warning
13% - 10%	Loss of consciousness
< 10%	Suffocation, death in a few minutes

N.B.: the oxygen concentration and the related effects on human beings can greatly vary from a human to another. Slight differences on the percentages and effects can be found in the literature ([3], [4], [5], [6]).

3. APPLICABLE REGULATIONS

French regulations do not deal clearly with oxygen deficiency risk. In the Labor Code, two short articles address specifically the work in confined spaces. They require checking the air quality before entering the area and ventilating the area. Several more general articles explain some requirements regarding the ventilation and the air exchange at the workplace.

Some higher-level dictates of the French Labor Code have more impact on worker's safety. Those dictates (such as "avoid risks", "evaluate risks" or "replace what is dangerous by what is not or less dangerous") are universal but implies everyday safety rules at CSG.

Nevertheless, French public administration published guide books (such as [3]) detailing some guidelines to work safely in confined spaces.

Interesting information coming from French public administration is about Seveso directive. That European directive implies to study the risks generated by an industrial activity on the environment (including people, other facilities and so on) before beginning its exploitation. With a huge amount of liquefied inert gas, a risk of oxygen-deficient atmosphere outside the establishment can be found. To treat that topic, French public administration determines two thresholds: 18% as the limit of irreversible effects on people and 11% as the limit of immediately dangerous effects on people.

CSG safety regulation "REI" dedicates an entire article to oxygen-deficient atmosphere. That article takes up several risk mitigation measures explained at chapter 5 such as detection devices or oxygen mask.

4. SOURCES OF ANOXIA

There are two ways to create an oxygen-deficient atmosphere: by removing oxygen or by adding other gases which replace oxygen.

Anoxia is more about to occur in confined space but is not limited to confined space. In fact, if the worker is close to oxygen deficiency source, he can be affected by anoxia even in high-volume building or in open space.

The main phenomena to reduce oxygen are combustion and fermentation, which are uncommon on launch preparation activities. Fermentation generally produces carbon dioxide and consumes oxygen so that there is a double effect on human.

There are many gases likely to replace the oxygen in air. The release of those gases can be either voluntary release or accidental leak. The touchiest gases are the inert ones because their only effect is the anoxia in humans. The sole way to be aware of their presence is by gas concentration detection. Various inert gases are used during launch preparation operations, mainly nitrogen and helium. Other generally unreactive gases are used on the launch range and are dedicated to spacecraft electrical propulsion system such as xenon.

Other classes of gas can be source of anoxia. Toxic gases like carbon dioxide (used for instance in some fire extinguishing systems) or flammable gases like hydrogen can also replace the oxygen in air. As mentioned in [5], the other hazards (toxicity or fire) generated by those gases are generally faster than anoxia. Consequently, anoxia can be considered as covered by those risks.

Many situations can result in gas leakage or release. A leakage can occur due to a non-tight connector or due to a non-tight valve. Those leaks are accidental.

Gas release can result from normal working of a safety device such as a relief valve. A relief valve is generally used to avoid overpressure of a fluidic system by releasing the fluid in case of too high pressure. The relief valve can be either collected or not. For a nontoxic and nonflammable gas, the relief valve can emit inside the facilities and cause oxygen-deficient atmosphere.

In dedicated situations, an inert gas can be used to cool down a pressure vessel in order to control the temperature while pressurizing. Another purpose of that kind of gas is to inert facilities previously to hazardous fuelling operations. In those cases, the area must be cleared of people before beginning the hazardous operations.

As mentioned in [5], the risk of oxygen-deficient atmosphere is increased when using liquefied inert gases such as liquid nitrogen or liquid helium. At atmospheric pressure, the boiling point of nitrogen is -196°C (77K) and its liquid density at boiling point is around 800 kg/m^3. Its gas density is 4.6 kg/m^3 at boiling point and 1.18 kg/m3 in standard conditions (atmospheric pressure, 15°C).

So, its liquid/gas equivalent is 680 in volume. It means that each liter of liquid nitrogen will create 680 liters of gaseous nitrogen while evaporating. The helium liquid/gas equivalent is 740 in volume. That phenomenon increases significantly the area subject to oxygen deficiency risk. When involving high volume of liquid nitrogen (e.g.: 100 m^3), the oxygen deficiency risk zone can be extended outside the facility limits and possibly expose public.

5. RISK MITIGATION MEASURES AND SAFETY DEVICES

When addressing the risk prevention measures, the voluntary release and the accidental leak of gas must be distinguished.

To avoid accidental leak, the first risk mitigation measure is to design and to operate the fluidic systems in a safe way. The fluidic systems tightness must be ensured by design rules, by qualification tests and by acceptance tests. Also, the whole architecture of the fluidic systems must be FS/FS (Fail Safe/Fail Safe) regarding the risk of leak. At CSG, all the hazardous systems are analysed and assessed following the safety submission process in order to ensure a system as safe as possible.

Those basic requirements must be met for ground systems and for flight hardware, including the launch vehicle and the spacecraft. Furthermore, the FS/FS criterion must be checked for the whole ground/flight systems. If both requirements are not met, access for workers is subject to restrictive conditions. Because the gas leakage can be very fast and the oxygen concentration can drop quickly, the FS/FS criterion is applied only on preventive measures and the protection measures are not included in that criterion.

In case of voluntary release of inert gas, the area must be cleared beforehand and no access will be granted for workers until the oxygen concentration of the area get standard again.

Between the prevention measures and the protection measure, there are the oxygen detection devices. Their goal is to measure the oxygen concentration in air and to warn to workers in case of low oxygen rate. At Guiana Space Centre, the threshold is set up at 19% of oxygen. Those detection devices can be either mobile or fixed.

When entering an oxygen-deficient risk area, there must be at least two segregated detection devices. In a conservative approach, the leak of inert gas or the wrong access to an oxygen-deficient area is considered as the first failure. To be compliant with the FS/FS criterion, two redundant detection devices are requested.

In facilities where inert gases are used, a network of fixed real-time detection devices covering the whole area is set up. When the oxygen concentration decreases below the detection threshold, an alarm arises. The alarm triggers both optical and acoustical signals in order to warn immediately the workers about the current risk.

Like any safety system (fire detection, toxic gas detection, etc.), the alarms of fixed detection devices are reported and relayed to the safety officer's operation panel. Depending on the kind and the status of the ongoing operations, he/she can act to evacuate the workers out of the hazardous area and to inform the firefighters.

These fixed detection devices can be either real fixed detection devices or portable detection devices placed on site only when anoxia risk exists. The second case includes the work inside launch vehicle inter-stages. In this example, the mobile detection devices are set up right after the launch vehicle stage arrival in the integration building and are dropped off at the latest before the closure of launch vehicle inter-stage access doors.

Any person intending to access an area with anoxia risk must wear a portable oxygen detection device. If the fixed detection devices are not available or if they do not cover the given area, the worker must bear two portable devices. What the portable detection device measures must be as representative as possible of the air breathed by the worker. That is why portable device are worn as close as possible from the nose. For instance, one can fix the portable device at his helmet or around his neck.

If the workers must access an area which is hard to evacuate quickly and safely, some additional risk mitigation measures must be applied.

First, an external observer must stay outside this area during the whole presence of workers. The observer is equipped with a communication device in order to keep in touch with safety officer. The observer is not requested to rescue the workers inside the high-risk area. His/her role is to call safety/emergency officers.

Secondly, breathing equipment is available for the workers entering the area. This safety equipment can be either a self-contained breathing apparatus (SCBA) or an oxygen mask connected to the breathable air network.

An SCBA typically has three main components: a high-pressure air tank, a pressure regulator and a mask, mounted to a carrying frame.

The oxygen mask is dedicated to a given area and is earlier connected to the breathable air network of the facility by a long hose. The maximum allowable number of people inside the area is the number of available masks inside the area minus one.

The breathable air is produced by the air supply system. This system can be supplied by air compressors or by air bottles. In both cases, breathable air is analyzed periodically to ensure its quality. The main parameters of air quality are the concentrations of nitrogen, oxygen, water vapor, carbon monoxide and dioxide, oil and dust.
Besides the typical ventilation system, confined spaces are equipped with forced air ventilation systems which aim at reducing as much as possible the effect of leakage inside that space by injecting fresh air.

Any area subject to a low oxygen concentration must have recovered a standard oxygen concentration and must be checked by a firefighter before granting access for workers. The firefighter measures the oxygen concentration at several points of the area, using a long hose and a portable detection device in order to control the whole atmosphere.

Any operation is identified with a "permit to work" called at CSG the "FASO" (Operation Analysis & Follow-up File, *Fiche d'Analyse et de Suivi d'Opération* in French). That FASO contains the operations description and constraints and sum up also the safety requirements derived from the safety risk analysis, including the ones concerning oxygen-deficient atmosphere. An operation cannot start without formal approval of the FASO by the safety officer, so as to allow him/her to check safety constraints with respect to other operations and to potentially add extra safety requirements.

The last barrier against life-threatening effects on workers due to an oxygen-deficient atmosphere is first aid and fast rescue. A 90-man firefighter's brigade is in charge of rescuing workers at CSG. They all are trained to standard rescue and followed a specific training about chemical and industrial hazards.

The fixed detection devices are not automatically relayed to the rescue center, like they are to safety officer's control panel. That feature is not mandatory because the arrival delay from rescue center to the area is too long to allow efficient measures by firefighters. Indeed, the best practices in case of oxygen deficiency are to evacuate the area and, if not possible, to put an oxygen mask and then evacuate quietly.

6. METHOD OF RISK ANALYSIS

After describing the risk mitigation measures that may be applied, we must determine which measures have to be applied depending on the risk case of oxygen deficiency. To do so, several cases are defined according to the following criteria:

C1. Presence of gas or liquid likely to create an oxygen-deficient atmosphere in the area (including inner parts of fluid systems)
C2. Presence of such a fluid in nominal situation in the area's atmosphere
C3. Compliance of fluidic systems to the above-mentioned design rules
C4. Easiness to evacuate the area

Depending on the value (yes/no) taken by these criteria, six cases are deduced in Tab. 2:

Table 2. Cases and criteria

	C1	C2	C3	C4
Case 0	No	N/A	N/A	N/A
Case 1a	Yes	No	Yes	Yes
Case 1b	Yes	No	No	Yes
Case 1c	Yes	No	Yes	No
Case 1d	Yes	No	No	No
Case 2	Yes	N/A	N/A	N/A

Taking into account the oxygen deficiency risk in the area induced by the operational conditions (criteria C1 to C4), the following associated risk mitigation measures must be applied in order to master this risk. Every oxygen deficiency risk case and its related risk mitigations measures are hereunder detailed:

Case 0:

- absence of gas or liquid likely to create an oxygen-deficient atmosphere in the area

⇨ No dedicated risk mitigation measure. Wearing a portable oxygen detection device is nice-to-have but not mandatory.

Case 1a:

- presence of gas or liquid likely to create an oxygen-deficient atmosphere in degraded case,
- system architecture and design compliant to above-mentioned design rules
- area easy to evacuate

⇨ Two redundant and independent oxygen detection devices are mandatory, following the above-mentioned rules.

Case 1b:

- presence of gas or liquid likely to create an oxygen-deficient atmosphere in degraded case,
- system architecture and design noncompliant to above-mentioned design rules
- area easy to evacuate

⇨ Two redundant and independent oxygen detection devices are mandatory, following the above-mentioned rules.

Case 1c:

- presence of gas or liquid likely to create an oxygen-deficient atmosphere in degraded case,
- system architecture and design compliant to above-mentioned design rules
- area hard to evacuate

⇨ Two redundant and independent oxygen detection devices are mandatory, following the above-mentioned rules.

- ⇨ Oxygen masks are available in the area.
- ⇨ An external observer stays outside the oxygen deficiency risk area and stays in touch with safety officer using a communication device.

Case 1d:

- presence of gas or liquid likely to create an oxygen-deficient atmosphere in degraded case,
- system architecture and design noncompliant to above-mentioned design rules
- area hard to evacuate

- ⇨ Two redundant and independent oxygen detection devices are mandatory, following the above-mentioned rules.
- ⇨ A firefighter is ready to rescue close to the area and has the first aid equipment.
- ⇨ Oxygen masks are available in the area.

Case 2:

- presence of gas or liquid creating an oxygen-deficient atmosphere in nominal case

- ⇨ No personnel are allowed in the area.
- ⇨ Fixed oxygen detection network, with alarm raised remotely

To each case, several risk mitigation measures are required. A difficult issue lies in how to define the area subject to oxygen deficiency and to the related risk mitigation measures. In the case of a tight closed room, the answer is quite easy even if the case of opening/closing the room access door must be studied.

A high-volume building with adequate ventilation and air renew can be seen as an open space. It does not mean that oxygen deficiency risk is non-existent. Even in open space an oxygen-deficient atmosphere can be met.

The whole volume of inner walls or closed rooms (subject to oxygen deficiency) is considered with a risk of oxygen-deficient atmosphere. Consequently, the risk mitigations measures must be applied in those whole areas.

For now, no method dedicated to limit the oxygen-deficiency risk zone at CSG is mature enough to be applied. An interesting track is to follow a similar reasoning to explosive atmosphere (ATEX) zones. Several ATEX standards explain how to define the limits of explosive atmosphere volumes, depending on the characteristics of the release source and on the environment (ventilation, etc.). That method must still be developed and a full check set must be done on every kind of hazardous operations or situations at CSG in order to assess by computation the consistency and the relevancy of the defined risk zones.

7. EXAMPLES AT CSG

Case 0 refers to any operation realized in a fluid-free area or in high-volume hall. At CSG, that kind of operations includes for instance operations in payload preparation facilities high bay and in launcher integration building without voluntary inert gas release.

Operations inside closed rooms, semi-closed spaces or confined spaces are included in case 1. As example for case 1a, we can mention the fluid system configuration in inner walls. In nominal case, there is no gas release, the system architecture is compliant with abovementioned criteria and the evacuation of the area is not difficult.

Case 1b refers to operations whose system architecture is not compliant with abovementioned rules, in particular the FS/FS criterion. For instance, during a ground-to-flight umbilical hardware connection, both interfaces are full of inert gas (helium). In order to avoid pollution, the gas goes as far as possible on both sides. The gas relies on a single barrier. The area being easy to evacuate, no extra risk mitigation measure is required regarding to case 1a.

Some launch vehicles operated from CSG are very wide like the 5.4m-wide Ariane 5's central core. Many operations are carried out with workers inside the inter-stage skirt, in particular to access the upper cryogenic engine and stage for flight preparation. Every fluid system of the launcher (engine, propellant feeding system, etc.) is full of inert gas, in order to avoid pollution. In degraded case (such as leakage), the atmosphere can become oxygen-deficient. In case of low oxygen concentration, workers must put the oxygen mask (previously disposed inside the inter-stage and connected to breathable air network) and then evacuate quietly the area. The external observer calls the safety officer for rescue and put his own oxygen mask. If needed, he can then help the workers to go out, without entering the oxygen-deficient area.

The most critical operations with personnel are situated in case 1d, because the safety architecture criteria are not met and the working area is hard to evacuate. Ariane 5's lower cryogenic stage is linked to ground segment (mainly for fuelling purpose) by ground-to-onboard links. Those links are located inside boxes which are mechanically closed during liftoff. As for other ground-to-flight-hardware connection, both sides are full of inert gas. Moreover, the worker's environment is difficult and it is hard to evacuate. That is why during the whole duration of this operation a firefighter is pre-positioned in the area and is ready to rescue the workers in case of worker's fainting.

In itself the case 2 does not involve workers, so that there is no oxygen deficiency risk for workers. Actually, the area is totally cleared before beginning that kind of operations. At CSG, an important operation related to case 2 occurs during the Ariane 5's launch countdown. Before the entrance of liquid hydrogen in the launch zone (bunker, mast), the related rooms and areas are inerted using nitrogen. According to the abovementioned rules, the whole area (the launch zone) must be totally cleared. No personnel are allowed to enter these launch pad areas since that moment, even in degraded cases.

Before granting the access (e.g., after liftoff or due to aborted launch) to these areas, the oxygen concentration must retrieve its nominal value of 20.9%.

8. CONCLUSION

Oxygen-deficient atmosphere can lead to asphyxia and anoxia in a very fast timeframe. That hazard is very vicious because no other warning or alarm except gas detection system can detect and warn the worker of the current danger. At CSG, two people lost life twenty years ago due to that hazard. They are the only (but still too many) victims of launch activities in French Guiana since more than fifty years of launches. Safety officers often update the risk mitigation measures policy in order to manage at best that risk and to be suited to new operating mode.

That article is not a breakthrough but aims at describing a simple but efficient method to identify that risk and to choose the associated risk mitigations measures. The hidden goal of that article remains to create a debate and discussion about that topic so as to continuously improve the safety at CSG and on other launch ranges and to ensure safety for all.

9. REFERENCES

1. Bijlenga, P. (2009). Pendant combien de temps le cerveau peut-il rester sans oxygène ?, Hôpitaux Universitaires de Genève, Switzerland, retrieved February 15, 2016, from http://www.rts.ch/decouverte/sante-et-medecine/corps-humain/4642357-pendant-combien-de-temps-le-cerveau-peut-il-rester-sans-oxygene-sans-sequelles.html.

2. Leverve, X. & Guignier, M. Conséquences métaboliques de l'hypoxie tissulaire: intérêt de la mesure des paramètres biochimiques, CHU de Grenoble, France, retrieved February 17, 2016 from www.icar-web.fr/IMG/pdf/1-09.pdf

3. Group of authors (2015). Espaces confinés, Institut National de Recherche et de Sécurité, Paris, France.

4. Gaffney M. (2014). Oxygen Deficiency Hazards (ODH), System Classification and Controls, Brookhaven National Laboratory, Upton, NY, United States.

5. Group of authors. Oxygen Deficiency, Linde AG, Pullach, Germany.

6. Group of authors (2014). Safetygram 17: Dangers of oxygen-deficient atmospheres, Air Products and Chemicals, Inc., United States.

THE PAYLOAD SAFETY HANDBOOK: A SELF -SUPPORTING TOOL TO ENSURE THE RESPECT OF REGULATIONS APPLICABLE TO SPACECRAFT AT THE GUIANA SPACE CENTER

Loriane Bourjac (1)

(1) *CNES, Centre Spatial Guyanais, BP n°726 – 97387 Kourou Cedex, French Guiana, Email: loriane.bourjac@cnes.fr*

ABSTRACT

The Payload Safety Handbook [1] has been recently drawn up and transmitted to spacecraft's manufacturers and satellites customers involved in AIT (Assembly, Integration and Test) launch campaigns at Guiana Space Center (CSG), the spaceport of Europe.
This document supersedes the former *Ground Safety Regulations*, no more effective, due to the evolution of the laws, particularly the French Space Law, known as LOS [2] and the Order regulating the operations of the CSG facilities, known as REI [3].
It defines and compiles the principles and rules applicable to the design, qualification, production and implementation of payloads to be launched from CSG.

1. INTRODUCTION

For onboard equipments, the Payload Safety Handbook (PSH), just like the former *Ground Safety Regulations*, defines and compiles the principles and rules applicable to design, qualification, production and implementation at CSG.

For ground equipments, the applicable regulations can vary, depending on the country where the payload is prepared and launched. At CSG, the respect of relevant French and European regulations is required for all activities. However, considering the diversity of the clients and manufacturers' origins, international laws and standards are obviously applicable also.

After reminding the general objectives of the range safety at CSG, the paper will present:

- The safety submission process, including the requirements in terms of documentation,
- The standard equivalence and non- conformances processing,
- A recall of the main relevant responsibilities at each phase of the safety submission process

2. GENERAL OBJECTIVES OF RANGE SAFETY AT CSG

The purpose of the range safety procedures is to ensure the protection of people, goods and the environment.

For this purpose, the PSH defines, in relation to the CSG ground safety department the general principles, the design rules, the operational rules, the principles for range safety submissions.
Designers, manufacturers and operators must abide by these rules which are designed to reduce the specific risks related to the implementation of a payload to the CSG.

These specific risks are notably related to nominal campaign operations at the CSG facilities, known as EPCU (Payload Preparation and Filling Facilities, *Ensemble de Préparation des Charges Utiles* in French) as well as non-nominal campaign operations or accidental situations involving the payload (for example, aborted launch).

The nominal and usual campaign operations at the CSG facilities are the following:

- handling of pyrotechnic articles,
- integration of solid-propellant engines,
- pressurisation of tanks or vessels,
- filling tanks with toxic or hypergolic propellants,
- switching on hazardous electrical circuits,
- handling of payloads fuelled with liquid propellant,
- transferring payloads

The satellite customers, for each risk-related activity it intends to perform on the ground have to:

- Identify and assesses the risks according to the two categories defined by the severity of the potential damage (catastrophic or critical consequences),
- Implement a risk-reduction programme where appropriate,
- Supply the CSG ground safety department, as early as possible a written file containing a demonstration of compliance with the specific stipulations and rules applicable to payloads.

Tab.1 hereunder shows the two categories, defined by the severity of the potential damage:

Table 1. Classes of risks relative to activities on the ground

Risk classes	Definition of damage
Risk of catastrophic consequences	• Loss of human life, immediately or later • Permanent disability • Irreversible harm to public health
Risk of serious consequences	• Serious injury to people not leading to loss of human life, nor to permanent disability • Reversible harm to public health • Significant damage to property: - total or partial destruction of public or private property - total or partial destruction of a critical facility for the launch operation • Significant damage to the environment

Accordingly, the functional requirements aim at eliminating or reducing the occurrence of an event by designing payloads so that as few operators as possible are exposed to risk during nominal or exceptional implementation operations (for example, when returning to a safe state). It also aims at reducing the severity of any event by remote monitoring of the configuration and characteristics of systems classified at risk.

Anyway, all the systems, in particular those which could cause risks, have to undergo a safety submission process.

3. PAYLOAD SAFETY SUBMISSION PROCESS

The payload safety submission process, which is iterative and continuous, enables the satellite customer to demonstrate to the CSG ground safety department that it fully complies with the stipulations and requirements set out in the PSH.

It consists in four main phases:

- A possible phase 0, in case of advanced feasibility for new platforms using innovative technologies,
- A phase 1, for design validation,
- A phase 2, for qualification and manufacturing validation,
- A phase 3, for implementation validation.

Phase 0 – Feasibility

The feasibility dossier shall include as a minimum:

- A list of risk-related systems with descriptions,
- An overview of the technical choices envisaged,
- A list of the risks related to the flight system and corresponding specific ground segment equipment,
- A list of the laws, regulations and standards applicable to the project (where appropriate)

Phase 1 – Design

The design dossier shall include as a minimum:

- A detailed description of risk-related systems, their command and control circuits and their corresponding ground segment equipment,
- The planned levels and frequency bands of the transmitters and receivers, and the transmission characteristics (spectrum, power, etc.).
  - A preliminary risk analysis, including in particular the feared events, the classification of risk-related systems, and the reliability data by which to assess the compliance of system design with [2],
- The qualification plans for the component parts of risk-related systems.

Phase 2 – Qualification and production

The qualification dossier shall include as a minimum:

- The results of partial or overall qualification tests of risk-related systems,
- The plan for partial or overall acceptance test plans for risk-related systems,
- Any particular study or engineering notes by which to judge the characteristics of systems (for example, rupture analysis),
- A document defining the interfaces between the payload and the corresponding equipment and the various facilities at the CSG,
- In-depth risk analyses of flight and ground systems:
  - the technical data of the range safety analyses,
  - tests performed for risk-related systems,
  - any risk-reduction measures implemented,
- An analysis of compliance with the range safety requirements of the launcher on the ground, for configuring and reprotecting payloads in the launch area.

Phase 3 – Implementation

The Phase 3 dossier shall include:

- The satellite operations plan (POS) containing in particular the sequence of all operations to be performed at the CSG,
- The procedures for performing risk-related activities, in particular the procedures relating to the activities for ensuring safety and the emergency procedures to be followed in case of incident. These procedures must contain the following information:
  - an identification of risk-related activities,
  - a consideration of the specific features of the CSG (sites, means, designations, etc.),
  - the details, for each step, of the number and function of the essential personnel in the danger areas,

  - a detailed list of the equipment and products used,
  - a step-by-step description of the procedure for returning to safety status,
  - an indication of the duration of these activities, including those for returning to safety status,
- The results of formal acceptance tests of flight components of risk-related systems (tanks, batteries, etc.)
- The certificates following tests and/or periodic checks of ground segment equipment,
- The medical fitness certificates for the personnel working on certain risk-related systems, especially those emitting ionizing radiation or containing toxic products,
- The certificates authorizing personnel to handle pyrotechnic devices, or for any other operation requiring a special authorization,
- The licences for the possession and use of equipment subject to administrative authorization (for example, object emitting ionizing radiation),
- The items necessary for checking that the customer resources and operations comply with the regulations applicable to the CSG (for example explosive atmosphere, chemicals, restrictions on personnel, amount of propellant, etc.),
- The items necessary for checking the integrity of flight or ground systems implemented on site (for example results of flight airtightness tests, leak tests on fuelling carts, etc.).

The procedure for submissions includes a demonstration that risk has been controlled. This demonstration may be more or less simplified depending on whether the payload is of an entirely new type, or part of a family with great similarities. It may depend also on the risk- related systems integrated in the payload.

Tab.2 hereunder precise the safety submission cycle depending on the payload's types:

Table 2. Safety submission cycle

Safety submission	Phase 0	Phase 1	Phase 2	Phase 3	Validity
Specific S/C	X	X	X	X	3 years from the closure
Generic platform	*If necessary (new technology)*	X	X	X	3 years from the closure
S/C based on known generic platform		*If necessary (new hazard systems)*	X	X	Done for each campaign

The various phases closure objectives are:

- Phase 0: before the end of feasibility cycle
- Phases 1 & 2: before arrival at CSG
- Phase 3: before actual operation

4. STANDARDS EQUIVALENCE

At CSG, the respect of relevant French and European regulations is required for all activities.
However, for ground equipment, the applicable regulations can vary, depending on the country where the payload is prepared and launched.
So, considering the diversity of the clients and manufacturers' origins (US, Japan, India, South America…), international laws and standards are obviously applicable also.

Concerning this point, the new approach described in the PSH is the introduction of standards equivalence. Indeed, for all ground support equipment (GSE) designed and operated in accordance with the national laws and consensus standards of their country of origin, but not certified in EU reference table, particular analyses must be realized.
The objective is to define the safety minimal requirements taken into account by the client in his declared reference table, and as a consequence, to guarantee the conformity in comparison with EU reference table.

Considering the number of reference tables and the diversity of domains and systems (pyrotechnic, static electricity, fluids, laser, explosive atmosphere…), this ambitious work is still in progress, in partnership with international spacecraft projects.
In some domains (lifting devices, pressurized and electrical systems…), technical and theoretical analyses have already been realized and equivalence tables are available.

It is the case for example with the COPVs (composite overwrapped pressure vessels). Depending on the suppliers, most of them are qualified based on US Military Standards [4]. The technical comparison between the European and the US standards has been realized few years ago and concluded to the same level of safety requirements. Then, all equipment qualified according to [4] requirement, is completely recognized as compliant for implementation at CSG.

5. PROCESSING NON-CONFORMANCES

In the event that it proves impossible to comply with one or more of the stipulations of the PSH, the satellite customer shall address a formal request for waiver to the Director of CNES/CSG, with a dossier supporting the request which should detail and justify:

- The impossibility, in the case considered, of taking all the necessary steps to ensure the establishment, maintenance or restoration of compliance with the safety rules,
- The measures taken to comply as closely as possible with the safety rules,
- The level of residual risk resulting from the non-conformance,

An exceptional waiver, depending in particular on the conditions of the operational environment at the time, can be granted.

6. MAIN DESIGN AND OPERATIONAL RULES AT CSG

The satellite customer is responsible for:

- The respect of range Safety requirements for the design, qualification, production and commissioning of the payload,
- The complete description of the payload during the CSG Safety submission process, including those systems or sub-systems which involve a risk,
- Supplying, within the deadlines, the documents requested by the CSG Safety department,
- The training and qualification of its operators at their workstations,
- The performing of risk-related operations according to operations procedures that have been approved by the CSG Safety department.

The CSG Safety department is responsible for:

- Verifying, the submissions provided by the satellite customer and within certain time limits, to ensure that range Safety requirements have been satisfied,
- Defining the technical means necessary for the supervision of risk-related activities and in particular:
  - video resources needed for monitoring activities on the site,
  - detection resources to be implemented (toxicity, anoxia, fire, etc.),
  - the means of communication for keeping in touch with the personnel concerned,
  - the means of disseminating information or broadcasting warning messages (evacuation alert, fallback order, etc.).
- Defining the means and conditions necessary for safe operations, and especially:
  - fire-fighting resources, both fixed (the fire hydrant system known as RIA, *Réseau d'Incendie Armé* in French, fire extinguishers, etc.) and mobile (Paris Fire Brigade known as BSPP, *Brigade des Sapeurs-Pompiers de Paris* in French, etc.) and alerting the medical services,
  - watchman service and access control,
  - individual protection for operational personnel (PPE, etc.),
  - weather surveillance,
  - assigning a Range Safety official to the site,
  - the evacuation of affected areas during risk-related operations,
  - the compatibility of activities carried out in parallel,
- Training the satellite customer's personnel and raising their awareness of Range Safety,
- Verifying the compliance of installations and in particular the proper functioning of the safety equipment before any risk-related operation,
- The management of accidental situations.

For these reasons, a risk-related activity may not start unless:

- Operational procedures have been approved by the Range Safety department,
- The means and conditions necessary for performing operations are met (weather, status of Range Safety resources),
- The performance of the risk-related operation is compatible with other ongoing operations,
- Agreement has been received from the Range Safety department.

7. CONCLUSION

The PSH has been drawn up for spacecraft's manufacturers and satellite customers involved in AIT launch campaigns at CSG, to help them with payload safety submissions, by developing requirements of [3].

It is self-supporting, regarding applicable regulation in terms of safety. It defines and compiles the principles and rules applicable to the design, qualification, production and implementation of payloads to be launched from CSG, including all good lessons learnt from decades of experience with worldwide customers and manufacturers.

It also permits to define the safety approach in terms of design and qualification for GSE which are based on foreign standard requirements.

To conclude, we can write that the main objective of this document is to clarify all the safety constraints applicable on the launch range, so as to ensure safety for all at CSG.

8. REFERENCES

1. Payload Safety Handbook (PSH) – Ed/rev 01/01, CSG-NT-SBU-16687-CNES

2. Space Operation Act N°2008-518 of 3 June 2008 concerning space operations (known as LOS)

3. Order issued by the CEO of CNES N°2010-1 of 9 December 2010 concerning the regulations governing the use of Guiana Space Center facilities (known as REI)

4. Military Standard, Standard general requirements for safe design and operation of pressurized missile and space systems, MIL- STD-1522A

ENSURING PAYLOAD SAFETY ON MISSIONS INVOLVING SPECIAL PARTNERSHIPS

Calvert A. Staubus, P.E., CSP [(1)], Rachel C. Willenbring [(2)], Michael D. Blankenship [(3)]

[(1)] National Aeronautics and Space Administration (NASA) Expendable Launch Vehicle (ELV) Payload Safety Program Manager, John F. Kennedy Space Center mail code SA-G Kennedy Space Center, FL 32899, calvert.a.stabus@nasa.gov
[(2)] NASA ELV Payload Safety Program Engineer, John F. Kennedy Space Center mail code SA-G Kennedy Space Center, FL 32899, rachel.c.willenbring@nasa.gov
[(3)] NASA Technical Writing Specialist, John F. Kennedy Space Center mail code SA-G Kennedy Space Center, FL 32899, michael.d.blankenship@nasa.gov

ABSTRACT

The National Aeronautics and Space Administration (NASA) Expendable Launch Vehicle (ELV) payload space flight missions involve cooperative work between NASA and partners including spacecraft (or payload) contractors, universities, nonprofit research centers, Agency payload organizations, Range Safety organizations, Agency launch service organizations, and launch vehicle contractors. The role of NASA's Safety and Mission Assurance (SMA) Directorate is typically fairly straightforward, but when a mission's partnerships become more complex to realize cost and science benefits (e.g., multi-agency payloads or cooperative international missions), the task of ensuring payload safety becomes much more challenging. This paper discusses lessons learned from NASA safety professionals working multiple-agency missions and offers suggestions to help fellow safety professionals working multiple-agency missions.

1. INTRODUCTION

The National Aeronautics and Space Administration (NASA) routinely collaborates with partners including spacecraft (or payload) contractors, universities, nonprofit research centers, Agency payload organizations, Range Safety organizations, Agency launch service organizations, and launch vehicle contractors. NASA depends on the Expendable Launch Vehicle (ELV) Payload Safety Program to manage the Agency's interests in safety and mission success for these joint efforts.

NASA's ELV Payload Safety Program was originally developed in response to events that occurred during the Cloud-Aerosol LIDAR and Infrared Pathfinder Satellite Observation (CALIPSO) mission. CALIPSO can be seen in *Figure 1*. CALIPSO was a joint science mission between NASA Langley Research Center (LaRC), Goddard Space Flight Center (GSFC), and Centre National d'Etudes Spatiales (CNES). This multi-agency mission between NASA and CNES was also a dual payload mission since CALIPSO was launched with CLOUDSAT, another cloud and aerosol observer satellite. CloudSat used a Cloud Profiling Radar (CPR) instrument, a 94-GHz nadir-looking radar, to measure the power backscattered by clouds as a function of distance from the radar. The design of the CPR was developed jointly by NASA Jet Propulsion Laboratory (JPL) and the Canadian Space Agency (CSA).

Figure 1. CALIPSO spacecraft

CALIPSO demonstrates well the importance and challenges of performing cross-agency safety engineering and safety coordination. As the mission progressed, concerns were raised involving design requirements that threatened to delay the mission.

One of the biggest challenges of joint missions is meeting multiple sets of safety requirements, especially those that cross agencies. However, in this particular case, the disagreement primarily existed within NASA. Alcatel Space Industries manufactured the off-the-shelf, hydrazine-fueled Proteus propulsion bus used by CALIPSO. CNES provided the spacecraft bus as part of their in-kind contribution to the joint mission. Even though NASA JPL had used the Proteus propulsion bus for a previous oceanography mission (Jason-1) without a safety concern, GSFC Safety and Mission Assurance (SMA) believed that the threaded fittings in the Proteus propulsion system did not meet NASA fault tolerance design guidelines. SMA at

JPL and other NASA Centers did not necessarily agree with GSFC SMA's position.

The existing payload safety review and approval process at the time was not sufficiently prepared to handle GSFC SMA's safety concern with CALIPSO's Proteus bus. A safety concern that should have been addressed at a much lower level of decision-making was left unnoticed until late in processing, and the NASA Chief of SMA, Center Directors, Air Force Range Safety, CALIPSO Project Manager, and other upper level managers were forced to intervene to resolve the issue. Ultimately the NASA Engineering and Safety Center (NESC) was brought in to independently assess the Proteus propulsion bus threaded fittings safety concern, and the NESC generated a report with recommended safety requirements to aid in the safe processing of CALIPSO (RP-04-01/03-001-E) and future Proteus bus missions.

2. NASA'S ELV PAYLOAD SAFETY PROGRAM

NASA recognized an opportunity for improvement following CALIPSO and established NASA's ELV Payload Safety Program in May 2008 to ensure the safety of personnel and resources from hazards associated with NASA payloads flying on unmanned ELVs. The Program does not cover safety during flight operations or payloads going to the International Space Station. The Program seeks to improve NASA safety review and approval process and delineates clear roles and responsibilities for those involved in ensuring payload safety. The Program also improves and formalizes NASA's payload safety review and approval processes, establishes formal decision-making authorities, and clarifies a process for handling dissenting opinions.

NASA's ELV Payload Safety Program consists of a full-time ELV Payload Safety Manager with a safety engineer support contractor and a part-time Agency Team. Agency Team members are payload safety managers located at GSFC, NASA Headquarters, JPL, Kennedy Space Center (KSC), and Wallops Flight Facility (WFF). The Agency Team helps ensure consistent application of payload safety requirements across the Agency for NASA ELV payloads.

NASA's ELV Payload Safety Program has two primary documents. The first, a NASA Procedural Requirements document (NPR) 8715.7, Expendable Launch Vehicle Payload Safety Program, provides the payload safety review and approval process and related roles and responsibilities. The second is a NASA Standard, NASA-STD 8719.24, NASA Expendable Launch Vehicle Payload Safety Requirements, written jointly with U.S. Air Force Range Safety since most NASA payloads are launched from the Air Force's Eastern and Western Ranges. A joint Air Force/NASA team tailored AFSPCMAN 91-710, Range Safety User Requirements, and incorporated applicable NASA safety requirements to form NASA-STD 8719.24.

3. METHODS USED TO FACILITATE SAFE AND SUCCESSFUL NASA ELV PAYLAOD MISSIONS

While multi-agency and international cooperative missions offer challenges, the benefits justify the effort. International cooperation is a cornerstone principle of NASA's activities and has been part of NASA since its inception. The 2010 National Space Policy of the United States of America lists as one of the goals to expand international cooperation on mutually beneficial space activities. The benefits from leveraging partner funding and capabilities through multi-agency missions are numerous. Cooperative research objectives between partners can accomplished efficiently with access to an increased number of talented engineers and scientists. Additionally, the overall cost of research can be shared and reduced while facilitating wider distribution of research and science data. It is easy to see why multi-agency missions will continue, especially in years of smaller agency budgets.

Agencies wishing to benefit from cooperative missions have several hurdles to overcome to ensure safety and mission success. While participants in joint partnership have individual goals, it is important to realize that all members have the same basic objective: a safe and successful launch of the payload. In working toward that goal, the safety requirements used by all members of the joint effort are similar. It is the coordination and communication of the payload safety review and approval process and the resolution of non-similar safety requirements that require more time and effort to resolve.

Recognizing this, partners need to understand that performing payload safety for joint missions requires a different approach than missions involving a single agency. Most importantly, partners need to recognize that time, as much as an additional 50% or more, is required to overcome the challenges associated with joint missions. Challenges include:

- Confirming that all safety concerns have been addressed. With varying levels of experience in joint missions, it may take time to define and explain roles and responsibilities and educate about payload safety requirements and safety review processes.

- Coordinating between multiple Centers, contractors, and agencies and facilitating necessary safety discussions.
- Multiple lines of management may need to review and approve mission processes.
- Communication between partners is not always easy to accomplish when international partners work cooperatively, and basic communication can become difficult when language barriers exist.
- Small misunderstandings can halt progress, and upper-level managers of partner organizations working toward milestone goals do not want these items to become show-stoppers that delay launch operations.
- There are financial and legal aspects to consider, and establishing contracts and agreements takes time.

NASA ELV Payload Safety has implemented processes to avoid these common pitfalls.

3.1 Payload Safety Working Group

The standing Payload Safety Review Panel used by Shuttle Program was expensive to maintain but worked well for a program with funding to support it. NASA ELV Payload projects do not have funding to support a standing panel, so a more cost-effective, mission-specific Payload Safety Working Group (PSWG) is formed for each NASA ELV payload project. The PSWG is made up of safety engineers representing the various organizations involved in the mission. These typically include the payload project system safety engineer, the payload contractor safety engineer, the launch site range safety engineer, the launch vehicle contractor system safety engineer, the payload processing facility safety engineer, and the NASA Launch Services Program system safety engineer who often chairs the PSWG. *Figure 2* shows the typical members of a PSWG in blue with green indicating specialists that may be called in to assist when needed.

Each mission generally has four safety review milestones in which the PSWG is involved. Additional review meetings may be held depending on the complexity and hazards related to the mission. It is not uncommon to hold several meetings to complete one of these safety reviews. The safety review process is kicked off with a Payload Safety Introductory Briefing (PSIB) held early in the preliminary design phase. The PSIB is followed by Safety Review I occurring around the Preliminary Design Review (PDR), Safety Review II held around the Critical Design Review (CDR), and Safety Review III completed prior to shipping the payload to the launch site processing area.

Figure 2. Payload Safety Working Group

The payload organization provides safety data packages to the PSWG for review in accordance with the mission-specific tailored NASA-STD 8719.24 Annex. Hazards are identified and abated to the satisfaction of the PSWG members. When the PSWG has issues or needs assistance, the ELV Payload Safety Program Agency Team is available to step in and help. Hazards are identified and documented along with agreed-to abatement actions on Hazard Reports (NASA Form NF 1825). Open safety verification actions are tracked until closed on a Safety Verification Tracking Log (SVTL).

NASA's ELV payload safety review and approval process finishes with the signing and distribution of the Mission's Certificate of ELV Payload Safety Compliance. From this point forward, local safety professionals, most of whom were on or represented on the PSWG, ensure safety during payload processing operations in their jurisdictions and the SVTL is closed out.

3.2 Trilateral Safety and Mission Assurance Payload Safety Task Force

The European Space Agency (ESA), Japanese Aerospace Exploration Agency (JAXA), and NASA established a Trilateral SMA in 2008 to enhance the success of our cooperative programs. At the first Trilateral SMA Conference in April 2008 (TRISMAC 2008), a recommendation was made to establish a Payload Safety Task Force (PSTF) to compare payload safety requirements used by the agencies. NASA hosted the first PSTF meeting in April 2009 at the Kennedy Space Center. Since then, PSTF meetings (mostly telephone conferences) have occurred. The U.S. Air Force participated in some of these meetings. To date the meetings have focused on general payload safety and fault tolerances, propulsion and pressure systems safety requirements, pyrotechnics safety requirements, and payload safety review processes.

In general, the payload safety requirements used by ESA, JAXA and NASA, including the U.S. Air Force, were similar. A hazard is a hazard regardless of where you are in the world or in space. Fault tolerance requirements are largely the same once agreement is reached on hazard level classification (catastrophic, critical, or marginal). Hazardous energies and materials used in the spacecraft (payload) and payload processing are recognized, addressed, and abated in similar fashions. Each agency has a comprehensive safety review and approval process with some slight differences. NASA uses a mission-specific PSWG safety review approach composed of the safety engineers involved in the mission. JAXA uses a standing System Safety Review Panel (SSRP) approach where the SSRP is staffed by SMA engineers and experts. Both safety review and approval processes achieve similar results.

4. LESSONS LEARNED FROM MULTI-AGENCY MISSIONS

Many of the lessons learned by NASA's experience are applicable to any organization participating in joint missions. Addressing the following common obstacles allows these organizations to optimize the benefits of cooperative missions.

4.1 Early Planning

As in any mission, planning is needed to allow for a viable safety review and approval process. For multi-agency missions, expect (and plan) to spend more time on planning. The mission safety professional must initiate planning, coordination, and communication activities with their mission partners involved in safety early in the project life cycle.

The mission safety engineer should try to be involved in the development of the mission contracts and cooperative agreements to ensure appropriate safety program requirements will be included and implemented. If this is not possible, the safety engineer should review and understand the existing contracts and cooperative agreements. Mission safety professionals should consider writing additional agreements to aid in the safety review process when needed or feasible. These safety agreements should define roles, responsibilities, processes, and schedule. An agreement could even be used to bring in third party support or expertise to augment the safety team and the safety review process.

Plan to establish temporary agreements allowing contractors to converse and communicate with international team members early in the process. For the United States and often other nations performing international work, contractors are required to have a signed Technical Assistance Agreements (TAA) approved by the U.S. Department of State. Without agreement documents, contractors are not authorized to talk directly to international team members. Processing the TAA approvals may require six months to one year, so early submittal of the TAA request is recommended. Additionally, most space agencies and companies require nondisclosure agreements signed by team members and management.

By establishing your safety review team and review processes, obtaining required facility access training, and obtaining necessary TAAs to comply with any local launch site requirements, these common causes of delays can be avoided.

4.2 Export Control Considerations

International Trafficking Arms Regulations (ITAR) must also be managed. Meeting ITAR regulations involves:

- Understanding governing regulations
- Training safety team members
- Identifying possible avenues for ITAR violations
- Establishing contacts and dialog with ITAR experts and legal representatives
- Applying early for export licenses

There are four organizations that work internationally to prevent and control the export of anything used for making or delivery of nuclear, chemical, or biological weapons. Missile Technology Control Regime operates as one of those four organizations staffed by multiple countries from the United Nations. Rocket technology falls under the category of missile delivery systems. Specifically, ITAR states the restrictions on missile technology. The U.S. Department of State's responsibilities includes ITAR.

ITAR expands to twenty-five categories. Training personnel on ITAR specifics concerning their work is extremely important. Companies and agencies often create ITAR education for their personnel. The NASA Export Control Program further educates, regulates, and monitors ITAR applying to NASA programs and projects. Projects with international involvement or having plans to transfer information internationally require an export license outlining and documenting specific information for transfer.

Allowing access to missile technology information in any media violates ITAR. Many methods exist for export other than transfer of an actual physical item. Telephone calls, data transmission, casual conversation, and presentations are considered export of information. ITAR violations may result in harsh consequences including company fines, individual fines, and jail sentences. Verifying nationality status for members in the partnership is crucial to prevent ITAR violations. Enacting an export control plan within mission teams helps prevent accidental violation of regulations. Laziness and ignorance, not only malicious intent, could lead to illegal technology transfer. Proper ITAR planning and vigilance will allow for ITAR and payload safety compliance while maintaining mission schedule.

The following ITAR-related actions are recommended in planning an international project:

- Read governing regulations, and effectively train personnel
- Identify possible avenues for ITAR violations
- When possible, employ an export control expert or consult an expert
- Apply early for export licenses from state department or governing authority
- Keep information concerning regulated items on secure, controlled data storage devices
- Contact lawyers or export control experts when questions arise in the duration of the mission

4.3 Coordinating, Communicating, and Documenting

Multi-agency missions require additional coordinating and communicating to complete the safety review process for the payload. Frequently it is necessary to perform increased documentation control as well.

Face-to-face meetings are very beneficial when communicating with other nationalities. Approximately 70% of communication is from body language. In-person meetings allow for body language and clearer verbal communication thus aiding in bridging the communication and cultural differences. However, face-to-face meetings with another space agency that require international travel are often infeasible. NASA uses videoconferences with good results with international partners. For all meetings, additional administrative planning and coordination time is necessary to connect with the desired parties. Emails and written correspondence are used regularly, but they may not be the most effective way to resolve safety concerns.

Documentation control in cooperative projects is also imperative. Only designated individuals should control and update documentation that is stored in a central repository available to appropriate partners. Personnel access to current versions of mission safety related documents must be controlled while being available to safety representatives for individual review. Recording meeting minutes and making them available for the attendees of inter-agency meetings has proven to be an effective way of providing clearer understanding of actions, roles, and responsibilities.

5. CONCLUSION

Over the years, NASA has gained valuable payload safety experience related to multi-agency missions through developing joint requirements with the U.S. Air Force (NASA-STD 8719.24), working with the Trilateral SMA PSTF, working multi-U.S. agency missions, and working multi-national agency missions with other space agencies. NASA has observed that payload safety engineering professionals of all agencies tend to have the same primary objective – safe payload processing and launch leading toward mission success. However, meeting this objective takes more time—more time for planning, for coordinating safety reviews, understanding contracts and agreements, for communications and cultural differences, for establishing trusted and respected safety relationships, and to allow for additional training or to bring in a third party consultant. It is our passion for safety and our respect of our fellow safety professional that motivates us to give the extra time it takes to ensure safe and successful multi-agency missions.

6. REFERENCES

1. NASA STD 8719.24 NASA Expendable Launch Vehicle Payload Safety Requirements

2. NPR 8715.7 NASA Expendable Launch Vehicle Payload Safety Program

3. NSTS/ISS 18798 Interpretations of NSTS/ISS Payload Safety Requirements

4. Japanese Safety Requirements for ELV payload paper for IAASS

5. Trilateral document on ELV payload safety by NASA, JAXA, and ESA

6. NASA Small Spacecraft Technology State of the Art Report, Mission Design Division
Ames Research Center, Moffett Field California
December 2015

7. Export Control Training Kennedy Space Center March 2016

8. Will Conn, February 29, 2016 Interview, Global Precipitation Measurement (GPM) Project Safety Manager, Goddard Space Flight Center

9. Missile Technology Control Regime
http://mtcr.info/

10. NASA Office of International and Interagency Relations, Export Control and Interagency Liaison Division Export Control Program,
http://oiir.hq.nasa.gov/ecild.html

11. US Department of State Directorate of Defense Trade Controls
http://pmddtc.state.gov/compliance/index.html

PROCEEDINGS OF THE 8TH INTERNATIONAL SPACE SAFETY CONFERENCE

SESSION 08 DESIGNING SAFETY - I

FLAT H FRANGIBLE JOINT EVOLUTION

Thomas E. Diegelman[1], Todd J. Hinkel[2], Andrew Benjamin[3], Brian V. Rochon[4], Christopher W. Brown[5]

[1] *NASA Johnson Space Center,2101 NASA Parkway, Mail Code NE311,Houston Texas 77058, Thomas.e.diegelman@nasa.gov*
[2] *NASA Johnson Space Center,2101 NASA Parkway,Mail Code EP311,Houston Texas 77058, Todd.hinkel-1@nasa.gov*
[3] *NASA Johnson Space Center,2101 NASA Parkway,Mail Code EP311,Houston Texas 77058, Andrew.l.benjamin@nasa.gov*
[4] *NASA Johnson Space Center,2101 NASA Parkway,Mail Code NC311,Houston Texas 77058, Brian.v.Rochon@nasa.gov*
[5] *NASA Johnson Space Center,2101 NASA Parkway,Mail Code EP311,Houston Texas 77058, Christopher.w.brown@nasa.gov*

1. INTRODUCTION

Space vehicle staging and separation events require pyrotechnic devices. They are single-use mechanisms that cannot be tested, nor can failure-tolerant performance be demonstrated in actual flight articles prior to flight use. This necessitates the implementation of a robust design and test approach coupled with a fully redundant, failure-tolerant explosive mechanism to ensure that the system functions even in the event of a single failure.

Historically, NASA has followed the single failure--tolerant (SFT) design philosophy for all human-rated spacecraft, including the Space Shuttle Program. Following the end of this program, aerospace companies proposed building the next generation human-rated vehicles with off-the-shelf, non-redundant, zero-failure-tolerant (ZFT) separation systems.

Currently, spacecraft and launch vehicle providers for both the Orion and Commercial Crew Programs (CCPs) plan to deviate from the heritage safety approach and NASA's SFT human rating requirements. Both programs' partners have base-lined ZFT frangible joints for vehicle staging and fairing separation. These joints are commercially available from pyrotechnic vendors. Non-human-rated missions have flown them numerous times. The joints are relatively easy to integrate structurally within the spacecraft. In addition, the separation event is debris free, and the resultant pyro shock is lower than that of other design solutions.

It is, however, a serious deficiency to lack failure tolerance. When used for critical applications on human-rated vehicles, a single failure could potentially lead to loss of crew (LOC) or loss of mission (LOM)).

The Engineering and Safety & Mission Assurance directorates within the NASA Johnson Space Center took action to address this safety issue by initiating a project to develop a fully redundant, SFT frangible joint design, known as the Flat H. Critical to the ability to retrofit on launch vehicles being developed, the SFT mechanisms must fit within the same three-dimensional envelope as current designs as well as meet structural loads requirements. There is increased mass associated with the redundant design, and the goal is to minimize the weight impact as much as possible. These requirements presented significant challenges, both technically and financially; these challenges will be explored in this paper.

Perhaps greater than the technical issues confronted during this design process, were the financial considerations. These were a significant part of the story of this design and development plan. Insufficient financial and labor resources were formidable barriers to completing this project.

Nevertheless, JSC personnel successfully conducted several test series at JSC with very useful results. The many lessons learned drove design improvements, performance efficiency, and increased functional reliability.

This paper examines the significant technical and financial challenges that these requirements posed to the project team. It discusses the evolution of the SFT frangible joint design, including optimization, testing, and successful partnering of the Johnson Space Center (JSC) engineering and JSC safety organizations, to enhance the flight safety margin for America's next generation of human-rated space vehicles.

2. FRANGIBLE JOINT DESCRIPTION

The primary function of a frangible joint is to separate launch vehicle stages or payload fairings. It offers a debris-free separation with significantly lower shock than a Linear Shaped Charge (LSC) separation system, and it weighs less than discrete release devices, such as separation nuts and explosive bolts. Figure 1 depicts where frangible joints are often used on a launch vehicle. Current unmanned spacecraft flight designs implement ignition redundancy by detonating the ends of the separation segments with zero-failure tolerance in the single cord between the detonators. Past attempts to achieve pyrotechnic failure tolerance focused on maintaining the existing exterior joint geometry and only altered the design interior by placing two explosive cords in the same relative location where there previously had been only one. This option was minimally successful. In contrast, the design described in this document is fully redundant and has two independent explosive trains.

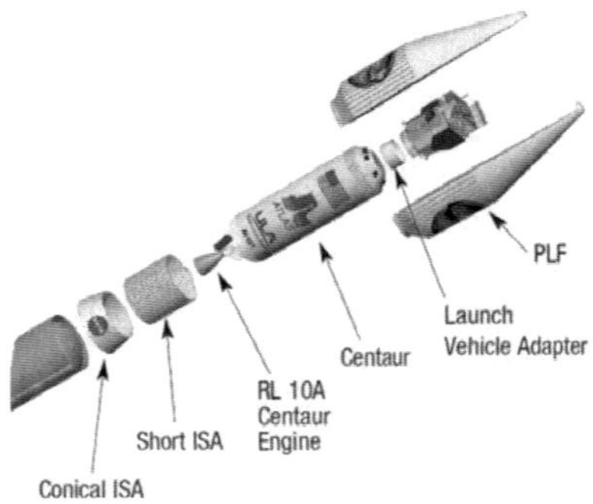

Figure 1: Spacecraft Separation Systems Using Frangible Joints

The frangible joint design has an explosive cord centered inside a steel tube with an elastomeric charge holder. This is the expansion tube assembly (XTA). The XTA serves the dual purpose of transferring the explosive cord energy to the joint while also containing the pyrotechnic combustion by-products. Detonation of the explosive cord, also known as a mild detonating fuse (MDF), sends a shock wave through the structure and expands the confinement tube. Both the shock and expansion force contribute to the separation.

In Figure 2, the cross-sectional shape of the tube has two flat sides and the cord centered between the flats. Detonating the cord returns the tube to a round shape and fractures the structure in its expansion path.

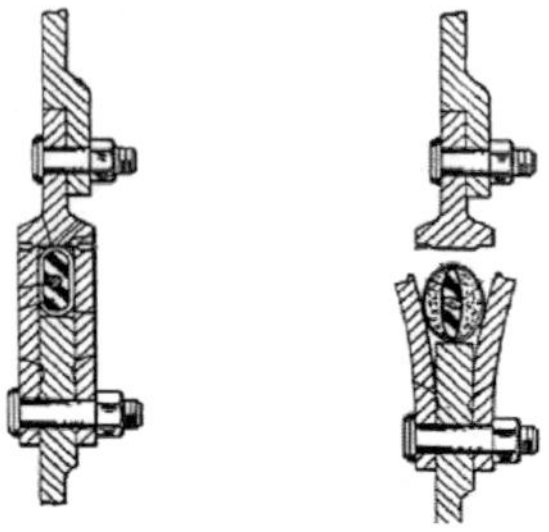

Figure 2: Cross Section of a Single XTA Frangible Joint

Figure 3 illustrates the before and after of a typical XTA.

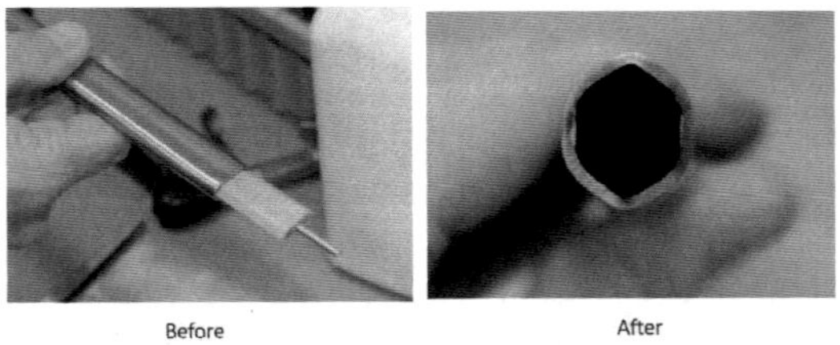

Figure 3: Typical Expanding Tube Assembly

Figure 4 depicts the XTA fracturing the separation plane when activated.

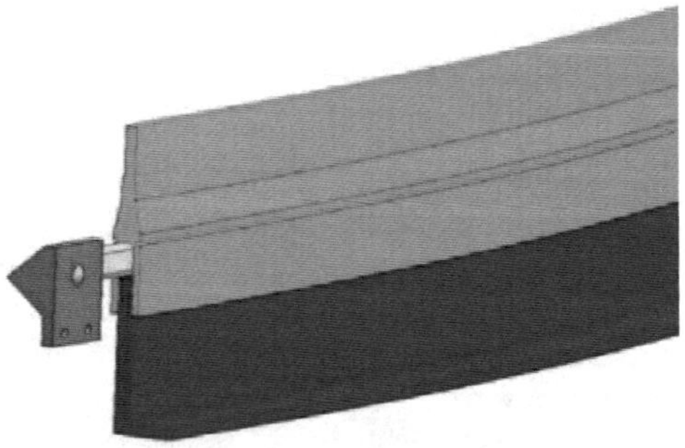

Figure 4a: Pre-Separation Frangible Joint

Figure 4b: Post-Separation Frangible Joint

3. FRANGIBLE JOINT APPLICATION

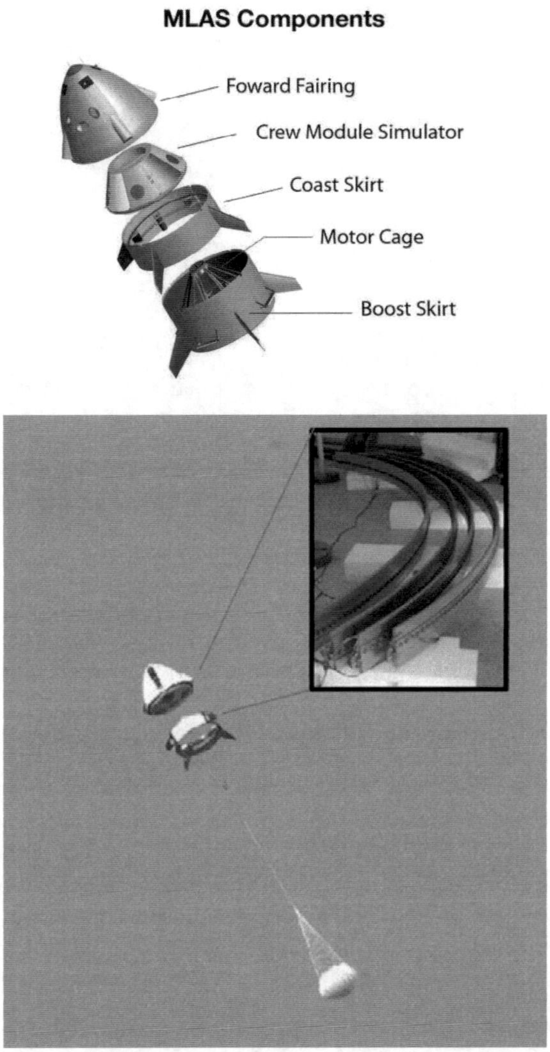

Figure 5 illustrates a typical fairing separation system.

Figure 5: Max Launch Abort System (MLAS) Frangible Application.
These separation designs are ZFT, and the Orion Program and the CCP have designated them for use in highly critical operations where failure to function, or premature function, leads to LOC/LOM outcomes.

To date, no NASA human-rated program has selected the use of ZFT frangible joints. The Shuttle Program for emergency crew egress selected and flew XTA sub-assemblies installed in the side hatch and overhead window jettison assemblies to break a series of bolts. They were not required to sever a continuous joint segment, as currently proposed for the Orion and CCP vehicles. These SSP designs were aligned with the NASA heritage SFT design philosophy, using redundant XTAs to maximize the best chance for system function, which is in stark contrast to the ZFT XTA proposed for use in the next generation of spacecraft.

4. PYROTECHNIC COMPONENT FAILURES

NASA has documented several cases, specifically relevant to frangible joints, in which ZFT pyrotechnic devices have failed. The root causes of some component failures have been resolved; however, other incidents remain unresolved, even after extensive investigations. Due to following best practices and employing functionally redundant pyrotechnic trains in human spacecraft, it is possible that more flight failures occurred than were discovered and documented. Several ZFT failure cases are discussed below as illustrative examples.

Case Study 1: Shuttle Overhead Window Jettison Qualification Test

The Space Shuttle Orbiter overhead window assembly contained explosives to allow ground egress in case the crew was not able to exit normally through the side hatch upon landing. The inner window contained dual expansion tube assemblies, while the outer window had dual MDF lines (see Figure 6).

After a successful jettison of the outer window during a qualification test, inspection revealed that one of the MDF lines had not fired as planned. The root cause of the MDF failure remains unknown. The implementation of redundancy enabled complete functionality of the system.

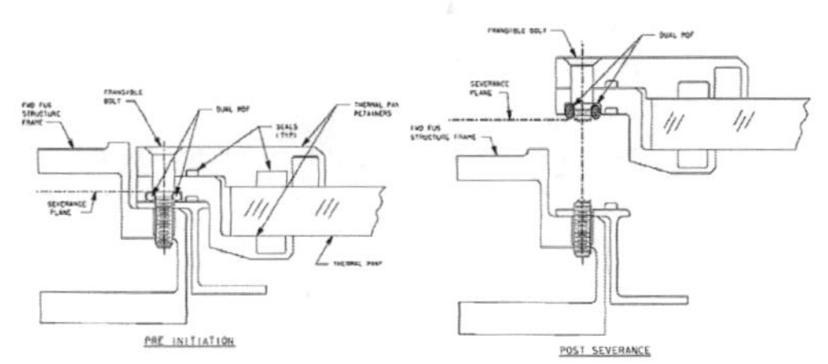

Figure 6: Shuttle Overhead Window Jettison System

Case Study 2: Shuttle Flexible Confined Detonating Cord (FCDC) Failure

During safing operations following the retirement of the Shuttle Orbiters, the Program management made the decision to fire these lines in place due to the extensive efforts required for complete removal. As a result, technicians removed all pyrotechnic devices from the vehicles, with the exception of the window egress system FCDC lines and the exterior T-handle initiator. They also installed swell cap function indicators on the end of each line.

Upon firing, one of the caps did not swell as expected, indicating that the line did not fully function. Further investigation revealed a fault between an elbow fitting and the swell cap located at the center console (see Figure 7).

Although the root cause of the failure was not determined, the leading theory is that repeated and excessive bending of the lines during cockpit refurbishment degraded the FCDC.

System functionality was achieved specifically due to SFT redundant design implementation.

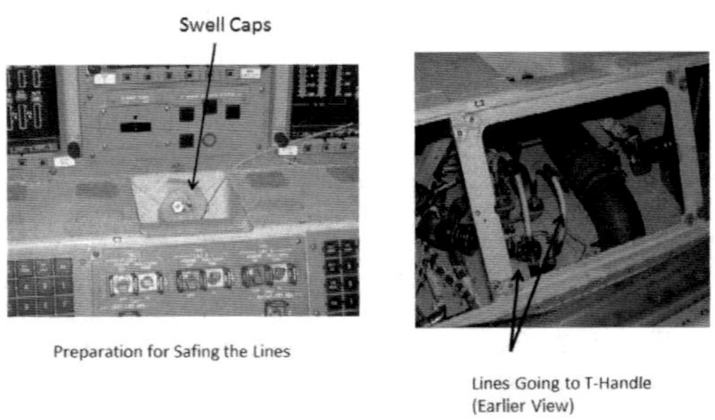

Figure 7: Shuttle FCDC Lines Being Configured for Safing

5. FLAT-H CONCEPTUAL DESIGN AND OPERATION

Figure 8 illustrates how the Flat-H design is configured. Both break plates have shelves extending out close to the center of the flat surfaces of the XTAs. When one or both XTAs function, the expansion rotates the corresponding break plate shelves. The rotation of the shelves severs the ligaments in both break plates, separating the Flat H joint. A combination of tension and torque at the break plate ligaments causes the fracture.

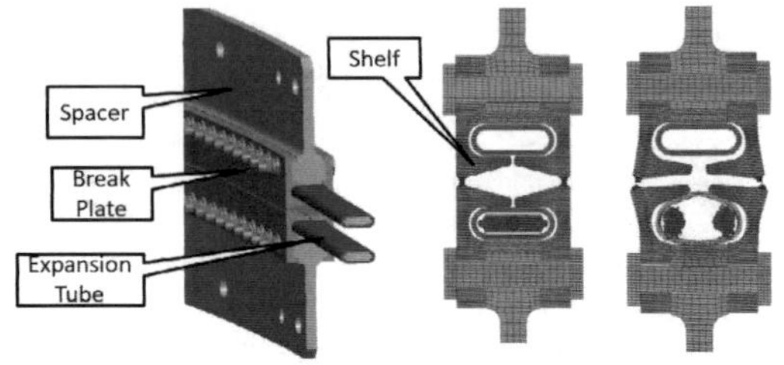

Figure 8: Flat H Redundant Frangible Joint with Two Expanding Tubes

Our new design has two XTAs unlike the earlier designs, and the XTAs are positioned horizontally, as opposed to the vertical positions as shown in Figure 2. This "flat" position makes a more compact assembly for structural support and stability. The Flat-H assembly is comprised of two XTAs that join the adjacent structures, along with two break plates, which sandwich the XTAs and spacers with fasteners, thereby holding the assembly together. Thus, SFT status can be achieved.

6. PROTOTYPE TESTING: FRANGIBLE JOINT LESSONS LEARNED

In early 2014, the Johnson Space Center (JSC) Pyrotechnics Group began the fabrication and preliminary evaluation of the Flat-H design. Requirements dictated that the design could be retrofitted on existing vehicle configurations and that it could accommodate the flight loads for the Orion Service Module fairing separation. The early Project Objectives were:

- Developing concepts into fabrication drawings;
- Performing approximately 15 firings at JSC;
- Developing hardware for further work to perform testing at White Sands Test Facility (WSTF) using high speed data acquisition equipment; and,
- Conducting stress analysis and load testing.

The schedule was success-oriented and provided a limited window of opportunity to leverage WSTF assets. Since the schedule for testing opportunities was prohibitive of a full typical design review cycle, rapid pace events occurred. No delays for dynamic analysis, no predictive models, and no time for failure analysis, if needed, were possible. Thus, the empirical design development approach was based exclusively on trial-and-error, professional experience, and intuition. Data collection and review supported the design refinement process.

The JSC tests resulted in successful separation using a single XTA, and demonstrated that the design could sustain compression and tensile structural loads. Unfortunately, when testing a similar configuration at WSTF the joint failed to separate. The team believes movement in the spacers and flexing of the bolts that hold the joint together caused the failures. With the unfavorable test results at WSTF, design momentum was temporarily lost.

A tremendous wealth of data from the WSTF testing required assessment, reduction and analysis to proceed. The analysis revealed exactly what happened to the Flat-H during the testing. Data analysis determined that there was relative motion between the joint components where there should have been none. Energy from the XTA firing was lost due to this motion, and this energy loss prevented the joint from breaking fully. Using the WSTF failure test data, JSC Pyrotechnic group completed reconstruction of the analysis models. The analysis revealed that modifications to the design would correct the failure. Consequently, there was excellent correlation between the actual WSTF test data and the JSC analytic model data. The results of this initial test phase motivated further investment in the Flat H concept.

This further investment was accompanied by an augmentation of project goals, pursuing closure of the initial test matrix with the common goal of serving all three human spaceflight programs – Orion, Space Launch System (SLS), and Commercial Crew Programs. Note this is a significantly larger role than originally envisioned for this project.

The funding of the new model validation approach increased design credibility, test predictions, and an understanding of component interactions for risk mitigation. In addition, it enabled team members to create purposeful analytic models, employing test data and prior corporate test knowledge of frangible joint technology.

The potential on-ramp of three programs provides for the utilization of the technology, needing only an opportunity to pass performance tests at JSC Energy Systems Test Area (ESTA) and WSTF. It also enables us to take the calculated steps to prepare for the next set of configurations and tests, including assessing load cases with new geometry and curved assemblies.

7. 2016 FLAT-H FORWARD

The project leadership anticipates returning to the Human Exploration Operations Mission Directorate (HEOMD) armed with the necessary data that warrants the investment into the next project phase that develops a frangible joint specific for Orion vehicle needs. This will reduce the LOC risk for future flights namely Orion's EM-3 mission and beyond where there are crew onboard. Early in the development process, the project may have to add the rolling of three, 5-foot joint segments as an activity to demonstrate the feasibility of the manufacturing process that emulates the Orion vehicle's geometry. Additional funding will be required, and a four-month timeline extension to procure curved geometry sections for combined load tests and the statistical data base enhancement. The estimated start date will be fiscal year 2017.

The JSC Pyrotechnics Group mitigated the flaw in the design, proposed a new test results based design configuration that precluded relative motion in testing, and successfully fractured the joint along its length using lower core load MDF than initially predicted. These are very significant strides made on the partial funding and in a matter of a few months.

Further investment has increased confidence in the SFT technology readiness state for full-scale development. Also, it gives the project an opportunity to demonstrate low cost additional functional tests at ESTA, as well as structural load carrying capability. This two-pronged data-driven approach entails modeling and analysis coupled with non-destructive and destructive testing. The data rich test suite attains the ultimate goal of the creation of an analytic joint model that can be used to assess a multitude of configurations prior to entering the destructive test validation phase of any vehicle or geometry specific joint design.

8. NEW FLAT-H DESIGN APPROACH

Unlike the first approach trial-and-error, test-only methodology, this new project plan includes modeling and analysis coupled with validation testing. It will save schedule time and provide good data correlation. Exploiting LS-DYNA simulation should yield the highest fidelity output in both 2D and 3D configurations yet available.

The modeling and analysis tasks to be completed include:
1. Construct models to recreate the results from previous ESTA testing.
2. Reconstruct models to recreate the results from previous WSTF testing.
3. Build models to evaluate enhanced designs
4. Perform static loads analysis using appropriate models.

The rationale for requesting both 2D and 3D models is that 2D evaluation is limited in certain areas. One rationale specifically concerns the assembly bolt response to firing. The 2D model treats the through bolt as a solid, unmoving beam, whereas the 3D model more accurately shows the flexure of the fastener. That bending equates to energy loss, which is then unavailable to break the joint. Both model configurations will enhance our understanding of how the joint behaves. The initial WSTF test cycle provided evidence of the distinction between the modeling techniques. Prior to those tests, only 2D analyses were performed. These did not provide the predictive results of energy loss due to component slippage. However, that failure phenomenon was validated via analysis utilizing a 3D model.

The ultimate goal is to create a model that can provide predictive results for varying joint geometries and configurations. The optimization process may illustrate that the functional margin can be increased significantly. The analytical results should show where energy is lost or misdirected. Once discovered, it can be mitigated. The outcome will be a configuration that routinely breaks the joint using minimized explosive cord loads. That would provide a functional margin and the potential to lower the pyro shock transferred to the vehicle.

In mid-2016, the JSC Pyrotechnics Group will conduct tests at ESTA and WSTF. These tests are described in detail below.

The ESTA firings will be conducted in pairs, where two tests will be performed on one test article. The test article will have the primary and secondary expanding tube assemblies (XTAs) installed in the joint, but only one XTA will be functioned at a time. The first firing will show positive separation of the joint. The unfired XTA will then be functioned. The second test configuration is to show that the XTA, when fired and when not using the resultant energy to fracture the joint, remains contained within the joint half. This is a critical design feature needed to prevent the highly detrimental release of excess Foreign Object Debris (FOD). Two separate joint subscale article will be tested.

The Project will conduct the structural load testing. This will require the evaluation of joint sections in both tension and compression. The results should show a positive margin in either case and should also align with the LS-DYNA model predictions.

Following the JSC test campaign, the intent is to return to WSTF, where the Pyrotechnic group will conduct two additional test firings. The newly designed test articles will be configured exactly as they were for the ESTA tests. However, only one XTA will be functioned to effect separation. Eliminating the second XTA functioning precludes the potential of damaging expensive hardware test assets, due to excessive energy release.

9. FUTURE FLAT-H FRANGIBLE JOINT TESTS

In fall 2016, a series of functional tests at the WSTF will be scheduled to understand the complex separation dynamics of the new joint design. The WSTF test data will provide substantial detail regarding the effects, which will support enhanced analysis modeling verification. Several tests will be identified to sufficiently establish the feasibility of a new design. This test plan will accompany a Memorandum of Understanding (MOU) between the NASA Engineering and Safety Center (NESC), the JSC Pyrotechnics Group, and JSC Safety to leverage single-use instrumentation assets, including digital image correlation, photon Doppler velocimetry (PDV), and other cameras as needed, at WSTF.

10. CONCLUSION

The pyrotechnic industry relies upon single-lot control, quality assurance, and rigorous testing to develop confidence in SFT device designs. However, even in ideal situations where these practices are fully implemented, process escapes still occur during production. For frangible joints, which rely on subscale, instead of flightlike test articles, assurance in reliability diminishes. While SFT is the standard for pyrotechnic devices used in human-rated vehicles, such devices must become commercially available, just like the ZFT devices utilized on non-crewed vehicles.

The causes of frangible joint pyrotechnic failures have proven elusive. Both of the failure cases presented in this paper lack full explanations. Given the majority of failures have no determined root cause, there may be more problems lurking than testing has exposed.

Substantial effort has been invested to completely understand the failure modes and design sensitivities of frangible joint devices. Even with a comprehensive understanding of frangible joint technologies, human errors, such as improper installation, poor handling methods, deficient procedures, inadequate manufacturing methods, and substandard materials and faulty workmanship, are inevitably a risk to crew safety, and must be controlled and effects mitigated. The SFT frangible joint achieves redundancy and increased reliability, and crew safety through pyrotechnically failure tolerant design utilizing a single fracture plane.

PRESSURE SUITS FOR SUBORBITAL SPACEFLIGHT MANDATORY OR NOT?

Christian Lüthen, M.D.(1); Neil Jaschinski, Dipl. Ing. (FH)(2)

(1) Erasmus MC – University Hospital Rotterdam, Dept. of Cardio-Thoracic Anesthesia, 's Gravendijkwal 230, 3015 CE Rotterdam, The Netherlands, c.luthen@erasmusmc.nl
(2) Applied Rocket Technology, Reginastraat 5 2645 NM Delfgauw, The Netherlands, neil@spacesuits.eu

ABSTRACT

This paper will present the human physiological basis concerning survival at higher altitudes and in the vacuum of space. Based upon those the paper will discuss the necessity for pressure suits for suborbital spaceflight passengers and pilots. The paper will also focus on design implications for pressure suits for spaceflight participants of varying age and health. The authors will enable and encourage the discussion whether or not pressure suits should be mandatory for suborbital human spaceflight.

1. INTRODUCTION

With the latest successes in rocket reusability (SpaceX "Falcon 9", Blue Origin "New Shepard"), we are getting closer to get more people into space than ever before. Horizontally landing spacecrafts like Virgin Galactic's "SpaceShip One" passed the Kármán line (100 km; 328.000 ft) during early test flights, competitors like XCOR's "Lynx" are getting ready to start test flights as well. In the beginning, suborbital scenic flights, and later point to point flights, will play a major roll. The current status of passenger space flight is in the same situation as air traffic in its early stages in the early 20th century. The use by some few, in an experimental state, rapidly lead to public transportation for everybody. This development was not without risk and this time the risk is even higher because we are leaving the Earth's atmosphere and entering space as an inherently hostile environment for the survival of humans. The question is: Can we lower the risk by using pressure suits as mandatory equipment for space travelers? The paper will discuss this question according to suborbital and near space flights, as this will be the first trips a wider range of people are able to afford.

2. HISTORICAL BACKGROUND

As soon as human mankind found out how to fly, going higher and higher was one goal and very soon, the first individuals faced hypoxia, a medical state in which the human body suffers from the lack of oxygen supply.
One example is James Glaisher. On 5th September 1862 he passed out at an altitude above 8.500 m (27.900 ft) [1] but recovered after the balloon descended to a lower altitude. The physiological explanation is based on the partial pressure of oxygen in the ambient air. Normally, the human body is adapted to an altitude slightly above sea level. With an air pressure of 1013,25 mbar (14,69 psi) at sea level, the partial pressure of oxygen in the air is 212,2 mbar (3,07 psi) [2]. The body itself tries to adapt to the lack of oxygen on short term by increasing respiratory frequency and tidal volume, and on long term by increasing the hemoglobin level (oxygen carriers) of the blood. Since flying leads to fast changes in altitude and therefore to a decreased oxygen supply, the body can only adapt on short term. This is only possible up to a certain altitude and for a certain amount of time without getting symptoms of hypoxia.

In an attempt to solve this dilemma, additional oxygen can be applied to raise the ambient partial pressure to the pressure of sea level. This is done on a regular base in military planes, in emergency situations in an airliner or for high mountain activities. But there is a maximum limit for this technique. When the altitude is so high that ambient pressure of the air is equal to the partial pressure of oxygen at sea level, than 100 % oxygen is needed. This equals to an altitude of approx. 12.100 km (39.700 ft). Above this altitude and breathing pure oxygen, we face the problem that the oxygen supply to the human body is not enough to survive.

The solution was utilizing a pressurized cabin for balloons. On 27th May 1931, Auguste Piccard and Paul Kipfer reached an altitude of 15.781 m (51.775 ft) with a hydrogen balloon with a pressurized cabin. Later they set an altitude record of 23.000 m (75.459 ft) [3].

On January 30th 1934, a Russian high altitude attempt failed. The balloon Osoaviakhim-1 with a crew of three men reached an altitude of 22.000 m (72.000 ft) for several hours. This balloon also used a pressurized cabin. During descent, the balloon lost buoyancy. The crew tried to bail out but failed. They tried to open the cabin hatch which was secured with 24 bolts, but the crew was able to remove only seven [4].

A pressurized cabin is now standard for passenger planes flying at higher altitudes. This gives the passenger more comfort during flight. The pressure inside the cabin varies over the flight. At beginning, the pressure inside of the cabin is slightly higher than outside ambient pressure at the airport. This will make the body of the plane stiffer and the plane can be build lighter. During ascent, the pressure is reduced to a pressure below an altitude of approx. 2.400 m (7.900 ft). This prevents most cases of hypoxia. Newer planes have an even lower cabin altitude. But all planes regularly fly below the 12.000 m (39.400 ft) altitude, even with pressurized cabins, because in the case of losing cabin pressure only oxygen mask can be used, and these masks are only supporting an oxygen concentration of 40 to 60 % [5]. There are only some few exceptions such as the Airbus A380 with 13.100 m (43.000 ft) [6] and the Concorde with 18.000 m (60.000 ft). For both planes special procedures are used to handle a loss of cabin pressure above 12.000 m (39.400 ft). The Airbus A380 uses an automatic descent mode to a safe altitude to protect passengers from a long exposure to a low pressure environment. The Concorde used the same technique with additional countermeasures. The size of the windows of the Concorde was smaller as the usual plane window, in order to reduce the speed of decompression, and the pilots used positive pressure oxygen masks. That means, that pure oxygen was pressed into the lungs of the pilots to establish a higher pressure locally in the lungs of the pilots to get enough oxygen into the body even above 12.100 m (40.000 ft). These masks had to be put on within 10 – 15 seconds, which is the time of useful consciousness at this altitude.

The main problem of this emergency procedure is the descent time to an altitude with a safe pressure. In the case of a suborbital flight once the engines shut down, the flight path is parabolic. This is the time frame of the flight were the passenger will have the experience of weightlessness in Zero-G, it takes approximately 4 minutes for a trip to an altitude of 100 km (328.000 ft) (depending on how long the boosting phase takes). In the case of an emergency, the initiated flight path can not be aborted. It takes the whole rocket fuel of the craft to get into this path and it will need the same amount to shorten it. Also if an emergency appears within the boosting phase of the flight and the engine is shut down earlier, it will end in a parabolic flight path, though of course shorter than the maximum possible. Therefore the design of the spacecraft has to take this into account and the resulting question is: Is it possible to design a pressure cabin that cannot lose cabin pressure in the case of an emergency?

The cabin needs two basic functions: a hatch to enter the cabin (windows to look out) and some kind of pressure regulation valve to keep cabin pressure constant and to equalize the cabin pressure to ambient pressure after landing. There were accidents in the past with both systems: Apollo 1 and Soyuz 11.

During a ground test of Apollo 1 a fire occurred in the capsule. The capsule hatch was designed in a way that the pressure difference between cabin and the outside (with a higher pressure in the inside) pressed the hatch more into the sealing and therefore generated a tighter sealing. The hatch had to be opened inwardly. Under normal condition this secures the sealing, in the case of a fire inside of the cabin it prevents to open the hatch. The fire increased the pressure of the cabin until the cabin busted in the lower part close to the heat shield. Technicians outside of the capsule were not able to open the door during this time.

With both an inward or outward hatch design, the cabin pressure has to be equalized before opening the hatch. Even the smallest pressure difference will cause a high mechanical load to the hatch because of the big surface area of the hatch. In the case that the locking mechanism can be opened, an outward opening door would pop out like a Champagne cork, putting persons outside to extreme risk of injury. An inward opening hatch would just stay in place until someone strong enough can open it a little bit that air can pass through a gap to equalize the pressure. Planes with a pressure cabin usually have a pressure indicator at the doors to visualize the state of cabin pressure for persons outside of the plane to prevent any faulty use.

The pressure equalization is done with a valve. Also during normal operation, there must be a valve to keep cabin pressure constant. In a plane, air in- and outtake is regulated by pressure sensors and valves. In a space ship, oxygen from a tank is supplied and carbon dioxide is removed by either flushing it out of the cabin, or by chemical binding to different type of lime (soda, barium or lithium lime). The equalization/regulation valve can fail as it happened during Soyuz 11 mission.

The spacecraft used for Soyuz 11 mission was a Soyuz 7K-OKS. This type was too small for the crew of three cosmonauts dressed in pressure suits. Therefore the decision was taken to fly without the bulky pressure suits. The whole 24 day mission went fine until the return to earth. During the separation of the service and descent modules, the explosive bolts of the locking mechanism were fired simultaneous instead of sequentially.

This caused a malfunction of the pressure equalization valve. Instead of opening when reaching an altitude with enough ambient pressure, it already opened in an altitude of 168 km (551.000 ft). This caused a drop of cabin pressure. After 212s the cabin pressure went down to zero [7]. Cardiac arrest (heart stops beating) of the cosmonauts occurred within 40 seconds of the onset of the pressure loss. Ground crew tried to reanimate the crew with cardiopulmonary resuscitation after landing, but the time span was too long to be successful. If we take the time line of a Soyuz descent, the separation is done 26 minutes before landing. This was the moment the malfunction of the valve happened. 15 minute before landing or 9 minutes after the accident, the parachutes are deployed in an altitude of approx 10 km (32.800 ft), still too high for a sufficient supply of oxygen with masks.

If we take a look to the speed of decompression (212 s) and the speed of cardiac arrest (40 s) and compare this with the time frame of a suborbital flight (approx. 240 s), than the period of flying too high for oxygen masks is still too long. A solution must be found to bypass the leak of pressure (oxygen).

3. HYPOXIA

Hypoxia means an insufficient supply of the body (or parts of the body) with oxygen. For a suborbital flight the risk of rapid onset hypoxia for the whole body is given. The symptoms are:

- ataxia
- confusion / disorientation / hallucinations
- behavioral change
- euphoria
- severe headaches
- reduced level of consciousness
- papilloedema
- breathlessness pallor
- tachycardia, later bradycardia
- low blood pressure

If not treated hypoxia is fatal.

Some of the symptoms like behavioral change and reduced level of consciousness, including euphoria, impede people from realizing the first signs of hypoxia and therefore they will not initiate the necessary emergency procedures. Therefore these procedures must be activated automatically by the spacecraft or its subsystems. The time gap between exposed to high altitude oxygen concentrations and first symptoms of hypoxia is called TUC (**t**ime of **u**seful **c**onsciousness). This is the time the person still knows what she or he is doing and able to correctly respond to it. The list below gives an overview of time vs. altitude for an average trained military pilot [8]. These times of course vary with age and health status, and would be shorter in untrained passengers:

Flight Level	Altitude [m]	TUC
FL 150	4.500	30 min +
FL 180	5.500	20 – 30 min
FL 220	6.700	5 – 10 min
FL 250	7.600	3 – 6 min
FL 280	8.500	2:30 – 3 min
FL 300	9.100	1 – 3 min
FL 350	10.700	30 – 60 s
FL 400	12.200	15 – 20 s
FL 430	13.100	9 – 15 s
FL 500 and above	> 15.200	6 – 9 s

Going suborbital above the Kármán-Line (100 km; 328.000 ft) is far above any flight level, and we have to take the shortest TUC into account. This time is short in a manner that it could be difficult to put on an oxygen mask in the right manner (especially for non-trained civilians) and the mask will only work if the pressure is maintained high enough (below FL 400).

4. COMMERCIAL SPACEFLIGHT APPROACH

Trying to build the safest spacecraft possible will result in extra weight and development time, which will demand higher development expenses and therefore will result in a higher ticket prize. The higher ticket prize lead to fewer customers and even with the safest spacecraft malfunctions still can occur.

One solution could be to take enough bottled air or oxygen with you to maintain cabin pressure for a limited frame of time for a maximum size of leak of the cabin. This of course will only work for very small leaks, and as we learned from the Soyuz 11 accident, even the small volume of the cabin was evacuated very quickly, leading to fatalities. To reduce tank size, Liquid Oxygen (LOX) could be used with a powerful evaporator. Of course fire prevention has to be taken into account. From a design point of view, the question would be: Does the gas tank weights and costs less than a pressure suit system for each passenger? The weight of the tank increases with the hole size which has to be compensated, the weight of a pressure suit system not. The weight of the pressure suit only increases with the low pressure duration time but not with the leak size, since the suit can work under zero pressure.

Another point of view is the marketing aspect. The customers want to have a nice flight without getting scared about things that might go wrong. With a pressure suit, they are wrapped in a constant reminder all measures for safety have been taken and they do not need to be scared. After all: doesn't a space suit belong to an astronaut? The typical press picture depicts an astronaut wearing a space suit (or pressure suit). The main problems of a pressure suit nowadays are the financial price, and the price the customer has to pay in reduced freedom during and before flight. A pressure suit limits the field of view caused by the helmet and the freedom of movement of the limbs. Depending on the used pressure inside of the suit during an emergency, pre-breathing has to be done before the flight to reduce the risk of decompression sickness (DCS). Pilots of U2 and SR-71 planes did regular flights to altitudes where they have to wear pressure suits in the case of bailouts. The pre-breathing time is approx. 30 to 60 minutes, which could be too long for an impatient customer. This has to be done because the suit has a pressure of only 233 mbar (3,38 psi) [9]. The low pressure was chosen to maintain suit flexibility. The weight of the suit was 14 kg (31 lb). A pressure of 650 mbar (9,4 psi) or above wouldn't need any pre breathing to prevent DCS but comes with a stiffer suit with the actual suit design.

5. ONBOARD FIRE

Fire on a spacecraft is a more complex situation than in a normal passenger plane. As we already know there is no possibility for a shortcut to make an emergency landing on the nearest airport. A plane is able to flush

the cabin with a constant amount of fresh air to remove smoke and toxic gases. The only way to get rid of those gases is evacuating the atmosphere of the spacecraft. This will also remove the oxygen supply for the fire. The life support system of Apollo 1 reacted to the fire by supplying more oxygen. This was a counterproductive reaction. For fire safety reasons, an oxygen reduced atmosphere within the spacecraft would be better since the capability of materials catching fire and keep the reaction running needs a certain level (16%) of oxygen [10]. If this is established, passengers have to wear oxygen masks. These masks must be designed in a way, that the spacecraft atmosphere doesn't accumulate oxygen. Wearing a spacesuit with a closed helmet would create a safe breathing micro- environment, which also would be easier to use and more comfortable than having an oxygen mask pressed into the face.

6. SUMMARY / CONCLUSION

It is difficult to predict the failure-free rate of future spacecrafts. A look into the past showed that accidents always can happen. Modern technologies like computer simulations help to minimize the risks but there will be always an uncertainty until enough flights are done. The challenge / risk to the human body in a perfect suborbital flight make supporting systems mandatory. If we add emergency measurements to these systems, a pressure suit is the easiest of all solutions. The benefits of lower risk of hypoxia and fire protection overweight the drawbacks of additional costs, slight discomfort and reduced freedom (movement, visual). Also the suit itself gives the passenger a real astronaut feeling. Based on this reflection the authors come to the conclusion that a pressure suit should be mandatory for suborbital spaceflights.

7. ABOUT THE AUTHORS

7.1 Christian Lüthen

Works as a consultant at the Department of Cardio- Thoracic Anesthesia at Erasmus MC – University Hospital Rotterdam, The Netherlands. He is also the Team Lead of the Biomedical Team of the Austrian Space Forum (Oesterreichisches Weltraum Forum, OeWF), involved in Mars Analogue Missions, (including the analogue spacesuit system of the OeWF).in deserts, caves and high altitude mountain settings. He is an alumnus of the International Space University.

7.2 Neil Jaschinski

Works as an engineer of material sciences in different fields of applications including aerospace

8. REFERENCES

1. Appletons' annual cyclopaedia and register of important events of the year: 1862. New York: D. Appleton & Company. 1863. p. 186.

2. Respiratory Protection Handbook: William H. Revoir 1997. p. 47.

3. Dr. Erich Tilgenkamp - Reisen in ungewöhnliche Räume - Eine autorisierte Biographie - Verlag neues Leben Berlin 1956.

4. Russian Spacesuits: 2003. Springer: Isaak P. Abramov, A. Ingemar Skoog

5. http://www.easyoxygen.com.au/should-i-use-a-nasal-cannula-or-oxygen-mask-with-my-oxygen-concentrator/.

6. "Exemption No. 8695". Renton, Washington: Federal Aviation Authority. 2006-03-24. Retrieved 2008-10-02.

7. "The Partnership: A History of the Apollo-Soyuz Test Project". NASA. Archived from the original on 23 August 2007. Retrieved 20 October 2007.

8. Mark Wolff (2006-01-06). "Cabin Decompression and Hypoxia". theairlinepilots.com. Retrieved 2008-09-01

9. Dressing for Altitude U.S. Aviation Pressure Suits–Wiley Post to Space Shuttle: Dennis R. Jenkins p. 329.

10. Bounding the Spacecraft Atmosphere Design Space for Future Exploration Missions. Kevin E. Lange, Alan T. Perka, Bruce E. Duffield and Frank F. Jeng. Jacobs Sverdrup ESC Group. Houston, Texas. NASA/CR—2005–213689.

OVERVIEW OF NON-IONIZING RADIATION SAFETY OPERATIONS ON THE INTERNATIONAL SPACE STATION

John Flores-Mclaughlin(1), James Runnells(2), Sam Ghalayini(3), Ramona Gaza(3)

(1) *University of Houston, 4800 Calhoun Rd, Houston, TX USA 77004, Email: john.flores-mclaughlin@nasa.gov*
(2) *HX5, LLC - Johnson Engineering and Technical Contract (JETS), 2224 Bay Area Blvd, Houston, TX USA 77058, Email: james.runnells@jacobs.com*
(3) *Lockheed Martin Information Systems & Global Solutions, 2101 NASA Parkway, Houston TX USA 77058, Email: sam.ghalayini@nasa.gov, ramona.gaza-1@nasa.gov*

ABSTRACT

Non-ionizing radiation (NIR) refers to any type of electromagnetic radiation that does not carry enough energy per quantum to ionize atoms or molecules directly. NASA human spaceflight standards for Laser, Incoherent Light and RF electromagnetic field human protection standards utilize methodology and thresholds from common industry guidelines.

The unique space operations environment and international partner arrangement of the International Space Station (ISS) presents multiple qualifiers in the implementation of NIR safety guidelines. While terrestrial regulatory guidelines are usually applied comprehensively, functional modifications must be made to accommodate the uniqueness of training and health of spaceflight crew as well as the operational logistics present in spaceflight. Additionally, equivalent safety considerations are made where differing jurisdictional safety guidelines are utilized in the process of manufacturing and verifying the safety of hardware. The combination of these qualities requires a proactive and highly adaptive safety oversight that is very unique compared to a traditional non-ionizing radiation safety program.

1. OVERVIEW

Non-ionizing exposure sources that may cause adverse biological effects to humans may be grouped into electromagnetic field (radio-frequency), broadband light and laser sources (Fig. 1). Exposure requirements for the International Space Station are outlined in NASA SSP 50005 [1]. Additionally, Electromagnetic field may induce currents that adversely effect hardware and subsequently create risk of harm to humans.

NASA electromagnetic exposures limits to humans are established from the guideline *IEEE C95.1 Standard for Safety Levels with Respect to Human Exposure to Radio Frequency Electromagnetic Fields* with modification to exclude consideration of small children and those who are ill. The requirement protects against the increase of 1 degree Celsius in any 1 gram of tissue. Other potential effects such as cancer are not yet conclusively supported by regulatory entities. ISS requirements for electromagnetic field induced hardware effects are created from hardware specific testing and captured in SSP 30237 [2], and SSP 30243 [3]. These documents provide information on what is acceptable with respect to ElectroMagnetic Interference (EMI) and

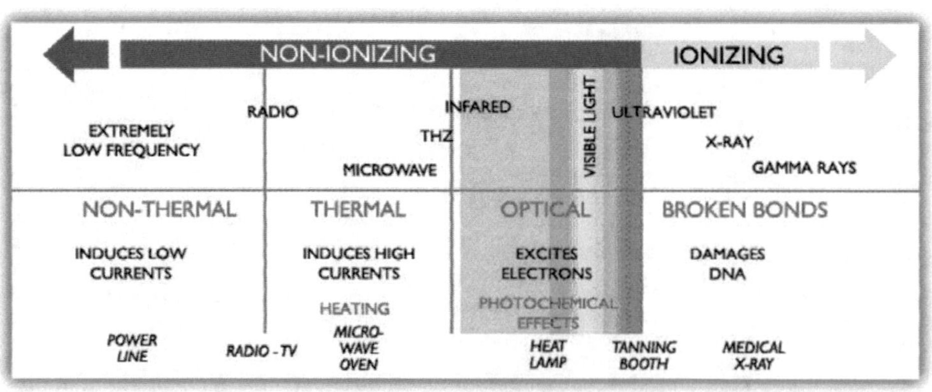

Figure 1. Categorization of ionizing and non-ionizing radiation.

Table 1 – Characteristics of RF Transmitters with No ISS Hardware Electromagnetic Effects (EME) Concerns[1]

Frequency Range	RS03 – 10dB	Power Density from RS03 – 10dB
14kHz to 200MHz	1.58V/m (124dBμV/m)	0.0066 (W/m^2)
200 MHz to 8 GHz	19 V/m (145.6dBμV/m)	0.955 (W/m^2)
8GHz to 10 GHz	6.3 V/m (136dBμV/m)	0.106 (W/m^2)
10 GHz to 13.7 GHz	(linear)	(linear)
13.7 GHz to 15.2 GHz	79 V/m (158dBμV/m)	16.58 W/m^2)

[1] Assuming that the measured point is 1 [m] away from radiation source. (SSP30237)

Table 2 – Characteristics of RF Transmitters with No ISS Crew Radiation Concerns[2]

Frequency Range	Maximum Radiated Power	Maximum Contact Current
110 kHz – 450 MHz (b)	< 7 watts	<100mA
450 MHz -1500 MHz (b)	7 wattsx450/frequency(MHz)	<100mA
> 1500Mhz (b)	Specific Absorption Rate < 0.4W/kg	<100mA

[2] Assuming no hardware failure scenario will allow deviation below 110 kHz output

Table 1. Low power radio-frequency exposure limits. Taken from NASA RF Memo OE-14-002.

ElectroMagnetic Compatibility (EMC) when operating on the ISS (IVA or EVA).

Broadband light exposure in space my result from both hardware and the sun. NASA broadband exposure limits are established from publications from the American Conference of Governmental Industrial Hygienists (ACGIH) and the International Commission on Non- Ionizing Radiation Protection (ICNIRP). With external consultation, NASA adopted requirements have reduced factors of safety intended for the broader general population. Broadband light exposure may result in injury from multiple pathways including to the skin and/ or eye potentially resulting in multiple requirements being enforced.

Laser exposure guidelines are adopted directly from the *American National Standard for Safe Use of Lasers* (ANSI Z136.1). Due to the rapid progression of semiconductor technology, the consideration of laser emissions has evolved from an original combined consideration with broadband light. With the increase of laser optical communication and Light Detection & Ranging (LIDAR) applications, it is expected that new thresholds and interpretations will be discerned.

2. INTENTIONAL RADIO FREQUENCY TRANSMITTERS

There are two areas of concern for intentional RF Transmitters operated on the ISS: hardware interference or damage, and crew injury/illness. Due to the large number of low powered transmitters flown to ISS, the JSC Safety and Mission Assurance/Program Risk Office (OE) has captured the relevant RF levels for the safe use of Intentional RF NASA Memo OE-14-002 [4]. The levels documented therein address both safe ISS hardware limits and safe crew limits for incidental or intentional contact with intentional RF transmitters (including housings/enclosures containing RF transmitter) and antennas. Table 1 is provided from the OE-14-002 memo and defines what the acceptable low power RF limits are for a RF transmitter. These limits allow an RF Transmitter to operate (intentionally or unintentionally) without impacting the ISS hardware or the crew. Levels above these thresholds will require further analysis and potentially the need for additional hardware and operational inhibits. The levels provided in OE-14-002 are for worst case RF output levels form an intentional RF transmitter, including over the operating temperature and frequency range, and including antenna gain. The "worst case" RF output power is a critical caveat for projects to address in their safety analysis. Many out of the box RF transmitters offer settable power and frequency levels. These are often selected by the project by means of software or firmware settings supporting the RF transmitter operation. While a project may not intend to operate, or set, a RF transmitter to a level higher than what is deemed safe, the actual setting and control of the RF output power can be considered a safety control for excessive RF interference or crew injury. Therefore, establishing the worst case output power for the RF transmitter, even if it is not intended to operate on the ISS (such as for deployable hardware), is a significant step for the safe flight and use of intentional RF transmitter on or around the ISS.

3. BROADBAND LIGHT

There are numerous windows providing viewing and sunlight exposure opportunities for ISS crew. The hazard of primary significance is ultraviolet (UV) radiation which is present in high amounts due to the lack of atmosphere surrounding the ISS. Fortunately, the polycarbonate structure of the U.S. segment windows attenuates the UV component significantly to below levels experienced on Earth. However, visible and infrared portions of the spectrum are still present at significant levels (Fig. 2).

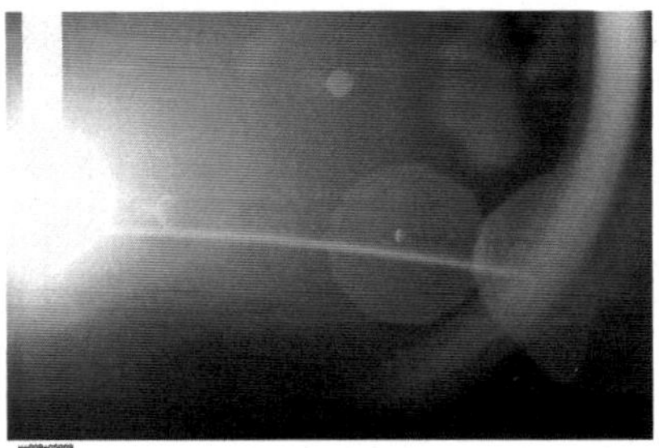

Figure 2. View from the ISS Lab Module windows. Credit: NASA

Aversion reflex to bright light akin to ground solar exposure provides adequate protection from acute injury. Additional protection in the form of sunglasses with special lenses (SPACE LENS™) designed by Adidas are provided to NASA, ESA, JAXA and CSA crew members. Extra dark inserts may be provided to crew members with operational needs to look at the sun directly.

Hardware that emits broadband light comes in many forms such as indicator lights and illumination lamps. Given the large amount of low power illuminating devices that are manifested, a luminance action level is provided for visible output of 10,000 nits or 10,000 Candela per meter-squared. Hardware with visible light output below this action level are considered safe and do not need to submit to subsequent analysis per NASA SSP 50005 requirements [1].

Visible light emitting hardware that exceeds the action level requirements, emits ultraviolet or infrared light will have to submit a thorough exposure analysis (Fig. 3) or inhibit exposure (Fig. 4)

Figure 3. VEGGIE - plant growth payload. Non-hazardous red, green and blue LED sources. The image is saturated, the sunglasses are for comfort. Credit: NASA

Figure 4. Microgravity Science Glovebox (MSG). UV-C source that is structurally contained and no hazard to crew. Credit: NASA

4. LASERS

Applications of laser sources are rapidly progressing in the space environment. Optical communication, LIDAR, environmental gas monitoring and fluorescent stimulation are just a few of the applications implemented on the ISS. The range of hazard severity spans from negligible to catastrophic to the human eye.

Internal laser emissions exceeding exposure limits on the ISS must be inhibited to prevent exposure. However, structural inhibits such as lasers covers and/or shields, are often not possible for external operations such as LIDAR and optical communication. In these scenarios comprehensive planning is undertaken to provide operational procedures that ensure crew safety. Much of this preparation may include communication with ground based authorities (Fig. 5) inclusive of different regulatory agencies and ISS international partners.

Figure 5. Laser light show with ISS laser tracking coordinated by Kennedy Space Center. Successful coordination to confirm safety to ISS crew. Credit: NASA

5. SUMMARY

Operational application of non-ionizing radiation safety requirements for ISS requires significant consideration of the diversity and quantity of hardware manifested. The overall goal of the NIR radiation safety program at JSC is to aid utilization of ISS while streamlining the existing complex safety verification process. In order to perform this task adequately due diligence is required to manage the dynamic ISS operations environment.

6. REFERENCES

1. NASA SSP 50005, International Space Station Flight Crew Integration Standard, Jun. 30, 2006.

2. NASA SSP 30237, Space Station Electromagnetic Emission and Susceptibility Requirements, Jan. 17, 2012.

3. NASA SSP 30243, Space Station Requirements for Electromagnetic Compatibility, Dec. 13, 2011.

4. NASA Memorandum, OE-14-002. Documentation of Intentional Radio Frequency Transmitter Hazards, Feb. 4, 2014.

PROCEEDINGS OF THE 8TH INTERNATIONAL SPACE SAFETY CONFERENCE

SESSION 09 SAFETY CRITICAL SOFTWARE

STPA FOR SPACE SOFTWARE DEPENDABILITY AND SAFETY

Carlos H N Lahoz(1)(3), Synara R G Medeiros(2)

(1) *Instituto de Aeronatica e Espaco IAE. Praca Mal Eduardo Gomes 50. Sao Jose dos Campos (SP) Brasil 122208 904*
lahozchnl@iae.cta.br
(2) *Embraer S.A. Avenida Brigadeiro Faria Lima, 2170. Sao Jose dos Campos (SP) Brasil12227-901*
synara.medeiros@embraer.com.br
(3) *Massachusetts Institute of Technology MIT. 77, Massachusetts Avenue. Cambridge (MA) U.S.A. 02139-4307*
lahoz@mit.edu

ABSTRACT

More and more Systems-Theoretic Process Analysis (STPA) is being applied as a system safety approach in areas like aviation, automobile and medical. Traditional techniques, such as FTA, FMECA or CCA are difficult to incorporate non-linear relationships including feedback, as well as not covers the interactions between hardware, software and human. Nowadays, due to the space systems to be using more and more software and human-machine interaction (pilot, ground system and support decision tools), is urgent that a new approach as STPA must be evaluated and applied. This new initiative started in the ESA Software Product Assurance Workshop in 2015, when a first study about STPA applied to the ESA product assurance related to dependability and safety (D&S) analysis for software was made.

1. INTRODUCTION

This paper present how Systems-Theoretic Process Analysis (STPA) is can be used in space domain mainly for software intensive system, in the context of dependability and safety analysis defined in ESA standards [1] [2]. The STPA is a new hazard analysis approach based on the STAMP (System-Theoretic Accident Model and Process) causality model. STAMP, as presented in Figure 1, is a model of accident based on three basic concepts: safety constraints, a safety control structure, and process models. STAMP treats accidents as a control problem, not a failure problem.

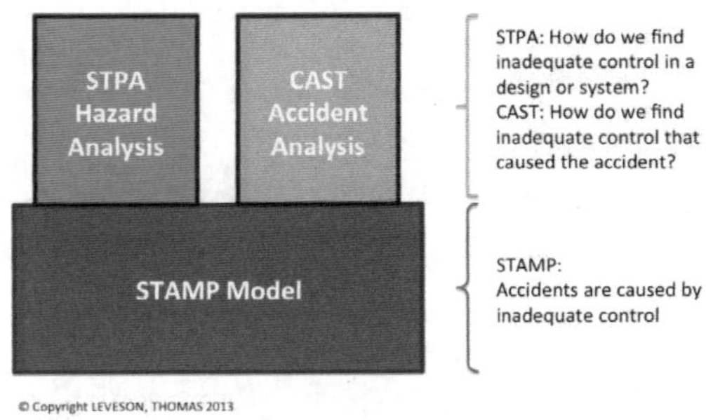

Figure 1. STAMP causality model.

In the STAMP model, accidents are defined as an "undesired or unplanned event that results in a loss, including loss of human life or injury, property damage, environmental pollution, mission loss, etc." and the hazards are "a system state or set of conditions, that, together with a particular set of worst-case environmental conditions, will lead to an accident" [3].

These high-level hazards are then turned into high-level safety constraints that the designers and engineers use in their system engineering processes, which are akin to design requirements. Then, a safety control structure is generated through a hierarchical, functional control diagram. Captures components failure accidents, unsafe interactions among components, complex behaviours between human-software, design errors, flawed requirements related to accidents.

According [3] in STAMP system is viewed as interrelated components kept in a state of dynamic equilibrium by feedback control loops. Systems are not treated as static but dynamic processes that are continually adapting to achieve their ends and to react to changes in themselves and their environment. Another goal of STAMP is to assist in understanding why the accidents occur and to use that understanding to create new and better ways to prevent losses. Thus, was created a process called CAST (Casual Analysis based on STAMP) to assist in performing this kind of analysis. Accident analysis in CAST looks for understanding the dynamic process that led the loss. The accident process is documented by showing the sociotechnical safety control structure for the system involved and the safety constraints that were violated at each level of this control structure and why.

Nowadays, STPA is used in several areas, such as aviation, space, ground transportation, medicine, defense, environment, nuclear, software and security. STPA hazard analysis was demonstrated in the H-II Transfer Vehicle (HTV) [4] and a new Crew Return Vehicle (CRV) manned Spacecraft [5], both initiatives from JAXA. Another study has been conducted using STPA complementary to SFTA and SFMECA in Brazilian rocket software [6].

In next section an overview about how to apply STPA is discussed. After that, is showed the ECSS approach for safety and dependability (D&S) for software. Following is presented how the STPA outcomes can be used in the ECSS D&S for software and the final considerations.

2. THE STPA

The STPA is a new hazard analysis technique with the main goal is to identify scenarios leading to identified hazards and thus to losses so they can be eliminated or controlled. It can be used at any stage of the system life cycle, and the process analysis is divided into discrete steps that reduce the analytical burden on the safety engineers and provides a structured process for hazards analysis. STPA can be used at any stage of the system life cycle, and the process analysis is divided into discrete steps that reduce the analytical burden on the safety engineers and provides a structured process for hazards analysis. STPA views the system as a collection of hierarchical control structures.

According [7] the assessment begins with establishing the System Engineering foundation associated with safety, identifying system accidents and its respective hazards and translating them into top-level system safety constraints. Next, a basic control structure is defined. A control structure diagram depicts the components of the system and the paths of control and feedback. Using the control structure diagram as a guide for conducting the step 1 of this analysis, where each Control Action (CA) is assessed for potential contribution to hazards. Thus, is identified the Unsafe Control Actions (UCA) are used to refine system safety constraints. Finally, step 2 if to identify the causes of the UCAs potentially hazardous, looking for how these unsafe actions could occur using scenarios.

2.1 Establish the System Engineering Foundation

STPA starts from the basic early system engineering activities associated with safety: define accidents or losses to be considered in development, identify the hazards associated with these accidents, and specify safety requirements (constraints). After that, a high- level functional control structure model should be draw. The STPA analysis will use this control structure. In the STAMP/STPA model, accidents are defined as an undesired or unplanned event that results in a loss, including loss of human life or injury, property damage, environmental pollution, mission loss and the hazards are a system state or set of conditions, that, together with a particular set of worst-case environmental conditions, will lead to an accident [3].

To exemplify the STPA application in this work is used a case study presented in [6]. This study is based on a hazard analysis of a Sequence of Flight Events (SFE) of rocket control system that comprises two (engines) stages: 1E and 2E. In this case, the possible kind of accident is that *The on-board computer (OBC) was not able to perform the SFE causing loss of mission*. One of the high-level system hazard is *Missed the 1E separation*.

In STAMP/STPA there is a set of safety constraints related to the system components - physical, human, and social - that enforces the safety property. Accidents occur when the interactions violate the safety constraints, that is, appropriate constraints are not imposed on the interactions. This concept of constraints on behaviour plays an important role in STAMP/STPA. Then, once the high-level system hazards are identified, they can be translated into safety requirements or constraints [8], like in the Table 1. This process is very simple but important because it translates the hazards into the requirements and constraints engineers and system designers need in their standard engineering processes.

Table 1. Examples of hazards and theirs related safety constraints.

Hazard	***Safety Constraint***
H1= Missed the communication with the inertial sensor	**SC1**=Inertial sensor communication must be work until the end of 2E flight
H2= The liftoff was not detected	**SC2**= The command NAV ON must be detected before the take off
H3= Missed the 2E ignition	**SC3**= The ignition of the 2E must be commanded before 1E propellants are exhausted
H4=Missed 1E separation	**SC4**= The 1E must be detached after 2E ignition

These high-level hazards are then turned into high-level safety constraints that the designers and engineers use in their system engineering processes, which are similar to design requirements. Then, a safety control structure is generated through a hierarchical, functional control diagram. To complete the system engineering fundamentals, using in STAMP/STPA, along with safety constraints and hierarchical safety structures, is process models.

The Figure 2 presents a generic model of the control structure, where the Controller (human or automated) contains a Control Algorithm for deciding what Control Actions are provided to ensure that the safety constraints are being maintained in the Controlled Process. The Control Algorithm uses a Process Model, with the current state of the system it is controlling, to help make this decision. The Feedback is for provide the inputs necessary to keep the Controller consistent with the actual state of the Controlled Process and close the loop.

Controller (automated or human)

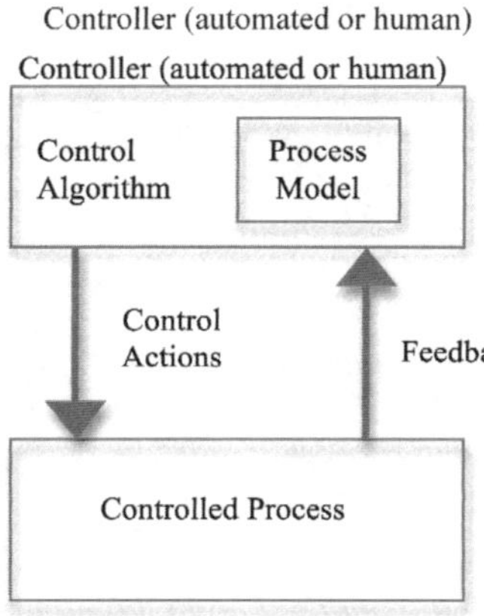

Figure 2. Safety Control Structure.

As an example of the Safety Control Structure of the mentioned rocket, Figure 3, the Controller is the OBC and the Controlled Process is the Rocket, including all stage engines. The Control Algorithm is this specific case is the routines to command 2E ignition, command 1E separation by AD channel, acquire flight time, etc. The contents of the Process Model are the flight time, the actuator stage (1 or 2), acceleration, etc. The Control Action is set the event (actuator channel) detect acceleration, command the ignition, etc. The feedback is the acceleration that measures the acceleration improvement after the separation and the flight time that acquires the time from the computer clock.

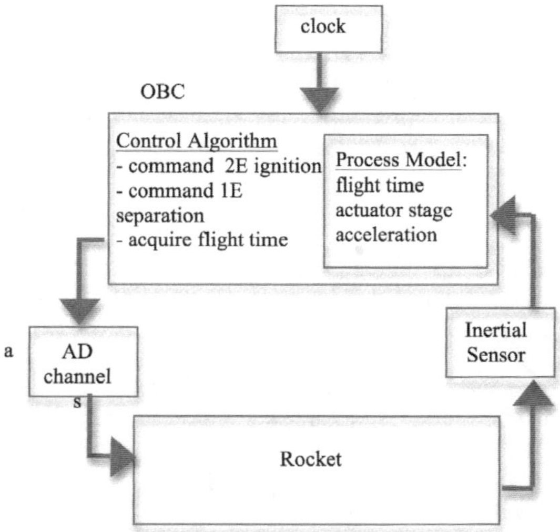

Figure 3. OBC Safety Control Structure

Also, according to [3] when STPA is used for safety-guided design only the system level requirements and constraints may be available at the beginning of the process. In the latter case, these requirements and constraints are refined and traced to individual system components as he iterative design and analysis process proceeds.

2.2 Identify UCAs

Once the entire System Engineering Foundation is established, the identification of unsafe control actions begins. This is known as STPA Step 1. This step starts with analysing the safety control structure diagram to identify the potential that the inadequate control could lead the system to a hazard state. Hazard states are result from inadequate control or enforcement of the safety constraints, which can occur due to four types of unsafe control [8]:

1) A control action for safety is not provided
2) A control action for safety is incorrectly provided
3) A control action for safety is provided in wrong time or order
4) A control action for safety is stopped too soon or applied too long (for a continuous or non-discrete control action)

There is a fifth UCAs that can be compromised: a required control action is provided but is not followed, but this fifth possibility will be handled in step 2.

STPA suggests a structure phrase to compose the Unsafe Control Actions in four parts, as showed in the Figure 4.

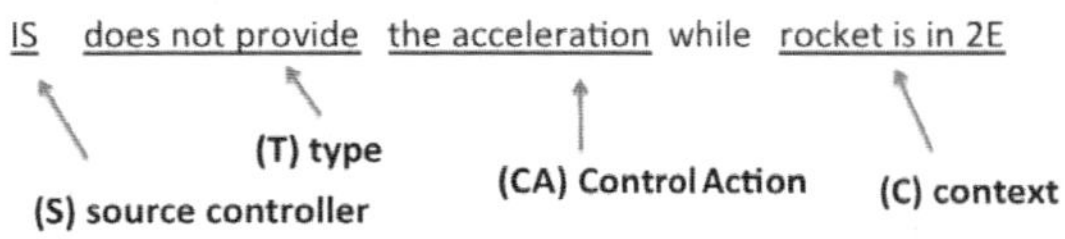

Figure 4. The UCAs composition phrase example.

Where:

(S) Source controller: the controller that can provide the control action: Inertial Sensor (IS).
(T) Type: whether the control action was provided or not provided.
(CA) Control Action: the controller's command that was provided / missing: acceleration.
(C) Context: conditions for the hazard to occur (system or environmental state in which command is provided): rocket in the second stage flight.

An excerpt of the explanation about the UCA's are presented in Table 2, but are by no means restricted to this format or meaning.

Table 2. Example of UCAs

Control Action	**1) Not Provide** *A required control action to maintain safety is not provided*	**2) Incorrectly provided** *An incorrect or unsafe control action is provided that induces a hazard or loss*
(CA.4) command the 1E separation	**(H4.UCA.1) OBC** do not command the 1E separation command while rocket is in the 2E	**(H4.UCA.2)** OBC command the 1E separation command while rocket is not in the 2E flight

Control Action	**3) Wrong timing or order** *A potentially correct or adequate control action is provided too early, too late, or out of sequence*	**4) Stop too soon or applied too long** *A correct control action is stopped to soon or applied to long causes a hazard*
(CA.3) command 2E ignition	**(H3.UCA.3) OBC** command too late the 2E ignition after exhaust the 1E propellants	**NA**

2.3 Identify scenarios of UCAs

The second step of STPA is to identify how the potentially hazardous control actions presented in step 1 could occur in order to provide mitigations. This step 2 is critical because it identifies additional safety requirements both on the Controller in the loop being analysed and on the other parts of the system, such as the Controlled Process, Control Actions and Feedback. It is also where information is generated to assist the designers in eliminating or mitigating the potential causes of the hazards.

The examination of the control loop is mainly to identify control flaws. This process of the analysis identifies scenarios that can lead to the UCAs, that is, the causes of the process model being incorrect. In addition, the analysis could identify a (safe) control action that should be given but not executed by the system yet. This is typically known as design features (controls) that could be used to protect the system against the identified UCAs scenarios. The design features that could be used to protect the system are not part of the analysis but usually are done by domain experts using the analysis results [8].

The step 2 could be split in two parts:
Step 2A: Identify causal factors that explain how it could happen. Potential causes of UCAs. (under the point of view of the Controller).
Step 2B: Identify causal factors that explain how control actions may not be followed or executed properly. Potential CAs not followed (starts under the point of view of the Controlled process).

Figure 5 shows one scenario of the sequence of flight events that cases of unsafe control actions from *OBC do not command the 1E separation command while rocket is in the 2E flight.*

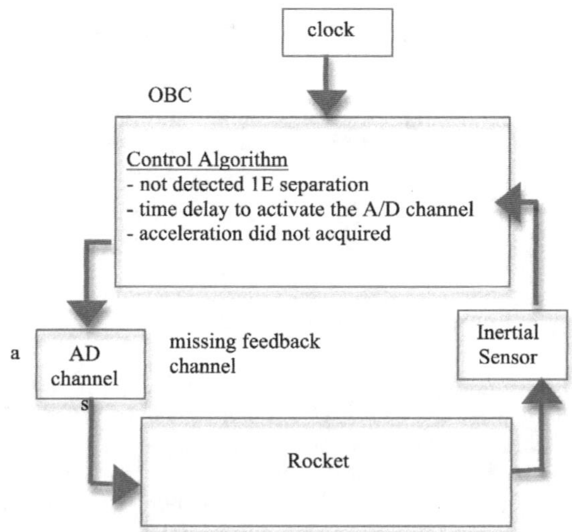

Figure 5 — One scenario of Unsafe Control Actions from "set actuation command".

One important unsafe control action identified in this case study example, is that there is no feedback by the actuation command from I/O channels. For example: information to the flight events of the 1E was physically separated after the activation of the respective digital channels (output). In this case, the software should read the data from 1E movable nozzle actuation channel (input), located in the 2E, to check if the value is zero. The zero value in this channel means that the 1E was physically separated.

Thus, the safety properties are controlled, imposing constraints on the behaviour and interaction of system components. As a safety constraint examples related to the hazard presented in this section are the *OBC must be detect the 1E body separation after the TF1ES (time frame 1E separation)*. Then, a new Safety Constraint table can be created after perform STPA steps 1 and 2, using the UCAs (and related hazardous scenarios) to refine the constraints that will become new requirements to mitigation and control the hazards. An extract of the refined Safety Constraint table is presented in Table 3.

Table 3. Examples of UCAs and theirs related safety constraints.

UCA	*Safety Constraint (Requirement)*
(H4.UCA.1) OBC do not command the 1E separation command while rocket is in the 2E flight	**H4.SC1.1**= the OBC must be command the 1E body separation during the TF1ES (time frame 1E separation)
	H4.SC1= the OBC must be detect the 1E body separation after the TF1ES
(H3.UCA.3) OBC command too late the 2E ignition after exhaust the 1E propellants	**H3.SC3**= the OBC must command the 2E ignition 5 seconds before exhaust the 1E propellants

3. ECSS APPROACH FOR SOFTWARE D&S

The ECSS [1] explains that the criticality classification of software products is based on the severity of failure consequences, which should be performed through the dependability and safety (D&S) analysis of the software. The main goal for performing this software analysis is to determine the criticality of the individual software components and then uses the results in the system level [9] [10].

In the results, should be applied engineering measures to reduce the number of critical software components (and the associated risk), addressing in particular with the additional failure modes identified at software design level and recommendations for system-level activities, like modifications in the architecture. Also, in this approach, the Hardware-Software Interaction Analysis (HSIA) must be performance in order to ensure that the software reacts in an acceptable way to hardware failure.

According to ECSS, dependability is the extent to which the fulfilment of a required function can be justifiably trusted, and safety is the state where an acceptable level of risk is not exceeded [11]. In [12] is mentioned that dependability is the discipline that addresses all aspects to ensure that the dependability performance (performance related to availability, reliability, maintainability and maintenance support) is met for the space product, including system functions implemented in software and the interaction between hardware and software. In particular it includes design rules and dependability analyses (e.g. worst case circuit performance, failure mode and effects, criticality). Also, the safety discipline addresses all aspects to ensure that all safety risks associated with the design, development, production and operations of space product are identified, assessed, minimized, controlled and finally accepted. In particular, related to the assessment of the risks based on qualitative and quantitative analyses (e.g. hazard, fault-tree, sneak).

Reference [10] requires that system-level safety requirements be defined during the early phases of the project. These requirements are derived from preliminary and advanced safety analyses, as well as from lesson learnt and safety requirements associated with similar previous missions. The system-level safety requirements include, or need to be translated into safety requirements for the software as well as, the software safety requirements discovered during this process, can be transformed in safety requirements for system.

3.1 Software RAMS

The ECSS proposed a D&S programme, that is a complete set of activities performed to ensure that dependability and safety requirements applicable to software are specified and complied with. Different types of space applications such as spacecraft, launchers, ground applications have specific software dependability and safety issues and each programme should be tailored. The programme focuses is on product assurance analyses and activities performed at software level, although system-level and software engineering activities related to software RAMS (reliability, availability, maintainability and safety) are addressed.

Software Reliability in this context is the property of software of being "free from faults" in space applications (failure-free operation). That means the ability of a computer program to perform its intended functions and operations in its environment, without any critical degradation of system performance, or even the complete mission loss.

Software Maintainability is the capability of the software to be retained or restored to a state in which it can perform a required function, when maintenance is performed. In satellite software component, for instance, the maintainability relates to the ease with which the software can be modified and put back into operation mainly when the system are in safe mode for a limited period of time.

Software Availability is the capability of the software to be in a state to perform a required function at a given instant of time or over a given time interval (providing services when needed). The availability is strongly related to the reliability and maintainability: frequent software failures and long periods of maintenance reduce the possibility of the system is available and operational at a certain time or for a certain period of time. In mission control systems or payload data ground systems, the dependability requirements are often expressed in terms of the availability of generated data. Safety is a system property. Software in itself cannot cause or prevent harm to human beings, system loss or damage to environment (disaster-free operation).

However, software is a main component of the system, and therefore contributes to its safety. The ECSS [10] defines "safety-critical" the software whose loss or degradation of its function, or its incorrect or inadvertent operation, can result in catastrophic or critical consequences.

3.2 Software related to System D&S

In the ECSS handbook for software D&S [2] is mentioned that the system safety requirements allocated to software mostly correspond to the need of eventually identifying safety-critical software components, through software criticality classification, and apply to them the suitable engineering and product assurance measures. It is explicitly required that no software error causes additional failures with hazardous effects, or propagate to cause the hazardous operation of interfacing hardware. The system-level dependability requirement for software specifies a degree of tolerance to hardware failures or software malfunctions, the detection, isolation, diagnosis of system failures and the system recovery (its restoration to an acceptable state), the requirement for the prevention of failures crossing interfaces with unacceptable consequences (failure propagation avoidance).

Also, the software D&S analysis shall be performed in the software products and using the results of system level D&S analyses, in order to determine the criticality of the individual software components. This analysis, including methods and techniques, SFMECA, SFTA and SCCA for instance, shall be agreed between supplier and customer. Based on the results of the software criticality analysis, engineering measures must be applied to reduce the number of critical software components and mitigate the risks associated with them. In order to control the critical software components related to the risk management, a cross-reference with the Project Risk List should be maintained.

The results of the software dependability and safety analysis for integration into the system level dependability and safety analyses should address in particular the additional failure modes identified at software design level and recommendations for system level activities. An additional Hardware-Software Interaction Analysis (HSIA) should be performed to identifying and specify requirements for the software behaviour in the event of hardware failure.

The HSIA is a system-level analysis that requires a significant support from software developers. According to [2], the situation where different software suppliers contribute to the development of a software product deserves special consideration. A typical case is the platform software of a satellite, which is split into basic software, developed by the supplier of the CPU hardware, and application software, developed by a software development organization. The ultimate responsibility for the production of the HSIA is with the system developer, i.e. Prime contractor. It is often the case, however, that the Prime organization has not enough visibility on the hardware and software features to generate a thorough and detailed HSIA.

The ECSS classifies four major groups of techniques to achieve software dependability and safety: fault prevention, removal, and tolerance, forecasting [2]. Fault prevention aims to prevent the introduction of faults in the final product. This is mainly achieved through the application of strict rules during the requirements definition, design and production of the software, such as, application of well established development methods, and rigorous design and coding standards. Fault removal is meant to detect faults and remove them from the software. Verification and validation are the main approach. Fault tolerance is intended to avoid that the presence of software faults leads to system failure, by tolerating the errors caused by faults and keeping the system functioning. It based on error detection and recovery, attained by different means, such exception handling or failure propagation avoidance. Fault forecasting aims to estimate the present number, the future incidence, and the likely consequences of faults. This estimation can be qualitative, using techniques such as FMEA and FTA, and quantitative, using statistical models.

4. STPA AND ECSS SOFTWARE D&S

The STPA is a new technique that has proven to be effective in establishing the safety requirements and constraints at the system level in many areas even the space, like the works presented in previous IAASS conferences related to JAXA H-II Transfer Vehicle (HTV) [7] [13][14], model-based specification technique on JAXA space Crew Return Vehicle design [15], safety spacecraft design [16], or in other proceedings and journals papers about design methodology to an outer planet exploration mission [17]. More recently, some initiatives are using STPA for identifying software safety in system context [18] [19][20][21][22].

The broad view of the STAMP/STPA, become this approach different from ECSS techniques to achieve software dependability and safety, STPA cannot be classified as a Fault Forecast technique or a Fault Prevention technique. The Control diagram and the analyses the Unsafe Control Actions, describing the Safety Constraints and after that, identifying the scenarios, becomes this approach much more comprehensive than the traditional techniques proposed by ECSS. However, STPA provides lots of evidence about the safety constraints of the system and software level that can be used to determine the criticality and severity of the system/software, as required by the ECSS D&S analysis. The results, like Control Algorithm (software functionalities) and its Control Actions can be used in the D&S analyses in the system level, in order to determine the criticality of the individual software components.

First, in the software level, it can use the Control Process to identify the software functionalities (Process Algorithms) and the variables involved in the software execution/monitoring (Process Model). For each Process Algorithm will be possible to evaluate e determine its criticality, taking into account their UCAs related. In the example presented in this study, the Control Algorithm *command the 1E separation* was classified in ECSS criticality category B, *Software that if not executed, or if not correctly executed, or whose anomalous behaviour can cause or contribute to a system failure resulting in Critical consequences* [1], and severity level 2, Loss of mission [10]. Second, in the system level, the idea is use the software criticality A, B or C to prioritize what software functionalities (Process Algorithms) must be developed new scenarios (system scenarios). In this way, it can be more practicable to select a minimum set of casual factors in order to better identify the Controller Process Model, analyse it in terms of control path, feedback path and process as well. Thus, the results could help to refine high-level safety constraints in more detailed safety requirements on components and scenarios for losses, improving the system design. In this particular case, new software functionality (new Control Algorithm) was identified to assure the separation of 1E, called *detect 1E separation*.

One of the possibilities of the STAP outcomes is put in tables combining strategic information, like the Control Algorithm (software functionality) with the UCAs and the related Criticality, as presented in Table 4. Or simply combining the Hazards with the Safety Constraints, as presented in the end of step 2.

Table 4. Example of CA, UCA and ECSS criticality

Control Algorithm	*UCA*	*Criticality*	*Severity*
command 1E separation	**(H4.UCA.1)** **(H4.UCA.2)**	B	2
command 2E ignition	**(H3.UCA.3)**	B	2

Thus, the idea is use the outcomes of STPA analysis, mainly the Control Algorithms, the CA, the UCAs, the safety constraints table and the scenarios as a input for ECSS D&S analysis. Due to the broad view that provides STPA analysis, it is possible to identify more than that safety issues. Attributes such as maintainability and availability can be identified by this analysis, even some functional requirements.

5. FINAL CONSIDERATIONS

This paper presented the STAMP/STPA as a new approach to deal with safety (and dependability, also), completely different from the methods and techniques mentioned in ECSS [2]. This technique can identify more casual factors for quality and safety losses, including problems with component interactions, software flaws, omissions and many other external influences, such as socio-technical influences, environmental and regulations.

The Control Algorithms and the UCAs are capable to identify another unsafe factors than traditional hazard analysis, like human behaviour, regulations constraints, and software-hardware interaction. STPA presented already some studies about how this approach helps to identify potential hazard causes in human controller by analysing patterns of mistakes caused by cognitive behaviours errors. [23], identified more casual factors for quality losses than FMEA or FTA, including component interactions, software flaws, and omissions and external noises [24] and besides to address misbehaviours due to software problems, and how it help address regulatory concerns. [25].

Another emergency property of STAMP/STPA is how to deal with system security [26][27][28]. The D&S analysis of ECSS does not cover this property yet. The STAMP/STPA claims that due to the great economic and strategic value in space system, it is mandatory that fault prevention and fault forecasting techniques paying attention and investing more time investigating how to preserve the space assets, like facilities and information systems against security violations and threats.

STPA should be understood as a paradigm shift in how to deal with the hazard analysis (and culture) of an organisation. The work begins with trying to view the system with a collection of interaction between human, hardware and software and looking for identify not only technical problems but also issues about human interactions with software, regulations and managerial problems, for instance. Due to a practical reasons, in a intermediate step, the outcomes of STPA can help to make a transition between ECSS D&S traditional techniques, recommended by [2], to this new and more holistic approach.

6. REFERENCES

1. European Cooperation for Space Standardization, ECSS-Q-ST-80C. (2009). Space product assurance - Software Product Assurance. The Netherlands.

2. European Cooperation for Space Standardization, ECSS-Q-HB-80-03A. (2012). Space product assurance - Software dependability and safety. The Netherlands.

3. Leveson, N.G. (2012). Engineering a Safer World: Systems Thinking Applied to Safety, MIT. Cambridge.

4. Ishimatsu, T., Leveson N.G., Thomas, J., Katahira M., Miyamoto, Y., Nakao H. (2012). Modeling and Hazard Analysis Using STPA. 4th Conference of the International Association for the Advancement of Space Safety (IAASS).

5. Nakao H., Katahira, M., Miyamoto, Y., Leveson, N. G. (2013). Safety Guided Design of Crew Return in Concept Design Phase Using STAMP/STPA, 6th Conference of the International Association for the Advancement of Space Safety (IAASS).

6. Lahoz, C.H.N., Abdala, M.A.D. (2013) Dependability Techniques Applied to Space Software – a Research Project Report, International System Safety Conference. Boston, USA.

7. Ishimatsu, T., Leveson, N., Thomas, J., Katahira, M., Miyamoto, Y., and Nakao, H. (2010). Modeling and Hazard Analysis using STPA. 4th Conference International Association for the Advancement of Space Safety (IAASS). Huntsville, USA.

8. An STPA Primer, Version 1 (2013). Massachusetts Institute of Technology MIT, System Engineering Research Laboratory. (updated June 2015).

9. European Cooperation for Space Standardization, ECSS-Q-ST-30 (2009). Space product assurance - Dependability. The Netherlands.

10. European Cooperation for Space Standardization, ECSS-Q-ST-40C (2009) Space product assurance - Safety. The Netherlands.

11. European Cooperation for Space Standardization, ECSS-S-ST-00-01C (2012). ECSS system. Glossary of terms. The Netherlands.

12. European Cooperation for Space Standardization, ECSS-S-ST-00C (2008). ECSS system. Description, implementation and general requirements. The Netherlands.

13. Ishimatsu, T., Leveson, N. G., Fleming, C., Katahira, M., Miyamoto, Y., Nakao, H. (2011). Multiple Controller Contributions to Hazards. 5th Conference of the International Association for the Advancement of Space Safety (IAASS).

14. Ishimatsu, T., Leveson, N. G., Thomas, J., Fleming, C.H., Katahira, M., Miyamoto, Y., Ryo, Ujiie, Nakao, H., Hoshino, N. (2014). Hazard Analysis of Complex Spacecraft using Systems-Theoretic Process Analysis. Journal of Spacecraft and Rockets (AIAA), p. 509-522.

15. Nakao, H., Katahira, M., Miyamoto, Y., Leveson, N.G. (2013). Safety Guided Design of Crew Return in Concept Design Phase Using STAMP/STPA. 6th Conference of the International Association for the Advancement of Space Safety (IAASS).

16. Fleming, C., Ishimatsu, T., Miyamoto, Y., Nakao, H., Katahira, M., Hoshino, N., Thomas, J., Leveson, N.G. (2011). Safety Guided Spacecraft Design using Model-Based Specifications. 5th I Conference of the International Association for the Advancement of Space Safety (IAASS). p.17-19, 2011

17. Owens, B.D., Herring, M.S., Dulac, N., Leveson, N.G. (2008). Application of a Safety-Driven Design Methodology to an Outer Planet Exploration Mission. Aerospace Conference, IEEE, p. 1 - 24.

18. Stringfellow, M., Leveson, N.G., Owens, B. (2010). Safety-Driven Design for Software-Intensive Aerospace and Automotive Systems. Proceedings of the IEEE, v. 98, n. 4, p. 515-525.

19. Xu, X., Zhong, D., Lu M., Bao, X. (2013). Application of STAMP-Based Safety Analysis on Navigation Software Development Management. IEEE International Conference on Computer Sciences and Applications (CSA), p. 214 - 218.

20. Asplund, F., Martin, J., El-khoury. (2012). Safety-Guided Design through System-Theoretic Process Analysis, Benefits and Difficulties. 30th International System Safety Conference (ISSC), 2012.

21. Abdulkhaleq, A., Wagner, S. (2014). A Software Safety Verification Method Based on System-Theoretic Process Analysis. In Computer Safety, Reliability, and Security (Safecomp). Lecture Notes in Computer Science, Springer International Publishing, Vol. 8696, pp. 401-412.

22. Abdulkhaleq, A., Wagner, S., Leveson, N. (2015) A Comprehensive Safety Engineering Approach for Software-Intensive Systems Based on STPA. Procedia Engineering, Volume 128, 2015, Pages 2-11, ISSN 1877-7058.

23. Hoshino. (2014). Applying Human Mental Model to STAMP/STP. STAMP Workshop. Cambridge, USA.

24. Goerges. (2013). The Application of STPA in Commercial Product Development to Identify Causal Factors for Quality TM Losses. STAMP Workshop Cambridge, USA.

25. Torok, Geddes. (2013). Systems Theoretic Process Analysis (STPA) Applied to a Nuclear Power Plant Control System. STAMP Workshop. Cambridge, USA.

26. Young, W., Leveson, N. (2013). Systems thinking for safety and security. Proceedings of the 29th Annual Computer Security Applications Conference (ACM).

27. Salim H.M. (2014). Cyber Safety: A Systems Thinking and Systems Theory Approach to Managing Cyber Security Risks. Master of Science in Engineering and Management, Massachusetts Institute of Technology. Cambridge, USA.

28. Young, W., Leveson, N.G. (2014). An integrated approach to safety and security based on systems theory. Communications of the ACM 57.2. p 31-35.

A FDIR IMPLEMENTATION BASED ON FUNCTIONAL AND SOFTWARE UNITS DESIGN

Olivier Boudillet[(1)], Jacques Magné[(1)]

[(1)] *AIRBUS Defence & Space, 51-61 route de Verneuil, 78133 Les Mureaux Cedex (France),*
Email: olivier.boudillet@astrium.eads.net, jacques.magne@astrium.eads.net

ABSTRACT

It is a very common requirement for space vehicle design to be one or two failures tolerant. This is for example the case for launchers where mission success is economically paramount, independently if the mission itself is crewed or uncrewed, where in the former case human safety is the key driver. This is of course the case for spaceships that have to dock to the International Space Station (ISS) in order to ensure the station crew safety.

This paper will first recall the objectives and illustrate the process to perform efficient Failure Detection, Isolation and Recovery (FDIR) activities for space vehicle design. It will then show how the vehicle function architecture can be formalised based on the Functional Units (FU) and Software Units (SU) concepts used by Airbus Defence & Space in Les Mureaux. This FU/SU method has been introduced as early as 1980 for Ariane5 launcher flight software design, matured on the 2 failures tolerant Automated Transfer Vehicle (ATV), further matured and tailored on the ORION European Space Module (ESM) design, and will be fully implemented in its latter improvement on the Ariane6 launcher.

The paper will then present how FU and SU can be formally described using SysML based modelling. The model contains all the functions (both hardware and software) of the system considered and the detailed way to correctly and safely operate the system in its nominal operations phases (called "virtual" modes). But the model defines as well all the RAMS based identified failures that can be monitored by the SUs that control the FUs and as well the detailed description of the sequences of command to be operated to recover the failure whenever implemented in the system. All the monitorings and corrective operations are associated with precisely defined recovery phases, named "configuration changes". Through this methodical description of both the nominal and recovery related operations, when they can be operated and in which sequence, the overall design of the system can be more easily reviewed and checked.

The model can be as well associated to an embedded software library that can be used to generate in an automated way the commanding sequences and the core software to execute these sequences according to current mode and configuration on the real hardware equipment or simulated ones.

Whether the system is a relatively simple one or a very complex one, the very same approach can be used to define it, check it and validate it, bringing the overall verification process to an even higher level than with a more traditional non model based approach.

1. INTRODUCTION

Even more than for their terrestrial counterparts, space vehicles and satellites must exhibit a high level of reliability for their mission: once launched from the Ground, their maintenance is either impossible (for a launcher whose mission duration is too short) or extremely costly (as for the 4 Space Shuttle missions STS-61, STS-82, STS-103 and STS-109 sent to service and extend the life of the Hubble space telescope, or as for the International Space Station (ISS) where parts can be replaced at regular pace). In this paper, we will propose a method to design the space vehicle flight software in order to implement an efficient software mechanism for hardware failure detections and their recovery. This approach has been based upon the Automated Transfer Vehicle (ATV) designed and built by AIRBUS Defence & Space for the European Space Agency (ESA) to service the ISS. This method has been further refined and is now used for Ariane6 Flight Software design, and for the NASA ORION European Service Module (ESM) functional design.

2. SPACE VEHICLE FDIR

FDIR stands for "Failure Detection, Isolation and Recovery". We will limit the scope of this study to hardware failures and not failure in the onboard Flight Software itself. The latter can be mitigated for mission or safety critical vehicles by designing segregated fault free safety software functions (as for the ATV, for which a specific safety critical category A software has been implemented on a segregated computer).

Figure 1. ATV4 at ISS vicinity

The first step in FDIR is failure detection of an incorrectly or non operating hardware equipment: this detection can be performed in the equipment itself (internal sensor and built in test mechanism), by an other equipment implementing sensor monitoring the first equipment, or deported to the Flight Software which can combine the monitoring of various sensors with its own knowledge of the overall mission phases.

The second step in FDIR is failure isolation, understood as the set of means that ensure failure containment, avoiding propagation of the failure to the rest of the system. Once again, this can be done locally at equipment level. This is generally the preferred solution when very short latency isolation is needed: e.g. use of a fuse to prevent over-current electrical damages, burst valves to prevent tank or engine destruction in case of fluidic overpressurisation. This is referred as "reflex FDIR".

When the isolation response time is less critical and may involve more than one equipment, it is generally preferred to design the hardware to offer commands than can be issued by the Flight Software to perform the isolation operation. This brings more flexibility both for the initial system design and its future evolutions when mission and equipment usage constraints are better defined or even evolving.

The third phase in FDIR is the failure location. A single equipment failure can possibly impact several FUs and therefore trigger several alarms when failure propagation cannot be avoided. It is important to analyse the root cause of the problem and organise the treatment of the failure in an orderly way between the FUs (as an output of the FDIR analysis) so to define a unique and predefined recovery sequence:

The forth phase in FDIR is failure recovery. As we have seen, almost all space vehicles must be designed to sustain the possible failure of some of their equipment during the mission without expecting any external repair service. The key for recovery is the use of redundant equipment. If a temporary failure can possibly be resolved by a switch off – switch on like operation (or reissue of the initialisation commanding sequence), a permanent equipment damage can only be mitigated through the use of distinct spare equipment.

Here as well, the recovery latency time is a key factor in the system design. A power distribution equipment must sustain its service with no disruption in case of internal failure: so the recovery may be an internally built up redundancy lane that automatically replaces the faulty nominal one without any external recovery command. On the other hand, the failure in any specific attitude control thruster may have an impact on pressurisation control, propellant lanes configuration and new thrusters selection in the thrusters commanding selection algorithm at Guidance Navigation and Control (GNC) software level. So the reconfiguration is to be managed at Flight Software level, possibly including the Ground Mission Control Center, to determine what the preferred recovery solution is. This may lead to mission continuation or mission abort decision.

So for permanent failure on key equipment, hardware redundancy is necessary with at least one duplication of the equipment. Note that recovery latency is one of the parameters to define the need for hot (both nominal and redundant equipment are operational), warm (redundant equipment is powered on but not fully configured) or cold redundancy (only the nominal equipment is operating, the redundant one being powered off). For very complex systems like the ATV, the most critical mission phase was the docking approach to the ISS and included cooperative optical navigation thanks to retroreflectors targets implemented close to the ISS docking port. Visual equipment were non only duplicated in number but also in technology since 2 identical videometers were used as primary sensors while being monitored by 2 telegoniometers to prevent a possible functional failure, in such case common to both videometers (e.g. unexpected out of validated operational condition happening during the real flight) . Basic electrical failures of primary navigation means (videometers in the case of the ATV) are easy to detect. It is common mode failures affecting both primary means w/o simple failure signature that requested the implementation of secondary means (telegononiometers in the case of the ATV) as an independent monitoring mean.

FDIR deals with failure management in order to improve the failure tolerance rate of the system well above the failure tolerance rate of any of its hardware component alone.

The FDIR management phases are depicted in Figure 2

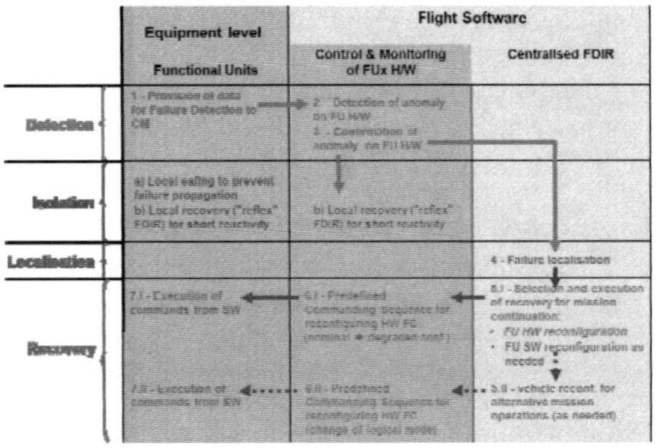

Figure 2. FDIR management phase concept

There is therefore a very strong link between FDIR activities and RAMS activities. The first uses failure propagation analysis from the latter to define the requirements towards the equipments (reflex mechanisms, observability and commandability) and the requirements toward the Flight Software which operate the FDIR.

3. THE FU AND SU APPROACH

At system level, the functional analysis defines the Main Functions of the overall vehicle, including the supply scheme definition, the production phase, the operational phase and the disposable phase. We will only consider here the Flight Segment in its operational phase.

3.1 FU and SU definitions

The acronym FU stands for "Functional Unit". The practical implementation of the vehicle functionalities is realised by means of FU which allow structuring the design of the vehicle equipment.
The FU definition processing concerns the flight mission phases but can also include the Ground pre- launch test phases of the Flight Segment. It is based on the refinement of the Main Functions towards hardware subsystems and controlling services at the centralised Flight Software level.

A Functional Unit is a set of active electrical and/or fluidic hardware, and/or software items, organized so as to provide a common set of services and capabilities to the vehicle and its mission, including the items providing common hardware resources as regards:
- Power supply and distribution;
- Data exchange on data bus.

Depending on the physical architecture of a vehicle, equipment inside a same Functional Unit may belong to a same subsystem or to different subsystems. But no equipment shall be shared by two or more FUs: otherwise services are to be defined between FUs.

In the ATV, the Flight Software (named FAS) was implemented on a Fault Tolerant Computer (FTC) connected to the various equipment through four 1553 busses as depicted in Figure3.

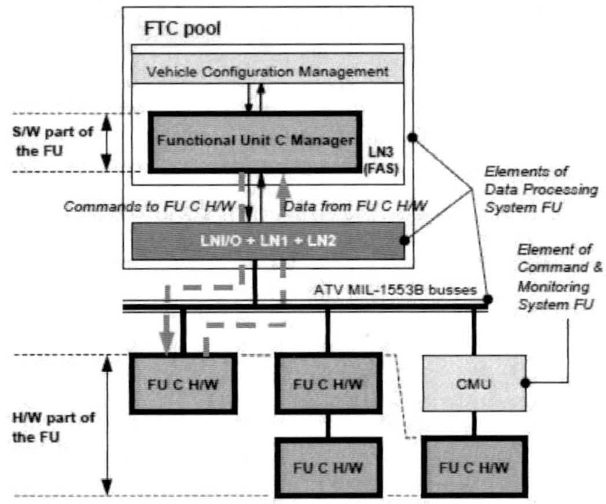

Figure 3. ATV FU mapping on vehicle HW and SW

As an example, 14 FU were defined for the quite complex ATV vehicle, as follows:
- COM: Communication - Defines the communication system for telecommand and telemetry with the Ground.
- CMS: Command and Monitoring System – manages data acquisition and command distribution in the service of ATV FUs and SUs
- PWS: Power System – Provides electrical power to the users by means of centralised distribution.
- GMS: GNC Measurement System - manages the supply of data from the GNC sensors (Gyros, telegoniometers, star trackers …)
- PRO: Propulsion - Provides the propellants and propulsive capabilities for orbit transfer, attitude control and rendezvous with ISS.
- DRS: Docking and Refuelling System – provides the docking related event detection and power and propellant exchange control with ISS.
- SGS: Solar Generation System – manages solar array deployment and rotation.
- A5I: ATV/A5 Separation and Interface system – manages the detection of separation with the launcher.
- TCC: Thermal Control Chain – manages the heaters and other means to ensure thermal control of the ATV versus power budget and balance.

- WGD: Water and Gas Delivery – manages the distribution of water and gas to and from the ISS.
- ECLS: Environmental Control and Life Support – manages the air distribution, illumination and fire detection within the pressurised part of the ATV
- EMA: External Monitoring Aids – provides visual elements to help the ISS to compute the ATV attitude and range.
- DPS: Data Processing System – features the fault tolerant computer which hosts the Flight Software and which ensures communication with the ATV equipment through 1553 buses.
- PFS: Proximity Flight Software – prevents collision with the ISS through segregated means (dedicated onboard computer, sensors and actuators).

To each FU is associated a Functional Unit Manager (FUM) which contains all the software services needed (if any) to operate the FU, including the functions to control the FU equipment through predefined sets of commands and the monitoring of the equipment measurements.

A Software Unit (SU) is a Functional Unit implemented exclusively in software, without hardware components.

Specific SU are defined to manage functionalities of the system that are implemented exclusively in the Software. The Flight Software then can use the FU Managers and the SU as dedicated bricks to operate each FU and perform higher level functions.

As an example, the ATV defined 5 SUs:
- GNC: Guidance Navigation and Control
- TMTC: TeleMetry and TeleCommand – processes data packets to and from the Ground and the ISS on the communication links of the ATV.
- MVM: Mission and Vehicle Management - consists first of the onboard capability of executing sequences of automatic operations during pre-defined phases as they have been scheduled and defined by the ATV CC as a set of commands stored in the On-board Mission Plan (OMP) and secondly to configure the vehicle according to predefined commands or commands issued by the onboard FDIR for a pre-defined set of failure alarms generated by the FUs and SUs.
- FCM: Flight Control Monitoring – performs, as part of the main Flight Software, trajectory monitoring to prevent collision with the ISS.
- FRMS: Reference Frame Services – services to convert ATV position between the various orbital frames (launch site, Earth or ISS based quaternions).

Figure 4 shows the proposed FU and SU architecture as it has been designed for the ATV.

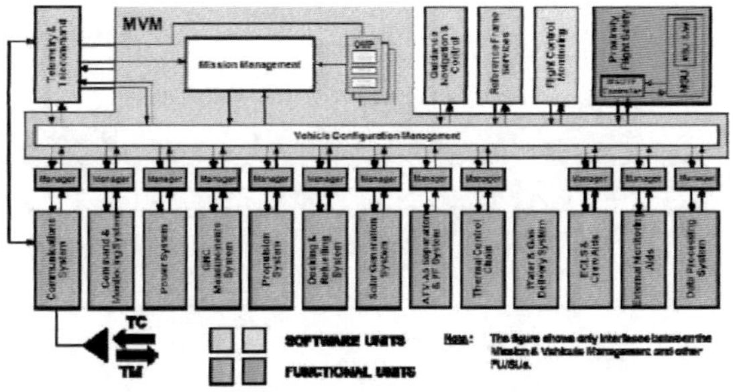

Figure 4. ATV FU / SU architecture

3.2 Modes and configurations concepts

The operation of FU is based on a concept that relies on the notion of Modes and Configurations which are representation of the physical and logical states taken by the FU to execute the mission and to recover from failure situation. This is introduced at two levels:

- at Functional Unit level → FU Modes and nominal / degraded Configurations;
- at overall Flight Software level → the vehicle modes.

The FU Modes define the stable working conditions of the Functional Units, corresponding to a given set of provided services and a given set of used services. The Mode of a FU describes the state taken by vehicle hardware and possibly software.
For Functional Units, there exist in general several Configurations corresponding to the same Mode, describing the Nominal Configuration and possible Degraded Configurations resulting from the occurrence of failures on Functional Chain equipment. For instance there may be the Configuration using the nominal hardware equipment and (an)other Configuration(s) using (a) backup hardware equipment(s).
The knowledge of the current Mode and nominal / degraded Configuration of a FU allows knowing precisely and unambiguously which support software items required by a FU should be active and which ones inactive: regulation loops, monitoring checks…
All possible combinations of FU Modes and Configurations (nominal or degraded) can be graphically represented by means of states (see Figure 5). The transitions between these graphically. represented states are operated, all along the mission, by the execution of the appropriate commanding sequences (sequence of unitary commands toward equipment) to activate, deactivate or change the internal Mode of a piece of FU equipment, via the system bus (or possibly some dedicated hardwired commands).

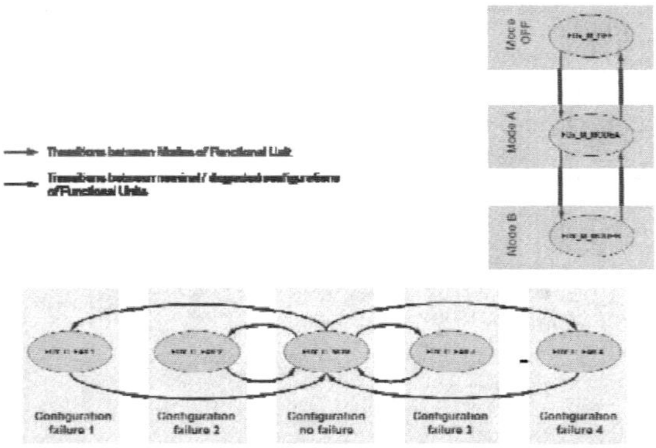

Figure 5. Modes and Configurations concept

There can be also Commanding Sequences for the update of data internally to an equipment or the change of the state of this equipment without impact on the FU Mode or Configuration itself.
Note that other FU software services such as regulation loops or failure monitoring must be performed in a coherent way with respect to the FU current Mode.

The benefits of such an implementation are the following:

- It streamlines the command and monitoring interfaces between the equipment items of a FU and the Flight Software.
- It streamlines the specification of the validity conditions vis-à-vis the state of the FU associated to the commanding sequences or to failure monitoring.
- It allows simplifying the demonstration that the Flight Software specification is correct versus the vehicle hardware behaviour and the qualification of the whole vehicle, thanks to the introduction of symmetries in the way nominal or degraded states are managed on-board

To ensure determinism in the execution of transitions between FU states, only one Command Sequence should be executed at the same time, to prevent conflicts between commands that could arise in case different Command Sequences would be executed in parallel.
Also, the execution of Command Sequences should not be interruptible, in order to prevent leaving the FU in an intermediary and unpredictable behaviour that would have not been qualified for working in steady-state regime.

As a complement to FU Modes, are defined vehicle Modes, in order to define the stable working conditions at overall vehicle level. Basically, these vehicle Modes are defined as combinations of Modes of the different FUs.
The rationales for this are the following:

- It allows taking into account the fact that the different FUs which constitute the vehicle do not work and are not operated in a completely segregated fashion from each other. Indeed, there exist strong couplings between the various functions of the vehicle (such as in the ATV between the Power Supply System, the Solar Generation System and the Thermal Control System FUs) so that the definition of vehicle level Modes allows to better take them into account.
- The identification of modes at vehicle level is performed based on the combinations of FU Modes which are used in the performance of mission operations.
- Such an approach helps simplifying the vehicle command and monitoring interfaces, as defining constraints for chaining FU Commanding Sequences at vehicle level each time a same transition has to be executed as part of mission operations.

4. FU/SU BASED FDIR DESIGN

At project CDR time, the "System Failure Modes - Effects and Critical Analysis" (FMECA) delivered by the RAMS team is used as input to define the FDIR algorithms for failure detection and then recovery.

Using the FU/SU approach, this is done through the following steps:

4.1 Elementary monitoring definition step

The measurements from all vehicle equipment are acquired by the onboard computer according to bus defined frame rate for each message. Some software validity flags may be associated to the measurements provided by equipment to prevent the further use by software of corrupted or missing data.

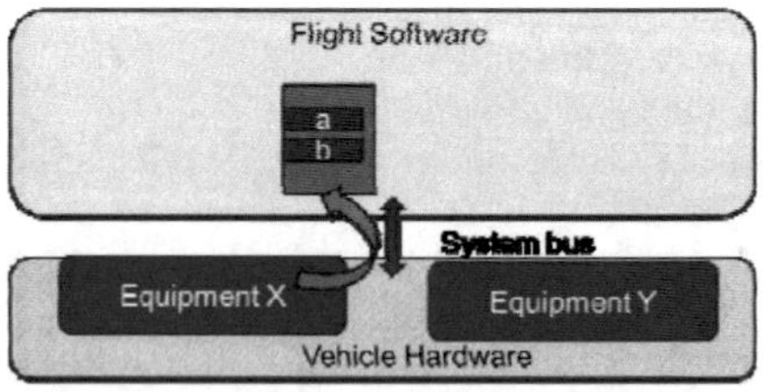

Figure 6. Measurement acquisition by Flight Software

For each FU, the acquired measurement/statuses are called "checked parameters" when they are associated to an "elementary monitoring" which evaluate the measurement or status versus an expected value or threshold and over a predefined number of consecutive occurrences, with possible measurement combination or filtering conditions (such as counters) in order to detect elementary failure symptoms of the FU.

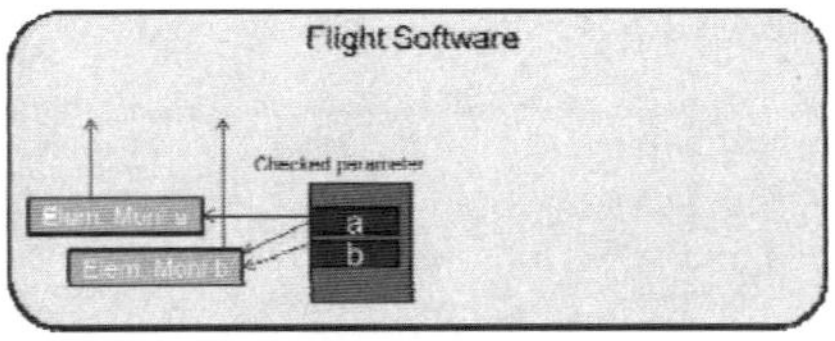

Figure 7. Elementary Monitoring

4.2 Failure alarm definition step

The combination of elementary monitorings produces a predefined alarm (associated to a single failure detection identifier) that is reported either to the FU local FDIR or the vehicle central FDIR (see Chapter 2). These monitorings only operate according to an activation condition which is defined versus the current Mode and Configuration of the FU (or possibly vehicle Mode and Configuration in some complex cases). So at any given time, the table of the activated failure detection monitorings is available.

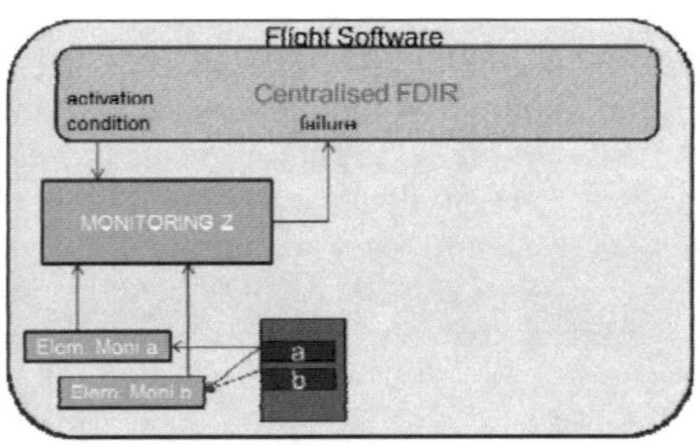

Figure 8. Identified failure detection

Further aggregation of elementary monitorings into the same failure alarm can be done when the these elementary monitorings have the same recovery(ies) over the different modes of the FU.

4.3 Recovery action definition step

Each identified failure has an associated Commanding Sequence for recovery (of course when permitted by the hardware redundancy). The recovery is related to a change in the FU Configuration that is performed by the execution of the defined Commanding Sequence to the vehicle hardware by the Flight Software. When several FU are involved, each FU defines its own Commanding Sequence as part of the overall recovery scheme. The recovery can be triggered either automatically at FU level when independent from vehicle Mode and Configuration, or according to the current vehicle mission phase and/or Control Center decision (see chapter 2) depending on latency requirements for recovering from the failure.

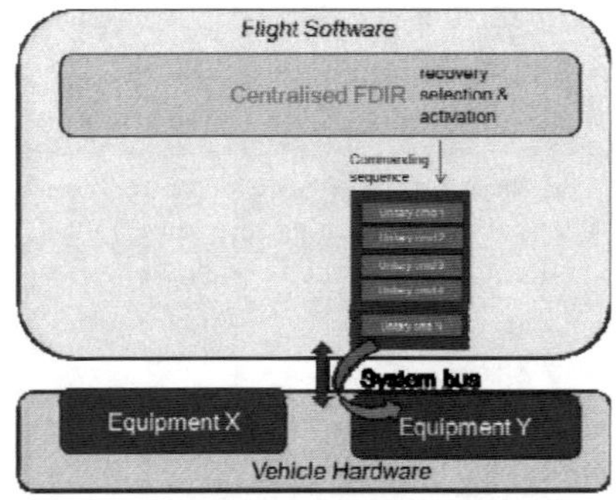

Figure 9. Identified failure recovery

A predefined table associates for each identified failure the condition and actions to be operated for recovery.

4.4 Modelling of the FU and SU

The process
- to collect the equipment measurements and unitary commands,
- to define the elementary monitoring algorithms, to define their association to FMECA based Failure Identifiers and aggregating them in failure alarms,
- to define the Modes and Configurations to activate the Failure detection, and finally
- to define and associate to each identified failure the Commanding Sequence to be operated for recovery and the associated Configuration change

can become very complex. So there is a great advantage to define for each FU a model that contains all the above mentioned data. The vehicle's model is then based on each FU sub-model. The model can be used to help to build the various tables clearly associating the various elements used for FDIR. Figure 10 shows which Commanding Sequence is defined for the recovery of a given failure according to the current FU mode.

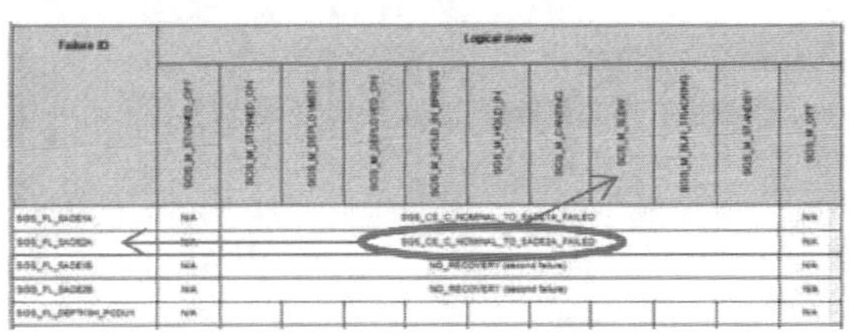

Figure 10. Failure recovery table

4.5 Software Generic Architecture

This Model Based System Engineering (MBSE) approach is used for Ariane6 software design at AIRBUS Defence & Space in Les Mureaux, where all FU and SU are defined within a SysML model. This model is the reference to produce a coherent set for the generation of the design justification file document, the Flight Software specification document, the FDIR synthesis document and for a partial Flight Software code generation (code for the overall software architecture, the Mode and Configuration automatons and the Commanding Sequences).

To implement the FU concept, a Generic Flight Software applicative architecture has been designed, based on the concept of a centralised Mission and Vehicle Management (MVM) implemented on an On Board Control Procedure (OBCP) that sends commands to the FU software parts. The FU software part is made up of finite state machines defining its behaviour in case of a command reception and a set of blocs implementing cyclical algorithms and monitorings whose executions are dependent of FU state. A specific ADA2012 library has been designed by AIRBUS Defence & Space to provide the general framework to execute the FU specific services software according to the Mode and Configuration states and transitions. More details can be found in [1].

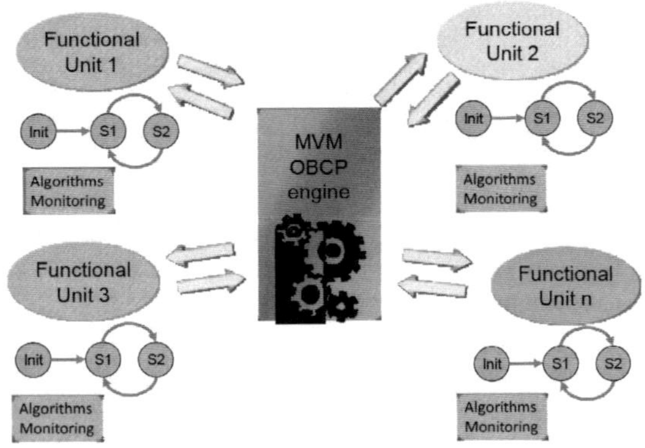

Figure 11. The software generic architecture

5. CONCLUSION

The return of experience from Ariane5 and the ATV both on FDIR and onboard Flight Software design is currently being used on the ORION ESM functional design and on the Ariane6 Flight Software design.

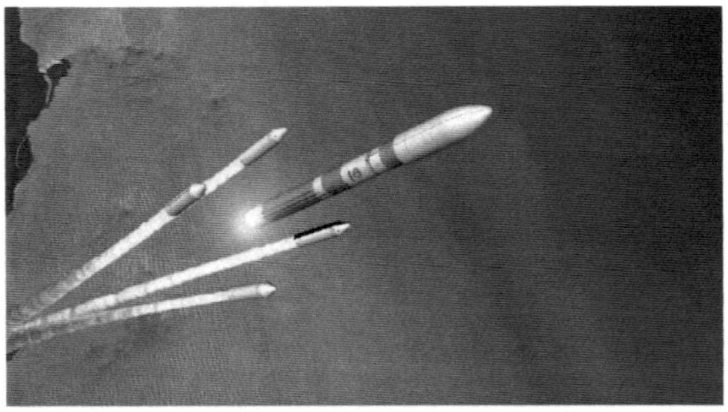

Figure 12. Ariane6 booster separation phase

The FU approach brings definitive benefits. It allows an efficient FDIR design approach, improves functional test coverage verification while avoiding costly duplication of tests. For Ariane6, it drives the design of the Flight Software and simplifies the management of mission phases thanks to strictly defined functional modes. This approach has also proved, for very complex orbital space vehicles, such as the ATV, to have greatly contributed to the overall mission safety. For ORION, it allows to present and explain the proposed ESM FDIR concepts to help NASA to design the ESM control in the vehicle Flight Software.

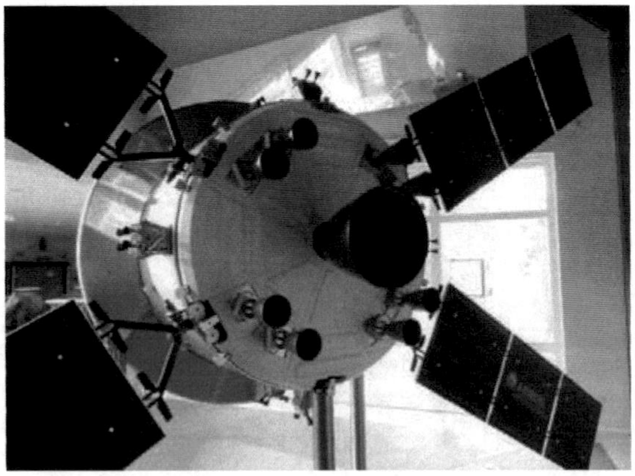

Figure 13. The ORION ESM

This approach will even be more profitable for even more complex orbital missions such as future space debris removal vehicles: the mission complexity is increased due to the need to perform non cooperative rendezvous with small debris, with extreme safety constraints to avoid any collision or dangerous disposal of the debris (human casualty on Earth after de- orbitation). This means the use of new and not yet "flight proven" technologies which increases safety concern and the addition of further backup equipment.

Figure 14. Urgent need for debris removal

6. REFERENCES

1. Emilien CARRE, Philippe GAST, Emmanuel HIRON, Alain LEBLANC, David LESENS, "Generic Software Architecture for Launchers", DASIA 2015 Conference Proceedings, EUROSPACE

SAFETY CHARACTERISTICS IN SYSTEM APPLICATION SOFTWARE FOR HUMAN RATED EXPLORATION MISSIONS

Edward J. Mango[(1)]

[(1)] *National Aeronautics and Space Administration, Kennedy Space Center, Florida, U.S.*

ABSTRACT

NASA and its industry and international partners are embarking on a bold and inspiring development effort to design and build an exploration class space system. The space system is made up of the Orion system, the Space Launch System (SLS) and the Ground Systems Development and Operations (GSDO) system. All are highly coupled together and dependent on each other for the combined safety of the space system. A key area of system safety focus needs to be in the ground and flight application software system (GFAS). In the development, certification and operations of GFAS, there are a series of safety characteristics that define the approach to ensure mission success. This paper will explore and examine the safety characteristics of the GFAS development.

The GFAS system integrates the flight software packages of the Orion and SLS with the ground systems and launch countdown sequencers through the 'agile' software development process. A unique approach is needed to develop the GFAS project capabilities within this agile process. NASA has defined the software development process through a set of standards. The standards were written during the infancy of the so-called industry 'agile development' movement and must be tailored to adapt to the highly integrated environment of human exploration systems. Safety of the space systems and the eventual crew on board is paramount during the preparation of the exploration flight systems. A series of software safety characteristics have been incorporated into the development and certification efforts to ensure readiness for use and compatibility with the space systems.

Three underlining factors in the exploration architecture require the GFAS system to be unique in its approach to ensure safety for the space systems, both the flight as well as the ground systems. The first are the missions themselves, which are exploration in nature, and go far beyond the comfort of low Earth orbit operations. The second is the current exploration system will launch only one mission per year even less during its developmental phases. Finally, the third is the partnered approach through the use of many different prime contractors, including commercial and international partners, to design and build the exploration systems. These three factors make the challenges to meet the mission preparations and the safety expectations extremely difficult to implement.

As NASA leads a team of partners in the exploration beyond earth's influence, it is a safety imperative that the application software used to test, checkout, prepare and launch the exploration systems put safety of the hardware and mission first. Software safety characteristics are built into the design and development process to enable the human rated systems to begin their missions safely and successfully. Exploration missions beyond Earth are inherently risky, however, with solid safety approaches in both hardware and software, the boldness of these missions can be realized for all on the home planet.

1. EXPLORATION SYSTEMS STRUCTURE

NASA's Exploration Systems Development is building the agency's crew vehicle, next generation rocket, and ground systems and operations to enable human exploration throughout deep space — a capability the world has not had for more than 40 years.

The Orion spacecraft, Space Launch System (SLS) and a modernized Kennedy Space Center spaceport will support missions to multiple deep space destinations extending beyond our Moon, to Mars and across our solar system. This innovative approach aligns with NASA's bold new mission to design and build the capability to extend human existence to deep space.

A program at the Johnson Space Center manages the Orion, a program at the Marshal Space Flight Center manages the SLS and a program at Kennedy Space Center manages the Ground Systems Development and Operations (GSDO). The GSDO program is broken down into major projects under the program. Command, Control and Communications (C3) is one of those projects. The Ground and Flight Application Software, GFAS, is a project within GSDO.

2. EXPLORATION MISSIONS

Human exploration missions are in the critical development phases. All three programs under ESD have completed their various Critical Design Reviews (CDRs). Each of the CDRs is at different levels of maturity with the details of their designs. In order to be ready for human flight in the 2021 timeframe a series of two highly integrated flights, test flights prior to a human test flight are required. The first flight test was the Exploration Flight Test -1 (EFT-1) in December 2014. The Orion capsule was launched on a Delta 4 V launch vehicle. The goal was to test the capsule structure as well as re-entry thermal and guidance. The second flight test is Exploration Mission – 1, (EM-1). This first fully integrated mission of NASA's Orion spacecraft and Space Launch System (SLS) will launch from a modernized Kennedy spaceport. The 70-ton SLS will send an uncrewed Orion into lunar distant retrograde orbit, a large orbit around the moon that is farther into space than any human spaceflight system has ever ventured. The mission will last 25 days and splashdown in the Pacific Ocean. Exploration mission – 2 (EM-2) will be the first "crewed mission" of Orion launched from Kennedy spaceport again using the Space Launch System. The flight will focus on checking out the mission systems, crew systems, and demonstrating the capability for the crew to operate this vehicle in deep space, near the moon, and return safely to earth. The mission is planned to be two weeks in duration. After the EM-2 flight, the integrated systems will be used to perform all of the future human exploration missions beyond Earth. Destinations and complexity will evolve as the missions are defined and developed. The entire architecture is based on capability not only on destination.

3. GROUND AND FLIGHT APPLICATION SOFTWARE

The overall construct of the C3 project includes the Launch Control System (LCS) as well as the GFAS system. The LCS comprises a software and hardware platform structure and includes a set of system software that operates and interfaces with each of the flight vehicles, as well as the various ground systems' interfaces, and the Launch Control Center (LCC). Consoles within the LCC Firing Room are used by both the flight vehicle and ground system engineers to monitor and command the hardware, perform the testing, checkout, and preparations including launch countdown of the flight vehicles. Each of the flight hardware vehicles has its own avionics and flight software that must interface with the LCS. The Orion spacecraft, the SLS Core vehicle and the Interim Cryogenic Propulsion Stage (ICPS) all have different approaches to their corresponding flight avionics systems and flight software. LCS must interface with each of these three different interface approaches as well as integrate the downlinked information and command uplink information all through a common set of equipment for the various Firing Room engineers. The LCS platform system software is similar to various operating systems, like Windows or Apple OS operating systems.

As with most command and control architecture, there needs to be applications on top of the platform of software structure. The GFAS system is the equivalent command and control applications on those various operating systems. The GFAS applications integrate the flight software packages of the Orion, the flight software of the SLS, and the ground control systems through the LCS. The launch countdown sequencers are also in the GFAS applications. All GFAS applications use the 'agile' software development process.

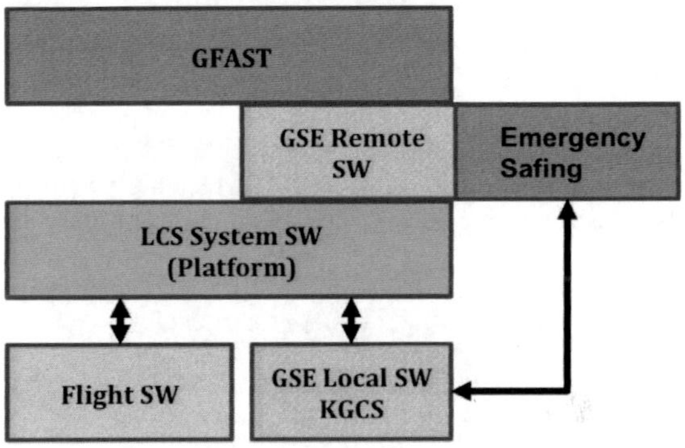

Figure 1 – GFAS Application Structure

The GFAS will develop the integrated firing room console applications and displays for pre and post launch activities to support flight and ground processing and integrated ground subsystem processing as required for Orion, Core Stage, Booster and ICPS. These software applications and displays will support integrated Ground Systems verification and validation, launch vehicle and spacecraft integration and launch processing. These displays and applications will also support off-line processing, recovery, and de-servicing of the Orion after splashdown.

The applications and displays include Graphical User Interface (GUI) and applications for Command Control Sequences, Prerequisite Logic, Reactive Control Logic, and Data Fusion.

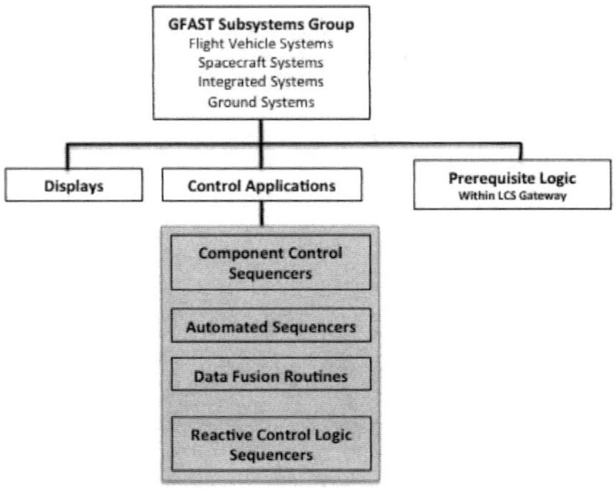

Figure 2 – GFAS Application Types

The GFAS software is used to satisfy integrated subsystem requirements, integrated ground to flight requirements, and integrated flight processing requirements. The GFAS team is made up of ten grouped subsystems comprising of approximately 70 software and processing engineers. Each subsystem team supports all the applications for the subsystem regardless of which flight vehicle they are supporting. For example, Avionics team supports the avionics installed in Orion, ICPS, SLS, Boosters and Engines. Below is the listing of each of the GFAS Teams.

Avionics
Cryo/Prop
Electrical
ECLSS/Hypers
Flt Controls
Mech
APU/HYD
Comm
Integration
Master

Figure 3 – GFAS Software Teams

The LCS system is developed through a series of Agile processes as well and is required to support Ground system verification and validation in late summer of 2016. The GFAS subsystems are expected to have verifications completed prior to first use of any applications used in flight hardware processing and testing. The first critical milestone for this effort is currently expected in Fall of 2017 against the Orion vehicle in the Multi-Purpose Processing Facility (MPPF). The Orion vehicle will go through a series of commodity loading activities and initial vehicle tests, prior to integration of the spacecraft hardware with the SLS in the Vertical Assembly Building (VAB).

4. SOFTWARE DEVELOPMENT PLAN

The GFAS project uses a form of "agile development" like many software application projects. Agile development is an alternative to classic project management techniques, in that it adapts quickly to unpredictable and multi- faceted requirement, design, verification and production situations. Agile is an alternative to a standard "waterfall" or traditional sequential project development scheme. For software applications, 'agile develop' is ideal when the vehicle hardware and interface hardware is still in design and development. Otherwise a classic project approach would require software applications to be developed after the hardware design is matured, qualified and even in production, thus causing an extensive and lengthy overall program development schedule. In the case of the ESD development, without agile development to many of the various software products, at least two or more years would be added to the overall program development, as the tasks would be sequential instead of paralleled. The Agile project development approach provides opportunity to assess and evaluate the direction of a project throughout the development lifecycle and adjust development priorities to match the assessment. This saves time during development. To achieve this project effort, work is accomplished via a series of cadences called 'sprints'. For GFAS, a series of 'sprints' are made up into a Drop. Each Drop has a specific set of actionable products to produce. Each 'sprint' within a Drop is two weeks in duration, with a total of six 'sprints' per twelve-week Drop. Thus the development can be measured in short increments. Changes, alterations, and issues that arise can be discussed and implemented within a Drop, and even from 'sprint' to 'sprint'. There are a total of fourteen Drops for the overall GFAS project development. In the Spring of 2016, GFAS will complete Drop 6 and start Drop 7. Additional Drops of requirement content, and software development activities can be added, if needed, due to flight hardware design modifications occur, or issues are found with ground or flight qualification activities. This approach allows the focus on safety and preparing the correct design for each mission.

The GFAS project control is through two separate activities. As with any project there are the three pillars of project management, cost, schedule and technical project management. The Technical Review Panel (TRP) is used to support the technical aspects of the GFAS project. The TRP, baselines or modifies software design requirements and implementation methods to include verification and validation. The TRP is the gateway between the driving NASA standards, specifications, and the flight/ground hardware design solutions and requirements. The TRP meets at least weekly to codify and accept the various GFAS team's detailed development plans. The second portion of the project control is the GFAS Drop Reviews. These Drop Reviews evaluate progress against the near term plan established for each twelve week Drop. Progress is measured in schedule performance, and work effort against the expectations for that Drop. The Drop Reviews, completed at the beginning, mid-point and end of each Drop help establish the cost and schedule impacts to the overall GFAS project effort.

The figure below illustrates the Agile process used within the GFAS project.

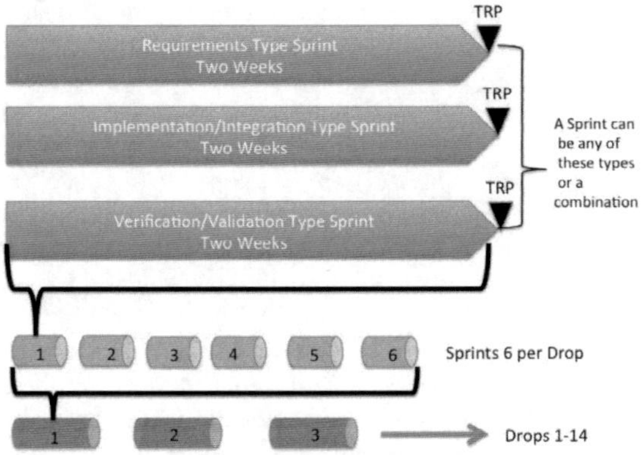

Figure 4 – GFAST Agile Process

The application software design lifecycle is made up of six phases for each set of the ten application's teams. These lifecycles are broken down by subsystem for the teams processing site locations which require an application software set of deliverables. For example, the electrical applications used during the MPPF processing may be totally different than those used during the Vertical Assembly Building (VAB) processing. The six phases include the Doc 45 Design Review, the Doc 90 Design Review, Implementation, Integration, Verification and finally Validation. Release of the application products for production use occurs after each of the lifecycle phases is completed for the software required for the site deliverable. During any specific 'sprint' or Drop, different subsystems may have various application products in the different phases. Therefore, not every application is at the same state of development and it's that flexibility that makes the overall software development approach 'agile'.

4.1 Doc 45 Software Design Review

The preliminary design development is commonly called the Doc 45 Design Review. This is the initial development of the requirements for the applications for specific subsystems. Since during this phase, many of the flight hardware design states are unknown or still not mature enough in their designs, the content for the Doc 45 Design Review is considered preliminary. At this phase the Software Requirements and Design Specifications (SRDS) are started. The SDRS will address the scope of the software applications, the types of applications, either Displays, Sequencers and/or Algorithms. In addition, the SRDS includes the approximate number of measurements and commands the applications are expected to process, the operational scenarios for the applications and any interface requirements to other applications will be documented along with the LCS platform needs or the flight hardware software interfaces. Each SRDS is brought forward to the TRP for approval.

4.2 Doc 90 Software Design Review

At the Doc 90 Design Review the software design package is ready for final review and the start of implementation. At this point the flight hardware requirements are established, through the Operational Maintenance Requirements Set (OMRS), the Launch Commit Criteria (LCC), software standards and specifications and the flight hardware subsystem critical design. A mapping of the software applications against each of the established requirements is completed and the resulting product is called the Requirements Traceability Verification Matrix (RTVM). The RTVM is critical to establish the proper links between the flight hardware requirements, the safety specifications, and the detail software applications and code for implementation. The RTVM is used throughout the rest of the software lifecycle including Verification and Validation. From the RTVM, a set of Design Verification Objectives and Design Validation Objectives (DVO's and DVaO's) are produced and baseline for use during verification and validation phases. The software Doc 90 Design package is accepted by the TRP as well as by the independent technical authorities as a check and balance of the software design.

4.3 Implementation

Implementation is the next phase in the software development process. Implementation is the actual display development and software coding of the applications, scripts, sequencers and control logic. For GFAS, the applications are coded in a C++ based Application Control Language (ACL). During this phase the applications are put under configuration control. Individual applications or components of an application are tested for syntax errors, logic gates, logic interfaces and also for its compatibility to the RTVM. In order to remain as agile as possible, a certain amount of implementation can be worked in parallel with the Doc 45 and Doc 90 design effort. This is necessary in the GFAS development lifecycle, as the flight hardware design and development is not often at a state for the ground software applications to proceed with fully accepted design details. This approach is done at risk both to the schedule but more importantly to the technical content of the software requirements and actual code. The TRP process and the GFAS project management pays particular close attention to the risk of performing this type of agile development effort. Once the implementation phase is near completion, there is a re-alignment of the software applications to the RTVM and the corresponding DVO's and DVaO's. This is critical in the extremely agile environment of the GFAS project to ensure all the requirements for usability are met, and safety remains as a focus for implementation at each step of the lifecycle.

4.4 Integration

The Integration phase provides the initial integration of the individual software applications to be compiled into a configuration managed Test Configuration Identifier Description (TCID). The complied action is the first time the application is incorporated into the LCS platform. Integration testing includes checks to assure that the applications have the appropriate measurement identifier information, and can properly communicate with other portions of the LCS systems. Any regression testing due to changes in the LCS platform operating system could also be performed during this phase against the application. The DVO's and DVaO's are incorporated into a set of software verification test procedures. A review and alignment of the SRDS, the RTVM, the application, and the verification procedures are completed by the TRP and accepted for readiness to begin the next step of the lifecycle, verification.

4.5 Verification

The Software Verification phase contains the verification activities needed to provide software requirements buyoff from quality and safety for the GFAS Displays and Applications. This testing is performed to ensure that all allocated functional requirements, including DVOs, are adequately and correctly satisfied. A Software Test Readiness Review (STRR) is conducted early in the phase to ensure the GFAS software team and the project stakeholders are all cognizant of the test plans, procedures and schedule to complete the appropriate testing. Verification Test execution is performed by a member of the GFAS team that was not directly involved in the development of the software. This provides a level of independent verification. In addition, the Safety and Mission Assurance personnel buy-off the test procedure steps, as well as the closeout of the procedure. A formal non-conformance process is also in place during the testing. Any non-conformance is formally documented and appropriate corrective action is put in place before the GFAS application set is considered complete through the verification test process.

4.6 Validation

The Validation phase is the final phase prior to formal release of the software applications to the operational TCID set. Once the verification testing is completed, and any non-conformances are formally disposition, the applications are ready for validation testing. Like verification testing, most validation testing is performed against a set of software models built by a separate software organization. The models are developed to simulate the flight hardware configurations. Most are not full physics based models, but are adequate to verify the application software. For validation testing, the stakeholder must accept the software for their use. In the case of processing for Orion, ICPS and SLS, the stakeholders are the operations and processing engineers and test controllers who will be performing the various checkout tests and processing on the flight hardware. In order to help ensure stakeholder needs were fully incorporated into the GFAS applications, the GFAST project appointed the GFAS "leads" from the operations engineering community at KSC. So the validation testing will include members of the GFAS teams as well as other operations engineers. A few of the validation tests will also be performed against the flight hardware test rigs at the hardware integration test rigs in Denver and Huntsville. Validation testing is executed in a similar manner to the verification testing with complete configuration control, S&MA oversight and non-conformance processing. Once the Validation testing is complete, the GFAS Application set is ready to use in an operational TCID and ready for a flight hardware Test Readiness Review (TRR).

5. SOFTWARE SAFETY ANALYSIS

Software projects under NASA's control are subject to the NASA Standard NPR 7150.2B. This Standard addresses the various top-level standards for software development, safety and configuration management. Individual software projects derive lower level software requirements and standards from the set of standards found in the NPR. Under the GFAS project there are a number of additional standards, which must be followed for proper development. They include the End To End Command and Control Software Development Plan, as well as the GSDO End To End Command and Control System and Software Safety Plan, and End-to-End Command and

Control Software Assurance Plan. Each of these is a lower level derived from the NPR 7150.2B.

There are a number of software safety characteristics built into the GFAS system. Throughout the six phases of the project development, software safety characteristics are embedded into the phases. Each of these characteristics adds a critical safety element in the overall structure of the GFAS software applications. Besides those characteristics embedded in the development phases, there are five overarching characteristics that help ensure GFAS is ready and capable of performing human spaceflight mission preparations

The first of these characteristics is the classification of GFAS software. Since GFAS software applications are used in the human spaceflight, GFAS must follow the elements of the Class A classification for NASA software according to NPR 7150.2B. The Class A classification is applicable to software systems and applications that are used to ensure safe and sustainable software for human rated space vehicles, during ground and flight operations. For GFAS, this represents all of the applications used to monitor, command and control the various elements of the flight hardware from initial power up at KSC through the completion of a launch, scrub recycle and recovery operations. The classification is applicable to all 1.5+ million lines of code, thousands of applications and displays across the ten GFAS subsystems.

Another key characteristic is the embedding of system safety engineers into the design and implementation process. The GFAS system safety engineers are tasked with ensuring that the subsystem software design is safe and reliable when used for its intended purpose. During the design and implementation of the flight programs, the flight system safety engineer is responsible for analyzing the overall subsystem design and the hardware component failures that can lead to subsystem level hazards. This is accomplished by performing a hazards analysis and a qualitative reliability analyses, or failure modes and effects analysis upon the subsystem design. If the analysis performed by the system safety engineer identifies concerns within the design, it is then the responsibility of the system safety engineer to raise the concerns through the flight system design review process. The GFAS software safety engineers review the SRDS both at the Doc 45 and the Doc 90 reviews for incorporation of the flight system hazards through the proper implementation of the OMRS into the software applications. Since the GFAS software is used to implement flight hardware activities and preparations for flight, many of the hazards and hazard analysis developed by the flight programs are translated into other operational requirements and are further implemented into the GFAS software code. Thus the system safety engineers who are assigned to the GFAS subsystem teams are focused on the proper implementation of those flight hardware hazards through the OMRS, the other operational requirements, into the software code.

Another characteristic is the review and concurrence the safety engineers perform on the software code itself to ensure the software applications meet the software standards as defined. These reviews ensure that applications for displays, command control sequences, prerequisite logic, reactive control logic, and data fusion are standardized. In addition, the software chief engineer must concur with the application design through the approval of any deviation against the software application standards.

Yet another overarching characteristic is the quality assurance effort required throughout the development process. Quality engineering supports the Doc 45 and Doc 90 reviews and once baselined, using the SRDS and the RVTM, the Quality Engineer ensures through each step in the development that the software application meets the RVTM. This is critical in the Integration, Verification and Validation steps in the development process. It is also critical that any non-conformance is addressed against the software design requirements (SRDS) and traceable back to the RTVM.

The final characteristic is the configuration control of the software application from its initial implementation through final validation and inclusion into the TCID used against flight hardware. There is a rigorous process that aligns with the standards defined in NPR 7150.2B. The GFAS team as a whole is accountable to configuration control of the software and the clear traceability from the requirements, through the implementation and into the verification and validation efforts. Ultimately, at the Design Certification Review (DCR), the GFAS project attests to the completeness and readiness of the GFAS applications to perform their intended purpose against the flight and ground hardware.

These five overarching characteristics are key to building a foundation of application software that will be safe to use against human rated flight hardware.

6. CONCLUSION

The ESD is in the midst of developing the first human rated beyond Earth space exploration capabilities in over 40 years. The capabilities being developed will bring human crews further into the exploration of our solar system than ever before. As the hardware and software capabilities are developed on the flight systems, a critical portion of the overall development is the design, development and implementation of ground applications to test, checkout, prepare and launch the flight systems. The GFAS project is made up of ten software subsystems and is closely aligned, to the flight hardware and software systems. The GFAS project follows a series of software requirements and standards. The implementation of these NASA Agency and GSDO Program requirements and standards is completed through a set of GFAS software safety characteristics to ensure safe and effective software applications. As the missions are designed, it is the solid safety approaches that will ensure the inherently risky missions can be successfully carried out for human exploration of our near space neighborhood.

Figure 5 – 2016, National Aeronautics and Space Administration, Space Launch System First Flight

7. REFERENCES

1. 2014, National Aeronautics and Space Administration, NASA Software Engineering Requirements, NASA NPR 7150.2

2. 2015, National Aeronautics and Space Administration, Ground Systems Development And Operations Program, *End To End Command And Control Software Development Plan, Volume 1*, K000064476-PLN Rev: B

3. 2016, National Aeronautics and Space Administration, Ground Systems Development And Operations Program, *End to End GSE/GFAST Subsystem Implementation Standards*, K0000254500- GEN

4. 2014, National Aeronautics and Space Administration, Ground Systems Development And Operations Program, *End-To-End Command and Control System and Software Safety Plan*, C3R-E2ECC-303 Rev: A

5. 2015, National Aeronautics and Space Administration, Ground Systems Development And Operations Program, *End-To-End Command and Control System Software Assurance Plan*, K0000136629-PLN Rev: C

PROCEEDINGS OF THE 8TH INTERNATIONAL SPACE SAFETY CONFERENCE

SESSION 10
LAUNCH SAFETY - I

LAUNCH AND REENTRY SAFETY OBJECTIVES

Jerold M. Haber[(1)]

[(1)] *ACTA Inc., 2790 Skypark Drive, Suite 310, Torrance, California, United States of America, haber@actainc.com*

ABSTRACT

The primary objective of Safety organizations has always been the protection of life. The earliest protection objectives focused exclusively on protection against fatalities. The initial risk measure employed was the probability the most exposed individual would be killed. Over time, the projection objectives have broadened. Many facilities now protect against multiple levels of injury (typically, severe injury or casualties, and fatalities). Most facilities have established some maximum tolerable societal, or collective, risk that is considered tolerable. Some facilities have placed limits on tolerable catastrophic risk as measured by the probability that large numbers of people may be simultaneously injured.

The 2010 publication of the Range Commanders Council guidance document, RCC 321, broadened protection objectives by defining the basis for protecting critical assets that are a part of the launch complex. These criteria ensured that the chance of damaging the critical infrastructure required for continued operations of the launch complex and the facilities required for emergency response to a mishap were sufficiently small.

For commercial launches the United States Federal Aviation Administration (FAA) Office of Commercial Space Transportation (AST) has promulgated financial responsibility standards based on the Maximum Probable Loss (MPL) that will result from launch operations. The MPL is a combined measure of potential damage and injury that may occur with a low probability (1E-7).

Individual ranges have imposed a variety of protection standards to protect unique resources including environmentally sensitive assets.

As launch operations become more common, once isolated launch complexes are surrounded by industry and communities they support. Zoning regulations rarely consider the constraints these developments may impose on launch operations. Moreover, current safety regulations are confined to protecting these communities from injuries resulting from launch operations.

This paper explores the issues of

- What should be protected?
- Why protection should be provided?
- Potential risk measures for assessing compatibility between launch operations and surrounding communities
- The need for ranges/launch complexes to have a voice in land use decisions for the regions surrounding them.

The purpose of this paper is to stimulate discussion so that the International Launch Safety Community addresses these issues before circumstances produce a *de facto* resolution.

1. INTRODUCTION

As a result of a number of spectacular missteps, society has developed concerns for balancing adverse consequences of new technology against the benefits to be derived from the technology. In many cases, when dealing with new technology significant segments of the public desire to avoid any risk. The Safety community must discover the proper balance between protection objectives and technology development.

In the early years, the Range Commander and the Range Safety organization at each range set protection standards with no accountability to outside agencies nor even the need to communicate the standards. Over time, many local, regional and national agencies[1] have published regulations to limit adverse consequences of human activi-

1 Relevant Federal agencies in the United States include the Environmental Protection Agency, Consumer Product Safety Commission, Federal Communications Commission, Federal Trade Commission, Nuclear Regulatory Commission, and Occupational Safety and Health Administration, Natural Resources Conservation Service, Federal Energy Regulatory Commission, Office of Infrastructure Protection, Bureau of Land Management, Bureau of Alcohol, Tobacco, Firearms and Explosives among others. Numerous such agencies also exist at the state, regional and local levels.

ties. While the focus of these agencies has typically not been launch and re-entry activities, their actions have set important precedents.

Agencies have a variety of protection objectives ranging from environmental protection and nuisance avoidance to protection of property and protecting people from injuries or fatalities.

2. WHAT SHOULD BE PROTECTED

Launch and reentry operations are relatively unique activities: They can place people at risk on land, at sea and in the air. A single operation may potentially hazard people and property on land, at sea, in the air and in orbit. Moreover, the regions at risk may have characteristic dimensions on the order of many hundreds to thousands of miles.

2.1. Protection of Life

While it is generally accepted that protecting human life is the highest priority, there is a divergence of opinion and practice regarding other protection objectives. Arguments in favour of only protecting against loss of life have included:

- It is a common metric of protection used internationally for a broad class of activities.
- It has been used as a measure of weapons effectiveness (lethality).
- Death is an unambiguous state.
- Death is so much more onerous than lesser injuries, let alone nuisance or "mere damage" to the environment or property.
- Many believe that characterizing other adverse consequences has too much uncertainty.

By contrast, the National Ranges in the United States and the Federal Aviation Administration's Office of Commercial Transportation have been moving in the direction of increasingly broader protection concerns. The Air Force Ranges were among the first to recognize the importance of addressing lesser injuries to people. The ranges added the consequence of casualty, serious injury or worse, as the primary consequence metric for several reasons including the following [1]:

- Using casualties as the criteria avoids the uncertainty associated with the promptness and effectiveness of medical treatment that can prevent serious injuries from becoming fatal.
- Casualty measures are necessary to provide a reasonable level of protection from all launch vehicle hazards, particularly toxic exposure.
- Serious injuries are onerous.

Supplemental criteria have been needed to address lesser injuries from toxic hazards to be consistent with the concerns of other regulatory agencies. Accidents with toxic releases may subject very large numbers of people to relatively low exposure levels. Although a single person subjected to such levels might be regarded as tolerable and the medical consequences may range from benign to minor irritation, adverse public reaction from affecting many people may significantly impact the ability to perform future launch operations.

In the United States the prevailing standard has become that launch facility operators must protect against casualties and may need to consider lesser injuries when large numbers of people can be affected from a single incident. At that point it was evident that protecting people from severe injuries was essential. The question was "Is that sufficient?"

By contrast, there are agencies around the world who still base their protection criteria on fatalities exclusively.

2.2. Catastrophic Risks

The baseline for protection described in the previous section was ensuring the maximally exposed individual would be protected against unacceptable probabilities of incurring the adverse outcome and that collectively the societal risk, as measured by the number of people who, statistically, might be subjected to the adverse outcome was limited. The discussion about large numbers of people sustaining lesser levels of injury from a single event raised a new question: What further limitations needed to be imposed on adverse consequences to many people from the same event?

While the question arose from considering toxic injuries from a launch and the international reaction to the disaster in Bhopal [2], the Range Safety community quickly realized that there were other classes of accidents that could produce multiple severe injuries or deaths. Examples of such accidents include a spent stage striking a ship, large debris striking an aircraft, and explosive debris falling in a public gathering, such as a stadium.

Bhopal caused the range safety community to recognize the need to address "High Consequence Events". These events may include outcomes that have a significant impact on continued range or launch operations, significant environmental impacts, impacts on relationships with other countries, and other long term or irreversible consequences. Signal events have a major impact on society as a result of a combination of dread and lack of visibility and understanding by the general public. Some examples of potential high consequence events include:

- Events that may produce significant dollar damage or large numbers of casualties,
- Events that damage critical assets or cultural treasures or natural wonders,
- Events that create a public perception of irresponsible action on the part of the range – whether or not any damage or injuries resulted,
- Events that damage the local economy, such as creating an atmosphere of fear in a tourist dependent community,
- Events that violate or appear to violate the rights of foreign nationals

These factors are addressed in the remaining narrative of this section. Both the US and the international community are more ambiguous with respect to catastrophic loss potential. These communities acknowledge the potential for high consequence events; they agree that catastrophic risk is a special concern. The community has not yet agreed upon what constitutes a catastrophe nor what actions are appropriate for protecting against high consequence events. Risk profiles and F-N curves have been employed by some to compute and display catastrophic potential property damage and injuries. Reference [1] (Section 5.5) summarizes some of the most pertinent work related to catastrophe aversion to injury; many of these are illustrated in Fig. 1.

An accident resulting in downing an aircraft or sinking a ship is clearly catastrophic and a feared outcome. Concerns about simultaneously injuring large numbers of people has resulted in criteria to protect against sinking a ship or downing an aircraft and catastrophe averse criteria such as displayed in Fig. 1. The most prominent examples of criteria to protect people from mishaps induced by launch and reentry failures are the RCC 321 provisional criteria for protecting occupants of ships and aircraft.

The US FAA Office of Commercial Space Transportation (AST) requirements for licensees to demonstrate financial responsibility was one of the drivers for assessing catastrophic loss potential from a launch. The FAA/AST requires

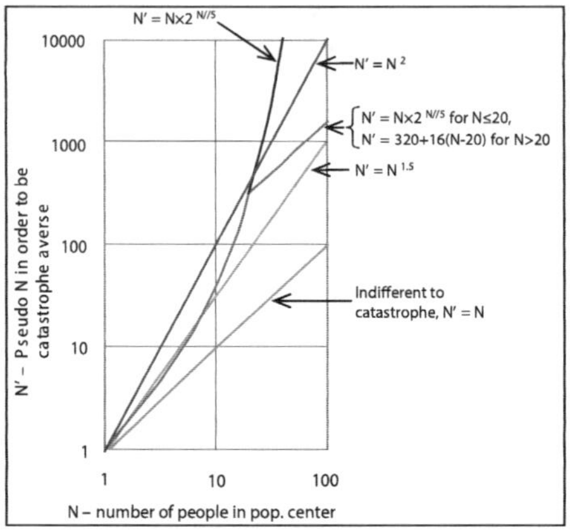

Figure 1. Catastrophe Aversive Criteria

computation of the Maximum Probable Loss (MPL) from a launch operation. The MPL is an estimate of the loss potential from property damage, death and injury at the probability of occurrence level of 1×10^{-7}. It is important to understand the extent of the applicability of this precedent. It is a requirement to assess the catastrophic potential of a launch. By assigning a value to death and injury it produces a single metric of catastrophe. However, MPL is to be used solely for the purpose of establishing insurance or bonding requirements. High values of MPL are not taken as a reason to limit the acceptability of a launch.

Implied by the MPL regulation is the assertion that it is acceptable to launch a vehicle that has a foreseeable potential to produce extremely large damages as long as the launch operator can demonstrate his capability to pay for the losses. (Clearly, this is an unintended implication of the regulation and not the intent of the regulators.) That leaves, however, the question of what property protection should be required.

As discussed in the next section, protection of critical assets for the ranges and for public health and safety has become a part of the safety guidance for U.S. ranges.

No guidelines have been developed for what cultural assets or natural wonders merit protection. Although these have been flagged as warranting attention in RCC 321, they are only addressed in the Environmental Protection Process, if at all.

What constitutes responsible behavior by the ranges? While there is no codified answer and public perception of responsibility frequently changes there are basic principles that decision makers must consider. These derive from considerations of what a prudent man would require. Important principles are the use of a documented safety management process and guidelines for adherence to the process. There must be well- defined measures and thresholds of acceptability. The process must uphold the standards of transparency, clarity, consistency, and reasonableness.

Decisions are derived from modeling of events and undesired outcomes. Models must be evaluated to ensure they capture the appropriate physics and, to the extent feasible, are validated by "real world" results.

Software implementing the models must be tested to ensure the models and logic are accurately implemented. Both models and software must be subjected for formal configuration control so that analyses are traceable and repeatable. Data required for safety analyses are often developed from empirical sources or other models. The range must ensure the validity of this input data and the appropriateness to the particular analyses.

Decision makers must rely on the information provided to them by analysts. Analysts must convey to the decision maker the basis of the analysis, sources of uncertainty, the magnitude of the uncertainty, and any known biases in the analysis. The analyst must interpret these in the light of safety and regulatory standards so that the decision maker understands the basis of the analyses and its limitations. Commonly, the decision maker is responsible for any subsequent communication of decisions to the public. It is essential that the analysts support the decision maker in formulating clear language for communicating such information. Although this may be complicated by concerns of protecting proprietary or classified information, such concerns must not be used as a "smoke-screen" to hide relevant disclosures.

Concerns about real, or perceived, potential impacts on tourism or other aspects of the local economies by launch or reentry operations must be addressed. In locations such as Hawaii with strong tourist economies, failure to sufficiently involve the public in frank discussions has at times precluded the ability to create a Space Port. A notable example of this occurred with the proposed Palima Point Space Port. Absent adequate communication, the local press raised the specter of accident releasing toxic materials devastating the local quality of life and economy. The public reacted. After the sensationalism there was no opportunity for reconciliation. By contrast, the cooperative environment created in Brevard County Florida between local officials including those responsible for emergency response has fostered the safe growth of Space lift operations there.

Protection of the rights of foreign nationals has had a mixed record. Nevertheless, an important milestone was established by the International Association for the Advancement of Space Safety (IAASS) [5] when they articulated the principle:

> *Ensure that citizens of all nations are equally protected from "unreasonable levels" of risk from overflight by missiles, launch vehicles and returning spacecraft.*

2.3 Protection of Property

Several factors in the United States point to the need to consider the protection of property. Early documentation of the role of range safety stated that the responsibility of range safety was "the protection of life and property". The MPL regulation discussed in the previous section was designed to ensure a launch operator was financially responsible; nevertheless, it points to the significance of protecting against large property damage losses.

The 2010 publication of RCC 321 added the requirement for protection of critical assets. Critical assets are defined as any resources requiring protection.
These normally include property or infrastructure that is essential to protect the public health and safety, maintain the minimum operations of the range, or protect the national security of foreign policy interests of the United States. These requirements result in the protection of property to ensure critical functionality.

Protection requirements have also been expressed in terms of mission rules determining if Flight Termination Systems should be required and the conditions under which they should be activated. Historically, the criteria have been based on protecting people. Flight termination limits have been defined so that a failed vehicle does not unduly endanger people. In addition, some ranges have required that the terminate limits be evaluated to ensure that a flight termination action, if taken, does not create unacceptable levels of risk to people. Should these requirements in some manner be extended to property protection? Recent discussions in the United States have addressed the possibility of including some measure of the conditional risk of damaging property given a failure of a booster to the criteria for assessing when a Flight Termination System should be required.

The Range Commanders Council (RCC) sponsored a study (expected to be completed soon) to define criteria for protecting public infrastructure. While the initial focus of the project is public infrastructure associated with renewable energy systems such as wind farms, solar arrays and their associated substations and transmission systems, RCC envisioned a far broader scope including:

- Transportation system infrastructure
- Critical infrastructure required to sustain life
- Energy infrastructure
- Hazardous waste infrastructure
- Information and communication infrastructure
- Public works
- Solid waste – generation, collection and management
- Sustainable urban infrastructure
- Water infrastructure
- Wastewater infrastructure

It is noteworthy, again, that the focus is on the potential for impairing the functionality of the infrastructure rather than the monetary cost of repair.

A question raised by some parties is whether protection of property is an appropriate responsibility for a range safety organization. Fatalities cannot be cured by financial payment; casualties may at best be remediated by financial compensation but not completely cured. By contrast, property damage (both direct and consequential) can, in most instances be restored by financial compensation.

There are, of course, exceptions to this assertion. Cultural or historical property cannot be replaced. Non- monetary impacts of property damage, such as disruption of family or community life, cannot be restored. These effects can, at best, only be offset by compensation.

2.4 Environmental Protection

In the United States, customary practice has been to build launch complexes in relatively remote regions. Such locations initially pose few restrictions from the perspective of protecting people. However, such remote locations often have environmentally sensitive regions either in the immediate vicinity of the launch complex or about likely flight paths. Papahanaumokuakea Marine National Monument in the vicinity of the Pacific Missile Range is one example of this type of situation. The California Channel Islands contain important geological formations and sensitive species that must be protected during launches from Vandenberg Air Force Base. In the vicinity of NASA Wallops Flight Facility are several such areas including Wallops Island National Wildlife Refuge, Chincoteague National Wildlife Refuge, and Assateague Island National Seashore. In the United States, the need to manage environmental impacts is the rule rather than the exception.

Normally performing launch vehicles and malfunctioning vehicles can generate a variety of environmental threats:

- Impacting inert or explosive debris
- Toxic gases or fluids
- Sonic boom and acoustical insults

Evaluation of these threats is in the United States typically performed as part of Programmatic Environmental Assessments or, on occasion, Environmental Assessments for a facility. As such, guidelines for what is acceptable are derived from a variety of National, State, and Local Agencies. This regulatory environment has resulted in significant variations throughout the nation.

Should the consideration of environmental impacts, degradation of quality of life, and impairment of natural heritage be the responsibility of safety organizations? If so, should the safety organization be involved in the evaluation of what should be protected or should their role be limited to executing a policy defined by an Environmental organization?

3. REASONS FOR PROTECTION

References [3] and [1] highlight major considerations in assessing what should be tolerable risk levels for range safety. The following summarizes some of the key points.

As noted previously, adverse outcomes that may be the subject of protection criteria include: threat to life (with severities of threat ranging from quality of life considerations to fatalities), threat to property from all launch related hazards (damage to facilities and their contents), threats to operations (direct and indirect, including economic loss and interference with ability to perform a mission) and threat to the environment (wildlife, plants, habitat, and natural formations).

There are many papers addressing the acceptance of risk; most neglect the underlying motivations. These include:

- Satisfying legal or regulatory mandates
- Protection of life and property
- Addressing societal expectations (world, national and regional)
- Protection of decision makers and their organizations

This last item may be the driving factor in many instances and includes: concerns about liability and litigation, personal threat of loss of position, fines or other adverse outcomes.

A variety of factors are commonly considered when assessing what constitutes tolerable levels of risk. These include precedents set by other regulators, historical risk from the class of activities, historical risk from comparable activities and background levels of risk. Note that all of the items listed except the last one point to levels of risk that are tolerable. When the risk induced by a new activity is small in comparison to the background risk, it is arguably tolerable in the sense that falls within the band of variability of the background risk.

4. CANDIDATE RISK MEASURES

A diverse set of possible candidates for protection have been considered in this paper. It is clear that by their very nature the risk measures employed for protection will not all be the same. This section will present some candidate approaches to risk measures to provide protection.

Historically, protection of life has addressed two primary concepts: individual risk and societal risk. The most common level of consequences considered are fatality, casualty (major injury plus fatality), and minor injury in that order. Individual risk is typically measured by the probability of the maximally exposed individual suffering a particular consequence level. Several measures are commonly employed for societal risk: In the United States, the most common measures are the expected number of individuals who are expected to suffer a particular adverse consequence and risk profiles (a curve depicting the probability that N or more individuals will suffer a particular adverse outcome). In Europe, the first point of a risk profile (the probability that at least one individual will suffer a particular adverse outcome) is commonly used to characterize societal risk.

Exclusion regions for the protection of people are commonly defined in terms of tolerable impact probabilities for particular classes of debris (for aircraft and ships the classes are debris capable of causing a casualty and debris capable of causing a catastrophe). Recent refinements related have replaced impact probability for casualty causing debris with probability of casualty to include the effect of other hazards.

Two approaches have been used for property protection. The first is based on the premise that risk is expressed in terms of measures of adverse consequences. Dollar value of damage, loss of service, and interruption of the supply chain allows all of these consequences to be treated on a common basis. This measure requires consequence models for every class of building at risk and for building contents. Such models are not generally available in the literature. Typically, they require a combination of two submodels – the first model characterizes the extent of the damage and the second model translates the damage level into an economic loss.

The second approach is to use impact probability for some class of debris. This approach avoids the added modelling requirement; it only requires definition of hazardous debris. This measure by itself has the further limit that it does not directly address any hazard except inert debris. It may be generalized by defining hazard thresholds for each hazard of concern. For example, for explosive hazards this might be overpressures exceeding some threshold. The alternative measure would then be the probability that any of the credible hazards exceed their threshold values at the receptor.

5. LAND USE PLANNING FOR RANGES AND THEIR NEIGHBORHOODS

While launch complexes are typically build in areas that are initially remote from population centers [4], by virtue of that isolation they are frequently near ecologically risk habitats. The compatibility of the launch complex and the nearby habitats is commonly rationalized as having minimal impact because the complex, when first established, is expected to support minimal traffic. Of course, if the facility is successful traffic levels grow. Although initially the facility was isolated, launch vehicle processing and integration and other related services eventually attract supporting industry and communities to house the worker base. Access to these facilities requires improving the highway, rail, shipping and air access. Growing ranges require facilities for processing, tracking and flight termination result in an increased demand for power. Moreover, such transportation and power infrastructure may well serve a variety of communities not in the immediate vicinity of the range.

Growth in the number of launch operations and their complexity increases the possibility of conflict between the ranges ability to support launches and to ensure adequate protection to the surrounding populations, facilities and infrastructure. Compatible land use typically involves multiple agencies at various levels of government from local to Federal.

In some regions of the United States these concerns about land use have resulted in litigation brought by land holders to protect their ability to develop their real estate unimpeded. In recent times, Federal agencies have created specialized offices to address particular aspects of land use compatibility with the missions of Federal Ranges. Nevertheless, there does not exist a central forum for each range and its surrounding communities to plan together to ensure both community and range needs are met. Addressing such needs would pose significant challenges: Missions to be supported are becoming ever more complex and typically involve progressively larger boosters. Building and development are governed by many agencies ranging from local municipalities, special purpose districts to national and international agencies.

Societal expectations for who and what must be protected have been constantly evolving. When the United States built railroads spanning the continent, it was *de facto* accepted that people would die in the course of the development and that natural obstacles were to be overcome. There was no consideration of ecological sensitivity, impact on the previous occupants of the land nor impact on wildlife. By contrast, these same types of issues are increasingly in the forefront in addressing the land use today. A recent book review in Science [6] summarizes the thesis of a prominent biologist published in his recent book [7] that still more extreme measures – turning half of the planet into protected reserves – is required to preclude another mass extinction of animal and plant species.

With a multitude of public and private interests, how should launch complex operators effectively engage to ensure compatible land use and acceptable levels of constraints on launch operations?

6. SUMMARY AND CONCLUSION

As the number of space launches and space launch operators have increased, launch and re-entry operations risk management has evolved with respect to transparency, commonality, and protection objectives. Early decisions were based exclusively on protection against fatalities; the criteria for such decisions were closely guarded. Consequently, there was little to ensure consistency in the application of criteria between operations or among launch facilities.

Evolving standards began to address hazards other than inert debris, lesser injuries, and adverse consequences other than injuring people. This process is clearly illustrated in the United States by the movement from individual range standards to a common guideline (RCC 321) and the evolution of RCC 321. RCC 321 [1] has been a major step in codifying protection standards, generating greater consistency among launching agencies, and proposing standards for Transparency, Clarity, Consistency and Reasonableness of risk models and acceptability criteria. Although RCC 321 is a guidance document for US Ranges, it has become a master template for such risk management around the world.

In response to evolving circumstances in the United States, RCC 321 has evolved to include guidance for individual, societal, and catastrophic risk. Moreover, as circumstances have dictated the need to broaden the scope of protection, the standard has incrementally addressed new assets at risk.

As noted in the beginning of the paper, space-lift operations pose a potential threat to large regions of the globe. Although the examples cited are oriented toward the United States, the implications are international. Space faring nations may launch from their own soil or from more desirable locations in other nations. Even launches within a nation may overfly foreign assets in the extended launch region. The regions at risk during return from orbit are, almost certainly, within territories of nations other than that of the launch operator.

Added considerations within international implications are the growing interdependencies on common infrastructure among the nations of the world. Property damage to a component of infrastructure may have impacts the immediate region in which damage occurred.

The questions before us are:

- As safety professionals what must be protected?
  - People
  - Critical Assets
  - Infrastructure
  - Cultural Resources
  - Sensitive Environmental Habitats
- When we provide protection, how do we ensure that system level consequences are addressed in addition to damage to a unit component?
  - For injury to people we consider individual risk and societal risk
  - What are the appropriate measures if transportation systems, power systems, and utilities are placed at risk?
  - What are the appropriate measures for protecting natural and cultural resources?
- What risk measures, tolerable values, and severity of adverse consequences should be adopted for each item to be protected?
- How do we establish appropriate forums to include stakeholders and launch operators in defining land use and operational rules for launch and re-entry?

7. REFERENCES

1. Range Commanders Council Range Safety Group Risk Committee (2010). Common Risk Criteria for National Test Ranges Supplement, RCC 321-10, U.S. Army White Sands Missile Range, New Mexico.

2. Broughton, E. (2005). The Bhopal disaster and its aftermath: a review. Environmental Health 2005 4:6.

3. Haber, J. and R. Nyman (2014). Range Safety Training – Risk Acceptability, ACTA, Inc., Torrance, California.

4. Sgobba, T., F. Allahdadi, I. Rongier, and P. Wilde (2013). Safety Design for Space Operations. Elsevier Butterworth-Heinemann.

5. Bahr, N., T. Sgobba, R. Jakhu, K. Schrogl, J. Haber, P. Wilde, P. Kirkpatrick, J. Trinchero, and H. Baccini. (2007). An ICAO for Space?, IAASS.

6. Griffiths, C. G. (March 18, 2006). Back to Nature in Science (Vol 351, Issue 6279), p. 1271.

7. Wilson, E. O. (2016). Half-Earth, Our Planet's Fight for Life, Liveright Publishing Company, New York.

IMPROVED RANGE SAFETY METHODOLOGIES FOR LONG-DURATION HEAVY-LIFT BALLOON MISSIONS OVER POPULATED REGIONS

George M. Lloyd(1), Kevin Benn(1), Jerry Haber(1), Danielle Franklin(2)

(1) ACTA Inc., 2790 Skypark Drive Suite 310, Torrance, California 90505-5345, lloyd@actainc.com
(2) Pacific Missile Range Facility, PO Box 128, Kekaha, HI 96752

ABSTRACT

This paper reviews the development of a new probabilistic risk methodology that has been developed to address concerns raised by long-duration fatality level exposures to aircraft and ground-level receptors from long/ultra-long duration balloon (U/LDB) flights. The methodology addresses risks during the launch and ascent to float phase, and for any float manoeuvres and planned descent and recovery operations. It is computationally feasible in that it allows for T-1 day or even day-of-launch risk assessments to be generated within a few hours so as to accommodate the most favourable launch conditions. This capability minimizes risks to people and infrastructure while simultaneously expediting the launch go/no-go decision process because deterministic limit line exclusion areas are replaced with risk-informed information. The methodology has been successfully applied to NASA's LDSD 2015 Campaign, notional elements of which are used as a demonstration.

1. INTRODUCTION

Range safety methodologies for scientific and military testing are typically designed for rocket-borne missions that complete within minutes, if not sooner. One result of high velocity flight on a linear trajectory is that incidental exposure to planned debris can be readily identified and mitigated. Moreover, the potential exposure of aircraft and ground receptors to unplanned debris can be quantified by accounting for debris aerodynamic properties and atmospheric uncertainties and propagating ensembles of representative break-up fragments to receptors such as aircraft, ships, and people. For both planned and unplanned debris, the dispersions arising from vehicle path and falling fragment uncertainties can be simulated using well developed aerodynamic modelling techniques. The process is often very computationally expensive, but the calculations can be done well in advance while still accounting for much of the launch-day uncertainties.

A category of lift vehicles with a long history that is gaining popularity for deployment of large payloads to the stratosphere and lower mesosphere is the long/ultra-long duration balloon (U/LDB), an example being NASA's zero pressure difference helium balloons. This class of balloons has several advantages for both scientific and military testing as well as commercial applications [2, 13, 14–16, 19].

For example, balloons are typically less expensive than rockets and enjoy the flexibility of being deployed from mobile launch facilities. Because control and FTS systems are simpler, flight readiness is simpler to verify, and because helium is inert and the balloon lift assembly does not contain propellants, balloons permit quick turn-around schedules. The absence of propellants also limits ground and near-ground catastrophic hazards that are possible from explosive motor failures. Balloons can be designed to lift sensitive payloads of over 3500 (kg). Because balloons can ascend and loiter into the lower mesosphere, they can serve as stable platforms for long-duration high-altitude missions lasting several hours or days. With sufficient gas reserve and ballast (typically 6–8% of system mass) so that this neutrally-stable vehicle can response to diurnal changes, projected loitering times are expected to become hundreds of days in the case of ULDB's [5, 8].

Although launches are routinely conducted without incident, the un-powered nature of U/LBD's and their large hazard area for planned debris coupled with the ascent dispersion and their large payload capacity means that large populations can be placed at risk. Moreover, unplanned incidents have a high probability of being catastrophic and resulting in fatalities. Stated in other words, the risks from balloons derive from a small inventory of large, *high ballistic* coefficient fragments. In contrast to balloons, for rockets much of traditional flight safety methodology as drawn from RCC 321-10 and other standards revolves around quantifying the <u>descent</u> hazards from *low ballistic* coefficient wind-dispersed fragments which intrude into aircraft corridors and onto ground receptors.

For balloon launches the converse is true. Much of the uncertainty is dictated by wind-driven dispersion during <u>ascent</u>, while the footprints from planned and unplanned debris (such as large recovery parachutes and gondolas) during descent are moderated by the simpler vehicle and the lack of significant forward velocity and ΔV's [11]. Because much of the wind uncertainty for the ascent phase cannot be resolved until day-of-launch for U/LDB's, there is a much greater need for real-time risk assessment methodologies for both planned and unplanned debris in order to minimize any catastrophic risks to people and infrastructure. With these distinctions between the two risk management problems in mind, this paper illustrates the development of a new probabilistic risk methodology that has been developed to address the differences by quantifying the long-duration fatality-likely exposures to aircraft and ground-level receptors to U/LDB flights.

The methodology addresses risks during to launch operations during the ascent to float phase, and for any float manoeuvres and planned descent and recovery. It is computationally feasible, allowing for day-of-launch risk assessments to be generated within a few hours so as to accommodate the most favourable launch conditions. Since break-up state vectors and associated debris are physically propagated to ground receptors, the methodology avoids the undesirable smoothing which can occur with kernel density estimator methods [7] and expedites launch decisions by avoiding the need for "hard" exclusion corridors for ground risk. The methodology has been successfully applied to NASA's LDSD 2015 Campaign, notional elements of which will be used as a demonstration.

2. DESCRIPTION OF PROBLEM

References [1] and [2] provide background on historical and recent developments in U/LDB's. These balloons have continued to increase in size and life capacity such that payloads on the order of 3000 (kg) can be lofted to 40–45 (km), and above 50 (km) with smaller payloads.

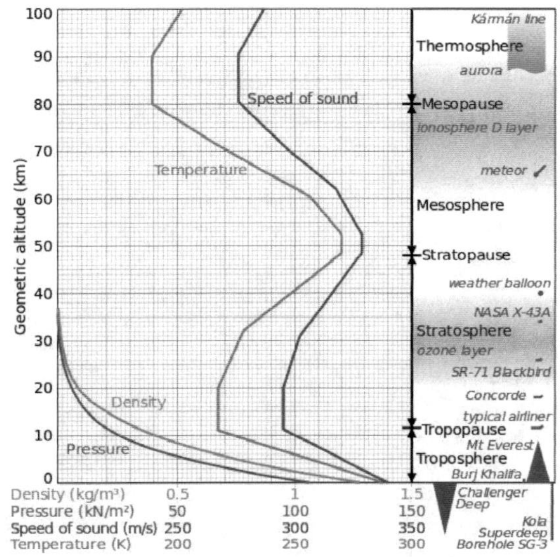

Figure 1. Atmosphere profile relevant to U/LDB operations. Earth balloons regularly ascend to densities of 0.033 (kg/m3). LDSD tests are conducted between 30 and 50 (km).

These altitudes can be viewed in the context of the physical atmosphere depicted in Figure 1. The ability to inexpensively lift tons of payload above 99.7% to 99.9% of the atmosphere and permitting its safe recovery while minimizing launch debris is significant.

Figure 2 shows a typical altitude flight profile for these missions—in this case from the NASA-sponsored "Big 60" mission launched on August 25, 2002, Lynn Lake, Manitoba, Canada. Not atypically, the actual launch occurred after several weeks of weather delays. The balloon—designed to have an ultimate lift capacity of 750 kg) — carried instrumentation weighing 690 (kg) for the study of cosmic rays. As Figure 2 shows, the balloon climbed to a peak altitude of 49.4 (km). The mission was terminated normally after approximately 23 hours of flight time.

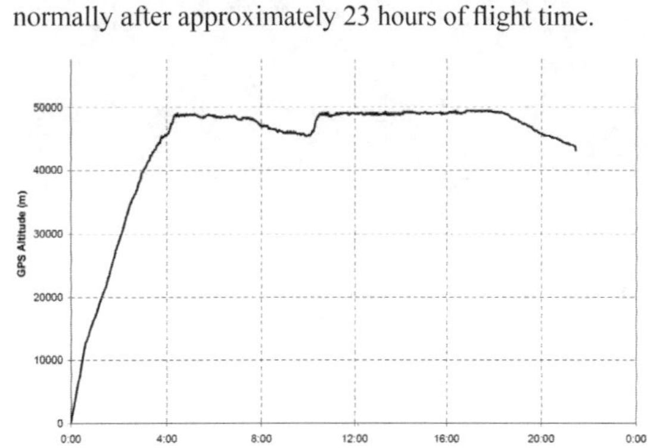

Figure 2. Balloon Flight Profile, 1.7 (Mm^3) balloon, carrying 690 (kg) of instrumentation [2]. Horizontal axis is in hours.

A more interesting payload was provided by the LDSD mission, which is the focal subject of this paper. This is a NASA effort to develop inflatable pressure vessels called Supersonic Inflatable Aerodynamic Decelerators (SIADs) for delivering future payload and manned missions to the Martian surface. These drag devices are attached to the outer rim of an atmospheric re-entry vehicle and inflate at Mach 3.5 or greater, in order to decelerate the vehicle to Mach 2 where it becomes safe to deploy a supersonic parachute. The scientific program developed by NASA to test this concept in the earth's upper atmosphere—where conditions mirror those on Mars—is given in Refs. [2–3, 6].

Stratospheric tests of the LDSD were conducted in 2014 and 2015 at the Pacific Missile Range Facility operated by the U.S. Navy on Kauai, Hawaii, as shown in Figure 3 [3, 4].

The mission revolves around a U/LDB provided by NASA Wallops Flight Facility and the Columbia Scientific Balloon Facility. This zero-pressure balloon is used to lift a solid-rocket powered test vehicle (TV) to an altitude of about 120,000 (ft) [37 (km)]. Once within earth's stratosphere and having transited over a suitable location and with the TV correctly oriented, the LDSD payload is dropped and then boosted by the solid rocket to supersonic speeds as discussed above.

Figure 3. Overview of PMRF and the Western Shore of Kauai [17]. Launch site in blue box.

Once boosted to "Martian" re-entry velocities, the tests of the deployment and functioning of the SIAD's commences, followed by recovery of the balloon and test vehicle in the Pacific Ocean. The mission sequence is shown in Figure 4. This paper is only concerned with the risk assessment from balloon launch to stratospheric float, conditioned on malfunctions during this phase.

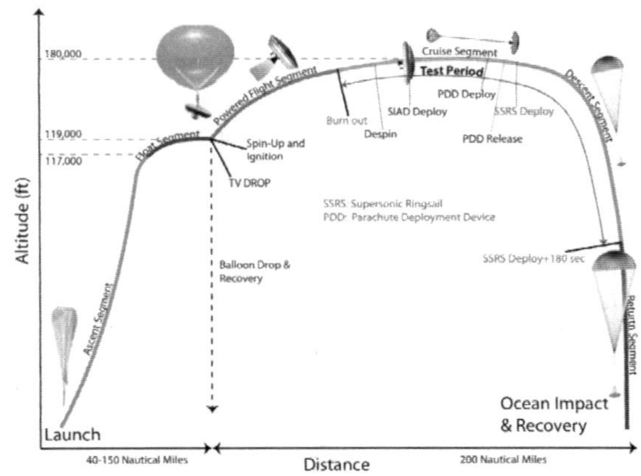

Figure 4. Flight profile for the high-altitude LDSD tests in Earth's stratosphere [4].

Risk assessment considerations differ from standard rocket launches, as discussed in the introduction [9]. Risk assessment for rockets generally proceeds from the standpoint of a nominal trajectory and planned debris, with guidance and performance uncertainties well described and constrained over the entire flight through an ensemble of trajectories. Malfunction trajectories and debris are incorporated on top of this basic methodology.

The only analogue to planned debris occurs primarily during the ascent phase, with carefully timed ballast releases which are intended to "boost" ascent where local temperature and pressure act to retard upward acceleration. Apart from the ballast releases, which are engineered to produce no risk to people or aircraft, there the only risk arises from unplanned debris, should a balloon malfunction occur, an inadvertent release of the TV, or the need to remedy other conditions with a flight termination action. These scenarios are typified by at most a handful of extremely heavy and/or large debris. Because of the large ballistic coefficients associated with the debris, the descent footprint is associated with much less uncertainty than that during ascent.

More severely, the balloon equivalent of a nominal trajectory does not exist. Balloon ascent trajectories depend on the current weather and atmospheric profiles, solar conditions, and shading from clouds. During a typical launch window period the ensemble of possible ascent trajectories can meander and "sprawl" to essentially carpet a vast area. In contrast to constrained ensembles of nominal and off-nominal trajectories for rockets, the character of an ensemble of ascent trajectories appears much more like a random walk, with the variance growing the longer the mission, and the longer the mission window. In order to respond to mission-terminate needs, whether due to balloon or equipment malfunctions, ballast exhaustion or weather anomalies, flight termination systems must be active constantly to protect civilian airspace and limit ground exposure and exhibit high reliability.

In the case of the LDSD, under *ideal* launch conditions the expected ascent track will carry it directly off-shore, or off-shore immediately after in-shore surface winds are breached [4]. Figure 5 illustrates the ground tracks when such ideal conditions are present. Under these "ideal" ascent conditions it is evident that the ascent ground track is over land for a limited period of time. With the general north-easterly ascent paths shown in Figure 5 the only exposure to population and population centers occurs in the beach areas. Experience at this location has indicated that under normal weather conditions the uncertainty associated with ascent trajectories is not significantly increased once weather forecasts are within a T-1 day window. As a general rule, therefore, launch go/no-go decisions are made on the basis of T-1 day forecasts.

The ground tracks shown in Figure 5 can be considered "ideal." The movement of weather fronts and anomalies in seasonal wind patterns, among many other factors, combine to produce ascent trajectories which are likely to over-fly population centers. Because the over-flight can affect many population centers, and the over-flight tracks can be disparate over a launch window, the strict use of hazard containment boundaries can lead to mission delays even when actual risks are tolerable.

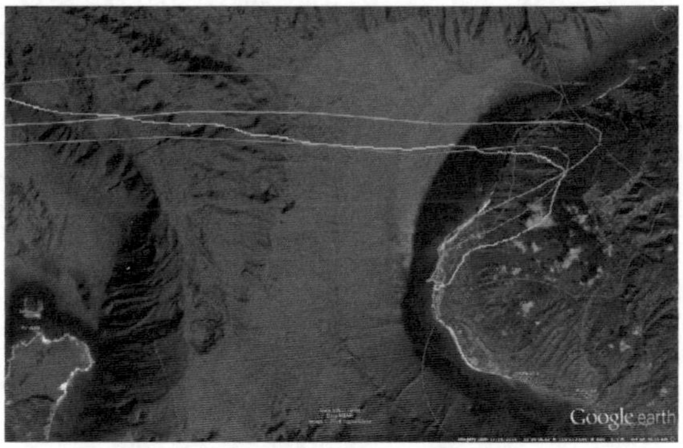

Figure 5. LDSD balloon ground tracks.

3. PROBABILISTIC RISK ANALYSIS USING META-RRAT

The time "parameterization" requirement identified in the previous section was implemented using a tool called *Meta-RRAT*, a risk analysis program for complex mission scenarios developed by ACTA Inc. *Meta-RRAT* is designed to allow the organization and execution of one or more sets of risk analyses which have been parameterized by a number of possible mission parameters, such as failure and break-up modes.

Meta-RRAT essentially functions as a high-level scripting language in which an arbitrary number of single risk analysis can be performed, and the risks combined in various ways such as weighting by individual analysis or union-combining to ascertain the maximum risk.

Meta-RRAT analyses include the following three types of mission scenarios. First, it may be used when there exists very limited data to describe the vehicle performance and malfunction behaviours. The "simplification" is that the vehicle malfunction probabilities and associated response are not realistically modelled. The input break-up state vectors represent only very general trajectory misbehaviour and the probabilities of each type of malfunction are significantly over-estimated (perhaps 100% for each). In this type of situation it is often desirable to consider the maximum risk from any type of malfunction. This can be particularly useful early in mission planning to provide indications of potential risk issues.

The second situation is when a wide range of trajectories is possible during mission planning. An example is the launch of an interceptor toward a target. During planning there is typically a large engagement volume where the intercept can occur. When the mission actually occurs, between the times of target and interceptor launch, the intercept location is determined with uncertainty much smaller than the whole intercept region. During mission planning it is therefore appropriate to study the range of trajectories that reach the volume. By performing a *Meta-RRAT* analysis to examine the risks of all these trajectories a range safety office can approve launch for any engagements within the planned intercept box.

A third situation is to perform basic sensitivity analyses in order to compute uncertainty on E_c and for risk profiles. Here, several of the internal computations and results of each baseline analysis are randomly adjusted to account for uncertainty in the risk models. With a standard input set for a given vehicle, the risk engine will produce a single mission casualty expectation, E_c, or risk profile. This value is a point estimate of E_c or a risk profile. In order to obtain the uncertainty about this best estimate value requires a set of results where each corresponds to a randomly perturbed input data set.

The LDSD mission constitutes a fourth application, one which seems undocumented in the literature. One can think of the balloon state vectors at failure as defining a rising and expanding "tube" which envelopes the ascent space. The state vectors are based on the balloon ascent dynamics and the T-1 day wind forecasts. Break-up times are randomly sampled over the duration of the ascent phase. The resulting "termination points" are mapped to debris which is then propagated to the ground, using the same forecasted wind field, and used for hazard assessment. Both the break-up list parameters and the wind field incorporate uncertainties.

This new methodology yields time-dependent estimates of the individual and collective risks, and the "evolution" of the debris footprints on the ground. This information was invaluable in order to unambiguously establish the periods in the mission which contributed the most to the risk, and whether, for example, the relative risks during those times in comparison to the total mission risk was excessively concentrated. The next section will illustrate this fouth methodology.

4. EXAMPLE ANALYSIS

The first step in the risk assessment commingles the forecasted wind profile obtained from the Global Forecast System (GFS) with the JPL wind uncertainty model in a form usable by *Meta-RRAT* [4, 11]. GFS is a weather forecast model produced by the National Centers for Environmental Prediction (NCEP). It is run four times per day, with forecasts up to sixteen days ahead. JPL has empirically found that (in the absence of weather systems or local anomalies) that ascent uncertainty is not greatly improved once inside a twenty four hour window to launch. An example of a standard GFS forecast obtained for T-1 day is shown in Figure 6.

As can be inferred from Figure 6, only a mean forecast is provided. An algorithm was developed to incorporate the JPL 3-σ uncertainty model for the PMRF location in order to generate the appropriate covariance matrices to model local wind ensembles, with typical results for each step shown in Figure 7 and Figure 8.

P(mb)	HT(FT)	DRCT	SPD(KTS)	T(C)	DP(C)
1000	422	356	6	22	17
975	1140	356	6	21	16
950	1875	355	6	19	15
925	2624	353	7	16	14
900	3389	352	7	15	12
875	4172	358	7	14	8
850	4974	8	7	13	4
825	5797	15	7	12	3
800	6642	17	8	11	3
750	8401	6	10	8	3
700	10261	325	8	5	-3
650	12242	289	11	4	-15
600	14367	290	14	2	-26
550	16649	289	21	-2	-31
		(lines omitted)			
100	54006	297	48	-72	-82
70	61000	294	15	-66	-84
50	67790	78	4	-61	-87
30	78288	77	20	-56	-88
20	86844	95	22	-51	-89
10	101974	117	19	-40	
7	110035	122	9	-36	
5	117797	112	15	-30	
3	129982	96	30	-21	
2	139885	95	45	-17	
1	156996	93	59	-15	
END					

Figure 6. Example of an "as received" GFS forecast.

Summary Listing of Processed GFS Data

No	P (Pa)	H (m)	DRT	x (m/sec)	T (C)	3σ EW (m/s)	3σ NS (m/s)
1	100000.0	128.6	356.0	3.08	22.00	6.82	4.63
2	97500.0	347.5	356.0	3.08	21.00	6.01	4.74
3	95000.0	571.5	355.0	3.08	19.00	5.20	4.84
4	92500.0	799.8	353.0	3.60	16.00	4.39	4.95
5	90000.0	1033.0	352.0	3.60	15.00	5.02	5.40
6	87500.0	1271.6	358.0	3.60	14.00	5.64	5.85
7	85000.0	1516.1	8.0	3.60	13.00	6.27	6.29
8	82500.0	1766.9	15.0	3.60	12.00	6.63	6.21
9	80000.0	2024.5	17.0	4.11	11.00	6.99	6.13
10	75000.0	2560.6	6.0	5.14	8.00	7.70	5.97
11	70000.0	3127.6	325.0	4.11	5.00	8.42	5.81
12	65000.0	3731.4	289.0	5.65	4.00	8.25	6.05
13	60000.0	4379.1	290.0	7.20	2.00	8.08	6.29
14	55000.0	5074.6	289.0	10.79	-2.00	7.90	6.52
15	50000.0	5824.4	286.0	15.42	-7.00	7.73	6.76
16	45000.0	6636.4	291.0	18.50	-13.00	8.62	7.10
17	40000.0	7523.4	305.0	21.07	-19.00	9.51	7.44
18	37500.0	8001.3	309.0	24.16	-22.00	9.70	8.30
19	35000.0	8505.4	309.0	26.73	-26.00	9.88	9.17
20	32500.0	9037.9	306.0	28.78	-30.00	10.07	10.04
21	30000.0	9602.1	303.0	29.81	-35.00	10.26	10.90
22	27500.0	10202.6	304.0	29.81	-40.00	11.82	11.36
23	25000.0	10845.1	314.0	29.30	-46.00	13.38	11.81
24	22500.0	11537.3	332.0	31.87	-51.00	13.63	11.07
25	20000.0	12292.3	343.0	35.98	-57.00	13.89	10.32

Figure 7. Processed GFS T-1 day forecast, including 3-σ uncertainty, used for wind covariance matrices.

The analogue to the standard nominal trajectory ensemble—namely an ensemble of ascent trajectories—was simulated by JPL, using models for the balloon, environment, wind forecasts, along with uncertainty models for each aspect of the physics. The simulations were performed with SINBAD (Scientific Balloon Analysis Model), as discussed in Refs. [3, 4, 18]. The resulting model was used to simulate ascent trajectories to failure, resulting in a file consisting of several thousand randomly spaced (in time) points called termination points, defined by the time since launch and the predicted altitude at failure. Each termination point is a balloon state vector at conditional balloon failure, defined by the geodetic position, altitude, and the three components of the mean "free stream" velocity. The reliability of the balloon was considered constant over the ascent, with a value of about 6%, consistent with historical data for these types of balloons [10, 12].

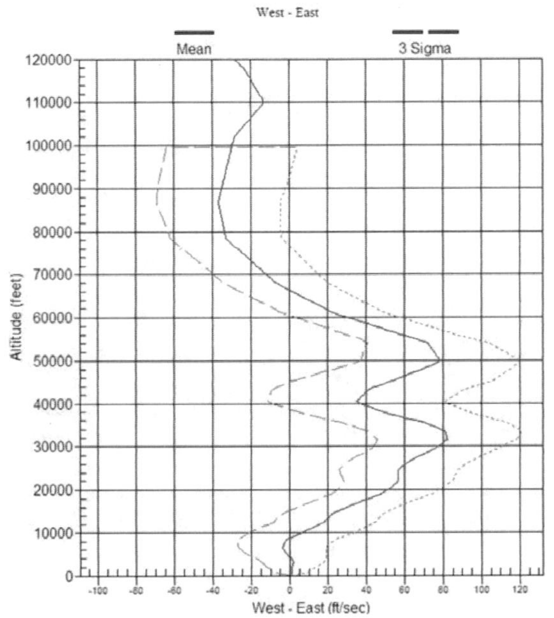

Figure 8. Final processed GFS forecast with 3-σ uncertainty in RRAT (only the W–E component is shown here).

The resulting ensemble of termination points was segregated into a set of contiguous windows, five minute each in length. Depending on final termination time, these parameters resulting in between 30 and 40 separate risk analyses. A program was developed to aggregate, process, and partition the data into the appropriate input format.

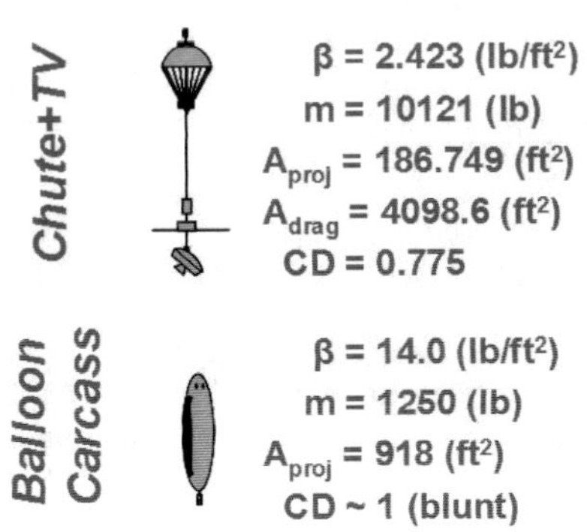

Figure 9. Basic modelling parameters for the recovery parachute and test vehicle, and the balloon carcass.

Figure 10 shows probability of impact contours when all the termination states are considered for the entire ascent, in this case thirty five minutes. One can infer from this depiction that the launch initially proceeds off-shore and then drifts south and, at some time, turns eastward along the coast. While individual population centers can be interrogated for total mission risk, neither such information nor Figure 10 gives any indication regarding how much time the balloon spends over any location nor where in time the risk is concentrated. Situations can and do occur when the balloon loiters as it ascends, "building up overflight risk", or conversely when it sweeps rapidly over large populations and/or critical infrastructure.

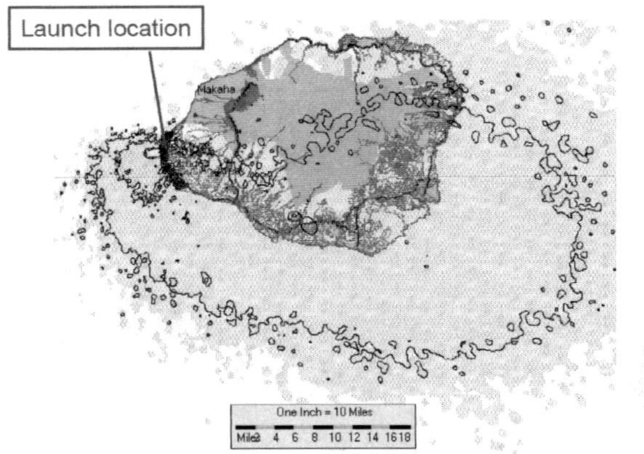

Figure 10. Probability of impact contours for all SINBAD Monte Carlo termination states during ascent, calculated by Meta-RRAT.

A reason why such information may be of great value for go/no-go decisions can be ascertained by examining the total mission risks for this particular case, as shown below.

MISSION E_C = 1.73E-04 = (173 / 1 million)
MISSION E_F = 1.57E-04 = (157 / 1 million)

These aggregated risks are typically the chief concern for rocket missions, given a rocket's short flight time. Note that as mentioned earlier there is little distinction between the fatality risk and casualty risk for balloons, since most of the debris is very massive or, in the case of the balloon carcass, has great hazard area (see Figure 9).

Mission decisions from a risk management perspective normally are considered relative to mandatory criteria published in RCC 321-10 [20], and for these risks the acceptance criteria for the general public are as follows:

RCC Criteria: E_c = 0.000100 (100 / 1 million)
RCC Criteria: E_f = 0.000030 (30 /1 million)

Since E_f and E_c are essentially the same, the acceptance is driven by fatality risk, which in this case is clearly excessive (157 versus 30). For this example, the exposure of the population centers along the southern coast and the populated south intuitively contribute to the accumulated risk. However Figure 10 provides no information at all regarding when during the ascent the risk is concentrated.

This type of crucial information is provided by the new methodology based on time-parameterization of the mission using *Meta-RRAT*. The actual time-dependent risks from which the mission E_f and E_c were obtained are shown in Figure 11. Figure 11 indicates that risk does not begin accumulating until well after five minutes into the launch.

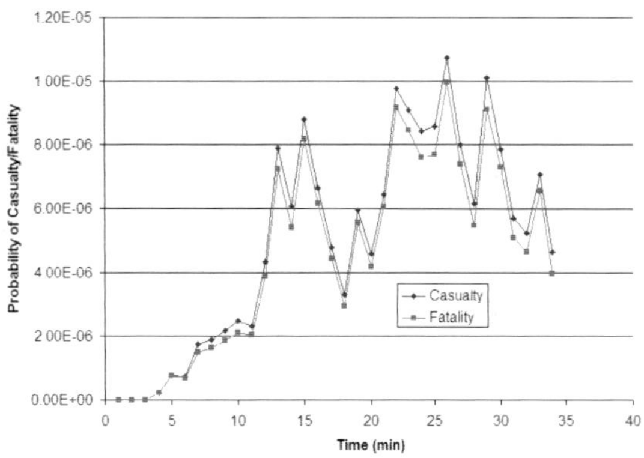

Figure 11. Time-dependent casualty and fatality risks obtained using a five minute window.

The methodology also enables quantitative comparisons between ascent situations under different weather patterns to be made, by examining temporal trends, as shown in Figure 12. Here, two different days (green and magenta) exhibit maximum risk during the initial launch phase, while the third (in blue) has a similar maximum risk but a much slower decay rate.

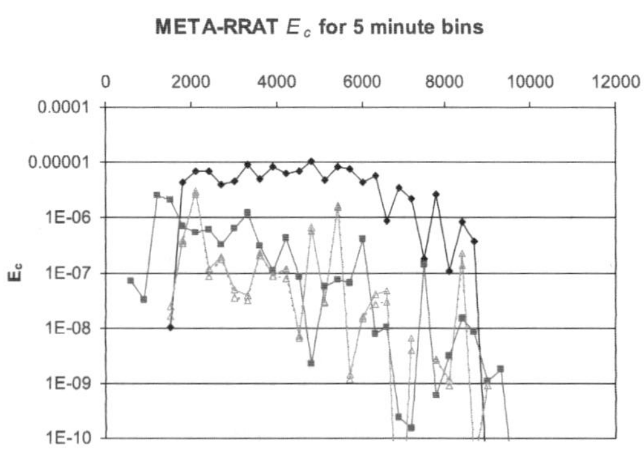

Figure 12. Comparison of risk evolution across ascent time for three different mission scenarios.

Fatality information at this level of time/spatial fidelity is particularly important in identifying potential situations which could "fly under the radar" on a total-mission basis but which could pose a catastrophic hazard, if only for a relatively short amount of time. Such situations could and have been observed to occur when the time-dependent footprints sweep across large crowds, such as invited observers and popular beach areas, or critical government and public infrastructure. It is also important to realise the potential for balloon failures (leaks, for example) to occur at any time, necessitating flight termination decisions to be instigated immediately. Knowledge of the time-evolution of the risk footprint provides additional information for screening the best time or times for bringing down the balloon.

5. COMMENTS AND CONCLUSIONS

Risk analysis for long duration balloon missions is seen to be characterized by mission times that can exceed several hours or days, a feature that is coupled with ascent uncertainties which manifest themselves as trajectories which exhibit features of a "random walk", namely spatial variance increasing as some function of time. These two features stand in marked contrast to more typical rocket-based missions, whose durations are limited to hundreds of seconds (and usually much less), and which have impact distributions dominated by descent uncertainty, especially for the smaller unplanned debris. And, unlike the same for rockets, balloon risk analysis must be conducted within a T-1 day window, in order to use accurate wind and environmental forecasts, since that dominates ascent uncertainty.

The methodology described in the paper builds on traditional and accepted practices refined for shorter missions. What it accomplishes is a time-parameterization of the risk over a lengthy mission, so that launch phase and subsequent FTS decisions can be based on detailed population and critical infrastructure exposures throughout the mission. This methodology moves the framework from a hazard containment approach to a risk-informed one, and has helped avoid excessive conservatism associated with the former.

The methodology uses an ensemble of termination points which are binned into discrete time windows. The failure modes associated with the termination points are mapped to break-up state vectors, and these are propagated to the ground through the correlated wind field in order to produce probability of impact and casualty/fatality probability distributions.

Apart from the generic risk modeling questions which arise in any mission, the primary question which the new analysis raises is the choice of window width. For an LDSD-type mission, with a maximum mission time expected not to exceed four hours, a window width of five minutes was found to yield excellent characterization of time-risk profiles consistent with distribution of population centers. This recommendation is based on analysis of several different day-of-launch weather profiles.

It should be noted that this recommendation is conditioned on the ensemble wind and overall weather conditions for this mission locale, namely Kauai, and for well-behaved weather patterns. Application to other locales will need to verify the best choice of window time.

The methodology was implemented in Meta-RRAT and optimized to enable throughput of day-of-launch winds into time and location dependent go/no-go decision criteria within an hour.

6. ACKNOWLEDGEMENTS

The assistance of JPL's LDSD 2015 Campaign team, including discussions with Dr. Rob Manning, Mark Ivanov, and Mark Adler, were extremely helpful in refining, cross-validating, and applying the methodology presented herein. Their assistance is greatly appreciated.

7. REFERENCES

1. Jones, W. V. (2005). *Evolution of Scientific Research Ballooning*, 29th International Cosmic Ray Conference, 10, (173–184).

2. Rainwater, E. L. & Smith, M. S. (2004). *Ultra high altitude balloons for medium-to-large payloads*, Advances in Space Research, 33(10), 1648–1652.

3. Clark, I. G., Adler, M., Manning, R., (2015). *Summary of the First High-Altitude, Supersonic Flight Dynamics Test for the Low-Density Supersonic Decelerator Project*, in proceedings: 23rd AIAA Aerodynamic Decelerator Systems Technology Conference Daytona Beach, Florida, March 30–April 2.

4. Randolph, T. M., Mullenax, R., Schwantes, C., Sell, S. W., Ball, D. R. J., (2015). *The First Balloon Flight of the Low Density Supersonic Decelerator Technology Demonstration Mission*, in proceedings: 2015 IEEE Aerospace Conference (New Technologies and Instruments for Scientific Balloon Missions), Big Sky, Montana, March 07–14.

5. Israel, M., et al. (2010). *NASA Stratospheric Balloons Science at the edge of Space Report of the Scientific Ballooning Assessment Group*, January.

6. Clark, I., (2016). *Summary of the Second, High-Altitude Supersonic Flight Dynamics Test for the LDSD Project*, in proceedings: 2016 IEEE Aerospace Conference, Big Sky, Montana, March 05–12.

7. Larson, E., Lloyd, G, (2010). *Application of Kernel Density Estimation to Impact Probability Density Determination for Risk Analysis*, Paper AIAA–2010–1541, 48th AIAA Aerospace Sciences Meeting Including the New Horizons Forum and Aerospace Exposition, Orlando, Florida.

8. NASA Range Safety Program, 2006 Annual Report, Wallops Flight Facility (WFF).

9. Range Safety Manual (RSM–2002), Goddard Space Flight Center (GSFC) / Wallops Flight Facility (WFF), Suborbital and Special Orbital Projects Directorate, June 28, 2002.

10. Noor, A. K., Venneri, S. L., (1997). *Future Aeronautical and Space Systems*, Progress in Aeronautics and Aeronautics, pg. 240.

11. Palumbo, R., Morani, G., & Corraro, F., (2011). *Balloon Trajectory Prediction Methodologies for the Unmanned Space Vehicles Programme*, in proceedings: 20th Symposium on European Rocket and Balloon Programmes and Related Research, Hyére, France, October.

12. Rotter, J., Marquez, M., and Orr, G., (2007). *Zero-Pressure Balloon Reliability Factors*, AIAA Balloon Systems Conference, Balloon Systems Conferences, Paper AIAA–2007–2614.

13. Hazlewood, K., (1981). *The Development of Plastic*, Zero Pressure Balloon Design Since 1945, Advances in Space Research, 11(1), (157–161).

14. Zhang, Y., Liu, D., (2015). *Influences of Initial Launch Conditions on Flight Performance of High Altitude Balloon Ascending Process*, Advances in Space Research, 56(4), (605–618).

15. Pankine, A., Li, Z., Parsons, D., Purucker, M., Weinstock, E., Wiscombe, W., Nock, K., (2009). *Stratospheric Satellites for Earth Observations*, BAMS, American Meteor-ological Society, August, (1–11).

16. Cathey, H. M., Pierce, D. L., (2007). *The Development of the NASA Ultra-Long Duration Balloon*, Paper C3P3, NASA Science Technology Conference.

17. *National Aeronautics and Space Administration Low Density Supersonic Decelerator Technology Demonstration Mission Pacific Missile Range Facility, Supplemental Environmental Assessment*, February, 2015.

18. Pankine, A. A., Heun, M. K., Nguyen, N., Schlaifer, R. S. (2005). *NAVAJO: Advanced Software Tool for Balloon Performance Simulation*, Paper AIAA 2005–7411, in proceedings: AIAA 5th Aviation, Technology, Integration, and Operations Conference (ATIO), 26–28 September, Arlington, Virginia.

19. Beyers, N. J., Miers, B. T. (1970). *Measurements from a Zero-pressure Balloon in the Stratopause* (48 km), J. Geophysical Research, 75, (3513–3522).

20. RCC 321-10, (2010). *Common Risk Criteria Standards for National Test Ranges, Range Commanders Council Range Safety Group.*

NEAR RANGE SAFETY ANALYSIS FOR A REUSABLE LAUNCHER CONCEPT BASED ON TOSS-BACK

A. Martinez Torio[1], J.M. Bahu[1], D. Delorme[1], V. Guenard[1], H. Poussin[2]

[1] *CNES Launchers Directorate, 52, rue Jacques Hillairet, 75012 Paris (France),*
Email: alexandra.martineztorio@cnes.fr, jean-marc.bahu@cnes.fr, david.delorme@cnes.fr, vanessa.guenard@cnes.fr

[2] *CNES Guiana Space Center, Centre Spatial Guyanais, 97387 Kourou (French Guiana),*
Email: herve.poussin@cnes.fr

ABSTRACT

The last decade saw the emergence of new private actors stiffening space launch market competition. In order to face this competition, in addition to the development of Ariane 6 low-cost launcher, European agencies and industry are investigating the introduction of low-cost propulsion technologies and partly-reusable launcher concepts. This paper deals with one of the concepts among those which are traded-off by CNES Launchers Directorate: a LOX/CH4 two-stage partially reusable vehicle, with a main stage return phase performed by a Toss-Back maneuver.

In particular, this article provides an overview of the various flight safety issues associated to such a concept for the French Guiana protection. Then, considering that this concept is a brand new one with respect to those operated in French Guiana, and without anticipating on any formal acceptability in terms of compliance to the French Space Operations Act, a list of possible risk reduction measures has been identified.

1. INTRODUCTION

This article presents an overview of the various flight safety issues associated to a LOX/CH4 two-stage partially reusable launcher by Toss-Back, operated from Guiana Space Center.

First a representative model of debris footprint in case of launcher neutralization, adapted to such launcher characteristics, has been developed. In order to illustrate the accuracy of such a model, a dedicated restitution of the footprint observed after recent LOX-hydrocarbure launcher failures over the world has been led.

Then, a near range safety assessment for both the ascent and the return trajectory of the aforementioned concept has been performed. The successive steps of this analysis were: a RAMS analysis of the launcher possible failures, an identification of the worst case scenarios in terms of explosion and neutralization events for both trajectories, and an assessment of the debris fallout footprint extents produced.

Finally, considering that such a concept is a brand new one with respect to those operated in French Guiana, and without anticipating on any formal acceptability in terms of conformance to the French Space Operations Act, a list of possible risk reduction measures to be considered has been identified: the introduction of launch azimuth constraints for the ascent phase, the definition of adapted neutralization criteria and neutralization means in order to limit the size of the fallout footprints, the increase of the distance between landing site and the protected areas (barge-platform trade-off), the improvement of the risky subsystems reliability, etc.

2. CONTEXT

The last decade saw the emergence of new private actors stiffening space launch market competition. In order to face this competition, in addition to the development of Ariane 6 low-cost launcher, European agencies and industry are investigating different launcher conception approaches, following two main guidelines:

- Launcher intrinsic simplification: aiming at reducing the number of engine and propellant types used and at maximizing synergies between boosters and central body and between lower and upper stages.

The introduction of hydrocarbon-based propellants and low-cost propulsion technologies allows a simpler engine conception and thus, lower development and production costs.

Taking into account the higher technological maturity level of Methane with respect to Kerosene in Europe, LOX/CH4 seems to be a promising propellant for a reusable stage. In this frame, a technological demonstrator, called PROMETHEE (Precursor Reusable Oxygen METHane cost Effective Engine), aims at consolidating the future LOX/CH4 low-cost reusable engine [1].

- Launcher reusability: limited to lower stages (boosters or central body) due to accessibility constraints. This strategy is being studied for different types of return maneuvers:

- Fly-back: consisting of a maneuver performed on a winged aircraft powered vehicle to return to the launch site [2].

- Toss-Back (or Boost-Back): consisting of a maneuver performed, after the booster/1st stage jettisoning, by the main propulsion system, which must be reignitable and throttable. This maneuver enables to obtain a return ballistic trajectory with an impact point near the landing site, to brake and reduce dynamic pressure and to land vertically either on the launch site or on a barge [2].

  An alternative maneuver called Dead-Leaf would consist of a winged version of the Toss-Back. After the first boost, which purpose is to cancel the horizontal velocity vector, the return phase is performed by gliding back (without using propulsion), in order to land horizontally on the landing pad [2].

  Another return strategy derived from the Toss-back maneuver is called ballistic arc: consisting of a free ballistic arc trajectory after booster stage separation and a final landing on a barge.

Since 20 years, CNES Launchers Directorate, in cooperation with European agencies and industry, has studied these different return strategies in the frame of projects such as Flex (Future Launcher Experiment), Aldebaran, Bargouzin, FLPP, etc. Nevertheless a revisit of such studies and a complete trade-off remain necessary as market targets and launcher technologies have evolved.

Therefore, the aforementioned launcher concepts are currently being traded-off by CNES in the framework of a national program and in the framework of the ESA program LEE (Launcher Evolution Element). By on- ground and, eventually, in-flight demonstration, the critical technologies will be matured, and experience on recovery, refurbishment and reuse of a reusable stage will be acquired.

Among these launcher concepts, this paper deals in particular with a LOX/CH4 two-stage partially reusable vehicle, with a main stage return phase close to Ariane 5 launch pad, performed by a Toss-Back maneuver.

Lower stage (reusable): C400 composed of 7 engines. Thrust level 1000kN approx.

Dimensions: 30m height, 5.4m of diameter

Upper stage: C91 composed of 1 engine derived from the lower stage.

Figure 1 LOX/CH4 two-stage partially reusable launcher configuration and staging

The ascent and return trajectories, as well as its staging (C400-C91) have been optimized in order to achieve a 5t performance for an 800km SSO mission.

The resulting return trajectory consists of 3 boosted maneuvers: a first boost after stage separation to reverse the velocity vector and to point it to the landing site, a second one to brake and reduce the impact of general loads and dynamic pressure during the fallout and a third boost controls the final vertical landing.

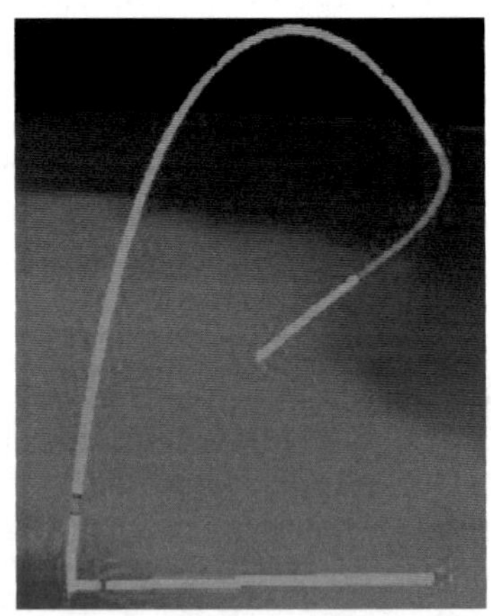

Figure 2 Return trajectory

Since a partly-reusable launcher, based on first stage Toss-back, is a brand new concept with respect to those operated from the Guiana Space Center to date, and so as to make a judgment on its feasibility, a preliminary dedicated study has been led in order to give rise to the safety issues involved.

3. FRENCH SPACE OPERATIONS ACT

Since 2010, the French Space Operations Act (FSOA – LOS) governs the space activities carried out by private entities. Its purpose is to set up a coherent national regime of authorization and verification of space operations in order to protect people, properties, public health and the environment (including space orbit), in accordance with the following principles, issued from ratified UN treaties [3]:

- Obligation for the French Government to authorize and to check national space activities carried out by non- governmental entities (Outer Space Treaty 1967)
- Absolute liability for damage caused on the surface of the Earth or in the airspace (Liability Convention 1972)
- Obligation to register the Space Object when France is the appropriate Launching State (Convention on registration of Objects launched into Outer Space 1975)

Particularly, Guiana Space Center Safety Regulations Decree (REI-GSC) sets the safety rules applicable to GSC exploitation: on ground transportation of space objects, tests and operations performed at GSC, as well as near range in-flight operations [3].

4. NEAR RANGE FLIGHT SAFETY

For a given mission, in order to verify its compliance with respect to the REI requirements, near-range flight safety studies should be carried out in order to ensure the safety of life and property in French Guiana from thermal flow, overpressure, projection of debris and toxicity caused by Launch System operations and failures.

In particular, the near range safety aims at equipping the launch system with intervention means (operated by the GSC) in order to neutralize the launcher in case of hazard for the population and property.

The safety neutralization criteria are established according to a deterministic approach, where no casualty risk is tolerated. Besides, the launch system should respect a Fail-Safe/Fail-Safe criterion, i.e. all safety means involved should be tolerant to a double system failure without entailing any danger.

In order to achieve this objective, protection limits (Fig. 3) are defined in order to preserve protected areas from different kinds of risks: mechanical, toxic, thermal flow and overpressure.

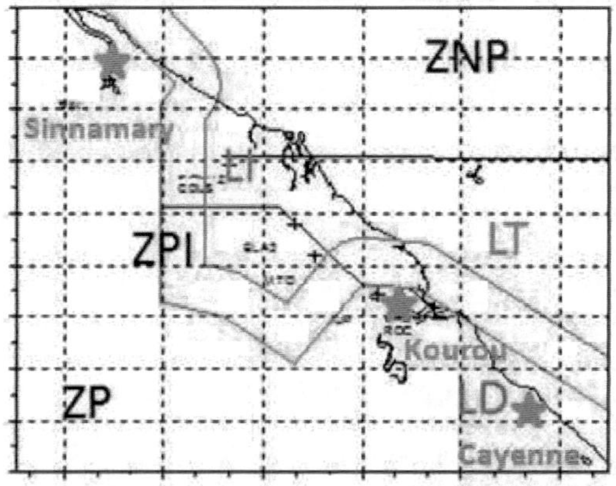

Figure 3 Toxicity and impact protection limits of GSC[5]

5. NEAR RANGE FLIGHT SAFETY PRELIMINARY STUDY FOR REUSABLE LAUNCHERS

The ultimate objective of the present study is to identify the safety issues involved and to define a list of possible risk reduction measures, without anticipating on any formal acceptability of such a concept in terms of conformance to the French Space Operations Act

The results will contribute to trade-offs on:

- The required safety systems and neutralization criteria (possible combination of):
  - Neutralization based on destruction by activation of pyrotechnic cords on receipt of an order from the ground as decided by the safety operators.
  - Neutralization based on propulsion cutoff commanded by the safety operators from the ground.
  - On-board subsystem (commanding neutralization based on destruction by activation of pyrotechnic cords or propulsion cutoff).
- The landing site:
  - Close or not to the launch pad: usually closer to the protected areas.
  - Alternatively, a sea barge: further from populated areas
- The return strategy:
  - Toss-Back, or
  - Fly-Back

5.1 Flight Safety Issues

Safety issues related to the space center:

The European spaceport, the Guiana Space Center, is placed in French Guiana, close to the equator, between two main cities, Kourou and Sinnamary. The advantage of having an uninhabited territory to the east allows nominal stages fallback over open sea, nullifying the casualty risk of a launch.

Nevertheless, the proximity of the cities of Kourou, Sinnamary and Cayenne (Fig. 3), implies the need of a neutralization system and a safety strategy in order to protect population and properties in case of trajectory deviation.

Safety issues related to a Toss-Back mission:

- Ascent phase:

The trajectory optimization process maximizes the targeted performance while minimizing its cost in terms of propellant.

A return phase requires the enchainment of ballistic phases and boosted maneuvers in order to reverse the velocity vector of the separated stage, to approach its impact point to the final landing site, to reduce the dynamic pressure and to finally brake in order to vertically land at almost zero velocity.

So as to minimize the propellant cost of the return maneuver, the distance of the impact point and the landing site at stage separation should be reduced.

The resulting optimized ascent trajectory is characterized by a steep verticality. As a consequence, compared to a standard expendable SSO mission, the launcher reaches higher altitudes before its impact point clears off the launch pad area. In case of accidental explosion or neutralization, the impact footprint will be larger and will remain closer to the launch site, and thus, the protected areas, becoming a potential hazard for property and population.

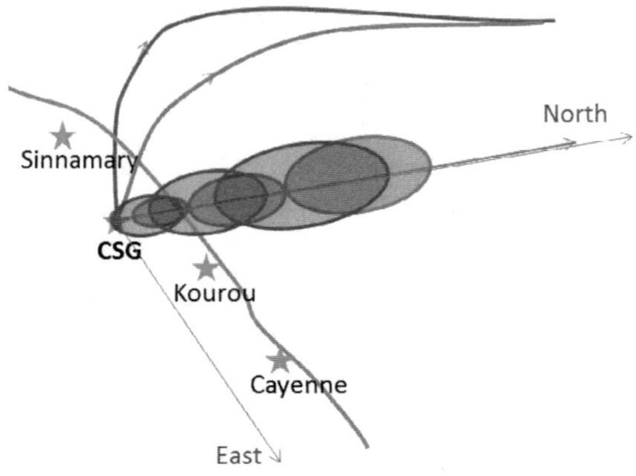

Figure 4 Qualitative representation of debris impact footprints during the ascent trajectory of a Toss-Back mission (blue) compared to a standard trajectory (red)

- Return phase:

On the other hand, a nominal return trajectory naturally flies land areas during its descent to the landing site. Without any specific constraint, in case of accidental explosion or neutralization, the stage debris may fall over protected areas. Especially at the beginning of the return maneuver (end of first return boost), where the launcher is at higher altitudes, debris impact footprint extents will be larger.

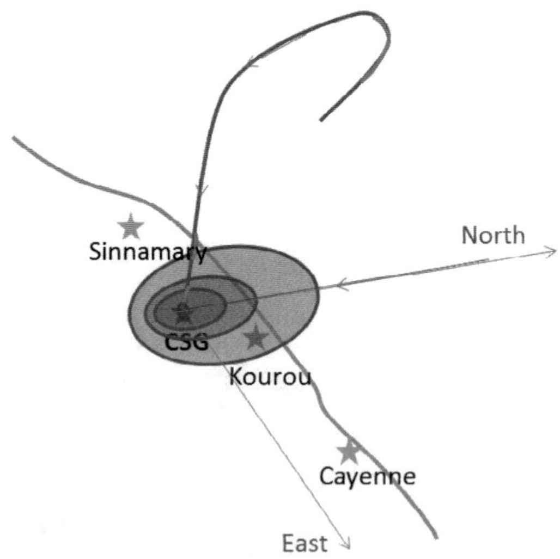

Figure 5 Qualitative representation of debris impact footprints at return boost extinctions

By controlling the ascent trajectory vertical profile, the launch azimuth and the number and duration of the boosts needed for the return maneuver, these safety issues can be mitigated.

However, these solutions would imply a reduction of performance. Therefore, an iterative loop of launcher concept, trajectory design and staging would potentially be needed, in order to take into account the safety constraints.

5.2 RAMS Analysis

In order to properly assess the casualty risks and to determine the worst case scenarios, an exhaustive hazard assessment study is required.

This analysis consists of a description of the possible hazards involved during each phase of the ascent and return trajectories, in both nominal and degraded cases, regardless of the origin of the failure.

- Ascent phase:

  During the atmospheric ascent phase, in case of failure at GNC chain or propulsive level, a deviation of the trajectory can be expected. In this case, neutralization (pyrotechnical cords activation or propulsion cut-off) can be commanded by the safety operators from the ground or automatically triggered onboard.

  Some feared events, like an engine explosion or a fire of the stage, as well as a structural rupture, can lead to a launcher explosion.

- Return phase:

  In case of a failure at GNC or propulsion level, that may lead to a deviation of the trajectory, during the boosted phases, a commanded or an automatic neutralization, based on explosion or on propulsion cutoff, would be required.

  Nevertheless, a neutralization system based on explosion may not be feasible regarding the impact footprint size at the end of 1st boost (when the altitude of the trajectory reaches its higher point) or afterwards as the impact point flies over the protected areas. Therefore, a neutralization based on propulsion cutoff appears to be the most suitable option.

  An engine explosion or a fire of the stage, as well as a structural rupture (especially at maximal dynamic pressure) can lead to a launcher explosion on its nominal trajectory.

  During the ballistic phase, at the moment of the atmospheric reentry, the mechanical and thermal (heat flux over its maximum level) conditions are also likely to lead to a structural rupture of the stage, with an explosion as a consequence.

  Nevertheless, the most critical instants of the return maneuver are the boosted phases and in particular, the ignitions and extinctions of the stage.

In case of neutralization on a deviated trajectory or explosion on a nominal trajectory, a verification of the debris impact footprint with respect to the protected areas is required.

The neutralization procedure can be based on explosion by activation of the pyrotechnical cords or through engine cutoff. The latter case can result in an explosion during the atmospheric fallout of the stage, as a consequence of the loss of structural integrity at propellant tank level (caused by the general loads suffered during fallout) leading to the mixture of fuel and oxidizer in a non-confined space. These scenarios cover the case of stage explosion once impacting on ground.

The representativeness of both scenarios following a propulsion cutoff depends on the structural stability and integrity of the vehicle, as well as the dynamic loads encountered during the fallout.

Regardless of the neutralization system (through explosion by activation of the pyrotechnical cords or through engine cutoff), it appears difficult, in the frame of preliminary analysis, to exclude accidental explosion of the vehicle, which is the worst hazard to be managed.

5.3 Fragmentation and Explosion Models Customization

Once the worst case scenarios have been identified, the hypotheses of the study are established, and the safety models for fragmentation and explosion can be customized.

Two sets of models have been prepared: a "conservative" one, taking into account worst-case scenario hypotheses and margins to cover uncertainties on the launcher architecture and failure scenarios; and a "realistic" one, more representative of a LOX/CH4 reusable launcher but potentially not sizing with respect to all scenarios of failures. This latter "realistic" model was created by fitting its results (debris footprints) empirically to recent LOX-hydrocarbure launcher failures observations.

As a first step, in a preliminary study, the analysis is restricted to mechanical effects (toxicity is not considered).

5.4 Near Range Safety Analysis of the Trajectories

In order to verify the feasibility of a Toss-Back maneuver operated from the Guiana Space Center, the safety impact of the mission should be analyzed for the ascent and return trajectories. The purpose of this preliminary study has been to quantitatively assess the safety issues identified in chapter 3.1 and to evaluate the robustness of the current safety means and procedures set in GSC against the identified safety constraints. If required, a list of possible risk reduction measures will be proposed.

Ascent phase:

For the ascent phase, the methodology consists on identifying the launch azimuth constraints for a given verticality profile, aiming at determining the acceptable SSO range in order to ensure the protection of populated areas.

The work plan consists in using the launch azimuth as a parameter sweep from 55°/East to 95°/East for SSO missions. The same analyses have been carried out for a non-verticalized trajectory profile with a northern launch azimuth of 90°/East, so as to quantify the verticalization impact on the results.

Theses analyses have been led for explosion-based and engine cutoff neutralization scenarios, in order to make a judgment on the suitability of the neutralization system, for both the "realistic" and the "conservative" explosion and fragmentation models.

- Nominal trajectory

  As a first step, the nominal trajectories have been analyzed. The acceptability of a trajectory depends on the respect of the protected areas by the debris footprint extents resulting from a failure leading to an explosion or from neutralization at any time of the nominal trajectory.

  It has been noticed that the impact footprints extents in case of propulsion cutoff are smaller since the altitudes, at which an eventual explosion would arise, after a ballistic fallout phase, are certainly lower.

  Nonetheless, since it appears difficult, in the frame of preliminary analysis, to exclude an accidental explosion of the vehicle, the scenario of explosion is considered to be the most dimensioning case.

  Explosion scenario:

  For the "realistic" model, all the trajectories seem to be compliant with the respect of protected areas.

Fig. 6 depicts the impact footprints corresponding to different explosions along the trajectory for 55°/East (dark blue), 75°/East (orange) and 95°/East (pink). The different debris footprints do not represent a risk with respect to the protected areas at any moment. Thus, the complete range of launch azimuths is compliant with the safety requirements.

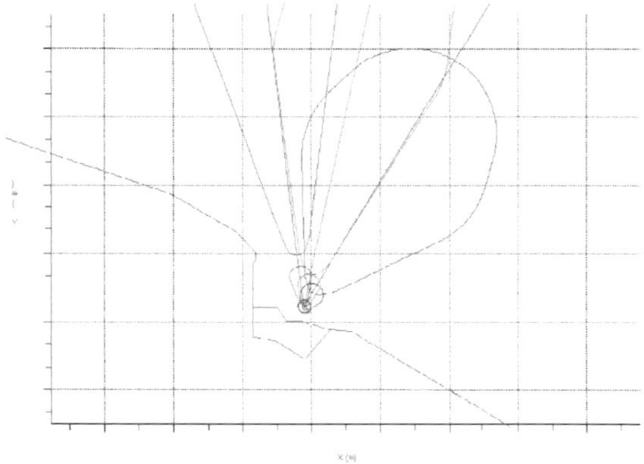

Figure 6 Debris impact footprints in case of explosion, for 55°/East (dark blue), 75°/East (orange) and 95°/East (pink), for the "realistic" model

For the "conservative" model, none of the trajectories are acceptable.

Fig. 7 depicts the impact footprints corresponding to different explosions along the trajectory for the complete range of azimuths. At each launch azimuth, there is a critical time interval where the debris footprints may impact over the protected areas.

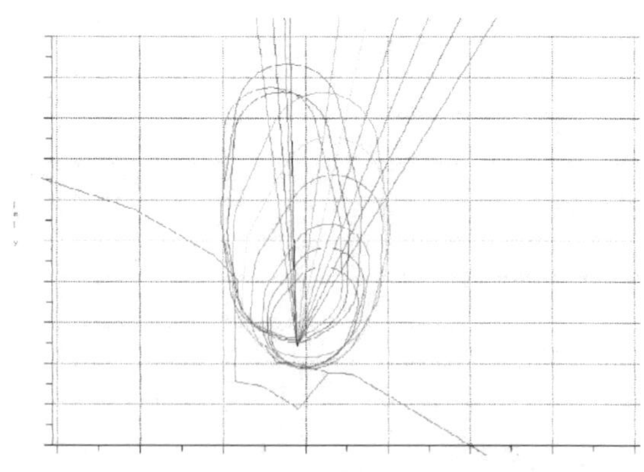

Figure 7 Debris impact footprints in case of explosion, for 55°/East (purple), 60°/East (dark blue), 65°/East (light blue), 70°/East (green), 75°/East (yellow), 80°/East (orange), 90°/East (red), 92°/East (dark red) and 95°/East (grey), for the "conservative" model

Compared to a non verticalized trajectory, in these trajectories the launcher reaches higher altitudes before its impact point clears off the launch pad area. As expected and depicted in Fig. 8, at the same time of explosion, the debris impact footprint of a verticalized trajectory is larger and remains closer to the launch pad area, whereas a non verticalized trajectory produces longer impact footprints as a result of the greater horizontal projection of the velocity vector.

For the "conservative" model, a non verticalized trajectory with launch azimuth of 90°/East is therefore acceptable in nominal case.

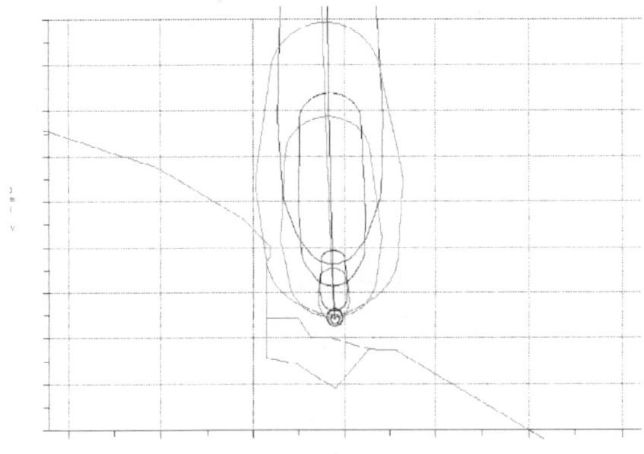

Figure 8 Debris impact footprints in case of explosion, for a verticalized trajectory at 90°/East (red) and a non-verticalized trajectory 90°/East (black), for the "conservative" model

- Deviated trajectories

Then, worst-case deviated trajectories are treated. In this case, the compliance of the trajectory depends on the time margin between the moment the launcher is declared non-nominal and the last instant of the deviated trajectory in which a neutralization represents no danger with respect to the protected areas. This time should be enough to allow an intervention by the ground operators or by an onboard algorithm.

Explosion scenario:

For the "realistic" model, Eastern trajectories appear to be more critical. Only trajectories which launch azimuth sweeps from 70°/East to 95°/East seem to be compliant with the respect of protected areas.

Therefore, for Eastern trajectories, and thus GTO missions, a decrease of the trajectory verticality with respect to the SSO reference trajectory is required in order to be able to perform a Toss-Back maneuver without implying any danger to the population of Kourou.

The results obtained through the "conservative" model show that none of the analyzed launch azimuths are acceptable with respect to the safety requirements. However, the "realistic" one shows that the Northern trajectories may be admissible.

Hence, so as to foreseen a verticalized Toss-Back mission, operated from the GSC, a refining of the existing explosion and fragmentation models has to be led when launcher architecture and failure scenarios will be better defined, in order to reduce the potential conservatisms and to obtain results closer to the observations.

Propulsion cutoff scenario:

Impact footprints extents in case of propulsion cutoff are less important since the altitudes, at which an eventual explosion would arise, after a ballistic fallout phase, are certainly lower.

Therefore, for the "realistic" model, all the trajectories seem to be compliant with the respect of protected areas.

The choice of a neutralization system based on engine cutoff seems to be more favorable provided that the structural stability of the vehicle is so as to avoid any rupture at the beginning of the ballistic fallout.

A neutralization based on propulsion cutoff could avoid considering an explosion on a deviated trajectory, and thus, all launch azimuths could become acceptable.

Return phase:

The compliance of the return trajectory with respect to the safety requirements depends on the respect of the protected areas by the debris footprint extents resulting from an eventual failure leading to an explosion or from neutralization at any time of the descent.

The objective of this preliminary analysis is to have an idea of the debris footprint size and evolution during the return trajectory to the landing site.

Special attention is paid to the most critical instants, from a RAMS point of view, along the descent: boost ignitions and extinctions.

Theses analyses have been led for explosion-based and engine cutoff neutralization scenarios, in order to make a judgment on the suitability of the neutralization system, for the "realistic" model of fragmentation and explosion. Since the "conservative" model was not acceptable for the ascent phase, it has not been considered for the analysis of the return phase.

The scenario of accidental explosion is also considered.

<u>Explosion scenario:</u>

Debris impact footprint due to accidental explosions or neutralization based on explosion, at the end of the boost 1 are not compliant with the near range flight safety constraints.

Fig. 11 depicts the impact footprint corresponding to an explosion at the first boost extinction time, where the debris footprint impacts over the protected areas.

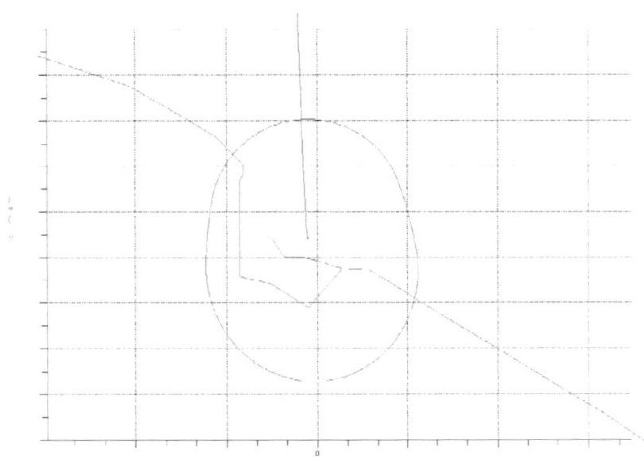

Figure 9 Debris impact footprints at boost 1 extinction

However, the size of the debris fallout footprints caused during boosts 2 and 3 seem to be acceptable with respect to the safety limits around the landing site.

<u>Propulsion cutoff scenario:</u>

Impact footprints extents in case of propulsion cutoff are less important since the altitudes, at which an eventual explosion would arise, after a ballistic fallout phase, are certainly lower.

A Monte-Carlo campaign, modeling the uncertainty on the instant of explosion during the stage fallout has been performed in order to calculate the envelope of the debris fallout footprints.

Even in case the stage only explodes once impacting on ground, the size of the fallout footprint at boost 1 extinction is not compliant with the flight safety requirements.

Nevertheless, boosts 2 and 3 seem to still be acceptable.

In any case, the choice of a neutralization system based on engine cutoff seems to be more favorable provided that the structural strength of the vehicle is so as to avoid any rupture at the beginning of the ballistic fallout.

Nonetheless, since it appears difficult, in the frame of preliminary analysis, to dismiss the hypothesis of an accidental explosion of the vehicle. The scenario of explosion is considered to be the most sizing case.

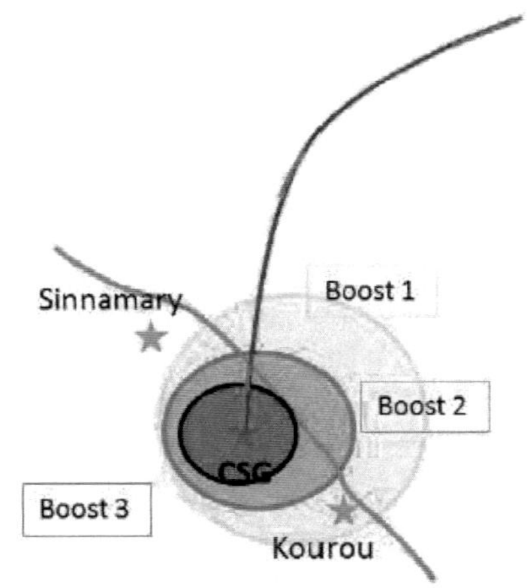

Figure 10 Debris impact footprint at boost 1 extinction falls back over protected areas. Boost 2 and 3 seem to be compliant with respect to the safety requirements

In order to make boost 1 acceptable with respect to the near range flight safety constraints, the distance between protected areas and the debris impact point at the end of the boost 1 should be increased. This can be done by shortening the first boost duration. An earlier extinction will prevent the debris fallout footprint from approaching the protected areas.

Such a boost 1 shortening will imply a need for trajectory re-optimization of the rest of the return phase, entailing an impact on the required performance, as boosts 2 and 3 will need to approach the vehicle from a further distance and will use more propellant.

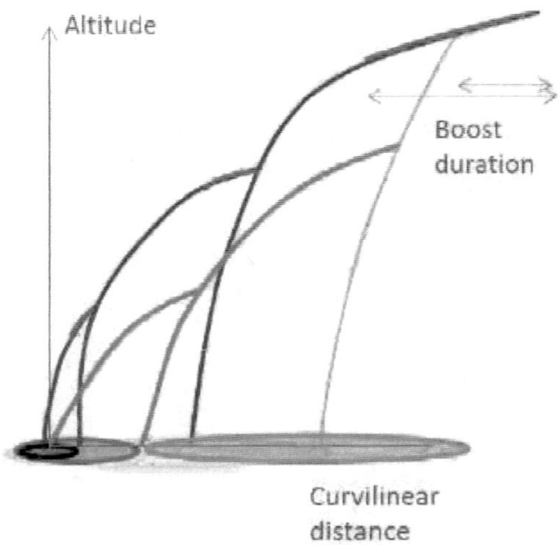

Figure 11 First boost duration shortening distances the impact footprint at extinction from the protected areas. Nevertheless, the re-optimization of boosts 2 and 3 implies a reduction of performance

Another solution would consist on distancing the landing site from the launch site and thus, from the protected areas. Landing on a sea barge would solve the aforementioned safety issues for the return phase. However, ground safety operations would be less efficient.

6. CONCLUSIONS

This article has presented an overview of the various flight safety issues associated to a LOX/CH4 two-stage partially reusable vehicle, with a main stage return phase performed by a Toss-Back maneuver, operated from the Guiana Space Center, for a 800km SSO mission.

A representative "realistic" model of debris footprint in case of launcher neutralization has been developed. Then, using this "realistic" model and a more "conservative" model taking into account uncertainties on launcher architecture and failure scenarios, a near range safety assessment for both the ascent and the return trajectory of the aforementioned concept has been performed, for explosion-based and engine cutoff neutralization scenarios.

The results obtained through the "conservative" model are not acceptable with respect to the safety requirements while the results are positive for most of the trajectories with the "realistic" model. Hence, an effort will be needed to refine the explosion and fragmentation models once the launcher will be better defined, in order to reduce the potential conservatisms and to obtain results closer to the observations.

It has been noticed that the impact footprints extents in case of propulsion cutoff are less important since the altitudes, at which an eventual explosion would arise, after a ballistic fallout phase, are certainly lower. Nonetheless, it appears difficult, in the frame of preliminary analysis, to exclude an accidental explosion of the vehicle.

The choice of a neutralization system based on engine cutoff seems to be more favorable provided that the structural strength of the vehicle is so as to avoid any rupture at the beginning of the ballistic fallout. A neutralization based of propulsion cutoff would avoid considering an explosion of a deviated trajectory.

For the ascent phase, the launch azimuth constraints have been identified. Eastern trajectories appear to be more critical. In order to make them acceptable, a decrease of the trajectory verticality with respect to the SSO reference would be required.

For the return phase, the boost duration constraints have been determined. By shortening the first boost duration, the trajectory will be compliant. Nevertheless, boosts 2 and 3 will need to be re-optimized in order to approach the vehicle from a further distance, as the impact point at the new first boost extinction would be further with respect to the initial one.

However, these solutions imply a reduction of performance.

Another solution would consist on distancing the landing site from the launch site and thus, from the protected areas (e.g. landing on a sea barge).

In order to determine a compromise between performance and safety compliance, iterative loops of launcher concept, staging and trajectory are to be performed taking into account the aforementioned constraints.

In order to consolidate the results of this preliminary study, once the launcher architecture design will be matured and failure scenarios will be better defined, an analysis of the safety impact of deviated trajectories during the return phase will be performed.

7. REFERENCES

1. Bonhomme, C., Palerm, S., Petitot, S., Louaas, E., Tchou-kien, D., Chopinet, J.N., Danous, P., Dreyer, S. (2015). The Future of Liquid Propulsion: French Activities'. IAC-15-C4.1.4

2. Baiocco, P., Bonnal, C. (2015). The Future for Technology Demonstration for Reusable Launchers. Planet. AA-5620, Acta Astronautica 120 (2016) 43–58

3. French Space Operations Act June 3rd 2008 - n°2008-518

4. Lazare B., The French Space Operations Act: Technical Regulations, Acta Astronautica, 92, Issue 2, 209-212, December 2013

5. REI: Decree regulating the operations of the Guiana Space centre facilities.

COMMON CAUSE FAILURE MODELING IN SPACE LAUNCH VEHICLES

Frank Hark[(1)], Rob Ring[(1)], Steven D. Novack[(1)], Paul Britton[(2)]

[(1)] *Bastion Technologies Incorporated, 17625 El Camino Real #330, TX 77058, USA,*
Emails: frank.hark@nasa.gov, Robert.r.ring@nasa.gov, steven.d.novack@nasa.gov,

[(2)] *NASA Marshall Space Flight Center, Huntsville, AL 35812, USA , Email: paul.t.britton@nasa.gov*

ABSTRACT

Common Cause Failures (CCFs) are a known and documented phenomenon that defeats system redundancy. CCFs are a set of dependent type of failures that can be caused for example by system environments, manufacturing, transportation, storage, maintenance, and assembly. Since there are many factors that contribute to CCFs, they can be reduced, but are difficult to eliminate entirely. Furthermore, failure databases sometimes fail to differentiate between independent and dependent CCF. Because common cause failure data is limited in the aerospace industry, the Probabilistic Risk Assessment (PRA) Team at Bastion Technology Inc. is estimating CCF risk using generic data collected by the Nuclear Regulatory Commission (NRC). Consequently, common cause risk estimates based on this database, when applied to other industry applications, are highly uncertain. Therefore, it is important to account for a range of values for independent and CCF risk and to communicate the uncertainty to decision makers.

There is an existing methodology for reducing CCF risk during design, which includes a checklist of 40+ factors grouped into eight categories. Using this checklist, an approach to produce a beta factor estimate is being investigated that quantitatively relates these factors. In this example, the checklist will be tailored to space launch vehicles, a quantitative approach will be described, and an example of the method will be presented.

1. INTRODUCTION

CCF is a known and documented phenomenon that can occur due to coupling factors that result in multiple dependent failures of identical components in a redundant design configuration. Consequently, if not understood, identified, and mitigated these factors limit the benefit of system redundancy as a design approach to achieve high reliability. Because of their extremely high cost, low launch rate, and national reputation, the public expects and demands reliable launch vehicle operation and mission success. To achieve high reliability, design engineers employ functional redundancy in the design to achieve reliability goals. The success of this design approach requires steps be taken during system design and throughout the system lifecycle to limit and reduce CCF. An important step toward this end is to implement a deliberate and documented procedure throughout the design, development, and operational life of the system to understand and mitigate coupling factors that can result in CCFs. Failure to actively pursue these steps may result in highly redundant configurations with added cost, complexity, and weight that also fail to achieve their reliability goals. Furthermore, reliability prediction methodologies that do not address CCFs significantly misrepresent the true reliability of a system that relies on redundancy.

Another related issue that is a source of uncertainty in predicting design reliability of launch vehicles is sparse data. Also, problem reporting databases, when they are implemented and maintained, typically do not record operating time and other information that makes it difficult to accurately predict system reliability. Even less common are documented proximate and root cause analyses to identify whether coupling factors may have contributed to the failure or precursor to failure. Our experience in reviewing failures and anomalies suggests that CCFs have often not been identified as such.

A generic common cause failure database maintained by the Nuclear Regulatory Agency is being used by the Bastion Probabilistic Risk Assessment (PRA) team to estimate the risk of CCF for launch vehicles. Without investigating details of the specific system, these generic CCF factors may grossly under estimate the magnitude of the risk if these estimates are not adjusted to reflect significant differences in other industry applications [1]. An accepted methodology for reducing CCF is a CCF checklist to help PRA analysts identify common cause coupling factors and to use the insights gained to improve the quantitative CCF estimate [2]. The checklist is an aid to judging the overall susceptibility of the system to CCF of coupling mechanisms with specific qualities of the system. As technology advances and new system designs achieve higher levels of reliability, it is imperative that procedures to reduce common cause are implemented early in the development cycle.

If steps are not taken to actively reduce CCFs through process, training, and design, the occurrence of CCFs may be significantly higher than the estimates documented in the NRC for the nuclear industry. According to a published paper [1], other industries have in fact experienced significantly higher occurrence of common cause failure.

There exists a need to link the magnitude of CCF risk to system qualities and to communicate the qualities that affect CCF. The systems analysis methodology detailed in this paper produces CCF estimates, based on an examination of the qualitative standards specific to the industry or system being analyzed as a first step to modeling CCF risk when no data on CCF exists. Before discussing the method, it is necessary to provide a review of the basic CCF calculation to show how this method may improve this CCF risk estimate compared to an unexamined application of generic data from the NRC.

2. BASIC COMMON CAUSE CALCULATION

As defined in the NASA PRA procedures guide [2] Section 7.3:

> *A common cause failure event is defined as the failure (or unavailable state_ of more than one component due to a shared cause during the system mission. Viewed in this fashion, CCFs are inseparable from the class of dependent failures and the distinction is mainly based on the level of treatment and choice of modeling approach in reliability analysis.*

There are a number of different methods for modeling CCFs and estimating the effect on system reliability such as the Multiple Greek Letter, the Alpha Model, and the Beta-Factor model. The beta-factor model is one of the simplest methods for modeling the impact of CCF and will be the focus of discussion in this paper.

To show how the beta factor estimates CCF risk, it is applied to a simple redundant system with two identical components in parallel in which at least one out of two is required to work. It is assumed that the total failure probability consists of independent and common cause failures. In this example, the total failure rate is equal to the sum of the independent failure rate and the common cause failure rate.

$$\lambda_T = \lambda_I + \lambda_{cc}$$

While assuming that $\lambda_{cc} << l_T$.

A factor, β, is defined as the fraction of the total failure rate due to common cause. From this relationship, the independent failure rate (λ_I) can be calculated.

$$\beta = \frac{\lambda_{cc}}{\lambda_T}$$

$$\lambda_{cc} = \beta\lambda_T$$

$$\lambda_I = (1-\beta)\,\lambda_T$$

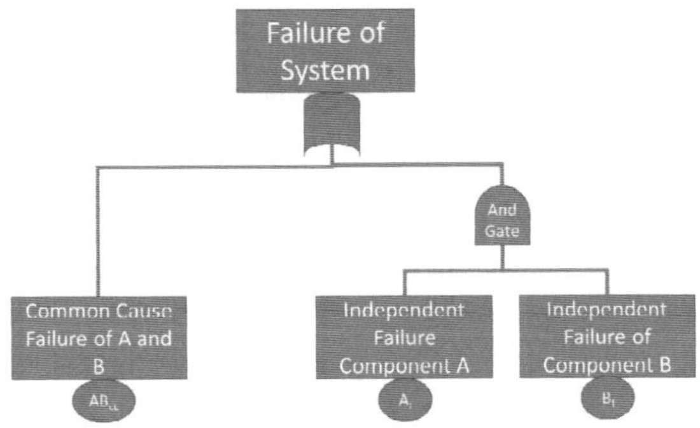

Figure 1: System Failure calculation including CCF

The total failure probability (or failure rate) can be taken from failure databases or testing. By assuming or estimating a β the failure probability of the parallel system is calculated as:

Failure of System = $(\lambda_I t)^2 + (\lambda_c t)$ =

$$= [(1-\beta)\,\lambda_T t]^2 + \beta\,\lambda_T t$$

Assume λ_T = 1E-3 (failures/hr), t = 1 hour, β = 0.1

= [1E-03*(1E-03)*1]2 + 0.1*1E-03*1

The first term is the independent failure contribution and second term is the CCF contribution, which is equal to:

=8.1E-03 + 1.0E-01.

Even when λ_{cc} is much less than λ_I (and β is less than 0.1) it shows that the CCF risk estimate dominates system risk. In addition, the β is the main driver of the overall system risk estimate.

Again, most databases only track total failures and make no distinction between independent and CCF. The NRC estimates for a range of generic component groups vary CCF β from 5-10% of total failures. From the same study cited earlier [1], failure data showed CCFs in some industries as high as 33% of total failures. This shows the importance of estimating a β that accounts for specific qualities of the system being modeled for CCF.

We have established the basics of estimating CCF risk and have shown how important the β is to the CCF estimate. We now will review factors that contribute to this phenomenon.

3. FACTORS THAT CONTRIBUTE TO COMMON CAUSE FAILURE

There are a set of primary factors that are significant contributors to CCFs and are described by eight areas that effect CCF coupling mechanisms:

1. Separation/segregation
2. Diversity/ redundancy
3. Complexity/maturity of design/experience
4. Use of assessments/ analysis and feedback data
5. Procedures/ human interface (e.g. maintenance/ testing)
6. Competence/ training/ safety culture
7. Environmental control (e.g., temperature, humidity, personnel access)
8. Environmental testing

1 Separation/segregation:
This first group is the most often quoted factor associated with CCFs. Redundant subsystems in close proximity can be susceptible to the same faults. Keeping systems segregated helps limit these common cause coupling factors. A famous example is a fire or a leakage that causes corrosion. There is a famous example of an early flight of the DC-10 which failed to separate multiple redundant hydraulic controls is well documented in the literature. [Wetherholt paper]:

> *An actual example demonstrating single physical point failure is the case of United Airlines Flight 232 which was flying from Denver, Colorado to Chicago-O'Hare. On July 19, 1989 on the DC-10, the number 2 engine (on the tail of the plane) experienced a failure which threw shrapnel into the hydraulic lines passing through a 10 inch wide channel in the tail. All three redundant hydraulic systems lost fluid, leading to loss of flight control surface actuation.*

This should make it clear that the more separation/segregation of redundant components can be achieved the less this factor influences CCFs. However, there are limits to how much separation can be achieved — separated for systems that are tightly constrained by volume limitations, such as in aerospace applications.

2 Diversity/ Redundancy:
The more diverse the better in regards to the design and maintenance of redundant subsystems. It is important to examine whether redundant systems were developed from separate requirements, by distinctly different design groups, with independent testing and design verification teams. Common cause coupling is decreased by adding more diversity.

3 Complexity/maturity/experience:
A complex design with many components makes discovering coupling factors more difficult. The maturity of the design, along with testing off-nominal situations that stress the system, will bring coupling factors to light. The amount of experience designers/operators/maintainers have with a redundant system will also affect the common cause coupling factors and the overall susceptibility to CCF.

Analysis and feedback data: The more analysis methods, data tracking, and feedback to the program that is done the better the control/reduction of CCF will be for a given system. Some of the questions to be addressed in this category are: Has a detailed Failure Modes and Effects Analysis been completed? Is there an active problem reporting and analysis program for the system? Has a comprehensive reliability allocation and risk assessment been completed? Is root cause analysis performed after the occurrence of a system fault or failure? Are these analyses and data effectively communicated to the design engineers and management, and are these insights used to make the appropriate changes in the design, operational procedures, and training programs? Analysis and data feedback effectively used to improve the system can reduce common cause failure coupling factors.

4 Procedures/human interface:
This aspect of CCF covers operation and maintenance as regarding human interfaces. Some of the questions to be addressed in this category are: Are there specific maintenance procedures for diverse/redundant systems/components that suggest a staggered or non-staggered approach? Are maintenance actions and faults documented and investigated with respect to other redundant systems/ components. Do maintenance manuals address possible common-cause coupling factors, for example assuring components are separated in case of a cable/connections re-routing. Is diverse equipment maintained by different shifts of staff? Are procedures updated after failure investigations? Is personnel training modified and updated following failure investigation findings?

5 Competence/training/culture:
The safety culture plays a role in reducing CCFs. Some of the questions to be addressed in this category are: Have designers been trained to understand CCF? Do designers have a variety of technical background and experience? Does the industry put an emphasis on increasing reliabil-

ity and safety? Have installers and maintenance personnel been trained to understand CCF? Have maintenance personnel been trained to understand CCF? Are maintenance personnel periodically retrained and updated on the results of failure assessments?

6 Environmental control:
The environment itself can cause or initiate CCF mechanisms. Some of the questions to be addressed in this category are: Is the environment adequately controlled and regulated? Will the system encounter extreme environments? How will the system respond to localized, rare, extreme events? (E.g. earthquakes, flooding, loss of power from grid, etc.)

7 Environmental testing:
Some of the questions to be addressed in this category are: Has the system been tested for the environmental conditions in which it operates (e.g. extended low or high temperatures, salt spray, vibration, etc.)? Has the system response been tested for electro-magnetic radiation?

Now that the eight areas that affect CCFs have been described, the next step is to present the methodology for developing a quantitative estimate of β.

4. BASIC METHODLOGY FOR ESTIMATING BETA FACTOR REALATED TO EIGHT COMMON CAUSE CATEGORIES

The first step in estimating the common cause beta factor is to assume a Maximum Common Cause Value (MCCV). One of three possible values must be selected (10%, 20%, or 30%). MCCV is based on judgment and experience and represents the maximum industry beta value based on a number of root cause considerations particular to the industry, such as its safety culture, management effectiveness, budget and schedule constraints, training effectiveness, maintenance program, and industry failure history. For example, in an industry that has a strong safety culture compared to other industries, one might select the lowest level of MCCV = 10%. Other industries that have a poor safety culture might have the highest MCCV = 30%.
The next step is assess each of the eight common cause susceptibility categories and for each, assign a Susceptibility Score of 1, 5, or 10 corresponding to the susceptibility category of Low, Medium, or High, respectively. The total Common Cause Score (CCS) is a sum of the products calculated by multiplying the number of categories (N_{low}, N_{medium}, N_{high}) assigned to each score by its Susceptibility Score.

$$CCS = 1 \cdot N_{low} + 5 \cdot N_{medium} + 10 \cdot N_{high}$$

The maximum possible CCS score with 8 high categories is T = 80. Then the CCF Beta is calculated as follows:

$$CCF\ Beta = \frac{CCF}{T} \times MCCV$$

For Table 1 given below with all low Susceptibility Scores, CCS=8, T=80, MCCV=30%

$$CCF\ Beta = \frac{8}{80} \times .30 = 3\%,$$

Table 1: basic evaluation for CCF with all low values

Area	Low	Medium	High
Separation/segregation	x		
Diversity/ Redundancy	x		
Complexity/maturity/experience	x		
Analysis and feedback data	x		
Procedures/human interface	x		
Competence/training/culture	x		
Environmental control	x		
Environmental testing	x		
sum of x's	8	0	0
Scoring	1	5	10
x * scoring	8	0	0
			CCS
	total of x * scoring		8

This, incidentally would be the minimum CCF Beta this method allows using this assumed MCCV.

5. INDUSTRY EXAMPLE #1: STRONG SAFETY CULTURE

Now that the basis of the methodology has been described, we next apply it to an industry with a strong safety culture to illustrate how the result compares to the generic database values.

MCCV for this industry is assumed to be 20% since this industry makes an effort to reduce common cause failures.

Separation/segregation: This industry has successfully applied this design principle by physically separating redundant components; therefore. This is judged this factor to be low for common cause.

Diversity/ Redundancy: This example industry uses functional redundancy and diversity in the design, and also ensures diversity in maintenance procedures (for example, by having written maintenance procedures that require different maintenance operators checking out redundant subsystems, or by staggering the testing of redundant subsystems). In general, this industry is aware of CCFs. However, most designs in this example industry are 40 or more years old, and since not much was known about CCF during early design, this is judged to be medium.

Complexity/maturity/experience: By its nature Example #1 industry is complex. It is also decades old with many years of operating experience and is subject to strict government oversight and regulation. This is judged to be medium.

Analysis and feedback data: Example #1 industry also collects and maintains extensive data on CCF multiple plants and many years of operating experience as required by its government's regulatory agency. This is judged to be a low contributor to CCF.

Procedures/human interface: The regulatory agency overseeing this industry's design and maintenance procedures expend resources to minimize CCFs. The agency also oversees industry operators to ensure procedures meet exceedingly well defined standards. This is judged to be a low contributor.

Competence/training/culture: The oversight agency regulating this industry requires operators to have specific training. In addition, the agency monitors and strictly enforces standards.

Environmental control: The plant operating environment within this industry is well controlled. This would be assumed to be low factor for CCF. However, other environmental factors include extreme weather and natural disasters (such as Earth quakes and floods) that raises this to medium.

Environmental testing: This industry expends significant effort in testing for environmental conditions such as rare weather and disaster events (i.e. earthquakes) for the design basis risk. Procedures and failure studies are conducted and published industry wide to disseminate information. This is judged to be a low contributor to CCFs.

These results are summarized in Table 2 below

$$CCF\ Beta = \frac{CCF}{T} \times MCCV$$

So for the table given below, CCS=20, T=80, and MCCV=20%.

Table 2: CCF Example #1 Evaluation

Area	Low	Medium	High
Separation/segregation	x		
Diversity/ Redundancy		x	
Complexity/maturity/experience		x	
Analysis and feedback data	x		
Procedures/human interface	x		
Competence/training/culture	x		
Environmental control		x	
Environmental testing	x		
sum of x's	5	3	0
Scoring	1	5	10
x * scoring	5	15	0
			CCS
	total of x * scoring		20

$$CCF\ Beta = \frac{20}{80} \times .20 = 5\%,$$

This methods estimates the generic CCF beta for this example industry at 5%.

6. EXAMPLE #2: POOR SAFETY CULTURE

In this hypothetical example, the industry has a poor safety record, operations are very hazardous, the working environmental conditions are harsh, and government oversight is not as intense as in example #1. For this industry, MCCV is assumed to be 30% since it has not made a significant effort to reduce common cause failures.

Separation/segregation: The physical working environment is somewhat limited, but relatively large compared to other systems, such as spacecraft. With careful design consideration, redundant systems can be adequately separated/segregated. However, the industry lacks awareness of CCF and historically has valued cost savings more than safety. Therefore this is deemed, for current industry standards, as a high contributor to CCFs.

Diversity/ Redundancy: Where system redundancy is present the focus is on demanding production goals rather than safe operations. Without more specific information or design details, this is judged to be medium.

Complexity/maturity/experience: This industry has made significant advances in technology and performance in the last two decades without associated increases in safety measures. With a poor record of compliance with respect to government regulations and a limited number of on-site government inspectors compared, this category is a high contributor to CCFs.

Analysis and feedback data: While some effort is made, no specific government regulation or oversight is made for detailed analysis or use of feedback data. In addition, competition between companies means data is not shared among industry participants. Lastly, different companies have different levels of analysis. This is judged to be high contributor to CCFs.

Procedures/human interface: This industry requires extensive human interface and experience. Many processes should be well documented and require many work intensive steps. This makes this a high contributor to CCFs.

Competence/training/culture: The standards in this industry require a high level of competence. Training requirements vary among companies but still maintain a certain level. Culture varies among the different companies. This is judged to be a medium contributor to CCFs.

Environmental control: Depending upon the operating location, environmental conditions range from mild to extreme. Typically, the environment is harsh and working conditions are extremely difficult. This is assumed to be a medium contributor to CCFs.

Environmental testing: The harsh environments should require more testing for hardware and system processes. This one is estimated to be a medium contributor to CCFs.

These results are summarized in the Table 3 below

$$CCF\ Beta = \frac{CCF}{T} \times MCCV$$

So for the table given below, CCS=60, T=80, and MCCV=30%.

Table 3: Example #2 CCF Example Evaluation

Area	Low	Medium	High
Separation/segregation			x
Diversity/ Redundancy		x	
Complexity/maturity/experience			x
Analysis and feedback data			x
Procedures/human interface			x
Competence/training/culture		x	
Environmental control		x	
Environmental testing		x	
sum of x's	0	4	4
Scoring	1	5	10
x * scoring	0	20	40
			CCS
	total of x * scoring		60

$$CCF\ Beta = \frac{60}{80} \times .30 = 23\%,$$

This methods estimates the generic CCF beta this example industry at 23%.

7. SUMMARY AND CONCLUSIONS

Systems require ever higher levels of reliability and in many cases this is achieved by increase redundancy. Without care, increasing redundancy may lead to increased CCFs coupling thereby reducing the benefit of increased system reliability. Until more data is collected and analyzed it will be difficult to judge just how high CCF risk is. However, using generic data, which are specifically designed to reduce common cause mechanisms, leads one to suspect that other industries may grossly underestimate CCF risk and also significantly overestimate system reliability.

We have reviewed the β factor CCF model and shown how it effects overall system reliability and how important that it in reliability predictions. We have provided a methodology that relates eight areas of susceptibility to common cause coupling factors to the β estimate. We then assessed the common cause β for two example industries: Example #1 Strong Safety Culture and Example #2 Poor Safety Culture and in both cases estimated reasonable β values based on data assumed to be known through an examination of eight factors that contribute to common cause failure.

Much work has been done to detail the factors that affect CCFs. By using this proposed method there is a better justification for generic CCF based on these factors that may lead to more realistic or credible CCF risk and overall system reliability when system specific data is not available. These results can then be used to highlight system design details that could investigated to reduce the system's susceptibility to CCF before data can be collected and investigate specifically to reduce CCF.

This method serves two purposes: 1) allows an analyst to better model CCF with factors specific to a system that doesn't have CCF data, and; 2) better communicate CCF coupling mechanisms to system designers, operators and maintainers. The more effort to educate industries about CCF coupling mechanisms, the more systems can be made that reduce CCF in the future.

8. REFERENCES

1. Hauge, S., Hoem A.S., Hokstad P., Habrekke, S, Lundteigen, M.A., Common Cause Failures in Safety Instrumented Systems, SINTEF Technology and Society, May 20, 2015.

2. Stamatelatos, M., Probabilistic Risk Assessment Procedures Guide for NASA Managers and Practitioners version 1.1, Office Of Safety and Mission Assurance, NASA Headquarters, August, 2002.

3. Procedures for Treating Common Cause Failures in Safety and Reliability Studies, NUREG/ CR-4780

4. Wetherholt, J, Heimann, T.J. Common Cause Failure Modes, MSFC SMA, 2011

PROCEEDINGS OF THE 8TH INTERNATIONAL SPACE SAFETY CONFERENCE

SESSION 11 OPERATIONS SAFETY

YOU CAN'T REACH FOR THE STARS IF YOU ARE TRIPPING OVER THE GROUND! (PREVENTING SLIPS, TRIPS, AND FALLS)

Darcy Miller[1], Mark Raysich[2], Mary Kirkland[3]

[1] *National Aeronautics and Space Administration, Kennedy Space Center, SA-E2, Kennedy Space Center, FL, USA 32899, Darcy.H.Miller@nasa.gov*
[2] *A-P-T Research, Inc, Kennedy Space Center, FL, USA 32899, Mark.S.Raysich@nasa.gov*
[3] *Integrated Mission Support Services, Kennedy Space Center, FL, USA 32899, Mary.K.Kirkland@nasa.gov*

ABSTRACT

Although there are very few mishaps related to ground, vehicle or payload processing at the Kennedy Space Center (KSC), employees have experienced a significant number of injuries due to slips, trips, and falls outside of performing flight processing operations. Slips, trips, and falls are major causes of occupational injuries at KSC, the National Aeronautics and Space Administration (NASA), and in general industry.

To help KSC employees avoid these injuries, and allow them to be fully productive, KSC launched an initiative in 2013 to reduce slips, trips, and falls. This initiative is based on a four-part model focusing on DATA analysis, HAZARD awareness, PREVENTIVE methods, and BALANCE.

1. INTRODUCTION

1.1 Define problem

The Kennedy Space Center (KSC) is the primary launch complex for NASA human space flight. Safety is built into every process and procedure of all vehicle and payload operations as well as ground support operations. This has resulted in very few mishaps related to ground, vehicle or payload processing. However, KSC employees have experienced a significant number of injuries due to slips, trips, and falls (STF) outside of performing flight processing operations. This is the most significant contributor to injuries at KSC, throughout NASA and in general industry in the United States.
Falls account for over eight million hospital emergency room visits each year, representing the leading cause of emergency room visits (21.3%). [1] Slips, trips, and falls cause 15% of all accidental deaths, which is second only to motor vehicles. [2] In addition to the cost and loss of productivity, these injuries can lead to a lifetime of pain.

1.2 Traditional Efforts

Traditional efforts to reduce STF injuries have focused on the work environment. Decades of research and countless dollars have been expended throughout the United States to identify and reduce walking/working surface hazards, including flooring materials, uneven surfaces, surface transitions, and weather-related hazards such as rain and ice.

1.3 Vision, Mission, and Goals

While traditional methods are still being deployed, a new initiative has been developed that includes a focus on the employee in reducing STF injuries.

Vision: No one should suffer the pain or cost of an injury due to a slip, trip, or fall

Mission: To contribute globally to the reduction of injuries due to slips, trips, and falls

Goal: Develop an effective Slips, Trips, and Falls Reduction Initiative that includes innovative solutions within a comprehensive program designed to be replicable across other NASA Centers, general industry, and outside of work.
This initiative looks to strengthen our safety culture through changes in attitudes, actions and an increased sense of personal responsibility.

The STF initiative focuses on four parts:
- Case data analysis
- Hazard awareness
- Preventive and situational awareness
- Balance initiative

2. KENNEDY SPACE CENTER SLIPS, TRIPS, AND FALLS REDUCTION INITIATIVE

2.1 Slips, Trips, and Falls Initiative Model

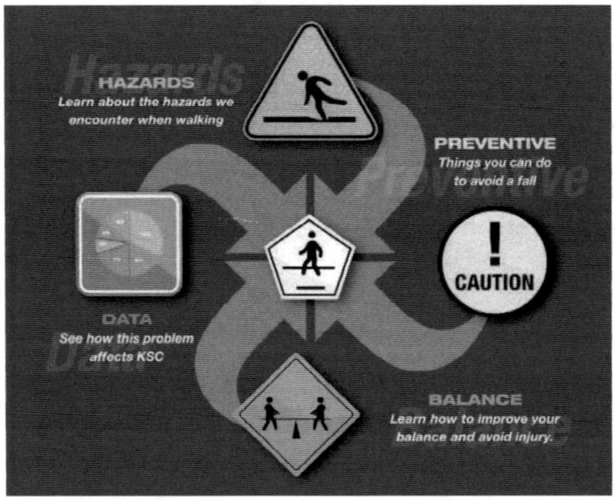

Figure 1: Slips, Trips, and Falls Initiative 4-Part Model

The KSC Slips, Trips, and Falls Initiative model was developed to address the problem in four inter-related ways. The KSC team decided to continue the traditional efforts to ensure a safe work environment while also adding a new focus on employee's responsibilities.

The first two parts, DATA ANALYSIS and HAZARD AWARENESS, continue to build upon traditional efforts. Data analysis has been expanded to a greater level of detail. Hazards are still mitigated as much as possible, but more emphasis is placed on educating employees to be aware of existing hazards.

The second two parts of the model, PREVENTION and BALANCE, are more innovative. They build on existing information but also give employees methods to reduce their risk of injuries from STFs. Updated prevention techniques are being taught to make employees aware of their responsibilities in reducing their likelihood of an STF injury. These techniques focus areas employees can control, such as their actions and decisions. Finally, employees are being offered hands-on training to improve their balance and physical abilities to reduce the likelihood of injury.

2.2 Timeline (How we did this)

In 2013, the KSC Institutional Safety and Mission Assurance (S&MA) organization began focusing on the STF problem by developing a detailed trending analysis of case data, which clearly defined the problem. At that time, a partnership was formed between the S&MA organization and the on-site physical rehabilitation group, RehabWorks [3].

S&MA developed the Slips, Trips, and Falls 4-Part Model in 2014. S&MA and RehabWorks also began conducting STF prevention classes and balance classes, with over 1600 employees trained to date.

Two Balance Zones equipped with traditional balance equipment and electronic balance training devices opened in 2015. An initial test group was formed to establish balance baselines, and testing and training protocols.

3. DATA ANALYSIS

KSC has maintained a long-standing practice of trending the events and exposures leading to the Center's Occupational Safety and Health Administration (OSHA) reportable injuries. Slips, trips, and falls are among the top contributors leading to these occupational injuries. In the fall of 2013, the S&MA Directorate launched an initiative to reduce these injuries. This began with finite analysis of not only OSHA reportable STF injuries, but all injuries that were the result of STFs. These included injuries that required only first aid treatment and minor injuries that resulted in medical evaluation only by KSC on-site health care professionals.

The analysis was expanded to first break down each injury by the category of a slip, a trip, or a fall from other factors. Then each category was analyzed for the mechanism that led to the injury. Additionally, each injury was mapped by location to identify "hot spots" of STF injuries.

This effort led to a greater understanding of the types and causes of STF injuries at KSC. The analysis first showed that STF injuries resulted in the most severe OSHA reportable injuries, and were often responsible for employees missing significant time away from work, or being restricted from performing some of the duties associated with their job. This meant that not only were there direct medical costs associated with these injuries, but that productivity was reduced because a member of the team was not present or was limited in their level of productivity.

As seen in Fig. 2, the analysis shows the most prevalent STF category to date is trips, followed by slips, and finally falls from other factors.

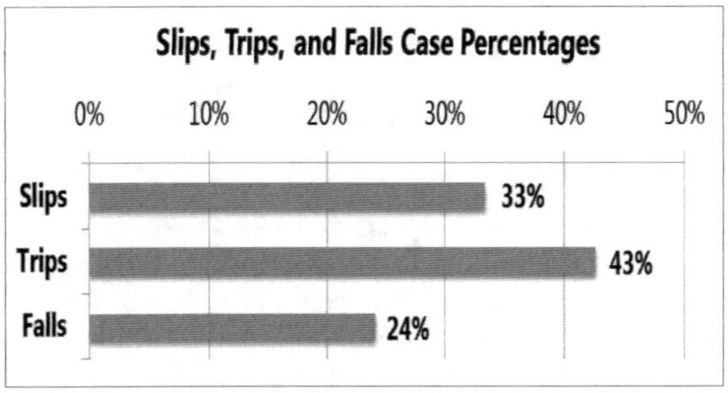

Figure 2: Breakdown of Slips, Trips and Falls at KSC

3.1 Slips

A slip occurs when there is too little friction or traction between your footwear and the floor surface. In most slip events where a worker is walking, the worker's heel on the front foot slips forward as the individual is transferring weight causing the worker to fall backwards
Fig. 3 shows the mechanisms involved in KSC slip injuries. The most prevalent mechanism was wet surfaces. Many slip injuries occurred where the employee was unable to identify exactly what caused the slip.

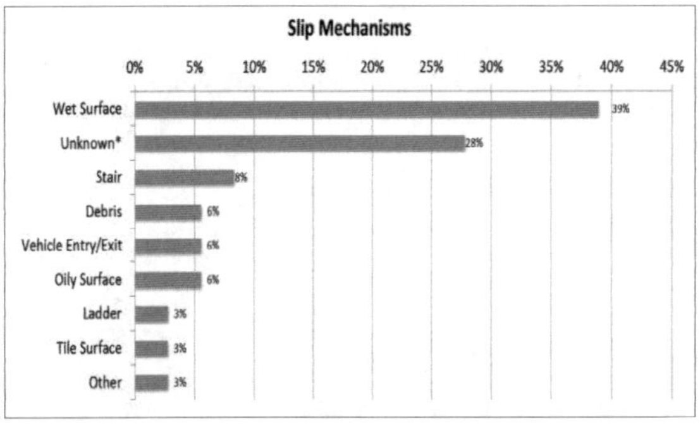

Figure 3: Mechanisms Involved in Slips at KSC

3.2 Trips

A trip occurs when your foot strikes an object resulting in a loss of balance. In a trip, your momentum causes your body to continue forward and commonly results in a forward fall.

Fig. 4 shows the mechanisms involved in KSC trip injuries. The most prevalent mechanism was colliding with miscellaneous objects such as pallets, hoses, or boxes. This was followed by stairs, debris, surface transitions and service carts as cases where the employee was unable to identify exactly what caused them to trip.

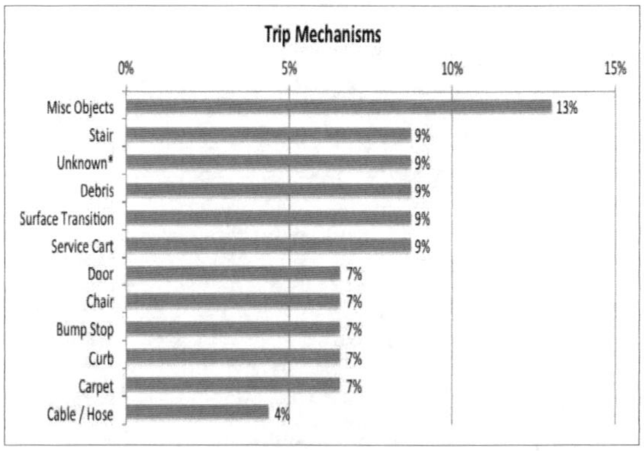

Figure 4: Mechanisms Involved in Trips at KSC

3.3 Falls from Other Factors

Falls can also be the result of personal factors such as balance, medications and health conditions.

Fig. 5 shows the mechanisms involved in KSC falls from other factors. The most prevalent mechanism was an employee's leg or ankle giving way. This was followed by unsafe acts and other personal factors.

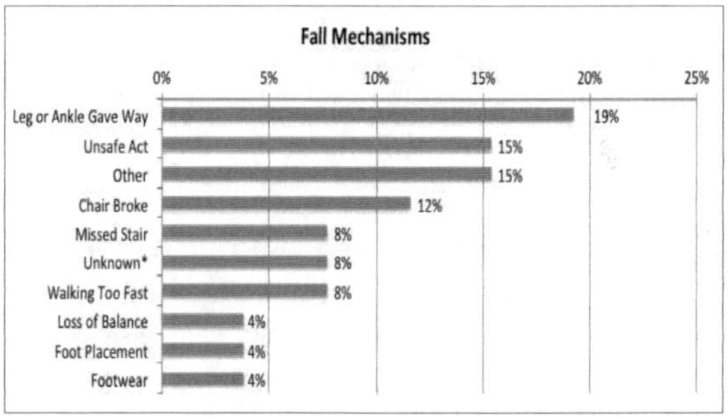

Figure 5: Mechanisms Causing Falls at KSC

4. HAZARD AWARENESS

Employees encounter various hazards when walking in the workplace. KSC is vigilant in ensuring employees have a safe work environment. This includes mitigating slips, trips and falls, but some infrastructure elements and work environments contain innate hazards that can lead to a slip, trip, or fall. This part of the model emphasized educating employees about these hazards and raising awareness to help them avoid these injuries. The first step is to identify the various hazards that can lead to slip, trip or fall injuries.

4.1 Slip Hazards

Slips can be caused by weather events that result in ice, snow or water collecting on a surface, but they can also be the result of surface contamination or inadequate floor coverings, materials or preparation.

Contaminated surfaces are broken down into two major categories: wet floor contamination and dry floor contamination. Examples of wet floor contamination are ice, water, mud, grease, oil, food, and soap. Examples of dry floor contamination are dusts, powders, granules, and small objects such as metal nuts and bolts. Examples of inadequate floor coverings, materials and floor preparation are loose or unsecured mats, sloped walking surfaces, ramps and gangplanks without anti- skid material. Even poor lighting can lead to a slip if the individual cannot see a surface transition.

4.2 Trip Hazards

Trip hazards include changes in elevation such as seams or irregularities in sidewalks, door thresholds, uncovered hoses, cables or extension cords, clutter or obstacles in walkways and work areas, open cabinet or desk drawers, and construction materials not stored or staged properly. Curbs and stairs cause many tripping injuries. Trips are also caused by unmarked steps or ramps, rumpled or rolled up carpets, and ramps.

4.3 Falls from Other Factors

In addition to the slip and trip hazards we are all exposed to every day, there are additional personal factors that can make an individual more prone to a fall. These include balance, medications or combinations of medications, vision, and musculoskeletal conditions. You are also at a higher risk to fall if you have a fall history or have a health condition such as diabetes, osteoporosis, or foot, ankle or leg injuries. Obesity, arthritis, and stress can also affect your balance.

5. PREVENTION METHODS

This portion of the initiative focuses on the personal responsibility of the employee. In addition to sharing injury data in the training sessions, an attempt is made to make this more personal while also emphasizing teamwork and looking out for each other at work.

5.1 Training

Once employees understand the difference between a slip, trip, and fall, they are taught some specific things they can do to prevent each of these types of injuries. This includes common-sense preventive methods, plus newer topics, such as distractions from cell phones. This is important because distracted walking cases in the U.S. have quadrupled in the past seven years [4]. People may be hesitant to report their use of a cell phone after an injury, so these cases may be underreported.

It may seem like common sense to wear proper shoes for the environment, but many STFs are caused by improper footwear. To reinforce the concept, employees are taught more details, such as how shoe surface area and traction can increase the chance of slipping or falling. Some shoes may look sturdy but have smooth leather soles and lack tread. Other shoes, like high heels, don't provide an adequate contact patch between the foot and the ground, especially at the heel.

Walking techniques are discussed such as stride, foot position, keeping hands free for balance, walking in designated areas, visibility, and carrying loads.

Keeping work areas clear, common obstacles in work areas, lighting, and wet surfaces are discussed to encourage employees to change their environments to reduce the likelihood of STFs.

Being aware of the walking areas is also emphasized, including slowing down and paying attention. One catch phrase is, "WWW - Watch Where You're Walking."

Since overall health can make a difference, lifestyle decisions such as, getting adequate rest, exercising, and keeping stress under control are also discussed.

Everyone acts unsafely at times, but the training emphasizes that these are controllable decisions that can affect risk of injury. Some examples include, standing on chairs or tables with wheels, or on items like furniture that are not intended for standing/climbing.

5.2 Make it Personal

The fact that STFs are the main source of injury at KSC, NASA, and in general industry is not enough to motivate most people to make changes. It is necessary to bring the message to a more personal level. None of the people who were injured at KSC expected to go home injured or go to the hospital that day. These injuries can have a far-reaching impact on each of us, our families, and our employers. It is important to discuss this in the training and spread the message through promotional materials.

A related catch-phrase is used, "Beat the Statistics, Don't Become One!"

5.3 It is Up to All of Us

The STF Initiative emphasizes that preventing accidents and injuries in the workplace starts with each one of us, individually. We need to be willing to make decisions and take actions that not only keep us safe, as an individual, but to keep our co-workers from getting injured.
One of the ways to help keep our co-workers safe is to report safety concerns, particularly those that may pose an STF hazard. At KSC, the various methods of reporting emergencies and safety concerns is widely publicized. It can be found at the top of the main KSC web page, and it is communicated in such occasions as the beginning of most training classes, and during new employee orientations. At NASA, being knowledgeable and willing to report problems is an important part of a strong safety culture.
All personnel need to work cooperatively to identify, report and correct unsafe conditions and behaviors. Ideally, if we see someone walking at risk, performing an unsafe act, or appearing to not see a hazard, we would step up and help them by speaking up. One action may help avoid an injury.

In the STF training, attendees are asked to make a commitment to themselves to develop an awareness of the potential walking hazards in their work and home lives and share what they learn with their coworkers and family.

The message can be summarized in this catch-phrases; "Remember at the End of the Day...Safety Is Up To You!"

6. BALANCE

The final part of the initiative provides a programmatic approach to reduce the risk of injuries due to STFs by focusing on three key areas: Education, hands-on training, and research.

Employees can take action by participating in the balance activities offered at KSC to understand personal risk factors, improve awareness, and train to improve balance.

When the S&MA organization and KSC RehabWorks teamed to tackle the problem of STFs, they didn't find a program that fit the needs of the KSC population, which has approximately 8000 employees with an average age of about 50 years. A great deal of research exists in the area of sports concussion management and geriatric fall rehabilitation. However there was very little research or programs found for the middle-aged population in an occupational setting.

6.1 Education

Healthcare professionals, including both athletic trainers, and health and wellness professionals, teach onsite classes to provide employees with self- assessment tools and awareness of personal risk factors that contribute to STFs. Presentations are also made available online.

An important piece of the initial training is to give employees an understanding of how their bodies stay in balance. Whether we are on Earth or in space, our bodies use information to maintain balance from three complex sensory input systems. Simply stated, these systems include vision, proprioception (unconsciously perceiving body position from sensory mechanisms throughout the body) and vestibular (sensory mechanisms in the inner ear which track the position of the head). All of the sensory input information is integrated in the brain to produce a motor response output. Through balance training, the speed of this response can mean the difference between falling and recovering when a trip hazard is encountered.

To help employees learn more about their own balance needs, the balance team developed a short Fall Risk Questionnaire and a longer Health History Form that relates to issues that may affect balance.

The Fall Risk Questionnaire is a rapid assessment of five risk factors that increase the chance of an STF injury (Fig. 6). If any apply to an employee, then they are directed how to take action. The five areas include: Age, medication, vision, recent injuries or surgeries, and systemic conditions.

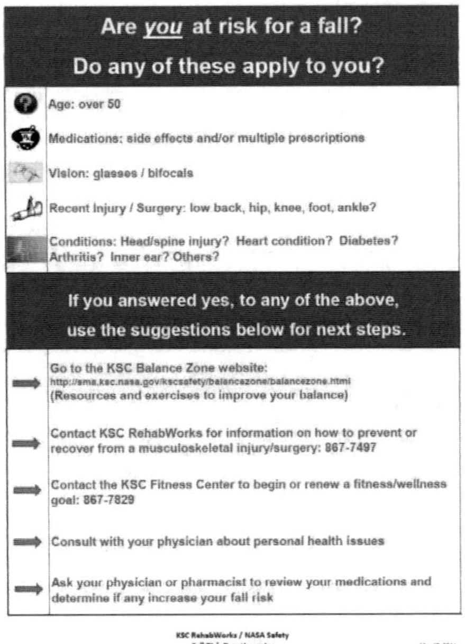

Are *you* at risk for a fall?
Do any of these apply to you?

- Age: over 50
- Medications: side effects and/or multiple prescriptions
- Vision: glasses / bifocals
- Recent Injury / Surgery: low back, hip, knee, foot, ankle?
- Conditions: Head/spine injury? Heart condition? Diabetes? Arthritis? Inner ear? Others?

If you answered yes, to any of the above, use the suggestions below for next steps.

- Go to the KSC Balance Zone website: http://sma.ksc.nasa.gov/kscsafety/balancezone/balancezone.html (Resources and exercises to improve your balance)
- Contact KSC RehabWorks for information on how to prevent or recover from a musculoskeletal injury/surgery: 867-7497
- Contact the KSC Fitness Center to begin or renew a fitness/wellness goal: 867-7829
- Consult with your physician about personal health issues
- Ask your physician or pharmacist to review your medications and determine if any increase your fall risk

KSC RehabWorks / NASA Safety
Fall Risk Questionnaire
May 15, 2014

Figure 6: Fall Risk Questionnaire

A more detailed Health History Form builds upon the Fall Risk Questionnaire to fully assess a person's risk of an STF injury.

- Musculoskeletal: weak muscles, arthritis/joint pain, low back, hip, or lower leg injury/surgery
- Systemic: High and low blood pressure/orthostatic hypotension, irregular heartbeat/arrhythmia, diabetic neuropathy (numbness) of the feet, pregnancy, obesity
- Visual: change in prescriptions, bifocals/trifocals, glaucoma, vestibular, inner ear damage, ear/nose/throat infection, Meniere's disease
- Medication side effects and interactions
- Neurological: Stroke, Parkinson's, Multiple Sclerosis, ALS

Through a Balance Initiative Speaker Series, subject matter experts were invited from KSC and within the local community to talk about issues that affect balance. They included an ophthalmologist, neurologist, pharmacist, gerontologist, and athletic trainer.

Promotion for the Balance Initiative is accomplished through KSC safety and health events, emails, newsletters, and an internal NASA website. The website includes videos showing balance drills, educational videos, and a video by Astronaut Cady Coleman discussing the importance of balance training [5].

Figure 7: Balance Team with Astronaut Cady Coleman

As a quick reference, a laminated card was developed to fit in the badge holders of KSC employees (see Fig. 8).

Informal on-line surveys were developed to learn about employees' interest in participating in the Balance Initiative and to gather feedback from those who participated.

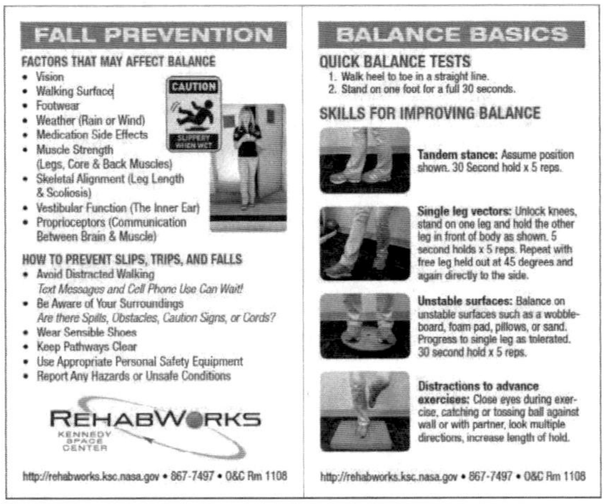

FALL PREVENTION

FACTORS THAT MAY AFFECT BALANCE
- Vision
- Walking Surface
- Footwear
- Weather (Rain or Wind)
- Medication Side Effects
- Muscle Strength (Legs, Core & Back Muscles)
- Skeletal Alignment (Leg Length & Scoliosis)
- Vestibular Function (The Inner Ear)
- Proprioceptors (Communication Between Brain & Muscle)

CAUTION
SLIPPERY WHEN WET

HOW TO PREVENT SLIPS, TRIPS, AND FALLS
- Avoid Distracted Walking
  Text Messages and Cell Phone Use Can Wait!
- Be Aware of Your Surroundings
  Are there Spills, Obstacles, Caution Signs, or Cords?
- Wear Sensible Shoes
- Keep Pathways Clear
- Use Appropriate Personal Safety Equipment
- Report Any Hazards or Unsafe Conditions

REHABWORKS
KENNEDY SPACE CENTER

http://rehabworks.ksc.nasa.gov • 867-7497 • O&C Rm 1108

BALANCE BASICS

QUICK BALANCE TESTS
1. Walk heel to toe in a straight line.
2. Stand on one foot for a full 30 seconds.

SKILLS FOR IMPROVING BALANCE

Tandem stance: Assume position shown. 30 Second hold x 5 reps.

Single leg vectors: Unlock knees, stand on one leg and hold the other leg in front of body as shown. 5 second holds x 5 reps. Repeat with free leg held out at 45 degrees and again directly to the side.

Unstable surfaces: Balance on unstable surfaces such as a wobble-board, foam pad, pillows, or sand. Progress to single leg as tolerated. 30 second hold x 5 reps.

Distractions to advance exercises: Close eyes during exercise, catching or tossing ball against wall or with partner, look multiple directions, increase length of hold.

http://rehabworks.ksc.nasa.gov • 867-7497 • O&C Rm 1108

Figure 8: Two-sided Balance Badge

6.2 Training

Balance Zones (BZ) are being established across KSC where employees can train through a combination of conventional drills, electronic gaming devices, and core stabilization exercises. Participants follow a progressive protocol designed to incorporate and improve the various components of balance.

BZ facilitators provide individualized, hands-on training sessions in the BZs. The facilitators include healthcare professionals and trained employee volunteers. Employees are required to use the "Buddy System" to ensure no one is alone in the unlikely event an injury would occur.

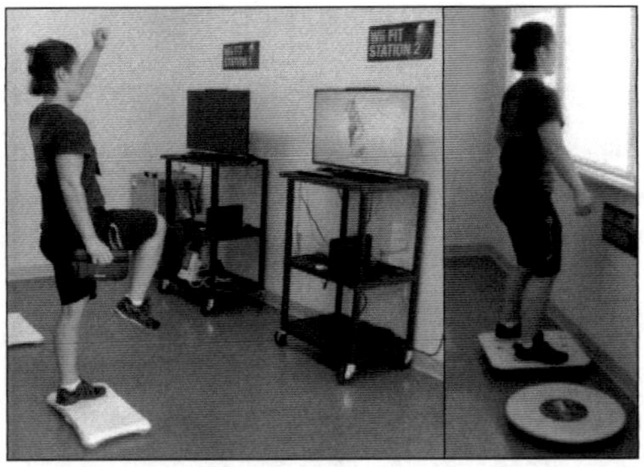

Figure 9: Balance Zone Training Session

6.3 Research

The KSC workforce represents an untapped data set of the 40 – 70 year old demographic for researchers of the musculoskeletal, visual, vestibular, and somatosensory systems' effect on balance related to STF injuries. In an effort to capture both the magnitude of the balance challenges to the KSC workforce, as well as the effectiveness of the balance training, a Biodex SD Balance System, a product of Biodex Medical Systems, is being used to assess and train volunteer KSC employees (see Fig. 10).

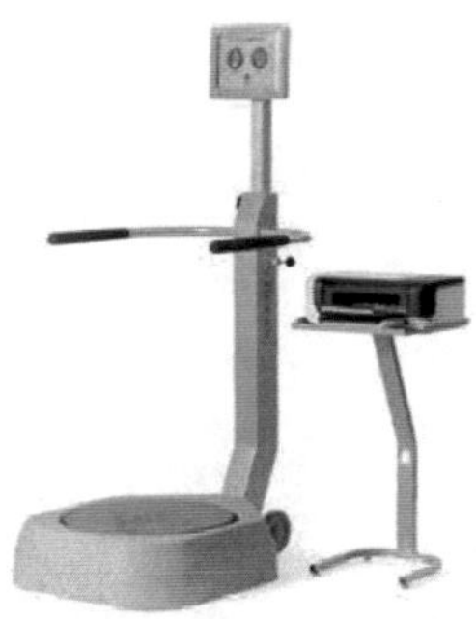

Figure 10: Biodex Balance System

Initially, a pilot study was conducted with 36 NASA Safety employees who completed pre and post testing of the Fall Risk Screening (FRS), which can identify a balance problem in under 2 minutes, and the Modified Clinical Test of Sensory Integration of Balance (m- CTSIB) Test, which helps determine which sensory system (visual, somatosensory, or vestibular) the patient relies on to maintain balance (see Fig. 11).

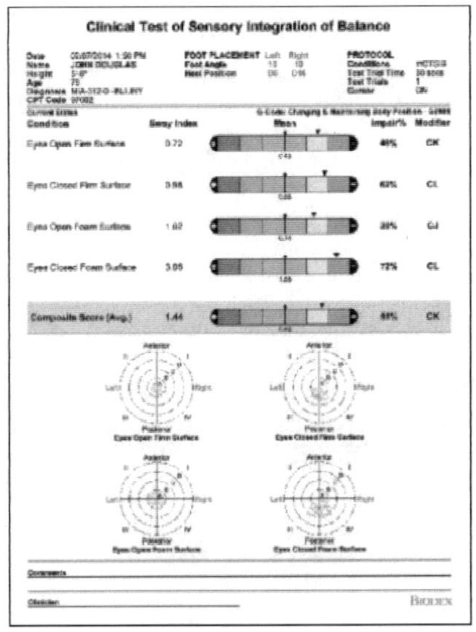

Figure 11: Clinical Test of Sensory Integration of Balance report

Additional research is being conducted by KSC RehabWorks entitled, *"Establishment of Normative Data for the Biodex Fall Risk Balance Test in Adults Aged 40-70 Years of Age,"* which will help fill the gap in normative data with regard to STF injuries in the middle aged worker. Upon completion of the study, the data will provide a baseline for future training and testing of the KSC workforce using the existing BZ facilities and equipment.

7. AWARENESS AND OUTREACH

To make the program successful, a paradigm shift must occur which is generally a slow and consistent process. To allow this to become part of employees' normal way of life, the team has chosen to reinforce the messages at many levels. These include training classes, a web site, videos, posters, employee newsletters, emails, and safety tips. The program began development in 2013 but was officially rolled out in March of 2015. However, it will be necessary to continually reemphasize the message in different ways over time.

8. PARTNERS

One of the most important parts of being successful in this new initiative is developing strong partnerships with people in organizations all around KSC and at the NASA Agency level. These include: KSC Wellness Program and Fitness Center, KSC Aerospace Medicine and Occupational Health, Indiana University Purdue University

Indianapolis (IUPUI), KSC Web Development, KSC Graphics, KSC Facility Management, KSC Public Affairs, NASA Safety Center, and KSC employees who volunteered for balance testing and training.

Over time, the support for the initiative has become very strong from senior management, at KSC and NASA Headquarters.

9. SUMMARY

The STF Initiative has evolved since 2013 and will be ongoing for years to come. It has gained support from leadership and employees. To make this program sustainable, employees are being empowered to take personal responsibility for preventing STF injuries at work and at home. Over time, KSC's Safety Culture has been shifting from an expectation that the employer will remove all hazards, to a culture where employees will take action to reduce their risks of injuries. This is being done through changes in attitudes, actions, and an increased sense of personal responsibility.

10. REFERENCES

1. National Floor Safety Institute, "Quick Facts," accessed 3/28/2016, https://nfsi.org/nfsi-research/quick-facts/

2. Occupational Safety & Health Administration, *"Walking/Working Surfaces,"* Accessed on 3/28/2016, https://www.osha.gov/SLTC/walkingworkingsurfaces/index.html

3. Kennedy Space Center RehabWorks, Integrated Mission Support Services https://mescforyou.ksc.nasa.gov/RehabWorks

4. The Advisory Board Company, "Walking While Texting is Bad for Your Health", Accessed on 3-31-2016, https://www.advisory.com/daily-briefing/2012/07/31/distracted-walking-injuries-quadruple

5. NASA Video, "NASA Astronaut Cady Coleman Visits KSC Balance Zone Team," Accessed March 18, 2016, https://www.youtube.com/watch?v=8UbafiFEs7M

PROBABILISTIC SURVIVABILITY VERSUS TIME MODELING)

James J. Joyner, Sr.

ABSTRACT

This presentation documents Kennedy Space Center's Independent Assessment work completed on three assessments for the Ground Systems Development and Operations (GSDO) Program to assist the Chief Safety and Mission Assurance Officer during key programmatic reviews and provided the GSDO Program with analyses of how egress time affects the likelihood of astronaut and ground worker survival during an emergency.

For each assessment, a team developed probability distributions for hazard scenarios to address statistical uncertainty, resulting in survivability plots over time. The first assessment developed a mathematical model of probabilistic survivability versus time to reach a safe location using an ideal Emergency Egress System at Launch Complex 39B (LC-39B); the second used the first model to evaluate and compare various egress systems under consideration at LC-39B. The third used a modified LC-39B model to determine if a specific hazard decreased survivability more rapidly than other events during flight hardware processing in Kennedy's Vehicle Assembly Building.

1. PROBABILISTIC SURVIVABILITY VERSUS TIME MODELING

The KSC Independent Assessment (IA) Team conducts assessments as requested by a Customer, providing the Customer with objective, non-advocacy recommendations and solutions. IA Team customers include the Office of Safety and Mission Assurance (OSMA), the KSC Center Director, Program Managers, Chief Safety and Mission Assurance Officers, and other first line Directors at KSC. Assessments performed by the team cover a variety of subjects ranging from systemic process evaluations and technical assessments to process improvement. Assessments exclude any criminal investigation.

This article summarizes three KSC assessments in which Survivability versus Time Models were developed as a decision-evaluation tool. The first assessment developed a mathematical model of Survivability versus Time for an emergency egress system at Launch Complex 39B (LC-39B). The second assessment used the first model to evaluate and compare various emergency egress systems under consideration at LC-39B. The third assessment used a modified LC-39B model to determine if a specific hazard(s) decreased survivability more rapidly than other events in the Vehicle Assembly Building (VAB). Probability distributions were developed for hazard scenarios to address statistical uncertainty, resulting in survivability plots over time. Composite survivability plots encompassing multiple hazard scenarios were also developed. These assessments produced a set of plots that acted as a decision-informing tool for the Ground Systems Development and Operations (GSDO) Chief Safety and Mission Assurance Officer (CSO) and GSDO Program Manager during key programmatic reviews.

In 2012, the GSDO CSO requested that KSC IA perform an assessment of the LC-39B Emergency Egress Systems. If an emergency situation (e.g., fire, imminent explosion) developed with the Orion Crew Module or Space Launch System (SLS) vehicle at LC 39B during launch countdown, an Emergency Egress System (EES) would quickly transport four astronauts to safety. The intent of this first assessment (KSC IA 1207, "Crew Emergency Egress Survivability") was to determine if survivability as a function of time to reach a safe location could be used to develop a Figure of Merit (see Fig 1) to enlighten EES design trade studies. By-products included a list of scenarios leading to emergency crew egress and initial estimates of the likely crew survivability.

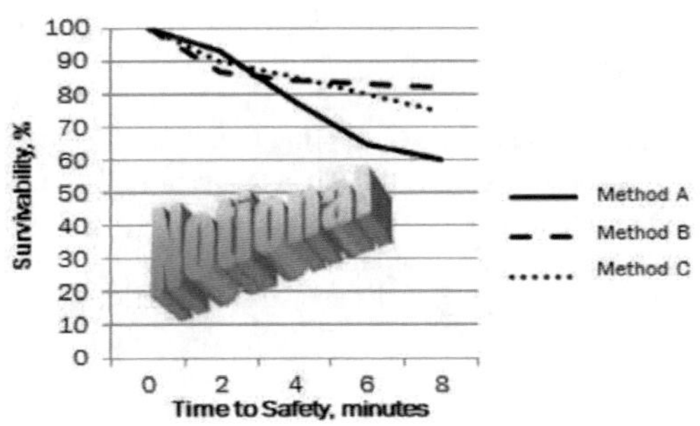

Figure 1 A Figure of Merit

To accomplish this task, the IA Team started by defining terms, groundrules, and assumptions early in the process. Death was defined as 0% survival if any of the crew members died before reaching a safe location. The IA Team made no assumption as to the nature of the Ideal EES (e.g., rail, slidewire, elevator) or a destination of a safe location (an area where the crew are free from the effects of a hazard). The evaluation started once all four crew members were on the Crew Access Arm, noting survivability at specific time intervals until a safe location was reached.

The Fault Tree Analysis (FTA) Method was used to determine which Hazard Scenarios would require an emergency egress. The FTA Method resulted in the simplified Fault Tree (soon in in Fig 2), which enabled the IA Team to examine all paths to establish credible scenarios.

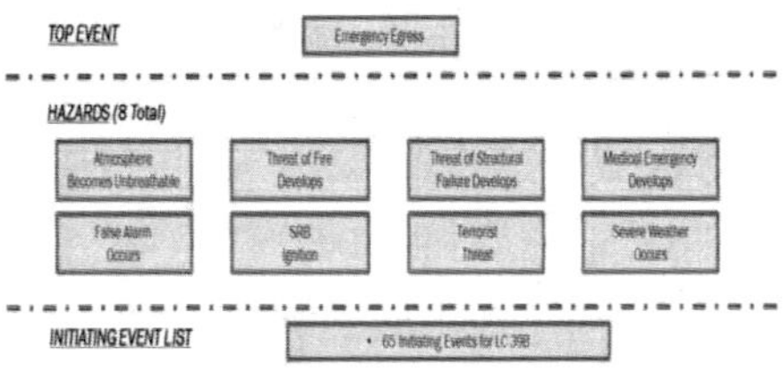

Figure 2 Simplified FTA

For each credible initiating event, the Likelihood of Occurrence and the Probability of Survival (given the initiating event occurred) were determined. The "Most Likely" value for the Likelihood of Occurrence was determined through data analysis and review of historical information or, in the absence of numerical data, via expert elicitation. Uncertainties in the Most Likely value were bound by Maximum and Minimum limits determined using the Error Factor method described in the NASA PRA Guidebook.

The "Most Likely" survivability value was bound by Minimum ("Bad Day" or everything working against crew surviving the hazard) and Maximum ("Good Day" or optimal conditions/not severe as expected) limits. These Maximum, Most Likely, and Minimum values establish the survivability distribution. All three survivability values were determined via team consensus, interviews, or consequence rating from hazard reports, or by a combination of these methods.
To calculate the probability of survival for all credible initiating events:

- P_E = likelihood of event occurring. [input in Failure Space
- $P_{S|E}$ = probability of surviving if event occurs. [input in Success Space]

Since a Failure Space Distribution should not be multiplied by a Success Space Distribution, $P_{S|E}$ needs to be converted into Failure Space, so $P_{D|E}$ = probability of dying, calculated by Eq. 1:

$$P_{D|E} = 1 - P_{S|E} \tag{1}$$

P_D = probability of dying due to this event, which is calculated by Eq. 2:

$$P_D = P_E * P_{D|E} = P_E * (1 - P_{S|E}) \tag{2}$$

Converting back into Success Space (Survivability):
P_S = probability of surviving due to this event, which is calculated by Eq. 3:

$$P_S = 1 - P_D = (1 - (P_E * (1 - P_{S|E})) \tag{3}$$

$P_{S\ all}$ = probability of surviving the occurrence of all Initiating Events (assumes events are independent), which is calculated by Eq. 4:

$$P_{Sall} = \Pi(P_{Si}) = P_{S1} * P_{S2} * ... * P_{S65} \tag{4}$$

The @RISK software was used calculate the survivability distributions. The Latin Hypercube sampling method was chosen to run probabilistic simulations at 50,000 iterations in order to calculate the output for each time interval. The first assessment produced an Ideal EES model with Most Likely value bounded by a Maximum and Minimum limit. Based on the composite survivability, there is a soft "knee" in the curves at eight minutes (ten minutes after the egress order was given). As a secondary effort, the team also developed survivability estimates assuming the existing launch pad elevator would be used after significant hardening changes. Use of the upgraded launch pad elevator as the EES decreases survivability somewhat compared to the Ideal EES since it involves taking the flight crew past the hazard for some scenarios. The second LC-39B assessment (KSC IA 1304, "Crew Emergency Egress Comparison") evaluated each emergency egress system under consideration (e.g., rail, slide-wire, elevator) based on the astronauts' survivability as a function of time to reach a safe location (a bunker inside the blast danger area or outside the blast danger area). The GSDO Program provided the IA Team with a study that outlined each EES route, time, distance, and safe location. For each EES design option, the IA Team considered 65 initiating events, rescored the survivable for each credible initiating event at each time interval to account for the characteristics of that EES transporting the astronauts to the safe location, and then calculated a P_{Sall} for that EES. .

The @RISK software calculated P_{Sall} for each EES and produced the curves shown in the figure of merit. Notional composite graphs of the Most Likely values for the seven egress methods assessed are shown in Fig 3:

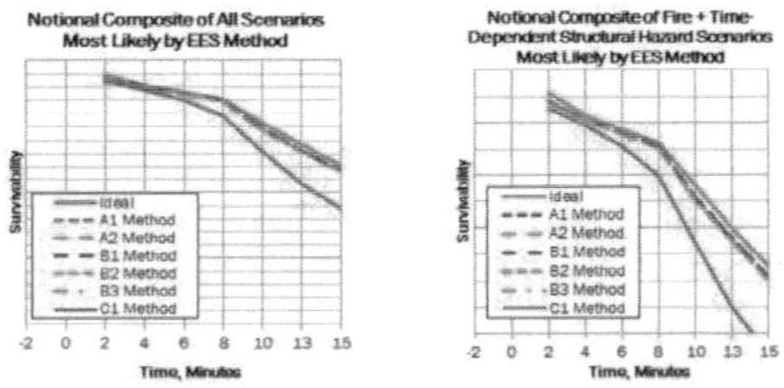

Figure 3 Notional Composite Graphs of Most Likely Values

Results from both assessments were briefed to the stakeholders, GSDO CSO and the GSDO Program Manager. The GSDO Program held various checkpoint reviews leading up to an EES design concept down-select at the GSDO Program Control Board. Method B was selected as the EES design for the LC- 39B.

The third KSC assessment in which a Survivability versus Time Model was developed started in 2013 (KSC IA 1308, "VAB Emergency Egress Survivability Assessment"). As before, the GSDO CSO requested an assessment of the VAB egress routes using the same methodology developed for the LC 39B Assessment. For the VAB Assessment, the IA Team was asked to determine if specific hazard scenarios encountered in processing the SLS and Orion flight hardware reflected a survivability that decreased more rapidly than the other event(s). The VAB assessment evaluated multiple workers (~14 to 90 people) egressing from multiple locations from VAB High Bay 3, compared to the LC 39B Assessments, which evaluated four astronauts moving as a unit using a single egress route. Assembly and testing in the VAB occurs over several months and was divided into seven different processing phases, each with different time durations, spanning from the start of solid rocket motor stacking operations to SLS/Orion rollout to the LC 39B. Each processing phase has a different number of workers in multiple work locations. Due to the size of High Bay 3 and the distribution of workers, the High Bay was divided into eight different work zones (see Fig 4). Egress paths to reach an exit varied by work zone between approximately 30 to 180 feet. The Customer requested the time to reach an exit from each work zone/processing phase be assessed at eight different time intervals.

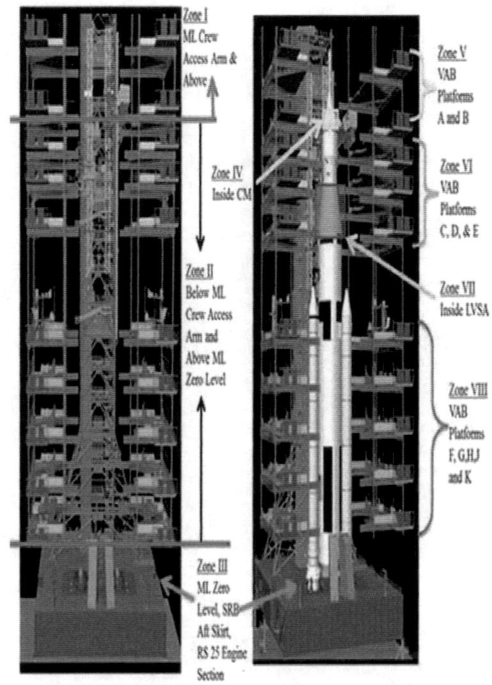

Figure 4 VAB Zones

To determine survivability for multiple personnel at various locations for a specific time, an Aggregate Survival Level was calculated as a weighted average based on manloading and the Survival Level assigned to each Zone.

- The Aggregate Survival Level formula for an individual Initiating Event during one Phase and at one time interval is Eq. 5:

$$P_{(SAggregate|E)} = \Sigma_{i=1}^{8} [(\frac{Headcountzone\ i}{Total\ Headcount}) * P(S_{zone\ i\ |E})] \quad (5)$$

As outlined in the LC 39B assessment, then P_S for all Zones, one Phase, one time interval and Initiating Event is Eq. 6:

$$P_S = 1 - P_D = (1 - (P_E * (1 - P_{(SAggregate|E)}) \quad (6)$$

Thus, the VAB Emergency Egress Analysis formula for the probability of surviving all individual Initiating Events at the same time is calculated in success space by Eq. 7:

$$P_{SAll} = \Pi(P_{Si}) = P_{S1} * P_{S2} * ... * P_{S78} \quad (7)$$

Thus, a composite scenario was developed, denoted P_{SAll}, which represents the probability of surviving all initiating events for all Phase, each time interval, and for all

Zones. The @RISK software calculated PSAll and produced curves for the Maximum, Most Likely, and Minimum. Fig 5 represents a notational composite survivability versus time graph for the VAB Assessment.

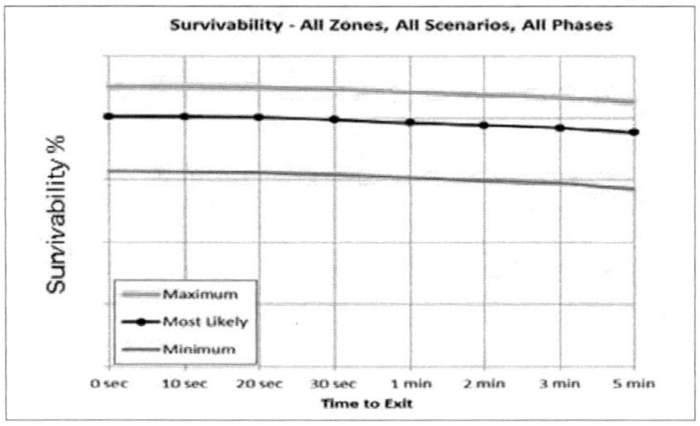

Figure 5 Notional Composite Survivability versus Time Graph

2. SUMMARY

Based on the composite survivability versus time graphs from the first two assessments, there was a soft "knee" in the Figure of Merit graphs at eight minutes (ten minutes after egress ordered). Thus, the graphs illustrated to the decision makers that the final emergency egress design selected should have the capability of transporting the flight crew from the top of LC 39B to a safe location in eight minutes or less. Results for the third assessment were dominated by hazards that were classified as instantaneous in nature (e.g. stacking mishaps) and therefore had no effect on survivability vs time to egress the Vehicle Assembly Building (VAB). VAB emergency scenarios that degraded over time (e.g. fire) produced survivability vs time graphs that were line with aerospace industry norms.

MAJOR INTEREST OF PAPER

Major Interest of the paper: Space Transportation and Safety, Operational Hazards, Emergencies, Egress, Survival, Decision Making, Probability Distribution Functions, Decision Support Systems, Safety Management, Fault Trees, Figure of Merit.

POINT OF CONTACT

James J. Joyner, Sr
Mail Code: SA-G
John F Kennedy Space Center
Kennedy Space Center, FL 32899
Telephone number: USA (321) 867-8313
Email Address: James.J.Joyner@nasa.gov

ADDITIONAL INFORMATION

For more details on the three assessments and the findings, you can read full report "Probabilistic Survivability Versus Time Modeling", NASA/TP-2015-218876 at url address: http://ntrs.nasa.gov or http://hdl.handle.net/2060/20150021266.

PROCEEDINGS OF THE 8TH INTERNATIONAL SPACE SAFETY CONFERENCE

SESSION 12 PROBABILISTIC RISK ASSESSMENT

SPACE MISSION PRAs

D. L. DeMott

SAIC, 2450 NASA Parkway, Houston, Texas 77058, USA,
e-mail: diana.l.demott@saic.com, diana.l.demott@nasa.gov

ABSTRACT

There is general consensus that *"Probabilistic risk assessment (PRA) is a systematic and comprehensive methodology to evaluate risks associated with complex engineered technological entities to estimate risk by computing real numbers to determine what can go wrong, how likely is it, and what are its consequences"* (Various versions of this statement appear in books, NRC NUREGs, and NASA documents).

PRA methodology describes what information is needed and how to organize and analyze this data to provide meaningful results. How an individual or organization plans to use the results of their risk assessment will often dictate the format and scope of the PRA. Risk can be defined in multiple ways including: risk to the public, risk to an individual, risk of facility shutdown, economic risks or risk to the environment. Previous PRA performed for space based missions identified the risks as Loss of Crew (LOC) and/or Loss of Mission (LOM). Potential failures for different phases (launch, ascent, low earth orbit, entry, and landing) of the Mission were included.

Many tools and techniques developed for earth based commercial PRAs are not easily applied on space based PRAs due to the dynamics forces and various extreme environments involved. Systems and equipment need to meet the goals and requirements for complex operations under different environmental parameters, and these same systems often incorporate new technologies with limited reliability data available. Other differences include multiple interfacing organizations and support systems that contribute to determining a reasonable prediction of mission safety and risk. Early identification of what functions and support are needed to understand mission risk helps define the goals and scope needed to support PRA development. Examples of differing considerations for a space based mission PRA include:

- Requirements for manned flight: including government agencies, mission control and flight operations.
- Defining scope of the planned PRA: when does the "mission" start, when does it end, and does the PRA cover all aspect of flight and mission?
- How to address the different environments and environmental stresses the equipment must be designed to operate within.
- Space based applications often use new equipment designs and technologies which may have limited or no specific equipment or system reliability data available.
- Integration of supporting organizational risks into a comprehensive PRA that identifies the overall risk for a successful mission.
- Determining what needs to be included, and at what level of detail to decide the "right" balance to acceptably identify risk.

Although the methodology for performing the probabilistic risk assessment may be the same as other industries, the implementation and performance of the overall Risk Assessment involved in identifying an overall risk for a Mission includes a variety of additional factors and nuances specific to the aerospace industry with its complex and dynamic technologies which really are rocket science.

1. INTRODUCTION

Sending a vehicle into space is not a "safe" venture, there are a variety of inherent risks. Understanding those risks is the first step in determining how to address the issue of what's "acceptable", what is avoidable, what can be eliminated and what can be mitigated.

Probabilistic Risk Assessment (PRA) is used to identify, characterize, and quantify failure probabilities to determine an overall probabilistic risk value. The methodology identifies failure events and risk contributors (such as: system, equipment, procedures, operational activities, human actions, environmental effect) that could result in the undesired event or failure. Risk is defined as a feasible unfavorable outcome considering both magnitude (severity) and likelihood of occurrence. The basis for identifying risk is responding to three questions: (1) what can go wrong, (2) what is the likelihood of the potential events or scenarios and (3) what consequences could result from these events or scenarios?

Most commercial industry PRAs determine a worst case accident or event and develop this scenario as a starting point. PRAs for spacecraft generally identify the major risks as Loss of Crew (LOC) or Loss of Mission (LOM) as the starting point, and cover the potential failures for all phases of the Mission, from prelaunch to landing. Aside from the differences in scope for a Mission based PRA, there are also a number of other factors that are different when using PRA methodology for assessing the risk of launching a crewed spacecraft on a mission.

2. WHAT'S DIFFERENT FOR A SPACE BASED MISSION PRA

In the past, the most common use for PRA has been for nuclear power plants due to regulatory requirements. Although NASA used PRA by the end of the Shuttle program [1] [2], application of the methodology for space vehicles is not well known. Since the commercial application for nuclear is more common, the following discussion uses it as a basis for comparison of the similarities and differences with a space based mission.

Basic PRA concepts remain the same: design capabilities generally means that the systems and equipment successfully operates within specified parameters based on identified requirements, analyzing the "risk" involves the determination of what can cause the design to fail and the consequences of that failure. Looking at risk involves identifying failure scenarios and determining the capabilities and reliability of the system or equipment based on the failure scenario.

Tools or techniques developed for commercial nuclear power PRAs that are not as easily applied on space based PRAs include:

- Identification of initiating events in commercial reactors are now well-developed based on years of past experience and the number of operating facilities. For space based applications, the identification of initiating events are more mission and equipment specific with advances in new technologies causing limited actual experience and hardware knowledge to draw from.
- With years of operational experience and accumulated data available, commercial nuclear PRAs have actual data and experience to clarify and quantify critical events and probability of occurrence. In the case of new designs and technologies with limited or no test data available, the application of simulation techniques to provide reasonable assumptions for development of the risk analysis may provide the needed information until actual data can be obtained.
- The application of new technologies means that designs are still in flux and different designs can be proposed to provide improvement in a specific area. When these happen during vehicle development, these changes can be input into the model to determine if the change will affect other areas including risk and safety. Providing this information during the decision making process can save in reworks and revisions in the future.

2.1. Surface similarities

Figure 1 provides a pictorial illustration of how elements of a PRA interrelate and provide a comprehensive representation of complex systems being analyzed. Not all of the elements shown are used in every PRA. Both current commercial industry PRAs and space based PRAs use the elements shown. A variety of inputs are fed into the fault and event trees to arrive at the predicted risk number.

Areas considered in the performance of a probabilistic risk assessment also include:

- Scope of Analysis which includes: level of model detail, assumptions, what's important, design life, likelihood of occurrence, and start / stop points for the assessment.
- Identification of potential failures / initiating events. (What could go wrong)
- Defining: top events, risk definitions, undesirable consequences, and environmental factors influencing equipment and risk scenarios. (Modeled consequences)
- Types and quality of data input into the model.
- Human performance and reliability inputs consider experience and training to ensure acceptable operations.
- Event trees: complexity and level of detail. (Determine what is important and how much detail is needed to express the desired question or information)
- Fault trees: complexity and level of detail. (Determine what is important and how much detail is needed to express the desired question or information)

2.2. Fundamental Differences

The following table provides a comparison of various input and process needs for a PRA and shows how the same elements could be interpreted for different industries. Since the most common use for PRA in the past was by the commercial nuclear power industry, it is used to provide a basis for comparison between how PRA has been im-

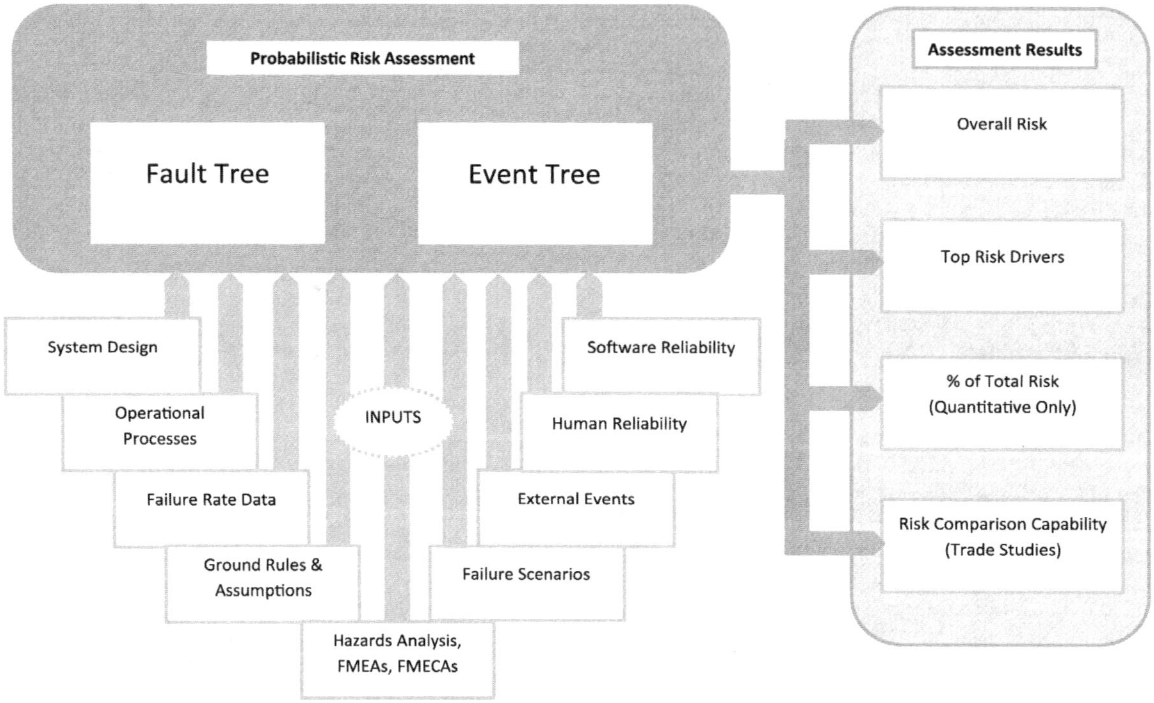

Figure 1. Elements of a PRA

plemented in the past and what new or different considerations arise when applying the methodology to aerospace programs. Some of these differences identified include:

- Goals for using PRA: there may be differences of uses of results in the space industry (subtle difference in use for commercial purposes versus experimental or scientific goals) and how regulatory bodies (such as the NRC) would use the final product.
- Variations in scope of assessments: pre-launch or launch as start, landing or crew is recovered; everything needed to get into space and return or not considering any safety support equipment or only parts of stack; abort or no abort components
- How the dynamic aspect of space flight affect a PRA Model: changing environments, changing requirements, changes in responsibilities, changes in capabilities, changes in support systems, changes in what's important
- Multiple assemblies and components with specific purposes (pre-launch preparations, rocket engines, abort systems, emergency egress systems, life support, landing systems, rescue support systems, etc.) are only needed for specific times and purposes.

3. APPLICATION / IMPLEMENTATION FOR SPACE MISSIONS

PRA techniques have generally been used after systems, equipment and processes have been built and operational information is available. These PRAs determined risk of as-built systems or a facilities and provided a basis for comparison for future changes to ensure risk was not increased or to show the benefit of change in reducing risk.

Current applications for the methodology have expanded to design phases to identify potential risks and what design elements contribute to the risk. This provides opportunities to understand the risk elements better and potentially improve the design by eliminating or mitigating identified risk drivers (what aspect, component, system or equipment failure increases the possibility for not meeting success). By considering these aspects in the design phase, costly re-work could be avoided and a safer design established. Design comparisons can be made by developing a model of the new design and comparing the risk value to the old design.

When using PRA as a safety tool in space based designs and Missions, being aware that there are some basic dif-

Table 1: Comparison of Implementing Parameters

	Nuclear Power Utility	Crewed Space Vehicles
Requirement	IPE*	NASA requirement for Crewed Missions
Scope	Identify and quantify plant vulnerabilities to hardware failures and human faults in design and operation	Mission success elements may include one or all of the following: launch facilities, launch vehicle, crew capsule, mission control centers, recovery operations, etc.
Event Trees	Identification of initiating events in commercial reactors are now well-developed based on years of past experience and the number of operating facilities.	Identification of initiating events are more mission and equipment specific; mission phases; dynamic situations
Fault Trees	Systems and equipment used in commercial reactors based on years of experience and numerous operating facilities.	Limited actual experience and hardware knowledge to draw from; few flights, design and mission differences between programs (ex.: Apollo, Gemini, Shuttle)
Data	Years of operational experience and accumulated data available, actual data and experience to clarify and quantify critical events and probability of occurrence	New or modified designs & technologies have limited or no test data available. Data often consists of "similar to", based on compilation of subcomponents, expert elicitation and the use of simulations to provide needed information until actual data can be obtained
Environment	Earth based location subject to weather and earthquake	• Launch: pressures, temperature, acceleration stresses • Ascent: rapidly changing pressures, temperature, acceleration stresses, MMOD • In-Space: MMOD, pressures, radiation, temperature, weightlessness, vacuum • Entry & Descent: pressures, temperature, deceleration stresses • Landing/Recovery: various potential weather and terrain/ water conditions
Operating Organization & Interfaces	Generally single utility generate power safely	Multiple organizations providing: launch site & launch control, range safety, vehicle design, rocket design, flight operations, crew medical, communications and navigation support, rescue and recovery
Operations	• Power plant operators • Technical support personnel	• Flight Crew • Flight Operations personnel • Mission Control personnel • Launch Control personnel • Ground Support personnel • Technical Support personnel
Systems / Equipment	• Generate and transmit electricity • Containment of nuclear material • Emergency power generation • Coolant supply system • Control system • HVAC	• Launch Pad and emergency egress equipment for on Pad • Launch Control • Pad abort System • Crew vehicle: life support (air, water, temp control, pressure, etc.), communications, guidance and navigation, control systems, power systems, safety systems, thermal shell, etc. • Launch Systems: Booster rockets, fuel tanks, control instrumentation, etc. • Mission Control • Communications systems • Landing equipment: parachutes, thermal shielding, structural integrity, etc. • Recovery Equipment
Timeframes	20 years of operation	• Generally days for crewed vehicle missions / years for ISS • Mission phases can last from minutes to days

*NRC required that each nuclear power plant in the US perform an independent plant evaluation (IPE) to identify and quantify plant vulnerabilities to hardware failures and human faults in design and operation.

ferences that should be considered upfront will provide a more reasonable and useable product. Table 1 outlines some of the basic differences to consider. The following information provides some additional details for application and implementation of PRA techniques for space based missions.

3.1 Identifying the Scope for Space Missions

Scope of a PRA is dependent on what task is required, there may be different levels of effort based on what needs to be analyzed. Examples of these types of different levels include:

- Vendor supplied equipment: only model the equipment supplied based on the equipment requirements.
- Contractor for vehicle module: Group assembling the space vehicle would model the vehicle components and interfaces as defined in their specifications.
- Responsible for Mission success: Those responsible for the entire program would need to integrate all the various elements needed to launch and operate the mission successfully.

For a quantitative PRA, the final value is dependent of the scope and detail of the model. The level of detail is especially important since failure data needs to be available to support the quantification of the model. Level of detail could stop at the system level if system failure data is available, or continue to the equipment and often component levels.

3.2 Developing Event Trees for Space Missions

An Event Tree for a space mission may include identified initiating events and define the different paths that could occur as the scenario progresses. These events covering the entire mission could include items such as:

- Launch is successful,
- Ascent to low earth orbit is successful,
- Transit to target is successful,
- Transit back to Earth is successful,
- Entry into atmosphere is successful
- Landing is successful.

Each of these top level events would add a node to the tree with a split of probabilities branching for success or failure incorporating the effects of functioning or failed systems as the event occurs and provides an overall success or failure probability to identify and quantify risk.

3.3 Developing Fault Trees for Space Missions

Starting with the basic failure scenario, often based on engineering analysis of system behavior in specified conditions, a Fault Tree Analysis (FTA) can provide a system level model to understand how the systems/equipment could fail. A logical relationship between elements using a top down, logical failure analysis starting with an undesired system event and identifying system-level failures, and more basic events such as component-level failures that contribute to system failure. Contributing elements are combined using Boolean logic with failure data for basic events. The result is a pictorial logic diagram which shows the initiating event at the top level with subsequent levels illustrating the logical relationship of supporting equipment, subsystems and components. This would include interrelationships and interactions between various systems and their supporting systems and equipment to identify risk contributions between systems. Support systems could include: electrical power distribution and generation, command and control logic, instrumentation, structures, and software.

3.4 Populating Data fields for Space Mission Equipment, Components, Operations and Phenomenological Events

Each of the basic events identified in the model require data. This can be expressed for equipment or components as the failure rate, the number of failures per unit time for a specified component or element, or a demand failure rate, how often the system or component fails to function on demand. These failure rates are used in the basic events to input into the Boolean logic calculation for the fault or event trees to provide risk values. For phenomenological events, consequences can be expressed numerically and when combined with the likelihood of occurrence will provide a numerical risk which can be included in fault or event trees. Types of data typically used in PRA models for space missions may include:

- Initiating event frequencies and uncertainties
- Component failure rates or failure probabilities and uncertainties
- Common cause failure parameters and uncertainties
- Human error (HRA) rates and uncertainties for: Crew, Flight Operations or Launch Control Services
- Event or phenomena probabilities such as: Micro-Meteoroid and Orbital Debris (MMOD) strike, or radiation
- Simulator and computerized model simulations data
- Expert Elicitation data

Issues that can arise in the selection of data for space based missions include: component failure rate data for common components may not take into account the different environment the component will be exposed to during operations, many of the systems are unique or new technology which has limited or no available data on failure rates, and simulations are dependent on how accurate their information and assumptions are.

3.5 Analyzing the Results

In any complex system there are a multitude of ways for things to go wrong. Understanding where failures are most likely to occur and be most significant is the first step in managing risks. The underlying reason for per-

forming a risk assessment is to identify risk contributors and determine what can be done to eliminate or mitigate the cause. A PRA can provide this information with the added benefit of being able to provide the percentage of risk attributed to each risk driver. Having this information available allows management to work on the highest risk contributors and assess how successful the solutions are by comparing the risk value of the proposed solution with the original risk.
Using the same concept, a specific system fault tree can be used to determine how design or equipment changes affect the risk by comparing the new design (reflected in modifications to the fault tree model) to the old design. As with any quality PRA that provides an integrated model of the entire system or program, interrelationships and dependencies become more clear and potential failure consequences better understood.

Other uses for a space based PRA would include having the ability to determine the differences in the risk between different Mission parameters (more days, fewer low Earth orbits, etc.) to determine how these changes affect the overall Mission risk.

4. CONCLUSION

The value in using PRA methodologies for space provides a way to bring all the disparate entities, programs and technologies into an organized and integrated model showing how they work together. Although the methodology for performing the probabilistic risk assessment may be the same as other industries, the implementation and performance of the overall Risk Assessment involved in identifying an overall risk for a mission includes a variety of additional factors and nuances specific to the aerospace industry with its complex and dynamic technologies which really are rocket science.

5. REFERENCES

1. NASA/SP-2011-3421, *Probabilistic Risk Assessment Procedures Guide for NASA Managers and Practitioners.* www.hq.nasa.gov/office/codeq/doctree/SP20113421.pdf

2. Hamlin, T., Canga, M., Boyer, R. and Thigpen, E., *2009 Space Shuttle Probabilistic Risk Assessment Overview*, 2010, ntrs.nasa.gov/archive/nasa/casi.ntrs.nasa.gov/20100005659.pdf

3. Safie, F., Stutts, R., Huang, Z., *Reliability and Probabilistic Risk Assessment - How They Play Together*, ntrs.nasa.gov/archive/nasa/casi.ntrs.nasa.gov/20150002964.pdf, Annual Reliability and Maintainability Symposium

4. Stamatelatos, Michael, *Probabilistic Risk Assessment: What Is It And Why Is It Worth Performing It?*, NASA Office of Safety and Mission Assurance, 2000, www.hq.nasa.gov/office/codeq/qnews/pra.pdf

D. L. DeMott – Sr. HRA/PRA Analyst
SAIC
2450 NASA Parkway
Houston, Texas 77058 USA

e-mail: diana.l.demott@saic.com,
diana.l.demott@nasa.gov

BS Nuclear Engineering and MBA, and began working career in the nuclear power industry with positions that included: systems engineer, licensing engineer, system safety and PRA specialist. Tasks included: trend analysis, root cause analysis, FMEA, project management, risk assessment and prioritization, auditing, development of specifications and requirements, equipment reliability and probabilistic risk assessment (PRA). Moving out of the nuclear arena, worked as a consultant for clients such as GE and CSX providing services in the area of quality improvement (QI) techniques and facilitation. Moving to Houston and entered the Aerospace industry as a Senior PRA Analyst working on NASA's Johnson Space Center (JSC) Safety & Mission Assurance (S&MA) contract with tasks that included review and development of fault and event tree models based on mission and system design, development of human reliability assessments for Shuttle and development of PRA models in 2012 for a conceptual Mars Mission. Currently works for SAIC supporting an integrated ORION mission PRA and developing HRA assessments for flight crews and flight operations personnel expected to operate the next manned NASA missions. Member of SRE and an AIAA standards committee member.

RESILIENCE ENGINEERING FOR SPACE MISSIONS' SAFETY ASSESSMENT

Riccardo Patriarca[(1)]*, Giulio Di Gravio[(1)], Francesco Costantino[(1)]

[(1)] *Sapienza University of Rome, Department of Mechanical and Aerospace Engineering,*
Via Eudossiana 18 – 00184 Rome (Italy)
[riccardo.patriarca; giulio.digravio; francesco.costantino]@uniroma1.it
**Corresponding author*

ABSTRACT

New challenges constantly affect space safety, mainly due to the complex technological, and procedural evolution of recent years. The human interaction with these factors became steadily more complex, too. Therefore, the traditional linear thinking and the static cause-effect relationship between factors may fail to represent current and future systems' safety needs adequately. A new trend emerges in safety thinking: focusing on non-linear interactions and tight couplings among functions. In order to enhance the probability of success in future space missions, it is thus necessary to go beyond the traditional risk and safety assessment techniques. In a top-down assessment, it becomes crucial focusing on the system taken as a whole, not on parts taken as separate. In line with this approach, this paper highlights the relevance of new systemic methods, in particular the Functional Resonance Analysis Method (FRAM). FRAM recognizes the importance of paying attention both to "things that go wrong" and "things that go right", in line with Safety-II perspective, evolving traditional Safety-I. According to Resilience Engineering concepts, FRAM therefore aims to describe how things happen rather than consider only the linear aetiology of a failure. This core concept recognizes the limits of decomposition and causality, in favour of a more complex principle, called functional resonance. It is therefore more important to understand system dynamics and performance variability rather than modelling individual technological, human or organization failures. FRAM permits thus considering inter-related patterns of events rather than simple causal sequences, in order to enhance traditional safety assessment.

This paper highlights the possibility of adopting FRAM for safety assessment for future space mission, where high availability is mandatory. This paper also evolves the traditional qualitative FRAM structure, paving the way to a probabilistic FRAM that could better fit with space safety assessment needs. By Monte Carlo simulations, this innovative FRAM model highlights potential critical patterns of functions and proactively defines critical couplings, as shown in a preliminary case study relative to the Apollo lunar missions.

1. INTRODUCTION

Accident analysis and risk assessment methods significantly evolved from their initial structure, usually to cope with "new" types of accidents or to adapt to the evolutions in technology.

Starting from the technical methods, based on the reliability and probability theory, after the Three Mile Island accident in 1979, in late 1980s the focus shifts to human factors. Due to the accidents of Challenger and Chernobyl, both in 1986, it appears necessary to introduce also organizational factors and develop thus specific organisational methods and models [1]. In these latter, probabilistic safety assessment parameters introduce the dependence on organizational factors [2].

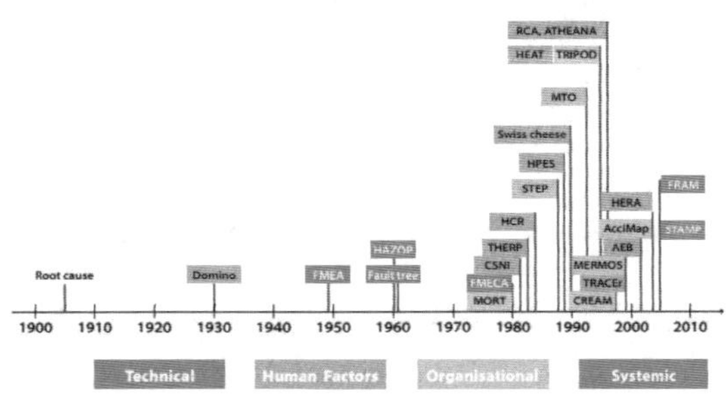

Figure 1. Accident analysis and risk assessment methods and models evolution [3].

Failure Mode and Effect Analysis (FMEA) represents one of the first risk assessment methods, firstly used in 1960s during the Apollo missions [4]. Nowadays it is a common method in several different industries, paving the way also to the development in 1974 of MIL-STD-1629A for Failure Modes, Effects and Criticality Analysis (FMECA). It represents an evolution of FMEA that introduces a criticality analysis. Both the methods are widely applied in the space context, because they allow identifying the product or process failure modes, estimating the risk associated with specific causes and prioritizing mitigating actions [5].

Note, however, that FMEA and FMECA do not take into account the role of human factor, which has been proved being crucial in modern systems. Humans, not only as the flight crew, but also as designers, manufacturers, and ground support, are considered part of the spacecraft system. HFE (Human Factors Engineering) aims to address the role of human performance in a complex system. For instance, the development of HFPFMEA (Human Factor Process FMEA) allows identifying criticalities of human/machine interfaces, potential errors and related consequences [6]. The core assumptions of HFPFMEA consist of imaging that any human error can be controlled by building barriers, adding controls to detect and correct it before it may lead to an undesirable outcome [7].

In the 1970s, Probability Risk Assessment (PRA) gained particular interest in the field of safety management cause of it successfully application to the WASH-1400 "Reactor Safety Study" [8], which established it as the standard approach in the safety-assessment of modern nuclear power plants. PRA is a system scenario-based risk assessment that uses a combination of Fault Trees Analysis (FTA), Event Tree Analysis (ETA), event sequence diagrams, probability and statistical data to analyse the risk of a system, a process or an activity [9]. At NASA, during Apollo program, pessimistic risk estimates discouraged the agency to adopt quantitative risk analysis. Due mainly to the extremely conservative approach and the absence of dedicated software, NASA preferred to adopt qualitative estimates based on FMEA/FMECA, at least until the Challenger accident in the 1986 [10]. Since then, the optimistic perceptions about risk dramatically changed and caused NASA moving towards quantitative analysis and adopting progressively PRA, for space shuttle, for some unmanned space missions and for the space station, in late 1980s and early 1990s. The PRA helps to obtain the probability of a major accident as a function of the probabilities of component and subsystem failures. Based on these results, the decision-maker uses PRA as one of the decision support systems to identify improvements in space missions' safety [11]. The advantages of adopting PRA consists of covering also human factors, and, at least in theory, the organizational factor, too. This idea relies on the certainty that organizational failures are representable in the same way of technical failures. However, modern socio-technical systems are characterized by an increasing complexity in terms of interactions among technical, human and organizational factors, which invalidates this assumption.

2. THE NEED OF SYSTEMIC METHODS

Several safety problems emerge more due to coincidences that happen among the hidden variability of technological, organizational and human factors (or due to a combination of them) rather than to evident failures or incorrect human actions. The inherent complexity of space systems requires an evolution of accident analysis and risk assessment methods, to address properly the interactions between the agents of the system, i.e. human, technical, procedural and organizational functions.

On this path, it appears clear how system description become extremely elaborate, requiring many details. Furthermore, it may be possible that some principles of functioning will be partly unknown, considering the dynamic behaviour of the system and the interdependences among agents, too. These observations reflect the nature of the so-called intractable systems, i.e. systems that cannot completely described or specified. Traditional accident analysis and risk assessment methods decompose systems in order to find the cause- effect link behind each failure and are built under the hypothesis of tractable systems, following the so-called Safety-I.

Nowadays, in complex systems, these hypotheses are no longer universally valid: systems' complexity does not allow a proper description and thus a significant decomposition. Safety focuses thus on the ability to succeed under varying conditions, managing performance variability (Safety-II), rather than constraining it (Safety-I). Resilience Engineering represents a reliable solution for taking into account the systems' intractable nature, i.e. tight couplings and dynamic interactions among functions.

Resilience Engineering aims to model performance variability in a dynamic way, acknowledging the need of analysing the system as a whole, rather than decomposing it. It considers physiological factors (both fundamental, e.g. perception and vigilance, and higher level, e.g. ingenuity, adaptability), organisational factors, social factors, contextual factors and the effects of domain's unpredictability. Resilience Engineering aims to represent the variability of the system in order to identify the factors that may affect system's performance, positively and especially adversely.

On this path, new systemic methods such as STAMP [12] (System Theoretic Accident Modelling and Processes) and FRAM [13] (Functional Resonance Analysis Method) better handle the levels of complexity of current system rather than traditional causality models, as proved in several contexts. This paper evolves the purely qualitative structure of FRAM in order to extend its applicability in the space context.

3. THE FUNCTIONAL RESONANCE ANALYSIS METHOD

Although FRAM is a relatively recent safety assessment method, it has been widely adopted, especially in the aviation context, both for risk assessment [14], [15], and for accident analysis [16], [17]. In addition, several applications are related to different sectors, (e.g.) nuclear power plant [18], oil refinery [19], railway security [20]. FRAM is also important to assess critically the differences among work as done and work as imagined in the industrial context [21], [22].

3.1 The FRAM principles

The traditional FRAM relies on four principles [13]:

- Equivalence of failures and successes. Failure and successes both arise from everyday work variability, which allows both things go right, working as they are imagined to do, and things go wrong. This latter is possible due either to a single system component failure or to a system failure with no single cause recognisable.
- Principle of approximate adjustments. According to the specific situation, people as individuals or as a group and organizations adjust their work, modifying their everyday performance to match the partly intractable and underspecified working conditions of the large-scale socio-technical systems. This principle envisages considering also the changes in everyday work, due to abnormal or unpredictable events.
- Principle of emergence. This principle acknowledges that it is not ever possible to identify the aetiology of a specific safety event or risk. In this context, an event is emergent, as opposed to traditional resultant events. The emergency is due to specific combinations of time and space conditions, which can lead to unwanted and unpredictable events. Furthermore, these events' combinations may be transient and hidden, not leaving any traces.
- Functional resonance. The functional resonance represents the detectable signal emerging from the unintended interaction of the normal variabilities of multiple signals. This resonance is not completely stochastic, because the signals variability is not completely random but it is subject to certain regularities, i.e. recognizable short cut or heuristic, that characterize different types of functions, as shown (e.g.) in Figure 2.

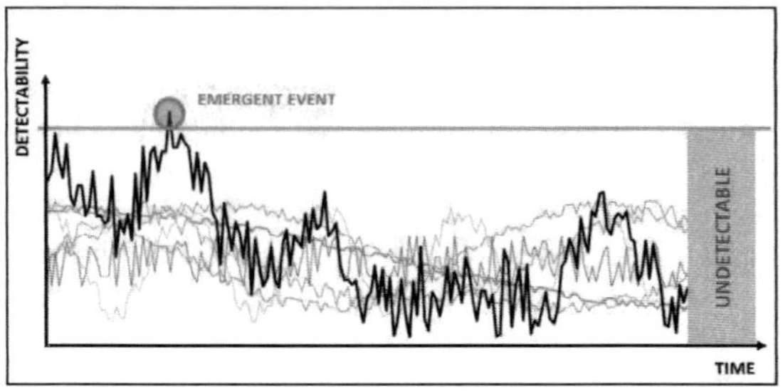

Figure 2. Example of functional resonance (the black line) arising from multiple signals (the blue lines).

Performance variability depends on the nature of the functions, i.e. technological, human and organizational ones. The technological functions usually represent all the machineries' work and, for this reason, their variability is generally considered low and negligible with respect to the other classes' functions. For a space equipment, this variability can be assumed as the probability of failure of the specific machinery [23].

On the other hand, human functions consist of the actions taken by individual or small groups of individuals. Differently from technological functions, human functions change rapidly, with high frequency and high amplitude. The changes generally depend on physiological, psychological status and working conditions.

Lastly, organizational functions differ from human functions because they are functions of the system per se, rather than of the people who work in the system. Their variability is low in frequency but high in amplitude. This concept is in line with the low frequency of changes about (e.g.) rules, regulations or policies, which however foresee big differences in the change (and thus a high amplitude).

3.2 The FRAM building steps

Hollnagel (2012) proposed also four steps to apply FRAM for safety assessment in a complex system. Although this paper proposes an evolution of the traditional FRAM, it is useful to follow the same traditional building steps, detailing the evolutions in methodology. Note also that beside the four steps, there is a preliminary step, i.e. the Step-0, which consists of clearly defining the scope of the analysis. More specifically, one has to decide preliminarily if the FRAM is built for the purpose of risk assessment or accident analysis. This paper focuses on risk assessment, in order to identify proactively couplings and interactions among system's components.

3.2.1 Step 1: Identification and description of system's functions

Functions aim to describe everyday system's functioning, not simple every component individual activities. A FRAM function represents the activities required to produce a certain outcome. Six different aspects can characterize each function:

- Input (I): that which is used or transformed to produce the output.
- Output (O): that which is produced by the function. It can be either an entity or a state change and constitutes links to subsequent functions.
- Precondition (P): system conditions that must be fulfilled before the function can be carried out.
- Resource (R): that which is needed or consumed by the function to process input (e.g. matter, energy, hardware, software, and operators).
- Control (C): that which supervises or adjust a function. It can be plans, procedures or guidelines.
- Time (T): temporal constraints of the function, with regard to both duration and time of execution.

The six aspects are traditionally placed at the corners of a hexagon, which represents the function (see Figure 3).

It is useful to follow the traditional classification of functions in two classes: foreground and background. The background functions represent the system's components not analysed in detail in the FRAM analysis and therefore they have only one input or one output. On the other hand, the foreground functions represent the core of the analysis, and thus have to be exploited as much as possible, potentially with respect to all the six aspects.

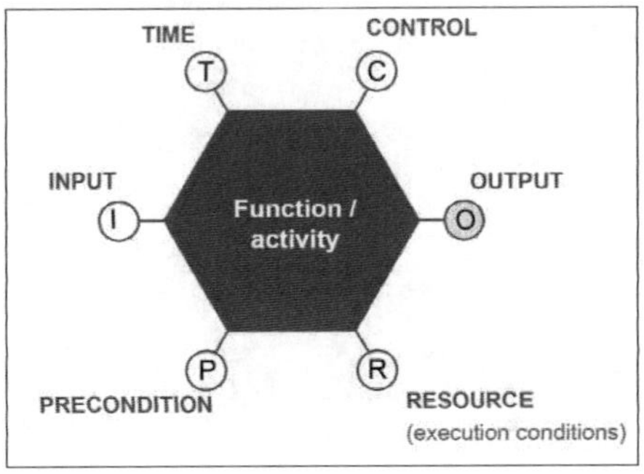

Figure 3. Graphical representation of a FRAM function.

3.2.2 Step 2: Identification of performance variability

Understanding performance variability is a crucial point of a FRAM analysis. Performance variability describes how functions can couple and how this can lead to unexpected outcomes. For this purpose, it is mandatory a wise and accurate description.

In line with the traditional FRAM approach, it is possible to define performance variability through different manifestations: the phenotypes (PH), i.e. speed, distance, sequence, object, force, duration, direction, timing, which describe the endogenous and exogenous causal factors [13].

Table 1 shows an example of variability classification for each phenotype. Table 1 shows also a numerical score for each performance variability state. This score represents a solution to combine the linguistic terms with a deterministic value, in order to express the effects on system performance of a function's variability. The Subject Matter Experts (SMEs) involved in the analysis should assign the numerical scores according to the requirements of the specific mission (or phase of mission). The higher the score, the more critical the output variability.

According to this choice, an output may occur, (e.g.) in terms of time, either at nominal time, or too early or too late. In this case, it may have no impact on mission success and safety, or a low impact on mission success not compromising safety, or even have impact both on mission success and safety. This latter possibility represents the most critical situation, in which an output is produced too late (or even too early) that it is dangerous for the mission safety. Imagine for example the timing of a decision about land or abort a specific phase of a mission, subsequently to output of a visibility test. Exceeding the nominal time may cause critical consequences for both direct linked functions, (e.g.) spacecraft structure and transient linked ones, (e.g.) crew's physiological status. Although, in the first case, the variability may be easily detectable by traditional causality methods, in the second case, it may be hardly recognizable and may require a systemic analysis.

Table 1. Variability score for each phenotype.

PH=1	Possible Output variability with regard to time			
Time Variability	Nominal	Too late or too early - no impact on mission success and safety	Too late or too early - low impact on mission success	Too late or too early - impact on mission success and safety
Score	1	2	3	4
PH=2	**Possible Output variability with regard to precision**			
Precision Variability	Nominal	Acceptable - no impact on mission success and safety	Imprecise - low impact on mission success	Imprecise - impact on mission success and safety
Score	1	2	3	4
PH=3	**Possible Output variability with regard to speed**			
Speed Variability	Nominal	Too slow or too fast - no impact on mission success and safety	Too slow or too fast - low impact on mission success	Too slow or too fast - impact on mission success and safety
Score	1	2	3	4
PH=4	**Possible Output variability with regard to object**			
Object Variability	Nominal		Detectable Failed	Not Detectable Failed
Score	1	2	3	4
PH=5	**Possible Output variability with regard to direction**			
Direction Variability	Nominal	Not nominal direction - no impact on mission success and safety	Imprecise direction - low impact on mission success	Wrong direction - impact on mission success and safety
Score	1	2	3	4
PH=6	**Possible Output variability with regard to force**			
Force Variability	Nominal	Too much or too little - no impact on mission success and safety	Too much or too litlle - low impact on mission success	Too much or too little - impact on mission success and safety
Score	1	2	3	4

By a rule block based on IF-THEN logic, it is possible to combine, for each j-th output, the variability in terms of each phenotype $V_{j,PH}$, where PH=1, 2,...6. This rule block, called here PRB (Phenotypes' Rule Block), allows obtaining the output variability OV_j represented by one of the five dimensions described in Table 2.

Table 2. Output variability.

OV_j	Possible Output variability			
Output Variability	Nominal	Negligibly Variable	Varibile with impact on mission success	Varibile with impact on mission safety
Score	1	2	3	4

The rule block's dimension depends on the considered number of phenotypes and on their dimensions. In this case, based on Table 1, the rule block has 3072 rows. Figure 4 shows an example of these rules. The output variability is then represented easily combining its phenotypes' values.

In a practical sense, however, the main problem at defining the variability scores arises from the awareness that a static behaviour for a complex system component may not reflect properly its real status. For example, even though an instrument with a low failure rate generates a generally nominal output, it may have rare errors that cause potential delays on transmissions, resulting thus in a late output (in terms of time phenotype). This idea acquires an even more relevant role for organizational and human functions, due to the more unpredictable effects of everyday work. Defining a static variability score may be an error that compromises a proper representation of the everyday variability.

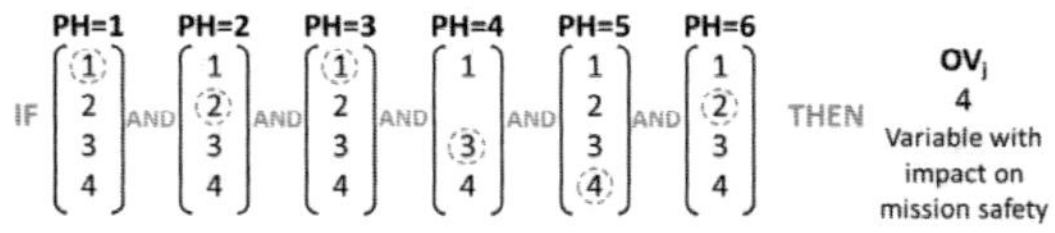

Figure 4. Example of PRB rules' combination.

Figure 5 shows a potential discrete distribution, arising from the description in terms of precision of a technological function. Technical datasheet, if any, and simulating tests and scenario analysis can help to draw in detail the variability of each function. It is thus possible to adopt Monte Carlo simulation in order to evaluate multiple combinations of performance variability and their effect on the output, once assigned the variability in terms of each phenotype.

3.2.3 Step 3: Aggregation of variability

This step focuses on examining how the potential variability of each function can become resonant and lead to unexpected results, following the functional resonance principle for each upstream-downstream coupling.
The variability of a function results as a combination of the function variability and the variability deriving from the output of the upstream functions, depending on the function type and the linked aspects' type.

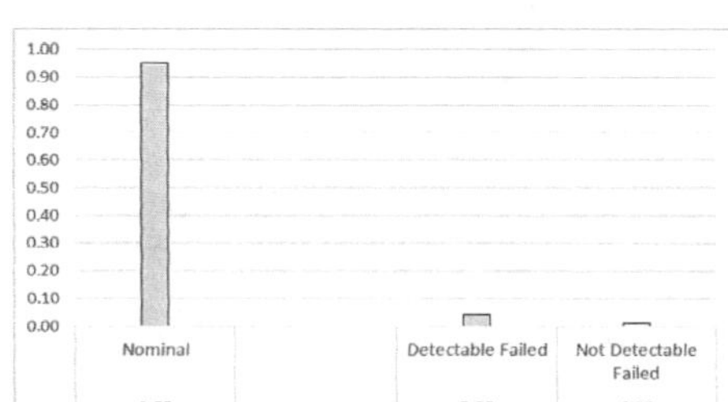

Figure 5. Example of a probability distribution function describing the performance variability.

For a generic function, it is possible to define qualitatively the effects of a coupling. For example, an upstream output that represents a precondition for the downstream function may cause a loss of time if it arrives too late, and then amplify the variability. On the other hand, the same output may damp the variability, if on time, or may result in a false start and amplify variability, if too early. Also in terms of precision, the same output may cause misunderstanding and amplify variability, if imprecise or even a loss of time to eliminate potential disambiguation (e.g. a readback). This research sketches the damping or amplifying effects of each coupling by a specific index, i.e. the Damping/Amplification factor DA_{ij}, of the upstream output j and the downstream function , as shown in Table 3.

Table 3. Output variability.

DA_{ij}	Possible Coupling variability			
Damping Amplification Factor	Damping Effect	No Effect	Amplifying Effect	Critical Amplifying Effect
Score	1	2	3	4

By an additional rule block, called here CRB (Coupling Rule Block), it is thus possible to combine the output variability OV_j with the DA_{ij} and obtain a synthetic index, which address how critical is each coupling, i.e. the Variability Priority Number VPN_{ij}. This latter has the same structure of OV_j, as shown in Table 4. The outcome of this step is a sequence - one for each Monte Carlo iteration - of VPN_{ij}, which can be summarized as a probability distribution function of the scores shown in Table 4. Figure 5 sketches the PRB and CRB, and how they allow evaluating VPN_{ij}.

Table 4. Coupling variability.

VPN_{ij}	Possible Coupling variability			
Coupling Variability	Insignificant	Minor	Moderate	Critical
Score	1	2	3	4

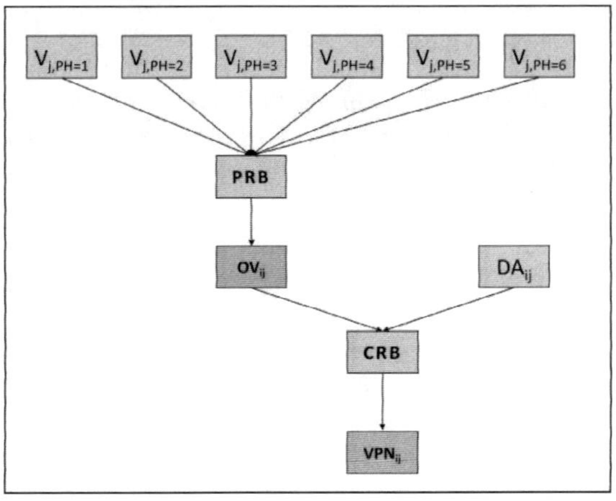

Figure 6. Input and output of PRB and CRB.

3.2.4 Step 4: Managing the variability

The evolved step 3 defines the variability of each coupling taking into account the upstream variability and the upstream-downstream link, i.e. looking at VPN_{ij}. The fourth step aims to identify the more critical couplings, in order to reduce the complexity of system's representation.

For this purpose, a coupling is defined critical if its cumulative distribution over a specific threshold VPN^* is major than a specific confidence level P^*. Furthermore, rather than focusing on single couplings, it would be possible to define critical paths, considering multiple critical couplings which relate the same functions. In this case, for each critical coupling, it is necessary to identify if other couplings, backwards or afterwards, are critical too. The functional resonance emerges relating multiple upstream-downstream functions. This analysis defines the priority of intervention, addressing where the investigation should start, as shown in the following preliminary case study.

4. CASE STUDY: APOLLO LUNAR LANDING

The Apollo program has revolutionised the space exploration as well man's perspective in the universe. The program, bringing unprecedented results to humanity, realized both technological innovations and cultural implications, with a global interest in the space field and, more generally in the entire scientific world [24]. This paper shows a preliminary application of the proposed innovative FRAM methodology to highlight the potentialities of the model. More specifically, it models the performance variability by the FRAM, for specific operations of the powered descent to lunar surface of Mission G, i.e. the operations leading to lunar touchdown, starting with the visibility phase. The analysis relies on the MSC Internal Note NO. S-PA-8N- 021 A [25], technical reports of Apollo missions, and Apollo flight journal [26].

4.1 Description of powered descent

The MSC Internal Note NO. S-PA-8N-021 A [25], accurately describes the sequence of events for the descent phase of the G mission, the data flow and the decision logic. In a general perspective, the Mission Control Center – Houston (MCC-H) is responsible for detecting insidious, slow drift malfunction and advising the crew of its decision. The crew involved in this phase, i.e. the Lunar Modul Pilot (LMP) and the Commander (CDR), is responsible for detecting errors that require immediate action. Here follows a brief description of the events' sequence.

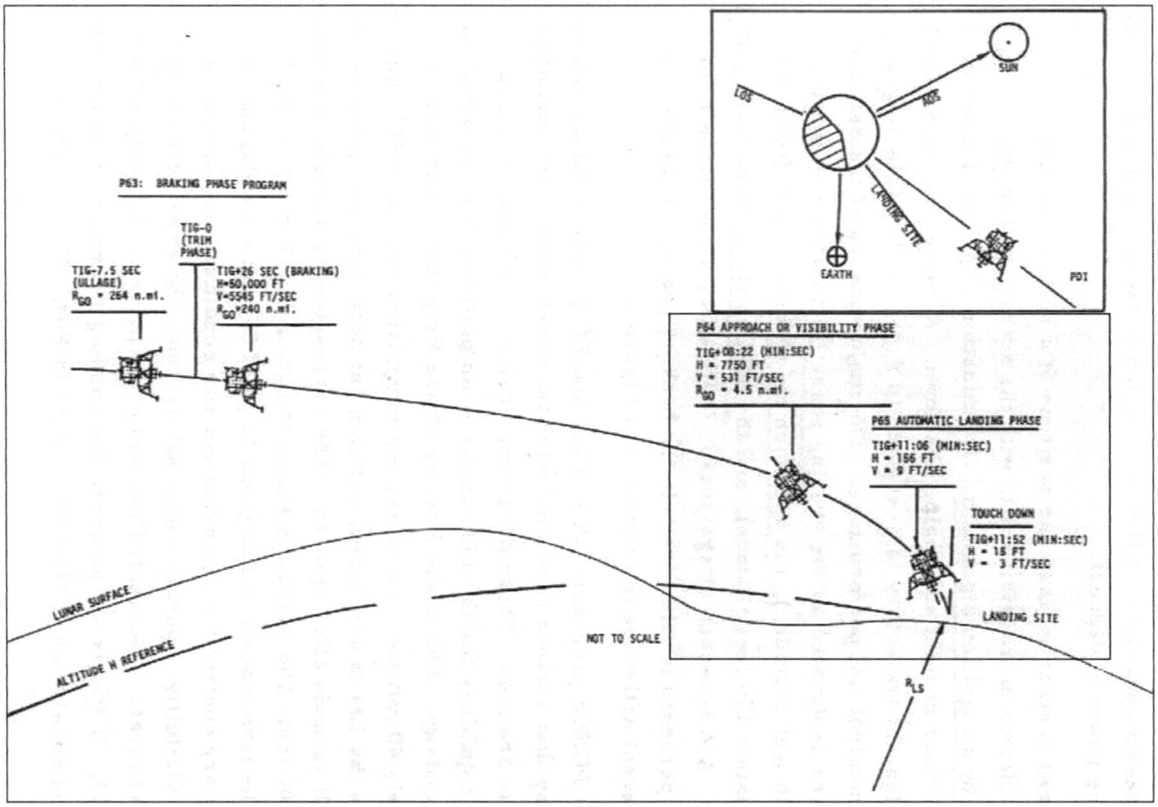

Figure 7. LM powered descent events, where the yellow box highlights the focus of this case study.

Powered descent begins near perilune of a 60-nautical mile by 50000-foot Hohmann transfer orbit, and terminates with lunar contact and engine shutdown.

The major phases of the descent are the braking phase, which terminates at approx. 7700 feet altitude, the approach or visibility phase, which terminates at 150 feet altitude, and the landing phase, terminating at the touchdown.

Specific PGNCS programs (P63, P64, P65) control the trajectory in terms of velocity, position, acceleration. P63 is the braking Phase program and initiates approximately 40 minutes before DPS (Descent Propulsion System) ignition. It is active until 8 minutes 20 seconds after ignition. The P64 starts automatically at the end of P63 and exercises trajectory control in order to achieve the aim conditions and providing landing site visibility.

During a nominal descent, the crew will select the P66 (Landing Phase Program) prior the completion of P64. If P66 is not entered, P65 (Landing Program) would be automatically activated to ensure a safe landing. Furthermore, to obtain manual control of the DPS throttle and digital autopilot, the crew may use P67 (Manual Landing Program).

Following LM contact with the lunar surface, CDR will disarm the DPS engine and decide if continue with lunar surface operations.

4.2 Applying FRAM Step1

The first step consists of identifying the system's functions, describing them univocally and discovering their interactions, through the definition of the six aspects. Based on [25], once individuated the main phases of the process (see Figure 9), it has been possible to define the system through 50 functions, following the FRAM structure (see Figure 8). Note that the grey hexagons indicate the background functions.

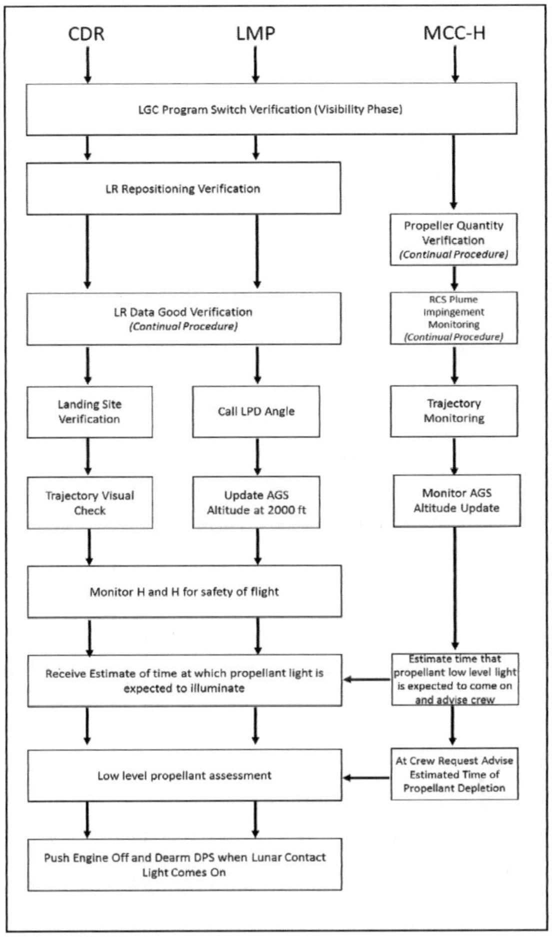

Figure 8. Phases of the analysed process.

4.3 Applying FRAM Step 2

Following the process described in §3.2.2., once defined the functions, it is possible to define their variability in terms of phenotypes, and assign then the relative score, which allow defining a probabilistic distribution function. Following the PRB, it is then possible to obtain the output variability OV_j. Table 5 shows an example for the function "Velocity reasonability test".

Table 5. Example of a FRAM function with variability exploited.

Name of function	*Velocity reasonability test*
Description	Part of LR Data Good Verification (Continual Procedure) which could lead to a LR self check. The crew may continue without the velocity update as long as the visual assessment in adequate and the remaining LR-PGNCS altitude difference is acceptable [25].
Variability	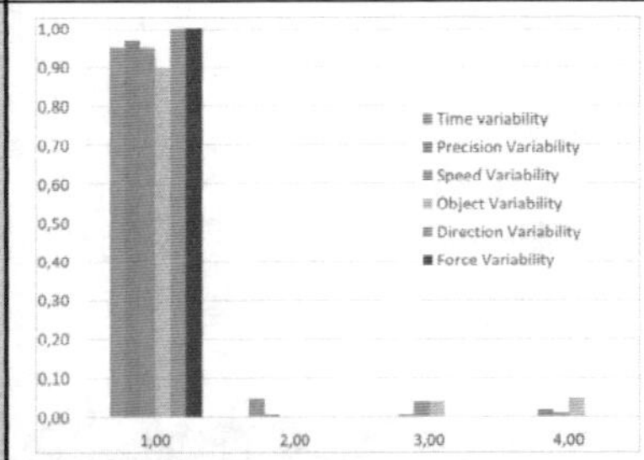
Aspect	**Description of Aspect**
Input	LR Altitude Light flashing
	Velocity Fail Light flashing
Output	Mission Continued to Landing Site Verification
	No LR Update
Precondition	LR Altitude Light steady
	Velocity Fail Light steady
	P64 activated
Resource	Velocity Fail Light working
	LR Altitude Light working
	CDR
	LMP
Control	LR self check
Score	Mission continued to LR Data Good verification

4.4 Applying FRAM Step 3

Once built the map through the definition of the system's functions and their couplings (Step 1), and defined the variability of each output (Step 2), it is necessary to evaluate how these couplings interact, leading to a potential criticality. For each Monte Carlo simulation, it is thus necessary to consider the output variability by the PRB and then the effect on system due to the coupling, by DA_{ij}. Table 6 shows some examples of couplings, highlighting the upstream/downstream link and its variability.

4.5 Applying FRAM Step 4

The application of Step 3 allows defining the variability of each coupling, which in the step 4, will be evaluated in order to define the more critical ones. For this purpose, this research considers $VPN^* = 2$, i.e. minor variability, and a confidence level $P^* = 0.99$.

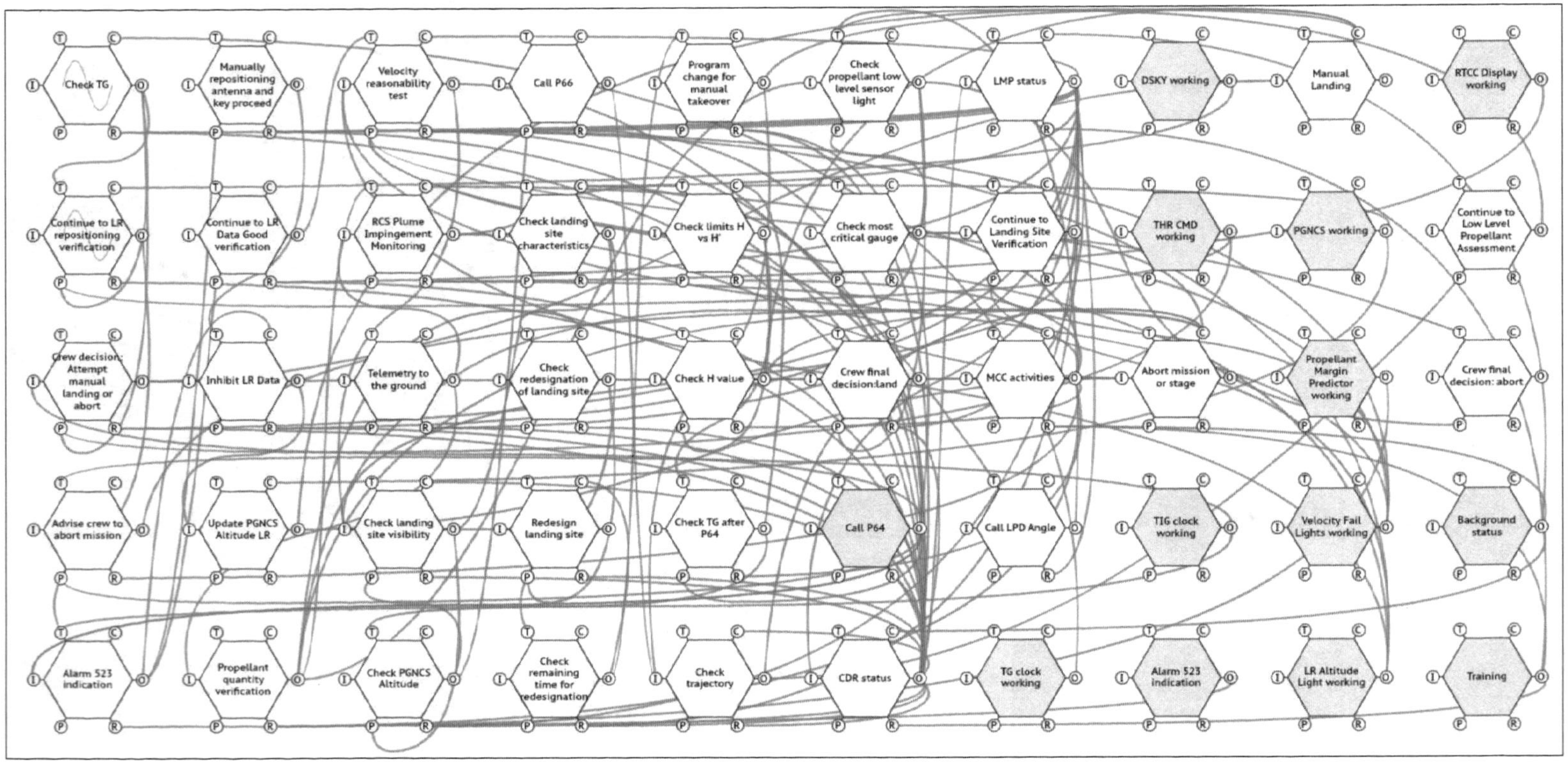

Figure 9. FRAM functions of the lunar powered descent.

Table 6. Some examples of aggregation of multiple upstream/downstream functions in a Monte Carlo iteration

Downstream function		Upstream function			DA_{ij}
Name of function	Aspect	Name of function	Description of Aspect	$V_{j,PH}$	
Velocity reasonability test	Precondition	Call P64	P64 activated	$V_{j,1}=2; V_{j,2}=1; V_{j,3}=1; V_{j,4}=1; V_{j,5}=1; V_{j,6}=1; OV_j=1$	1
	Time	Continue to LR repositioning verification	Mission continued to LR Data Good verification	$V_{j,1}=1; V_{j,2}=2; V_{j,3}=1; V_{j,4}=1; V_{j,5}=1; V_{j,6}=1; OV_j=2$	2
	Resource	Velocity Fail Light functioning	Velocity Fail Light working .	$V_{j,1}=2; V_{j,2}=1; V_{j,3}=1; V_{j,4}=3; V_{j,5}=1; V_{j,6}=1; OV_j=3$	3

This means that a coupling will be considered critical only if its variability in (at least) 1% of Monte Carlo iterations is greater than *VPN**. Figure 10 shows an example of a critical path, which highlights the critical role of propellant (and a reliable verification of its level) to achieve a safe landing. As confirmed from the critical path, it is relevant to highlight the importance of telemetry to ground and communication between the crew and MCC-H. Besides the technical failure of lights, it is necessary to consider the importance of the limits on the gauges of fuel. All these observations lead to a possible decision of landing or abort mission.

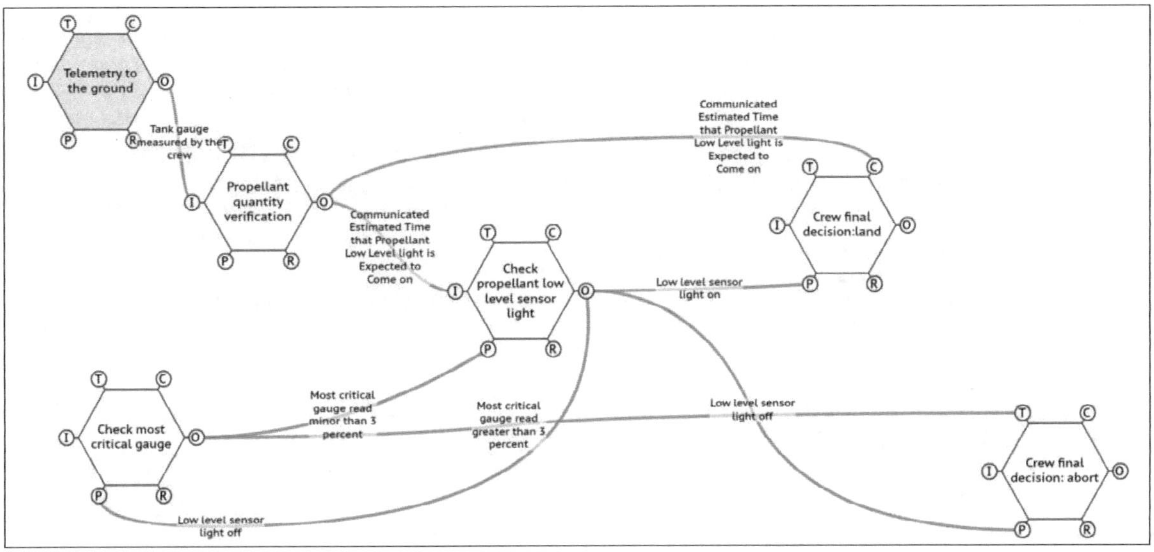

Figure 10. An example of a FRAM critical path.

5. CONCLUSION

FRAM allows a structured analysis of a complex system where complexity arises from transient cause-effect links and non-linear interactions among agents.

FRAM aims at individuating the variability of the interactions among two or more system's components, rather than the single probability of failure. FRAM, incorporating the principles of system theory and Resilience Engineering, allows taking into account both the organizational, human, procedural and technological aspects. This paper proposes an evolution of the traditional FRAM model, on the basis of a semi- quantitative structure and Monte Carlo iterations, which could be applied for risk assessment of a space mission. Although the case study in this paper is based on raw data, it depicts the possibility to highlight different risk sources, at organizational, human and technological level, (e.g.) training, human factors, emergency procedure effectiveness, budget and schedule pressure, requirements definition and validation, critical system redundancy.

The potential of the method will arise in case the semi-quantitative structure would rely on reliable dataset (e.g.) of failure probability and human training simulation results. Furthermore, for a potential application to assess risks of a complex space mission, it will be interesting to explore the possibility of combining it with the well-established PRA [27].

Furthermore, FRAM allows defining the effects of changes, or implementations on system safety. Once identified the critical couplings and paths, it would allow estimating the effects of a system's component in terms of variability of the system as a whole.

Further research directions may allow developing a multi-layer structure, in order to obtain a more manageable representation of the system. For example, in a three-layer representation, the third layer would be the FRAM traditional one, the second one would represent sub-processes, and the first one would represent the agents' and their responsibilities for the process. In this case, each layer would guide the risk assessment process with different implications for the analysis.

6. ACKNOWLEDGEMENTS

The authors developed a VBA code to execute the quantitative analysis conducted in this study and they interface it with the FRAM Model Visualizer (FMV), a software written and developed by Rees Hill, which the authors greatly acknowledge for his contribution to the development of traditional FRAM. The authors want to thank also ing. Fabrizio Bernardini, FBIS, for his support during the research.

7. REFERENCES

1. J. L. Hall, "Columbia and Challenger: Organizational failure at NASA," *Space Policy*, vol. 19, no. 4, pp. 239– 247, 2003.

2. G. Apostolakis and J.-S. Wu, "A structured approach to the assessment of the quality culture in nuclear installations," no. April, pp. 24– 28, 1995.

3. EUROCONTROL, "A White Paper on Resilience Engineering for ATM," pp. 1– 16,2009.

4. NASA, "Procedure for Failure Mode, Effects, and Criticality Analysis (FMECA), N70-76197," Washington DC, 1966.

5. L. Lipol and J. Haq, "Risk analysis method: FMEA/ FMECA in the organizations.," *Int. J. Basic Appl. Sci.* , vol. 11, no. 05, pp. 74– 83, 2011.

6. NASA, "JSC report 29867 - Human Reliability Analysis (HRA) Final Report," Johnson Space Center, 2002.

7. A. Hobbs, J. O'Hara, B. Adelstein, and C. Null, "Defining reliability and robustness from a human factors perspective," in *Space Safety is No Accident - The 7th Conference* , 2015, pp. 277–283.

8. Atomic Energy Commission, "Reactor safety study: an assessment of accident risks in US Commercial power plants (WASH-1400)," Washington DC, 1975.

9. M. Stamatelatos, "Pr obabilistic Risk Assessment Procedures Guide for NASA Managers and Practitioners," *Office* , no. December, p. 323, 2002.

10. National Research Council, "Post –Challenger Evaluation of Space Shuttle Risk Assessment and Management," Washington DC, 1988.

11. E. Paté-Cornell and R. Dillon, "Probabilistic risk analysis for the NASA space shuttle: a brief history and current work," *Reliab. Eng. Syst. Saf.* , vol. 74, no. 3, pp. 345– 352, 2001.

12. N. Leveson, "A new accident model for engineering safer systems," *Safety Science* , vol. 42, no. 4. pp. 237– 270, 2004.

13. E. Hollnagel, FRAM: *The Functional Resonance Analysis Method: Modelling Complex Sociotechnical Systems* . Ashgate, 2012.

14. T. Sawaragi, Y. Horiguchi, and A. Hina, "Safety Analysis of Systemic Accidents Triggered by Performance Deviation," in *2006 SICE-ICASE International Joint Conference* , 2006, pp. 1778–1781.

15. I. A. Herrera, E. Hollnagel, and S. Håbrekke, "Proposing safety performance indicators for helicopter offshore on the Norwegian Continental Shelf," *PSAM 10 - Tenth Conference on Probabilistic Safety Assessment and Management* . p. 10, Jun-2010.

16. E. Hollnagel, S. Pruchnicki, R. Woltjer, and S. Etcher, "Analysis of Comair flight 5191 with the functional resonance accident model," *8th International Symposium of the Australian Aviation Psychology Association* . p. 8 pages, 2008.

17. P. V. R. De Carvalho, "The use of Functional Resonance Analysis Method (FRAM) in a midair collision to understand some characteristics of the air traffic management system resilience," *Reliab. Eng. Syst. Saf.* , vol. 96, no. 11, pp. 1482–1498, Nov. 2011.

18. K. Lundblad and J. Speziali, "FRAM as a risk assessment method for nuclear fuel transportation," *Int. Confererence Work. Saf.* , no. October, 2008.

19. G. . Shirali, V. Ebrahipour, and L. Mohammad salahi, "Proactive Risk Assessment to Identify Emergent Risks using Functional Resonance Analysis Method (FRAM): A Case Study in an Oil Process Unit," *Iran Occup. Heal.* , vol. 10, no. 6, pp. 33– 46, May 2013.

20. R. Steen and T. Aven, "A risk perspective suitable for resilience engineering," *Saf. Sci.* , vol. 49, no. 2, pp. 292– 297, Feb. 2011.

21. A. G. Amorim and C. M. N. A. Pereira, "Improvisation at Workplace and Accident Causation - An Exploratory Study," *Procedia Manuf.* , vol. 3, no. Ahfe, pp. 1804– 1811, 2015.

22. R. Clay-Williams, J. Hounsgaard, and E. Hollnagel, "Where the rubber meets the road: using FRAM to align work-as-imagined with work-as-done when implementing clinical guidelines," *Implement. Sci.* , vol. 10, no. 1, p. 125, Aug. 2015.

23. NASA, "Saturn Component Failure Rates and Failure Rate Modifiers," Marshall Space Flight Center, Alabama, 1971.

24. R. C. J. Seamans, "Project Apollo - The Tough Decisions," Washington, DC, 2005.

25. NASA, "Apollo Mission Technique s - Mission G Lunar Descent (revision A)," Houston, Texas, 1969.

26. D. Woods, "Apollo Flight Journal," 2016. [Online]. Available: http://history.nasa.gov/afj/. [Accessed: 01-Mar-2016].

27. M. V. Frank, "Choosing among safety improvement strategies: a discussion with example of risk assessment and multi-criteria decision approaches for NASA," *Reliab. Eng. Syst. Saf.* , vol. 49, no. 3, pp. 311–324, 1995.

INTERNATIONAL COLLABORATION FOR HTV PRA ANALYSIS

Masami Miki[(1)], Satomi Takada [(1)], Takashi Goto[(1)], Koji Oga[(1)], Toru Yoshihara[(2)], Hiraku Kudo[(2)], Norimitsu Kamimori[(2)], Jinfeng Ni[(3)], John Yasensky[(3)], Philip Mortillaro[(4)]

[(1)] *Japan Manned Space Systems Corporation (JAMSS), Tsukuba Center Inc, D, 2-1-6, Sengen, Tsukuba, Ibaraki, 305- 0047, Japan, Email:miki.masami@jamss.co.jp*
[(2)] *Japan Aerospace Exploration Agency (JAXA), Tsukuba Space Center 2-1-1, Sengen, Tsukuba-shi, Ibaraki, 305-8505 Japan*
[(3)] *MAPI Contract, 16441 Space Center Blvd B-100, Houston, TX 77058 United States*
[(4)] *National Aeronautics & Space Administration(NASA), Lyndon B. Johnson Space Center Houston, Texas 77058 United States*

ABSTRACT

Japan participates in the International Space Station (ISS) program, and Japan Aerospace Exploration Agency (JAXA) provides an unmanned orbital carrier, H-II Transfer Vehicle (HTV). The initial HTV flight was launched in 2009. So far, five HTVs have been flown to ISS successfully.

In JAXA, hazard analysis has been conducted for all HTV flights. There has been no major incident so far and safety of the ISS crew and vehicle is assured by the adequate hazard analysis. In addition to this qualitative analysis, JAXA has conducted Probabilistic Risk Assessment (PRA) for the HTV to get better safety assurance in the design/manufacturing phase.

NASA also separately developed an HTV PRA model, mainly for calculating ISS generic risks, and for trade studies involving the HTV. However, NASA has several uncertainties in developing the HTV PRA model because NASA did not have enough data regarding the HTV. For this reason, among visiting vehicle PRA analysis in NASA, the analysis of the HTV has greater uncertainty than that of other visiting vehicles.

Therefore, NASA and JAXA started their joint efforts to update their models driven by the request from NASA. The outline of joint work is the followings:

(1) Share failure rate data with NASA
The original NASA HTV model uses failure rates of analogue components, while JAXA has unique failure rates for the HTV components. Therefore, JAXA shares failure rates with NASA.
(2) Compare the original NASA model and JAXA model
NASA/JAXA compare the original JAXA model/results with those of NASA in order to identify items/portions which should be modified in the NASA model for accuracy.
(3) Study and improve NASA model
This paper will explain the differences of the original JAXA and NASA PRA models, and how to improve the NASA model and the results.

1. INTRODUCTION OF THE HTV

Japan participates in the ISS program providing the Japanese Experimental Module (JEM) and the HTV. The HTV is an unmanned orbital carrier that can carry both pressurized and unpressurized cargo to the ISS. The HTV is composed of five sections; the Pressurized Carrier, Unpressurized Carrier, Avionics Module, Propulsion Module, and Exposed Pallet. In addition, a Proximity Communication System (PROX) is installed into the rack in the JEM system, a Hardware Command Panel (HCP) is in the US segment, and Laser Radar Reflectors (LRR) are on the bottom side of JEM.

The HTV is launched by the H-IIB transfer vehicle, aboard with up to 6,000kg of system and experiment racks, and supplies. When the HTV approaches close to the ISS, the Space Station Remote Manipulator System (SSRMS) grapples the HTV and berths it to the ISS. While the HTV is berthed, the ISS crew can enter in it to unload cargo such as food, clothes, and experiment equipment. Then, the HTV is loaded with the wasted materials. After the transfer, the HTV is deberthed from the ISS and reenters the atmosphere. The HTV initial flight vehicle was launched in 2009. So far, five HTVs have been flown to ISS successfully.

In JAXA, hazard analysis has been conducted for all HTV flights according to the ISS safety requirements. There has been no major incident so far and safety of the ISS crew and vehicle is assured by the adequate hazard analysis, controls and verification. In addition to this qualitative analysis, JAXA has conducted Probabilistic

Risk Assessment (PRA) for the HTV to get better safety assurance in the design/manufacturing phase.

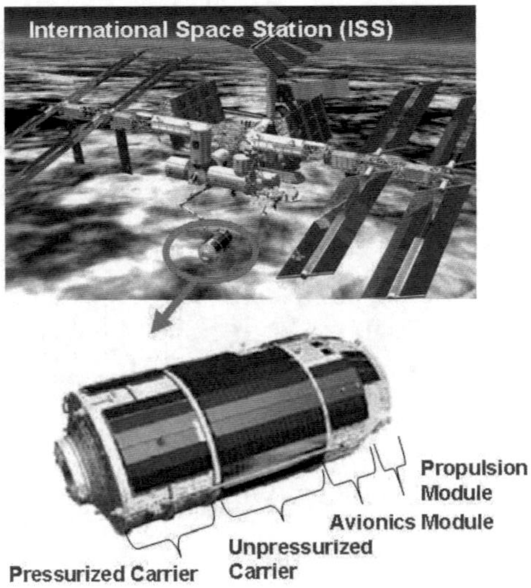

Figure 1. H-II Transfer Vehicle (HTV)

2. MOTIVATION FOR INTERNATIONAL COLLABORATION FOR HTV PRA

While JAXA performed PRA for the HTV individually, NASA also separately developed the HTV PRA model mainly for calculating ISS generic risks, and for trade studies involving the HTV. However, NASA has several uncertainties in developing the HTV PRA model because NASA did not have enough data regarding the HTV. For this reason, among visiting vehicle PRA analysis in NASA, the analysis of the HTV has greater uncertainty than other visiting vehicles.

Therefore, NASA and JAXA started their joint efforts to update their models driven by the request from NASA.

3. HISTORY OF NASA-JAXA JOINT WORK

First trigger of this joint work was driven by the request from NASA on Multilateral Safety & Mission Assurance Control Board (MSMACB), which is the board for the international coordination of the ISS program about Safety & Mission Assurance. In September of 2015, JAXA started to share their HTV PRA data with NASA according to the discussion at MSMACB. After that, NASA and JAXA reviewed each other's HTV PRA models. Based upon their comments/recommendation, NASA and JAXA had two Technical Interchange Meetings (TIMs) to update their models in March 2016 and April 2016.

4. COMPARISON OF ASSUMPTIONS FOR EACH MODEL

The NASA HTV PRA model and the JAXA HTV PRA model have their own assumptions. "Tab 1" shows the summary of comparison of assumptions for each model.

Table 1. Comparison of assumptions

	JAXA model	NASA model
How to build the model	HR based	FR based
Failure mode	Specific failure modes are included.	Specific failure modes are NOT included.
Scope (Flight phase)	• "Proximity operation phase" • "R-bar approach phase" • "Berthing Box phase"	• From "250m at Rbar phase" up to "Completion of HTV berthing" • "Departure phase"
Scope (H/Ws and operations)	• ISS systems are NOT modeled • Ground Control is modeled	• ISS systems are modeled • Ground Control is NOT modeled
End State	• "Collision" • "Abort"	• "Collision" • "Loss of port" • "No berthing" • "Stuck SSRMS"

4.1 How to build the model

The JAXA HTV PRA model was built based on Hazard Reports (HRs). The HTV has 17 HRs regarding ISS safety, and JAXA PRA model focuses on the HTV HR "HTV-0008 Collision with ISS", which assesses HTV collision with ISS due to the HTV failures. In the JAXA model, initiators of Event Trees (ETs) were picked up from hazard causes, therefore, the JAXA model has totally 25 ETs which are the same number as hazard causes in HRs. On the other hand, the NASA HTV PRA model was built based on Flight Rules (FRs). In the NASA model, initiator of ET was HTV beginning approach to ISS, followed by events involving systems/components failures, therefore, the NASA model has just one ET.

4.2 Failure model

As for the JAXA HTV PRA model, specific failure modes are modelled. For example, as for the valve failure, both OPEN failure and CLOSE failure of a valve are included in the model. On the other hand, the NASA HTV PRA model does NOT have specific failure mode like the JAXA model. That is, in the NASA model, just a component failure of valve is included without distinguishing OPEN failure and CLOSE failure.

4.3 Scope for flight phase

Scope of the JAXA HTV PRA model: (1) Proximity operation phase, (2) R-bar approach phase, and (3) Berthing Box phase.

"Fig. 2" shows Proximity phase. Integrated operation starts from 90 minutes prior to the AI manoeuvres. At the integrated operation, the HTV starts to communication with ISS directly and the HTV receives ISS GPS navigation data through the PROX. The AI manoeuvre is executed by ground command. After the AI manoeuvre, the HTV automatically approaches to the RI point which is 500 m below from ISS with relative GPS navigation.

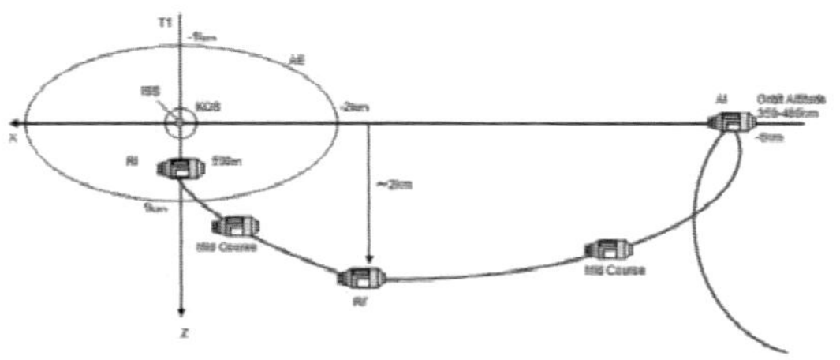

Figure 2 Proximity Phase

"Fig. 3" shows R-bar approach phase. After the arrival to RI point, the HTV starts relative navigation using the laser radar. The HTV ascent R-bar and stops at the berthing box to be captured by SSRMS.

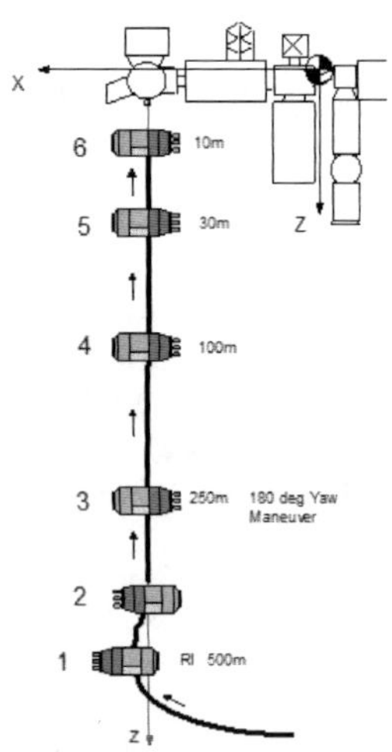

Figure 3 R-bar approach phase

"Fig. 4" shows Berthing Box phase. For SSRMS berthing operations, the HTV remains within the Berthing Box up to 99 seconds. At that point, the HTV is in safing mode that propulsion system is not activated (free drift mode) by the command. And SSRMS is close to the HTV.

On the other hand, the scope of the NASA model are (1) From "250m at R-bar phase" to "Completion of the HTV berthing", and (2) "Departure phase".
The NASA model starts 250m hold point at R-bar approach phase (See "Fig.3"). The NASA model continues through SSRMS grappling up to the completion of the HTV berthing to the ISS.

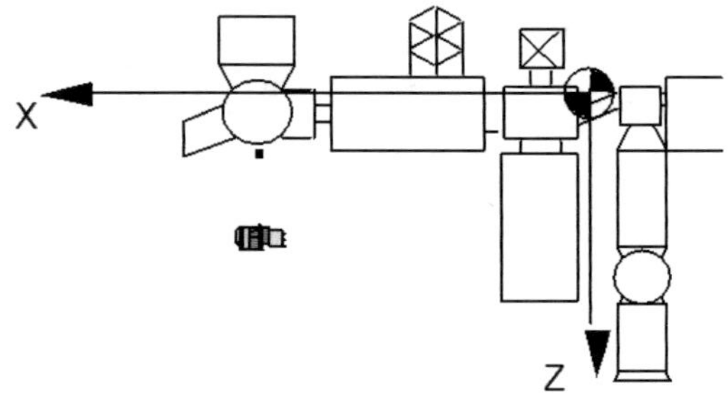

Figure 4 Berthing Box

"Fig. 5" shows berthing operation. After capturing FRGF of the HTV, SSRMS transfer the HTV to ISS Node2/Node1, and berthing the HTV to Nadir Port of ISS Node2/Node1. After the berthing, SSRMS separates from FRGF of the HTV. If SSRMS has any trouble at capturing or separating, the HTV uses the FRGF separation plane and separates from SSRMS.

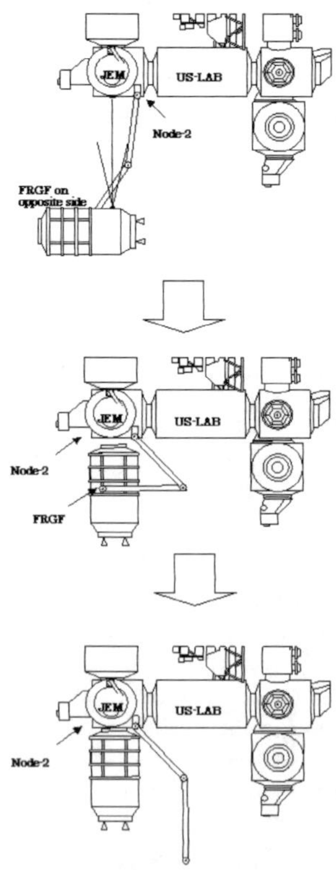

Figure 5 Berthing Operation

The NASA model also covers the departure phase.
"Fig. 6" shows the departure phase. The Departure Phase is defined as the period from the inactivation of utility lines prior to the start of deberthing operations, up to termination of the Integrated Operation. This phase includes the separation of the HTV from ISS, release of the HTV from SSRMS, and departure from ISS for reentry.

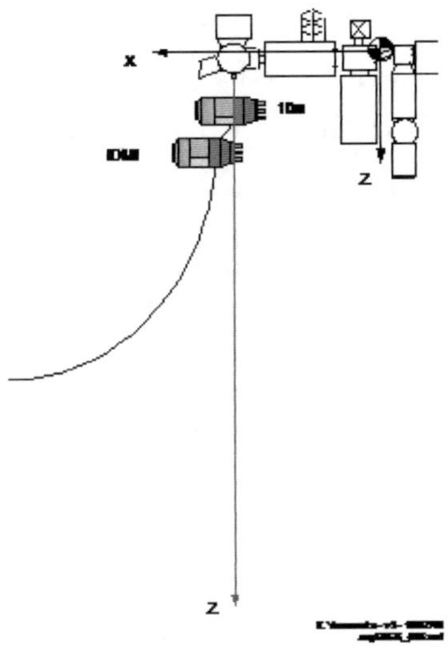

Figure 6 Departure phase

4.4 Scope for H/Ws and operations

In the JAXA HTV PRA model, HTV ground operations are modelled, which means monitoring, commanding and recovering from ground operators are in scope, while, ISS systems such as SSRMS, Common Berthing Mechanism (CBM) are NOT modelled because JAXA does not have detailed information about these items.

On the other hand, in the NASA model, HTV ground operations are NOT modelled, while ISS systems are in scope.

4.5 End State

In the JAXA HTV PRA model, End States in the ETs are (1) Collision and (2) Abort. "Abort" means that the HTV can abort successfully, while "Collision" means that the HTV collides with the ISS due to failure of abort. Because of the reason mentioned in section 4.4, JAXA could not build the models regarding ISS systems, therefore, End State of JAXA model are "Abort" and "Collision", focusing on the HTV systems.
While, NASA model has five End States: (1) Collision, (2) Loss of port, (3) No berthing, (4) Stuck SSRMS. "Loss of port" means that the HTV is not able to undock from the ISS due to CBM failure. "No berthing" means that the HTV cannot berth to the ISS, which includes abort of the HTV during the approach phase. "Stuck SSRMS" means that the HTV is not able to ungrapple from SSRMS due to SSRMS failure.

5. UPDATE OF NASA MODEL

The NASA HTV PRA model has been updated based on discussion and collaboration with JAXA.

5.1 Sharing of JAXA HTV PRA data

JAXA shared their HTV PRA data such as failure rates, Fault Trees (FTs), ETs with NASA according to the discussion at MSMACB. Based on the data, NASA updated their model. Main updates for the NASA original model are the followings:

- The NASA original model had the common cause failure (CCF) of Valve Drive Electronics (VDE). NASA removed CCF of VDE in the model because VDE2 is different components from VDE1, which was indicated in the JAXA model.
- NASA applied the information of basic events such as failure rates in JAXA model to NASA model.

5.2 Updating NASA model based on JAXA comments

JAXA reviewed the NASA original model with their detailed references such as HTV HRs, design review documents and provided comments, recommendation to NASA. Main updates based on the JAXA comments are the followings:

- The NASA original model did not have FTs regarding Flight-Releasable Grapple Fixture (FRGF) and associated electrical components on the HTV. JAXA commented that failure of FRGF and associated components should be considered during SSRMS grappling phase because the HTV could not be released and SSRMS would be stuck due to FRGF separation failure. NASA agreed and updated their model based on JAXA comments.
- The NASA original model had two Power Control Units and the common cause failure (CCF) of PCU, however, JAXA commented that the second PCU and the CCF should be removed because the HTV has only one PCU. NASA agreed and removed the second PCU and CCF of PCU in the model.

5.3 Results of new model

"Fig. 7" shows the comparison of the NASA original model and new model for each End State. "Tab. 2" shows the comparison of cut set for each End State.

Likelihood of "Collision" and "No berth" of new model are slightly higher than those of the original. As for the difference of likelihood of "Collision", the following reason is estimated according to the cut sets:

- applying the failure rate of JAXA basic event seems to be the dominant reason because each model has CCFs for dominant cut sets and differences caused by modifying FTs and ETs do not appear.

As for the difference of likelihood of "No berth", the following reason is estimated according to the cut sets:
Updating FTs regarding the HTV free drift mode seems to be the dominant reason because cut sets regarding this FT (Human error) appear in the new model while they do not in the original model.

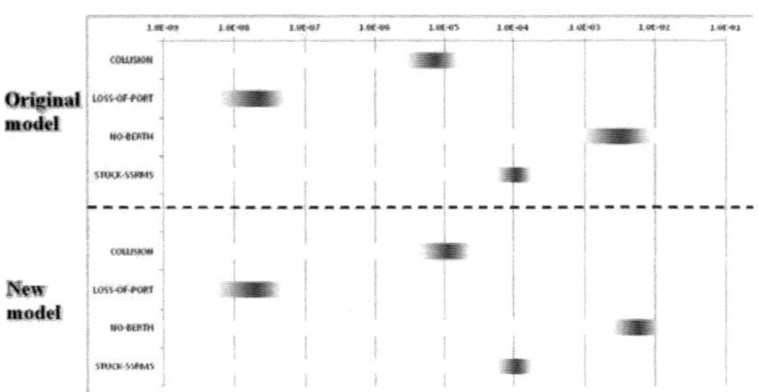

Figure 7 Comparison of NASA original model and new model

Table 2. Comparison of Cut sets for NASA each model

End State	*Cut Set %*	*Prob./ Freq.*	*Description*
COLLISION (Original model)	18.5	1.24E-06	HTV propulsion ox and ful injector common cause failure
	18.3	1.23E-06	HTV Main Bus Unit comman cause
	10.5	7.0E-07	Common Cause failure of VDE
	6.3	4.3E-07	Common Cause Failure of BIU1-A & B
	18.5	1.2E-06	Common Cause of all Latch Valves
COLLISION (NEW model)	36.4	3.8E-06	Common Cause of all Latch Valves
	8.04	8.4E-07	Common Cause Failure of BIU1-A & B
	8.04	8.4E-07	Common cause failure of CDH MCU
	3.87	4.1E-07	Failure of VDE 2 electronics
			Common Cause of HTV Thrusters
	2.87	3.0E-07	HTV Main Bus Unit common cause failure

End State	*Cut Set %*	*Prob./ Freq.*	*Description*
Loss of port (Original model)	100.	2.06E-008	HEP crew attempt to demate HTV from Node 2 fails
			ISS CBM fails during unberth, derived from the integrated model
Loss of port (NEW model)	Same Cut set as the Original model		

End State	*Cut Set %*	*Prob./ Freq.*	*Description*
No Berth (Original model)	38.6	1.1E-03	Common Cause of HTV Thrusters
	2.42	6.9E-05	SSRMS LOM during Capture (0.5 hours)
	2.21	6.3E-05	ISS CBM fails during berth, derived from the integrated model
	1.80	5.1E-05	Loss of S-Band
	1.08	3.1E-05	Electrical actuator fails to function
No Berth (NEW model)	37.39	2.0E-03	Crew fails to command control actuators
	21.22	1.1E-03	Common Cause of HTV Thrusters
	6.23	3.3E-04	Common cause failure event for Rendezvous Sensors
	1.29	6.9E-05	SSRMS LOM during Capture (0.5 hours)
	1.29	6.8E-05	ISS CBM fails during berth, derived from the integrated model

6. UPDATE OF JAXA MODEL

The JAXA HTV PRA model has also been updated based on discussion with NASA.

6.1 Common Cause Failure

Thanks to NASA reviews and comments, JAXA found that CCFs were not modelled correctly in FTs. (Almost all the CCFs of H/W failure were not modelled in the original JAXA model). Therefore, JAXA model has been updated by adding CCFs correctly.

6.2 Updating human error probability

In the JAXA original model, a constant value (1.0 E-03) was applied to the human error probability (HEP). During this collaboration with NASA, NASA provided CREAM HRA CALCULATOR (CHRAC) to JAXA.

CHRAC is a calculation tool for human error by using CREAM. By using CHRAC, HEP can be evaluated for various environments, operations and so forth. JAXA updated HEP by using CHRAC and applied the value to the model.

6.3 Results of new model

"Fig. 8" shows the comparison of the JAXA original model and the new model for each End State. "Tab. 3" shows the comparison of cut set for each End State.

Likelihood of "Collision during proximity phase" and "Collision during R-bar phase" of the new model are one-order-higher than those of the original, while, likelihood of "Collision during Berthing Box" of the new model is one-order-lower than that of the original model. As for the difference of likelihood of "Collision during proximity phase" and "Collision during R-bar phase", the following reason is estimated according to the cut sets:

- Adding CCFs correctly seems to be the dominant reason because CCFs appear in the new model while they do not in the original model.

As for the difference of likelihood of "Collision during Berthing Box", the following reason is estimated according to the cut sets:

- Applying CREAM to HEP and updating FTs seems to be the dominant reason because H/W failure of PDB1 appear in the new model for the top cut set while it does not in the original model.

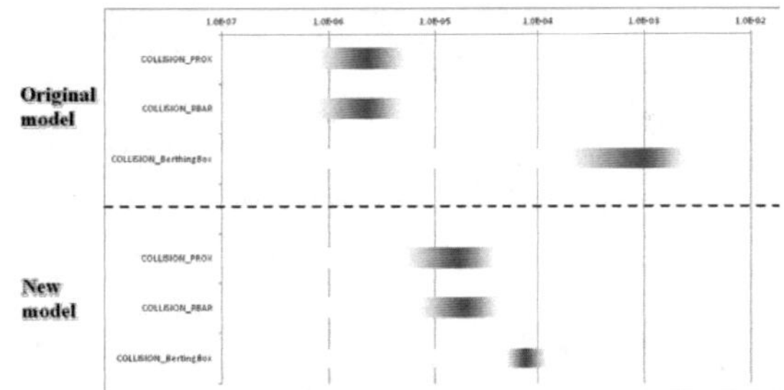

Figure 8 Comparison of JAXA original model and new model

Table 3. Comparison of Cut sets for JAXA each model

End State	Cut Set %	Prob./ Freq.	Description
COLLISION @PROX (Original model)	73.4	1.7E-06	false output of CDP-A
	20.5	4.7E-07	false output of BMIU1
	5.3	1.2E-07	false output from HCP
	0.1	3.3E-09	Mis command by Ground
			False GPSR output from SIGI
	0.1	3.3E-009	Mis operation of cause 2.3
			False GPSR output from SIGI
COLLISION @PROX (NEW model)	81.3	1.3E-05	CCF of SIGI
	5.28	8.5E-07	false output of CDP-A
	4.2	6.8E-07	CCF of CPU
	2.5	4.0E-07	false output of BBP-A
	1.8	2.9E-07	false output of P-TRX-A

End State	Cut Set %	Prob./ Freq.	Description
COLLISION @R-Bar (Original model)	73.1	1.7E-06	false output of CDP-A
	20.4	4.7E-07	false output of BMIU1
	5.3	1.2E-07	false output from HCP
	0.2	4.1E-09	I-TRX-A failure
			HUMAN-ERROR14
	0.1	3.3E-09	Mis operation Cause2.3Rbar
			False GPSR output from SIGI
COLLISION @R-Bar (NEW model)	70.1	1.3E-05	CCF of SIGI
	14.1	2.6E-06	CCF of IOC
	4.6	8.5E-07	false output of CDP-A
	3.6	6.8E-07	CCF of CPU
	2.1	4.0E-07	false output of BBP-A

End State	Cut Set %	Prob./ Freq.	Description
COLLISION @Berthing Box (Original model)	99.9	1.0E-03	Automated relief sequence is enabled Cause3.4.
	0.10	1.0E-06	mis-operation of Cause 3.4.3
			External Disturbance Bumping
	0.00	3.3E-09	Mis operation of Cause3.3
			CCF of SIGI
	0.00	1.2E-09	HCP H/W failure / HCP failure
			External Disturbance Bumping
	0.00	1.0E-09	Mis command for Inhibit 1 Cause 3.6.2
			Mis command for Inhibit 2 Cause 3.6.2
			Mis command for Inhibit 3 Cause 3.6.2
COLLISION @Berthing Box (NEW model)	37.2	2.8E-05	H/W failure of PDB1
	30.4	2.3E-05	Automated relief sequence is enabled Cause3.4.
			Automated relief sequence cannot be monitored
	8.1	6.2E-06	HCP H/W failure / HCP failure
	4.2	3.2E-06	CCF of BBP(PROX2)
	3.9	2.9E-06	H/W failure of VRVA

7. COMPARISON OF NASA AND JAXA MODEL

As section 4 mentioned, due to the difference of assumptions of the NASA and JAXA models, the results of each model cannot be compared simply. However, both NASA and JAXA models were modelled regarding "collision during R-bar phase and Berthing Box phase", therefore, comparison of the NASA and JAXA new models was conducted by focusing on the above specific duration and End State. "Fig. 9" shows the comparison of new models (As the reference, the result of original model are also shown). Likelihood of "Collision of JAXA new model" is one- order-higher than that of "Collision of NASA new model". As for this difference, the following reason is estimated according to the cut sets:

- Modelling of VRV relief failure during free drift phase seems to be the dominant reason; JAXA model includes VRV relief failure during capture phase based on HR, while, NASA model does not include

VRV relief failure during capture phase because it is not listed in the flight rules .

If VRA relief failure is removed from JAXA model, likelihood of collision of JAXA model is equivalent to that of NASA. Therefore, It can be said that both new models has been updated having accuracy to some extent.

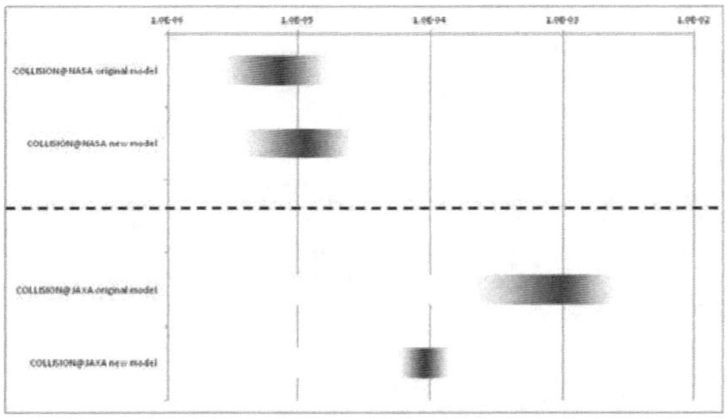

Figure 9 comparison of new models

8. CONCLUSION AND FUTURE WORK

Thanks to the international collaboration, Updating the NASA and JAXA models has been completed. As a result, NASA and JAXA will apply new models to the trade study performed previously by NASA and study the similarities and differences.

Also, JAXA is developing new transfer vehicle: the HTV-X. JAXA will study whether they can apply PRA to the HTV-X or not. If PRA is applied to the HTV-X, NASA and JAXA will start international collaboration to build aPRA model for the HTV-X.

Throughout this international collaboration, NASA and JAXA achieved great and fruitful success . Therefore, it is thought to be best for NASA and JAXA to continue the collaboration.

9. REFERENCES

1. T. Fukatsu, N. Takeuchi, T. Seike, K. Oga and Y. Sato, JEM Probabilistic Safety Assessment, PSAM 6, Vol.1, Elsevier Science Ltd., pp.949- 955 (2002).

2. N. Takeuchi, T. Sato, T. Seike, K. Oga, S. Takada and Y. Sato, PSA for Depressurization Hazard in ISS module, PSAM7, Vol. X. (2004)

3. S. Takada, T. Fukatsu, T. Sato, K. Oga, S. Yoshikawa, M. Miki and Y. Sato, Probabilistic Safety Assessment directed to Manned Space System Operation, 1st IAASS (2005)

4. S. Takada, T. Fukatsu, T. Sato , M. Miki, K. Oga, S. Yoshikawa and Y. Sato, Improvement of JEM Operation by PSA, PSAM8 (2006)

5. S. Takada, N. Takeuchi, M. Miki, Y. Sato, Effective Probabilistic Risk Assessment Application for Japanese Manned Space Systems, PSAM9 (2008)

6. S. Takada, M. Miki, Y. Sato, Probabilistic Risk Assessment for H-II Transfer Vehicle, 3rd IAASS (2008)

7. MAPI contract, H-II Transfer Vehicle Analysis Notebook (2016)

PROCEEDINGS OF THE 8TH INTERNATIONAL SPACE SAFETY CONFERENCE

SESSION 13
COMMERCIAL SPACEFLIGHT - II

SABRE – ENABLING COMMERCIAL SINGLE STAGE TO ORBIT, SAFELY

Dr Andy Quinn MSc PhD MRAeS CEng [(1)], Richard Varvill, Founder/Technical Director [(2)]

[(1)] *Saturn Safety Management Systems Ltd, Bath, BA2 9BR, UK*
[(2)] *Reaction Engines Ltd, Culham Science Centre, Abingdon, OX14 3DB, UK*

ABSTRACT

Single Stage to Orbit (SSTO) has been the 'Holy Grail' since the beginning of astronautical thinking. However, the technical solution has been out of reach due to costs, technical feasibility of designs, and low Technology Readiness Levels. SSTO spaceplanes can take off from a spaceport's runway and reach Low Earth Orbit with their design payload for delivery, the spaceplane could then recover the same payload or a previously deployed payload and return safely to the same spaceport for a glide approach and landing. The vehicle will then be ready for its next mission after a few days of maintenance and payload integration. The main enabling feature of such a spaceplane is an engine that can provide the required performance in terms of thrust, weight and specific impulse with typically an air-breathing engine envisaged as a requirement to achieve this. The Synergetic Air-Breathing Rocket Engine (SABRE) is such an engine. In the SABRE engine, the Air- Breathing system is employed from take-off up to 25km altitude at Mach 5.5 before transitioning to the separate rocket system for the remainder of the ascent into orbit. SABRE engines are usable in systems for space access and hypersonics and in this paper we explore SABRE within a SSTO spaceplane concept.

This concept has 2 engine nacelles on the wingtips and this allows for excellent abort capabilities, whether aborting on the runway before take-off, back to base or 'aborting to orbit/once around'. As the air-breathing and rocket engines are separate systems (except for the nozzles) separate development programmes can be accomplished with several thousand engine tests planned prior to achieving full certification. As part of the certification process a demonstrator flight test vehicle would prove the airworthiness/spaceworthiness of the SABRE; even this aspect can be split into high altitudeatmospheric flight tests (for the Air-Breathing engine) and later flight testing of the rocket engines to prove the transitioning aspects and rocket mode. This design approach provides not only redundancy, hence reduced vehicle loss rates, but more capabilities to abort the mission safely in the event of anomalies. This paper will provide an overview of the design features of SABRE as integrated into a spaceplane concept (such as Skylon) along with abort strategies, safe design philosophy and loss of vehicle/abort targets.

1. INTRODUCTION

For 30 years there has been activity in the United Kingdom to realise the vision of a single stage to orbit launch system using combined cycle engines that work in both air- breathing and pure rocket modes.
Figure 1: HOTOL Concept (BAE)

This activity started in the 1980's with the British Aerospace/Rolls-Royce Horizontal Take-Off and Landing (HOTOL) project using the Rolls-Royce RB545 engine invented by Alan Bond. Whilst the project was not pursued for various reasons, the HOTOL study had established that the use of combined cycle engine with an aircraft like airframe would be a technically realistic proposition to insert a payload of up to 7 tonnes in orbit.

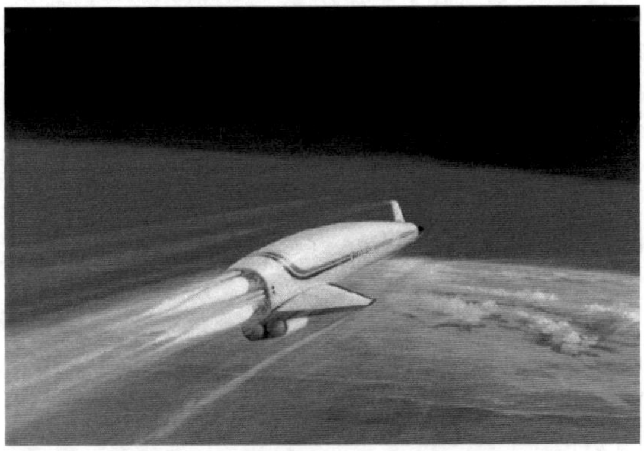

Figure 1: HOTOL Concept (BAE)

In 1989 Alan Bond formed Reaction Engines Limited (REL) together with Richard Varvill and John Scott-Scott. The HOTOL concept was progressed to overcome identified technical problems, from which the Synergetic Air Breathing Rocket Engine (SABRE) propulsion system was created. REL developed various SABRE configurations over the following 20 years culminating in the successful demonstration of the key heat- exchanger technology in 2013. This demonstrated that the intake air could be cooled to the required (minus) -150° Celsius within a fraction of a second.

In parallel with the SABRE design, REL developed a spaceplane concept called Skylon in order to explore the potential to apply SABRE in space access systems. This vehicle would be scalable during the early concept phase with the goal of inserting 15 tonnes to Low Earth Orbit (LEO) as its reference mission.

2. SABRE

SABRE is an enabler for orbital spaceplanes as well as other markets such as Hypersonic Transportation and Two-Stage To Orbit satellite launchers (see section 3.2).

The air-breathing part of the engine includes a helium closed loop system as part of a Brayton Cycle, which means the mass fraction required to reach orbit, for an SSTO, is 22% compared with 13% for an equivalent pure rocket system.

The SABRE 3 version is shown in Figure 2:

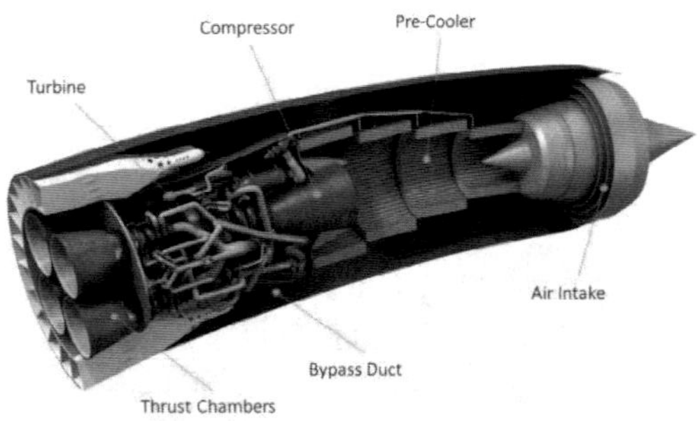

Figure 2: SABRE 3 (Reaction Engines)

2.1 Technology Development

2.1.1 Pre-Cooler

The pre-cooler demonstration included one heat-exchanger section mounted on the front of a Rolls-Royce Viper engine shown below in Figure 3:

Figure 3: Pre-cooler Tests (Reaction Engines)

The pre-cooler development has progressed since the 2013 demonstration and REL now has in-house manufacturing capability with the addition of a high temperature vacuum furnace for brazing. A typical pre-cooler heat exchanger subsystem for a large engine would have over 1 million leak-tight joints with over a 1000km of 1mm diameter tubing (20 microns thick) [1] and this would enable a total heat transfer of >400MW.

2.1.2 Nozzle

The current SABRE 4 engine requires a novel design of the rocket engine's thrust chamber and nozzle to allow operation in both air- breathing and rocket modes, as well as having a smooth transition between the two. The Advanced Nozzle experiment has been explaining the feasibility of this concept and represents a significant technology development effort towards the SABRE engine. The test engine incorporates several new technologies, including a 3D printed, actively cooled propellant injector (Figure 4).

Figure 4: 3D printed injector system for the Advanced Nozzle Project

The test engine has been successfully fired over 30 times since commissioning in spring 2015, Figure 5. The aerodynamic data collected from the firings is being used to validate in-house computational modelling and verifying the flow stability and expansion efficiency.

Figure 5: Image during testing of the SABRE Advanced Nozzle at Airborne Engineering Ltd., Westcott, UK

2.2 Technology Timeline

The main focus for REL is on SABRE and particularly on the air-breathing novel technologies. The aim is to have an engine undertaking ground tests by 2020.

2.3 Space Access Applications

Once proven and incrementally developed, SABRE will be ready for use in space access vehicles (section 3 details the Skylon Spaceplane as an example); these will be developed by suitable vehicle manufacturers i.e. REL will not develop any space access vehicles.

The proven SABRE will also be available to integrate into non-orbital vehicles i.e. suborbital hypersonic vehicles, suborbital vehicles deploying (orbital) satellite systems etc.

Having a reliable and safe engine such as SABRE will provide good confidence for developing Commercial Human-Rated spaceplanes as well. This will open up the access to space for commercial ventures such as Bigelow Aerospace (space hotels) and other Space Stations.

2.4 Other Applications

SABRE technologies also have ground-based uses and the technology can be exploited appropriately. Examples of this include the heat exchanger manufacturing and frost control technology.

3. SKYLON SPACEPLANE CONCEPT

The Skylon concept is a SSTO winged spaceplane designed to give routine low cost access to space and is useful in exploring the impact of using SABRE engines in space access systems. At a gross take-off weight of 325 tonnes of which 270 tonnes is propellant the vehicle is capable of placing 15 tonnes into an equatorial Low Earth Orbit (LEO). The vehicle configuration consists of a slender fuselage containing the propellant tankage and payload bay with delta wings located midway along the fuselage carrying the SABRE engines in axisymmetric nacelles on the wingtips. The vehicle takes off and lands horizontally on its own undercarriage. The fuselage is constructed as a multilayer structure consisting of aeroshell, insulation, structure and tankage.

Figure 6: Skylon (Reaction Engines)

3.1 Skylon Mission

The system performance of the Skylon D1 with its SABRE 4 engine has been extensively modelled with inhouse mission analysis software to track the impact of any changes as the SABRE engine design evolves. Although the required performance for Skylon is specified as a 15 tonnes payload to a 300km circular orbit from an equatorial launch site, the modelling uses a "standard mission" from a launch site at latitude 5.2 degrees; corresponding to the CSG spaceport at Kourou, French Guiana. The altitude versus time graph for the powered ascent trajectory until Main Engine Cut off is shown in Figure 7 below. This trajectory leaves Skylon in a 90 km by 300 km elliptical orbit, which is circularised at apogee using the orbital manoeuvring engines.

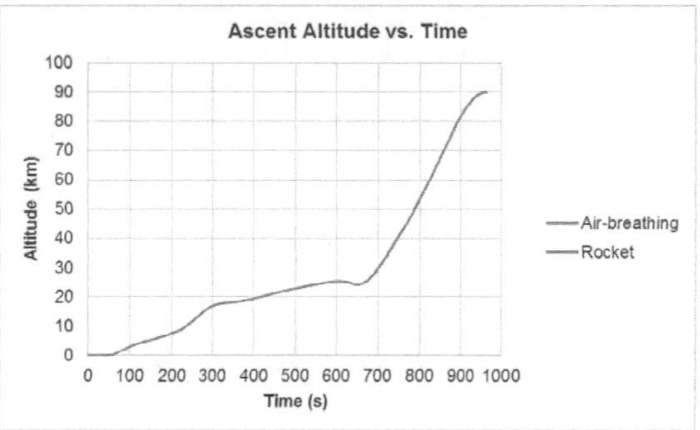

Figure 7: Skylon Standard Mission Ascent Altitude Vs. Time (Reaction Engines)

The mission analysis studies also confirmed the sensitivity of take-off thrust to weight ratio performance was insensitive to moderate changes (over the 0.35 to 0.5 range). This gives some scope to alter the vehicle take off mass after the engine thrust has been fixed during the detailed design phase. On orbit, Skylon is able to deploy a 15T payload to Low Earth Orbit. The payload includes a reusable Skylon Upper Stage (SUS) system. The SUS provides a means to insert payloads to higher orbits such as Geostationary Transfer Orbit (GTO). After delivering the payload to the GTO the SUS would then be able return back and rendezvous with Skylon; hence its reusability.

The return trajectories confirmed Skylon's flexibility of touchdown point with a wide range of downrange and cross range availability. The vehicle can recover to any spaceport with compatible latitude at least 6 times per day from any orbit, and can recover to an Equatorial spaceport from a low inclination orbit (less than 40 degrees) on any pass.

4. SAFETY BY DESIGN

Space safety standards dictate a Design for Minimum Risk (DMR) philosophy. This includes deriving Fault Tolerance, Safe-Life and Fail Safe criteria. The DMR philosophy within Advisory Circular AC 437.55-1 [2] details the following Safety Precedence Sequence:

- Eliminate hazards (by design or operation)
- Incorporate safety devices
- Provide warning devices
- Develop and implement procedures and training

Notice that in space the key term is 'hazard'. Hazards are analysed and then, as part of Probabilistic Risk Analysis (PRA), cumulative assessment is made in order to determine whether the Target Level of Safety has been met i.e. top-down analysis. In aviation, although the term hazard is recognised, the focus is on lower level failure conditions associated with failure modes i.e. in order to meet safety objectives such as 1x10-9 per flying hour for catastrophic events i.e. bottom- up analysis.

Hence REL is adopting a top-down safety target approach (this is also the approach taken by NASA). The REL Loss of Vehicle (LOV) target is 1 in 2000 per mission (all causes).

This is much safer than the best of the current space launch systems (Soyuz family) with a 97% success rate i.e. 1 in 33. REL has noted lessons from previous space programmes such as Shuttle where initial predictions for LOV were in the order of 1 in 100,000 per mission (which were then realistically reduced to 1 in 100 by the in-service phase; actual achieved rate at retirement being nearer 1 in 70).

The risk budget will be apportioned between the Skylon systems and it is estimated that the apportioned level of safety for the SABRE engines is estimated in the order of 1 in 10,000 per mission (for a catastrophic event from an engine technical fault).

Additionally, as the vehicle is a spaceplane with 2 engine nacelles, there is much more scope to abort safely. During the Air- Breathing and transition to rocket phases of flight Skylon can abort by returning to the launch site. During the rocket phase of flight Skylon can abort to a Transoceanic Abort Landing (TAL), Abort Once Around (AOA) or Abort to Orbit (ATO).

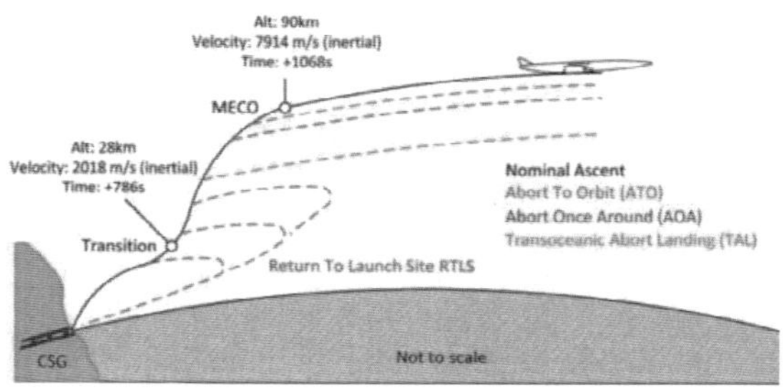

Figure 8: Spaceplane Abort Capability

REL system safety analysis will follow best practice from aviation and space by influencing the design from the beginning with key high level safety requirements (from standards and rationalised accordingly) and providing derived safety requirements from analyses.

4.1 Key Safety Requirements

For an SSTO and in particular spaceplanes, it is essential to understand both space and aviation requirements and thereby identifying relevant key safety & functional requirements that should be achieved in the design. For instance, a space requirement is that any Inadvertent Failure Modes that could result in a catastrophic outcome should have 3 Inhibits. This means 3 separate and independent inhibits which could be hardware (physical switches/guards etc.), software (latches) or combination thereof. The IAASS Space Safety Manual [3] is based on the NASA and European ECSS standards and rationalised/consolidated into one manual; in relation to 'Functions Resulting in Catastrophic Hazards' the following requirement is stated:

"A system function whose inadvertent operation could result in a catastrophic hazard shall be controlled by a minimum of three independent inhibits, whenever the hazard potential exists. One of these inhibits shall preclude operation by a radio frequency (RF) command or the RF link shall be encrypted. In addition, the ground return for the function circuit must be interrupted by one of the independent inhibits. At least two of the three required inhibits shall be monitored."

In relation to Skylon & SABRE this important safety requirement is already defined and as the programme develops, these inhibits will be detailed and then form part of the Validation & Verification Plan.

A key safety requirement is that catastrophic hazards should be controlled such that no combination of two failures or operator errors can result in a catastrophic event (for the unmanned system this would relate to ground control, and for a manned system the 'pilot in command' has limited control functions as unmanned spaceplanes, such as Skylon, cannot be 'flown'). In relation to SABRE this would concern the Engine Control System failure modes for example and ensuring adequate Design Assurance Levels along with appropriate fail-safe and redundancy (as detailed by derived safety requirements). In terms of functional requirements, a system whose loss of function could result in a catastrophic outcome should be two fault- tolerant.

Another requirement is that a spaceplane shall be able to abort its nominal powered ascent at any point and return safety to a landing site following a single engine failure (as well as other hazardous/ critical system failures) i.e. return to base directly or per *Figure 8* via TAL/AOA/ATO as required.

5. SABRE CERTIFICATION

There is no international agreed process on the safety certification of space systems, each nation doing what it regards as sufficient to meet its obligations under the Outer Space Treaty. Currently NASA has the 1100-Series certification requirements which appear to contain a well-structured set of high level requirements with applicable means of compliance from previous NASA Standards:

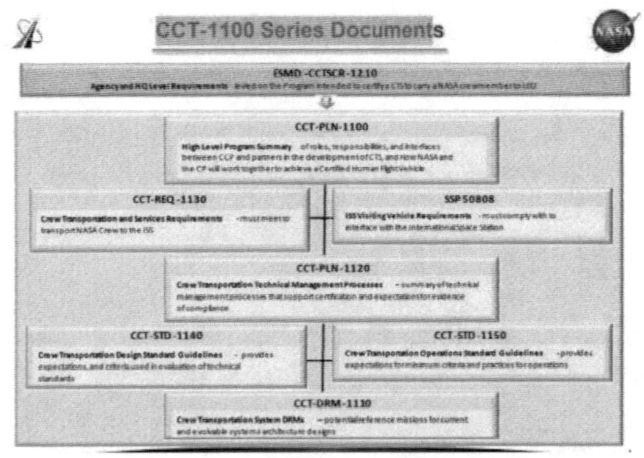

Figure 9: NASA High Level Requirements Set

In addition to obtaining the NASA certification approval, space launch systems (such as Space-X) must then meet the FAA- AST Launch Licensing requirements.

For aircraft-based systems, under the terms of the Chicago Convention on International Civil Aviation, any launch system that at some point relies on wings for lift will require certification from the civil aviation authority of the country from which they operate or over-fly.

Spaceplanes would fly through the atmosphere (i.e. Skylon using an air-breathing engine) and then continue the trajectory under rocket power to reach orbit. Throughout the ascent the vehicle has a good abort capability (per Figure 8). During the atmospheric phase which includes aborting back to the launch site, spaceplanes such as Skylon act as an airplane i.e. they use the wings for lift and therefore meet the ICAO definition of an aircraft/aeroplane [4]:

"An aircraft is any machine that can derive support in the atmosphere from the reactions of the air other than the reactions of the air against the Earth's surface"

Hence spaceplanes within Europe (and systems including SABRE) should adopt a certification approach by a competent Authority. The certification basis (airworthiness requirements) would require all phases of flight to be covered and therefore this would include the space segment with associated space system requirements; hence this would require the Authority to consult with ESA specialists. This then requires a unique approach integrating air and space requirements with appropriate means of compliance and guidance material.

In regards to SABRE, Reaction Engines are currently working with the UK CAA as well as ESA to derive such requirements that can be verified and validated throughout the engine development (EASA are currently not resourced for commercial space-related activities).

The primary focus is (experimental) certification of SABRE which will be carried out through an incremental approach with sub- system tests (as part of qualification), through to ground engine tests and finally to flight test engines. However, to derive appropriate engine requirements one needs an understanding of higher level (space and aviation) requirements, including those that may be imposed by the Authorities.

REL envisage a rationalised high-level risk and performance based requirements approach from the Authorities; thus leaving the designer to arrive at the product (solution) based on agreed means of compliance in accordance with agreed standards. This would be similar to the NASA approach but perhaps leaner and taking cognisance of the changing European certification approach:

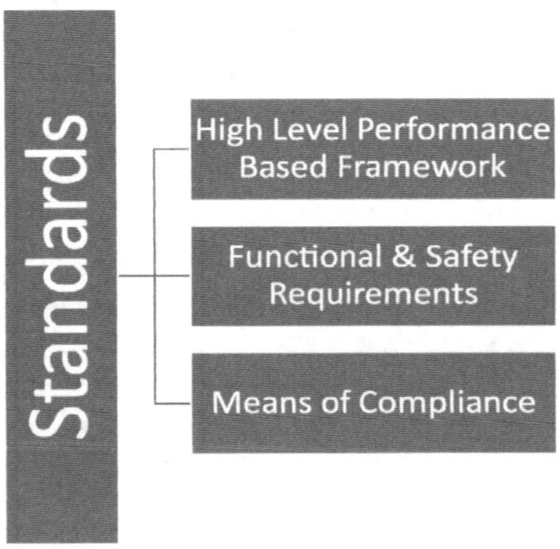

Figure 10: High Level Performance Based Requirements

The SABRE propulsion system has many functional requirements that are similar to aviation engines (in relation to the Air- Breathing phase) and hence CS-E is considered a good starting point. As indicated in *Figure 11* below, REL's initial findings are that roughly a third of CS-E airworthiness codes apply, a third do not apply (reverse thrust, etc.) and a third of requirements require Special Conditions (meaning a re-write of the intent of the requirement for SABRE). These lower level airworthiness (and relevant spaceworthinesss) requirements will be linked with the higher level functional, performance and safety requirements. Additionally Space standards (ECSS) will provide additional requirements for SABRE, particularly in rocket mode and also for the remainder of the flight (orbit through to re-entry).

The V&V programme will provide appropriate test and analysis evidence to demonstrate the means of compliance in order to attain certification (in the first instance experimental approval for flight demonstration).

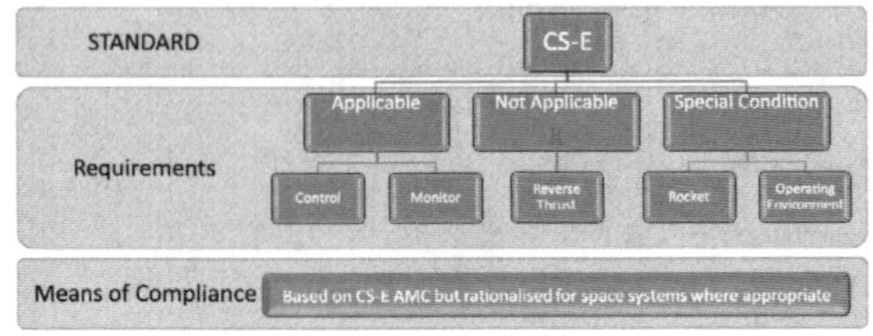

Figure 11: CS-E Requirements

6. SUMMARY

This paper has discussed an SSTO spaceplane concept which can take off from a spaceport's runway, deliver a 15 Tonne payload to Low Earth Orbit, then recover the same payload or a previously deployed payload and return safely to the same spaceport. The spaceplane could then be ready for its next mission after a few days of maintenance and payload integration. The main enabling feature is SABRE, encompassing an innovative air breathing engine to Mach 5.5 at 25km and a separate rocket engine to propel the vehicle to orbit. The aim is to have an air-breathing engine on test by 2020. The abort capability is an important part of the safety case.

Spaceplanes are predicted to be much safer than current vertical launch systems. The LOV target for spaceplanes (such as Skylon) is in the order of 1 in 2000 per mission for all causes.

A rationalised European-based certification programme should be derived with the Authorities; this is likely to be a high-level performance and risk-based regulatory framework focusing on functional and safety requirements, thereby allowing the designer to develop the product (vehicle/engine) whilst demonstrating means of compliance to appropriate standards.

7. REFERENCES

1. R. Varvill "Technological Evolution of Propulsion Systems in the UK", Culpepper Lecture, presented at the 20th AIAA International Space Planes and Hypersonic Systems and Technologies Conference, July 2015

2. FAA-AST, Advisory Circular AC 437.55-1

3. IAASS-ISSB-S-1700-Rev-B, Space Safety Standard, Commercial Human-Rated System, Requirement 201.2.2

4. ICAO – Annex 1 to the Convention on Civil Aviation, Annex 6 Part I. Montreal, ICAO. Available to order from www.icao.int

8. ACRONYMS/ABBREVIATIONS

Acronym	Meaning
AOA	Abort Once Around
ATO	Abort to Orbit
CAA	Civil Aviation Authority
DMR	Design for Minimum Risk
EASA	European Aviation Safety Agency
ESA	European Space Agency
FAA-AST	Federal Aviation Administration – Office for Commercial Space Transportation
FMECA	Failure Modes Effects & Criticality Analysis
FTA	Fault Tree Analysis
GTO	Geostationary Transfer Orbit
HOTOL	Horizontal Take-Off and Landing
LEO	Low Earth Orbit
LOV	Loss of Vehicle
NASA	National Aeronautics & Space Administration
REL	Reaction Engines Limited
SMS	Safety Management System
SPP	Safety Program Plan
SQEP	Suitably Qualified Experienced Personnel
SSTO	Single Stage To Orbit
SUS	Skylon Upper Stage
TAL	Transoceanic Abort Landing
TSTO	Two Stage To Orbit
V&V	Validation & Verification

OPERATIONAL SAFETY CONSIDERATIONS FOR RAPID TURNAROUND PRIVATE SUBORBITAL FLIGHT PROVIDERS

Justin Karl, Ph.D.[(1)], Jennifer Thropp, Ph.D.[(2)]

Embry-Riddle Aeronautical University, 600 S. Clyde Morris Blvd., Daytona Beach, Florida, 32114, USA
E-mail: [(1)]karlj@erau.edu, [(2)]thropppj@erau.edu

ABSTRACT

Evidenced by recent successes of private suborbital spaceflight providers, the era of for-hire launches catering to researchers and casual tourists is finally near. A major reason for the rapid success of these providers is the relative autonomy afforded them by funding self-reliance and limited interference by regulating agencies. When their vehicles are proven, these development companies and their derivatives will transition into flight operators, and they expect to continue at a similar pace, leading to internal predictions of high launch volume and rapid turnaround.

While their designs are relatively simplistic and have a short mission duration, these flights will be transporting human occupants above the Karman line on rocket-powered vehicles with unprecedented frequency. Moreover, the operations will be conducted in many cases at private launch sites, or at leased sites where commercial interests and economic development play a role in decision-making.

All of these factors beg the following questions: (1) What are the regulations that the majority of these providers will be subject to, and how should enforcement be executed?, (2) What specific modifications to current risk mitigation methods will these non-traditional operators be making?, and (3) What safety and mission assurance plans have these companies already put in place?

This effort explores these questions, with an intent to extract and characterize the core issues. Information direct from several leading private suborbital spaceflight companies is disseminated and analyzed against that of larger orbital launch providers.

1. INTRODUCTION

In the emergent suborbital reusable launch vehicle (sRLV) sector, a number of factors are contributing to difficulty in determining ultimately how profitable, successful, and safe vehicles will be. Often compared to the early aviation industry, differing approaches from a multitude of designers hold one thing in common: a goal of aircraft-like turnaround times for sRLV's.

In each case, the roles of the designers are mixed, with companies like Blue Origin and XCOR designing, manufacturing, and ultimately conducting commercial flight operations of the vehicles. Individual lifecycles are long, and decoupling of manufacturing, operations, and maintenance is not currently planned. In the orbital launch sector, larger vehicles are often managed by a single company, though aided by contractors and having a very slow refurbishment process if any reusability exists at all. Some experts have cited that sRLV companies working like this is problematic, as civil aviation designers do not conduct commercial flight operations [1].

Considering that parallels with aviation may be limited, further problems arise when it is shown that most space-related organizational, operational, assessment, and safety practices have evolved over decades to support the slower-paced orbital launch providers. Additionally, historically ambiguous handling of suborbital flights in space law [2] means that regulating space vehicles of this nature is not as simple as applying the same methods utilized in cases of large orbital rockets and their associated missions.

Though generally perceived as a negative aspect of sRLV development and operation, the fact that the industry does not explicitly adhere to either aviation or traditional spaceflight methods also affords a degree of flexibility. It can be reasoned that a set of best practices can be instituted that offer individual derivations from multiple sources, each selected from where most appropriate.

2. OVERVIEW OF sRLV's

Suborbital vehicles have been utilized for many decades for defense and research purposes. Only within the last 15 years has a push for reusable versions that could offer cheap entry-level space access been present. Driven by a perceived high-growth market [3], several companies began developing vehicles that could take passengers or payloads above the Karman line.

2.1 Vehicle Design and Performance

While not strictly novel in any technical sense, the vehicles have performance envelopes that would be considered severe by any measure. Regimes include supersonic flight and entry [4] with high aerodynamic loads present. During a nominal boost phase, high acceleration, acoustic, and vibratory loads [5] are expected. All vehicle designs utilize hybrid or liquid rocket motors, with inherent risk in the high stresses and temperatures involved.

Compared to larger orbital vehicles, some metrics vary a great deal, with ultimate downrange speeds (thus driving eventual entry speeds) for orbital vehicles being much higher. Additionally, propellant mass fractions for large vehicles are generally in the range of 0.9, while suborbital vehicles near those of commercial long-range aircraft [6], with 0.6 being a conservative figure.

A quick comparison of suborbital and orbital vehicle flight characteristics is available in Tab. 1.

Table 1: Vehicle mission characteristics

Vehicle	MECO time (s)	Flight Duration	Typical Downrange Velocity
NS	2m33s	~11min	0 km/s
Lynx	3m25s	~25min	0.1km/s
F9-D	2m55s	Weeks	7.5km/s
Ant-Cyg	3m54s	Weeks	7.5km/s
NS: Blue Origin New Shepard, Lynx: XCOR Lynx Mk. II, F9-D: SpaceX Falcon9/Dragon, Ant-Cyg: Orbital ATK Antares/Cygnus			

However, the launch and boost phases share some similarities, as they are both under rocket-powered ascent and the burn time for a suborbital flight is similar to that of the first stage of a larger commercial rocket like the SpaceX Falcon 9 [7]. Effectively, sRLV operators will need to develop competencies such that they can repeat one of the most failure-prone phases of spaceflight with unprecedentedly short turnaround times.

2.2 Launch and Turnaround Rates

Aided by the relatively low technical complexity and driven by models which predict profitability increases with launch frequency [8], suborbital providers are working to make turnaround times per vehicle much lower than any other reusable launch system. Historically, large vehicles like ULA's Delta and Atlas launcher families have typically flown between 5 and 10 times [9] per year. Addressing partial reusability in large vehicles, the Space Shuttle Orbiter flew with each requiring a refurbishment process that took on the order of months [10], giving the program a few flights per year with a fleet of 4 vehicles at any given time. SpaceX believes that optimization in their processes and the ability to recover the majority of first stages could have them launching Falcon 9 every 2 to 3 weeks [11].

In the first months of 2016, Blue Origin demonstrated their ability to refurbish and re-fly their boosters and capsules on the order of weeks. Considering the timelines posted in their Payload User's Guide [5] and their intention to fly multiple vehicles, it can be surmised that when testing is complete, their ultimate launch schedule might have a New Shepard flight on a near-daily basis. XCOR makes even bolder claims, with an expressed intention to make the Lynx spacecraft serviceable in under two hours [12] and capable of 4 flights per day.

2.3 Vehicle and Operations Costs

With launch frequency being a primary driver and suborbital vehicles having simpler designs with low propellant mass fractions, it is clear that the ratio of operations cost to vehicle hardware cost will be higher than those of larger vehicles. Low propellant mass fraction translates to less necessity for high-performance materials, and simpler design similarly reduces cost in the development phase. The need for maintenance increases, and maintaining vehicles with such high performance on a fast turnaround schedule should imply higher cost [13]. The danger that arises in this scenario is that operational cost creep might drive measures that lower time or cost but increase risk. Tab. 2 gives an overview of several expendable, partially reusable, and reusable vehicles. Note that the Space Shuttle figures are for post-Challenger operations and reflect increased hazard mitigation expenditures, and analysts predict commercial aircraft-like ratios [14] for the suborbital spaceflight industry.

Table 2: Spacecraft and aircraft vehicle cost ratios.

Vehicle	Hardware cost (%)	Operations cost (%)
Titan IV	64%	36%
Space Shuttle	14%	86%
B777-200ER	26%	74%

3. OPERATIONAL CONSIDERATIONS FOR SUBORBITAL FLIGHT PROVIDERS

A number of issues specific to rapid turnaround suborbital flight providers are generated from the high performance nature of the vehicles meeting the requirement to launch frequently. These issues have a direct impact on risk, and must be carefully managed.

3.1 Vehicle Health Management

To control hazards associated with the vehicle itself, vehicle health over long periods must be taken into account during both the design and operations phases, with design ultimately impacting the ability for operations to rapidly refurbish the vehicle. High factors of safety, redundancy, avoidance of complex or failure-prone components, and embedding sensors and avionics for health monitoring are recommended to designers [15] by FAA/AST with the clear intention of reducing risk.

Beyond the vehicle itself, a robust integrated health management (IHM) system must be implemented across the entire operational infrastructure and address the constraints of a high launch frequency system. Key components must be verifiable as ready for flight with minimal testing via a built-in test (BIT) method. Hazards associated with maintenance worker errors or unintentional damage during testing and refurbishment are reduced. In cases where BIT is not possible, easy accessibility to key systems must be considered. Both risk and timelines can be reduced when components with suspected wear are line replaceable.

An IHM system that utilizes avionics to constantly monitor sensors and quantify parameters during flight and ground segments can reduce human subjectivity in the turnaround process [16], directing operations to the components or systems in need of testing or replacement. This is especially critical in situations where complex systems exist in difficult arrangements, as shown in Fig. 1. An added benefit for rapid turnaround operations is that is that an IHM can recognize aggregate damage types like fatigue or creep by analysis of data gathered over many flights [17].

Figure 1: A suborbital vehicle propulsion cluster with many components. (Courtesy XCOR)

3.2 Crew and Personnel Management

Building on the assertion that a data-driven IHM system must be present throughout the entire operations process, it is critical that the correct automated data reduction methods be similarly utilized. Operations and maintenance personnel should rely on anomaly or damage detection software to inform them in place of performing testing or analysis themselves. Essentially, vehicles and processes should be built to be as automated as possible, thus reducing the personnel workload and avoiding time lost to unnecessary manual operations on a vehicle. With sensor redundancy and a reliable system for anomaly detection (SAFD), an IHM can be built that reduces cost, time, and risk all at once by decreasing unnecessary action and removing opportunity for human error.

One final issue very recently highlighted [18] involves risk arising from situations where human error could lead to significant issues or loss—the ability of the pilot to perform his or her job effectively. In the case of piloted vehicles, workload is high and the physical strains of high-g launch and entry can further decrease the ability to perform [19]. It is absolutely necessary that operators implement programs of training, scheduled rest, and workload sharing to mitigate risks associated with pilot error leading to a vehicle loss. Additionally, procedures should be developed such that no single erroneous action leads to an emergency, effectively implementing fault tolerance with regard to human factors considerations.

4. REGULATORY CHALLENGES

With suborbital vehicles themselves technologically situated between aircraft and orbital launchers, they likewise fall into a regulation gap between aircraft and orbital spacecraft. International space law has not clearly defined suborbital flight in either discussions at the United Nations Committee on the Peaceful Uses of Outer Space (COPUOS) nor is it mentioned in the Outer Space Treaty [20] adhered to by spacefaring nations since 1967. Recognizing no clear path, there have been notions that perhaps they should be regulated as high-performance aircraft. In fact, Scaled Composites model 316P (SpaceShipOne) initially flew under existing experimental aircraft regulation.

After the first demonstrations of private suborbital technologies, the Commercial Space Launch Act (CLSA) was amended (see CSLAA) to regulate all US commercial human spaceflight, both orbital and suborbital. The FAA/AST became the authority for certification, licensing and operation, with guidelines and regulations issued with particular relevance to suborbital operators [21]. These regulations are currently viewed as only a basic framework, with the assumed intention that it makes it easier for the market to develop. Risk can thus be prevalent in early operators, as suborbital-specific safety issues are not comprehensively addressed.

Additional difficulty and opportunity for risk intrusion is encountered in handling of hybrid systems in which a rocket-powered vehicle is launched from a carrier aircraft. In this case, the carrier aircraft and suborbital vehicle are subject to different licensing and regulation standards, even if significant system similarities exist. A more difficult problem still is encountered when proposing a jet-powered aircraft that includes a rocket propulsion option for trans-atmospheric flight.

A final hurdle to mitigating risk through regulation is encountered in launch location licensing. Specifically, some arrangements like those of the all-private Blue Origin facility are not subject to FAA regulation [22]. Furthermore, a vehicle launching from a temporary or mobile site as proposed by Armadillo Aerospace, is not only not subjected to FAA regulation [23], but also has immunity from a number of occupational safety regulations, as well.

5. CONCLUSIONS

Suborbital spaceflight presents significant risk due to the nature of the flight profile and performance level of the vehicles. Compounding this risk is the non-standardized development and operations procedures that commercial suborbital flight providers currently engage in. In the absence of well-defined regulations that guide the practices of these entities, both risk categories can go unchecked.

As vehicles are flown with increasing frequency the opportunity to fill in regulatory gaps should be seized. In the United States, the FAA/AST has built frameworks that should evolve to be more stringent and engaging. Designers of sRLV's should be subject to deeper scrutiny during the vehicle certification process to ensure that design requirements for reusable vehicles are met or exceeded, and that proper accommodations for integrated health monitoring are built in to the design.

Similarly, more stringent operating and spaceport licensing requirements should be imposed to verify that flight, training, and ground operations include best practices like pilot workload sharing, health monitoring, built-in testing, and worn parts replaceability. Policies should exist to verify that companies never sacrifice risk mitigation in favor of decreased turnaround times. Exclusive, private operators and vehicles that do not require ground infrastructure should not be exempt from regulation.

In summary, an industry with a patchwork of borrowed practices and such a high growth rate requires a culture of safety through all phases of vehicle lifecycle. In the absence of a professional authority or set of specific standards, increased involvement of regulatory agencies may be the best way to mitigate the unknowns that increase risk.

6. REFERENCES

1. Quinn, A., Bond, S.J. & Maropoulos, P. (2011). New safety model for the commercial human spaceflight industry. Paper presented at the Fifth IAASS Conference: A Safer Space for a Safer World, October 17-19 2011, Versailles, France.

2. Moro-Aguilar, R. (2015). National Regulation of Private Suborbital Flights: A Fresh View, 10 FIU L. Rev. 679.

3. Young, A. (2015). The Twenty-First Century Commercial Space Imperative. Springer International, Basel, Switzerland.

4. XCOR (2014). Lynx Payload User's Guide v5. XCOR Aerospace, Inc., Mojave, CA, USA.

5. Blue Origin (2015). New Shepard Payload User's Guide for Research and Education Missions. Blue Origin Texas, LLC, Van Horn, TX, USA.

6. Jane, F. T., Jackson, P., Munson, K. & Peacock, L. (2009). Jane's All the World's Aircraft 2009-2010. McGraw-Hill, New York, NY, USA.

7. SpaceX (2015). Falcon 9 Launch Vehicle Payload User's Guide v2. Space Exploration Technologies, Hawthorne, CA, USA.

8. The Tauri Group (2012). Suborbital Reusable Vehicles: A Ten-Year Forecast of Market Demand. The Tauri Group, Alexandria, VA, USA.

9. Angelo, J. (2014). Encyclopedia of Space and Astronomy. Infobase Publishing, New York, NY, USA.

10. Erickson, L. (2010). Space Flight: History and Technology, and Operations. Rowman and Littlefield, Lanham, MD, USA.

11. Foust, J. (2016). SpaceX seeks to accelerate Falcon 9 production and launch rates this year. SpaceNews, retrieved from http://spacenews.com/spacex-seeks-to-accelerate-falcon-9-production-and-launch-rates-this-year/ on 04-15-2016.

12. XCOR Aerospace. (2016). XCOR Aerospace Frequently Asked Questions. XCOR Aerospace Corporation, retrieved from http://www.xcor.com/faq/ on 04-15-2016.

13. Hammond, W. (1999). Space Transportation: A Systems Approach to Analysis and Design. AIAA, Reston, VA, USA.

14. FAA/AST. (2012). The U.S. Commercial Suborbital Industry: A Space Renaissance in the Making. US DOT, FAA, Washington, DC, USA.

15. FAA/AST. (2012). Reusable Launch Vehicle Critical Structures Guide v2.3. US DOT, FAA, Washington, DC, USA.

16. Swearingen, K., et al. (2007). An Open System Architecture for Condition Based Maintenance Overview. Paper Presented at the IEEE Aerospace Conference, March 3-10 2007, Big Sky, MT, USA.

17. Zagrai, A., et al. (2012). Active structural health monitoring during sub-orbital space flight. Journal of the Acoustical Society of America, Vol 132:1964.

18. NTSB. (2015). Report: In-Flight Breakup During Test Flight, Scaled Composites SpaceShipTwo, N339SS, Near Koehn Dry Lake, California. US NTSB, Washington, DC, USA.

19. Aerospace Medical Association Commercial Spaceflight Working Group. (2011). Suborbital commercial spaceflight crewmember medical issues. Journal of Aviation, Space, and Environmental Medicine 82.4:475.

20. Masson-Zwaan, T. & Moro-Aguilar, R. (2013). Regulating Private Human Suborbital Flight at the International and European Level: Tendencies and Suggestions, 92 ACTA ASTRONAUTICA 243, 243-54.

21. United States Code of Federal Regulations, FAA Final Rule – CSLAA. (2006). Human Space Flight Requirements for Crew and Space Flight Participants. US DOT, FAA, Washington, DC, USA.

22. FAA/AST. (2006). Legal Interpretation Findings Letter- Meredith, Blue Origin. US DOT, FAA, Washington, DC, USA.

23. FAA/AST. (2006). Legal Interpretation Findings Letter- Milburn, Armadillo Aerospace. US DOT, FAA, Washington, DC, USA.

DEVELOPMENT OF A COMMERCIALLY AVAILABLE PRESSURE SUIT FOR SUBORBITAL FLIGHT

Erik Seedhouse, Ph.D.

Embry-Riddle Aeronautical University. Bldg. 6, College of Aviation Room 314,
Daytona Beach Campus Embry-Riddle Aeronautical University.
600 S. Clyde Morris Boulevard Daytona Beach, FL 32114-3900
seedhoue@erau.edu

1. BACKGROUND

On October 31, 2014, the crash of Virgin Galactic's SpaceShipTwo (SS2) became the first commercial human spacecraft accident investigation led by the National Transportation Safety Board (NTSB). The spacecraft was at an altitude of about 15,000m and traveling at Mach 1 when the accident occurred. The extreme aerodynamic forces induced structural failure and explosive decompression of the cabin. Both pilots were wearing flight helmets with oxygen masks, but not pressurized flight suits. The crew was immediately exposed to –57°C, an atmospheric pressure of 15% sea level, and had less than ten seconds before hypoxia onset. The co-pilot did not survive [1].

Ever since the suborbital flight of SpaceShipOne in 2004, the pilots and co-pilots of these spacecraft have never worn pressure suits because, with a pressurized cabin, it was thought a pressure suit was unnecessary for trips into the stratosphere. The stratosphere begins at an altitude of approximately 10 km and extends to 50 km (space starts at 100 km). For over half a century, this has been the flight domain of the U-2 aircraft which has an operational ceiling above 21,000m and pilots of this aircraft have always worn pressure suits and helmets.

The SS2 flights were significantly higher than those of the U-2 aircraft, but significantly lower than the Mercury suborbital flights. Neither the Air Force nor NASA ever considered attempting human flight in these environments without a pressurized flight suit but, until the SS2 crash, Virgin Galactic maintained that pressure suits would be unnecessary. In fact, just two weeks before the SS2 crash, Virgin Galactic's CEO George Whitesides told *The Telegraph*, "We think the safest thing is to not have people in pressure suits but to have them in flight suits and then in a cabin which protects them and allows them the freedom of microgravity because these people will be able to get out of their seats and float around the cabin". Since the SS2 crash the industry is rethinking their perspective and it is likely commercial suborbital flights will require the use of pressure suits. But existing commercial suborbital spacesuit options are limited, since commercially available space suits are extremely expensive (several hundred thousands of dollars), heavy, and have limited mobility capabilities. But Final Frontier Design (FFD) is currently developing a space suit [2] that is more adjustable, more mobile, less expensive and lighter weight than current options. The suit is currently undergoing user testing and documentation for Flight Safety Approval from the FAA-AST, and is slated for engineering evaluations through Critical Design Review (CDR) for orbital space flight, via a NASA 2014 Space Act Agreement (SAA) and NASA Flight Opportunities Program (FOP) grant

2. METHODOLOGY/APPROACH

Historically, pressure suits have been designed to address requirements associated with individual applications. This will hold true for commercial suborbital flights for which the suit ensemble must address myriad requirements, some of which will represent a significant departure from traditional aeronautical/aerospace programs. This includes wide- ranging temperature and pressure profiles (in the event of a contingency), mobility/range of motion requirements and other factors unique to supersonic free-fall from extremely high altitudes (in the event of a contingency). The proposed approach in this study is to develop and test the FFD suit for commercial suborbital flight against and in parallel with a current U-2 suit which will be modified to the requirements of suborbital flight. Since the U-2 suit has been through extensive qualification and testing for use in high altitude operations, many of the requirements for suborbital spaceflight have been met, so the intent is to prove out the FFD suit and its capabilities against the U-2 suit.

Embry-Riddle Aeronautical University (ERAU) is in possession of a U-2 suit [3] and a FFD Demonstrator Suit (DS). The DS was developed to evaluate the capabilities of a full pressure suit for commercial suborbital spaceflight. FFD is uniquely qualified in the design and manufacture of commercial pressure/space suits, having pioneered this highly specialized field for the past several years in cooperation with a NASA Space Act Agreement (SAA).

3. PROPOSED RESEARCH

The main objective of this project is to develop two space suits for suborbital missions. Note: this research involves human subjects. In this application we propose to test the FFD DS against the established U- 2 suit across the following factors [4, 5]:

Category	Requirement	
Factors affecting performance on every mission	1. Comfort (unpressurized) 2. Visibility 3. Mobility (unpressurized)	4. Ease of donning/doffing 5. Tactility 6. Oxygen breathing system
Factors occurring frequently but not on every mission	1. Valsalva 2. Altitude protection 3. Mobility (pressurized)	4. Comfort (pressurized) 5. Leakage (pressurized) 6. Structural integrity
Factors occurring under emergency conditions	1. Ejection G forces 2. Parachute compatibility 3. Wind loading protection 4. Flotation provision	5. Impact and penetration protection 6. Water immersion protection 7. Explosive decompression

A. Retention of oxygen pressure and carbon dioxide washout

1.1 *Leak Rate Test: performed with mannequin inside suit.*
The ventilating outlet on the suit torsos will be blocked and the suits connected to leak test instrumentation. The leak test instrumentation will measure the pressure in the suits, a flowmeter will measure air entering the suits, and a manometer will measure the pressure at the flowmeter [6].

1.2 *Carbon Dioxide (CO_2) Washout Test: one test performed with mannequin inside suit. One test with human.*
The objective of this test is to characterize workloads and flow rates at which CO_2 is washed away from the mannequin's oronasal area. Data will be used to define acceptable workloads and flow rates for spacecraft simulator-based ground testing and inflight testing with the suits. Data will also be used toward building a database of CO_2 washout test data that can be used to validate analysis models, and inform future FFD space suit helmet and ventilation flow path designs. Each subject will perform at four metabolic levels; resting, 1000, 2000, and 3000 BTU/hr, at three air flow rates: 4, 5 and 6 ACFM. Each subject will perform testing on two different days so data can be compared for day to day variations. Gas will be collected from a fixed sensor located above the subject's head and from two sensors places on either side of an oral/nasal mask. Resting metabolic rate will be obtained with subject in the donning stand. 1000 BTU/hr will be generated by having each subject operate a hand ergometer. 2000 and 3000 BTU/hr will be produced by walking on a treadmill. Standard air flow for pressurized suited events is 6 ACFM.

B. Assess the work envelop during a simulated suborbital spaceflight.

2.1 *Functional Arm Reach Test: performed with subject inside suit.*
The radii of the maximum functional arm-reach envelope [7] will be measured at intervals in each of 12 vertical planes using an arm-reach measuring device. Data will be obtained on the volume of the grasping reach envelope in shirt sleeves, suited and unpressurized, and suited and pressurized to 5 psig.

2.2 *Vision Tests: performed with subject inside suit.*
Visor optical characteristics will be determined using the U-2 and FFD suit visor's inherent refractive power, calculated from radius curvature supplied by USAF and FFD. Critical and noncritical areas of the visor will be checked in eight areas according to MIL-V-27446 (USAF). Vertical and horizontal prismatic deviations will be investigated. Each subject will be tested for unaided visual acuity using a Snellen Eye Sight Chart the screen set for 6 m letter size--right eye 20/20+, left eye 20/20+, both eyes 20/20+. Field of vision studies will be conducted with the subject in the suit unpressurized with the visor up and pressurized, visor closed.

2.3 *Dexterity and Movement Tests: tests to be performed with subject inside suit.*

2.3.1 *Dexterity: suit*
Each test subject shall be able to bend over (without squatting), pick up a length of 8 mm to 10 mm diameter rope, pass it round his/her waist and tie a double knot in front of them.

2.3.2 *Dexterity: suit, lifejacket and parachute harness*
Each test subject wearing the suit system and a lifejacket shall be able to bend over (without squatting), pick up a length of 8 mm to 10 mm diameter rope, pass it round his/her waist and tie a double knot in front of them.

2.4 *Movement: suit: all tests to be performed with subject inside suit.*

2.4.1 *Movement:*
Each subject shall be able to complete the following series of movements wearing the suit system:

(i) Kneel on both knees, lean forward and place both hands on the floor 450 mm in front of the knees
(ii) Position hands at chest level, palms out, reach directly overhead, interlock thumbs, extend arms fully
(iii) Kneel on the right knee, place left foot on the floor with the left knee bent 90°. Touch toe of left foot with the thumb of right hand
(iv) Extend arms fully in front of body, lock thumbs together, twist upper body 90° to the left and right
(v) Stand with feet a shoulder width apart, arms at side. Raise the arms until they are parallel to the floor in front of the body. Squat down as far as possible

2.4.2 *Movement: suit, lifejacket and parachute harness*
Each subject shall be able to complete the following series of movements wearing the suit system, lifejacket and parachute harness:
(vi) Kneel on both knees, lean forward and place both hands on the floor 450 mm in front of the knees
(vii) Position hands at chest level, palms out, reach directly overhead, interlock thumbs, extend arms fully
(viii) Kneel on the right knee, place left foot on the floor with the left knee bent 90°. Touch toe of left foot with the thumb of right hand
(ix) Extend arms fully in front of body, lock thumbs together, twist upper body 90° to the left and right
(x) Stand with feet a shoulder width apart, arms at side. Raise the arms until they are parallel to the floor in front of the body. Squat down as far as possible

C. Emergency Bailout

3.1 *Emergency Egress Background: performed by two subjects*
To effect egress from a suborbital spacecraft the astronaut will need to transition from the seated to the standing position, and climb out of the vehicle [8, 9]. Since this action may occur at high altitude, a unique set of requirements comprise the thermal considerations since a large range of thermal regimes exist for suborbital missions – from as high as 40° C at ground level during launch in the summer, to as low as -60° C at ~100,000m. Another requirement is the effect of windblast when attempting to egress from a vehicle traveling at Mach speeds.
After egress the subject will deploy a parachute. The parachute system includes a drogue, main and reserve chute. The drogue chute will automatically deploy if a wrist-worn sensor senses 6 Gs or more for 6 continuous seconds. The drogue chute will also be activated manually during free-fall. The main chute will be manually deployed nominally, and the reserve chute has an automatic opening device, designed to deploy if the main chute does not open.

3.2 *All-weather chamber testing: one test performed with mannequin. One test performed with human.*

3.2.1 *Cold Test*
Testing will be performed in an environmental chamber. Each mannequin/subject will spend 60 minutes in the suit with the chamber at -60° C during two ventilation conditions: 13 cfm 5 mm Hg water vapor pressure and 8 cfm and 5mm Hg water vapor pressure. Skin temperature (taken at 5 points) and heart rate will be measured and recorded continuously throughout the test. Body heat storage (Qs) will be computed from the equation:

$$Qs = \frac{0.831 \text{ x } t_b \text{ x weight}}{\text{Body surface area}} \ (\text{kcal/m}^2)$$

These tests will demonstrate that the suit provides thermal protection from extreme cold.

3.2.2 *Heat Test: one test performed with mannequin. Second test performed with human.*
Testing will be performed in an environmental chamber. Subjects will spend 60 minutes in the suit with the chamber at 40° C during two ventilation conditions: 13 cfm 5 mm Hg water vapor pressure and 8 cfm and 5mm Hg water vapor pressure. A second test will assess the effectiveness of the Veskimo® Cooling device that the subject/mannequin will wear for 60 minutes while inside the environmental chamber at 40°C. Skin temperature (taken at 5 points) and heart rate will be measured and recorded continuously throughout the test. Body heat storage (Qs) will be computed from the equation:

$$Qs = \frac{0.831 \text{ x } t_b \text{ x weight}}{\text{Body surface area}} \ (\text{kcal/m}^2)$$

Sweat loss will be obtained by weighing the subject nude before and after the test. These tests will demonstrate that the suit provides the necessary thermal protection from extreme heat.

3.3 *Vertical wind tunnel tests: performed by each subject.*
To develop confidence freefalling in the pressure

suit, each subject will perform three vertical wind tunnel tests at the iFly Indoor Skydiving facility in Orlando to acclimatize subjects to pressurized and unpressurized free-fall in the suit, and to practice free- falling in a relatively closed environment. In the first test subjects will practice emergency egress, in the second they will practice upset maneuvers and in the third they will practice unstable exit, falling backwards and spinning. These tests will prove the suit's functionality during free-fall [8, 9].

3.4 *High altitude chamber test: one test performed with mannequin. One test performed with human.*
Each suited subject will perform an altitude chamber test to an equivalent altitude of 100,000 feet. The atmospheric and thermal profiles of the mission – ascent, capsule depressurization and suit pressurization, climbing out of the capsule, and rapid descent – will be simulated.

3.5 *Altitude bailout contingency test* [9].
This test will require each subject to bail out at 12,000 feet wearing the suit and parachute harness. This test will be performed to demonstrate the subject's comfort in a pressure suit at high altitude equivalent.

D. Flotation Test [10]

The DS and U-2 suit will be subjected to two flotation tests:

(i) Each subject will be dressed in full assembly with helmet with visor closed and enter the water from a 3 meter platform. Each subject will board a life- raft.
(ii) Each subject will be dressed in suit with helmet with visor open and enter the water from a 3 meter platform. Water will be allowed to enter the helmet so as to fill the bladder. When the suit is filled, each subject will tread water with their head out of the water for 3 minutes. Each subject will board a life-raft.

4.2 *Cold immersion testing.*
Testing of the pressure suit system will be performed in a cold water flume. Subjects will spend 60 minutes in the suit with the water at 40 degrees F. These tests will demonstrate the suit provides the necessary thermal protection from extreme cold while immersed in water.

4.3 *Assess the water bailout contingency*
Each test subject shall enter the water wearing suit plus lifejacket and shall swim for 25m ± 5 m and attempt to board a life-raft without assistance. This attempt shall be carried out in a test pool filled to a depth that prevents a test subject from pushing off from the bottom of the pool. The test subject may have as many attempts at this action as required within a time limit of 5 min.

E. Neutral Buoyancy Testing: performed by mannequin and subjects in Neutral Buoyancy Facility (NBF)

Neutral buoyancy has been the preferred method of evaluating microgravity EVA tasks since its development in the 1960s for the Gemini program. It consists of balancing the mass/volume of the spacesuit so that the mass of the suit matches the mass of the water it displaces. The suit is neutral; not floating or sinking, in the water column. This allows subjects to evaluate tasks while they float in simulated microgravity. Some of the benefits of neutral buoyancy are the ability to closely match microgravity conditions, and 6-degrees-of-freedom of motion (three translation axes and three rotation axes). The NBF represents a higher level of risk than lab testing or the Zero-G aircraft since the issues with suited testing are combined with the hazards of scuba diving, but the NBF is manned by 3 safety divers,

Two sets of unmanned NBF evaluations will be completed prior to manned use. The first will be completed on the pool deck and serve to dry run test procedures, verify nominal operation of the suit, and test off-nominal flow conditions. The second test – the mannequin test – will allow a full demonstration of test procedures. In this test the suit will be taken up and down the water column at typical ascent/descent rates to confirm the Environmental Control System (ECS) response to the new suit. If all ECS parameters are nominal, and test will proceed to manned testing.

5.1. *Manned NBF testing*
This will demonstrate the ability to interface with the NBF systems and weigh-out the suit, and to determine the subject's ability to use the suit underwater. Operational time underwater will be limited to 30 minutes until cycle testing is completed (a drink bag will be incorporated into the suit). Each subject will attempt the following tasks:

(i) Ingress/egress of a mockup hatch
(ii) Translating 10 ft across NBF carrying weighted bag
(iii) Translating 10 ft across NBF no weight
(iv) Manipulating medium-sized tools
(v) Articulate foot restraint
(vi) Performing two-handed tool operations

F. Microgravity Testing

Microgravity testing on a Zero-G flight will be conducted to determine ease of egress and operation of flight controls. Each suited subject will attempt to:

(i) Egress the Lynx seat in the suit
(ii) Attempt to perform translation and body stabilization tasks.
(iii) Attempt attaching seat belts once inside the seat.

G. Integrated Test

Each subject will operate a Cessna 172 while wearing the FFD suit and follow a test card with 18 flight profiles. The test will be performed over a 180-minute period including approximately 150 minutes while inside the suit and 130 minutes under pressure. The air flow rate will be 22-30 LPM throughout the test. Communications between the pilot and co-pilot will be via normal aircraft intercom. Each phase of flight will be assessed for visibility performance, arms, head, torso, legs mobility, smell, sweat, noise, communication clarity, overall comfort, and reaction time for the given task. Other parameters, including time for each phase and pressure, heart rate, temperature, level of assistance if any, and task gain level from low to high will be recorded. Gloved hand mobility and dexterity tests will be conducted during each phase of flight.

- (i) Start data recorder
- (ii) Rotate at 160 kt
- (iii) Climb out at 25° angle of attack
- (iv) Climb at full power
- (v) Keep ball centered
- (vi) Use checklists
- (vii) Check engine gauges
- (viii) Trim hands off level flight
- (ix) Operate (simulated) RCS
- (x) Perform spiral glide
- (xi) Record ground speed from GPS
- (xii) Note stall speed and Stall warning
- (xiii) Perform pitch up and pitch down maneuver
- (xiv) Recover altitude and speed
- (xv) Record rate of descent
- (xvi) Inspect airframe

4. REFERENCES

1. http://www.ntsb.gov/news/events/Documents/2015_spaceship2_BMG_abstract.pdf

2. Southern TC., Moiseev NA., Iturmendi M. Operational Testing of a Novel IVA Space Suit. 45th International Conference on Environmental Systems. Bellevue, 12 – 16 July 2015.

3. Jenkins, K. *Dressing for Altitude*, U.S. National Aeronautics and Space Administration, Washington DC, 2012

4. Ripps, T.B, Garcia, J.M., Macleod, S.M., Jacobs, S.E., "The Demonstrator Suit: Evaluating a Full Pressure Suit through Manned and Unmanned Testing." *41st International Conference on Environmental Systems*, July, 2011. AIAA 2011-5106

5. Barry, D.M., Bassick, J.W. "NASA Space Shuttle Advanced Crew Escape Suit Development." *25th International Conference on Environmental Systems*, July, 1995. SAE 951545

6. Korona, F. A., Jason Norcross. "Z-1 CO_2 Washout Test Report." *Internal NASA report*, CTSD-ADV-1024. November 2012. NASA JSC.

7. Tufts, D.B., Jacobs, S.E., Barry, D.M., "The Demonstrator Suit: Incorporating Waist, Hip and Shoulder Mobility in a Full Pressure Suit." *40th International Conference on Environmental Systems*, July, 2010. AIAA 2010-6178

8. Ryan, C., "The Pre-Astronauts. Manned Ballooning on the Threshold of Space." Bluejacket books, Naval Institute Press, Annapolis, MD. 1995.

9. "Summary Report: Findings of the Red Bull Stratos Scientific Summit." California Science Center, Los Angeles, CA. January 23, 2013

10. NPR 8705.2B, NASA Procedural Requirements, Human-Rating Requirements

PROCEEDINGS OF THE 8TH INTERNATIONAL SPACE SAFETY CONFERENCE

SESSION 14
LAUNCH SAFETY - II

POTENTIAL USES OF CONSEQUENCE ANALYSES FOR RANGE SAFETY

Paul D. Wilde Ph.D., P.E.

Federal Aviation Administration, 800 Independence Ave., Washington, DC 20591 USA, Email: paul.wilde@faa.gov

ABSTRACT

This paper summarizes FAA regulations regarding the use of a Flight Safety System (FSS) to terminate a launch vehicle's flight to prevent the hazardous effects of the resulting debris impacts from reaching any populated area and ensure that the launch satisfies the public risk criteria." This paper investigates how consequence analysis can be used to help ensure that safety interventions, such as FSS activation, prevent "high consequence" events and introduce reasonable conditional risks when implemented. This paper also summarizes the US consensus standards on conditional risk management (in RCC 321) and proposes reasonable definitions for the currently undefined terms such as high consequence hazards and acceptable conditional collective risks. This paper suggests that third party casualty consequences of aviation accidents can be the foundation for reasonable criteria to determine the circumstances that warrant activation of a FSS or other safety intervention.

1. FAA SAFETY MANAGEMENT SYSTEM

The US Federal Aviation Administration (FAA) Office of Commercial Space Transportation (AST) serves to protect the public health and safety, the safety of property, and the national security and foreign policy interests of the United States during commercial launch and reentry activities, and to encourage, facilitate, and promote US commercial space transportation. AST achieves the public safety aspects of its mission through the application of a regulatory framework for commercial space transportation and through safety promotion activities, which together make up AST's safety management system (SMS).

AST is statutorily directed by Commercial Space Launch Act of 1984 (CSLA) as amended to regulate commercial space transportation (CST) *"only to the extent necessary...to protect the public health and safety, safety of property, and national security and foreign policy interests of the United States."* Title 14, Chapter III of the Code of Federal Regulations (CFR) defines the regulatory lifecycle for CST, as shown in Fig. 1, by specifying the procedures and public safety requirements for each regulatory phase, and establishing the associated public risk thresholds for CST activities. The CSLA and the CST regulations form the foundation of AST's SMS by providing a systematic approach to identifying, assessing, and controlling risks associated with CST operations. This paper describes a situation involving the top part of the SMS lifecycle depicted in Fig. 1: where innovative CST activities challenged the current regulations, produced lessons learned that built on stakeholder engagement, and resulted in a more refined approach. Specifically in this case, innovative launch operations such as the return to launch site of the Falcon 9 first stage, led the FAA to appeal to foundational range safety principles and apply state-of-the-art techniques in the evaluation of a waiver request.

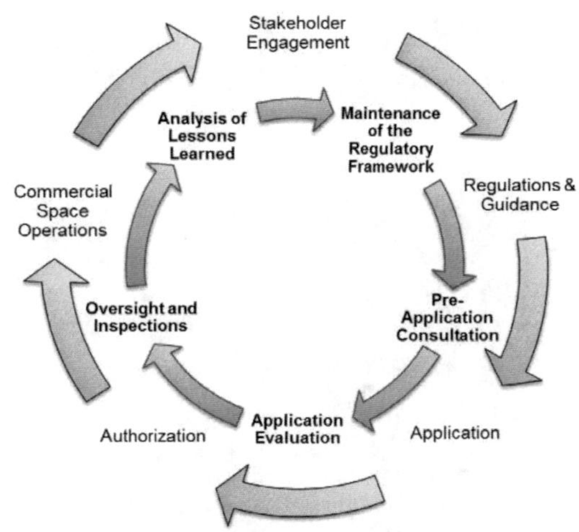

Figure 1. Safety Management System Lifecycle

2. OVERVIEW OF FAA APPROACH

This section explains in simple terms what AST finds "necessary" to protect public health and safety during CST. To understand what is "necessary" to protect the public safety requires answers to two critical and fundamental questions: how safe is safe enough for the public, and how are public risks "managed."

Webster's defines safety as "the condition of being free from undergoing or causing hurt, injury, or loss," and risk as "the possibility of loss, injury, disadvantage, or destruction." Absolute public safety is an abstract ideal that equates to zero risk or complete protection from harm.

Absolute safety during CST is possible in rare cases where physical limits isolate the public from harm. However, the high energy and complex nature of launch and reentry means that hardware, software, or human failures can readily lead to serious public consequences (serious injury or loss). Since absolute safety during CST is often unattainable, particularly for orbital launch and re-entry operations, the protection of public safety necessarily involves "risk management."

The fundamentals of public risk management are the same across all industries. Risk management involves a logical and systematic process to identify hazards and control the risk they pose. Hazard identification uses data on the planned performance and foreseeable malfunctions to identify scenarios that could threaten the public. Risk controls for any scenario must address at least one of three key elements of risk, illustrated in Fig. 2: (1) the probability of a dangerous event (such as a rocket crash), (2) the size of the danger area (such as the area destroyed by a rocket crash), and (3) the nature of the public exposure (such as the population density and sheltering provided in an area where a rocket could crash). Thus, public risks reflect the possibility of dangerous events that could produce serious public consequences.

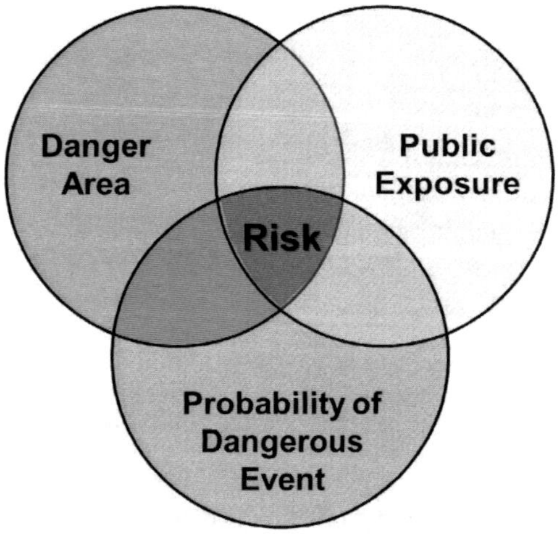

Figure 2. Key Elements of Risk

To illustrate these concepts, consider for example, that the public risk from an intact rocket impact may be controlled by reducing the probability of an intact impact, reducing the area destroyed by the impact (e.g. by ensuring the propellants are dispersed before impact), or by evacuating the public from the area that could be hit. (Note that an intact impact of a launch vehicle with substantial quantities of propellant onboard generally produces an explosion and creates a much larger danger area than purely inert debris impacts.) As a result, there are a number of ways to potentially mitigate public risk, and the level of effort required to demonstrate appropriate risk management is naturally linked to the size and complexity of the system, as well as the nature of the public exposure. As explained in the next section, a Flight Safety System (FSS) is an important means to mitigate public risks from large orbital rocket launches by reducing the danger area (ensuring propellants are dispersed before impact) and reducing the probability of impact in public areas.

One fundamental tenet of risk management is that acceptable risk levels are set with an understanding of the consequences of a hazard and the likelihood of its occurrence. Of course, no serious public consequences from launch or reentry are truly acceptable, in that no responsible authority would regard such an event as routine or permissible. AST seeks to maintain a level of public safety where adverse public consequences remain rare events by enforcing well-defined regulatory risk tolerability criteria supported by multiple lines of logic.[1,2] While the consequences of the hazards of CST operations can be substantial, the acceptable risk levels contained in AST's regulations are only a small fraction of the "normal background risk…accepted in the course of normal day-to-day activities."[3] For example, AST's risk limits equate to less than 1% of the annual risk accepted by US pedestrians on an individual and collective basis. The risk criteria set by AST in the commercial space regulations for licensed operations are the same for each mission. These limits were set as a means to manage the risk for current and expected future operations, consistent with the goal that adverse public events remain rare. AST periodically re-evaluates the risk criteria to account for the frequency of CST activities; the demonstrated safety record and benefits provided; technological capabilities and maturity of the industry; risks tolerated in other industries, and common perceptions of CST risks.

No US space launch has created a public casualty. AST seeks to maintain a level of public safety where adverse public consequences remain rare events. AST enforces public safety requirements that were initially developed and implemented by the US Air Force and NASA. Building upon this experience and approach, the FAA has implemented a comprehensive and interdependent set of public safety regulations that include specific operating requirements, specific safety requirements for critical systems, and the application of a public risk management process that uses quantitative analyses. A primary purpose of the public risk management process applied to launch and reentry is to facilitate informed decisions regarding the operating parameters and vehicle design features that are necessary to limit the predicted public risks to pre-defined criteria. AST uses a *risk informed* process to systematically identify, reduce, monitor, and ensure acceptable public risks. Compliance with the applicable regulations constitutes what is "necessary" to protect the public during CST in general.

3. FLIGHT SAFETY SYSTEMS (FSS)

As mentioned in the previous section, a FSS is an important and rather unique means to mitigate public risks from large orbital rocket launches. The FAA formally defined (in 14 CFR §417.3) a FSS as "the system that provides a means of control during flight for preventing a hazard from a launch vehicle, including any payload hazard, from reaching any populated or other protected area in the event of a launch vehicle failure." In plain language, a FSS provides a means to terminate flight (e.g. by terminating thrust or by triggering an explosive charge to destruct4 the vehicle) to prevent the hazard effects of an errant vehicle from reaching protected areas. Note that an important function of a FSS is the dispersion of propellants prior to impact with the Earth's surface.

As illustrated in Fig. 3, a traditional FSS has included hardware, software, and flight safety crew. A FSS must include an ability to track the vehicle trajectory and to terminate the flight if the vehicle experiences a malfunction that violates pre-defined mission rules. The FAA currently requires that any vehicle that employs a FSS must use tracking data sources that are "independent of one another, and at least one source must be independent of any vehicle guidance system." In the past, the decision to terminate flight was made by a human being, referred to here as the Mission Flight Control Officer (MFCO). Whether the decision to terminate flight is made by a MFCO or by an automated (i.e. compute) system, the terminate decision is made using mission rules, which generally involve "flight safety limits" (also known as destruct lines) and debris "Impact Limit Lines (ILLs)."

In simple terms, the flight safety limits are lines on a map that designate when the flight safety system should be triggered if the vehicle "tracking icon"[5] crosses them as illustrated in Fig. 3. The FAA requires (in § 417.213) that *"a flight safety analysis must identify the location of populated or other protected areas, and establish flight safety limits that define when a flight safety system must terminate a launch vehicle's flight to prevent the hazardous effects of the resulting debris impacts from reaching any populated or other protected area and ensure that the launch satisfies the public risk criteria."* The FAA furthermore requires an analysis to "establish designated impact limit lines to bound the area where debris with a ballistic coefficient of three or more is allowed to impact if the flight safety system functions properly." Although § 417.107(c) requires a launch operator's flight safety analysis to account for any inert debris impact with a mean expected kinetic energy at impact greater than or equal to 11 ft-lbs., impact kinetic energy was deemed an impractical metric for establishing impact limit lines because kinetic energy at impact can vary significantly depending on wind conditions, and impact limit lines that vary with wind condition are impractical. Thus, ballistic coefficient was deemed a better metric than impact kinetic energy to establish the debris that needs to be accounted for in the determination of valid flight safety limits.

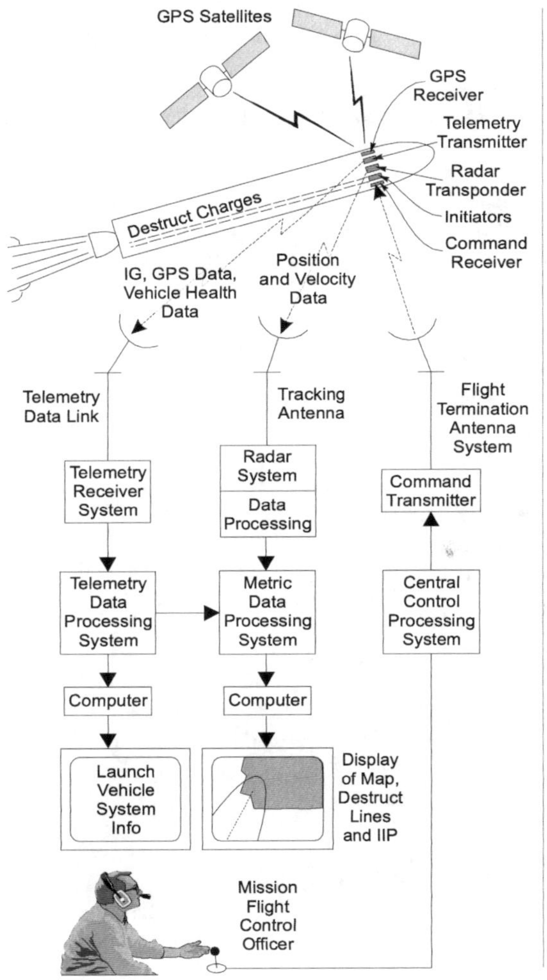

Figure 3. Elements of a Flight Safety System.

As explained further in the next section, the FAA regulations for flight safety limits, which are central to the FSS activation criteria, were intended to (1) capture the current practice of the USAF at the time the regulation was developed,[6] (2) provide a clear and consistent basis to establish impact limit lines that determine if an accident as defined by § 401.5 occurred,[7] and (3) help prevent a high consequence to the public given FSS activation.

4. FSS ACTIVATION CRITERIA

As alluded to in the previous section, and explained further in this section, the main purposes of a FSS are to (1) prevent high consequence events as a result of launch vehicle failures, and (2) ensure that each launch operation satisfies the public risk criteria. For example, §417.107(a) requires a launch operator to employ a FSS if either (1) a failure "would have a high consequence to the public" in the vicinity of the launch site, or (2) "if the absence of a flight safety system would significantly increase the accumulated risk from debris impacts" in the downrange area.

A FSS is typically necessary to ensure that an orbital launch satisfies the public risk criteria given (1) the relatively high probability of failure for Expendable Launch Vehicles (ELVs), particularly for new ELVs compared to certified aircraft, [8,9] and (2) the potential for a high consequence event given a failure, especially during the first stage of flight when large quantities of propellant are onboard. The demonstrated flight experience of even the most reliable ELVs, such as the Delta II, reveal failure probabilities on the order of 1%, and the demonstrated flight history shows that new vehicles developed by experienced manufacturers have a POF near 0.3 for initial launches. New ELVs from inexperienced developers have demonstrated failure probabilities about twice high. In contrast, certified commercial transport aircraft have demonstrated accident rates on the order to one in ten-million per flight, roughly five orders of magnitude lower than the most reliable ELVs.

Thus, public safety for large orbital launch vehicles has traditionally been protected with the use of a highly reliable FSS and quantitative risk analysis (QRA) to ensure any "residual risks" are acceptable. The term residual risk refers here to public risk from a launch or reentry that is not mitigated by a FSS.

Residual risk assessments must be conducted using tools that are both validated (fulfills the requirements of the task) and verified (correctly executes the function). Valid public risk analyses must use accurate data and scientific methods that are mathematically valid, with reasonably conservative assumptions applied in areas where significant uncertainty exists. Valid assessments usually use computational tools, which must be verified by comparison to available empirical data (e.g. from a launch failure, planned event or lab tests/experiments), or other validated tools, such as physics-based models.

An important example of residual risk that is central to this paper is the risk to the public posed by a launch vehicle failure that results in activation of the FSS when the vehicle violates a valid flight safety limit and the FSS works as designed. Successful activation of a FSS triggered by a flight safety limit violation can pose a significant residual risk because debris with a ballistic coefficient less than three pounds per square foot may be capable of producing a casualty. The fact is that kinetic energy at impact is better correlated with the potential for a debris piece to cause a casualty than ballistic coefficient.[10] However, impact kinetic energy was deemed an impractical metric for establishing flight safety limits (aka impact limit lines) because kinetic energy at impact can vary significantly depending on wind conditions,[11] and impact limit lines that vary with wind condition are impractical to implement. Thus, the FAA deemed that ballistic coefficient a better metric than impact kinetic energy to establish the debris that needed to be accounted for in establishing flight safety limits. In adopting the 3 psf ballistics coefficient standard, the FAA recognized that ballistic coefficient is not well correlated with the probability of a casualty producing impact.[12] There are significant probabilities that impacts with debris with a ballistic coefficient less than 3 psf might produce a casualty and that debris impacts with a ballistic coefficient greater than 3 psf might not produce a casualty. The population potentially exposed to an impact (e.g. whether in the open or sheltered in buildings, etc.), as well as the shape and impact orientation of debris, in addition to its energy and other characteristics, all influence whether or not an impact is likely to produce a casualty. Hence, the FAA requires an analysis of the residual public risks in addition to the establishment of impact limit lines. Note that while successful activation of a FSS triggered by a flight safety limit violation is usually a highly effective mitigation against high consequence events, the current flight safety limit requirements do not explicitly ensure protection against high consequence events.

Another example of a residual public risk from ELV launches involves downrange overflight of populated areas. For example, launches to the International Space Station (ISS) from Cape Canaveral Air Force Station, or from the Guiana Space Center (near Kourou, French Guiana) typically overfly portions of Europe while the upper-stage is thrusting prior to orbital insertion. A failure during downrange overflight, such as a thrust termination or an on-trajectory explosion, would result in debris impacts over a large area and could produce debris impacts in populated areas. During downrange overflight, the activation of a FSS cannot completely prevent the possibility of debris impacts in populated areas. Thus, downrange overflight usually involves a significant residual risk to the public, and in some cases, activation of a FSS could actually increase pubic risks.

There are several reasons that the public risks from downrange overflight are often below the acceptable risk criteria, which AST applies equally to foreign and domestic populations. First, the probability of a failure that could produce debris impacts in populated areas is often very low because downrange overflight usually involves a short dwell time of the Instantaneous Impact Point (IIP)[13] over populated areas; usually there is only a period of a few seconds where a failure could produce debris impacts in a populated area during downrange flight because the IIP moves very rapidly as the vehicle approaches orbital insertion. Second, the lower stages of the vehicle are usually jettisoned into the ocean long before the IIP reaches populated area during downrange overflight. A failure of an upper-stage as the vehicle approaches orbital insertion generally produces much smaller danger areas (also known as casualty areas) compared to failures earlier in lower-stage flight because (1) the inert mass associated with an upper-stage is generally much lower than a lower-stage, (2) propellants often disperse naturally following such high speed and high altitude break-ups, and (3) the inert debris may be reduced by ablation following a failure where the vehicle speed exceeds Mach 10.

The foregoing paragraph explains why activation of a FSS during downrange overflight may be unnecessary, and why the FAA only requires and ELV to be equipped with a FSS during downrange overflight "if the absence of a flight safety system would significantly increase the accumulated risk from debris impacts." For example, this regulation generally means that a FSS, or an alternative mitigation, is necessary to prevent a capsule with full propellant tanks from surviving to impact (and thus producing a relatively large danger area due to the ensuing explosion). A specific example is that SpaceX designed the thermal protection system of the Dragon capsule so that a launch failure during downrange overflight would result in break-up and demise, and thus mitigate the risk from the potential for the capsule to survive intact to impact.[14]

5. CONSEQUENCE ANALYSIS

Before delving into the particulars of how conditional risk analysis, also known as consequence analysis, can be useful in ensuring public safety, it is important to understand the essential difference between risk and consequence. As explained above, public risks reflect the probability of dangerous events that could produce negative public consequences. Whereas the risk from a launch or re-entry accident is quantified as the product of probability of the accident and the average (i.e. mean) consequence of the accident, a consequence analysis examines the outcome of an event independent of the probability of that event. Thus, consequence analysis is central to and embedded in risk analysis. Mathematically, the only difference between risk and consequence analyses is that a consequence analysis assumes that the event occurs: the probability of the event, e.g. a failure of some kind, is assumed to be one.

An examination of empirical data on the outcome of commercial transport aircraft departures helps illustrate the difference between risk and consequence. Empirical data for the 30-year period between 1984 and 2013, which are presented in more detail later in this paper, showed that the probability of an airline accident was about 2.9E-7 per departure: there were an average of about three airline accidents for every ten million departures in that period. The same empirical data set showed that an average airline accident between 1984 and 2013 produced about one (actually 0.9) ground fatality; in other words an airline accident in the US had a mean consequence of one fatality for people on the ground in that 30 year period. Therefore, the collective risk to the public from an airline departure was empirically demonstrated to be about 2.6E-7 expected fatalities between 1984 and 2013 (i.e. 2.9E-7 probability of an accident per departure multiplied by 0.9 fatality per accident). In simple terms, empirical data showed that about three ground fatalities resulted on average from every ten million airline departures in the 30 year period from 1984 to 2013.

As explained in the previous section, the traditional ELV approach, which was codified by the FAA Chapter 14 of the Code of Federal Regulations CFR in Part 417, uses a FTS designed and operated to protect populated (or otherwise protected) areas from debris impacts that exceed a ballistic coefficient of 3 psf together with risk analyses to demonstrate compliance with consensus requirements that limit the residual collective and individual risks to the public.[15] A drawback to that traditional ELV approach, which does NOT entail an examination of the conditional risks to the public given a FSS activation event occurs, is that it could unnecessarily limit occupied vehicle flights and produce elevated risks for vehicle occupants even if the threat to the public is extremely low. In addition, the traditional flight safety limit approach does not explicitly ensure protection against high consequence events as explained in the previous section.

Previous work done by the RCC Risk Committee outlined the steps involved in an innovative QRA approach that could provide a better balance between crew and public risk than the traditional ELV approach: "conditional risk management" for events that may involve a "safety intervention." The term "safety intervention" is used to encompass the entire range of risk mitigating actions that may be proposed for either ELVs or Reusable Launch Vehicles (RLVs), whether occupied or not. Activation of a FSS is an example of a common safety intervention for

ELVs, and a contingency abort to an alternative landing site is an important safety intervention for an RLV.
Figure outlines a systematic QRA approach to manage risks associated with safety interventions developed by the RCC Risk Committee. The extraordinary contributions of Mr. Jerry Haber of ACTA Inc in the development of the conditional risk management process as a whole and the basis for the abbreviated description that follows here are especially gratefully acknowledged. The conditional risk management approach was intended to supplement the current risk management requirements of RCC 321-10.

The conditional risk management process was designed to ensure that (1) safety interventions will be implemented to address "high consequence" events, and (2) implementation of safety interventions will produce reasonable conditional risks. The sequence of steps shown in Figure was designed to assure completeness and avoid unnecessary efforts whenever possible. The steps in this process are numbered for easy reference in the following description. Figure has two termination points: step 12 concludes the flow when it has been demonstrated that the risks from high consequence hazards are *de minimis;* step 13 is the result of a complete analysis to demonstrate that all safety interventions pose acceptable conditional risks.

1. "Hazard identification" refers to the initial step in a risk assessment in which hazards and possible consequences are identified and eliminated if possible. "High consequences hazards" may include outcomes that have a significant impact on continued operations, significant environmental impacts, impacts on relationships with other countries, and other long term or irreversible consequences.

2. If there are no identifiable high consequence hazards the process flow skips to Step 7.

3. When the potential for high consequence hazards have been identified, an assessment should be made of the chance of the high consequence result occurring given the safety intervention is implemented.

4. If it can be shown that the probability of a high consequence given the safety intervention is remote, the process flow skips to Step 7.

5. Quantitative or qualitative methods may be used to assess the high consequence risk potential. Safety personnel must be prepared to show that the assessment procedure has a traceable, defensible rationale.

6. Evaluation of the acceptability of risk must consider severity of the consequences and probability of occurrence. This assessment of the "residual risk" must account for all foreseeable outcomes, including safety interventions, and compare the results to acceptable risk criteria defined by appropriate authorities as described in the RCC 321 Standard in §2.3.4. If the public risks that include high consequences safety interventions are not acceptable, modify the intervention or the mission (Step11) and repeat the process.

7. Having ruled out significant high consequence hazard induced risks, the next step is to assess whether the probability of using the risk mitigation during the mission is negligibly small. When that can be established, the conditional risks are de minimis.

8. Acceptable safety mitigations should normally be expected to reduce the risk relative to no mitigation. While extenuating circumstances might warrant an exception, this is standard decision theory. If the mitigation does not reduce the risk, modify the intervention or the mission (Step11) and repeat the process.

9. To assure that the risk induced by the intervention is reasonable, two conditional risk measures should be calculated: the conditional collective risk expected given the mitigation and the expected value of the individual risk to the maximally exposed individual given the mitigation. Expected values of the two risk measures are used as the risk measure rather the peak values to produce more stable, consistent measures as the basis for decision making. The set of events over which averaging occurs will depend on the nature of the safety intervention. The general principles to define the groups for averaging are to group: (a) events that result from different instances of the same safety intervention (e.g. different outcomes from a FSS activation in response to a failure at a given time in flight), and (b) events that hazard substantially the same population centers (e.g. failures at various flight times that would trigger a safety intervention that would lead to debris impacts on a given population).

10. The conditional risks are reviewed for acceptability. If they are not acceptable, modify the intervention or the mission (Step11) and repeat the process. This is where any definitions of acceptable conditional risks are applied.

11. When some set of conditional risks have been determined to be unacceptable, the mission or the risk mitigations must be revised and the revised mission/risk mitigations must be reassessed to assure that it is now acceptable.

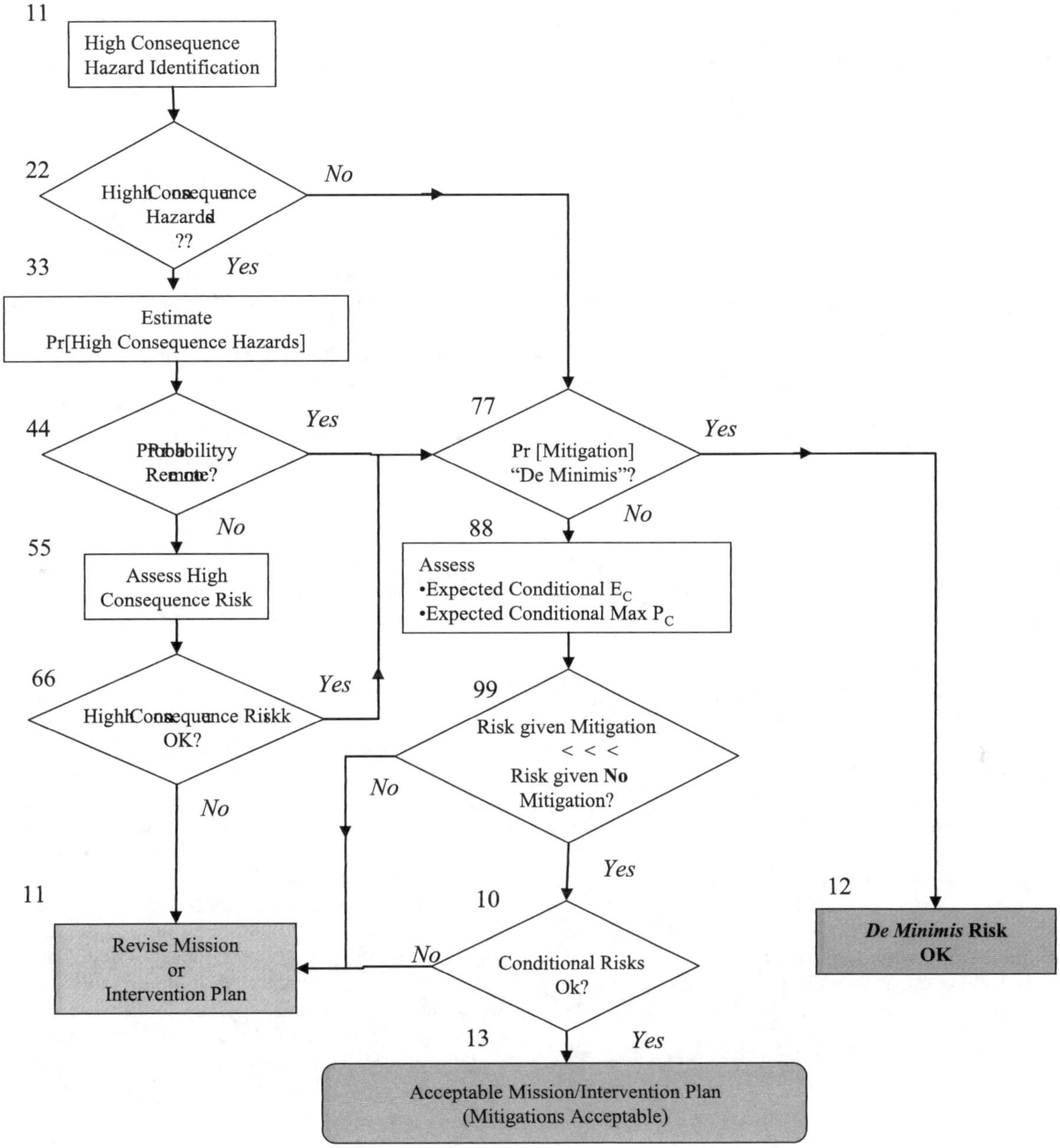

Figure 4. Conditional Risk Management Approach for Safety Interventions

12. When it has been shown that the conditional risks are de minimis, no further evaluation is required.

13. When the conditional risks given the mitigation are shown to be acceptable, the original risk analysis should be reviewed and, if necessary, revised to assure that mission risks are acceptable.

6. KEY DEFINITIONS

The conditional risk management process outlined in Fig. 4 includes several undefined terms and criteria, such as a "high consequence hazards" and calls for acceptable conditional risk criteria in step 10 and a de minimis conditional risk threshold in step 7. This section proposes a means to define several of these key terms and an acceptable conditional collective risk criterion.

Using data on the conditional risks posed by conventional aviation as a basis for quantitative definitions of key terms involved in conditional risk management appears reasonable and logical based on RCC policy goals and the historical intent that US space operations should be no more dangerous than conventional airplanes flying overhead.

Conditional risk management criteria derived from data on the conditional risks posed by conventional aviation would achieve the RCC policy goal that the uninvolved "public must not be exposed, individual and collectively, to a risk level greater than the background risk in comparable involuntary activities, and the risk of a catastrophic mishap should be mitigated." In this context "the RCC considers 'comparable involuntary activities' as those where the risk arises from manmade activities that (1) are subject to government regulations or are otherwise controlled by a government agency, (2) are of vital interest to the host nation, and (3) impose a risk involuntary risk of serious injury or worse on the public." [RCC 321-10 page 2-2]

Civil aviation and commercial space transportation in the US obviously meet the three conditions to be considered comparable involuntary activities in that both are manmade activities that

1. Are subject to government regulations or are otherwise controlled by a government agency, and

2. Are of vital interest to the US[16], and

3. Impose involuntary risk of serious injury or worse on the public.

Many organizations have identified the risk posed by conventional aircraft as an important benchmark for the acceptable risks posed by launch vehicles, including Congress,[17] the USAF, and an American National Standard. As previously mentioned, the FAA launch safety regulations have their genesis in USAF Range safety requirements. The USAF launch safety requirements were documented in EWR 127-1, which stated the governing principle that *"to provide for the public safety, the Ranges, using a Range Safety Program, shall ensure that the launch and flight of launch vehicles and payloads present no greater risk to the general public than that imposed by the over-flight of conventional aircraft."*[18] In addition, an American National Standard endorsed the same governing principle: "during the launch and flight phase of commercial space vehicle operations, the safety risk for the general public should be no more hazardous than that caused by other hazardous human activities (e.g., general aviation over flight)."[19]

Clearly civil aviation is a comparable involuntary activity and thus a legitimate benchmark for evaluating acceptable risks from commercial space activities. Thus, it is rational and reasonable to derive conditional risk limits to ensure that commercial space activities *"will be no more dangerous than conventional airplanes flying overhead."*

The NTSB aviation accident database provides a straightforward empirical basis to estimate the conditional risks to people on the ground from the flight of US civil aircraft. The FAA/AST recently updated its analysis of data from the NTSB on injuries, both minor and serious as defined in Federal regulations[20], and fatalities for people on the ground from civil aviation accidents for the 30-year period between 1984 and 2013. In general, the term "casualties" refers to serious injuries or worse, including fatalities, in the US flight safety community. In this analysis, the number of casualties was computed by adding the number of serious injuries, as defined in 49 CFR 830.2, to the number of fatalities reported by the NTSB. The results of this analysis, shown in Tables 1 and 2, indicate that the average number of casualties (serious injuries or worse) and fatalities on the ground from civil aviation accidents are 0.06 and 0.02, respectively. General aviation, which are relatively small aircraft compared to commercial airliners produce far fewer ground casualties per aircraft accident as one might expect.[21] For example, the data in Table 2 shows that the average airline crash produces one ground fatality, whereas one ground fatality occurs on average after 100 general aviation accidents. These data also indicate that crashes of larger vehicles generate a larger fraction of fatalities to casualties on the ground: about two ground casualties per fatality for airline accidents, and about three casualties per fatality for general aviation accidents.

A quantitative and comprehensive definition of a "high consequence hazard" is elusive because one should obviously include hazards that could result in long term or irreversible consequences, such major environmental impacts, and negative impacts on the national security or foreign policy interests of the launching state. In addition, the sparse set of data available for orbital launches makes probability estimates difficult and uncertain: whereas an estimated two billion departures of US civil aircraft occurred in the 30 year period between 1984 and 2013, the total number of orbital launches worldwide is less than ten thousand. The uncertainty in launch vehicle failure probabilities increases the importance of the acceptable conditional collective risk criteria called for in step 10.

Individual risks from orbital launches and re-entries are generally extremely small given the large number of people exposed. Hence the critical need is for an acceptable conditional collective risk criterion to be applied in step 10.

Table 1. Ground Casualties Per Fatal Accident Based on 30 Years of NTSB Data (1984 through 2013)

Aviation Category	Ground Casualties Per Accident	95 Percentile Upper Bound
All US Civil (Part 91, 121 and 135)	0.06	0.17
Airlines (Part 121)	1.86	8.31
Scheduled (Part 135)	0.09	0.44
Unscheduled (Part 135)	0.07	0.27
General Aviation (Part 91)	0.04	0.15

Table 2. Ground Casualties Per Fatal Accident Based on 30 Years of NTSB Data (1984 through 2013)

Aviation Category	Ground Fatalities Per Accident	95 Percentile Upper Bound
All US Civil (Part 91, 121 and 135)	0.02	0.06
Airlines (Part 121)	0.91	3.85
Scheduled (Part 135)	0.02	0.14
Unscheduled (Part 135)	0.02	0.09
General Aviation (Part 91)	0.01	0.03

Given the significant differences in aviation and space transportation safety paradigms, the uncertainty inherent in casualty consequence predictions for space launch and re-entry missions compared to the empirical data shown in Tables 1 and 2, and the unique prevailing conditions in commercial space transportation, the author recommends an acceptable conditional collective risk criterion of 0.01 expected casualties for use in step 10, which is a half order of magnitude lower than the maximum that could be justified as *"no more dangerous than conventional airplanes flying overhead."*

Note that 0.01 was used previously by the US Government as a conditional risk threshold, although in a different context and as a limit for the maximum individual risk. Specifically, the US military used a conditional risk of about 0.01 probability of fatality to define minimum separation distances between potential sources of high speed fragments (propelled by accidental explosions) and areas with exposed personnel. The Department of Defense quantity-distance limits were originally designed to limit conditional individual risk of fatality to 0.01 from inert debris fragment impacts.

An acceptable conditional collective risk criterion of 0.01 expected casualties is consistent with recent decisions made to protect public safety during commercial space operations. One example, involving downrange overflight of the Dragon was described briefly above. Specifically, a safety mitigation was implemented (an intentional weak spot in the thermal protection system) so that a launch failure would not result in Conditional Expected Casualties (CEC) above 0.01 due to an intact impact of the capsule.[22] More recently, in assessing the potential public safety impacts associated with debris with ballistic coefficients outside of the impact limit lines for the SpaceX Falcon-9 stage 1 return to launch site as part of the Orbcomm2 mission, the FAA applied state-of-the-art techniques to examine the CEC of a failure that could generate debris outside of the impact limit lines.[23] The analysis conducted by the FAA and 45SW/SELR demonstrated that the consequence of events that could produce debris (with ballistic coefficient above 3 psf) outside of the impact limit lines for a small portion of the stage 1 fly back operations (where the concern exists) was below this threshold (0.01 CEC), even with input data corresponding to the worst case weather conditions. Thus, the FAA determined that a waiver to the requirements of § 417.213 (a) and § 417.213 (d) would not jeopardize public health and safety or the safety of property. This waiver implies that a CEC below 0.01 does not constitute a "high consequence" event in the context of §417.107(a), which requires that FSS be employed if a failure "would have a high consequence to the public" in the vicinity of the launch site.

7. CONCLUSIONS

Activation of a Flight Safety System (FSS) is an important safety intervention that has been traditionally used to mitigate public risks from large orbital rocket launches. A drawback to the traditional ELV approach, which is codified in FAA regulations 14 CFR Part 417, is that it does NOT entail conditional risk management for the public (given a FSS activation event occurs), and thus it could unnecessarily limit launch operations and produce elevated risks for vehicle occupants even if the threat to the public is extremely low. In addition, the traditional flight safety limit approach does not necessarily ensure protection against high consequence events as explained here.

A conditional risk analysis can be useful to evaluate when activation of a FSS or other safety intervention should be implemented, both in the vicinity of the launch site and during downrange overflight. This paper explained how a conditional risk management process can ensure that (1) safety interventions will be implemented to address "high consequence" events, and (2) implementation of safety interventions will produce reasonable conditional risks. Based on an examination of 30 years of empirical data available on US civil aviation accident consequences for people on the ground, with due consideration to the differences between aviation and space transportation, this paper proposed 0.01 expected casualties as an acceptable conditional collective risk criterion, which is a half order of magnitude lower than the maximum that could be justified as *"no more dangerous than conventional airplanes flying overhead."*

It is important to note that while the risk management process, including traditional QRAs and conditional risk analyses, may be useful in planning for contingency aborts, etc. for occupied launch or reentry vehicles, especially those under autonomous control, the design and operation of piloted vehicles should also account for longstanding rules and conventions governing emergency situations and decision-making theory as well. For example, in an in-flight emergency requiring immediate action, the pilot in command generally has been given the freedom to deviate from any requirement to the extent necessary to cope with the emergency, and later file a report to document the circumstances and rationale for that deviation. The history of aviation emergencies includes many cases where pilots have taken extraordinary measures to protect the public from a crash landing.

8. REFERENCES

1. See Wilde P., "Public Risk Criteria and Rationale for Commercial Launch and Reentry," 5th IAASS Conference, Versailles, France, October 2011.

2. The FAA recently proposed an update to the public risk criteria. See 42241 of Federal Register, Vol. 79, No. 139, July 21, 2014

3. See page 19605 of Federal Register, Vol. 64, No. 76, April 21, 1999.

4. The FAA has formally defined "destruct" in §417.3 as the act of terminating the flight of a launch vehicle flown with a flight safety system in a way that destroys the launch vehicle and *disperses or expends all remaining propellant* and renders remaining energy sources non-propulsive before the launch vehicle or any launch vehicle component or payload impacts the Earth's surface.

5. The FAA has formally defined "tracking icon" in §417.3 as the representation of a launch vehicle's instantaneous impact point (IIP), debris footprint, or other vehicle performance metric that is displayed to a flight safety crew during real-time tracking of the launch vehicle's flight.

6. "The Air Force and the FAA remain committed to the partnership outlined in the MOA and … developing common launch safety requirements and for coordinating the common requirements." See p. 49471 of Federal Register, Vol. 67, July 30, 2002.

7. The term "accident" includes any unplanned event occurring during the flight of a launch vehicle resulting in the impact of any component of an expendable launch vehicle outside the designated impact limit lines.

8. Fragola J.R., *Aerospace Failure Data Handbook: A Reference Guide to Understanding and Assessing Risk and Reliability of Aerospace Systems and Spacecraft Design*, Valador, Inc., 2010.

9. Wilde P, Morse E, Rosati P., Cather C, "Probability of Failure Analysis Standards and Guidelines for Expendable Launch Vehicles," 6th International Association for the Advancement of Space Safety Conference in Montreal, Canada – May 21-23, 2013

10. For example, §417.107(c) requires a launch safety analysis to account for any inert debris impact with a mean expected kinetic energy at impact greater than or equal to 11 ft-lbs.

11. *Flight Safety Analysis Handbook Version 1.0*, Federal Aviation Administration, Department of Transportation, September, 2011

12. 67 FR at 49464.

13. *Instantaneous impact point* means an impact point, following thrust termination of a launch vehicle, calculated in the absence of atmospheric drag effects

14 See page 78619 of Federal Register, Vol. 75, No. 233, December 6, 2010.

15. RCC 321-10 Supplement. *Common Risk Criteria Standards for National Test Ranges*. January 2010.

16. In 2004, Congress identified space transportation as "inherently risky." (see 49 USC Chapter 701, referred to as the Commercial Space Launch Act (CSLA), §70101 (a)(12), 12/2004). At the same time, Congress found that "a robust US space transportation industry is vital to the Nation's economic well-being and national security," (CSLA §70101, which gives reference Pub. L.106-405, Sec. 2, 11/1/2000, 114 Stat. 1751). The Major Range and Test Facility Bases (MRTFBs) have long been regarded as "national assets," and thus vital to the interests of the US.

17. See Wilde P., "Public Risk Criteria and Rationale for Commercial Launch and Reentry," 5th IAASS Conference, Versailles, France, October 2011.

18. Eastern and Western Range 127-1, Range Safety Requirements, 1998, see page 1-viii.

19. ANSI/AIAA S-061-1998, "Commercial Launch Safety," see Section 4.5

20. Department of Transportation, Federal Aviation Administration, *Title 49 Code of Federal Regulations (CFR) Parts 830.2, Notification And Reporting Of Aircraft Accidents Or Incidents*, Federal Register Vol. 53, p. 36982, Sept. 23, 1988, as amended at 60 FR 40112, Aug. 7, 1995.

21. This data suggests that the likelihood a GA accident per departure is about 30 times higher than airlines.

22. See page 78619 of Federal Register, Vol. 75, No. 233, December 6, 2010.

23. See page 1470 of Federal Register, Vol. 81, No. 7, December 12, 2016.

MANAGING A SAFE AND SUCCESSFUL MULTI-USER SPACEPORT

Taylor Dacko[1], Kirk Ketterer[2], Phillip Meade[3]

[1] *NASA Kennedy Space Center, FL, 32899, taylor.m.dacko@nasa.gov*
[2] *NASA Kennedy Space Center, FL 32899, kirk.a.ketterer@nasa.gov*
[3] *NASA, Kennedy Space Center, FL 32899, phillip.t.meade@nasa.gov*

ABSTRACT

Encouraged by the creation of the Office of Commercial Space Transportation within the U.S. Federal Aviation Administration (FAA) in 1984 and the Commercial Space Act of 1998, the National Aeronautics and Space Administration (NASA) now relies on an extensive network of support from commercial companies and organizations. At NASA's Kennedy Space Center (KSC), this collaboration opens competitive opportunities for launch providers, including repurposing underutilized Shuttle Program resources, constructing new facilities, and utilizing center services and laboratories. The resulting multi- user spaceport fosters diverse activity, though it engenders risk from hazards associated with various spaceflight processing activities. The KSC Safety & Mission Assurance (S&MA) Directorate, in coordination with the center's Spaceport Integration & Services and Center Planning & Development organizations, has developed a novel approach to protect NASA's workforce, critical assets, and the public from hazardous, space-related activity associated with KSC's multi-user spaceport.

For NASA KSC S&MA, the transformation to a multi-user spaceport required implementing methods to foster safe and successful commercial activity while resolving challenges involving:

- Retirement of the Space Shuttle program
- Co-location of multiple NASA programs
- Relationships between the NASA programs
- Complex relationships between NASA programs and commercial partner operations in exclusive-use facilities
- Complex relationships between NASA programs and commercial partner operations in shared-use facilities

NASA KSC S&MA challenges were met with long- term planning and solutions involving cooperation with the Spaceport Integration and Services Directorate. This directorate is responsible for managing active commercial partnerships with customer advocacy and services management, providing a dedicated and consistent level of support to a wide array of commercial operations. This paper explores these solutions, their relevance to the current commercial space industry, and the challenges that continue to drive improvement with a focus on areas of safety management and risk assessment that have been crucial in KSC's evolution into a multi-user spaceport. These solutions may be useful to government entities and private companies looking to partner with the commercial space industry.

1. BACKGROUND

The realm of outer space, from low-earth-orbit to extraterrestrial bodies and beyond, is always viewed with great potential, and national pursuit of these endeavors spurs both technical and economic growth. In the United States (US), the road to promote a commercial space industry was shaped by national policy as well as the high risk of the activity. The National Aeronautics and Space Act of 1958, and subsequent creation of the National Aeronautics and Space Administration (NASA), declared that this civilian agency will exercise control of aeronautical and space activities, but could enter into cooperative agreements with other entities to conduct its work [1]. The US Space Program historically was a joint undertaking of the federal government and private industry, but this relationship principally existed through the use of commercial contractors for federally managed programs. An increased presence of commercially produced vehicles and satellites through NASA's first few decades led to renewed interest in enabling commercial space enterprise. The Commercial Space Launch Act of 1984 recognized the contributions of private industry, and declared that the United States should encourage private sector launches [2]. The Act directed NASA and the United States to promote entrepreneurial activity in space, and facilitate the use of government-developed technology to encourage the private sector. This act created the Office of Commercial Space Transportation within the Federal Aviation Administration (FAA), and subsequent NASA Authorization Acts saw similar language incorporated to promote private sector launches.

1.1 Kennedy Space Center

The Kennedy Space Center (KSC), located on Merritt Island, FL, serves as NASA's primary launch center for its spaceflight programs. Containing over 140,000 acres of land, KSC has the capabilities to support transportation via land, air, rail, sea, and space. The location of KSC on the Eastern Seaboard allows for multiple launch capabilities, including horizontal and vertical, human-rated and unmanned launch vehicles. The KSC organizational structure is composed of multiple directorates, which provide planning and implementation services for ground operations and spaceflight programs. These programs manage and share a unique infrastructure of processing facilities, launch pads, testing facilities, and laboratories. KSC assets are utilized by multiple NASA programs and contractors, and mostly recently, commercial partners. The KSC Institution supports these programs by providing and managing the essential functions of the center. Among these organizations, the Spaceport Integration and Services directorate integrates and manages center services and customer support to spaceport users. KSC's Safety & Mission Assurance (S&MA) Directorate sustains and strengthens the success of KSC's organizational structure, serving as an independent and value-added partner ensuring the mission success of programs while protecting the safety and health of the public, program team members, and those assets that the US entrusts to NASA [3].

From concept development, to production, to operation and retirement, the lifecycle of many programs existed at KSC. The last major program lifecycle to reach retirement, the Space Shuttle program, left KSC with not only excess real and personal property capacity, but a large amount of legacy procedures and processes to maintain. The center, and S&MA sought to consolidate the legacy documentation from the Space Shuttle program. S&MA as an organization originated within the program and engineering directorates, reporting directly to program management and providing safety engineering services to the center. NASA reassessed this model after the tragedies of Space Shuttle Challenger, and Space Shuttle Columbia. The Rogers Commission Report, investigating the Challenger accident, and the Columbia Accident Investigation Board noted that independent authority did not exist to manage technical requirements that addressed hazards [4, 5]. NASA's Office of Safety and Mission Assurance realigned to provide independent Technical Authority to NASA programs through each center's safety organization. This Technical Authority allows the KSC S&MA organization to perform independent assessment of NASA activity, and manage programmatic and institutional risks for the hazards of a government-managed spaceport.

The Space Shuttle program remained the largest NASA program on KSC for several decades, up until the final launch of STS-135 in July 2011. The presence of purely commercial work at KSC had been historically small, in comparison to the sizeable NASA contractor workforce and amount of center resources and infrastructure. External partnerships, both domestic and international, have increased in the last three decades. In 1998, NASA introduced the Launch Services Program to procure launch vehicle services from commercial providers. Although these providers launched federal payloads with NASA oversight, the launch provider managed a larger portion of risk, with NASA safety and quality participating in the reviews. The International Space Station program brought international partners and payloads to KSC's Space Station Processing Facility, where S&MA provided safety oversight for both civil servant, contractor, and international partner workforces. Throughout all of these efforts, NASA retained the primary responsibility for operations and processing activities. KSC also increased partnerships with academic research, prominently in the center's Space Life Sciences Laboratory. NASA had dedicated the bulk of the center's facilities, services, and workforce to the successful completion of the Space Shuttle program's mission.

1.2 Multi-User Spaceport

The transition and retirement of the Space Shuttle program freed up a number of assets at KSC previously utilized by the Shuttle processing and launch flows. NASA's current manned exploration vehicle, the Space Launch System, will be processed and launched from KSC. KSC determined that the facilities and services required for current and future NASA programs left many existing assets underutilized or mothballed. Maintaining these facilities and services at their previous capacity would result in excess cost to the government. A strategic goal in NASA's 2007 Strategic Plan was to, "encourage the pursuit of appropriate partnerships with the emerging commercial space sector." [6] A solution, therefore, was to partner with non-NASA entities, including federal, state, commercial and academic organizations, to make use of these assets. The NASA Authorization Act of 2010 identified KSC as a multi-user launch complex, for both government and commercial programs [7]. NASA sought commercial partners through both open solicitation for NASA partnerships, and by public announcements of available facilities. NASA made certain facilities available for exclusive-use including the Launch Complex 39A, the Orbiter Processing Facilities, Shuttle Landing Facility, Hypergolic Maintenance Facility, and others, and also proposed sharing space within active NASA facilities, including the Vehicle Assembly Building, Launch Control Center and laboratories.

2. REQUIREMENTS DEVELOPMENT

This concept for a space center with multiple users required a thorough review of existing requirements to determine how commercial launch activity would be incorporated. Within S&MA, this started with general institutional safety requirements. The basis for the NASA Safety program, at an agency level, is defined in the NASA Procedural Requirements (NPR) document NPR 8715.3 NASA General Safety Program Requirements. This document implements a comprehensive safety program that defines institutional, program, system, operational, training, and discipline-specific safety programs [8]. Referenced NASA standards and technical requirements included explosives safety, pressure vessels, lifting devices and equipment, lightning safety, personal protective equipment, and other occupational safety policies (industrial health and fire protection, while referenced in the NPR, are maintained under separate documents at the center level). This document provides a framework for NASA centers to detail specific requirements for their center. At KSC, these further requirements are addressed via the KSC Procedural Requirements (KNPR) 8715.3 document. This KNPR addresses the governing requirements and demonstrates how KSC implements the NASA Safety program and NASA Safety standards at the center [9]. The document applies to all entities, including civil servants, contractors, and visitors, that operate at KSC.

S&MA developed the KNPR as a catch-all to safety programs and practices on the center. KSC S&MA was concerned that these NASA requirements could be burdensome to companies looking to conduct business at KSC. KSC previously developed S&MA requirements to accommodate large-scale federal programs. Industrial safety requirements in several areas exceeded those specified by the US Occupational Safety & Health Administration (OSHA), and other standards such as the Safety Standard for Explosives are supplemental to existing federal requirements. Many commercial companies interested in partnerships maintain existing safety & health programs. KSC did not intend to supplant these existing safety programs and introduce strain in complying with a new set of requirements. S&MA needed to understand how to bring these safety & health programs under the umbrella of KSC's existing safety culture. Multiple safety and health programs would need to operate in harmony, with NASA's safety program acting as the final authority if required. NASA sought to fully understand and define an acceptable level of risk from commercial activity to NASA personnel and programs. The S&MA organization formulated two questions to guide development of a solution to this problem. First, what are the minimum necessary requirements to develop an acceptable risk profile for commercial activity? Secondly, what facets of the safety program must be universally standard on center?

2.1 Evolving Rationale

To develop a framework for a minimum set of NASA requirements, the KSC S&MA organization looked externally to find rationale. If general industry operated within the boundaries of the center, then KSC would have to understand general industry. A key facet of how OSHA interprets fault among multiple parties in the same location is by its Multiple Employer Citation Policy [10]. In the event of an accident that involves multiple organizations, OSHA distinguishes these employers into four categories: creating, controlling, correcting, and exposing employers. The creating employer introduces the source of the hazard, the controlling employer manages the hazard, the correcting employer mitigates the hazard, and the exposing employer has employees that could be harmed by the hazard. This is useful in determining responsibility in an incident, however the citation policy only engages once the accident has occurred. It is prudent for a worksite manager to ensure hazards are appropriately controlled in order to prevent such a mishap from occurring. This categorization of employers still provided a useful framework to understand how NASA wanted to position itself as a spaceport manager. Considerations included how NASA or partners introduced hazards to the center, how hazards are controlled and mitigated, and how respective employees might be exposed to those hazards. To address the question of minimum necessary requirements, S&MA considered NASA to be an exposing employer. According to the OSHA policy, in the event of a mishap, an employer can be cited if it, "knew of the hazardous condition or failed to exercise reasonable diligence to discover the condition." [10] KSC requirements on agreements with commercial partners would reflect due diligence by S&MA to understand local hazards, and through coordination with a partner, work to control them.

Hazard communication between NASA, partners, the public, and other participants of the multi-user spaceport is a strong component of KSC's S&MA policy. Communication and coordination alone does not satisfy the principles of NPR 8715.3 with regards to risk evaluation and acceptance. Safety policy must withstand the rigors of hazards specific to aerospace, including launch and reentry activities. The NPR states that NASA will, "ensure the conduct of assessments of quantitative and qualitative safety risks to people, property, or equipment, and include recommendations to either reduce the risks or accept them." [8] This is flowed down to the KNPR 8715.3 in the assessment section, under Chapter 4: Operational

Requirements. Chapter 4 requires that organizations on KSC establish a process to review and accept risks, perform an assessment to evaluate hazards associated with operations, and establish controls for the hazards [9]. This applies to NASA as the sole employer identifying, assessing, and controlling its own risks. In assessing a third party, and without the full complement of NASA requirements, the center would have to be judicious in how it collected information through an agreement. The solution was to pare down the requirements of KNPR 8715.3 to the minimum safety standards necessary to regulate activity on center, while bolstering the capabilities of the operational assessment that identifies hazard controls.

2.2 Operational Assessment

The core of this new multi-use safety policy is NASA's operational assessment, derived from the original Chapter 4 in KNPR 8715.3. NASA's ability to control and manage hazards is only as strong as its ability to have information available for assessment of the existing hazards. As a baseline, NASA maintains the final authority to mitigate a hazard, should partner- managed controls not be sufficient to reduce the risk to KSC personnel or property. At the same time, NASA wants partners to keep their safety programs and requirements intact, and acknowledges different approaches in safety. A commercial partner already must comply with all federal, state and local requirements, including OSHA and other incorporated standards, by operating in the United States. NASA is not responsible for enforcing a commercial company's compliance to OSHA and other laws as they pertain to the company's own employees. Rather than levy the full slate of KSC requirements on a partner, a partner could demonstrate that their safety programs represent an equivalent level of safety to NASA requirements that either exceed federal laws or target specific aerospace hazards. This equivalence would be assessed against the risk to KSC personnel and activity governed by NASA. The more partner documentation and data NASA could review, the better NASA could understand gaps between partner and NASA safety requirements. A complete review of a partner's business and company documentation would require significant time and resources on behalf of both parties, and might impose an undue burden on companies wishing to conduct operations. NASA would have to be judicious in its selection of material to review, while ensuring appropriate breadth to adequately assess risk.

The modification of the operational assessment began with specifying particular documents that NASA required prior to a partner initiating operations. The first of these is a partner's concept of operations, which defines the full extent, at a high-level, of all proposed activity. The partner has discretion on what form this concept of operations takes, and could originate from an existing document or as a new product. Acknowledging partner recommendations, NASA retains the authority to ask for further information or depth for a full assessment. Along with a concept of operations, partners would include a description of all hazards associated with those operations, and the corresponding controls or mitigations for those hazards. This is meant to be encapsulating language that included activity covering launch processing and ground operations. General hazards include occupational hazards to personnel, pressurized systems, hazardous chemicals, ordnance, critical lifting, and others to be identified by the partner. The partner would also report all hazardous chemicals so that KSC could catalogue existing chemicals on center and ensure compliance with NASA's Safety Standard for Explosives [11]. To prevent cataloguing insignificant quantities, chemicals identified as hazardous by the OSHA General Industry Hazard Communication standard served to encapsulate the information NASA sought [12]. With this information as a basis, NASA could begin to craft a complete story of the proposed activity and the hazards brought on KSC property. Further requirements would address specific highly hazardous activity, including high-pressure operations, ordnance, and range operations, but these would be tailored from the existing NASA standards. Any coordination on this information would need to begin well in advance of operations, to ensure no delays in NASA receiving the relevant information.

2.3 The Safety Requirement Volumes

Once NASA S&MA understood the hazards posed by a partner's operations, it was necessary to ensure those operations were compatible amongst existing center programs. The concept of operations document includes information on all locations utilized by the partner at KSC. The locations made available for commercial use fell into two categories: facilities whose operation, maintenance, and usage are fully transferred to a partner; and shared areas within existing NASA-controlled facilities. At facilities operated and maintained by a partner, designated exclusive-use, NASA would have flexibility to allow partner safety programs to manage the facility, provided NASA had full awareness of these policies and deemed them compatible with existing facility policies. Facilities designated for exclusive-use also contained NASA systems that fed through or originated within those facilities, traces of the former Shuttle program facility architecture. These facilities would require routine access from KSC personnel for maintenance or inspection, and KSC S&MA needed oversight of safety controls at these sites. Within a NASA-controlled facility or shared-use, NASA maintains existing security, safety, and scheduling procedures to be complied with by all occupants, including partners. Assessing each agreement and each facility to tailor the KNPR 8715.3 would be a laborious process, as each facility had unique attributes and infrastructure designed to support flight hardware processing. S&MA decided to take the original KNPR and split it into three volumes: the existing document relabeled Volume 1 for civil servants and contractors, a new Volume 2 for partners within shared-use or NASA managed facilities, and Volume 3 for partners within exclusive-use facilities. These individual volumes could be placed within an agreement, and accommodate a majority of safety provisions and concerns, with additional tailoring as needed to accommodate the specific needs of a partnership.

3. SURVEILLANCE AND VERIFICATION

The information gathered from the operational assessment of Volumes 2 and 3 provides the foundation of KSC's risk assessment of partners, but NASA must perform its own due diligence to ensure the information is accurate and complete. NASA

develops a surveillance plan for each agreement to assess the partner's compliance to S&MA requirements in the agreement. These agreement requirements largely come from Volumes 2 or 3, since these documents are intended to exist on a majority of agreements. The plans document which requirements are targeted, and how often KSC S&MA performs surveillance. S&MA decided that to reduce the burden on partners and manage the resources of KSC S&MA, the surveillance frequency would only be performed on a monthly, quarterly, or annual basis. The frequency is based on current KSC construction surveillance practices, and S&MA constantly reassesses the frequency of evaluation. Whether a contractor or partner has a satisfactory record of compliance or not, the evaluator may increase or decrease the surveillance frequency.

Surveillance of KSC S&MA partnership requirements falls into two general types: documentation reviews, and site visits. Documentation reviews occur early in the partnership process, and initially reference the documents requested in the operational assessment. At a minimum this includes a review of the concept of operations, hazard and chemical lists, and a copy of the partner's safety & health plan (or program), but may also include operational or design documentation and risk analysis. These reviews provide both NASA and the partner with a baseline of activities associated with ground and flight operational schedules, safety policies and surveillance activity over the life of the agreement. It also provides sufficient lead time should a particular activity require additional controls or coordination with NASA. Generally, these reviews are conducted once, at the inception of the partnership, but will be subsequently reviewed as updates are available. These are particularly critical for flight safety analyses developed for commercial launches, since NASA may perform a review in addition to those required by range regulators. The second type, site visits, are the best opportunities to verify that the information in the documentation is current and adequate. Site visits are pre-coordinated with the partner on a non-interference basis, and surveillance is only performed on requirements identified in the agreement or associated KNPR 8715.3 Volume. Stop- work authority, a critical item in KSC's safety culture that allows personnel to call a halt to immediate unsafe activity, is always explicitly clear for any activity.

4. SPACEPORT INTEGRATION

The usefulness of surveillance relies on close coordination with NASA partnership management, which lies within two key center organizations. S&MA maintains relationships with KSC managers and stakeholders of partnerships for awareness of any concerns with regards to safety requirements and policies. To understand how these relationships are maintained, it is important to describe how KSC realigned its structure to better support the multi-user spaceport concept. The Center Planning and Development directorate leads the development of partnership agreements at KSC, and thus is responsible for early partner coordination, negotiations, and ensuring the development of an executable and enforceable agreement which includes all relevant safety requirements. The management of the multi-user spaceport, and the liaisons to partner operations fall under the responsibility of the Spaceport Integration and Services Directorate.

The Spaceport Integration and Services directorate is responsible for the overall planning and assimilation of Kennedy Space Center processing activities and the execution of center services across NASA projects and programs, other government agencies, and commercial partners. These functions are integrated to create consistently safe, innovative, responsive, and cost-effective solutions, driving the success of all spaceport customers. The primary goal of the directorate is to ensure that the institutional needs of all spaceport customers, including NASA programs and commercial partners, are met. This is done by providing a consolidated operations and scheduling function for the center which includes developing and maintaining a top-level operations master schedule that integrates commercial and government entities operating at KSC to ensure awareness and deconfliction of hazardous and other operations that have potential impact across the customer base.

The directorate also provides customer liaisons to customers of the KSC spaceport for the purposes of identifying, coordinating, deconflicting, scheduling, and managing KSC spaceport services. These customer liaisons interface directly with partners on a daily basis to develop and communicate requirements, identify creative solutions to issues, and integrate between the various users and stakeholders of the spaceport (including S&MA). These liaisons also manage and implement the process for partners to procure services from KSC, facilitate approval for any customer-proposed modifications, and coordinate customers' requests with KSC implementing organizations to provide a viable support plan to the customer.

Through close partnership with Spaceport Integration and Services, S&MA is able to maintain a good understanding of the nature of operations occurring at KSC, the level of interaction between spaceport users and their operations, and the types of issues being faced by the customers. This partnership is critical to helping both organizations be successful and in ensuring that the spaceport enables the safe and successful operations of all government and commercial activity.

4.1 KSC Technical Integration

In addition to working with stakeholders from other directorates at KSC for agreement development and implementation, S&MA also retains institutional and programmatic safety, quality, and reliability experts within its organization. During the early stages of agreement development, the subject matter experts (SMEs) are consulted for specific requirements on activities that are expected to be highly sensitive or highly hazardous, including: program schedules, explosive siting, lifting devices, pressure vessel and range safety priorities. During agreement implementation, NASA KSC SMEs are available for interpretations of technical safety standards and program safety priorities. They also assist in the development of targeted surveillance activity for their disciplines.

5. COMMERCIAL LAUNCH PROVIDERS

The authority of KSC safety requirements also required reevaluation for commercial launches licensed by the FAA. For NASA-managed or procured launches, safety analysis is performed under existing cooperative agreements with its neighbor, the Air Force's 45th Space Wing. The Webb-McNamara Agreement, established in 1963 to minimize the duplication of responsibilities from both NASA and Air Force launch operations, drove this cooperation [13]. However, Webb-McNamara did not account for purely commercial launches, and the review and licensing of launch providers by the FAA therefore creates another regulatory entity. A review in 2013 by the NASA Executive Council determined that for FAA-licensed launches, safety reviews could be conducted by the 45th Space-Wing or an FAA- approved entity. This creates more options for commercial providers, but moves away from the traditional NASA model for safety review and concurrence.

The flight safety analysis required in the application for an FAA launch license is reviewed by the FAA. With the FAA maintaining the licensing authority, NASA does not have mandated approval of a launch, as the FAA ensures protection of the public. As the manager of the spaceport where this launch activity would occur, though, NASA still holds a fundamental responsibility for protecting personnel and property. Using the previous risk-based approach from the Operational Assessment in the KNPR Volumes, NASA will request and review a minimum set of data from the flight safety analysis for assurance that risk is minimized. If the risk to NASA facilities exceeds criteria established in the NPR 8715.5A Range Flight Safety Program, NASA will work with the launch provider to better quantify or control the risk [14]. Likewise, processing and ground safety operations will be conducted according to the shared or exclusive-use facility requirements in KNPR 8715.3 Volumes 2 and 3, respectively. The ability to review data, request information, and implement controls through these document Volumes allows any risk- based safety concerns, regardless of origin, to be covered by the KNPR requirements. Even in situations where NASA does not have immediate approval authority, NASA must manage the risk to its spaceport. Through coordination and levying prudent requirements, NASA will fully understand launch hazards and ensure the risk is acceptable to the center. KSC stakeholders and S&MA will work with commercial launch providers to identify controls and modifications that will allow them to successfully execute an FAA-licensed launch.

6. CONCLUSION

The multi-user spaceport concept invites new perspectives, novel engineering, and cutting-edge technologies to the KSC. The success of commercial partnerships, and NASA missions, hinges on cooperation and shared awareness. This awareness is the core of both KSC's Spaceport Integration and Services policies and S&MA safety policies that govern the risks and challenges inherent to the multi- user concept. Strong and suitable policies allow KSC to support a wide variety of partnerships without any compromise to NASA KSC's safety culture The center acknowledges the new risks associated with commercial launch activity, and has adapted its structure, and requirements, to better promote these partnerships. In enabling the multi-user spaceport, KSC also addresses its mission as levied in NASA's Space Act to foster an environment for commercial space access. As this access to space will always offer great rewards at great risk, it is the S&MA organization's continued mission to provide a framework that allows these goals to be safely attained.

7. REFERENCES

1. National Aeronautics and Space Act of 1958, Pub. L. No. 85-568, 72 Stat. 426-438, July 29 1958, NASA History Program Office

2. Commercial Space Launch Act of 1998 (Previously Commercial Space Launch act of 1984), Title II – P.L. 105-303, October 28 1998, Office of Commercial Space Transportation Legislation & Policies

3. KDP-B-1041 Rev. G, John F. Kennedy Space Center Business Operating and Agreement for Safety and Mission Assurance Directorate, November 12 2015, NASA

4. Presidential Commission, *Report of the Presidential Commission on the Space Shuttle Challenger Accident* (Rogers Commission Report), June 6 1986, Washington DC, NASA History Program Office

5. Columbia Accident Investigation Board, *Columbia Accident Investigation Board Report*, August 2003, NASA History Program Office

6. FY 2007 Performance Highlights, February 1 2008, NASA

7. NASA Authorization Act of 2010, Pub. L. No. 111-267, 124 State. 2805, October 11 2010, NASA

8. NPR 8715.3C (change 9) NASA General Safety Program Requirements, February 8 2013, NASA

9. KNPR 8715.3 Rev. J-1 KSC Safety Procedural Requirements, June 28 2012, NASA

10. Occupational Safety and Health Administration (OSHA) Instruction, Multi-Employer Citation Policy, CPL 02-00-124, December 12 1999, US Department of Labor

11. NASA-STD 8719.12 (change 2) Safety Standard for Explosives, Propellants, and Pyrotechnics, December 12 2011, NASA

12. OSHA Regulations (Standards – 29 CFR 1910.1200) Hazard Communication, May 25 2012, US Department of Labor

13. KCA-1645, Rev. Basic *Agreement between The Department of Defense and The National Aeronautics & Space Administration Regarding Management of The Atlantic Missile Range of DoD and The Merritt Island Launch Area of NASA*, January 17 1963, Florida Space Development Council Library

14. NPR 8715.5A (change 2) Range Flight Safety Program, September 12 2012, NASA

LAUNCH SYSTEM HAZARD STUDY: METHODOLOGY AND LESSONS LEARNT AFTER 5 YEARS OF APPLICATION

D. Delorme[(1)], A. Biard[(2)]

[(1)] *CNES/DLA, 52 rue Jacques Hillairet Paris 75612 Cedex 12, david.delorme@cnes.fr*
[(2)] *CNES/DLA, 52 rue Jacques Hillairet Paris 75612 Cedex 12, arnaud.biard@cnes.fr*

ABSTRACT

Risk management for new launch vehicles issuing from different origins and original industrial organizations, was a great challenge. In order to assess risk for launch operations, a specific type of risk analyses from take-off along the flight is performed. This analysis is called Launch System Hazard Study.
The method is a classical one for high-risk industry, for which all the risks shall be assessed and mastered. These classical studies identify the hazard on ground and establish the risk mitigation to reduce the risk occurrence (prevention measures) and/or to minimize the effect (correction measures), like in particular the evacuation of the hazard zone. One of the main difference between the risk industry and the launch vehicle is that this latter is a moving source of hazard, which can be very large and reach many places on Earth or in orbit.

More precisely the hazard study is an analysis that includes a description of all the hazards related to the operation in nominal and accidental operating situations, whether their cause is internal or external. The study specifies the nature and scope of the possible consequences of all these operating situations. When dealing with elements of the launch vehicle which are returned or which fall-back and are liable to reach the ground, the study presents the components of these elements, stating their dimensions, masses and materials used. Moreover the study must present an exhaustive analysis of the causes and consequences, as well as the probabilities of the critical events. The mitigations are also defined, if necessary.

Since 2010, each new development of an European launch vehicle required a hazard analysis.
The lessons learnt consist in the adaptation of the methodology not only to the scale of the launch system evolution, but also to the different phases or type of development, from a complete launch system to an adaption for a specific mission. To allow this flexibility, the challenge was to simplify the analysis, without losing from the accuracy and the completeness point of view. The benefit of this approach is to have a more efficient methodology, more widely usable and fruitful all along the development, to anticipate and optimize the mitigation measures.

1. INTRODUCTION

French Space Operation Act (hereafter FSOA) and associated regulations came into force on 2010, December 10th. Their goal is to establish the legal safety of each space operator. They aim at ensuring the space activities' technical risks mastery, considering in particular third party risks. These third party risks are mainly the consequences of damages to persons, to public health, to belongings and to environment that can occur during the space operation. According to the regulatory frame, the hazard study is the method to be applied in order to identify and mitigate those risks.

CNES, French Space Agency, has 2 types of missions: the first one is to develop and maintain the expertise in the methodologies to identify and mitigate the risks. The second one consists in verifying for the French government the accuracy of the operator demonstration of the launch operation compliance with the regulations.

For European launch systems, the hazard study is begun as soon as possible in the development plan. Today, after several developments, some improvements due to lessons learnt have been implemented.

2. FSOA TECHNICAL REGULATION

In accordance with II.2°c) of article 1 of the FSOA decree of 9th June 2009 [A1], the launch operator carries out a study of the potential hazards involved in the planned space operations.

To do so, in accordance with Article 7 of decree concerning Technical Regulation [A3] (hereafter TR), the launch operator must perform a Launch System Hazard Study.

"This hazard study includes a description of all hazards related to the operation in nominal and accidental operating situations, whether their cause is internal or external.

The study specifies the nature and scope of the possible consequences of all these operating situations. When dealing with elements of the launch vehicle which fall back and are liable to reach the ground, the study presents the components of these elements, stating their dimensions, masses and materials used.

The launch operator must therefore:

- demonstrate compliance with the quantitative requirements (cf. article 20 of TR [A3]), with regard to the collective casualty risk;
- evaluate the effects of any accidents on public health and the environment.

This study must cover the following events, in the conditions stipulated in chapter 3 of [A3]:

- damages linked to fall-back of elements designed to separate from the launcher;
- damages linked to controlled or uncontrolled re-entry of launcher elements placed in earth orbit;
- damages linked to failure of the launch vehicle;
- collision with manned space objects, for which the orbital parameters are precisely known and available;
- damages linked to explosion of a stage in orbit;
- collision with a celestial body.

The study must present an exhaustive analysis of the causes and consequences, as well as the probabilities of the aforementioned critical events. The risk reduction measures such as to comply with the technical requirements of the Technical Regulation (Chapter 3 of [A3]) are listed in the risk management plans laid out in the article 9 of [A3]."

3. GENERAL PRINCIPLES OF THE HAZARD STUDY METHODOLOGY

In answer to TR article 7 the methology described hereafter has been proposed by CNES and is described throughout the "Guidelines of Good Practices" cf. Article 54 of [A3].
This study includes a description of all the hazards related to the operation in nominal and accidental operating situations, whether their cause is internal or external to the Launch System.
The study specifies the nature and scope of the possible consequences of all these operating situations. When dealing with elements of the launch vehicle which fall back and are liable to reach the ground, the study presents the components of these elements, stating their dimensions, masses and materials used.

The study must present an exhaustive analysis of the causes and consequences, as well as the probabilities of the critical events at the origin of these hazard scenarios. The mitigation measures are listed in the risk management plans.

Main topics of the hazard study are the following:

- ✓ Description of the space operations, procedures and systems;
- ✓ Hazards identification;
- ✓ Identification of the feared events and their effects;
- ✓ Eligibility criteria for risks;
- ✓ Risks management (mitigation);
- ✓ Synthesis and conclusion: Synthesis tables & Mitigation risks measures list.

Note: The feedback from previous launches is taken into account in each part of the hazard study.

Five years after the first hazard studies performed in the scope of the different European launch systems, the general methodology as above-mentioned described in the "Guide of Good Practices" is still relevant. Nevertheless some improvements have been implemented.

4. IMPROVEMENTS OF THE METHODOLOGY

The first launch system hazard studies performed after the FSOA came into force had some weaknesses to be improved.
The analysis table containing the justification of the acceptability of the risks was a bit complex and contained up to 22 columns. It generated the following disadvantages:

- The study was too theoretical and difficult to read and be understood by people who are not usually in charge of risk management
- The logic of acceptability was uneasy to follow, as the requirements come from the different TR articles are not only quantitative
- The traceability to the reference of the detailed technical analyses used as input were difficult to follow

This complexity had also the disadvantage to make uneasy and very time consuming the evolution of the study in case of evolution of launcher mission and hardware configuration. The hazard study format was also not very well adapted to assess the impact of a new development or a specific mission.

The objectives of the hazard studies improvements aimed at making the final document more synthetic and more readable, with the strong constraint to loose neither reliability, nor exhaustiveness.

The expected benefits are:

- An enlarged diffusion and a better appropriation of the safety issues by the different actors involved in the project, and not only a document dedicated to the safety experts;

- A better understanding of the overall logic of the risk management for each potential effect thanks to a higher synthetic level;
- An improvement for the traceability of the referencing of the detailed technical analyses which were performed in the scope of the management process of the risk mitigation;
- An easier assessment of the impact of a launcher definition or mission evolution, like hardware or trajectory change.
- A better appropriation of the hazard study by the project team for a better anticipation and piloting of the risks throughout the development program.

4.1 Description of the space operations, procedures and systems;

In the hazard study the general description of the overall launch system can be limited to the key elements linked to safety (i.e. the main hazard situations), as the reader can have access to a more general description in other launch system documentation, like the definition file or the justification file.

With respect to the safety aspects, the flight domain can be described by the successive flight phases:

- Close range zone: authorized envelope for launch vehicle movements during the first moments of flight. The close-range zone ends no later than the radio horizon or the range limit of the CNES/CSG remote-control and neutralization station, according to the Decree regulating the operation of the Guiana space Center facilities (DRO) cf. [A4]
- Far field zone: this phase succeeds the previous one following either the loss of the neutralization capacity (ground or in-board) of the launcher vehicle, either the moment at which the impact zone is tangent to the territorial waters of the first State encountered along the nominal trajectory (cf. TR article 18). It ends when the launch vehicle is in orbit.
- Fall-back of the element designed to separate to the launcher: during the close range or far field zone, such elements must not impinge on the territory, including the territorial waters, of any State, without its agreement. (cf. TR article 23).
- Orbital phase: Orbital flight and maneuvers of the launcher or elements of the launcher, including in-orbit life, passivation and disposal maneuvers (cf. TR article 21).
- Re-entry phase: phase which starts with the re-entry of the space objects in the terrestrial atmosphere and ends when the debris reach the ground, in the scope of a controlled or not controlled re-entry (cf. TR Article 20).

Regarding the hazard study the main parameters to described are:

- The flight sequential of the launcher events linked with those latter safety topics
- The ground track of the instantaneous impact point
- The nominal fallback areas of elements designed to separate from the launcher and deorbitation area if any.

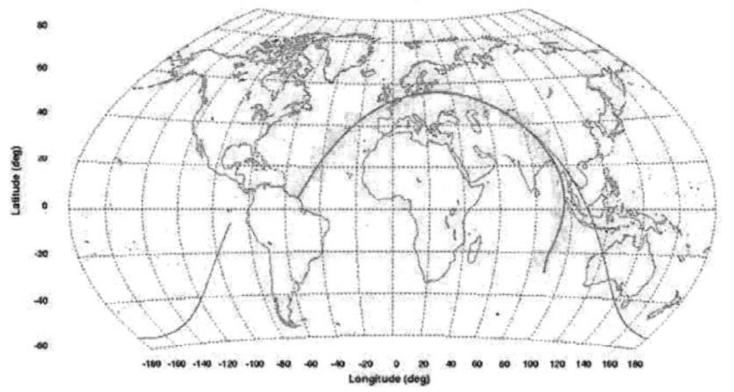

Figure 1: Ground track of impact point

An example of such flight sequential for an Ariane 5/ES mission is shown on Table 1.

Table 1 : Flight sequential example

	Event
Close range	Lift off
	Booster jettisoning : fallback in Atlantic
	Fairing jettisoning : fallback in Atlantic
	End of visibility of the neutralization remote control
Far field	Start of inhabited land overflying
	First stage jettisoning and fallback zone
	Ignition of upper stage 1st boost
Orbital flight	Start of orbital flight
	Cross of ISS orbit (altitude = 400 km)
	Shutdown of upper stage 1st boost
	Ignition of upper stage 1st boost
	Exit of A area (perigee > 2000km)
	Shutdown of upper stage 2nd boost
	Separation of first set of payload
	Separation of second set of payload
	Start of passivation of upper stage
	End of mission = stop of avionic

4.2 Hazard Sources identification

Two classes of danger source must be addressed: external and internal.
The external danger sources are linked to the launcher environment:

- Thunderbolts ;
- Winds on the launch pad;
- Winds in altitude (light fragment) ;
- Winds on coasts (toxicity impact) ;
- Rain / temperature drop / frost ;
- Excessive electromagnetic fields ;
- Outer space conditions (temperature, vacuum, …);

The external danger sources linked to humans are intentionally not considered in the hazard study:

- Malicious acts, sabotage or terrorism affecting the flight, but are subject to special measures in Guiana Space Center.
- Boats, planes, manned spacecraft or unmanned space objects are not considered as danger, but as "targets".
- The dangers associated with the preparation of launch are managed under other regulations (occupational safety, ICPE, ...).

Note: natural disasters (geological or hydrological or biological phenomenon, wildfire, severe turbulence and off-specification) were not selected because they are considered very unlikely in Guyana or irrelevant in our study

The internal danger sources are the launcher failures, which must be identified through usual dependability analyses, as launcher risk analysis and failure mode analysis, for the aim of achieving the exhaustiveness of the identification of danger sources.

4.3 Potential targets of aggression

4 types of potential targets of aggression are identified:

- Humans on Earth (land, air (planes), sea (boats)) and public health;
- Man-rated flights (International Space Station or other man-rated flights whose parameters are well known and available);
- Properties;
- The atmospheric environment (land, air, sea) and the outer space environment (specific requirements applicable for protected LEO and GEO zone cf. TR article 21, for celestial bodies, cf. TR article 26).

4.5 Acceptability criteria

For each effect the involved applicable articles of the TR or DRO are identified to define the relevant acceptability requirement.

These requirements depend of the considered flight phase, and are based on both quantitative and qualitative criteria, as:

- Qualitative failsafe/failsafe requirement at the launch system level for the close range zone and deterministic approach to exclude any hazard in the protected area, whether mechanical, thermal or toxic;
- Quantitative requirement for the collective risk of causing at least one casualty of 2.10-5 for the lift-off up to the orbital flight, and 2.10-5 for the reentry phase;

The chapter 3 of the TR contains other specific technical requirements for the launch operations. The acceptability criteria are specific to each article, and therefore the generic eligibility for Humans and Environment considered in the first hazard studies, led to a higher abstraction level that harmed the readability

As an improvement, the acceptability of the effects on the target is now strictly stated with regards to the criteria expressed in the concerned articles of the TR.

4.6 Identification of the feared events and their effects

The Feared Events (FE) are the phenomena which are potentially feared by the targets. The generic list of FE comes directly from the art. 7 of the TR (cf. §3).

From the launch system risk analysis and following a flight phase analysis, the different effects resulting from launcher failures which potentially lead to the FE are identified. All these cases are gathered in a synthesis table (see *Table 2*) containing the following information:

- Hazard source: origin of the hazard, coming from external environment, or from internal launcher failure;
- Feared Event: generic feared event coming from the art 7 of TR;
- Flight phase: part of the flight involved in the effect (close range, stage fallback, far field, orbital flight, reentry);
- Effect: wording of the effect
- Damage class: severity of the worst feared consequence on the target.

Table 2: format of the effect synthesis

Hazard sources	Feared Event	Flight Phase					Effect	Damage Class
		Close range	Stage Fallout	Far field	Orbital flight	Rentry		

For a dedicated mission, about twenty different effects are identified. As an illustration, the different effects analyzed for a particular mission of Ariane 5 are:

- Launcher fallout or explosion following an lightning impact on in-flight launcher in close range
- Launchpad destruction due to launcher drift following excessive ground wind at lift-off.
- Launcher fallout or explosion following excessive altitude wind in close range
- Nominal fallout of stage (boosters fairing, first stage)
- Launcher fallout or explosion after electrical failure due to radiation environment in far field or orbital flight
- Collision of the launcher or elements of the launcher with a plane
- Launcher fallout or explosion due to internal failure in close range
- Untimely separation of elements designed to separate from the launcher
- Non nominal fallout of elements designed to separate from the launcher
- Creation of maritime wrecks leading to potential collision with boat
- Launcher fallout or explosion due to internal failure in far field leading to fallout of debris on inhabited land
- Fallout of the upper composite following an under propulsion of the lower composite on inhabited land
- Fallout or explosion of the upper composite following an under propulsion resulting of a bad ignition of the first boost, on inhabited land
- Explosion of the upper composite following a bad ignition of the first boost leading to debris in orbit or collision of fragments with inhabited space stations
- Collision with the inhabited space station during the launch phase
- In orbit explosion due to internal failure of the upper composite before mission disposal leading to creation of in-orbit debris and chain reaction (Kessler syndrom) or loss of human life in inhabited space stations.
- Non controlled reentry of the upper composite or interference with the GEO protection region from its cemetery orbit

For each effect, a synthesis sheet is emitted (see *Table 3*) to summarize the treatment logic, the analyses which have been performed, the associated identified mitigation plan, implemented in development and to be performed in production, as well a clear status identifying the acceptability by the project or the identification of the analyses which remained to be done.

Table 3: effect synthesis sheet

Synthesis Sheet Template	
Flight phase	Concerned flight phase (cf. flight sequence)
Hazard	Identification of the origin of the Hazard
Target	Identification of the target
Scenario and effect	Description of the feared scenarios and events
Feared event Probability	Probability of causing a loss of human life, permanent invalidity or irreversible harm to public health, if the feared event occurs. This is the probability before and after the mitigation measure is effective
Reference of FE	Identification of the feared event
Regulation Requirement	Identification of the concerned requirement TR or DRO
Mitigation action	Implemented mitigation measures which allows to make acceptable the consequences of the Feared Events with respect to the regulatory requirements
Lessons learnt	Lessons learnt from other programs, when available
Project justification	Justification synthesis and reference of detailed notes performed by the development project team.
Exploitation actions	Identification, if any, of the activities to be done for each flight by the launch system operator in order to mitigate the associated risks
Status	Identification of what still to be done (project analyses and documentation) for the final demonstration of the acceptability in development and Acceptability status : green = acceptable risk, red = non acceptable risk

4.7 Mitigation

For each feared event, mitigation measures must be defined to make it acceptable.

In the close range zone, the key mitigation measure is the ability to neutralize the launcher from the ground by the safety operators. This neutralization is effective in near field phase.

There are two main types of mitigation measures:
- Preventive measures relying mainly on the launcher reliability;
- Corrective measures relying mainly on launcher neutralization.

All the mitigation measures identified in the effect synthesis sheets, cf. Table 3, are gathered in appendix of the hazard study. Hereafter some generic mitigation measures:
- Mitigation measures designed in development phase, as:
  - Definition of meteorological criteria for launch;
  - Design of neutralization chains to be remotely controlled by the safety operators;
  - Conception of dedicated tools and models, to be used in exploitation for the safety submissions files;
  - Definition and implementation of on- board automatic neutralization systems
  - Definition of test logic implemented in the flight software, in case of failure detection
  - Design of the passivation system;
  - …
- Mitigation measures to be applied in exploitation by the launch system operator:
  - Respect of meteorological criteria for launch;
  - Maintain of the system in qualified domain
  - Measures relative to mission analysis and inputs required in the safety submission (trajectory, flight corridor, fallback areas, tuning of predictive criteria, fragmentation model, casualty risk computation, injection accuracy, …)
  - Tuning of the flight software using mission analysis results
  - …

For each mitigation measure, the reference of the associated effect synthesis sheet is kept to ease the follow-one and traceability.

5. DIFFERENT TYPES OF HAZARD STUDIES

5.1 Delta hazard study

A typical hazard study covers a generic mission. For Ariane 5 program, the following dedicated hazard studies have been performed to cover the following generic missions:
- A5/ECA GTO dual launch;
- A5/ES ATV;
- A5/ES Galileo.

For these missions, the hazard study is based upon generic trajectories characterized by a qualified domain of key parameters, as launch azimuth, final orbit,… which fix the safety key events of the flight sequence. For each flight, the launch system operator provides safety submissions to check that the specific flight is still inside the generic domain and does not question the hazard study.

Some particular missions are outside the generic domain, due to system or hardware evolutions, leading to the need of additional development to demonstrate their airworthiness. This leads to redefine a specific flight sequence, trajectory, fallback areas, visibility from ground station, logic of de-orbitation… and therefore it is necessary to carry out a specific hazard study.

Such hazard study, emitted for few launches or even for one launch only, can be lighter than a generic hazard study, and can focus only on the difference between the specific mission and the generic ones. The main benefit of this approach is to highlight the safety and regulatory impacts and to manage them by particular mitigation measures. It allows also to reduce the amount of work necessary to conduct the analysis, which is performed by delta comparison with an existing generic hazard study.

The used methodology for these delta hazard studies follows the different steps:
- Identification of the generic mission which is closest to the specific mission, in terms of trajectories or hardware.
- Description of the particularities of the new mission in term of hardware, flight sequence and trajectories.
- For each effect identified in the synthesis table (cf. *Table 2*), assessment if the new mission has an impact or not on it.
- In case of impact, description of the demonstration provided by the project and identification of the specific mitigation measures that have to be done in development or in exploitation.

A new synthesis table is therefore established by the addition of 2 columns to the initial synthesis table (cf. *Table 2*), corresponding to the 2 latter bullets of the proposed methodology.

For Ariane 5, this methodology has been applied for the following specific missions, with a delta hazard study compared to the A5 / ECA GTO dual launch.

- A5/ECA GTO single launch
- A5/ECA GTO Demoflight (Commercial mission with additional experimental maneuvers performed after the payloads jettisoning for microgravity characterizations)

5.2 Preliminary hazard study

As a benefit of its simplification, the hazard study methodology is easier to be implemented. Consequently it appeared that this tool can be used very early in a development project. The systematic approach aiming at the exhaustiveness of the identification of the potential effects on humans and environment allows to identify all the technical analyses to be done in the scope of the demonstration of the compliance to the technical requirement of the chapter 3 of the TR. This approach is now implemented for all the Ariane future missions and complementary developments. The following preliminary hazard studies have been performed for the following missions:

- A5/ECA Upper Part Adaptation;
- A5/ECA mission with Bepi Colombo (mission toward Jupiter);
- A5/ECA mission with the James Webb Space Telescope;

As these missions used the A5/ECA version, the format of the hazard study is similar to the format described in §5.1 for the delta Hazard analysis.

6. SYNTHESIS

Since the French Space Operation Act and the associated regulations came into force in 2010, the methodology used to establish the launch system hazard studies has been improved in order:

- to ease its readability;
- to facilitate its drafting.

This improvement has been implemented without loss of exhaustiveness and with a better overview of the logic of risk mitigation, demonstration of the compliance to the regulatory requirements, through the synthetic sheets produced for each potential effect on humans or environment.

As a result, the diffusion has been enlarged and the hazard study is no more only a regulatory document, but is also more widely used as a project tool which summarizes the risk management of the safety feared events.
In this way the template of the hazard study is flexible and can be adapted to different types of situations, from a complete hazard study covering a generic mission domain, to a delta hazard study covering an evolution of the launch system by an assessment of the impact of each evolution all along the flight sequence. Preliminary hazard study, performed before the start of development, also allows to anticipate and optimize the mitigation measures, and to have a better follow-up of their implementation.

7. REFERENCES

A0. Loi N° 2008-518 du 3 juin 2008 relative aux opérations spatiales

A1. Décret n°2009-643 du 9 juin 2009 relatif aux autorisations délivrées en application de la loi n°2008-518 du 3 juin 2008 relative aux opérations spatiales

A2. Décret n°2009-644 du 9 juin 2009 modifiant le décret n°84-510 du 28 juin 1984 relatif au Centre national d'études spatiales

A3. Arrêté relatif à la Réglementation Technique en application de la Loi n°2008-518 du 3 juin 2008 relative aux opérations spatiales - 31/03/2011

A4. Arrêté portant Réglementation de l'Exploitation des Installations du Centre Spatial Guyanais - 09/12/2010

A5. F. Meyer Lassalle, G. de Blanchard, C. Aussilhou (2011) Launch system hazard study, methodology and application with 3 European launchers, IAASS, Montreal (2011)

THE USE OF AN ATMOSPHERIC MODEL FOR STUDYING THE GAS DISPERSION AT THE BRAZILIAN SPACE LAUNCH CENTER

Gilberto Fisch[(1)], Paulo Geovani Iriart[(2)], Daniel Andrade Schuch[(2)], Vinicius Couto Milanez[(1)], Carlos Augusto Teixeira de Moura[(3)]

[(1)] *Institute of Aeronautics and Space, Praça Marechal Eduardo Gomes, 50, CEP 12228-904, São José dos Campos, Brazil, Email: fisch.gilberto@gmail.com, vimcless@gmail.com*
[(2)] *Technical Institute of Aeronautics, Praça Marechal Eduardo Gomes, 50 CEP 12228-904, São José dos Campos, Brazil, Email: pgiriart@gmail.com*
[(3)] *Brazilian Air Force Major-Staff – Science&Technology Branch, Esplanada dos Ministérios - Bloco M - 6º andar - 70045-900 - Brasília - DF, Brazil, Email: catmoura2008@gmail.com*

ABSTRACT

This paper aims to use the atmospheric mesoscale model WRF coupled with its chemical module CHEM to study the simulation of the dispersion of exhausted gas released from solid propellant rockets launched from the Alcântara Launch Center (CLA), Brazil. For safety reasons, CLA needs better tools to simulate gases dispersion and properly define the exclusion zones. By now, there is a lack of studies devoted to pollutants dispersion in the Southern Hemisphere launch centers. The simulations were made for different representative cases of diurnal (daytime and nighttime) and seasonal (dry and wet seasons) scales. Observational data (radiosondes and wind tower data) was used to validate the wind field. The WRF-CHEM model was able to represent the plume dispersion and produces new suggestions for the safety procedure rules. New improvements are under development to represent the vertical dependence of the gas exhausted at the lower atmosphere as well as to include another environmental variable (HCl) in the equations of the WRF-CHEM.

1. INTRODUCTION

The Alcântara Space Center (CLA) is the Brazilian gate to the space. From this Center, the guided rocket Satellite Launcher Vehicle (VLS) and unguided sounding rockets (like VSB30) are launched. Especially for the sounding rockets, the winds are very important to determine the flight trajectory: it is computed as a sum of the internal (characteristics of the rocket) and external (winds) forces involved [1].

After the ignition the solid propellant rockets release a large amount of gases (mainly the major exhaust pollutants: Aluminum Oxide (Al_2O_3), Hydrogen Chloride (HCl) and Carbon Monoxide (CO)) from the combustion chambers, which results in the intended thrust. On the other hand, for safety reasons the evolution of the exhausted gases should be determined before the launch.

All the space centers has some method or tool to determine these gases dispersions. For instance, the US space centers (like NASA JFK/Cape Canaveral) use an operational model, which computes the gas dispersion based on the theory of the cloud rise [2]. This code is known as Rocket Exhaust Effluent Diffusion Model (REEDM) and it has been used to assess the environmental impact of aerospace activities. This dispersion model are based on Gaussian model concepts: the exhaust material (mixture amongst CO, CO_2, HCl and alumina) is assumed to be uniformly distributed in the vertical and to have a bivariate Gaussian distribution in the plane of the horizon at the point of cloud stabilization.

The model used at the European Spaceport of Kourou (French Guyana) is the SARRIM software [3], which is a multi-function atmospheric dispersion model dealing with the local and large-scale impact assessment for propellant and hypergolic rocket releases. It is an operational and fast running tool taking into account atmospheric stratifications inside the boundary layer using in situ data like radiosondes. This software allows the CNES safety department to organize the safety program and an environmental sampling monitoring survey at every launch.

The India Space Center (Satish Dhawan Space Center SDCS SHAR) uses a different approach: their scientists coupled a Hybrid Single-Particle Lagrangian Integrated Trajectory (HYSPLIT) model with an atmospheric mesoscale model (in this case, they have used the mesoscale meteorological model – MM5) to predict the dispersion of exhaust pollutant in the form of vapour and ground level concentrations [4].

The Brazilian space community is also addressing this problem for CLA since 2010. It has developed the model MSDEF (in portuguese, Modelo Simulador da Dispersão de Efluentes de Foguetes), which represents the solution for time-dependent advection-diffusion equation applying the Laplace transform considering the Atmospheric Boundary Layer as a multilayer system. This solution allows a time evolution description of the concentration field emitted from a source during a release lasting time "tr" and it takes into account deposition velocity, first-order chemical reaction, gravitational settling, precipitation scavenging, and plume rise effect. This solution is suitable for describing critical events relative to accidental release of toxic, flammable, or explosive substances [5]. An improvement of this model by coupling it to the CMAQ (Community Multi-scale Air Quality) is under way and some preliminary results were published by Nascimento et alli. [6]. Both studies addressed the CLA dispersion problem.

The objective of this work is to propose an alternative model or tool by the use of an atmospheric mesoscale model Weather Research and Forecasting (WRF) coupled with a Chemistry module (CHEM) in order to predict the dispersion of the gas exhausted by rockets at CLA.

2. DATA SET AND METHODOLOGY

CLA is located along the coast at the latitude 2° 19' S and the longitude 44° 22' W, and 40 m above sea-level (Fig. 1). The climate presents a precipitation regime split in wet period (from January up to June) being March and April the peak of the rain (higher than 300 mm/month), and a dry period (from July up to December) with precipitation lower than 15 mm/month (Fig. 2). The wind regime possesses a distinct behavior between the rainy and dry seasons. During the wet period the surface winds are weaker (typically around 4.0 – 5.0 m/s) than during the dry season (winds around 7.0 – 9.0 m/s). The wind regime at CLA is very complex as it is a superposition of the trade winds and the sea breeze developed due to the thermal contrast between ocean and continent [7]. This is due to the intensification of the sea breeze that presents its maximum influence during this time (particularly from September until November). The wind direction is from the NE at both periods.

Two observational data sets made by radiosondes have been used to validate the numerical outputs of the winds. The rawinsounding system used was from Vaisala Oy (Helsinki, Finland) MW11 using a sonde RS80-15G (with GPS capabilities). During the ascension phase, the GPS sensor sampled the balloon positions at every 2 s, and average winds for 50 m height interval were processed as the averaged wind profile for the layer. The data set used were obtained from an intensive field campaign from September 15 until 30, 2008 (representing the dry season) and from March 19 to 25, 2010 (representing the wet season).

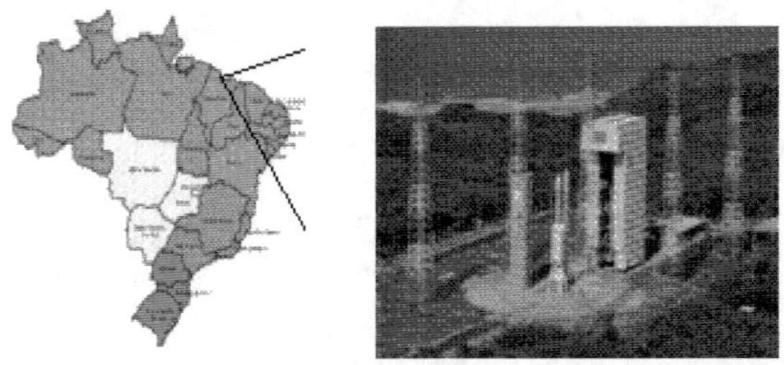

Figure 1. CLA geographic localization and a view of the launching pad.

For this work the mesoscale model, named WRF–ARW (Weather Research and Forecasting - Advanced Research WRF) version 3.4.1 was used as it is the "state-of-the-art" for regional models [8]. It also has some useful characteristics: open code, friendly user, several parametrizations available for radiation, boundary layer, convection, etc. that can be easily tested and changed to be locally adapted. The module chemistry (CHEM) has several options for the chemistry computation, and the Regional Atmospheric Chemistry Mechanism (RACM) with the library Kinect Pré-Processor (KPP) was used. The RACM has several equations to explicitly determine species of inorganic and organic contaminants, with a total of 237 chemical reactions. This software is relatively easy to install and operate at the launch center.

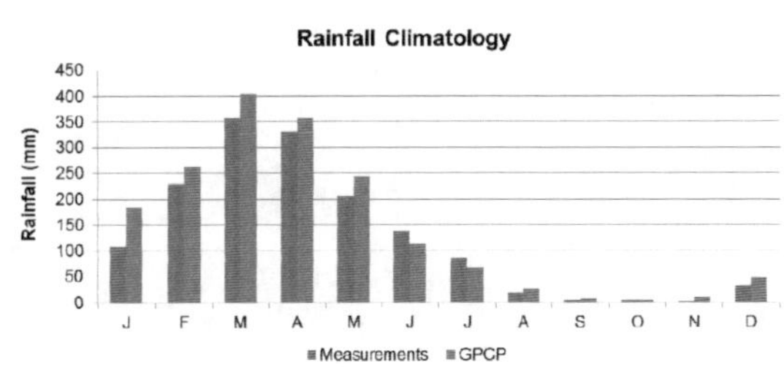

Figure 2. The rainfall climatology for CLA (GPCP: Global Precipitation Climatology Project).

3. RESULTS

The data used to initialize the model was the meteorological field from the National Centers for Environmental Prediction (NCEP) named GFS, with spatial resolution of 0.5° x 0.5° (approximately 55 x 55 km). The land use – land cover data was obtained at United States Geological Survey (USGS) (available at http://www2.mmm.ucar.edu/wrf/src/wps_files/geog.tar.gz), with spatial resolution 30". All the simulations were made with 42 vertical levels at the 3 nested grids centered at CLA (Fig. 3). The spatial resolution was 9, 3 and 1 km for the inner grid (D1). The simulation was started at 12 UTC and has a spin-up of 6 hours for the grids 1 and 2 . The legend for Fig. 4 is: EST represents the launch pad gate; MET is the area encompassing Meteorological Section and Technical Center; and VALC and VTAP are the Alcantara village and Tapireí settlement.

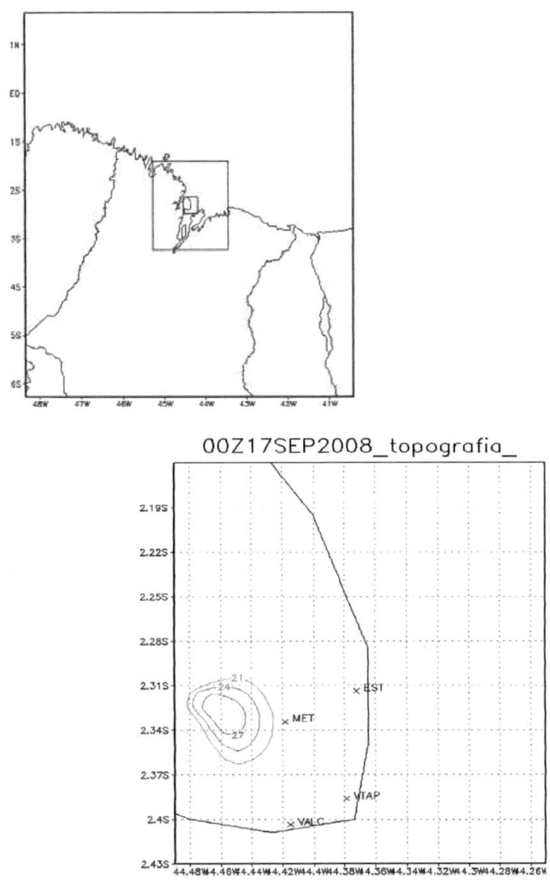

Figure 3. The 3 domains used for the numerical simulations. The D1 is showed in zoom.

Finally, the initial data assimilation (GFL or FNL) was tested. It should be mentioned that the GFL is an operational scheme (e.g., there are outputs from a larger model each 6 hours), and FNL is a better description of the meteorological fields, but it is not available rapidly. The results (not shown) did not present any significant difference between the 2 initial data-sets.

Fig. 4 presents the time series of the atmospheric flow (both wind speed and direction) associated with the concentration of the plume from the exhausted gas. One can clearly see the plume being dispersed and advected with the wind flow.

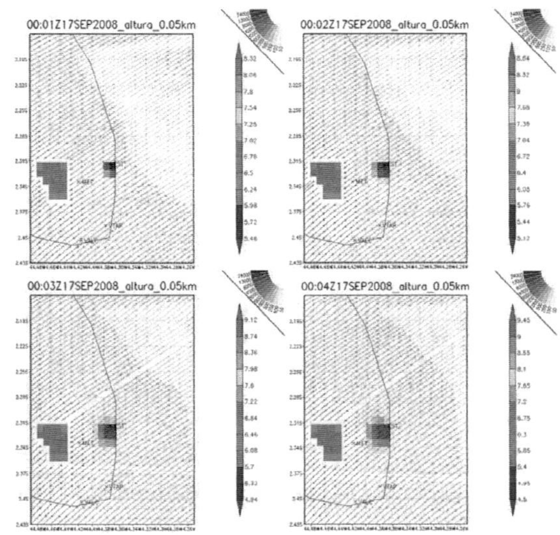

Figure 4. The time evolution of the CO concentration for the minutes 1 up to 4 after the gas release. Case Dry season 00 UTC.

In the following minutes (5 up to 8 min at Fig. 5 and 29 up to 32 min at Fig. 6) the plume dispersed, followed the atmospheric flow, and bypassed the hill, but did not reach the VALC.

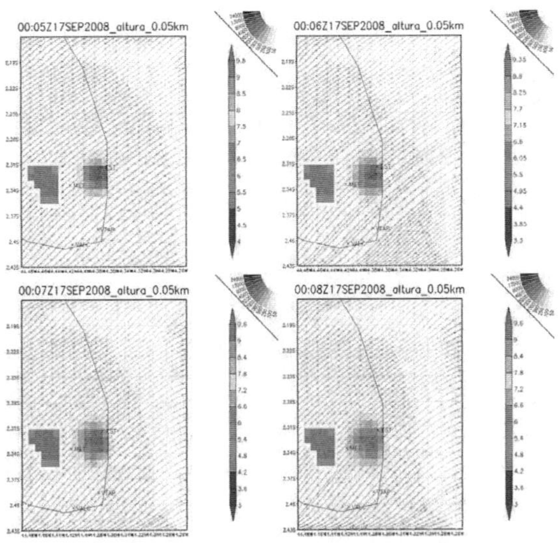

Figure 5. The time evolution of the CO concentration for the minutes 5 up to 8 after the gas release. Case Dry season 00 UTC.

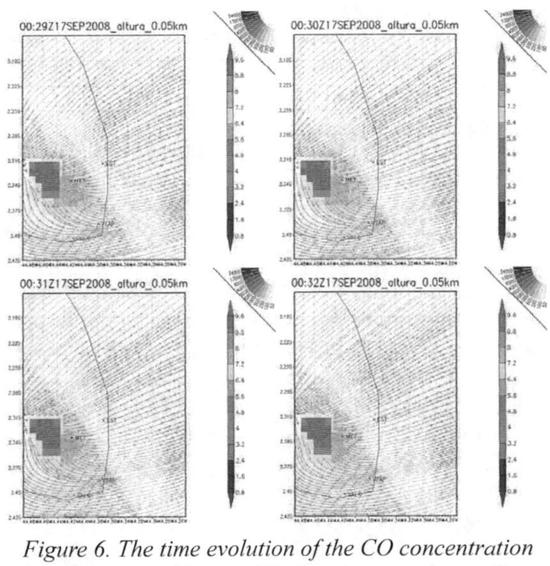

Figure 6. The time evolution of the CO concentration for the minutes 29 up to 32 after the gas release. Case Dry season 00 UTC.

However, for the Fig. 7, which represents the same emission released but with different meteorological initial conditions, especially for the wind direction, it can be seen that, for the same moment as Fig. 6, the plume reached the Alcantara village. That could represent some toxic hazard for the population.

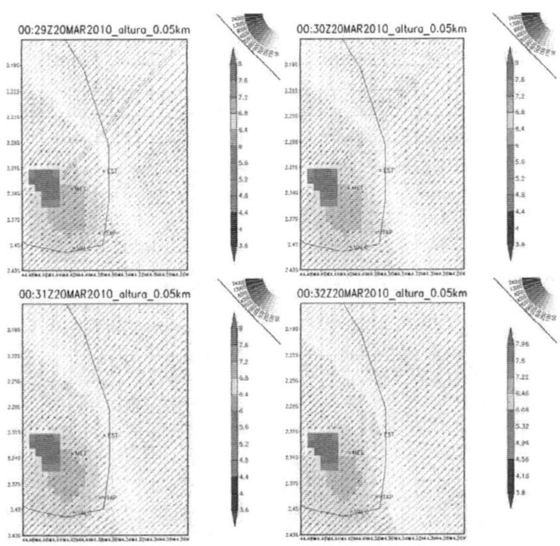

Figure 7. The time evolution of the CO concentration for the minutes 29 up to 32 after the gas release. Case Wet season 00 UTC.

All the simulations showed approximately the same patterns as the wind flow is very persistent (this is a characteristic of the trade winds). Typically, the simulations showed that the CO concentration (the variable used to represent the gases exhausted by the solid motors) at the launch pad is 2 order of magnitude higher than at the gate (1 km far) and 4 order of magnitude higher than that at Alcantara village (20 km far). It can reach 30000 ppm at the launching pad after H0 + 1 min (Fig. 8).

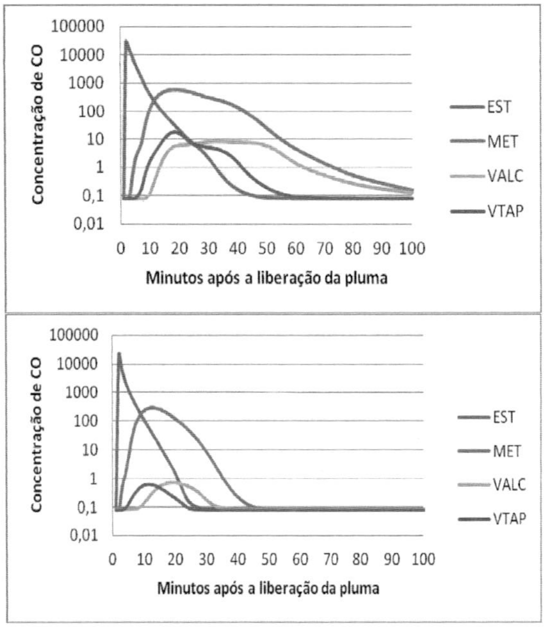

Figure 8. The CO concentration against the time after the release for 00 (first graphic) and 12(second one) UTC during dry period.

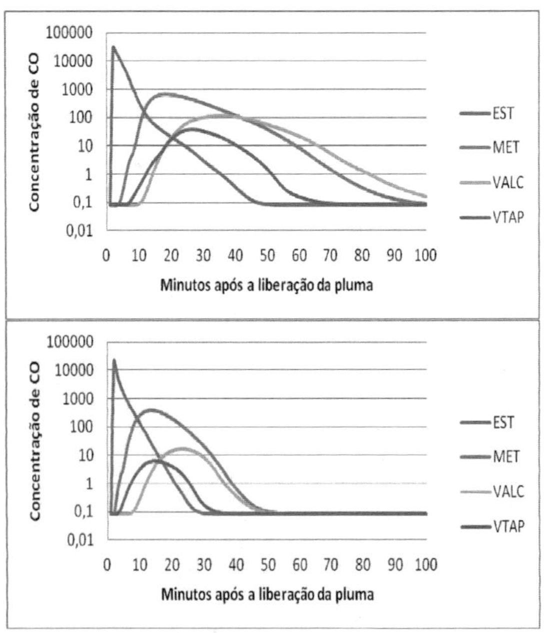

Figure 9. The CO concentration against the time after the release for 00 (first graphic) and 12(second one) UTC during wet period.

Considering the recommended safety tolerable values for human being, it was computed that the launch pad must stay isolated by 10-15 min before any other action for the complete dispersion. As the turbulent intensity is higher at 12 UTC (due to the daytime conditions), the total time for the complete dispersion of the plume is reduced (around 30-40 min) related to the nighttime conditions (70-90 min).

Table 1: Synthesis of the time dispersion needed for full dispersion

Period	*Time*	*Total time (min)*
Dry	*00 UTC*	*70*
	12 UTC	*30*
Wet	*00 UTC*	*90*
	12 UTC	*40*

4. CONCLUDING REMARKS

The dry season showed the fastest dispersion time, ranging from 30 up to 90 min. The plume dispersion shows strong sensitivity to the topography as well as the winds (especially the direction). The WRF-CHEM model was able to represent the plume dispersion and produces some new suggestions for the safety procedure rules (people must be protected for 10-15 min at launch pad and for 30 min at technical center/meteorological section).

This is an ongoing work that aims to improve this model configuration to include: 1) a vertical distribution (from the surface up to 2000 m height) of the exhausted gases due to the normal launching; and 2) small scale features (atmospheric turbulence generated by the surface change roughness associated with the steep escarpment) at the scale of 100 m. In the near future, this model should be operational for studying the gas released during the launches at CLA, even considering the chlorine compounds. This development is intended to be finished by late 2016 or earlier 2017.

5. ACKNOWLEDGMENTS

The authors would like to thanks the financial support from the Brazilian Organizations: CAPES – PRO-ESTRATÉGIA (Process number 2240/2012) and CNPq - National Council for Research (303720/2010-7). This work is part of the Master of Science Dissertation from Paulo Geovani Iriart presented in March 2015 and the ongoing PhD These from Daniel Andrade Schuch at ITA.

6. REFERENCES

1. Palmério, A.F. (2006). Introdução a Engenharia de Foguetes, ITA Mestrado Profissionalizante, São José dos Campos, Brazil, 142 p.

2. Bjorklund, J.R., Dumbauld, J.K., Cheney, C.S., & Geary, H.V. (1982). NASA Contractor Report 3646, 180 p. (Huntsville, AL.).

3. Cencetti, M., Veilleur, V., Albergel, A. & Olry, C. (2011). International Journal of Environment and Pollution, 44(1), 87-95, (DOI: 10.1504/IJEP.2011.038406).

4. Rajasekhar, M., Kumar, M.D., Subbananthan, T., Srivastava, V., Apparao, B., Rao, V.S. & Prasad, M. (2011). Proceedings Indo-US Conference-cum-Workshop on "Air Quality and Climate Research", ASCI Hyderabad, 8 p. (DOI: 10.13140/2.1.4008.1284).

5. Moreira, D.M., Trindade, L.B., Fisch, G., Moraes, M.R., Dorado, R.M. & Guedes, R.L. (2011). J. Aerosp. Technol. Manag., 3(1), 41-52.

6. Nascimento, E.G.S., Moreira, D.M., Fisch, G., Albuquerque, T.T.A. (2014). J. Aerosp. Technol. Manag., 6(2), 119-128.

7. Fisch, G. (2010). J. Aerosp. Technol. Manag., 2(1), 105-110.

8. Smararock, W.C., Klemp, J.B. & Dudhia, J. (2008). A description of the Advanced Research WRF – version 3, 113 p.

PROCEEDINGS OF THE 8TH INTERNATIONAL SPACE SAFETY CONFERENCE

SESSION 15 ORGANIZATION CULTURE - I

PURPOSE, PRINCIPLES, AND CHALLENGES OF THE NASA ENGINEERING AND SAFETY CENTER

Michael G. Gilbert, Ph.D.[(1)]

[(1)] *Principal Engineer, NASA Engineering and Safety Center NASA Langley Research Center*
11 Langley Blvd. MS 116 Hampton, VA, 23681 USA Email: m.g.gilbert@nasa.gov

ABSTRACT

NASA formed the NASA Engineering and Safety Center (NESC) in 2003 following the Space Shuttle Columbia accident. It is an Agency level, program-*independent* engineering resource supporting NASA's missions, programs, and projects. It functions to identify, resolve, and communicate engineering issues, risks, and, particularly, alternative technical opinions to NASA senior management. The goal is to help ensure fully informed, risk-based programmatic and operational decision-making processes. To date, the NESC has conducted or is actively working over 600 technical studies and projects, spread across all NASA Mission Directorates, and for various other U.S. Government and non-governmental agencies and organizations. Since inception, NESC human spaceflight-related activities, in particular, have transitioned from Shuttle Return-to- Flight and completion of the International Space Station (ISS) to ISS operations and Orion Multi-purpose Crew Vehicle (MPCV), Space Launch System (SLS), and Commercial Crew Program (CCP) vehicle design, integration, test, and certification. This transition has changed the character of NESC studies. For these development programs, the NESC must operate in a broader, system-level design and certification context as compared to the reactive, time-critical, hardware specific nature of flight operations support.

1. CREATION AND PURPOSE

The Space Shuttle *Columbia* accident occurred on the morning of February 2, 2003, in the skies over Texas. NASA immediately formed an accident investigation board, the Columbia Accident Investigation Board or CAIB. As the investigation progressed, the CAIB Chairman provided briefings and updates to NASA, the U.S. Congress, and the American public. On one such occasion in July 2003, speaking publically following a Congressional briefing, the CAIB Chairman stated

> *"The safety organization sits right beside the (Shuttle) person making the decision, but behind the safety organization there is nothing there, no people, money, engineering, expertise, analysis ... there is no 'there' there."*

NASA responded within weeks to this statement announcing in late July 2003 the formation of the NESC. The NESC was to be operational by the end of 2003, and provide the NASA Safety and Mission Assurance Director and the NASA Chief Engineer with the independent engineering capability the CAIB had pointed out publically was missing.

NASA's traditional safety philosophy of

- Strong in-line checks and balances
- Healthy tension between organization elements
- Value-added independent assessment

defines the NESC's role relative to the technical functions of the inline engineering organizations that support NASA programs and projects. Using its own funds, the NESC performs independent analyses and testing to bring challenging or confirming data to complex technical issues, and provides or supports alternative viewpoints on program or project technical decisions. If necessary, the NESC can and does make these alternative viewpoints known to the highest levels of NASA management, including the Administrator, to help make sure NASA's development and operational decisions are fully risk informed. The NESC vision is to be a high-performance, world-class technical resource routinely sought out by NASA's programs, projects, and organizations for its ability to provide timely, value- added solutions to difficult problems.

2. ORGANIZATION AND OPERATION

The NESC is organizationally part of the NASA Chief Engineer's Office supporting the Chief Engineer in his role as the Engineering Technical Authority for the Agency, reporting to the Administrator. The NESC is a distributed virtual organization, with the Director and management offices located at the NASA Langley Research Center (LaRC) and permanently assigned technical personnel located at each of the other NASA Centers. The NESC operates in a "tiger-team" model when conducting studies and assessments, forming a dedicated team of subject matter experts to address a well-defined technical issue.

2.1 Organization

The NESC is organized into five complimentary functional offices.

1) Principal Engineers Office, providing study leadership and management, particularly for longer term, multi-disciplinary assessments. The office is currently staffed by three Principal Engineers and one Associate Principal Engineer at LaRC, one Principal Engineer at NASA Johnson Space Center (JSC), and two Associate Principal Engineers, one located at NASA Marshall Space Flight Center (MSFC), and the other at NASA White Sands Test Facility (WSTF).

2) NESC Chief Engineers Office, one at each of the 10 NASA field centers to be the "eyes and ears" for the NESC and provide a conduit for alternative opinions. The NESC Chief Engineers track program and project technical progress and risks at their respective Centers, help identify and facilitate assignment of subject matter experts to support NESC teams, and manage pre- positioned resources to ensure ready and timely support. NESC Chief Engineers also lead assessments suited to their technical interests and capabilities.

3) NASA Technical Fellows Office, currently staffed by 18 NASA Technical Fellows, each recognized by NASA as a technical leader in their specific discipline. They, along with the NESC Chief Scientist and the NESC Chief Astronaut, each maintain a technical discipline team consisting of experienced subject matter experts from other NASA Centers, U.S. Government Agencies, academia, and industry. These technical teams are often the source of subject matter expertise supporting NESC studies. The NASA Technical Fellows often lead discipline-specific NESC assessments; however, recently the role of Technical Fellow has expanded to include an Agency-Level capability leadership role. The capability leadership role is to help ensure NASA has the necessary capability in each discipline, including people, facilities, and tools, for future missions and programs. As a result, each NASA Technical Fellow has one or more Deputies to assist with their responsibilities. Fig. 1 lists the 19 current technical discipline teams, with a new one for Sensors and Instrumentation in the process of being formed. Fig. 2 shows the typcial annual affiliation of subject matter experts participating on NESC Technical Discipline Teams.

4) The NESC Integration Office (NIO, formerly the Systems Engineering Office) provides technical integration and coordination between the NESC and the major NASA Program and Project Offices, and administers the NESC Review Board (NRB) review and approval processes. The technical integration function has become increasingly important and complex as NASA's current major development programs are completing design and beginning hardware build and test.

5) The Management and Technical Support Office (MTSO) provides financial, budgetary, and procurement support for the NESC. The ability of NESC study teams to form and respond quickly to urgent technical issues is dependent on the ready-access procurement instruments and pre-positioned resources managed by this office. The MTSO also provides contract technical editing, project logistics and coordination support, and audio and video services to support NESC assessments.

2.2 NESC Review Board

NESC Principal Engineers, NESC Center Chief Engineers, NESC assigned NASA Technical Fellows, the NESC Chief Scientist and Chief Astronaut, the Deputy Director for Safety, the Heads of the MTSO and NIO, and the Deputy Director and the Director of the NESC are voting members of the NRB. The NRB evaluates and approves each technical study and plan, monitors assessment progress, and reviews and approves the final report and recommendations to ensure technical rigor and accuracy. The NRB meets approximately weekly, and voting members attend in person or via web-video links.

The NRB decides on accepting new studies or assessments according to the following priorities

1. Technical support of projects in the flight phase
2. Technical support of projects in the design phase
3. Known problems not being addressed by any project
4. Work to avoid potential future problems
5. Work to improve a system

2.3 Technical Reports

In the CAIB final report, NASA was criticized for lack of technical reports supporting many Space Shuttle technical decisions and rationales. In many cases, the only source of documentation was view-graph presentation materials, which did not contain the technical data and analysis details to reconstruct conclusions or recommendations. During formulation of the NESC, this shortcoming was clearly noted, and the NESC adopted as a core value documenting all studies and assessments with technical, peer-reviewed final reports.

Final reports are required for any assessment containing NESC-performed independent analyses and/or testing, and must fully document models, analyses and assumptions, test articles, test procedures, test data, and conclusions and recommendations. Conclusions and recommendations are written in terms of Findings, Observations, and Recommendations (FORs). A finding is a relevant fact derived from analyses or tests, and an observation is a technical statement or opinion that could or should be of interest to the project requester, but not necessarily strictly derived from the tests or analyses. Recommendations are statements, directly traceable to one or more specific findings or observations, of the action or position the NESC recommends the project adopt. Each final report is peer-review internally and externally to the NESC, and receives final scrutiny and approval from the NRB before publication. Since in some cases assessment results must be delivered to the requester before the final report is complete, the NRB may also review and approve FORs separately from the report, with the expectation that they are final and will not change.

In some cases where the NESC does not perform independent analyses or tests, but instead reviews existing technical documentation or provides direct technical support to a project, resulting NESC opinions or recommendations may be documented as a "white paper", also approved by the NRB.

Depending on any distribution restrictions, for example due to export control laws or proprietary contents, the NESC publishes each report through the NASA publication system with appropriate availability. The NESC maintains a cadre of technical writers and configuration control specialists to assist with and manage the technical report process.

2.4 Individual and Team Recognition

The NESC recognizes that the individual subject matter experts, acting separately and as part of the assessment team, are the key to NESC operations. Without the ability to identify, recruit, and support highly skilled and capable technical experts external to the NESC, the NESC could not be effective, timely, or value-added. Because of this, the NESC has established its own annual recognition (awards) program to ensure that key individuals and teams are recognized for the impact their work makes. Awards are given for Leadership, Engineering Excellence, Administrative Excellence, and Group Achievement. The highest award, the Directors Award, is given to an individual that takes personal accountability and ownership of a controversial technical issue.

3. SIGNIFICANT MISSION SUPPORT

The technical studies conducted by the NESC are typically performed at the request of NASA management, program or project managers, project technical leads, and Center engineering organization leadership. However, the NESC also maintains an online presence for requests that bypass normal program or organizational reporting and chains-of-command to ensure alternative technical opinions can be made known and independently evaluated.

Fig. 3 shows the historical breakdown for the source of the assessment request, and Fig. 4 shows, by year, the number of assessments supporting each of the major NASA Mission Directorates. Evident in Fig. 4 is the long-term annual average number of assessments, slightly over 50, or about one per week. Since inception, the NESC has accepted 684 of 1122 assessment requests, completed 583, and has 101 either currently active or in final report preparation and closeout.

A comprehensive listing of all active and completed NESC assessments is beyond the limits of this paper; however, the list of current assessments below is representative of NESC mission support assessments.

- CubeSat Radiation Environments and ISS (International Space Station) Radiation Dose Data
- Additive Manufacturing Structural Integrity Initiative
- Burst Factor Assessment for Pressure Vessels
- (Electronic) Parts-level vs. Board-level and Box-level Screening Testing
- JWST (James Webb Space Telescope) Space Environment Launch Constraints
- Wear Resistant Titanium Bearing Technology for Spacesuits
- Independent Crew Vibration and Shock Requirements and Compliance Assessment
- Commercial Crew Aerodynamics Peer Review
- Review of the Orion-ESM (European Service Module) Interfaces

4. RISK REDUCTION AND EXTERNAL PROJECTS

The NESC, in addition to assessments addressing near-term mission needs, has undertaken several large-scale, risk-reduction projects for major NASA Programs that, for one reason or another, the program could not undertake on its own. These projects include the Max Launch Abort System (MLAS), the Composite Crew Module (CCM), and Shell Buckling Knockdown Factor (SBKF) projects, among others. Each of these projects required the design, development, and test of full-scale hardware over several years.

The NESC undertook the Max Launch Abort System project to develop and flight-test an alternative launch abort (escape) system design, as compared to NASA's traditional tower-based design used during Projects Mercury and Apollo. This design was risk-reduction for the Orion escape system, which at the time was experiencing development delays. This project defined a passively stable, aerodynamic faired, side-mounted rocket motor configuration as a potential drop-in replacement for the Orion Launch Abort System (LAS). With the configuration defined, the project designed and built full-scale flight hardware to confirm the stability characteristics during a pad-abort type scenario. The flight test vehicle launched on July 8, 2009 at the NASA Wallops Flight Test Facility, as shown in Fig. 5(a). Post- flight data analysis showed good agreement with pre- flight prediction [1].

The NESC performed a second risk-reduction project for Orion in the same period, investigating the potential use of composite materials for the Orion crew-module pressure vessel. The Composite Crew Module team designed, sized, and built a full-scale pressure vessel including hatches, mounting blocks, and access ports, Fig. 5b. The NESC tested the fully instrumented composite pressure vessel to failure in 2009 at the NASA LaRC [2].

A third risk-reduction project, still on going, is addressing shell buckling design-knockdown factors for orthogrid launch vehicle tanks. A knockdown factor reduces the analytical tank design stiffness in order to account for material and geometric imperfections and defects that can affect buckling loads and behaviour. This project has tested sub-scale and full-scale tank shell structures at the NASA MSFC. The objective is to validate buckling prediction methods for tank structures, and identify and reduce knockdown value conservatism, leading to lighter weight tank structures. Fig. 5(c) shows the full-scale test setup, along with equipment supporting a high-resolution, surface-deflection photogrammetry measurement system. This risk-reduction project will benefit future NASA launch vehicle developments projects [3].

The NESC, on occasion, also participates in or conducts technical or safety-related assessments for requestors outside of NASA. The reason for these requests vary, ranging from the need for total independence from another U.S. Government Agency, or because of the NESC capability to form subject matter expert teams quickly. Significant examples of NESC led outside assessments include

- Toyota Un-Intended Acceleration Study for the National Highway Transportation Safety Administration (NHTSA) [4], looking for latent hardware or software faults that could lead to sudden, un-commanded accelerations.
- Chilean Miners Rescue [5], in which the NESC over several days assembled an expert team of human factors, physiology, and structural designers to write requirements for the rescue capsule.

In each of these cases, the NESC assessment provided needed, un-biased or time-critical technical data on issues of national or international public concern.

5. FUTURE CHALLENGES

In recent years, the NESC's human spaceflight-focused activities have changed from Shuttle Return-to-Flight, completion of the ISS, and major operational science missions such as the Hubble Space Telescope. Today the focus is much more directed to the Orion MPCV, SLS, Ground Systems Development and Operations (GSDO), CCP, and JWST design, integration, test, and certification activities. For these programs, the identification and resolution of technical issues must come early enough in the development cycle to influence design decisions and/or minimize cascading system-level or architecture interactions. To do so, the NESC must engage directly with program engineering teams, but must also maintain independence in case of major technical disagreements or apparent compromises to safety or mission success requiring elevation to NASA senior management.

A lesson-learned from NESC engagement with the predecessor Constellation development program was that creating complex flight, guidance, control, avionics, structural, and/or aerodynamics models and simulations independent of program efforts requires long-lead times. Independently developed models allows the NESC to challenge modeling assumptions, implementation, and interpretation of results, which is not possible if the NESC must rely on the program-developed models because of time criticality. Consequently, the NESC has undertaken a long-term modeling and simulation effort to develop several key simulation capabilities in parallel to program efforts, specifically a comprehensive flight dynamics/controls/flexibility/slosh model of the SLS, and an entry/descent/landing model adaptable to either CCP provider. In this manner, the NESC models can be compared and validated with program or contractor models to ensure proper capture of system physics, but still be based on independent assumptions, tools, and techniques.

Farther in the future, Orion, SLS, CCP partners, and JWST will transition from development to flight operations. While an operational focus is more like the NESC's original role, the number of flight programs, the change in mission types, and, in the case of the CCP partners, commercial operations, will further challenge the NESC to adapt.

6. REFERENCES

1. Gilbert, M. G. (2014) The Max Launch Abort System - Concept, Flight Test, and Evolution. 7th IAASS Conference "Safety is No Accident", Friedrichshafen, Germany.

2. Kirsch, M. T. (2011) Composite Crew Module: Primary Structure, NASA TM-20110217185

3. Hilburger, M. W., Waters, W. A. Jr., and Haynie, W. T. Buckling Results from the 8-Foot-Diameter Orthogrid-Stiffened Cylinder Test Article TA01, NASA TP-2015-218785

4. Technical Support to the National Highway Traffic Safety Administration (NHTSA) on the Reported Toyota Motor Corporation (TMC) Unintended Acceleration (UA) Investigation, available at http://www.nhtsa.gov/UA

5. http://www.nasa.gov/feature/how-nasa-helped-the-33-chilean-miners

19 Technical Discipline Teams
Aerosciences
Avionics
Electrical Power
Flight Mechanics
Guidance, Navigation & Control
Human Factors
Human Spaceflight Operations
Life Support/Active Thermal
Loads and Dynamics
Materials
Mechanical Systems
Non-Destructive Evaluation
Passive Thermal
Propulsion
Robotic Spaceflight
Software
Space Environments
Structures
Systems Engineering

Figure 1. List of NESC Technical Discipline Teams

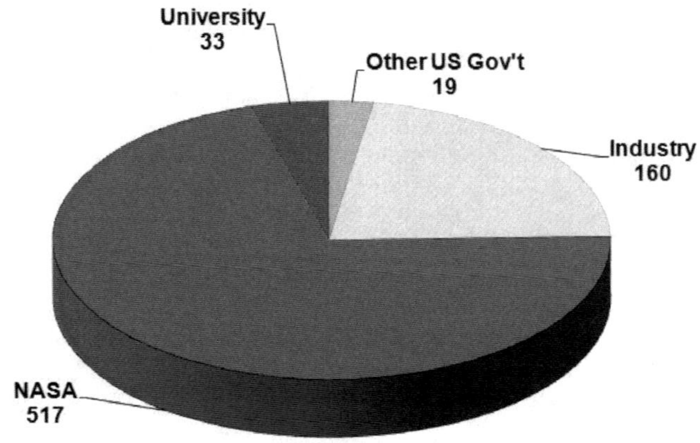

Figure 2. Affiliation of NESC Technical Team Membership, 2015

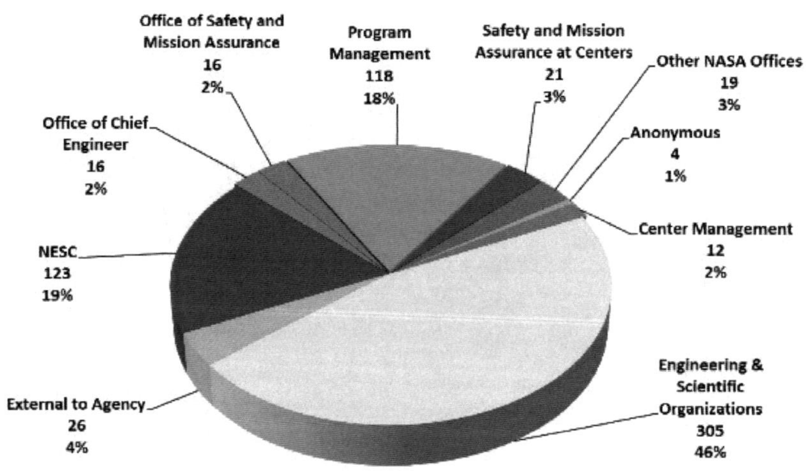

Figure 3. Historical Breakdown of Assessment Request Source

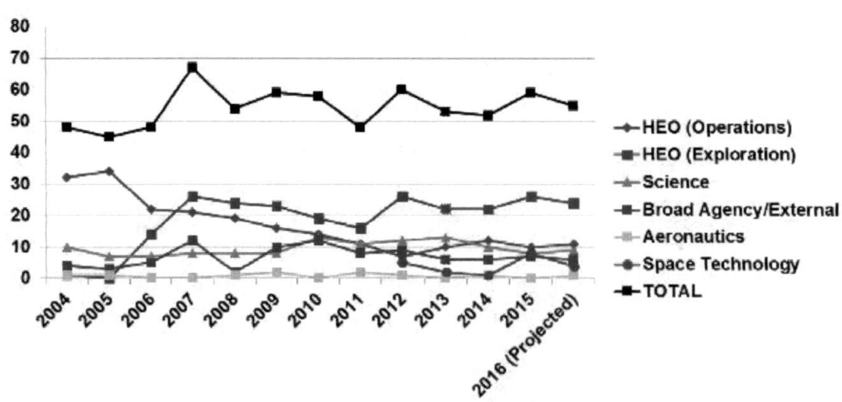

Figure 4. Historical Breakdown of NASA Mission Directorate Support

(a)

(b)

(c)

Figure 5. Major NESC Risk Reduction Projects (a) Max Launch Abort System, (b) Composite Crew Module, (c) Shell Buckling Knockdown Factor

OBSERVATIONS, REFLECTIONS, AND LESSONS LEARNED... FROM ~500 SSPCBS & ~1000 IMMTS... AND GOBS OF SORRS, FRRS, ETC.

George K Gafka[1]

[1] *NASA, Johnson Space Center, Houston, TX, USA, george.k.gafka@nasa.gov*

ABSTRACT

NASA established the Chief Safety and Mission Assurance Officer (CSO) role as part of the Agency's formal governance model following the Space Shuttle Columbia accident in an effort to formalize technical authority, provide a healthy tension against the pressures of cost and schedule, and strengthen the overall systems of checks and balances within NASA's Programs. Mr. Gafka was selected as the first CSO for the International Space Station (ISS) in 2005 and held that role for over 9 years. During his tenure as ISS CSO, Mr. Gafka oversaw Space Shuttle Return to Flight in 2005, continued assembly of ISS (including multiple new International Partner modules, new visiting vehicles, and new ground controls centers), the birth of US "commercial" cargo and crew transportation services, several historic and dangerous space walks, and some noteworthy major anomalies and close calls. In this paper/presentation, Mr. Gafka will share some of his most significant reflections and lessons-learned about conducting high-risk exploration and risk- informed decision making. The paper/presentation will not simply list a series of anomalies and subsequent corrective actions; but instead will focus on the underlying valuable lessons that tend to live deep in the context, the tone, the organization, the policy, the philosophy, and the culture surrounding both failures and amazing accomplishments. Mr. Gafka will discuss communication as a critical transaction, growing and sustaining excellence and resiliency, growing and sustaining curiosity and vigilance, establishing norms, using judgment, lessons learned, and imagination.

1. INTRODUCTION

NASA established the Chief Safety and Mission Assurance Officer (CSO) role as part of the Agency's formal governance model following the Space Shuttle Columbia accident in an effort to formalize technical authority, provide a healthy tension against the pressures of cost and schedule, and strengthen the overall systems of checks and balances within NASA's Programs. Mr. Gafka was selected as the first CSO for the International Space Station (ISS) in 2005 and held that role for over 9 years. As referenced in the title of this paper, during his tenure, Mr. Gafka held a formal seat for ~500 Space Station Program Control Boards (SSPCBs), ~1000 International Space Station Mission Management Team boards (IMMTs), and a significant number of both Space Station Program-level Stage Operations Readiness Reviews (SORRs) and Agency- level Flight Readiness Reviews (FRRs), where the Agency conducts its executive level readiness to commit to launch. This paper captures some of his most significant reflections and lessons learned during this historic time period in human spaceflight from the perspective of ISS CSO.

2. MORE THAN A COMMUNITY OF EXCELLENCE... A COMMUNITY OF CHOICE!

Today, many people refer to the phrase "communities of excellence." However, to have a robust Safety and Mission Assurance (S&MA) community, one must go even further and create a "community of choice." True, the S&MA community must have excellence, but that alone is not sufficient for success in high-risk exploration. S&MA is an assurance and influencing function at its core and, therefore, the S&MA community must be the "community of choice": the choice for the Program Manager who is struggling with a difficult risk trade decision, the choice for the nervous employee who is troubled by a safety concern, and even the choice for the best and brightest employees who are planning their career path. It is not enough to merely have a formal seat at the table or be a mandatory step in the process, and it is not enough to merely be knowledgeable about your technical area of expertise. People both internal and external to S&MA must understand the value of S&MA and genuinely want the expertise and services of the S&MA community. In order to have a truly robust S&MA community, one must go beyond excellence and become a "community of choice."

3. A MISSION FOCUSED ON TRANSACTIONS

It is very important to recognize that the "job-focused mission" of an S&MA professional within human spaceflight is a mission focused on intellectual transactions.

Much like a business must focus on the quality of financial transactions in order to achieve business success, the S&MA community must focus on the quality of intellectual transactions in order to achieve safety and mission success. There are many similarities here to explore, and also a few notable differences that pose even greater challenges for the S&MA professional.

For example, eBay, is an on-line auction website that seeks to be the market-leading community of choice for buyers and sellers to meet and transact. The infrastructure for these buyers and sellers to meet and transact is provided by eBay, and eBay makes a fee off each successful transaction. In a similar sense, by definition, S&MA processes and products are core infrastructure for "knowledge buyers" and "knowledge sellers" to meet and intellectually transact with respect to safety hazards, control strategies, residual risks, failure modes, etc. The knowledge buyers (decision makers) in this case need all the best risk information from the knowledge sellers (the people who know the information) in order to make a proper risk-informed decision. The S&MA processes and products are that infrastructure, which capture and organize the risk related information critical to enabling a successful intellectual transaction.

However, a few notable differences exist in this interesting analogy that pose an even greater challenge for the S&MA professional and those challenges must be understood in order to fully appreciate the S&MA task at hand. In the eBay case, there is one buyer meeting and transacting with one seller; however, in the case of S&MA for human spaceflight, the intellectual transactions typically comprise multiple knowledge buyers and sellers, each bringing their unique understanding, perspectives, and objectives to the complex culminating integrated decision or "sale." Furthermore, in the eBay case, both buyers and sellers are mutually attracted to each other in a fairly primal manner. The buyer wants something that the seller has and the seller wants money. In the case of S&MA for human spaceflight, all parties may not fully appreciate the value of fully utilizing the S&MA infrastructure as a facilitator and enabler of the decision or "sale." Some knowledge buyers and/or sellers may see the S&MA products and processes as overly burdensome or taxing as compared to the perceived value added by their use, and some knowledge buyers and/or sellers may think it is possible to have a "good enough" decision or "sale" without utilizing the S&MA products and processes, certainly not to their fullest potential. Finally, in the eBay case, eBay only provides the infrastructure for the transaction and makes no assertion with respect to the contents of the transaction. For example, if a buyer receives a product through eBay that is not per the pre- sale description, eBay had no role in assuring the correctness and validity of the contents of the transaction. In the case of S&MA for human spaceflight, not only does S&MA provide the infrastructure for the necessary risk-based intellectual transactions, but the correctness and validity of the contents of the transaction must be assured as well! This is a significant and challenging difference in scope.

4. GROWING AND SUSTAINING EXCELLENCE

As previously stated, excellence alone is not sufficient; however, it is still a critical piece of the S&MA foundation and very much required for success. S&MA professionals must be thoroughly knowledgeable of their respective technical discipline areas, the systems and architectures within their boundary of responsibility, as well as the various standards and requirements entrusted to their insight/oversight. Furthermore, the S&MA professional must understand the history and the "why" behind the spirit of their requirements, so as to deeply understand proper versus improper requirement tailoring as well as acceptable versus unacceptable deviations/waivers. A level of passion and resilience is required to grow and sustain this level of excellence, especially over long periods of time with little breaks in the intensity level.

5. GROWING AND SUSTAINING CURIOSITY AND VIGILANCE

It is clear that attitude is a major contributor to success. Much like growing and sustaining excellence, the S&MA professional must grow and sustain an attitude that exudes perpetual curiosity and vigilance. One must never get too comfortable when working in high-risk exploration, never get too arrogant that the risks are all identified and understood, and never get too complacent due to previous successes. Instead, one must scrub each risk decision with fresh energy and focused determination, continually scour the data to look for hints of unidentified trends and hidden risks, and always assume that they are not fully knowledgeable and that there is always more to learn/know. Similarly, a level of passion and resilience is required to grow and sustain this spirit of perpetual curiosity and vigilance, especially over long periods of time with little breaks in the intensity level.

6. ESTABLISHING NORMS

One of the more subtle, but important, things that organizations do is establish "norms"; that is, establish the organization's values, philosophies, protocols, ways of thinking, and just how things really get done. Many times, these norms are loosely understood rather than written down as, say, a requirement or work instruction. This is understandably so, as many times it would be impossible to do such as these norms often deal with fuzzy things like mission/vision definition, judgment, and wisdom. For human spaceflight, many age-old legacy norms are currently being challenged (and will continue to be challenged) as the role of the government versus commercial entities in human spaceflight, the required amount of government insight and oversight for crew safety, and the establishment of acceptable risk tolerances are all evolving at a rapid pace with much debate and pressure from all sides. For the S&MA professional, it is especially important to properly navigate the current dynamic nature of today's norms surrounding human space flight. In order to do that successfully, today's S&MA professional must develop a deep understanding of various legacy norms and the "why" behind them, in order to fully assess the new, innovative, and unconventional norms being proposed for trial use. It is important to not simply remain mindlessly shackled to the norms of the past, and yet it is equally important to recognize that many of our legacy norms were born out of tragedy and hard lessons learned. Therefore, new norms must be assessed openly, fairly, and thoroughly to ascertain whether the proposed new benefits can be achieved without undue risk and while retaining the spirit of important lessons learned from history.

7. "THE PLAN", USING JUDGMENT

In human spaceflight, it is very typical to spend a significant amount of effort trying to determine all the various things that could possibly go wrong and then developing plans to deal with such, in case of realizing those failures. In fact, beyond a "plan b", many times a "plan c" and "plan d" are also developed depending upon priority factors. However, there is also a very real programmatic cost to making significant investments in off-nominal planning. Much like everything else, a sensible balance must be achieved using good judgment. It also becomes very apparent that the nominal plan contains a complex set of interdependencies and, by the operational time frame, much investment has been made developing, assessing, and training to the nominal and selected off-nominal plans. This significant programmatic investment and the vast amount of time committed in training tends to make plans "sticky", meaning that there tends to be a mentality to stay the course and not deviate from the prescriptive playbook. When surprises arise that were not accounted for in pre- coordinated off-nominal planning, great care must be taken prior to executing recovery operations despite the desire to immediately conduct corrective actions… and yet, many times, the exact right thing to do is to throw the playbook away and scramble through the unknown. Judgment is key to this difficult decision process. Our history has many examples of our ability to agilely accomplish incredible things (i.e. performing an unplanned repair via Extra Vehicular Activity (EVA) to a solar array torn during deployment as shown in Figure 1); however, our history also has many examples of miscalculating or underappreciating very hazardous interdependencies when scrambling in the heat of the moment and totally off script (even with the best of intentions). Again, judgment is key when deciding whether or not the team is ready to "go for it."

Figure 1. Solar Array Repair via EVA

8. LESSONS LEARNED AND IMAGINATION

The term "lessons learned" gets a lot of airplay today, when in actuality the spirit of the term is rarely met. Many times an anomaly or failure is investigated and the proximate or likely contributing cause(s) are identified. Corrective actions and recurrence controls are then developed and implemented to address the original specific anomaly/failure… and all efforts typically stop there. This does not constitute "lessons learned." The spirit of "lessons learned" is taking certain specific or generic elements discovered in the investigation above, and then searching elsewhere for places where this new knowledge could help avert a similar issue from arising. It requires a programmatic investment and sometimes a lot of imagination to decide where best to look and just how far to take the search. Many times the acceptable answer is to only perform the very "local" investigation discussed above; however, if the investigation yields communication gaps, process holes, organizational problems, cultural deficiencies, or systemic disconnects, then taking the next steps to perform a more thorough "root cause" investigation and then spreading that outward via a true "lessons learned" campaign is warranted. Much like the ripples in the water of Figure 2, judgment must be used for each case-specific situation to determine the right size investment from an investigation perspective and from a true outward "lessons learned" perspective.

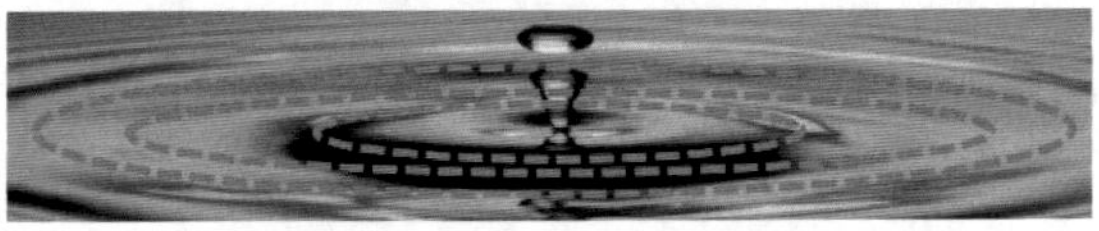

Figure 2. How far and wide should one look for potential areas of application?

9. SUMMARY

This paper is the result of much personal reflection from the ISS CSO perspective about a very complex, dynamic, and historic time period in human space flight. And in the end, these reflections lead back to the same general categories and questions of most such reflections regardless of industry or organization sector:

- *Who are we?*
- *What is our Mission?*
- *How do we fit in the bigger picture?*
- *What does success look like?*
- *How do we achieve success?*
- *How do we grow and sustain our position?*
- *What are the fuzzier things we must monitor?*

Success, especially in high-risk exploration, is many times very contingent upon understanding and addressing these questions and concepts. It is always necessary to grow and sustain the technical discipline and management excellence within any organization, such as human spaceflight S&MA; however, excellence alone is not sufficient. An assurance and influencing organization, such as human spaceflight S&MA, must become a "community of choice" within the largest possible boundary and sphere of influence… and must demonstrate a high level of leadership vision, judgment, and wisdom. An S&MA organization must be passionate, resilient, curious, and vigilant… and must use judgment to balance that necessary curiosity and critical imagination against the required investment and risk. More subtle than the more tangible S&MA products, processes, procedures, and work instructions, the S&MA professional must also have the keen eyesight to see into fuzzy areas such as underlying tone, organizational culture, and interpersonal/soft skills. Ultimately, success demands such.

EVOLVING UNITED STATES AIR FORCE SPACE SAFETY POLICY AND CULTURE TO MEET TODAY'S SPACE SAFETY CHALLENGES

Lt Col Steve Bogstie, Lt Col Beth Nayder, Maj James Souders

The views expressed in this presentation reflect those of the authors, and do not necessarily reflect the views or opinions of the Department of Defense or the U.S. Air Force.

1. INTRODUCTION

The United States Air Force (USAF) started to formally manage military space programs in 1959.[1] By 1962, the USAF published the USAF MISSILE/SPACE SYSTEMS SAFETY SURVEY GUIDE which outlined an accident prevention program[2]. The guide focused on detecting and correcting unsafe practices and conditions. "Most important are education, training, discipline, enforcement of appropriate regulations, and thorough investigation of missile/space accidents." These steps started the USAF Space Safety Program, which have continued to evolve over the past six plus decades.

Aerospace companies, developing a vast array of space launch systems and capabilities, have made space more accessible. This unprecedented growth spans not only the launch capabilities, but includes satellite design and manufacturing, and on-orbit operations. Safeguarding the American public and Department of Defense (DoD) space assets are our top priorities. Constant vigilance is required to adapt policy and guidance to meet the complexities associated with the growth in technology and capabilities.

Launches today are rarely one satellite on one launch vehicle owned by the same agency. We see increasing use of rideshare (secondary payloads hitching a ride on a primary launch mission for little to no launch costs) to reach orbit and the proliferation of small satellites. Unique examples include DoD hosted payload on a foreign satellite and a small satellite built by a university that happens to be a service academy. The Air Force (AF) is on the fore-front of space safety innovation and continues to update safety policy and inspire a culture to evolve with the technology and mission, while constantly improving mishap prevention.

2. AIR FORCE SPACE SAFETY POLICY

The USAF is a proud member of the Space Safety community. The Air Force Safety Center (HQ AFSEC), located at Kirtland Air Force Base, New Mexico, under the command of the Air Force Chief of Safety, develops, implements and oversees the Air Force Mishap Prevention Program. The purpose of the Air Force Mishap Prevention Program is to minimize the loss of Air Force resources and protect Air Force personnel from death, injuries or occupational illnesses by managing risks on and off-duty. HQ AFSEC develops safety programs, policies, goals, objectives and establishes guidelines to support and assess effectiveness of the Air Force Safety Management System (AFSMS).[3] The AFSMS is the core framework for the AF's Mishap Prevention Program.

As one of four operational safety divisions (Flight, Occupational, Weapons and Space) within HQ AFSEC, the Space Safety Division's mission is to preserve combat space power by anticipating, reducing and preventing mishaps. We develop, execute, and evaluate AF space mishap prevention programs that align our space safety programs to the AF's overall AFSMS structure. These safety mishap prevention programs are codified in Air Force Instruction (AFI) 91-202 and AFI 91-217.

Recognizing the safety communities best practices, and seizing on strengths of these established programs (e.g. Federal Aviation Administration (FAA)), the USAF adopted and incorporated the ideas of a Safety Management System into its guidance in 2015. AFSMS is now the foundation for all AF mishap prevention programs. The AFSMS vision is for the AF to be a world leader in safety management and provide care for our Airmen and our environment to meet our air, space and cyberspace missions. Accordingly, the AF is committed to the following three AFSMS priorities:

- Compliance. Comply with all safety and regulatory guidelines.
- Risk Reduction. Protect our assets, personnel and material by effectively identifying and managing risks.
- Continuous Improvement. Instill a culture that encourages and supports continuous improvement.

The AF operates responsibly with due regard for the safety of the general public, AF personnel, space support personnel, and public and government resources. The Space Safety program includes mishap prevention guidance across the life cycle of space systems and their unique support equipment, to include Design, Systems Development, Integration, Testing, Pre-Launch, Launch Operations, Range Operations, Orbital Operations and Ground-Based Space Systems.

The AF's Space Safety Process can improve operational effectiveness by managing or preventing close calls (events), reducing mishaps, and supporting mission assurance throughout the life cycle of a space system. It includes launch/range safety, orbital safety, ground-based space safety, system safety, software safety, risk management, and mishap investigation and reporting.[4] All aspects are considered when looking at risk reduction.

AFSMS utilizes the four pillars as a framework for structuring the AF mishap prevention programs and activities used to minimize risk and reduce the occurrence and cost of injuries, illnesses, fatalities and property damage. Space Safety Division efforts crosses all four pillars to protect the mission.

The first pillar in AFSMS is Policy and Leadership. Safety policy provides the framework to build a sound and proactive mishap prevention program. AF Safety Policy is established through AF Policy Directives, and implementation is directed through AFIs, Air Force Manuals (AFMANs) and Air Force Pamphlets (AFPAMs). These documents define the directives, policies and procedures, and organizational structures necessary to implement the AFSMS vision.[5] Active leadership involvement in the implementation and execution of the AFSMS supported at all levels of command is critical. Leadership implements the AFSMS by providing guidance and goals, establishing safety responsibility and accountability, applying risk management to all activities and promoting the AFSMS throughout the organization.[6]

The second pillar in AFSMS is Risk Management (RM). The Air Force's five-step deliberate RM process is the core of the Air Force safety and mishap preven-

tion program. AF units utilize RM to the maximum extent possible to identify and assess hazards from which mitigating controls are developed. Control measures are implemented and are continuously monitored and analyzed to assess their effectiveness. Should a control measure not meet expectations, new control measures will be implemented and the process continues. Streamlining the steps is essential in situations where risk decisions need to be made quickly and in real-time.[7]

The third pillar in AFSMS is Assurance. Safety assurance is the evaluation, review, mishap investigation and monitoring that assures commanders the elements of the mishap prevention program are being implemented, and guides continuous improvement efforts. Assurance programs measure whether organizations conform to standards and are making progress toward established goals.[8] The Space Safety Division conducts mishap safety investigations to determine root cause of a mishap and implements corrective actions.

The forth pillar is AFSMS is Promotion, Training and Education. This pillar ensures Airman are provided safety awareness information, organizations have ongoing embedded AFSMS unit training, and organizations implement effective risk control measures.[9]

The Space Safety Division applies AFSMS principals across the life-cycle of space systems and their unique support equipment, to include Design, Systems Development, Integration, Testing, Pre-Launch, Launch Operations, Range Operations, Orbital Operations and Ground-Based Space Systems to execute a robust space safety mishap prevention program.

There are a number of safety challenges that have recently arisen that affect various aspects of the space system life-cycle. Utilizing the four pillars of the AF-SMS; the AF Space Safety community is tackling these challenges head-on, while ensuring safe mission operations continue unhindered. For example, the Space Safety Division, along with other Space Safety stakeholders, are addressing policy for Design, Systems Development, Integration and Testing of space systems. Originally satellites were designed with a single user / single mission. The lines of safety authority and responsibility were clear and well defined. This practice has greatly changed over the course of the last decade. The AF is now processing satellite payloads which now have multiple users and multiple missions on-board a single spacecraft. Ensuring safety guidance properly aligns with these practices is critical to ensuring mission success.

Other evolving areas for safety policy to address are launch, orbital and re-entry operations. Advances in materials, power supply and electronic miniaturization has increased the number of small satellites (SmallSats) in orbit. In 2012 one-tenth of all cataloged Low Earth Orbit (LEO) objects fell into this category, and by 2017 we project nearly a quarter of the active LEO catalog will be CubeSats[10] or smaller. The proliferation of CubeSats and other small satellites is illustrated by the ORS-3 mission, which on 20 November 2013 launched 28 CubeSats and one small satellite into space. The mission held the record for the most objects deployed by a single launch – for one day. On the next day, a Russian Dnepr rocket launched 32 payloads, predominantly SmallSats, into space.

Additions to policy to try to capture all the different technologies, dependent resources and organizational relationships could grow exponential. The Department of Defense Space Test Program has some experience in handling these challenges. They adopted the mantra "Do No Harm" to your fellow mission partners.[11] While it does not take into account on-orbit risks of potential negative interactions between the different missions, but it provides a starting point.

With the Space Safety Division efforts to update policy with the perspective – "Do No Harm", the Air Force takes on greater mishap prevention responsibility. The program manager of a space mission must ensure, before launch, that the mission will do no harm to the space environment. The launch integrator must ensure that the satellite will do no harm to the launch vehicle or to other satellites "sharing a ride" on the same rocket. The launch commander must ensure the safety of the public from not only the launch itself, but also the reentry of any upper stages, reentry vehicles, and possibly, from the fly-back of reusable launch vehicle components. The on-orbit satellite operator is responsible for conducting on-orbit operations safety, with due regard not only for other satellites, but for the shared space environment. And that same operator must ensure the safe disposal of the mission at the end of its life. This is all part of our mission to Safeguard Combat Space Power via Mishap Prevention.

3. CULTURE CHANGES IN SUPPORT OF SPACE SAFETY

As technology evolves and the community expands, it important to identify opportunities for mishap prevention. Instituting cultural change is an effective way to prevent mishaps. The Space Safety Division is undertaking several initiatives to improve the space safety culture across the AF, utilizing the AFSAS framework.

The first is "Proactive Safety". Historically, we have practiced "active safety," which can be described as managing known hazards. The aviation community has embraced proactive safety, an effort to systematically and scientifically understand the gap between desired and actual flight performance. The goal is a safety culture that rewards the reporting of errors and hazards, versus one that either hides errors or that punishes those who err.

The AF started the Aviation Safety Action Program (ASAP) for voluntarily reporting threats and errors to flight operations. The Safety Center received hundreds of reports about threats and errors that never would have been exposed otherwise. Globally, over 105 airlines use ASAP-style initiatives that have yielded over 500,000 reports.[12] Though our community is smaller, expanding that openness of proactive safety to space is possible. We derive extremely important information from our mishap investigations, but at a high cost. ASAP-style initiative allows any person to report potential safety hazards before the lead to a mishap. This initiative is being expanded from a purely aviation focus to all the safety disciplines (Flight, Occupational, Weapons and Space) with the program name changing to Airman Safety Action Program.

Additionally, the Space Safety Division is looking to institute a space-centric Military Flight Operations Quality Assurance (MFOQA) program. MFOQA is a 12 year old aviation program that analyzes routine flight data to detect, measure and mitigate mishap precursors while protecting crew identity. A similar approach could be applied to space operations command and control. Where consoles data could be analyzed and risks identified in training, operations, or technical data. Implementing such a program for space would allow AF leaders to perform early intervention to correct adverse safety trends before they lead to mishaps.[13]

Next, the Space Safety Division is developing space specific training for mishap prevention and mishap investigation. This is a cultural change within the AF safety community; as space safety has always relied on other disciplines safety training courses to educate our safety personnel. This shift to space-specific mishap training will heighten safety awareness for all space professionals.

Finally, the Space Safety Division is reaching out to other government agencies to streamline and standardize mishap investigation policy and procedures. This collaboration has enabled better communications and effective policy implementation across the AF, FAA, the National Transportation Safety Board (NTSB), and the National Aeronautics and Space Administration (NASA). Collaboration between Federal Agencies has never been better; sharing lessons-learned, conducting joint training, and sharing mishap prevention strategies is raising the visibility of the AF's space safety efforts.

4. CONCLUSION

Just as important as it was over 50 years ago, "judicious application will be reflected in reduced mishaps and a higher degree of combat readiness." Recognizing the growing congestion in space and growing reliance upon its use, the Air Force formalized space safety as a mission operations discipline within the safety center in 2013. We protect the mission by providing policy, guidance, and oversight to preserve combat space power. In the past three years we have focused our efforts on updating policy that encourages prevention with the AF-SMS. The AF's Space Mishap Prevention is a continuous process that begins in the early acquisition phases of a Space Safety Program and includes reentry, recovery, and disposal activities for applicable systems. Supervisors at all levels execute established risk management (RM) practices/procedures throughout all phases of a space system's lifetime. More openness and expanding perspective will strengthen our culture. We will achieve Our Space Safety Vision: safe full-spectrum combat capability across the space domain.

5. REFERENCES

1. THE THREASHOLD OF SPACE: THE AIR FORCE IN THE NATIONAL SPACE PROGRAM, 1945-1959. USAF Historical Division Liaison Office. September 1960. http://www.afhso.af.mil/shared/media/document/AFD-110321-032.pdf

2. Directorate of Missile Safety, Deputy Inspector General for Safety, USAF. USAF MISSILE/SPACE SYSTEMS SAFETY SURVEY GUIDE. STUDY NO. 12 – 62.

3. AFI 91-202, The US Air Force Mishap Prevention Program, 24 JUNE 2015

4. AFI91-217, Space Safety and Mishap Prevention Program, 11 APRIL 2014

5. AFI 91-202, The US Air Force Mishap Prevention Program, 24 JUNE 2015

6. AFI 91-202, The US Air Force Mishap Prevention Program, 24 JUNE 2015

7. AFI 91-202, The US Air Force Mishap Prevention Program, 24 JUNE 2015

8. AFI 91-202, The US Air Force Mishap Prevention Program, 24 JUNE 2015

9. AFI 91-202, The US Air Force Mishap Prevention Program, 24 JUNE 2015

10 http://www.cubesat.org/

11 Andrew Read, Peter Chang, Barbara Braun. Rideshare Mission Assurance on Multi-Payload Missions. 29th Annual AIAA/USU Conference on Small Satellites. SSC15-VII-9

12. United States Air Force Safety Center Proactive Aviation Safety (ProSEF). http://www.afsec.af.mil/proactiveaviationsafety/index.asp

13. http://www.afsec.af.mil/proactiveaviationsafety/mfoqa/index.asp

LAUNCH SYSTEMS CONFORMITY TRAINING PROCESS AND ACADEMIC METHODOLOGY

C. Aussilhou & D. Miot [1]

[1] CNES, Launchers' Directorate, 52 rue Jacques Hillairet, 75012 Paris CEDEX France, caroline.aussilhou@cnes.fr

ABSTRACT

This article deals with "Organizational Culture and Safety" domain, and concerns the relationship between regulatory control, people and safety after a regulatory change. Beyond prevention of failures through launcher conformity, launch safety is integrating the mitigation of the consequences of a launch failure. The safety has to rely on technical requirements but also competent resources and thus trained people. After the reminder of the context of the implementation of the new law named FSOA [1] and the description of controller's new mission, this article explains the reasons of the implementation of a training adapted to the controller's new job and the concerns of a loss of technical skills. The reason was double: the information and the qualification needs. Indeed, the first difficulty met was the misunderstanding at interfaces on the new rights and duties of the controllers. This training was also included in the process of qualification of the operational functions of the conformity control. This training hinges on 3 key domains: the control plan, the launch system risk analysis and the mission analysis that relates to 3 jobs: controller, assessor and auditor. This training is adapted to be dispensed not only for the Controller but also for Launch System Operator, legal entity in charge of operating the launches.

1. INTRODUCTION AND REGULATORY CHANGE

FSOA (French Space Operation Act) [1] and associated regulations came into force on 2010, December 10th. Its goal is to establish the legal frame in terms of safety of each space actors, whether they belong to the public or the private sector. It aims at ensuring the space activities' technical risks mastery and control.

One of the responsibilities of CNES, French Space Agency is to guarantee the safety for activities conducted within or from the perimeter of the European Spaceport in Kourou, French Guiana, (called the CSG) but also, ensuring along the flight the protection of people, Public health, belongings and environment.

In this context, risk mastery for new launch vehicles issued from different origin and original industrial organizations, was a great challenge and should be supported by setting up a coherent national regime of authorization and control of space operations.

The basis of the regulations [3] was the experience gained on thirty years of Ariane launches, capitalized in the applicable CSG Launch range regulation [3].

This relation between Controllers and Operators has allowed to smoothly demonstrate the conformity of the nowadays launch systems regarding regulations. Although a specific set of tools, files and procedures has been introduced or adapted in order to comply with the legal framework, very few new technical requirements were introduced, except the requirement to conduct launch system hazard study in case of failure and therefore mitigate the consequences. This change is broadening the safety scope beyond prevention (and the conformity strategy) to protection and from the launcher system safety to the launch system safety.

In addition, this regulation change has modified the governance of the European space system through the clarifications of roles and responsibilities. CNES mission cannot be anymore only a quality control service to the Operator, but must shift towards a safety control of the launch system through conformity control, under the umbrella of French government. Launch System Operator is the first barrier of safety. It is necessary that the controller can be and remain a second effective barrier. The issue raised by the regulatory change, was: how to maintain a satisfying (or even raise) safety level after this regulatory change that increases the independence and distance from the fieldwork and data?

One important component of this new mission is the man power: to control means controllers! The job is changing which raises concerns but also opportunities. Several controllers were concerned to lose their engineering skills and expertise, having to control Operators doing their former job.

This paper presents the adaptation of the organization and the functioning implemented to constitute the core team control and above all the training put in place to qualify them. To summarize the specific point of view of this paper: "Behind the control, men and women!"

2. NEW LAUNCH SAFETY ASSESSMENT REQUIREMENTS

In accordance with II.2°c) of article 1 of the FSOA decree of 9th June 2009, Launch System Operator carries out a study of the potential hazards involved in the planned space operations.

To do so, in accordance with Article 7 of decree concerning Technical Regulation [2], Launch System Operator must perform a Launch System "Danger" Analysis, which is a specific Launch System Hazard Study.

This hazard study includes a description of all hazards related to the operation in nominal and accidental operating situations, whether their cause is internal or external to the launcher.

The study specifies the nature and scope of the possible consequences of all these operating situations. When dealing with elements of the launch vehicle which fall back and are liable to reach the ground, the study presents the components of these elements, stating their dimensions, masses and materials used.

Launch System Operator's new task must therefore:
- demonstrate compliance with the quantitative and qualitative risk assessment requirements (cf. article 20 of [2]), with regard to the risk of injury to individuals;
- evaluate the effects of any accidents on public health and the environment.

This study must cover the following events, in the conditions stipulated in chapter 3 of [2]:
- damages linked to fall-back of elements designed to separate from the launcher;
- damages linked to controlled or uncontrolled re-entry of launcher elements placed in earth orbit;
- damages linked to failure of the launch vehicle; collision with manned space objects, for which the orbital parameters are precisely known and available;
- damages linked to explosion of a stage in orbit;
- collision with a celestial body.

The controller new job is to check that these risk analysis studies are complete to identify all the possible risk reduction measures. Thus the priority major tool of the controller is the launch system risk analysis, like a scoreboard. For every technical fact analysis it is necessary to conclude on the acceptability of residual risks concerning the following feared events:
- Fallback of elements on Earth,
- Controlled or uncontrolled re-entry of launcher elements placed in earth orbit
- Launch vehicle failures,
- Creation of debris in orbit,
- Collisions in orbit.

and all the associated damages.

3. MANAGEMENT OF CHANGE & HUMAN RESOURCES

We were facing a problematic issue that is the management of change in high-risk systems. Change can bring opportunities (e.g. innovation) but also risks, well known in engineering, and concerns like "resistance to change". It is therefore a key process of safety management that can be deficient, contributing to cause an accident, such as experienced by NASA before Columbia shuttle loss [CAIB, 2003].

In this case, we face 3 types of change: an institutional change (the impact of a public policy with a new law), an organizational change (within CNES to establish the controller division) and a change of job and posture for the engineers.

Modifications and changes may occur at several levels of the sociotechnical system: competences and status of workers and management staff, procedures, structures, policy, culture and regulations. Important change may impact and cascade through 3 organizational dimensions [11] illustrated in Fig. 1: Organizational vertical and hierarchical dimension (Human resources from top management to field engineers and workers), Organizational transversal and inter-organizational dimension (interfaces between internal divisions and external entities) and over time (historical dimension).

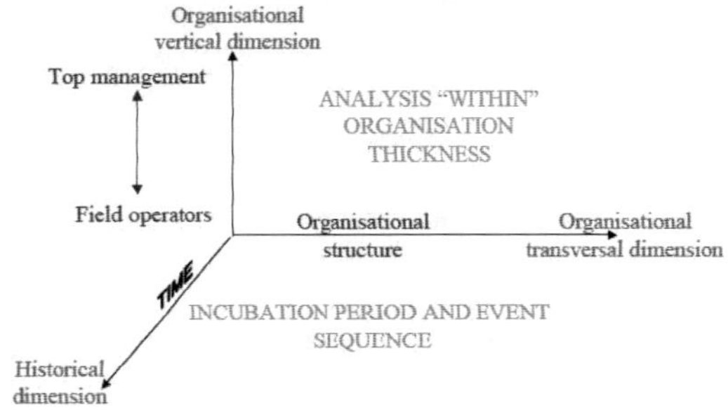

3 dimensions of « organisational analysis » of an event (Y. Dien, 2005)

Figure 1. 3 dimensions of "organizational analysis"

An important change could impact safety and production, in a positive or negative manner, at short or long term. After 5 years of experience (which is rather short still), it seems, no observable degradation has occurred yet.

Safety has to rely on technical requirements but also organizational and competent resources. The context of the implementation of this new mission in CNES Launchers Directorate had a direct impact on designated personnel for this mission and the conditions of carrying out this mission.

Careful thought has been put in place to define and characterize this new mission at the crossroads between safeguard, RAMS (Reliability, Availability, Maintenability and Safety) and product quality assurance activities.

It is also interesting to explain the management of change of this job. Above all, we focused on the position change for the staff involved in this new relationship with Launch System Operators. In CNES firm, they are development projects, Quality assurance and technical expertise functions, in a quite classical scheme of a firm: Expertise, Quality and projects.

In the past, some of Launchers Directorate staff were in charge of quality studies in an embedded project team with the Launch System Operators. It enabled them to have access to the data, to get the ground for their conformity judgment and to maintain their engineering skills.

Becoming an independent and external controller raised the concern of losing transparency on data and the engineering skills.

Indeed, it was a real transformation more difficult to manage by the experienced employees' than for those who began directly in this new job. To some extent, they view their job and added value as being the quality support of the Operator. Indeed, to illustrate, this was like a customer change: previously, the customer of this quality service was the Operator and then the customer is now directly the government.

After a phase of adaptation, it is back to normal again, but the relationship with the Operator was so, marked by this phase of learning longer than planned. The dialogue was essential during this phase of mutual understanding and explanations of the new requirements and the new ways of interacting and functioning. This change has many impacts to Launch System Operator with changes brought for him. There is no question of underestimating them but this is not the main subject of this article.

This experience could be compared with the one that was lived by the maintenance technicians of an electrical power operator who had to give their job to agents from the subcontractors in the nineties. Some agents of this operator then became controller of the activity where previously they were the actors in charge. This change leads to a loss of know-how. According to the situation, this new posture obliges them to be either a facilitator and to share their know- how, or policemen in the field who gives sanction and who are then marginalized by the checked agent, with no way to get information. This latter strategy undermined the younger competence acquisition process.

The existence of the triptych between the developer ESA (European Space Agency), Launch System Operator ARIANESPACE and the controller CNES (French Space Agency) had an impact on the process of implementation of this new profession and it is interesting to describe the impacts.

The developer and the Operator have both obligation to respect the law but it is essential that the developer prepares conformity demonstration files during development phase, mandatory files for Launch System Operator during exploitation phase.

4. MISSION DEFINITION & NEW JOB

The new mission is to verify compliance with safety requirements to protect people, properties, environment and public health. Before this new mission definition, safety concerns were taken into account with the goal of launch success. It could have been understood as: a safe launcher is a reliable launcher! Now it should be understood as, the launch safety is more than the reliability of the launcher. Indeed, the reliability of the launcher is the key factor to demonstrate conformity to the law. It is a tool used for prevention demonstration: a safe launcher is a reliable launcher, but launch hazard analysis should integrate also protection measures.

Implementation of this mission leads to apply RAMS (Dependability) approach with quality assurance methods such as the approach ADEQUA©, a root causes analysis method [11], but also implementation of audit of the Operator's processes and operation controls directly on the ground. The implementation of this new job set two years. During this period, around 2009 & 2010, it was an opportunity to establish useful guides and procedures in order to analyse all technical subjects which could jeopardize conformity to FSOA requirements. These guides have contributed strongly to grow the maturity of this new job and posture.

Furthermore, an another important component of this mission is the implementation of a project coordination type based on "empirical approach" revolving around practical knowledge and expert knowledge to allow cross checks in independent manner by using technical skills specific to space industry.

These technical skills are: Space debris, Planetary Protection, Modeling accidental physical phenomena, EMI/EMC, Dependability, quality control, Space Mechanics, mission analysis, orbit determination, trajectory skill, mechanical structures, thermic, liquid and solid propulsion, Pyrotechnics, guidance control, Aerodynamics, Flight program, board electrical System, Floor control Systems, Operations.

For technical experts, it is really necessary to distinguish well the work of technical expertise for the benefit of the development projects and the technical expertise for the benefit of conformity to the FSOA requirements.

That is why, intervention of technical experts is governed by a code of ethics to ensure impartiality of the advices. This is functioning quite well.

To summarize, controller job is a learned mixture between the coordination project and the technical expertise. Thus, control is made thanks to the following tandem: a non-specialized engineer with a strong culture of launch system and an engineer specialist in the domain concerned by the observed gap.

In order to assess risk for launch operations, controller mission was defined with the objective to maintain a control plan for each set of launch: a kind of contract between controllers and the ministry in charge of space.

Concerning this control plan, the basic method relies on the launch system risk analysis with Launch System Danger Analysis and the environmental studies.

This control plan will be described more in details in chapter 7.

This plan contains:

- all the milestones of the process of production and implementation where the controller has to participate and contribute,
- the type of check of files,
- the controls in operation and
- the audits.

Indeed, there are 3 main different type of control: files check, technical analysis either with complementary test of not, and control in the field.

That is why, this new controller job in the frame of FSOA, is the sum of 3 different functions as follow:

- The controller, the inspector who guaranties the respect of Launch System Operator's reference table on the field and the compliance of safety regulations by Launch System Operator,
- The assessor who analyses technical facts and who specifies complementary checks. It is also him who can propose recommendations of best practice in the demonstrations of conformity and who can propose modifications of regulations; modifications to specify a requirement more precisely or according to the case of benchmark, either to harden it either to soften it.
- The auditor who makes sure of the quality and efficiency of the processes of the developer.

The new job is thus lying on the following triptych: Controller, Assessor and Auditor; practically called "controller" to summarize.

Launch System Operator is the first barrier of safety. It is necessary that the controller can be a second effective barrier. The objective is to avoid impoverishing skills of the controller who would be too much taken away from the technical assessor job who has to be close to Launch System Operator to have a good anchoring to get necessary information. The independence between Launch System Operator function and Controller mission must be then weighted because a balance must be found. It is a question of getting good level of information, adequate and fed skills and capacity of an objective control.

5. TRAINING: TWO MAIN OBJECTIVES

After several years of implementation of this new mission at Launchers Directorate the constant concern was effectiveness. Several aspects should be improved; especially the visibility, awareness and recognition of this mission to enable all interfaces in interaction with controllers to better understand this new activity.

To improve the functioning on a daily basis, it was set up feedback experience meetings between Controllers and Launch System Operators.

The objectives are the following ones:

- To handle the subjects considered as non-blocking before the last launch. This decoupling allowing to concentrate during launch campaign on the blocking subjects.
- To review what did not work and to take measures to avoid that it reproduces. This meeting has to permit discussion and exchanges on the differences of situational understandings about a risk connected to a

waiver for example.

For example, in a file concerning safety margins of the a launcher structure, it was exchanged with the Operator the necessity of refining the calculations justifying the acceptability of a waiver when the reduction of the safety margins is significantly taken away from the value having allowed to pronounce the qualification, even if this margin is still comfortably positive.

Despite these exchanges, despite a day to day explanation, it is still difficult to be known and recognized because the job is new for the Operator and not fully understood. Furthermore, inside CNES, the French space agency, this controller's new job is also difficult to be known. The already existing job which has the same safety objective was the job of Range Safety Officer.

There is easy confusion between those two, between prevention of the causes for the one, the Controller, and reduction of the effects for the other one, the Range Safety Officer.

Furthermore, it was necessary to finalize a process of qualification to officialise the operational functions of the control in launch campaign.

Therefore, it was decided to set up training with two objectives:

- To help to discover this new job to all stakeholders within the CNES, the French space agency, and
- To complete qualification process to validate the controllers to different positions needed.

To summarize, implementation of this training had a double goal: information and qualification.

6. QUALIFICATION PROCESS

To better understand the second objective of this training, it is necessary to describe the qualification process which has been formalized in several steps with integration of a continuous feedback experience.

This qualification process was chosen with academic aspects, theoretical and practical aspects (day to day aspects) with a successful marriage between European quality production aspects and implementation aspects in French Guyana.

The controller needs to go through different steps of training to get his qualification for an operational position. These steps are from different types, like:

- Academic training like functional management, or management of critical situations,
- Methodological training concerning risk analysis method, Search for causes of anomaly,
- "Home-made" training developed in chapter 7, purpose of this paper,
- Training on the ground by companionship,
- Steering board to clearly assess the acceptance of the responsibility bound to this function.

Criteria of success for the qualification of the operational functions are criteria not only technical but also organizational and relational with specific life skills named « savoir-être » in French.

It is a question of giving evidence that the controller masters the regulations, the procedures to be unwound and the various actors in interface with an open, constructive dialogue and in the objective of safety, reliability, and risks mitigation management.

7. TRAINING CONTENT

The educational objectives are:

- To assimilate the stakes in this mission of the conformity control with feedback experience.
- To identify the role of the various actors of the conformity control.

Its content is distributed in various chapters as follows:

- Regulatory and legal framework
- Knowledge of the Launch System Operators s
- CNES Conformity control
- Preliminary Conformity control
- Control Plans for each type of launch system
- Conformity control during a campaign from Europe to lift-off
- Controls of operations workflow
- Evaluation tools of the risks of "Fallout effects" and "Debris in orbit"
- Conformity control by Safety Range in French Guyana

We are going to emphasize 3 important modules of this training.

7.1 Launch System Danger Analysis

A Launch System Danger Analysis is a specific Launch System Hazard Study.

We will not go through this method and refer to a previous paper [9] for that. The goal is here to emphasize on the main message delivered during this training.

For every significant technical fact analysis, it is asked to conclude on the acceptability of the residual risks (and associated damages) concerning the feared events as follows:

- Fallback of elements on Earth,
- Controlled or uncontrolled re-entry of launcher elements placed in earth orbit
- Launch vehicle failures,
- Creation of debris in orbit,
- Collisions in orbit.

Here below, there is Fig. 2 to illustrate the feared events and their effects of an upper stage for instance:

- damages linked to fallback of elements designed to separate from the launcher;
- damages linked to controlled re-entry of launcher elements placed in earth orbit;
- damages linked to uncontrolled re-entry of launcher elements placed in earth orbit;

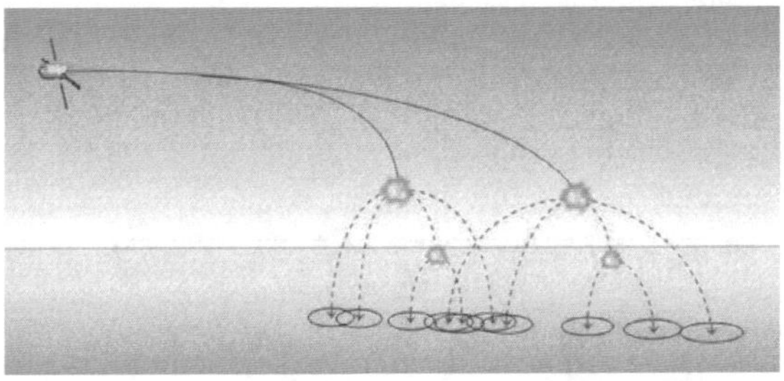

Figure 2. Degraded cases for controlled re-entry

Here below, there is Fig. 3 to illustrate the feared events and their effects of a launcher from lift off, in flight:

- damages linked to failure of the launch vehicle due to external danger sources;
- damages linked to failure of the launch vehicle due to internal danger sources;

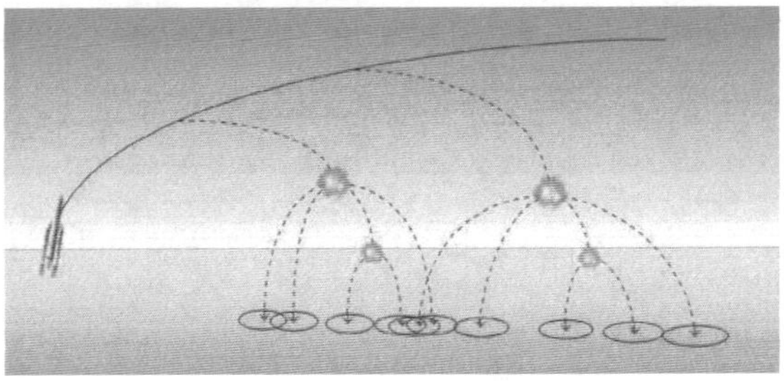

Figure 3. Degraded cases for launch phase

To complete these illustrations, the other feared events are:

- Collision with manned space objects, for which the orbital parameters are precisely known and available;
- Damages linked to explosion of a stage in orbit;
- Collision with a celestial body.

Launch System risk analysis is the reference method of work of the controller.

To analyse a technical fact, there are 2 possibilities. Either, we get rid of the causes by implementing risk reduction measures or by changing the equipment if the perimeter of the anomaly is known.

Either, we analyze the consequences of the failure of the affected product, the loss of the associated function, the effect on the stage, the impact on the launcher and finally the consequences on the Launch system; these consequences being listed in the most critical events.

To estimate the risks, it is thus necessary to manage to estimate the tandem "severity / probability" in a qualitative way most of the time to deduce from it if necessary risk reduction measures.

This approach to establish connection between the requirements of the regulations, the feared events of the Launch System Danger Analysis and the effects on Earth and in orbit is illustrated by the Fig. 4 below.

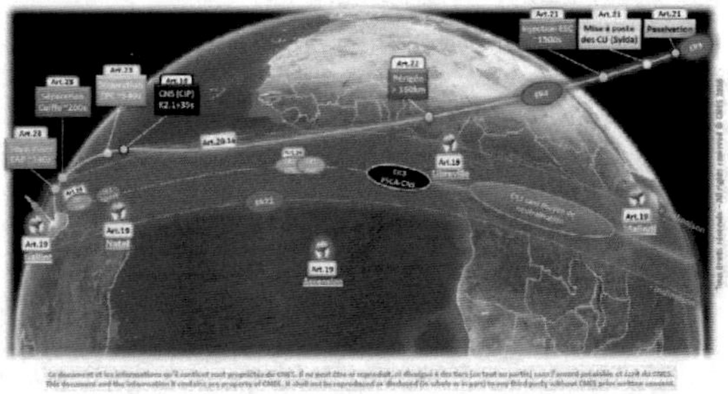

Figure 4. Connexion between requirements, launch system feared events and effects of Earth and in orbit

To build Launch System Risk analysis, a methodology tool is used, like a type of identity card for the risk. Here is below, an illustrated partial example (Fig. 5).

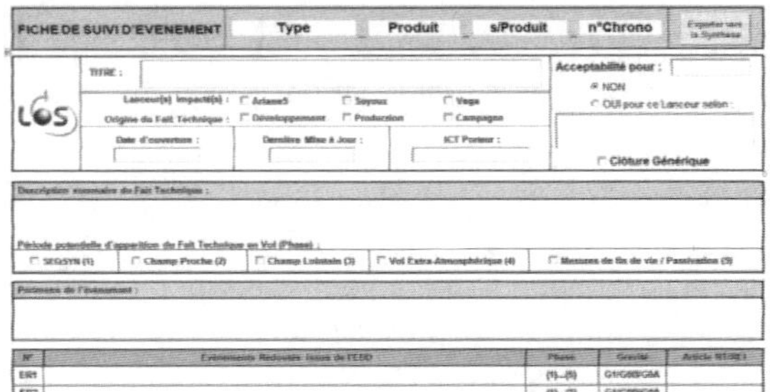

Figure 5. Methodology table to build Launch System Risk analysis

7.2 Checking Mission Analysis

In the work plan of the controller, there are two important phases:

- Control (cf. Article 16 of [1]) of production, integration and implementation, this is an invariant and recurring phase with a process linked to a long-term cycle for modifications because linked for example to obsolescence or performance improvements and ,
- Check (cf. Article 17 of [1]) of mission analysis, by definition adapted and specific in every mission asked by the satellite to the launcher and thus evolving according to a short-term cycle.

Control of production, integration and innocuousness of the implementation on the launcher covers the undergone technical facts because not planned and not wanted, as the anomalies or the alerts.

Mission analysis check deals with missionnisation file, wanted because decided for it. The check consists then in making sure that the specificities of the mission do not introduce new launch system risks and that we stay in launcher qualified domain of use.

7.3 The milestones of the control plan

The control plan describes precisely the modalities of the conformity control with technical regulations carried out by CNES for the government.

This document:

- Details the activities of the contributors to the Technical Conformity control to the everyday life with the type of intervention;
- Details the document for delivery according to the key points and the meetings in which they participate;
- Specifies the internal organization in CNES for the realization of these conformity controls as well as the interfaces between the various participants in CNES

This internal document is dedicated to CNES engineers involved in technical conformity control.

The activities of conformity control lean on the processes implemented by the Operator. So, all preparation activities of the mission, launcher manufacturing, launch system preparation, as well as cross-functional processes are identified within the control plan as:

- Mission Analysis
- Launcher production with its specific upper part,
- Operations and final preparation,
- Flight Exploitation,
- Configuration management,
- Technical facts and critical points management,
- Management, organization, risks assessment.

The control plan tailors for every milestone of every identified process, the activities in a global way without distinction of products, subsets or considered equipment's. Indeed, the levels of intervention of the controllers are modulated according to the identified risks and to the level of criticality of the event or the concerned activity. To favor the continuous-flow conformity control, Launch System Operator transmits files. These files will be cross-analyzed by the controllers who will watch to notify Launch System Operator of any risk of nonconformity. The final notice of conformity will be pronounced after examination of the final file delivered by Launch System Operator.

So, the adopted approach is naturally to tune on the quality processes of the operator. It is the least intrusive of the new control approach allowing not to destabilize the usual process well established by several decades of experience. Furthermore, it is the approach allowing not to miss any type of steps and thus to insure a strong cover of the potential risks.

In each of this step, the objective is either to get information to increase the acuteness of the analysis, or to proceed to a control by checking with regards to a technical criterion which leads to pronounce the state of conformity: compliant or not compliant.

In every step, control is made by samplings on the types of technical facts and by adopting Launch System Risk analysis approach.

So, the control plan is a subtle mixture balanced between the notion of technical evaluation and the notion of control, without forgetting the cross- functional dimension with audits. This joins well the 3 jobs described previously: Controller, Assessor and Auditor.

To confirm control plan relevancy and robustness, we led a check that every requirement of the regulations was covered by the control plan. This check concluded positively, since there is no sub-control or over control.

The adaptation of control plan is a constant concern for controllers, function of technical difficulties we could be facing, function of the maturity of the operator analyses; For example, one of the adjustment parameters is the studied technical facts sampling percentage.

8. DISCUSSION

With our long technical experience, we had to build from scratch a new job in the frame of European Space, Controller job. The result was satisfactory after nearly 3 years of experience even it is still a continuous process of improvement. That is why, this training was also profitable to the controllers to enrich the definition and the contents of this new job by building and by structuring this training. It was the first benefit of the phase of preparation of this training involving many actors of the conformity control.

This training benefits for all, provided that the leitmotiv is constantly the pedagogy, provided that the messages are clear and relevant to facilitate the understanding and the assimilation, provided that the good elements of language are used, with support of miscellaneous and modern tools without neglecting the density and quality of the technical basement.

The wealth of Fig. 4 illustrates perfectly this power of pedagogy.

This training was already provided 2 times in 2015 and 2016 within CNES. The feedback is interesting and the effects are already visible and obvious for all partners. Indeed, controllers meet fewer questioning on their role among those having attended this training course. It remains the case to those who did not follow it. It is thus, a training to be regularly provided to reach wide and new audiences.

To conclude, this feedback leads to an interesting and obvious point: this training course should be proposed to the Launch System Operators which remain the first safety barrier to mitigate launch risk.

In another frame, it was decided wisely that this training has to be given to the main European Launch System Operator in a size totally adapted to the Operator specific needs. This training is under definition at the present time in close cooperation with the main European Launch System Operator.

The specific demand of the Operator is the Launch Risk analysis method defined in a very practical and pedagogic way.

Indeed, his special need is to develop his culture of "protection" as regards accident, like crisis training to assess protection measures. Nowadays, the main obvious and practical focus point is prevention and quality to avoid accident, which is a very wise culture reliability risk oriented approach.

The objective should be to manage these two aspects of launch system risk mitigation approach: prevention culture and protection culture at launch system level.

The absence of accident in the European space world participates in the everyday acceptance of the risk which can become abstract because very rare and very far in French Guiana compared to concerns from European citizens on nuclear power plants after Fukushima (2011) and petrochemical (Seveso) plants after Toulouse (2001) or Buncefield (2005) disasters [12]. This success which we can only greet can lead to a reduction in vigilance or more exactly to a decrease of the awareness of the risk, even to the faith of risk disappearance.

This could lead in some situation that the controller has to be in the justification of its role and not in the reminder of its role due to the fact that launch system risk would not be credible any more given that the continual success can lead to forget it or tend to limit its effects. An easy deduction could lead to say that control process works well because the launch system success is the nowadays reference! But it would be a partial conclusion which also forgets the maturity of the Launch System Operator process well established in Europe and which showed its ability.

In front of this very comprehensive addiction to success, this training can then contribute to increase Launch System Operator vigilance by reminding the dangerousness of our job and a stronger societal requirement of risks mitigation. Indeed, confusion between reliability and safety is often likely to occur with more successes.

The important thing is then to share our own thoughts in the anticipation of the risk by a proactive approach thanks to the benchmark to improve continuously safety requirements and above all safety demonstrations.

9. CONCLUSION

Changes introduced by new legal frame of FSOA had significant consequences on CNES human resources. Defining Controller new mission within European space world was a fundamental task during past years. In particular this new CNES mission actually includes three complementary jobs: controller, assessor and auditor.

To go further in a continuous improvement process, it has been decided to build specific conformity training with two main goals: information and qualification.

This training focuses then on three domains: launch system danger analysis, mission analysis check and milestones of the control plan.

This change period was longer than expected because complex regarding several aspects (legality, process and technology), but it now gives quite acceptable satisfaction.

However, it is an effort of each moment. Indeed, we must demonstrate perseverance to convince maintaining and improving the level of safety risk for this major system like launch system. In Europe, fatal accident is so rare that it becomes almost abstract.

Furthermore, an important learning is especially that communication is fundamental within the tandem Controller / Operator introduced within FSOA new legal frame.

And it is thus very important to build up to itself the opportunities and the conditions of the technical dialogue during dedicated regular meeting but also not so formal meeting to share concerns in a very open way.

Following this communication effort, the training we first built for internal formation and qualification purposes is now adapted to be dispensed to Launch System Operators.

Above all, this training benefits for all, provided that the leitmotiv is constantly the pedagogy, provided that the messages are clear and relevant to facilitate understanding and assimilation, provided that good elements of language are used, with support of miscellaneous and modern tools without neglecting the density and quality of the technical basement.

This article is a testimony of our modest real-life experience of Controller new job after 5 years of regulatory change.
The feedback of this new job implementation seems interesting to share in terms of impacts of politic change, and specially in terms of management of all changes dimension: business, mission, organization, functioning. And of course the strong impact on human resources.
We hope this experience could be useful to anybody who has to face management of change in high-risk systems.

10. REFERENCES

1. FSOA: French Space Operation Act n° 2008-518 of 3rd June 2008

2. RT: Decree concerning Technical Regulation implementing decree N° 2009-643 of 9th June 2009 concerning license issued pursuant to FSOA

3. REI: Decree regulating the operation of the Guiana Space Centre facilities

4. IAC-09.E8.6.2 Advance in the implementation of the French Space Law on space operations in the launcher field (Philippe Clerc - Francois Cahuzac (CNES))

5. IAASS 2011 - Application of FSOA and development of space activities in the field of launchers (Francois Cahuzac, Arnaud Biard (CNES))

6. IAC-12, E7,5,13,x16006 Implementation of the French Space Operations Act for launchers, and contribution to the control of risks (Francois Cahuzac, Stéphane Louvel (CNES))

7. IAASS 2013 – Lessons learned from the application of the regulations associated to FSOA during the qualification and the first flights of Soyuz and Vega (Cahuzac François, Denoyers Jean-Yves (CNES))

8. IAASS 2013 – FSOA and associated regulations applied to future European launchers A5ME and Ariane6 (Francois Cahuzac, Arnaud Biard (CNES))

9. IAASS 2013 – Launch System Hazard Study, Methodology and Application with 3 European Launchers (Frederique Meyer-Lassalle, Guillaume de Blanchard, Caroline Aussilhou (CNES))

10. Méthode ADEQUA © - Groupe de Méthodologie Appliquée (GMA ©) - contact@adequa.info

11. 4th ESReDA Seminar - Risk Analysis and Management Across Industries - Managing sociotechnical changes: hard lessons learned from accidents and further challenges (Nicolas Dechy, Jean-Marie Rousseau (Institut de Radioprotection et de Sûreté Nucléaire (IRSN)), Michel Llory (senior researcher), France)

12. SPACE ACCESS 2011 - Safer access to space and hard lessons from soft sciences: organisational failures and challenges learned from space and technological disasters - John Kingston, The Noordwijk Risk Initiative Foundation, Yves Dien, EDF and Nicolas Dechy, IRSN, Direction de la Sûreté des Réacteurs, Service d'Etude des Facteurs Humains.

PROCEEDINGS OF THE 8TH INTERNATIONAL SPACE SAFETY CONFERENCE

SESSION 16 REGULATIONS & STANDARDS - II

NEED FOR SPACE GOVERNANCE FOR INDIA AND GLOBAL SPACE GOVERNANCE

Dr. Sanat Kaul

1. WHY STATE POLICY

Global space economy consisting of activities of governments and private companies grew to $314 billion in 2013. Private-sector commercial space activity is now growing at a faster pace, while governmental activity is slowing down. Between 2012-2013, commercial space products and services revenue grew 7%; commercial infrastructure and support industries grew by nearly 5%; while government spending decreased by almost 2%. Space investment is a major part of the infrastructure of communications – both telecommunications and broadcast – of weather and geological monitoring, and defense. Thus, commercial development of outer space is outpacing governmental activities in space. As private firms launch commercial space activities, the legal obligations and liability re of space-faring States proliferate as well

With the growth of number of States in space activities, many nations have enacted their owner space legislations. Many are enacting national space legislation, establishing governmental space regulatory institutions and giving them jurisdiction to license private actors and ensure compliance with regulatory requirements. In order to fulfil their international obligations based on space treaties ratified by them and to protect the citizens from harm, to protect their treasuries from liability, and to encourage and foster the development of commercial space activities, many states have their own rules and regulations on space issues generally harmonious with international treaties. This can be interpreted as filing a void in absence of an international regulatory regime addressing safety and navigation of aerospace vehicles. Though a number of commentators (including the present one) have urged the International Civil Aviation Organization [ICAO][1] to regulate the safety and navigation of aerospace vehicles, to date, this has not been done. Moreover, the world community has failed to draft a single multilateral treaty addressing space issues since 1979. ICAO and UNOOSA held for the first time a joint meeting in 2015 at the ICAO headquarters in Montreal to consider the issues of management of aerospace. This was the first step towards getting ICAO involved in issues of aerospace travel. There is a need for framing of Standard and Recommended Practices (SARPs) like ICAO is doing for aviation.

Five multilateral conventions, drafted in a dozen years, place numerous obligations upon States. Further, Space Law consists of a growing number of international multilateral and bilateral agreements and conventions, U.N. resolutions, decrees by international organizations, national legislation and regulations, and court decisions. These require States to adhere to principles of international law, assume responsibility and liability for activities in space (whether governmental or non-governmental), authorize and supervise the activities of their nationals in space, and notify and register their space objects. Among requirements imposed by the Outer Space Treaty of 1967 are the following:

- States must carry on space activities in accordance with principles of international law;
- States bear international responsibility for national activities in space and on the moon and celestial bodies, including activities of both governmental and non-governmental entities;
- States must authorize and supervise the activities of its nationals in space;
- States that (a) launch, (b) procure the launch, or (c) from whose territory or facility an object is launched, are internationally liable for damage to another or its national or juridical persons by such object in the air or in space;
- States on whose registry an object is launched must retain jurisdiction and control over the object and any personnel thereon;
- States must avoid harmful contamination and adverse environmental consequences from the introduction of extraterrestrial matter; if it believes an activity or experiment by it or its nationals in space would potentially harm or interfere with activities of other States in space, it must consult with such States before proceeding; and
- States must inform the UN Secretary General of the "nature, conduct, locations and results" of its activities in space.

Several of these provisions also are elaborated upon by the Liability Convention of 1972. Building on Article VII of the Outer Space Treaty, the Liability Convention imposes liability upon a launching State (i.e., the State that launches, procures the launch, or from whose territory or facility a space object is launched) to pay compensation for personal injury and property damage caused by its space objects on the surface of the Earth, or to aircraft. The Convention establishes a two- tier liability regime, providing that the "launching State" is absolutely liable for damage caused by its space objects on the surface of the Earth or to an aircraft in flight, and liable in negligence for damage caused to a space object of another State or to persons or property on board. Where there is more than one launching State, they shall be jointly and severally liable for the damage they cause. Hence, by ratifying or acceding to either the Outer Space Treaty of 1967, or the Liability Convention of 1972, the launching or launch- procuring State becomes potentially liable for damages caused by itself and its commercial launch sector. A ratifying State accepts absolute liability for damage on the ground or to aircraft in flight outside its territory when a launch takes place from its territory or facilities, or when it procures a launch from another State. A State incurs fault-based liability for damage caused in outer space. In addition to these multilateral conventions, additional legal obligations are imposed upon States through Customary international law, an array of United Nations Security Council and General Assembly Resolutions, and a growing body of "soft law."

Further, the role of ICAO is to harmonize State regulation of aircraft safety and navigation in – may apply to vehicles transporting space objects through air space. But to date, ICAO has promulgated no Standards and Recommended Practices governing aerospace vehicles or rockets, though in time, it may.

Considering the gap created by international treaties on space without proper international space legislation as done by ICAO for air, the international obligations and the liability exposure created thereby, as well as a desire to protect the health and safety of their citizens, their property and the environment, a growing number of States have promulgated national legislation to regulate commercial space activities. As one source notes, "Since a government can only act on the basis of laws or respective regulations, the establishment of national space laws is the most effective way of providing the State with the means to authorize and supervise non-governmental space activities." At least twenty-six States – about 14% of the members of the United Nations - regulate space activities. Among the States that have enacted national space legislation are Algeria, Argentina, Australia, Austria, Belgium, Brazil, Canada, Chile, the People's Republic of China [PRC], Colombia, France, Germany, Italy, Japan, Kazakhstan, Netherlands, Nigeria, Norway, Russian Federation, South Africa, the Republic of Korea [South Korea], Spain, Sweden, Ukraine, United Kingdom, United States, and Venezuela. Hong Kong also regulates space activities.

The United Nations Committee on the Peaceful Use of Outer Space [COPUOS] recommends that, "Space activities should require authorization by a competent national authority; the authorities and procedures, as well as the conditions for granting, modifying, suspending and revoking the authorization should be set out clearly to establish a predictable and reliable regulatory framework. ... The conditions for authorization should be consistent with the international obligations and commitments of States, in particular under the United Nations treaties on outer space..." Governmental oversight of space activities is essential to protect public safety, property, and the environment, and to fulfill State obligations under international law. Licensing becomes the bedrock of governmental regulation of commercial space activities.

2. INDIA: BACKGROUND TO SPACE ACTIVITIES[2]

India established Indian Space Research Organization (ISRO) in 1969. With self-reliance as a goal, the space programme was based on the premise of 'end to end' system concept. This premise enabled the Indian Space Programme to overcome numerous hurdles and challenges in leaning and experimenting with new technologies.

The Indian Space programme was started not by addressing India's military requirement but rather developmental needs. Satellite based instructional television was the beginning. The first was the Satellite Instructional Television Experiment (SITE) with telecommunication satellites. By 1970s, India had procured its own communication satellite 'Aryabhata', a 360 kg satellite named after an Indian astronomer. It was launched by Russian launch vehicle. India by then decided to have its indigenously designed space worthy satellite, conduct a series of complex operations in space, set up ground based receiving, transmitting and tracking stations, and also establish an infrastructure for fabrication of spacecraft systems.

The second phase of grand plan for indigenous 'end to end' system for development and manufacture started with satellite Aryabhata, led to Bhaskara series-India's first experimental geo-stationary satellite. Thereafter, came the development of Launch Vehicles such as SLV-3 and ASLV. By the late 1970s, India had already built a considerable infrastructure of laboratories and facilities and also initiated a 3-pronged programme- INSAT (Indian National Satellite) programme for communication satellites; IRS (Indian Remote Sensing Satellite) for remote sensing; and launch vehicle programmes for launching of Polar and Geo-synchronous satellites.

In the third phase starting in the year 2000, India had achieved the basic technological maturity required for developing a space system. However, India also announced around this time its augmentation satellite to compliment GPS in GNSS and further announced seven satellite based Indian Regional Navigation System known as IRNSS. This mini-constellation would be operational would be end of April 2016. It also started planning for an unmanned mission to Moon with Chandrayaan I, which was successful in 2008-09. It is now planning to go in for reusable launch vehicle.

Later in 2011, ISRO took upon a new challenge of a foray to Mars in the 2013. The Mars Orbiter Mission (MOM) was successfully launched in November, 2013 and has already completed more than 90% of its traverse to Mars. MOM has entered mars orbit and has started its experiments of imaging and measurements. But more significantly, MOM would establish the fact that India can successfully undertake long-duration planetary missions and has gained tremendous experience in this regard.

So far, the achievements of Indian space policies have been mainly due to government funding with ISRO as its implementer. The achievements can be summed up as:

- A full space infrastructure has been built up over half a century.
- Approval for 200 missions has been accorded by Indian government but 125 Missions have been accomplished – out of which 111 missions have been successful.
- Leading satellite capability that covers a wide variety of applications satellites. These can be grouped in five categories
  - i) satellite communications,
  - ii) Broadcasting satellites,
  - iii) Remote Sensing (Earth Observation),
  - iv) Location and Timing Satellites,
  - v) Metrological Satellites.
- Use of INSAT communications systems have resulted in the wide outreach of TV signals (from early 1980s onwards) to almost whole of the country. In addition, growth of large-scale DTH and VSAT data communication business has been achieved.
- The availability of low- priced and easily available IRS images (from about 20 IRS missions) has given a great thrust to use of images and geographical information techniques into many governance and national building activities, including disaster management.
- Weather and ocean services modeling have derived a great boost from the availability of INSAT and Oceansat images/data on a variety of ocean and atmospheric data – thus consolidating the scientific services of meteorological department and Earth Sciences.
- Forays in planetary missions have been made through Chandrayaan-1 and MOM-1 to establish the technological capability of Indian space to undertake far-reaching planetary exploration and also undertaking advanced scientific studies.
- Unique missions for astronomical observations – Astrosat and operational Positioning Services – through Indian Regional Navigational Satellite System (IRNSS) constellation of seven satellites already in orbit. Astrosat has multiple imaging methods: ultraviolet, soft x-ray and hard x-rays. This will involve expansion of space applications for ISRO.
- The next lunar mission Chandrayaan II by ISRO is being planned for which ISRO is developing a Rover, a lander and orbiter.
- ISRO has another major mission in mind: Aditya I, a satellite to look at the Sun.
- Global commercial operations of Indian space includes 43 commercial/foreign satellites launches; sale of IRS images and value- addition services and lastly more lucratively, transponder lease business in India. This has estimated to have resulted in revenue earnings of about INR 50 billion over the past 20+ years.

3. COMMERCIALIZATION OF SPACE APPS IN INDIA: ANTRIX CORPORATION 1992

By 1992, the government had realized that the potential of Commercial space activities and set up a wholly owned company called Antrix Corporation in that year to market Indian space capabilities developed by ISRO globally. Thus Indian Government entered the commercial space market in 1992 through a wholly owned company Antrix Corporation which is still attached to ISRO.

Antrix Corporation has access to specialized design, test and manufacturing facilities of ISRO and its staff of 16000 experienced scientists and engineers with in depth knowledge of the crucial parameters of the space industry. Antrix Corporation has focused on enhancing functionality and value to the customer. Antrix offers satellite systems and sub-systems with proven performance in space and incorporating advanced features of power & reliability and long life. The earth observation data services powered by a state of the art constellation of Indian remote sensing satellites (IRS) and a network of ground stations provide a valuable and assured resource for business to customers worldwide. Antrix facilitates the utilization of Indian Space assets in the field of telecommunications & broadcasting for a variety of services including TV feed,

DTH, VSAT, mobile communications and socially relevant services such as Telemedicine and Tele-education. A track record of successful space launches in a variety of earth orbits through India's PSLV and GSLV launch vehicles has made Antrix a choice for many customers.

Antrix Corporation has already achieved: -

- Successful supply of reliable satellite systems and sub-systems. Some of Antrix's better known customers are Hughes, Matra Marconi, World Space etc.
- Successful Commercial Satellite
- Execution of many support services to International Space Agencies. Some of the customers using Antrix services are World Space PANAMSAT, GE Americom, AFRISTAT.
- Successful launch of TecSar(Israel).
- Two satellites; one from France and another from Japan were launched in September 2012.
- Successful launch of five satellites, including French SPOT 7 satellite on 30 June 2014

Business Agreements

On 29 January 2014, Antrix Corporation Limited (Antrix), signed Launch Services Agreement with DMC International Imaging, the wholly owned subsidiary of Surrey Satellite Technology Limited (SSTL), United Kingdom (UK), for launch of three DMC-3 Earth Observation Satellites being built by SSTL, on-board ISRO's Polar Satellite Launch Vehicle (PSLV). On 5 February 2014, Antrix signed another Launch Services Agreement with ST Electronics (Satcom & Sensor Systems) Pte Ltd, Singapore, for launch of TeLEOS-1 Earth Observation Satellite, on-board PSLV. On September 29, 2014, Canada announced award of contract for launch of M3M (Maritime Monitoring and Messaging Micro- Satellite) communications satellite to Antrix at International Astronautical Congress at Toronto.

4. SIZE OF INDIAN COMMERCIAL SPACE INDUSTRY: NEED FOR PRIVATE SECTOR INITIATIVE

It is estimated that while the global space economy is worth \$314.17 billion which includes \$226 billion is commercial, the Indian Space industry is, however, worth only \$1.1 billion of which private commercial space economy consisting mainly of Space SME sector is only \$48 million. It is felt that Antrix Corporation, as a commercial arm of ISRO, will not be flexible enough to compete in the world space market. The need for small, lean, young, flexible organization is felt.

Some of the growth areas of Space based activities in India are disaster and flood management, Tele-medicine, Tele-education, vehicle tracking, women safety, management of land issues, remote sensing for minerals, urban development and creation of smart cities.

India has issued two policy documents so far: Satcom Policy in 1999 and Remote Sensing Policy in 2001 and 2011.

5. ANALYSIS OF SATCOM POLICY (1999) AND REMOTE SENSING DATA POLICY (2001 & 2011)

a) While stating the usual goals, the Satcom Policy of 1999 also mentioned that it would encourage and promote privatization of satellite communications in India by way of encouraging private sector investment in space industry and also attracting foreign investments. This proved to be true.

It also authorized leasing of capacity of Indian National satellites (INSAT) to non- government (Indian and foreign) parties on commercial terms. This was essential so that commercial services could ride on INSAT – envisaged to spur the DTH and VSAT market in India to a large extent. It also allowed Indian parties to provide services including TV uplinking through Indian satellites which helped opening up of a large variety of TV channels in India. It also opened up the operations from Indian soil using foreign satellites under certain conditions.

Further the Satcom Policy proved to be a great boost for DTH business, Vsat services, connectivity for education outreach across the country, reliable telemedicine connectivity, increased capacity leasing and a great growth in Indian ground equipment manufacturing. It was also envisaged that this policy should help position JVs for communication satellite ventures and also bring in a variety of new value-added services, especially for mobile phones.

The Satcom Policy gave preference to Indian Satellite Systems (ISS) in giving service licenses – thereby ensuring "protective cover" for INSAT for Indian services against global commercial systems. However, the Satcom Policy did not in any way prohibit the use of foreign satellite systems - which, after a due process, can be treated on par with ISS for service licensing in India. This has not happened mainly due to deficiencies of appropriate procedures in implementation. Further, it was decided to channelize all lead foreign satellites through ISRO/Antrix which brought in major bureaucratic hurdles for private persons.

The impact of SATCOM policy has been very positive. On top of this, announcement of "Digital India Policy" in 2014 has further added impetus. This has also led to e- commerce startups offering services of online shopping, to locating taxis, restaurants etc. Use of SMS alerts by banks, airlines has become common. Use of SMS to inform value of shares and other financial services are gaining momentum. But more important is the use of GIS in integrating information relating to land ownership and physical features (water bodies, roads, forests etc.). These can be super imposed on basic top sheets as has been done by the Government of Gujarat. The state government has also enacted a law for institutional framework for use of GIS. E- governance has now been made a policy by Government of India leading to growth of e-kiosks in small towns and villages. In the recent budget, funds have been allocated to lay out a National Optical Fibre Network Programme of 75000 kms to connect 250- 1000 villages. This is a part of the 'Digital India' strategy.

b) The Remote Sensing Data Policy (RSDP) 2001 & 2011 lays down rules for regulations for acquisition, dissemination of satellite images in India – earlier, RSDP-2001 and now RSDP-2011 governs how satellite images are to be acquired and distributed – allowing up to 1m images to be openly dissemination to users. The RSDP embeds the concept of "regulation" to address the dissemination for 1m images[3].

   The RSDP-2001 provided the basic "framework" for a comprehensive imaging policy. Remote sensing was identified as a "public good". The RSDP introduced the concept of "one- window" access to any image (Indian or foreign satellite) – which today appears to be against "free market" concept. Concept that RSDP- 2001 brought in was mainly to prevent access to 5.8m images that could become a security concern. The RSDP also required foreign satellite images to be routed through the national agency – National Remote Sensing Centre (then Agency), NRSC. Thus, even though the RSDP, 2001 clearly emanated from the competitive challenge of US 1m images against the Indian 5.8m IRS system in the Indian market – it was certainly a protective regime for IRS till it could also match with commercial 1m image availability from other sources.

   By 2005-06, India also launched 2.5m and 1m images but by then the larger proliferation of 1m images from US commercial satellites had also happened. Thus, the 5.8m thresh-hold of RSDP-2001 as "regime for non- discriminatory access" became irrelevant.

   With NRSC the "sole agency" for distributing images, it had become further monopolistic as it adopts IRS-centricity and pushed 2.5m and limited 1m images – thereby denying Indian users 0.3m level images for national development. At the same time, India was unable to match the resolution quality of US commercial systems (that have reached 0.3m level in global market) and had planned for a 0.5m imaging IRS in 2017 time- frame. Because US 1m images became widely popular in India as against 5.8m/2.5m images and very limited 1m images from IRS systems RSDP 2001 was a failure. Consequently, in RSDP-2011 the bar was lowered for "non- discriminatory access" to 1m while fully retaining all other aspects of RSDP-2001.

6. SHORTFALL OF ANTRIX CORPORATION

While ISRO together with Antrix has a monopoly for the trading of satellite bandwidth in India, less than 50 percent of commercial satellite demand in India is served today by Indian satellites, with the majority being provided by foreign satellite operators such as SES, Intelsat and AsiaSat through subleasing capacity to ISRO. ISRO in turn further leases it to Indian private companies while making for themselves undue profits. This situation is the result of a disparity between the lack of capacity available from ISRO and the strong growth in demand in recent years, driven by DTH TV broadcasting and monopoly rights with ISRO.

7. UPGRADING SATELLITE MANUFACTURING & LAUNCHING

The successful launch of GSLV- D05 using an indigenously developed in India a cryogenic engine in January, 2015 was a critical milestone for ISRO's communication satellite program. Routine operation of GSLV Mark 2 will help ISRO in achieving self- reliance for launching satellites now in the pipeline for Direct to Home television (DTH TV) and broadband communications. The development of the GSLV Mark 3 with its capacity of 4,000 to 5,000 kilograms should accelerate in the coming years after success with the indigenous cryo- stage, which has been a bottleneck in ISRO's geostationary launcher development program over the past 20 years. In future, therefore, India should be able to launch heavier satellites[4].

8. FUTURE OF COMMERCIAL SPACE APPS IN INDIA

Euroconsult has predicted in its recent study "India Satcom Markets 2014," an average growth of 8 percent per year in commercial bandwidth demand driven by DTH satellite pay-TV platforms, cable television, cellular backhaul and enterprise and government VSAT, or very small aperture terminal, communications networks has been predicted. To meet growing commercial demand, ISRO has not been able to provide enough satellites and instead has taken foreign satellites on lease for the past ten years especially for Ku- band. Although it considers those leases as gap-fillers until sufficient domestic capacity becomes available, it is highly likely that ISRO's dependence on foreign operators will continue, as it has not been able to launch more than one communication satellite per year in the past. Also, ISRO, being a government body has to reserve capacity for government users such as the Department of Telecommunications, All India Radio, Prasar Bharti and the military that account for more than one-third of its current bandwidth supply, therefore limiting the capacity available for commercial users. There is, therefore, a definite need for government to open satellite development and launch to private industries which is so far an ISRO preserve and thereby give ISRO competition on equal footing. As of today, ISRO has vendor base in private sector only.

Regulation still limits market growth, as all Ku-band leases have to go through ISRO/Antrix in a process that is often lengthy and cumbersome for operators, including strict price regulations, royalty fees and service-level license regulations by the Department of Telecommunications and the Ministry of Information and Broadcasting. Some of these regulatory restrictions are likely to be revised in the country's new satcom policy expected to be released shortly[5]. Expectations are high from satellite operators, equipment vendors and service providers alike.

9. ISRO TO SCALE UP OUTSOURCING TO PRIVATE INDUSTRIES

In a recent statement, a former of Chairman of Indian Space Research Organisation (ISRO) had stated that India would significantly scale up outsourcing to industries to meet the shortfall in demand and to fuel the quantum jump in the programmes being undertaken by it. He had also proposed to have a risk-sharing model, as more than 500 industries (micro, small, medium and large) already account for 60 percent production of the space agency's programme and their share would further go up. ISRO is witnessing a quantum jump in the production of rockets (Polar Satellite Launch Vehicle) and satellites, in the last two-three years.

At present, ISRO's suppliers are located in different parts of the country, and the idea now is to cut down the turnaround time and get the products on time. The ISRO has proposed to set up a specialized industrial park near the Sriharikota spaceport on the Andhra coast so that its industrial partners can set up their production there.

ISRO has also mooted an idea for industrial partners that they can work in consortium mode if they wish — like coming together of players in the field of electronics, production, metals and precision fabrication, among others. In spite of all said above, India still does not have a space policy or a space legislation.

10. NEED FOR GREATER PRIVATE INITIATIVE IN SPACE ACTIVITIES

Shortcomings in the Indian Space Ecosystem

The space ecosystem of India is driven by the Department of with upstream, downstream and commercialisation of space assets all done by the government institutions with limited support in supplies of products and service by the space industry. As Department of Space also acts as the regulator for all sectors of the industry, there exists a conflict of interest for it to promote private space industry. This, combined with industry entry barriers like high cost, high risk, lack of regulations, and sometimes long gestation periods, discourages the industries that supply to the space activities currently to be able to reach the turnkey solution provider in space capabilities and launch systems, and their ability to cater to international requirements. Antrix Corporation has the mandate to commercialise products and services including remote sensing data, launch contracts for foreign satellites, development of turnkey spacecraft, and others. Yet, given that Antrix and ISRO overlap–the former tends to function more like a quasi- government organisation rather than a private one that competes effectively in the international market. As private sector space enterprises grow, Antrix Corporation's powers becomes an issue to be studies for opening of the sector.

Capacity in Capturing the International Demand

With the extremely busy schedule of ISRO in the current five-year plan 12 with 58 missions, the ability of Antrix Corporation to leverage on the facilities and infrastructure developed in ISRO for AIT of satellites, manpower in consultancy will remain limited. The last of such (and the only instance of) satellite development taken up by ISRO via its commercial wing Antrix has been the satellites manufactured and shipped for two major European companies, W2M for Paris-based Eutelsat Communications (sub- contracted to India from European Aeronautic Defense and Space Company (EADS) subsidiary Astrium) and for the 13 14UK-based Avanti Screen Media. The failure of the W2M satellite had an effect on the international market sentiment for development of communication satellites by India.

However, it may be added as a background that ISRO has launched about 25 foreign satellites all with its workhorse rocket launcher PVSL without a single failure. It may be further added that PVSL has been continuously upgraded is now a very versatile vehicle which can do much more. GVSL, on the other hand, is using cryogenic engine and is using tested components of PVSL.

11. BOOSTING PRIVATE SECTOR IN SPACE INDUSTRY: NEED FOR SPACE POLICY AND SPACE LAW IN INDIA

As the geospatial market is growing and is around $315 billion, India has a lot of potential and capability to export is space products. Over 500 small scale industries provide a supplier base to ISRO which help make satellite components. As a result, ISRO, with its low development costs, can build a package which is nearly unbeatable in the market. ISRO's involvement of small and even big industries in India has already provided a solid technological foundation and eco- environment for private industry. However, the need now is to get beyond ISRO and Antrix corporation in India and let private sector get into commercial space activities on their own without the umbrella of Antrix corporation.

Government investments and polices with reference to space

The maturity and size of the space industry is heavily dependent on government investment and policies. The largest share of the global commercial space industry belongs to the United States due to higher investments by government and a well-defined space policy that promotes industrial participation. Futron's Space Competitiveness Report (2014) found that not only has the US been successful in leading with globally reputed space companies like Boeing, Lockheed Martin and SpaceX, but it is also leading in creating human capital. Much of this success can be attributed to its large investments in defence space, providing these companies to scale up their operations without largely affecting the cash flow.

12. INDIA'S OPPORTUNITY IN SPACE SECTOR

India's space budget has increased in size and is now one of the largest in the world; however, the lack of an active space industry at turnkey level might have an immense opportunity cost for India in manufacturing satellites and launch vehicles to service the global market. Further, not a single Indian company among the top space companies in the world and this needs to be addressed urgently through a policy push under the several grand schemes announced by the current government, such as 'Make in India' and 'Digital India' by declaring a space policy followed by space legislation. As India ramps up its space defence capabilities, the lack of a mature space industrial base will potentially hurt its ambitions. India counts itself among the top nations in the world in terms of government space investment, but is far behind when it comes to creating successful private industry that is globally reputed.

India's private sector has already proved its metal in software, automobile sectors etc. but has stayed away from space sector due to policy restrictions. The success of a privatized space industry will enhance capacities within the country and complement the government driven programme, which has been historically proven in advanced spacefaring countries such as the US. Capacity building in the private industry at a turnkey level for both upstream and downstream shall assist the economic development of the country by keeping up to the pace of requirement of the marketplace (e.g. Direct-to-Home TV, Broadband Internet), while reducing the inherent dependence on foreign assets. Only one among the seven DTH providers is leasing transponder from the INSAT system from Indian Satellite System and the primary reason for this disparity is the slow pace at which ISRO has added satellite transponders to the commercial market. The net effect is that the DTH providers are incurring higher transponder costs on foreign satellites when INSAT could have been an equally reliable, and more cost efficient, alternative.

India's cost advantage is well established. India's Moon and Mars missions have shown how India can do it cheaply. India has already complete one year of Mars orbital mission known as Mangalyaan which arrived at the red planet on 23rd September 2014 and its cost was around $73 million. Earlier India's Moon mission called Chandrayaan also completed its objective successfully and a second Moon mission has been planned for 2018 again at a very low cost. This cost advantage that India has now needs to be leveraged to enter space industry.

13. REFERENCES

1. ICAO was creation of the Chicago Convention, 1944. This Convention gave full details and powers of ICAO.

2. India's three leading Space Scientist Dr. Mukund Kaduswami Rao, Prof Sridhar Murthy and Dr. VS Ramamurthy have recently published a paper in which they have divided the Indian Space industry development in three phases: Future of Indian Space - Renewing Policy Dimensions, IAC14.E3.2.7 www.nias.res.in

3. A detailed analysis can be found in a paper "Perspective for a national GI policy" by Mukund Rao, KR Sridhar Murthy, produced for National Institute of Advance Studies Bangalore, available at www.nias.res.in

4. For full details of ISRO future activities, see "What Next for ISRO": the Economic Times, 19th Feb. 2015

5. India is still working on The Indian Telegraph Act of 1885, The Indian Wireless Telegraphy Act, 1933 and the Cable Television Networks (Regulation) Act, 1995. There is a dire for a new legislation on the subject.

POLICY CHALLENGES RELATED TO NANOSATELLITES

Matteo Emanuelli(1), Blake James Edwards(1), Matt Driedger(1), Justin Atchison(1), Jordan Sotudeh(1), Gabriel Lapilli(1), Maria Grulich(1), Laura Bettiol(1), Caroline Thro(1), Eren Gorur(1), Leehandi De Witt(1), Alon Davidi(1), Suman Gautam(1), Sirisha Bandla(1), Juan Gramajo(1), Milan Mijovic(1), Laura León Perez(1), Chantelle Dubois(1), Emmanuelle David(1), Meidad Pariente(2), Chris Johnson(3)

(1) Space Generation Advisory Council, Austria;
(2) Spacecialist, Israel;
(3) Secure World Foundation, USA

ABSTRACT

The emergence of nanosatellites, brought on by the miniaturisation of systems and the considerable decrease in development and launch costs, has allowed new actors such as start-ups, universities, NGO, and even amateurs supported by crowdfunding campaigns to launch and operate their own small satellites into space.

In October 2015, young professionals and students from around the world met in Jerusalem, Israel, for the annual Space Generation Congress (SGC). During SGC, amongst others, the Policy Working Group, supported by Secure World Foundation (SWF) was formed, with a focus on the Policy challenges related to nanosatellites. The group discussed and debated the implications of the current regulations and guidelines related to the growth of the nanosatellite market. Recommendations were made in consideration of the stakeholders with the intent of promoting a path for sustainable, cooperative use of space between existing satellite operators and new nanosatellite operators. This paper focuses on a detailed definition of the above-mentioned recommendations and best practices listed above to provide beneficial though non-binding best-practice for nanosatellite operators.

1. INTRODUCTION

Nanosatellites are part of a rapidly growing class of resident-space-objects. Specifically, nanosatellites have greatly reduced the traditional barriers to space access. As a result, it is not uncommon for high school students, start-up businesses, and even private amateurs to develop and launch simple nanosatellites. For example, the QB50 project is going to launch to launch with different means CubeSats built by university teams from all over the world in 2016[1]. There are other similar programs, such as the ESA "Fly Your Satellite!" program[2] and NASA's CubeSat Launch Initiative that aims to launch 50 CubeSats from 50 states in America in 5 years[3].

The ease at which nanosatellites can be developed offers a significant opportunity to promote innovation and increase the technology base of developing countries. It is essential that educational initiatives and small commercial ventures be given access to space through a reasonable framework of regulations. With sufficient access to space, estimates suggest that 3000 nano/microsatellites will be launched by 2022[4].

Figure 1 Nanosatellites released by the International Space Station (Credits: NASA)

2. DEFINITION

There is a variety of definitions of nanosatellites; but first of all, a nanosatellite is a satellite, meaning a 'space object'. According to the Liability Convention of 1972[5] : "The term 'space object' includes component parts of a space object as well as its launch vehicle and parts thereof"[6]. According to Bin Cheng, a space object is a manmade object that is launched or intended to be launched into outer space[7].

More precisely to scope of this paper, the definition adopted herein is that a nanosatellite is any artificial satellite with a mass less than or equal to 50 kg. At present, this relatively small size introduces challenges for long duration missions and for space catalogue maintenance and tracking. That is, it is not currently common for nanosatellites to have mission lifetimes greater than 3-5 years. Moreover, their small length scales imply small visual magnitudes or radar-cross- sections, making tracking nanosatellites difficult.

3. CURRENT PRACTICES AND POLICIES

Mass and dimension differences notwithstanding, nanosatellites are currently classified the same way as any other satellite. Simply put, "a satellite is a satellite".
They operate in the same orbital regions, share the same Radio Frequency (RF) spectrum, and they are expected to abide by the same national and international policies and laws[8]. This includes the laws and policies of 22 nations, as well as guidelines and recommendations from the European Code of Conduct (the latest UN draft of an International version is from 2014), the Inter-Agency Space Debris Coordination Committee (IADC), the International Telecommunications Union (ITU), and the UN Committee on the Peaceful Uses of Outer Space (UNCOPUOS). Most of these policies recommend deorbiting satellites within 25 years from the end of their mission, making no exception for nanosatellites – or any category of satellite. As recently as July 2015, it was reported that one fifth of CubeSats launched between 2003 and 2014 do not meet this guideline[9].

Nanosatellites, being designed and built economically - often from commercial off-the-shelf (COTS) components - do not have the redundancy or reliability of larger satellites. Consequently, nanosatellite missions are often designed to be completed within a short amount of time, rarely taking more than 3 years. Nanosatellites are typically launched as secondary payloads, and as such have little control over their orbital parameters, including elevation. As such, nanosatellites are often launched into orbits in which their operational lifetime makes up only a small fraction of their total orbital lifetime. Deorbiting is often not a priority, or even a feasibility for smaller nanosatellites. This is because deorbiting systems require a fairly large space allocation to provide a meaningful decrease in orbital lifetime, in addition to active attitude determination and control (ADC) systems. Nanosatellites are often made to form factors which are too small to fit these systems. Due to the lack of deorbiting systems, a short-lived (6 months – 3 years) satellite may remain in space for a long period of time (10+ years), posing risks to other satellite operators, who must actively avoid the inoperable nanosatellite.

Figure 2 Satellites integrated on the Dnepr vehicle prior to launch at 750 km altitude in June 2004 (Credits: GAUSS)

Satellite communications between the satellites, ground stations, and users on Earth, use radio frequencies to function. These frequencies are regulated ITU, a specialised United Nations agency. The ITU designates which spectrum bands may be used by various technologies and applications and which users are allowed to use specific frequencies.

For satellite communications, the application time for receiving one of these frequencies is fairly long, as it is essential that satellites are not rendered inoperable through interferences from other satellites. Satellites must share a limited number of frequencies, despite an increasing number of actors coming into the market.

4. MARKET PROJECTIONS

SpaceX, Facebook, and other new players are expected to enter, or expand their presence in, the low-Earth Orbit (LEO) environment, with over 2000 small satellites projected to be launched by 2020[10]. Although small satellite is a broader category than nanosatellites, it must be considered that these news spacecraft will end up in the same region as currently used by nanosatellites, increasing the spacecraft density in LEO.

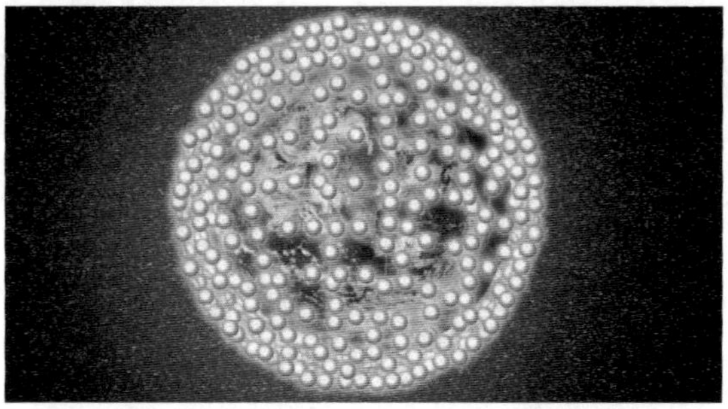

Figure 3 Many NewSpace companies, including SpaceX, Planet Labs and OneWeb, plan to create a mega-constellation of hundreds and sometimes thousands of smallsats into low Earth orbit. (Credits: SpaceNews/ Lance H. Marburger)

The commercial sector has begun to drastically increase its presence in the nanosatellite market, with 56% of nanosatellite launches expected to be commercial by 2016. Another shift in the nanosatellite market has been mission type, as by 2016 Earth observation and remote sensing missions will have increased from 12% to 52% of all launches[11]. These new players will place increased burdens on the spectrum bands and increase the risk of orbital debris.

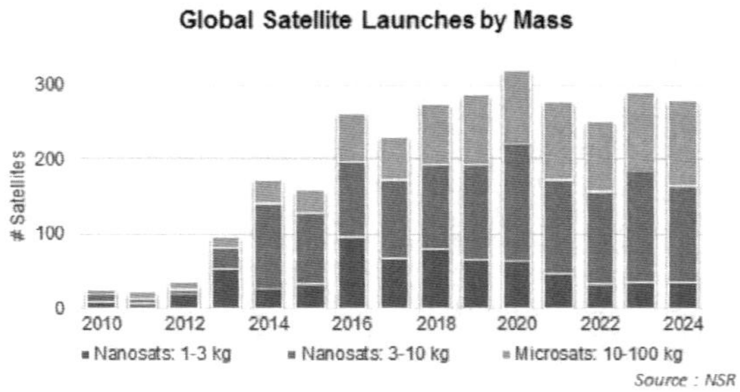

Figure 4 Global Satellite Launches by Mass[12]

Given that space is a shared resource, one might ask if these numerous, nanosatellites pose a disproportionate risk to the existing and future satellite communities. Do nanosatellites compromise the sustainability of space and the RF spectrum? Is there a best-practice that could benefit all space operators and ensure easy access to space? The 2015 SGC Policy Working group met to discuss this question and consider mitigating recommendations, if appropriate.

5. SPACE GENERATION CONGRESS 2015

In October 2015, young professionals and students from around the world met in Jerusalem, Israel, for the annual Space Generation Congress (SGC). During SGC, amongst others, the Policy Working Group, supported by Secure World Foundation (SWF) was formed, with a focus on the Policy challenges related to nanosatellites. The group was composed of 18 people from 13 different countries. During the 3 days of the Congress, the members of the group discussed and debated the implications of the current regulations and guidelines related to the growth of the nanosatellite market.
This section will discuss the recommendations made by the participants of the Space Generation Congress as providing detailed explanations to each point.

Figure 5 Picture of the Policy Working Group at the Space Generation Congress 2015 (Credits: SGAC/ Reinhard Tlustos)

6. PURPOSE

The purpose of this paper is to provide recommendations on ways to bring order to the currently unsustainable manner in which nanosatellites are utilizing LEO with respect to both debris mitigation and radio frequency usage. As previously mentioned, nanosatellites typically have operational lifetimes of less than 2 years. This is only a small fraction of their orbital lifetimes, during which the satellite is effectively space debris. Additionally, while these satellites are operational, they must be allocated radio spectrum bands for communication. These frequencies are a finite resource, which is unlikely to increase despite the increasing demand[13].

7. GUIDING PRINCIPLES

There are three guiding principles that were established in developing the recommendations. These principles served as the metrics by which recommendations were evaluated:

1. *Sustainability*: Space and radio spectrums are global resources that must be managed responsibly. Any practice, even if legal, ought to address its impact on the current and future state of access to space.
2. *Respect*: The community of existing satellite operators has developed policies and norms that any new satellite stakeholders should attempt to respect.
3. *Important Benefits*: Nanosatellites offer many clear benefits to the global community. They represent a significant means of encouraging innovation, education, science, and economic growth.

8. STAKEHOLDERS

In developing this report, the following stakeholders were identified and considered in developing recommendations:

- *Existing Satellite Operators* – Any recommendations ought not to risk existing satellites or capabilities.
- *Governments* – According to the Outer Space Treaty of 1967, governments are responsible for authorizing any satellite launches from their territory. This includes fault-based liability for on-orbit actions, intentional or unintentional. Any recommendations should reflect a government's rights and responsibilities.
- *Launch Vehicle Providers* – The launch vehicle provider accepts legal and financial burdens for their cargo. Additionally, there are technical constraints that limit the scope of deployment possibilities available for a given recommendation. For example, a finite number of upper-stage restarts may limit the number of separate deployment orbits available to a set of nanosatellites.
- *Insurance Companies* – Insurance companies have a financial stake in the deployment and survival of cargo. Any nanosatellite recommendation should not significantly increase the on-orbit risk.
- *Universities / Educational Institutions* – Many nanosatellites are developed and operated at universities and educational institutions globally. These are typically associated with student teams. The experience of a student team successfully completing a CubeSat is irreplaceable; however, typical limitations for a low-budget group of students must be acknowledged and, possibly, the expectations for adherence to complex practices should be lowered.
- *"New Space" Companies* – A variety of new business models are appearing. These capitalise on the short development time, economies of scale, and cheap launch-cost associated with nanosatellites. Inasmuch as technology is a critical international economic sector, the interests of New Space companies should be accommodated.
- *Developing Countries* – Nanosatellites are a first step for many developing countries to operate in space. It is appropriate to consider this access if introducing any guidelines. The guidelines ought not to introduce undue burden or restrict access to space.

9. RECOMMENDATIONS

Based on the previously mentioned guiding principles, and in consideration of the stakeholders, four actions are recommended here. The intent of these actions is to promote a path for sustainable, cooperative use of space between existing satellite operators and new nanosatellite operators. This path benefits all parties and, while non- binding, still sets forth a worthwhile best practice for satellite operators.

1. An appropriate, cognizant, international organisation should introduce dialog, initiate coordination, and possibly develop guidelines or best-practices for nanosatellite operators and launch organisations.
   a) Any specific agreement or guideline will likely be accepted only on the condition that it clearly benefits all of the stakeholders. Dialogue will be critical for ensuring that each party is satisfied.
   b) Given that the number of nanosatellites to be launched each year is dramatically increasing, this dialogue should be initiated soon, so that the conversation is prescriptive rather than reactive.
   c) It is recommended that one of the following organisations be considered as coordinator:
      i. United Nations
      ii. International Academy of Astronautics
      iii. International Astronautical Federation
      iv. International Institute of Space Law
      v. Inter-Agency Space Debris Coordination Committee.
      vi. International Association for Advancement of Space Safety

2. This organisation should consider proposing the establishment of open orbital regions for nanosatellites that would be constrained to an attitude up to 400km (LEO), likely below human spaceflight operations.

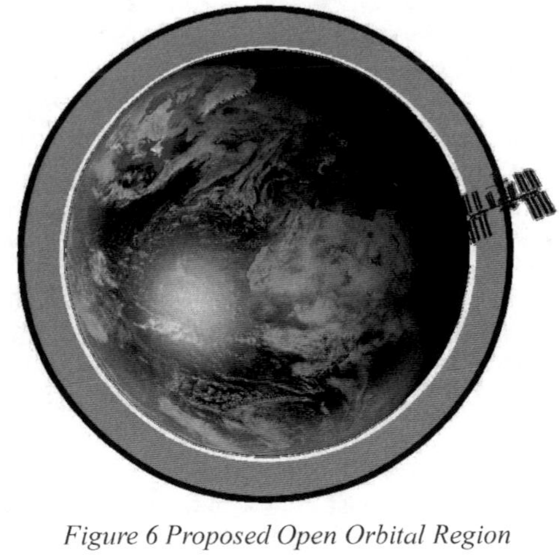

Figure 6 Proposed Open Orbital Region

3. The organisation should develop best practices to evaluate sustainability for the specified region(s).
   a) Specifically, this practice would ensure that short-lived satellites would have a commensurate orbit lifetime, to mitigate long- term, unsustainable debris issues.

b) Locating the nanosatellite's orbit regime below human spaceflight activities minimises any potential impact to the important human activities in LEO. In addition to the International Space Station, various countries are considering human space stations in their national plans. These plans ought to be taken into consideration.
c) The practice could make clear allowances for satellites to operate in any region of space. Such allowances might include consideration of the following:
   i. Satellite deorbit capability, including active and passive means
   ii. Anticipated satellite operational lifetime, including assessment of radiation hardening and redundancy
   iii. Satellite operator and intent (e.g., Developing country conducting scientific experiment)

4. The organisation should consider approaches to promote the small satellite launcher market. This enables or facilitates the dedicated regimes described in (2) above. Practically, this means:

   a. Where possible, governments and international organisations should facilitate and advocate for the sustainable regulation of new small satellite launch vehicle concepts.
   b. (4) is consistent with current trends that suggest a number of commercial options are entering the market.

10. RECOMMENDATIONS' RATIONALE

With the emerging nanosatellite market, an increasing number of well-known as well as new actors are developing and launching nanosatellites. Market projections foresee that more than 2,000 nanosatellites will be launched by 2020. A typical nanosatellite launched by a university or an amateur does not include a deorbiting mechanism or collision avoidance system that can avoid collisions with other objects in its orbit. This poses a great risk to the existing space assets, especially to the manned International Space Station or future outposts in LEO. Developing a new set of recommendations targeted to nanosatellites is considered a "must do" for the immediate future, in order to avoid collisions and mitigate the risk posed by space debris. The recommendations in this paper include the establishment of a specific orbital area, possibly below the International Space Station, where nanosatellites without the possibility of being controlled from the ground or performing orbital manoeuvres can be placed. This lower orbit means that the satellite will remain in orbit for 1-2 years at most, which is also the current typical operational time for a satellite of this class. Even in case of collisions, the debris caused by an impact in this lower orbit will deorbit in short time. On the contrary, in order to be authorised to launch a satellite in an orbit higher than the ISS, it will be necessary to provide the required documentation to demonstrate that the nanosatellite is able to perform orbit changes in due time and can avoid collisions. This means that the satellite has relevant ADC and propulsion systems onboard. In addition to this, the nanosatellite operator needs to prove that its craft will re-enter within 25 years, which will require a passive or active deorbiting system. In addition, the promotion of the small launcher market would make available new lines of rockets able to a more precise insertion in orbit for nanosatellites, avoiding piggyback launches to altitudes unsustainable for systems without proper control or deorbiting means.

11. CONCLUSIONS

The Policy Working Group at the Space Generation Congress 2015 discussed the implication of the raising number of nanosatellites against concern on the sustainability especially of the LEO regions. The group acknowledged the difficulties of tackling these issues especially taking into consideration the necessity of avoiding restrictions for current and above all, new players in the space sector, governmental, commercial, academic or amateur. The recommendations of this paper have been thought to preserve the rights of both old and new players to launch craft in space. Future works will consist in a more detailed definition of the proposed guidelines to propose to relevant organisations.

12. REFERENCES

1. Website: https://www.qb50.eu/index.php/project-description-obj

2. Website: http://www.esa.int/Education/CubeSats_and_Education_the_Fly_Your_Satellite!_programme

3. Website http://www.nasa.gov/directorates/heo/home/CubeSats_initiative#.VcqebDZRHv8

4. Space Works's Nano Microsatellite Market forecast 2016, http://spaceworksforecast.com/2016-market-forecast/

5. *Convention on International Liability for Damage Caused by Space Objects, entered into force* Oct. 9, 1973, 24 U.S.T. 2389, 961 U.N.T.S. 187.

6. Article 1 (d) of the Liability Convention

7. Bin Cheng, *International Responsibility and Liability for Launch Activities*, 20 AIR AND SPACE LAW, 1995, at 297

8. United Nations Office for Outer Space Affairs. "Compendium of space debris mitigation standards adopted by States and international organizations." Last Updated: 8 September 2015. http://www.unoosa.org/oosa/en/ourwork/topics/space-debris/compendium.html

9. Peter B Selding, 1 in 5 Cubesats Violates International Orbital Disposal Guidelines. 23 July 2015. http://spacenews.com/1-in-5-cubesats-violate-international-orbit-disposal-guidelines/

10. Website: http://www.economist.com/news/technology-quarterly/21603240-small-satellites-taking-advantage-smartphones-and-other-consumer-technologies

11. http://www.spaceworksforecast.com/docs/SpaceWorks_Nano_Microsatellite_Market_Assessment_2014.pdf

12. Carolyn Belle, MASS CHALLENGE FOR CUBESATS, Jul 6th, 2015. http://www.nsr.com/news-resources/the-bottom-line/mass-challenge-for-cubesats/

13. http://spacenews.com/itu-grapples-with-small-satellite-regulatory-challenge/

STATUS OF THE GOVERNING SPACE SAFETY TREATY ORGANIZATION

Leslie Ann Alford

A-P-T Research, Inc, 4950 Research Drive, Huntsville, AL 35805, Email: lalford@apt-research.com 256-701-7540, Fax: 256-837-7786

ABSTRACT

Cost and schedule drive industry in accomplishments in the market. Using standardization with the supporting infrastructure of policy and regulations forms part of the solution set. An overarching organization for consensus is needed to:

- Manage, document, and administer space safety agreements and policy,
- Create goals with implementation schedules, and
- Gain consensus for support of the regulatory structure.

Today, successful technology use is constrained to achieve airspace safety. Airspace regulations require evidence that safety critical systems are safe. However, regulatory guidance is controversial due to the cost of achieving that proof. If we do not design for safety but rather assess the risk, then for international consensus, we need agreement on risk determination methods and for identifying acceptable risk.

This dilemma has been solved for international airspace. To advance space safety, consistency with the International Civil Aviation Organization (ICAO) provides precedence.

1. WHAT IS ICAO?

ICAO is a subcommittee of the United Nations (UN) concerned with:

- Setting the standards and regulations of air navigation that in turn assist the global air transport system, and
- Fostering planning and development of international air transport [1].

Its strategic objectives are safety, security, infrastructure, and environment [2]. ICAO ensures the safe and orderly growth of international civil aviation throughout the world; encourages the arts of aircraft design and operation for peaceful purposes; and promotes safety of flight and the development of all aspects of civil aeronautics. It currently serves 191 nations, also called member states, and acts as a focal point for the entire aviation community.

ICAO standards require each member state to undertake and collaborate in securing the highest degree of uniformity in regulations, standards, procedures, and organization in relationship to aircraft, personnel, airways, and supporting services [1]. The resulting standards and recommended practices and procedures are the foundation for all member states. Their minimum standards are mandatory for international operations and have spawned international airspace regulations for safety including environmental protection. As a result, ICAO is the voice of all member states and facilitates global harmonization.

Using the path forged by ICAO provides a route to forming an evolving agency and future standards for international space safety. It gives precedence for establishing space safety regulations and environmental protection.

2. PROBLEMS STIMULATE STANDARDS DEVELOPMENT

The growth of air travel and the accompanying growth in the use of radio communications quickly reached the point of saturation, first in Europe and later on the eastern seaboard of the United States. This limited the growth in the market due to safety concerns.

These concerns led to an ICAO agreement establishing an increase in accuracy of systems in navigation. This accord resulted in the move to increase accuracy of aircraft interoperability and added additional radio frequencies for communication. This led the member states to sponsor development of aircraft minimum operational performance specifications (MOPS) for communication, navigation, and surveillance (CNS) and the associated ground support systems for air traffic management (ATM). Specifications for the minimum aviation-specific performance specifications (MASPS) for the components or avionics were developed as well [3].

This activity followed the ICAO infrastructure established earlier to support the standards and operations of ICAO. Figure 1 depicts the flow of airspace regulatory

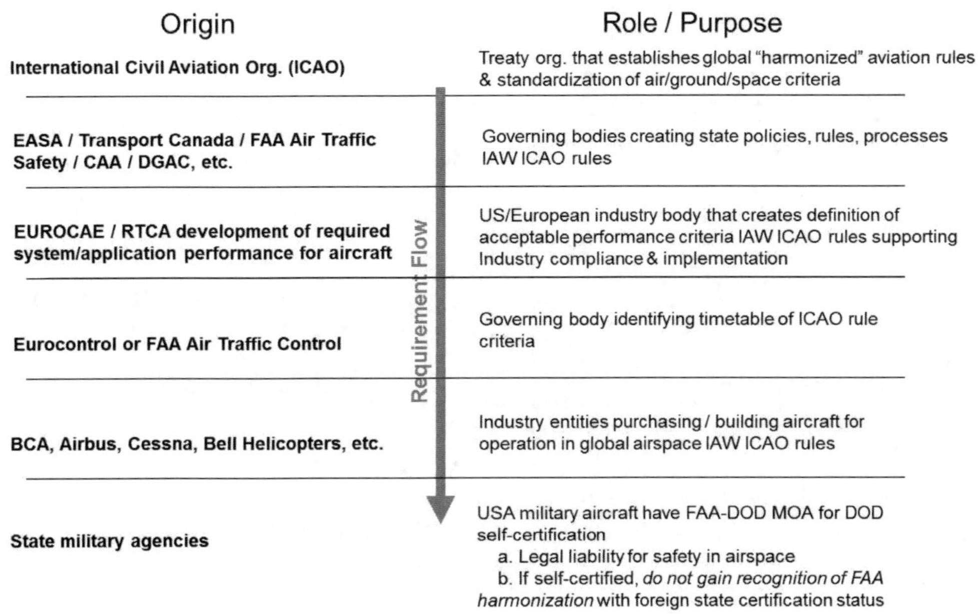

Figure 1. Airspace Regulatory Framework

development used for developing the CNS/ATM technical criteria.

The goal of the changes in CNS allow for more aircraft in the same airspace, flying closer together, communicating without radio interference, and increased situational awareness between aircraft. This also reduces the need for heavy involvement of air traffic control. While there are many additional aspects to the overall plans, the scheduled progress achieving the objectives of more aircraft and signals has been in implementation for several years now. And it impacts both civil and military entities world-wide.

This is a huge change for international airspace and is still in progress. The ICAO paradigm is impressive and powerful.

3. STANDARDS DEVELOPMENT INFRASTRUCTURE

In 1935, the Federal Aviation Agency (FAA) of the United States chartered the creation of a standards organization, the Radio Technical Commission for Aviation (RTCA). The objective of RTCA is to seek resolution of issues and challenges involving air transportation concepts, requirements, operational capabilities, and the associated use of technology and related considerations to aeronautical operations that affect the future ATM system [4].

Specific RTCA objectives include delivering consensus–based, objective and independent recommendations to the FAA [5]. Its industry membership is required to have balanced representation as depicted in *Figure 2*, promote transparency and accountability, and develop minimum performance standards.

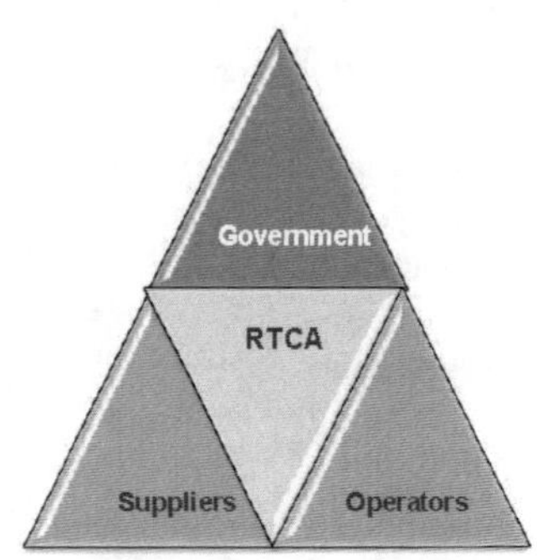

Figure 2. RTCA Balanced Representation

This expands the marketplace of solutions and provides anti-trust protection through industry engagement. Oversite, guidance, and integration is provided by the RTCA Program Management (PM) Committee formed of member companies who make available the expertise to support the development of the standards. The standards include Minimum Operational Performance Standards (MOPS) and Minimum Aviation System Performance Standards (MASPS), which are defined in Table 1 below.

Table 1. RTCA Chartered Documents

MOPS	These are standards for specific equipment useful to designers, manufacturers, installers, and users of the equipment. The word "equipment" used in MOPS includes all components and units necessary for the system to perform its intended function(s). MOPS provide the rationale for equipment characteristics and requirements stated; describe typical equipment applications and operational goals; and, establish the basis for required performance under the standard. Definitions and assumptions essential to proper understanding are provided, as well as installed equipment tests and operational performance characteristics for equipment installations.
MASPS	These specify characteristics useful to designers, installers, manufacturers, service providers, and users of systems intended for operational use within a defined airspace. If systems are global in nature, international applications are taken into consideration. MASPS describe the system, including subsystems and functions, and provide the rationale for system characteristics, operational goals, requirements, and typical applications. Definitions and assumptions are provided and include minimum system test procedures to verify system performance compliance (e.g., end-to-end performance verification).

The completed documents are recommendations to the FAA who can implement them as a reference via Regulation, Advisory Circular (AC), or by Technical Standard Order (TSO). Any adherence to an RTCA Standard is considered only one possible means of compliance. The creation of new standards or updating of existing standards is initiated by request from the FAA. The conditions driving creation or update of standards include:

- Experience using the standard or operating equipment,
- A forecast of change in the airspace environment,
- New or improved technology, and
- International activity.

In 1963, the European Organization for Civil Aviation Equipment (EUROCAE) was formed. Its membership is comprised of manufacturers, service providers, national and international aviation authorities, and users from Europe and elsewhere. EUROCAE works jointly with RTCA and with the Society of Automotive Engineers (SAE) for a single consensus in the development of system performance standards [6]. This working arrangement allows for one harmonized definition of a given technology. The EUROCAE documents are produced in context with the applicable ICAO standards to ensure global interoperability.

In November 2014, RTCA and EUROCAE signed a commitment to work toward harmonized standards. This formal alignment further strengthens the overarching infrastructure that allows a consistent, safe single sky for the world. Seventy percent of the published aeronautical standards are joint between the two organizations [7].

This highlights the cooperation and unification of the member states to work towards the goals of safe airspace through the creation, support, and implementation of the standards developed. This should be the goal of the international space safety member states.

4. REGULATORY REFERENCE TO STANDARDS

The next step in the process occurs when the Civil Aviation Agencies (CAA) acknowledge the standards by referencing them in regulatory writings. This is most commonly done by the FAA and the European Aviation Safety Agency (EASA) who accept a given standard as a means, but not the only means for compliance. Figure 3 below illustrates the CAA's role in the global infrastructure for a safe sky.

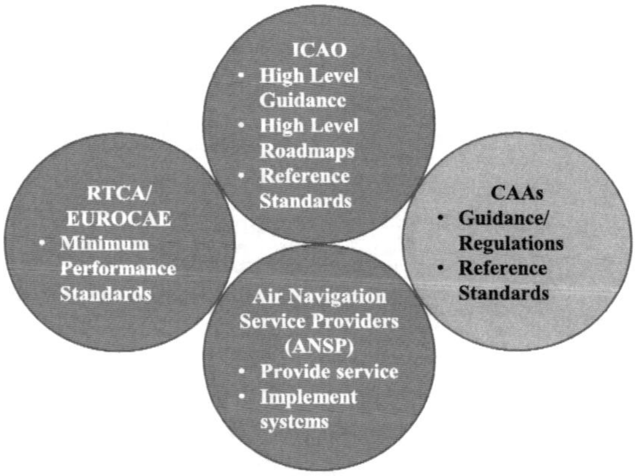

Figure 3. International Airspace Infrastructure for Safety

The civil authorities' roles are to ensure safety, set the regulations, and ensure compliance. An example of this is the recognition of RTCA/EUROCAE DO-178C/ED-12C [8] as an acceptable method for airborne software assurance in avionics. This was done through the FAA with Advisory Circular (AC) 20-115C [9] and by EASA with their Certification Specification (CS) Acceptable Means of Compliance (AMC) 20-115C [10].

This follow-through by the member states of ICAO allows air traffic control management to enforce the compliance of aircraft for airspace access in accordance with the recognized standards. The issues of international space can follow this same paradigm of issue resolution via a treaty level harmonization for technical standards development.

5. COMPONENT MARKET GROWTH RESULTS FROM STANDARDS

Establishment of standards enables marketplace growth for avionics components. The component manufacturers are part of the Air Navigation Service Providers (ANSP), the bottom bubble from Figure 3. Others in this category include airlines, aircraft operators, aircraft manufacturers, maintainers, and the ATM system. The ANSP are providing service and implementing systems, such as:

- Ensure safety through operation, repair, and maintenance
- Ensure efficiency, such as on-time flights, meeting development, and delivery schedules, etc.
- Build, maintain, or operate
- Provide services such as airports and navigation aids [4]

Services are a market and foster competition which keeps the avionics components at competitive levels. But this is not the end of the story.

Industry is fearful of overregulation and constantly pressures the civil authorities for a more risk-based approach trying to diverge from the technical proof of design and verification evidence for safety. Yet whatever is proposed must eventually include the other parts of the air safety infrastructure. And any damage or serious safety event results in liability. The burden of proof of safety is on the "applicant"—the entity requesting approval.

However, industry pressure has positive consequences. It brings about innovative approaches and technologies. One established alternative approach is afforded to the members' military aircraft and operations.

Dependent on a state's regulations, military may be required to fully meet established standards or they may be able to use a risk-based approach for safety. This particular deviation is established and an accepted part of the international air safety strategy [11].

6. A CASE HISTORY [12]

Some aircraft equipped with components addressing the Required Navigation Performance (RNP) have a system installed called Traffic Collision Avoidance System (TCAS). On July 1, 2002, two aircraft were on a collision course over Überlingen and Owingen, Germany. One was a Russian passenger jet with 69 people onboard and the other was a DHL Freight Transportation cargo jet with a crew of two. The air traffic control involved was handled by a Swiss traffic control service. All 71 people onboard were killed.

While this accident is well documented and can be found on the web, the incident included an international accident investigation, part of the air safety infrastructure, Air Traffic Control (ATC), standardized procedures for using a TCAS system and the responsibilities of the pilots and air traffic control. To use a TCAS system all parties must be trained; this is a required safety component based on take-off weight or seating capacity of the aircraft.

The TCAS system provides voice alarm warnings to dive or climb as needed. In this case, the Russian TCAS was stating climb, while a single air traffic controller was telling the pilot to dive. The Russian pilot did what he was told by the air traffic controller, in concert with the Russian Pilot's Guide and training. The DHL TCAS was instructing its pilot to dive… So there was a crash.

Some of the 21 investigative results are briefly summarized in Table 2 below:

Table 2. Synopsis of the Investigation Findings

<table>
<tr><th>Infrastructure Element</th><th>Issue Found</th><th>Finding/Recommendation</th></tr>
<tr><td>Development of TCAS systems to RTCA/DO-185 and EUROCAE/ED-143</td><td>Recommended upgrades for the system</td><td rowspan="2">1. ICAO was recommended to change international requirements requiring pilots to obey and follow TCAS resolution advisories (RAs), regardless of whether contrary ATC instruction is given prior to, during, or after the RAs are issued. Unless the situation is too dangerous to comply, the pilot flying should comply with the RA until TCAS indicates the airplane is clear of the conflict.
2. ICAO should ensure that rules and procedures regarding ACAS are uniform, clear and unambiguous. Compliance should be ensured in the ICAO Annexes 2, 6, the PANS and the Guidance Material. The procedure for pilots should include the following elements: In the event of an ACAS/TCAS Resolution Advisory (RA) to alter the flight path pilots shall: Respond immediately and maneuver as indicated, unless doing so would jeopardize the safety of the airplane.
Never maneuver in the opposite sense to an RA, nor maintain a vertical rate in the opposite sense to an RA.
3. ICAO should initiate the development of down-linking RAs to ATC, using such technologies as SSR Mode S and Automatic Dependent Surveillance-Broadcast (ADS-B)</td></tr>
<tr><td>Integration into the Air Traffic Control (ATC) procedures for TCAS systems internationally</td><td>Insufficient and did not correspond in all points with the system philosophy, therefore the regulation and procedures were not standardized and were incomplete</td></tr>
<tr><td>Pilot training to use a TCAS system</td><td>In question due to conflicting information in the Pilot's Guide</td><td>1. ICAO should ensure that a high level of acceptance and confidence of pilots in ACAS/TCAS (exists) by improving education and training.
2. Civil Aviation Authority (CAA) of the Russian Federation should ensure that the ACAS/TCAS training for flight crews is in accordance (with) the requirements and guidelines of [TCAS training procedures].
The following aspects should be covered: Flight simulators equipped with ACAS or inter-active Computer Based Training (CBT) should be utilized for training of ACAS/TCAs maneuvers.
ACAS event scenarios should be included in Crew Resource Management (CRM) and Line-Oriented Flight Training (LOFT) programs.
3. Civil Aviation Authority (CAA) of the Russian Federation should ensure that exposure to Crew Resource Management training within the airline industry is (improved). The use of flight simulators or appropriate synthetic training devices for Line-Orientated Flight Training should be promoted</td></tr>
<tr><td>TCAS Pilot's Guide</td><td>Conflicting information between using TCAS as an aid in one place and elsewhere, where TCAS was only to be followed–recommended update to remove the "aid" wording</td><td>Federal Aviation Administration (FAA) of the United States of America should ensure that the TCAS 2000 manufacturer re-phrases the TCAS 2000 Operating Manual to reflect the ACAS/TCAS system philosophy and the international ACAS/TCAS regulations and operating procedures in an unambiguous and consistent manner.</td></tr>
<tr><td>Air traffic controller training</td><td>In question due to the events</td><td>Federal Office for Civil Aviation Switzerland should ensure that air traffic controllers are imparted with initial and recurrent training covering the theoretical and practical (simulator) emergency procedures, particularly:
1. Recognition of potential air traffic conflict and maintenance of aircraft separation in accordance with the international standards.
2. A quick re-establishment of the minimum separation, in an event of having fallen below it.
3. The use of appropriate phraseology, with an emphasis on achieving maximum effects within minimum possible time/radio transmission</td></tr>
<tr><td>Air traffic controller procedures for coverage of flights requiring 2 or more people present at all times</td><td>Management and quality assurance of the air navigation service company did not ensure all night that workstations were continually manned; standard operating procedures had accepted less coverage at the workstations</td><td rowspan="2">Federal Office for Civil Aviation Switzerland should
1. Ensure that the ATC service issues/implements procedures to undertake maintenance work on the ATC Systems stipulating operational effects and available redundancies.
2. The ACC Zurich is manned with the minimum number of air traffic controllers as follows: There shall be at least two controllers on active duty at all times; at least two controllers to manage enroute sectors; i.e. one Radar Planner (RP) and one Radar Executive (RE).
When ACC is required to manage the approach services for Friedrichshafen and Altenrhein/St. Gallen, one additional controller shall be assigned to this task. Alternatively, this task should be taken over by APP Zurich.
Additional controllers shall be assigned to manage breaks.
3. Federal Office for Civil Aviation Switzerland should ensure that the ATC service provider issues and implements procedures to undertake maintenance work on the ATC Systems stipulating operational effects</td></tr>
<tr><td>Air traffic management system on the ground, including a ground based optical collision warning system, and a radar image processing system</td><td>Main radar image system was out for maintenance and the optical collision warning system was turned off</td></tr>
</table>

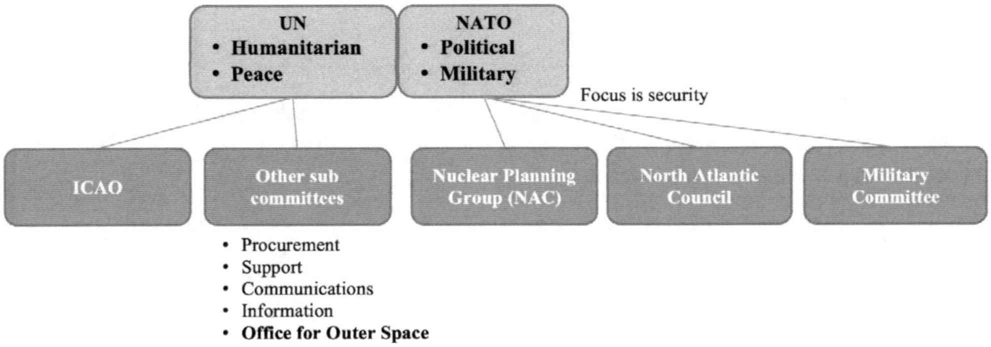

Figure 4. Context of the World's Treaty Organizations

The investigation done by the German Federal Bureau of Aircraft Accidents Investigation was comprehensive. The results were accepted by the appropriate parties and today, the recommendations have been implemented.

What is notable is that the ICAO infrastructure provided an orderly independent investigation, cooperation by all involved members, and correctional upgrades to the ICAO's standards, the TCAS system, Pilot's Manual, and respective training programs for all parties. This resulted in a closely monitored, safer airspace. Crisis management was not necessary.

7. INTERNATIONAL SPACE SAFETY FRAMEWORK

We know that there are two treaty organizations that lead to stabilization internationally. They are the United Nations (UN) and the Northern Atlantic Treaty Organization (NATO). The NATO treaty operates within the framework of the UN and is committed to the principle of collective defense setting the tone of solidarity [13]. The charters and interrelationships are shown in Figure 4.

There are 28 member countries in NATO and 192 member nations in the UN. Those that are in NATO are also in the UN. Each have liaison with the other that enhances the cooperation between the two organizations. This is an integral part of NATO's contribution to their approach to crisis management and operations.

7.1 Governing International Space Safety Organization

Space Safety fits into the existing treaty structure within the UN and is the responsibility of the Office of Outer Space Affairs (OOSA) [15]. OOSA is responsible for:

- Promoting international cooperation for peaceful use
- Space Object registering (asteroids, etc.)
- Global development and disaster management support
- Space weather and climate change
- Global navigation satellite systems
- Long term sustainability of outer space activities
- Treaty implementation

The principles of OOSA include free and peaceful use and exploration for all mankind. It has also established that space is not subject to claims of sovereignty by use or occupation and it is to be free of nuclear weapons. Member states have agreed to be responsible for national activities, liable for damage, and are to avoid harmful contamination. Thus space safety has a home in the UN and has governing treaties with the member states, shown in **Error! Reference source not found.**.

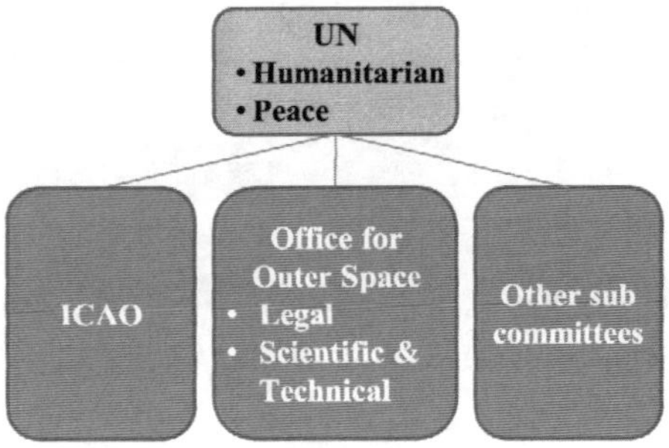

Figure 5.

7.2 Safety Governance

OOSA addresses governance of overarching issues for cooperation and addresses resolutions in its working groups on its areas of responsibility. This is powerful, stabilizing, and peaceful. Following the UN/ICAO model

identified in the case study, the working groups have proposed recommendations and in some cases have resulted in treaties such as the Nuclear Treaty, where the members declared that space would not be used for nuclear weapons. However, UN OOSA is in its infancy by comparison and the gravity of issues remain at the treaty level and in the domain of OOSA itself.
The overarching long-term development concerns for space include:

- Economy
- Society
- Accessibility
- Diplomacy

The UN OOSA is concerned with providing economic assistance to developing countries, including education; the evolution of benefits stemming from space activities; accessibility of space through strengthening an supporting national infrastructures and their capacity; and finally, increasing partnerships and international cooperation through diplomacy on space topics and activities.

7.3 Safety Guidance and Standards

Space debris management has already been pulled together in document for OOSA and forms an example of safety management policy and practice on space debris management. Space debris damages the vehicles and spacesuits, increases the cost of protection for missions, and destroys live satellites. The measurement, mitigation, and potential removal of debris are conducted by some participants in the space industry, but a systematic approach to prevention and removal is needed [14]. So at the treaty level the UN OOSA is building treaties for space policy through agreements, rules, and conventions [15].

7.4 Market Growth

The UN OOSA is still evolving and appears to be similar but different to the ICAO roadmap, as standards development is remaining at the treaty level. The future may include a market for third party products, similar to the aviation field, but it has not yet arrived. The investigatory governance and resolution remains at the treaty level as well.

8. SUMMARY

International space safety is in work within the UN. Setting policy at the international treaty level provides weight and import to achieving the goal of implementing and managing a safer space. It provides a forum for establishing relationships, cooperation, and policies.

Standards development provides participation of the global space science and technology community. As consensus based standards are created, they guide technological development. The effectiveness of the standards development bodies is measured in one sense by membership—members looking for a voice and a market.

9. REFERENCES

1. Fox, Mitchell, "What is ICAO – Why Should You Care?", Chief Flight Operations, ICAO, http://www.google.com/url?sa=t&rct=j&q=&esrc=s&source=web&cd=5&ved=0ahUKEwjOvYXSv5jMAhWI5iYKHRz-rsQFgg_MAQ&url=http%3A%2F%2Fwww.iaopa.org%2Fdoc%2F26th-assembly-presentations%2F2-what-is-icao.ppt&usg=AFQjCNEjogvigpAMX09pDM0X679ItOCyzQ&sig2=HfXf9VxfbVilrzEd_zKXRg

2. ICAO, Vision and Mission, http://www.icao.int/about-icao/Pages/Strategic-Objectives.aspx

3. Jenny, Margaret, RTCA Background on Standards for Lithium Batteries, National Transportation Board, April 2013, pp. 3-7, http://www.ntsb.gov/news/events/Documents/2013_Lithium_Batteries_FRM-Panel2b-Jenny.pdf

4. Iverson, Jennifer, RTCA & EUROCAE Standardization, 28 August 2013, pp. 3-30, http://www.google.com/url?sa=t&rct=j&q=&esrc=s&source=web&cd=4&sqi=2&ved=0ahUKEwiH5obFxJjMAhXBMyYKHW4uCYkQFggyMAM&url=http%3A%2F%2Fwww.aixm.aero%2Fgallery%2Fcontent%2Fpublic%2F2013_08_ATIEC%2FDay_2%2F11 - EUROCAE and RTCA Standardisation - Jennifer Iversen - EUROCAE-RTCA.ppt&usg=AFQjCNH13JUYOo8oTeHz4zeB18GFFhNMJg&sig2=Hohn0BDVUDDTSiRhJcws5Q&bvm=bv.119745492,d.eWE

5. FAA Order 1110.77V, RTCA, Inc. (Utilized as an Advisory Committee), 04/01/2015, pp. 1-5, http://www.faa.gov/regulations_policies/rulemaking/committees/documents/media/RTCA Charter Order1110.77V.pdf

6. SAE Press Release, SAE International and EUROCAE Sign Updated Memorandum of Cooperation, September 22, 2015,
https://www.sae.org/news/3150/https://www.google.com/url?sa=t&rct=j&q=&esrc=s&source=web&cd

=4&ved=0ahUKEwiDgpvfw5jMAhUJLyYKHdyAAfsQFggyMAM&url=https%3A%2F%2Fwww.eurocae.net%2Fstatic%2Fdownload%2F%3Fdownload%3DdocumentIdentifier%3BMoc1114%2Cdc%3Bdocuments%2Cdc%3Beurocae%2Cdc%3Bnet&usg=AFQjCNFFjW96y-ZeJlo2cEDu7DhTlFh99g&sig2=qGfcCOFWrus80OTE2ttYng&bvm=bv.119745492,d.eWE

7. Jenny, Margaret, "President's Message: RTCA-Results Through Collaboration in Aviation", http://www.rtca.org/content.asp?pl=49&sl=136&contentid=136

8. RTCA/DO-178C, Software Considerations in Airborne Systems and Equipment Certification, December 13, 2011; Documents may be purchased from RTCA, Inc., http://www.rtca.org

9. FAA AC 20-115C, Advisory Circular: Airborne Software Assurance, July 19, 2013, http://www.faa.gov/documentlibrary/media/advisory_circular/ac_20-115c.pdf

10. EASA, CS AMC20-115C, https://www.easa.europa.eu/system/files/dfu/Annex II - AMC 20-115C.pdf

11. NATO AOP-52, (EDITION 1) - GUIDANCE ON SOFTWARE SAFETY DESIGN AND ASSESSMENT OF MUNITION-RELATED COMPUTING SYSTEMS, NATO, http://www.hcrq.com/uploads/2/6/9/6/26962125/aop-52_-_guidance_on_software_safety_design_and_assessment_of_munition-related_computing_systems.pdf

12. AX001-1-2/02, Investigation Report, German Federal Bureau of Aircraft Accidents Investigation, May 2004, http://www.bfu-web.de/EN/Publications/Investigation Report/2002/Report_02_AX001-1-2_Ueberlingen_Report.pdf?__blob=publicationFile

13. NATO, http://www.nato.int/cps/en/natohq/topics_50321.htm

14. NASA, "The Threat of Orbital Debris and Protecting NASA Space Assets from Satellite Collision", 28 April 2009, http://images.spaceref.com/news/2009/ODMediaBriefing28Apr09-1.pdf

15. United Nations, Office for Outer Space Affairs, http://www.unoosa.org/oosa/en/ourwork/unispaceplus50/index.html

PROCEEDINGS OF THE 8TH INTERNATIONAL SPACE SAFETY CONFERENCE

SESSION 18
SPACE TRAFFIC CONTROL - I

PRELIMINARY STUDY ON INADVERTENT LASER ILLUMINATION HAZARDS POSED TO SATELLITE OPTICAL SENSORS

Patrick Shriver(1), Karen Yamamoto(1), Chad Cogburn(1), Chris Jones(1), Ken Miller(2)

(1) Metatech Corporation, 2340 Alamo Ave S.E. Ste. 300, Albuquerque, NM, 87106, U.S.A., Email:patrick.shriver@metatechcorp.com; karen.yamamoto@metatechcorp.com;chad.cogburn@metatechcorp.com; jones@metatechcorp.com
(2) Satellite Assessment Center, Air Force Research Laboratory, 3550 Aberdeen Ave. S.E., Kirtland AFB, NM, 87117,U.S.A., Email: afrl.rdte@us.af.mil

ABSTRACT

Hazards posed to space systems from inadvertent illuminations by outdoor laser activities are generally not well known, but will continue to grow as the numbers, applications, and intensities of lasers increase world-wide. The only known, proactive measures in place to protect against such hazards are those of the U.S. Department of Defense. Hence, satellite owners/operators are implicitly accepting the risks posed by the vast majority of lasers operating around the globe. To quantify the risks, a Probabilistic Risk Analysis methodology has been developed to calculate the statistical frequency of accidental illumination. In this paper, we report on a preliminary study aimed at characterizing the risk environment faced by satellites and examine hazards posed by common outdoor laser sources. Since the risks will only continue to grow over time, it is worthwhile to consider international measures of protection to ensure the safe interoperability between satellites and lasers.

1. INTRODUCTION

Outdoor laser activities can pose a hazard to satellites. Even if the laser is not intended to track satellites, an unblocked beam can still propagate through the atmosphere and accidentally sweep across a satellite while pointing in a desired direction or tracking the intended target. Although illumination does not always cause damage, damage can occur depending on the characteristics of the laser system and the susceptibility of the satellite to the laser beam.

There are no international policies/processes/standards in place to protect against inadvertent illumination hazards. The only known measures that proactively control the hazards of inadvertent illumination are those of the U.S. Department of Defense (DoD). Thus, all satellite owners/operators are implicitly accepting inadvertent illumination risks posed by the vast majority of lasers operating around the globe.

In this preliminary study, we have begun to characterize this risk environment faced by all satellites. In particular, we examine the hazards posed by common laser sources: atmospheric Light Detection and Ranging (LIDAR), weather monitoring ceilometers, Satellite Laser Ranging (SLR), and astronomical Laser Guide Stars (LGS). The potential for a laser source to accidentally damage a satellite optical sensor is evaluated through comparison of on-orbit irradiance calculations to optical detector material damage-onset thresholds. The probability of inadvertent illumination is calculated using a Probabilistic Risk Analysis (PRA) method created by the Air Force Research Laboratory (AFRL) Satellite Assessment Center (SatAC, AFRL/RDST).

Since the numbers, power levels, and applications of lasers will continue to grow world-wide, the risks will only continue to increase. As the risks become better understood with PRA, it may be mutually beneficial for the satellite and laser communities to consider some industry-wide, international measures that ensure the safe interoperability between satellites and lasers. With this paper, some of the issues and potential solutions from an international standards viewpoint are outlined to serve as a starting point for discussions within the satellite and laser communities.

2. INADVERTENT LASER ILLUMINATION

Unless fully blocked, an outdoor laser beam can propagate through the Earth's atmosphere and make it into space. Clouds may effectively block a beam by absorbing most, if not all, of the radiation, but transmission through clear, non-turbulent air typically reduces the amount of beam energy by only a small fraction for many laser wavelengths. Even some commercial hand-held lasers can be seen from orbit under ideal conditions. Although accomplished via an intentional, highly coordinated event, an amateur astronomy club pointed a 1 W laser at the International Space Station that was observed by an astronaut aboard the station (~350 km altitude) [1].

Yet, contrary to popular fiction, simply illuminating an object with a laser beam does not always cause damage. Instead, damage is a complex relation of the beam characteristics, dwell time, and illuminated component or material susceptibility. Thus, any laser does not necessarily pose a permanent damage hazard to all satellites. Illumination may lead to no effect at all or merely induce a temporary and recoverable effect (e.g., dazzling and blinding of optical sensors).

Optical sensors are typically the most susceptible satellite components to laser radiation. Of the elements in a sensor, the detector is generally regarded as the most susceptible component due to the sensor's high optical gain, which has a multiplier effect on the incident irradiance at the sensor aperture. A detector is also typically most susceptible to irradiance at a wavelength that is in-band to a sensor channel and the detector material's spectral response. It usually takes significantly higher irradiance levels at out-of-band wavelengths to cause damage (perhaps to other components in the optical train).

Imagers have optical gains commonly $\sim 10^9$ but may exceed 10^{10} for high fidelity systems. Given the high optical gains of imagers, damage depends on the detector material properties. Damage to a detector material can occur nearly instantaneously, as within the time of a single detector scan of a point on the ground (microseconds to tens of milliseconds) or a single laser pulse (nanoseconds to microseconds) [2], [3]. Hence, pulsed laser systems that yield high peak powers per pulse are of most concern for detector material damage.

As will be shown later, model computations of single pixel damage can occur for irradiance levels on the detector of $\sim 10^5$-10^{11} W/cm^2 (depending on the detector material, laser wavelength, and pulse width or dwell time). Thus, given the typical imaging sensor 10^9 optical gain, it is possible, under certain conditions, for pixel damage to occur with on-orbit irradiance levels incident at the sensor aperture of only $\sim 10^{-4}$ W/cm^2. With their high peak powers, many of the pulsed laser sources in this study can propagate on-orbit irradiances easily exceeding this incident level.

Experiments have shown that multi-pixel, wide area or full array damage can even occur at levels equivalent to or just a few times above single pixel damage thresholds [4], [5], [6]. From a brief review of pulsed laser experiments on silicon Charge Coupled Devices (CCDs), irradiance levels of $\sim 10^4$-10^{13} W/cm^2 have been observed to cause full array damage [2] - [7]. However, the tested CCDs were not space-qualified. Space-qualified imagers, hardened for the space environment, may have higher damage thresholds. Also, the experiments may be capturing damage due to electronic effects instead of absorption in the bulk material.

To ensure the protection of all space systems from inadvertent illumination and potential damage, DoD policy requires that DoD laser programs adhere to the Predictive Avoidance (PA) process established by the U.S. Strategic Command (USSTRATCOM) and executed by Laser Clearinghouse (LCH) [8]. This PA process is the only known, proactive measure in place to protect satellites against inadvertent illumination hazards. Although USSTRATCOM will consider U.S. non-DoD and foreign laser program requests for deconfliction, DoD laser programs are given priority. Since there are no international policies, standards, or processes for satellite and laser owners/operators to follow in order to mitigate inadvertent illumination risks, satellite owners/operators are implicitly accepting the risks posed by the vast majority of laser systems operating around the globe.

3. LASER APPLICATIONS AND TRENDS

From popular culture, lasers conjure up the fictional vision of weapons blasting away and easily obliterating their targets, but the reality is substantially different. The first practical use of lasers on the battlefield has been the support of conventional weapons. Laser designators have been fielded since the 1980s, which provide a reflected signal for precision-guided munitions [9]. Large chemical, 100s of kW to MW-class systems have demonstrated successes, but have not been fielded. Recent advancements, such as solid-state and fiber lasers, have made smaller, tactical weapons of the 10-100 kW-class viable [10]. The U.S. Navy has fielded a prototype tactical laser for ship defense [11].

Although the risks of inadvertent illumination from U.S. DoD lasers are controlled with the PA process, the U.S. is certainly not the only country developing and fielding military lasers. Hence, non-U.S. military lasers are a legitimate concern for inadvertent illumination hazards to space systems. However, military lasers are not yet globally prolific (as they are still slowly developing). Also, military lasers would only be used during infrequent, hostile situations or pre-planned tests.

Easily overlooked, lasers used for scientific purposes can also pose an inadvertent illumination hazard to space systems. In contrast to military lasers, scientific lasers are globally prolific and frequently operate; they will operate for as long and as frequently as possible in order to maximize the number of measurements. Scientific lasers are also typically pulsed with high peak powers, which, as mentioned in the previous section, is of particular concern for satellite optical sensors.

A brief survey of common types of scientific lasers and their networks was performed for this study. Table 1 summarizes the laser types considered here. Note that this survey was not exhaustive. Only information readily available in a convenient format obtained by on-line searches was used.

Table 1: Summary of Laser Types Included in this Study

Laser Type	# of Sources†	Wavelengths (nm)	Power (W)
LIDAR (pulsed)	250	266 – 10600	$10^2 – 10^8$ (peak)
Ceilometer (pulsed)	1993	855 – 1064	$10^1 – 10^2$ (peak)
LGS (CW and pulsed)	17	351 – 589	$10^0 – 10^1$ (CW, avg.) $10^1 – 10^4$ (peak)
SLR (pulsed)	50	532	$10^4 – 10^9$ (peak)

† does not include all lasers in the global population

3.1 LIDAR

Since the 1960s, LIDAR systems have been used to measure constituents in the atmosphere for weather, climate, and other environmental (e.g., pollution) monitoring [12]. Some of their applications include measuring aerosols, ozone, water vapor, temperature, and clouds. The basic principle is to send a laser pulse into the atmosphere where it interacts with and scatters off of aerosols, gasses, particulates, etc. The returned scatter signal is compared to the original outgoing signal to determine time of flight, composition, component density, and other atmosphere profile characteristics.

Today, these systems can be found across the globe in wide use by numerous institutions. The networks and numbers of sites are constantly changing as resources and technology advance. This analysis included approximately 250 separate LIDAR systems that are listed in Table 2. Some ground sites support more than one type of LIDAR, may belong to more than one network, or may operate transportable LIDAR systems. The information in Table 2 presents a subset of the full ground presence of LIDAR systems that may present a hazard to satellites. Other research stations (e.g., universities) that do not belong to the advertised networks are not included here.

The laser wavelengths used in atmospheric LIDAR applications range from the ultraviolet to the infrared. Although the operating wavelength for some of the sources was not readily available, approximately 61% of the wavelengths identified in this study are one of the first three laser Nd:YAG harmonics (16% at 355 nm, 29% at 532 nm, and 16% at 1064 nm). Other sources include a range of Raman wavelengths from 266-353 nm and 589-800 nm, and infrared wavelengths at 1.55 μm, 2.022 μm, and 10.6 μm.

Table 2: LIDAR Networks

Network	Website Reference	# Sites	Location
AD-Net	http://alg.umbc.edu/galion/	20	East Asia
ARM	http://alg.umbc.edu/galion/	7	US
CIS-LiNET	http://alg.umbc.edu/galion/	7	CIS
CORALNET	http://alg.umbc.edu/galion/	6	Canada
CREST	http://alg.umbc.edu/galion/	4	US & Puerto Rico
EARLINET	http://alg.umbc.edu/galion/	27	Europe
LALINET	http://lalinet.org/	9	S. America & Cuba
LEOnet	http://leo-net.eu/	15	worldwide
MPLNET	http://alg.umbc.edu/galion/	14	worldwide
NDACC	http://alg.umbc.edu/galion/	23	worldwide
PollyNet	http://polly.tropos.de/?p=home	8	worldwide
SPALINET	http://www.lidar.es/spalinet/en/home/	8	Spain
TOLNet	http://www.air.larc.nasa.gov/missions/TOLNet/	5	US

The LIDARs in this study have high peak power levels that range on the order from 10^2-10^8 W/pulse. These high peak powers are necessary to generate enough backscattered radiation to be detected on the ground. In the scientific community, there is a desire to extend atmospheric LIDAR analyses to higher altitudes, which is driving LIDAR technology trends towards using even higher peak powers.

3.2 Ceilometers

Ceilometers are automated laser systems that probe the atmosphere to determine cloud height and visibility conditions. They are self-contained systems that house the laser source, detection, data processing, and reporting systems. They are used in stand-alone applications and are also included in automated weather monitoring systems installed at airports around the globe, where they provide real-time cloud and visibility information to airport personnel and pilots.

Atmospheric LIDAR systems perform a similar function as ceilometers; however, ceilometers use lower power lasers that are generally safe to the unaided eye. While ceilometers are reported to be included at some atmospheric LIDAR network locations, ceilometer networks are operated and maintained on a national level for national meteorological services [13]. Ceilometers operate in the near-infrared region at specific wavelengths from 855-1064 nm. These wavelengths are covered by only a few different ceilometer models that operate at 6–10 kHz with pulse energies of less than 10 uJ and pulse widths from 50–135 ns.

3.3 SLR

Similar to LIDAR, SLR activities began in the 1960s and remain strongly active today. SLR activities take instantaneous range measurements of cooperative satellites with Laser Retro Reflectors (LRRs). The round-trip travel time of a laser pulse being transmitted, reflected, and received is measured to determine the distance traveled. These measurements have important implications for geodesy, fundamental physics, and Earth science [14].

SLR consists of a global network of ground-based observatories managed by the International Laser Ranging Service (ILRS). Currently, the network includes approximately 50 active ground stations with around 35 inactive and engineering stations distributed around the world. From 2006 to 2012, there was an increasing trend of ILRS tracking, which is only expected to increase further as more satellites with LRRs are launched [15]. The ILRS network of ground sites also includes lunar laser ranging systems; there were >100,000 measured lunar laser returns in 2008 [16].

ILRS guidance does require laser operators to ensure that the cooperative satellites with LRRs are not harmed during illumination. However, it is possible for other satellites passing in the foreground, background, or nearby the cooperative LRR satellite to be inadvertently illuminated and potentially damaged. Often, the ILRS operators will intentionally diverge their beams for better tracking; wider beams increases the chances that other satellites may be inadvertently illuminated, albeit also reducing the beam's intensity. For lunar ranging, ILRS systems are not required to deconflict their operations with satellites that may cross the beam's path as it propagates to the moon and is reflected back.

The majority of the ILRS systems operate at 532 nm. There is a general trend to achieve increasingly shorter pulse widths and higher pulse repetition frequencies. This trend allows for higher data rates, and therefore more data, to produce higher fidelity analysis results.

3.4 LGS

In astronomy, LGS systems have been in use for the past two decades and are used by observatories in conjunction with Adaptive Optics (AO) to negate atmospheric distortions. The LGS beam is used to create an artificial "star" in the night sky, from which the distortions are measured and minimized by the telescope's AO [17]. The first regular LGS operations began in the mid-1990s, and a study performed in 2008 showed that scientific productivity derived from LGS AO observations was on the rise due to the enhanced imaging capability [18]. About half dozen major civil observatories now regularly use LGS systems, which typically operate for several hours at a time and for many nights throughout the year.

There are two types of LGS systems. A sodium LGS is created by focusing a 589 nm laser in the sodium rich layer of the atmosphere at an altitude of about 90 km. A Rayleigh LGS is created by focusing a laser at a lower altitude to induce Rayleigh scattering. Common sources include 532 nm and 355 nm lasers that are easy to achieve and efficiently induce Rayleigh scattering. LGS sources identified in this study are both pulsed and CW with average powers ranging from a few to tens of watts. Pulsed sources operate from 5–13 kHz with pulse widths of 30–150 nsec.

4. POTENTIAL FOR SENSOR DAMAGE

In determining the potential of a laser source to cause damage, the amount of laser irradiance incident on a satellite can be estimated with Eqs. 1–3.

$$I_{inc} = \frac{4P_{eff}}{10^4 \cdot \pi D_{beam}^2} \tag{1}$$

$$P_{eff} = P_{src}T_{src}T_{atm} \tag{2}$$

$$D_{beam} = \frac{0.122\lambda RM^2}{D_{eff}} \tag{3}$$

where:

I_{inc} = irradiance incident on the satellite (W/cm^2)
P_{eff} = effective power of the laser source (W)
D_{beam} = beam diameter at the satellite range (m)
P_{src} = laser source peak power (W)
T_{src} = laser source transmission efficiency
T_{atm} = atmospheric transmission
λ = laser wavelength (μm)

R = satellite range (km)
M^2 = beam optical quality output at the aperture
D_{eff} = effective aperture beam diameter (cm)

Note that Eqs. 1–3 are approximations, but they provide a general rule-of-thumb to reasonably estimate the amount of incident irradiance. The actual irradiance may vary depending on the laser system and the exact conditions under which it is used. These equations and their assumptions are further described in [19].

Sensor optical gain will increase the irradiance incident on the detector as shown in Eq. 4.

$$I_{det} = K I_{inc} \qquad (4)$$

where:
I_{det} = irradiance incident on the detector (W/cm^2)
K = sensor optical gain

The results from Eqs. 1-4 for the laser sources in this study are compared against damage-onset thresholds for common satellite optical sensor detector materials. The "Bartoli" model for damage-onset is used to calculate the irradiance required to raise a single pixel of a detector material from its initial temperature to its melt temperature [20]. Nominal material parameters for the Bartoli calculation were obtained for four common detector materials: silicon (Si), indium-antimonide (InSb), indium-galium-arsenide (InGaAs), and mercury-cadmium-teluride (HgCdTe). For each detector material, a Bartoli damage-onset threshold curve was generated as a function of dwell time for individual laser wavelengths. Irradiance levels below these threshold curves are expected to be non-damaging, while irradiance levels above the curves are predicted to cause pixel melt.

Comparison plots of the Bartoli threshold curves and irradiance levels at wavelengths representing the four laser types presented in this analysis are shown in Figure 1 through Figure 5. Sources represented by dots in these figures indicate a pulsed source, while a line indicates the source is CW. For the CW source in Figure 3, typical dwell times for inadvertent illumination can range from 10 μsec to 10 msec. Also note that, due to the similarities in laser source characteristics, some sources have overlapping points on the plots; hence, while calculations were made for every laser source, the number of sources plotted may appear smaller than the actual number that was analyzed.

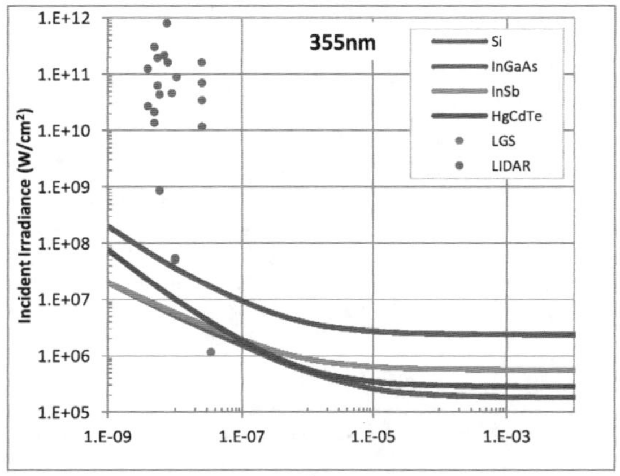

Figure 1: Damage-Onset at 355nm

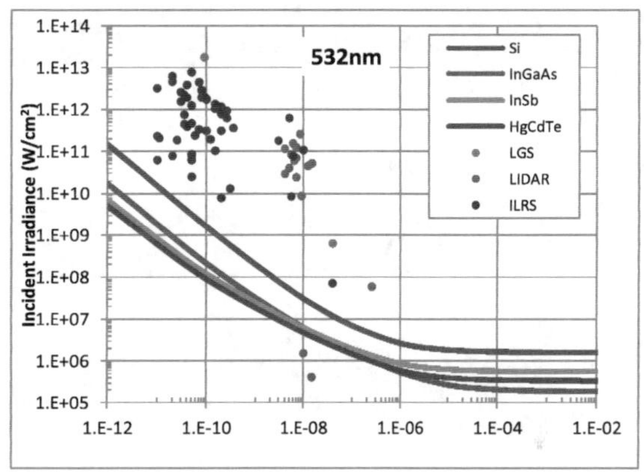

Figure 2: Damage-Onset at 532nm

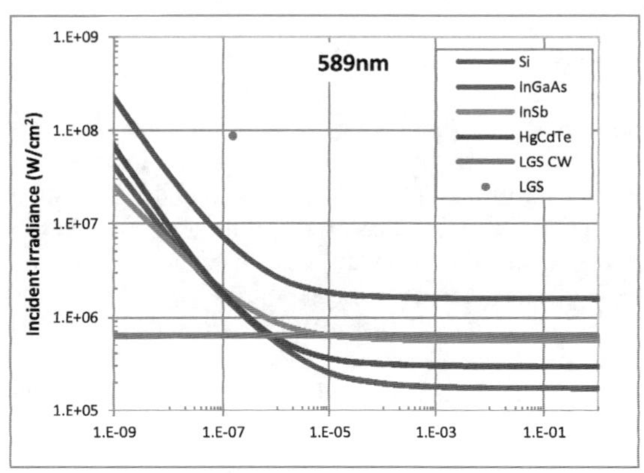

Figure 3: Damage-Onset at 589nm

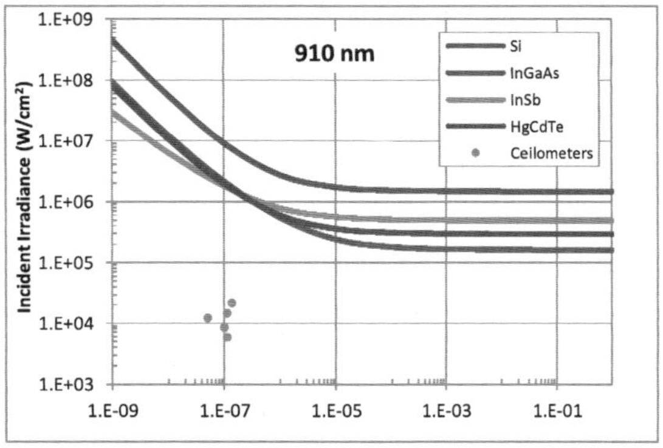

Figure 4: Damage-Onset at 910nm

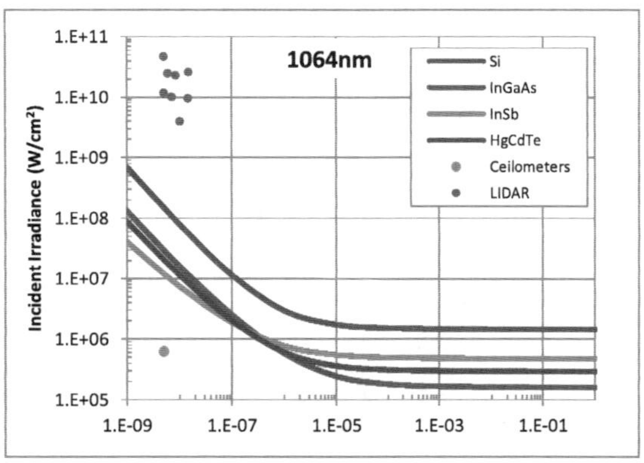

Figure 5: Damage-Onset at 1064nm

As seen in Figure 1 through Figure 5, most of the laser sources considered in this study have the potential to propagate an incident irradiance that exceeds the damage-onset thresholds for detector materials; only the ceilometers do not appear to be a cause for concern at this time. Many of the sources also exceed the damage-onset thresholds by multiple orders of magnitude, indicating a potential for wide area or even full array damage. In addition, the Bartoli model only accounts for thermal damage effects; at short dwell times (e.g., less than tens of nanoseconds), non-thermal damage mechanisms become important. Hence, particularly for the short pulses of ILRS, the damage threshold curves may be even lower than shown here.

Assumptions were made in this analysis that represent nominal and worst-case conditions. Hence, results may vary depending on the actual conditions during laser use. Here, it was assumed that there were no clouds to effectively block the beams or atmospheric scintillation to vary the beam irradiances. The combined product of atmosphere transmission and laser source throughput efficiency ($T_{atm}T_{src}$) was estimated to be 0.81. The slant range (R) was 625 km, which is the median perigee altitude for the satellite population considered here.

Furthermore, these comparisons only indicate the ability of the laser sources to propagate irradiances on orbit that exceed the damage-onset thresholds. Other statistical variations (e.g., such as those due to atmospheric scintillation or in experiment uncertainties from measured material parameters of the Bartoli model) are not addressed. Hence, these comparison plots do not account for the probability, or risk, that a laser source can cause damage.

5. INADVERTENT ILLUMINATION RISK

A series of coincident events must take place in order for a laser to inadvertently illuminate and potentially damage a satellite optical sensor. These events can be illustrated within the context of a typical sodium-LGS scenario. This scenario is notionally broken down into four main stochastic factors as depicted in Figure 6.

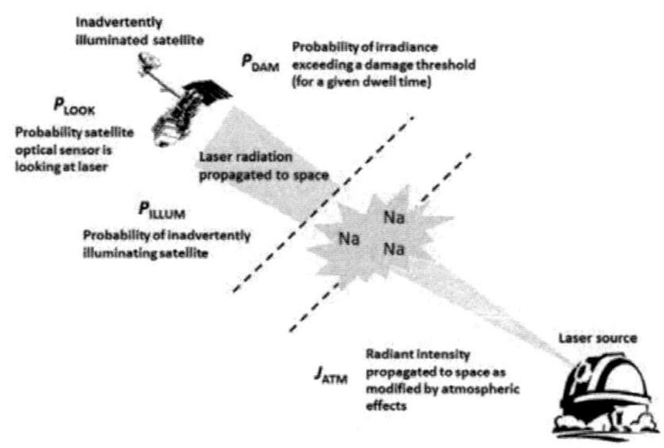

Figure 6. Stochastic Factors of Inadvertent Laser Illumination and Damage

Only about 2% up to 10% of the beam's radiation is absorbed in the sodium layer [21]. Since most of the beam can continue to propagate up to orbital altitudes, it is possible for a satellite passing overhead to be inadvertently exposed to the beam's radiation. However, the following factors must be considered: 1) the atmosphere will cause fluctuations in the beam size and intensity at orbital altitudes (J_{ATM}), 2) the satellite must happen to be passing overhead and through the location where the beam is pointing in the sky (P_{ILLUM}), 3) the optical sensor must be pointed at the laser source in order for the for the beam's radiation to impinge on the detector (P_{LOOK}), and 4) variances in the measured parameters for the Bartoli model yield uncertainties on the damage-onset threshold (P_{DAM}).

The AFRL SatAC has undertaken efforts to develop a PRA methodology to quantify and better understand the inadvertent illumination risks to satellites. This methodology uses a Monte Carlo approach to account for the factors illustrated in Figure 6. A more detailed description of the methodology can be found in [22]. With continued support from both AFRL leadership and the High Energy Joint Technology Office (HEL JTO), this PRA capability can address most laser applications.

However, this PRA approach is data-intensive and relies on numerous parameters related to the laser system, laser operations, atmospheric statistics, and space system designs and operations. As a preliminary study, we perform a less data-intensive analysis by using the PRA methodology to calculate only the probability of inadvertent illumination, i.e., only P_{ILLUM} as shown in Figure 6 but including the turbulence-induced beam broadening part of J_{ATM}. Thus, only key information pertaining to satellite orbits and laser sources is required.

For each of the four types of laser sources (LIDAR, ceilometers, SLR, LGS), PRA scenarios were defined using representative laser sites and the full satellite population. Each scenario simulated the operations of a representative laser and propagated the satellite orbits to determine the satellite positions relative to the beam. The satellite population and ephemerides were obtained by selecting all of the active satellites available on the Celestrak website as of October 2015, yielding a population of approximately 1300 satellites [23].

Simulating all laser sources in the global laser population was not possible due to the limited resources of this study. Hence, representative sites were chosen to reduce the required number of simulations. The numbers and locations of representative lasers simulated for each laser type were selected based upon the geographical distributions of laser site locations. For laser types with heavier concentrations of sources in a particular region, the regions of higher concentration were chosen for the representative sites. Laser CONcept of OPerations (CONOPs) for each ground site describe when (day or night), where (pointing direction), how often, and how long a laser operates. CONOPs information can be difficult to obtain, so assessments were made that may over-estimate the actual CONOPs.

During simulation, an illumination is counted when a satellite is determined to pass through the simulated laser beam cone. The resultant P_{ILLUM} is a mean value per laser operation, calculated from the outcomes of tens of millions or more of simulated laser operations performed for each scenario. This mean P_{ILLUM} is extrapolated to the number of laser operations within a year and the estimated number of laser sources that are operating around the globe. Since the illuminations are inadvertent, the P_{ILLUM} values for individual laser sources are treated as being independent when calculating the overall P_{ILLUM} for the global laser population.

Table 3 summarizes the results for these illumination-only PRA calculations. Note that this type of illumination-only analysis does not imply that any laser radiation has entered the optical system or has exceeded damage-onset thresholds. Instead, these results only pertain to the chance that one or more satellites can be inadvertently exposed to a laser beam, which, by itself, is not an indicator for damage.

As shown in Table 3, the probabilities of one or more satellites being inadvertently illuminated by a LIDAR, ceilometer, SLR, or LGS over the course of a year are extremely high. These high probabilities are due to the number of satellites in Earth orbit, the numbers of laser sources in the global laser population, and the high-activity laser CONOPs. Due to the conservative estimates made (particularly for laser CONOPs), these results should be interpreted as an upper limit on the actual probabilities. These results will be better refined in future work.

Table 3: Illumination-Only Probability (P_{ILLUM}) Results

	LIDAR	Ceilometer	SLR	LGS
Number of Representative Laser Sources Simulated	3	2	1	1
CONOPs for a Single Laser (hrs/day)	6	24	6	6
P_{ILLUM} for a Single Laser (1-year of CONOPs)	0.19 – 0.32	1.0†	0.91	0.12
Global Number of Lasers Found‡	250	250	50	17
P_{ILLUM} for Global Laser Population‡ (1-year of CONOPs)	1.0	1.0	1.0	0.89

† P_{ILLUM} for each of the two representative ceilometers that were simulated is 1 when accounting for a year of CONOPs

‡ does not include all lasers in the global population

6. LASER/SATELLITE INTEROPERABILITY

Given the high likelihood of one or more satellites being inadvertently illuminated, as suggested by the preliminary results in Table 3, it may be possible that satellites already have been or are now being inadvertently illuminated. If inadvertent illuminations have occurred, they may not have been publicly reported, noticed (perhaps due to negligible effects), or attributed to laser activities. Unless an inadvertent illumination results in a significant anomaly, it might be easy to overlook or dismiss as being due to natural phenomena.

For example, satellite optical sensors do commonly lose pixels on orbit due to the background radiation environment. Even a few damaged pixels do not necessarily impair the space mission. Image processing techniques are used in common practice to compensate for damaged pixels without significantly degrading the quality of an image, and image analysts are often not even aware of damaged pixels in an image.

A literature survey revealed average detector pixel loss rates on the order of ~0.01% to 4% per year [24], [25], [26], [27], [28]. Since losing a pixel is typically not a significant impact to the space mission, the exact causes of pixel loss are not usually investigated in detail or reported. Hence, at this time, it cannot be ruled out that some of the reported pixel losses might also be due to inadvertent laser illuminations.

Even though the illumination-only results of Table 3 are an upper bound, it is conceivably only a matter of time before a satellite is inadvertently illuminated by a laser operating somewhere around the globe. Although simply illuminating a satellite does not necessarily result in instantaneous and catastrophic damage, the potential for existing LIDAR, SLR, and LGS sources to damage satellite optical sensors (as indicated by Figure 1 through Figure 5) are a genuine cause for concern.

The numbers, applications, and power levels of lasers operating world-wide will increase over time with advances in technology. At the same time, the numbers, applications, and sensitivities of satellites to laser illumination hazards will also increase. As notionally represented by the graph of Figure 7, the risks of inadvertent illumination and damage will continue to rise, but the rate at which these risks are increasing is currently unknown.

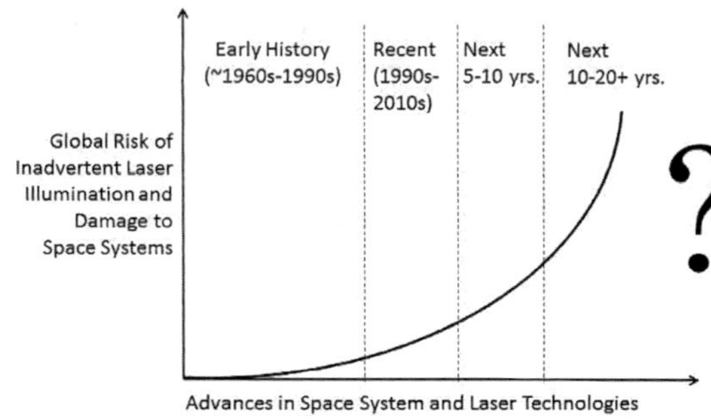

Figure 7. Increasing Risks of Inadvertent Illumination and Damage

As the risk environment becomes characterized with additional PRA studies, it is desirable to consider proactive measures before a significant mishap event occurs, potentially resulting in serious damage to a satellite or even an international incident. Given the importance of both space and laser technologies to our modern, global society, an appropriate balance of international measures is needed that ensure the safe interoperability between satellites and lasers without over-regulation. In many cases, the space and laser actors will be one and the same.

Certain elements of the existing DoD PA process could be considered, but simply scaling up the existing DoD process would likely be unworkable on an international scale. The DoD process only implements controls that regulate laser activities, placing most of the burden of mitigating the risk on laser programs. Further, a key obstacle would be the transnational sharing and protection of detailed information between national, commercial, and private entities required in performing deconfliction. Any information of a sensitive nature would entail international politics, security matters, and proprietary concerns that would have to be addressed.

There are measures that satellite owners/operators can take to mitigate these risks individually and as a community. In passively accepting inadvertent illumination risks, satellite owners/operators might consider insuring against potential, but unlikely, extreme pixel loss conditions that may seriously degrade images or render a detector unusable. Insurance solutions could be financial, covering some fractional cost up to the full cost of the satellite. Materiel solutions may also be possible for some satellite programs; for example, fully redundant "backup" spacecraft readily available for launch to replace/augment capability when necessary. Such financial and materiel solutions are certainly not anything new; they are regularly employed by the space industry to handle other kinds of risks.

Active measures of protection against inadvertent illumination could leverage readily available technology and design techniques. Optical filters can reject laser radiation, up to some exposure level, from common laser wavelengths; filters are commercially available and cost effective. Masking detector array electronics is important to protect against a single illumination rendering a detector array completely useless.

Similar to existing standards for hardening against space environment hazards, a consistent set of standards for protection against inadvertent illuminations could be promulgated within the space community. The standards would specify measures of protection up to a given exposure level within designated open-use bands, which would correspond to an accepted laser low intensity level. The open-use wavelengths would be another type of spectrum allocation as is currently done to prevent interference of radio frequency communications. In open-use wavelengths for lasers, portions of the electromagnetic spectrum would be set aside for lasers of sufficiently low intensity to operate without substantial satellite deconfliction requirements.

Some commonly used wavelengths to consider for the open-use bands could include the following:

- 0.532 μm and 0.589 μm: Both of these wavelengths appear widely worldwide and include LGS and SLR systems.
- 1.025 μm - 1.075 μm: Lasers in this wavelength band are used for many purposes including LIDAR and military applications.
- 1.52 μm - 1.58 μm: Many of the lasers involved in laser communications exist in this band.

Operationally, an international system could be established to collect and publish Notices to Space Operators (NOTSOs) of laser activities. NOTSOs would be analogous to the current Notices to Airmen (NOTAMs) that are used in the aviation industry to alert aircraft pilots of potential hazards that may be encountered en-route or in a particular area. Certain outdoor laser activities are already required by aviation authorities to produce NOTAMs so that pilots may avoid or at least be aware of potential laser hazards.

NOTSOs would serve a similar function by warning space operators of potentially hazardous laser activities. The same basic information as contained in NOTAMs should at least be included in NOTSOs, namely, where and when a potentially hazardous laser may be operating. A few additional details on the laser system (e.g., wavelength, power, aperture diameter, beam quality, pointing directions) may be stated in the NOTSO so that space operators can determine the degree to which the laser poses a hazard to their specific satellite.

7. CONCLUSIONS AND RECOMMENDATIONS

In this preliminary study, we have found that common laser sources have the ability to propagate irradiances on orbit that exceed optical detector material damage-onset thresholds. Also, the chance that these laser sources can inadvertently illuminate one or more satellites is extremely high with probabilities ranging from 0.89 to 1.0 over the course of a year. Although they are likely to be upper bounds on the actual risks (due to assumptions made), these results suggest that some satellites may have been or are being accidentally illuminated, and possibly at levels sufficient to cause pixel damage to optical detectors. However, unless a significant anomaly is identified, it can be easy to overlook or dismiss pixel losses as being caused by natural phenomena instead of from laser activities.

Regardless of whether or not inadvertent illuminations have occurred, the risks will continue to grow with advancements in both satellite and laser technologies. This situation will be further exacerbated with accelerating trends to utilize lasers onboard satellites (e.g., space-based laser communications [29] and LIDARs [30]). From the information provided in this paper, we hope to encourage discussion on a path forward towards the establishment of international measures of protection for laser activities.

As described here, possible measures to consider are derived from existing practices that include passive acceptance and insurance against the risks, promulgation and implementation of design standards, allocation of spectral bands, and operational notifications of laser activities. Given the importance of both space and laser technologies to our modern society, an appropriate balance of protection measures is needed without over-regulation. This effort will require the involvement of both satellite and laser communities. Since the risks will only continue to grow, it is worthwhile to consider measures of protection in order to ensure the safe interoperability between satellites and lasers.

8. ACKNOWLEDGEMENTS

The views expressed herein are solely those of the authors and are not the official stance of the U.S. government. Contracted efforts were performed under contract to the AFRL SatAC with support also provided by the HEL JTO.

9. REFERENCES

1. N. Atkinson, "Amateur Astronomers Flash the Space Station," 2012. [Online]. Available: http://www.universetoday.com/93987/amateurastronomers-flash-the-space-station/.

2. T. Lei, W. Jianhua, L. Jun and H. Qun, "Study on Damage Effect and Threshold of High Energy 1.06 Micron-Wavelength Long-Pulse Laser to Photo Detector," in *IEEE 2012 16th International Electromagnetic Launch Technology*, 2012.

3. S. Junfeng, L. Yang, W. Tingfeng and G. Jin, "Damage Effect of Charged Coupled Device with Multiple-Pulse Picosecond Laser," *Acta Armamentarii*, vol. 35, no. 9, pp. 1408-1413, 2014.

4. S. Hongbin, S. Xueju, Z. Bing, M. Shaojuan, J. Nan and L. Gang, "Experimental Study of 532 nm Pulse Laser Irradiating CCD," *High Power Laser and Particle Beams*, vol. 21, no. 10, pp. 1449- 1454, 2009.

5. Z. Zhiwu, Z. Zhen, C. Xiang'ai, H. Lianglin and L. Zejin, "Damage Phenomenon and Probability of CCD Detectors Under Single-Laser-Pulse Irradiation," *Infrared and Laser Engineering*, vol. 42, no. 1, pp. 113-118, 2013.

6. L. Hua, W. Xi, N. Jinsong, B. Jintian, L. Peng and H. Xiangnan, "Influence of Pulse Width on Damage Effects of CCD Detector Induced by Laser," *Infrared and Laser Engineering*, vol. 42, no. S2, pp. 403-406, 2013.

7. S. Junfeng, G. Jin, W. Tingfeng and W. Ming, "Experimental Research on Femtosecond Laser Damaging Array CCD Cameras," in *2nd International Symposium on Laser Interaction with Matter (LIMIS 2012)*, Proceedings of SPIE, vol. 8796, 2013.

8. U.S. Department of Defense, *Illumination of Objects in Space by Lasers, Department of Defense Instruction*, DODI 3100.11, March 2000.

9. A. Sijan, "Development of Highly Compact and Low Power Consumption Athermal Military Designators," in *Proceedings of SPIE Electro-Optical and Infrared Systems: Technology and Applications IX*, 2012.

10. D. Kiel, "Is This the Time for a High-Energy Laser Weapon Program?," *SPIE Optical Engineering*, vol. 52, no. 2, pp. 021008-1 - 021008-4, 2013.

11. P. Tucker, "Navy Tests Super Precise Laser Weapon in Persian Gulf," 2014. [Online]. Available: http://www.defenseone.com/technology/2014/12/navy-tests-super-precise-laser-weapon-persian-gulf/100929/.

12. NASA Langley Research Center, *Airborne UV Differential Absorption Lidar (DIAL): Measuring Ozone and Aerosols in the Atmosphere*, 2000.

13. W. Thomas, "Ceilometer and Lidar networks for aerosol detection," 19 March 2013. [Online]. Available: https://www.wmo.int/pages/prog/arep/gaw/documents/GAW-2013-Thomas-DWD-Ceilometer.pdf.

14. NASA Goddard Space Flight Center, "International Laser Ranging Service 2009-2010 Report," C. Noll & M. Pealman (eds.), Greenbelt, MD, 2012.

15. M. R. Pearlman, J. J. Degnan and J. M. Bosworth, "The International Laser Ranging Service," *Advances in Space Research*, vol. 30, no. 2, pp. 135-143, 2002.

16. C. Barache, "Lunar Laser Ranging Observations from 1969 to May 2013," [Online]. Available: http://polac.obspm.fr/llrdatae.html.

17. J. W. Hardy, Adaptive Optics for Astronomical Telescopes, Oxford University Press, 448pp., 1998.

18. M. C. Liu, "LGS AO Science Impact: Present and Future Perspectives," in *Proceedings of the SPIE Adaptive Optics Systems*, 2008.

19. P. Shriver, C. Cogburn, C. Jones, M. Campbell and K. Miller, "Considerations for a Laser/Satellite Interoperability Plan," in *International Laser Safety Conference Program and Proceedings*, Laser Institute of America, Paper #703, 185-194, 2015.

20. F. Bartoli, L. Esterowitz, R. Allen and M. Kruer, "A Generalized Thermal Model for Laser Damage in Infrared Detectors," *Journal of Applied Physics*, vol. 47, no. 7, pp. 2875-2881, 1976.

21. A. Quirrenbach, "Adaptive Optics with Laser Guide Stars: Basic Concepts and Limitations," in *Laser Guide Star Adaptive Optics for Astronomy*, NATO ASI Series, Netherlands, Kluwer Academic Publishers, vol. 551, Chapter 2, 2000, pp. 23-50.

22. P. Shriver, S. Gay, T. Gustafson, C. Cogburn, J. Davis, J. Montoya and S. Bump, "Probabilistic Risk Analysis Methodology on Inadvertent Laser Illumination of Satellite Optical Systems," AIAA *Journal of Spacecraft and Rockets*, vol. 51, no. 6, pp. 1994-2007, 2014.

23. T. S. Kelso, "Celestrak," Center for Space Standards and Innovation, [Online]. Available: http://celestrak.com/. [Accessed 2015].

24. J. L. Hora, B. M. Patten, G. G. Fazio and W. J. Glaccum, "The Effects of Cosmic Rays and Solar Flares on the IRAC Detectors: the First Two Years of Inflight Operation," in *Proceedings of SPIE High Energy, Optical, and Infrared Detectors for Astronomy II*, vol. 6276, 2006.

25. M. Buchwitz, H. Bovensmann, M. Reuter, K. Gerilowski and J. P. Burrows, "Carbon Monitoring Satellite - CarbonSat," Mission Requirements Document, version 3.2, IUP-CSMRD-0001, 31 May 2010.

26. S. Le Mouelic, Y. Langevin and S. Erard, "A new data reduction approach for the Clementine NIR data set: Application to Aristillus, Aristarchus and Kepler," *Journal of Geophysical Research*, vol. 104, no. E2, pp. 3833-3843, 25 February 1999.

27. J.-P. Halain, D. Berghmans, J.-M. Defise, E. Renotte, T. Thibert, E. Mazy, P. Rochus, B. Nicula, A. De Groof, D. Seaton and U. Schuhle, "First light of SWAP on-board PROBA2," *Proceedings of SPIE: Space Telescopes and Instrumentation 2010: Ultraviolet to Gamma Ray*, vol. 7732, 29 July 2010.

28. S. G. Garcia, F. Schreier, G. Lichtenberg and S. Slijkhuis, "Near infrared nadir retrieval of vertical column densities: methodology and application to SCIAMACHY," *Atmospheric Measurement Techniques*, vol. 4, pp. 2633-2657, 8 December 2011.

29. M. Toyoshima, "Trends in Laser Comminications in Space," *Space Japan Review*, pp. vol. 70, 1-6, 2010.

30. J. B. Abshire, *NASA's Space Lidar Measurements of the Earth and Planets: A Brief Overview*, 2011

THE IMPACT OF NEW TRENDS IN SATELLITE LAUNCHES ON THE ORBITAL DEBRIS ENVIRONMENT

Arif Göktuğ Karacalıoğlu(1), Jan Stupl(2)

(1) work performed while Associate Researcher, STC / NASA Ames Research Center
NASA Ames Research Center, Moffett Field, United States, E-Mail: goktug.karacalioglu@community.isunet.edu
(2) Research Scientist, SGT Inc. / NASA Ames Research Center
NASA Ames Research Center, Moffett Field, United States, E-Mail: jan.stupl@nasa.gov

ABSTRACT

The main goal of this study is to examine the impact of new trends in satellite launch activities on the orbital debris environment and collision risk.

As a foundation for the study, we developed a deployment scenario for satellites and associated rocket bodies based on publicly announced future missions. The upcoming orbital injection technologies, such as the new launch vehicles dedicated for small spacecraft and propulsive interstages, are also considered in this scenario. We then used a simulation tool developed in-house to propagate the objects within this scenario using variable-sized time-steps as small as one second to detect conjunctions between objects.

The simulation makes it possible to follow the short- and long-term effects of a particular satellite or constellation in the space environment. Likewise, the effects of changes in the debris environment on a particular satellite or constellation can be evaluated. It is our hope that the results of this paper and further utilization of the developed simulation tool will assist in the investigation of more accurate deorbiting metrics to replace the generic 25-year disposal guidelines, as well as to guide future launches toward more sustainable and safe orbits.

1. INTRODUCTION

Starting from the launch of the first artificial satellite in 1957, spaceborne technology has become an indispensable part of our lives. More than 7,000 satellites have been launched into Earth orbit since then, contributing significantly to our understanding of space and Earth, and making our life easier [1]. Unfortunately, these systems, similar to any other system on Earth, do not have an infinite lifetime, therefore they will stop functioning once they are out of fuel or one of their systems fails. Currently, only around 1,400 satellites in orbit are operational [2].

During the early space age, it was not considered what would happen to the satellites once they become nonoperational. No measures were implemented to retrieve or dispose them. This resulted in an unnecessary accumulation of retired spacecraft in Earth orbits. Not only satellites contributed to the increase of the orbital density, but also the upper stage engines, which carried the satellites to orbit, fragments from engine exhausts and many other human-made objects.

Today, the space community is aware of the orbital debris and the problems it causes. A worldwide system of ground-based radars, telescopes, along with space-based sensors is utilized for tracking and cataloging orbital objects. The SpaceTrack database is maintained by the Joint Space Operations Center (JSpOC), part of the U.S. Strategic Command. Conjunction warnings are provided by the JSpOC to space operators in order for them to execute collision avoidance maneuvers. Before a launch vehicle lifts off, its trajectory is checked against the trajectories of orbital objects to avoid any collisions. [3]

Unfortunately the number of fragments in specific orbits is so high that accidental collisions are unavoidably producing more debris pieces than the rate of debris to fall naturally back to Earth due to atmospheric drag. A 1978 article [4] by Donald Kessler and Burton Cour-Palais discussed, for the first time, this potential of orbital debris becoming self-perpetuating; this phenomenon has been addressed as the "Kessler syndrome" in the literature since then.

Beside the technical challenge of cleaning up orbital debris, one of the main reasons that cause the hesitation and abstention to initiate the implementation of mitigation and remediation operations are the economic uncertainties. Although the space community agrees that space debris is threatening the orbital assets, decision makers still do not have a clear understanding of whether the additional cost originated from orbital debris on space operations exceeds the cost of implementing active or passive remediation projects.

There are four major reasons which cause the increase of the number of artificial objects in Earth orbit:

- Injection of new objects into the orbit
- Explosion of on-orbit objects (due to thermal effects
- on residual propellants and pressurants)
- Accidental collisions between on-orbit objects
- Intentional destruction of on-orbit objects

Until recent years, the majority of the orbital break-ups (i.e. more than 200 break-up events to date) were caused by explosions; these explosions were the prevalent cause of orbital debris [5]. However, their effect has become less pronounced due to passivation techniques which eliminate stored energy (i.e. using all the fuel), thus reducing the chance of a breakup [6]. Today, the primary driver of orbital debris is the accidental collisions between objects located in highly crowded orbits. Even simulations assuming no future launches pinpoint a cascading effect. They show that the number of objects and collisions will increase in the absence of remediation activities (i.e. removal of five objects with high mass and collision probability per year from orbit) [7].

However as a no-launch scenario is unrealistic, the satellite industry is likely to grow with more and more start-up companies having been established around the globe. Currently several consulting companies prepare forecasts to estimate the number of satellite launches [8]. However these forecasts do not take into account the orbital parameters for these satellites, nor do they include the other debris-creating factors (i.e. rocket bodies) which are critical for the accuracy of an orbital debris simulation. For this reason, we aim to create our own deployment scenario and then use our simulation tool to propagate their orbits into the future and examine the impact of new trends in satellite launch activities on the orbital debris environment and collision risk.

Section 2 and Section 3 explain our approach to create the scenarios for satellites and rocket bodies, respectively.

2. SATELLITE LAUNCHES

Since the 1960s, the annual number of payloads launched into Earth orbit was around 100-150 without large deviations. Fig. 1 represents this using the data from SpaceTrack database. The blue lines in the graph represent the number of payloads injected into Low Earth Orbit, LEO, (i.e. below 2000 km apogee) in the respective year and the red lines indicate the payloads located at higher altitudes. Due to the past stability of injection rates, most of the state-of-the-art long-term orbital debris projections record the launch activities 8- or 10-years before the start of their simulation and repeat that launch cycle consecutively for the entire simulation period.

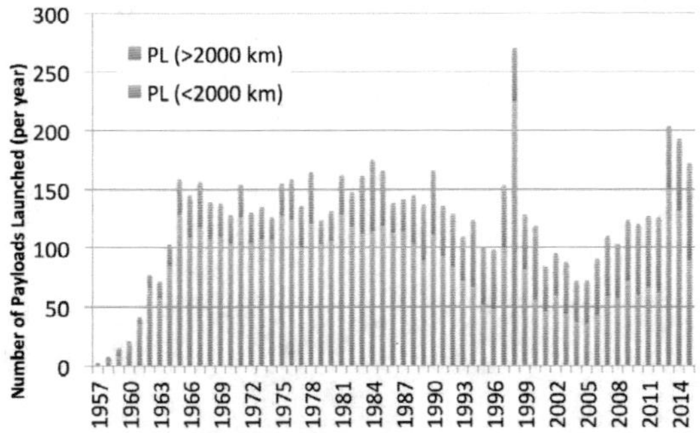

Figure 2-1 - Number of Payloads Injected into Orbit

In recent years, the satellite market has been undergoing a major evolution with new space companies replacing the traditional approach of deploying a few large, complex and costly satellites with a multitude of smaller, less complex and cheaper satellites. This new approach creates a sharp increase in the number of launched satellites and so the historic trends are no longer representative. The early effects of this change can already be observed in the right-most three bars of Fig. 1 representing the years 2013, 2014 and 2015. However, according to our research, the actual boom in small satellite market is likely to happen in the upcoming years.

The only way to make more realistic future predictions in such an emergent, thus unstable, environment is to put together the numbers from an up-to-date market analysis and try to estimate the trend. Since the orbital debris is more of a critical problem in LEO compared to higher altitudes, and since the recent changes in the satellite market is predominantly related with LEO missions, our study is dedicated to develop a scenario for LEO injections.

To develop this scenario, we systematically gather available data on future launches and collect it in a database. We aim to build a database that covers all the publicly available launch related information regarding the companies which intend to launch satellites into LEO between 2016-2030. These companies and/or constellations include, but are not limited to: Blacksky, CICERO, EROS, Landmapper, Leosat, Northstar, O3b, OmniEarth, OneWeb, OuterNet, PlanetIQ, Planet Labs, Radarsat, Terra Bella (formerly Skybox), SpaceX and Spire.

Data is gathered either through direct contact with the company or from online resources (i.e. company press releases and published interviews). Collected data includes: statements on the number of launches for each year between 2016-2030; the target orbits the constellation will be distributed to; and spacecraft mass and area. Whenever data are not available, estimations are made considering the constellation's purpose and company's previous missions, if any. The database also takes into account possible newcomers into commercial Earth observation and telecommunication markets (as additions), as well as the replenishment launches (as extrapolations) of the current and upcoming constellations to keep them operational. We are aware that it is unlikely that all of these companies will survive, however our model assumes a thriving "New Space" economy which would be a worst-case debris scenario.

Table 2-1 - Constellation Launch Data (2016-2020)
(values indicated as "u" refer to satellites in the respective CubeSat form factor; underlined values are our own estimates)

Constellation	Apogee (km)	Perigee (km)	Incl. (deg)	Mass (kg)	Area (m^2)	2016	2017	2018	2019	2020
Commercial Remote Sensing & Weather Tracking										
Landmapper-BC (Astro Digital	600	600	SSO	6u	6u	2	4	4		
Landmapper-HD (Astro Digital)			SSO	16u	16u	2	6	6	6	
GRUS (Axelspace)	675	675	SSO	80	0.4		3	10	10	10
BlackSky Global	450	450	40-55	50	0.8	6	18	18	18	
World View (Digital Globe)	620	620	98	2800	2.5	1				
Digital Globe & Taqnia Space								3	3	
CICERO (GeoOptics)	650	650	SSO	104	1	6	6	12	*4*	*4*
HOPSat (Hera Systems)			SSO	12u	12u	9	10	10	10	9
HyspecIQ	500	500	SSO	600	1.4			*2*		
EROS (ImageSat)	500	500	SSO	350	1		1			*1*
Radarsat Costellation Mission (MDA)	592.7	592.7	SSO	1400	2			3		
NorthStar (NorStar Space Data Inc.)			SSO	750	0.15			10	10	10
OmniEarth	680	680	98	100	0.5				*18*	
PlanetIQ	800	800	72	25	0.01	2	10	6		6
Planet Labs	400	400	0	3u	3u	250	*75*	*75*	*75*	*75*
Satellogic	500	500	SSO	35	0.18	6	19	*50*	*50*	*50*
Spire	550	550	0	50	0.03	50	50	*50*	*50*	*50*
Terra Bella	600	600	0	120	0.4	2	5	5		*8*
Generation 3 (UrtheCast)			0	100	0.5				8	8
Other Remote Sensing & Weather Sat.			0	40	0.15			*25*	50	80
Commercial Telecom										
Iridium NEXT	780	780	86.4	50	0.2	32	40			
LeoSat	1430	1430	0	100	0.5				54	54
O3b	8062	8062	0.1	700	1.5		4	4	*8*	*8*
OneWeb	1200	1200	0	150	0.7			320	330	100
OuterNet	200	200	0	1u	1u	*10*	*12*			
SpaceX	1100	1100	0	200	0.8				300	300
Other Telecom Satellites			0	100	0.5				50	80
Non-Commercial Satellites										
All non-commercial			0	1500	1.5	*115*	*125*	*130*	*135*	*140*
					TOTAL:	**493**	**388**	**743**	**1189**	**993**

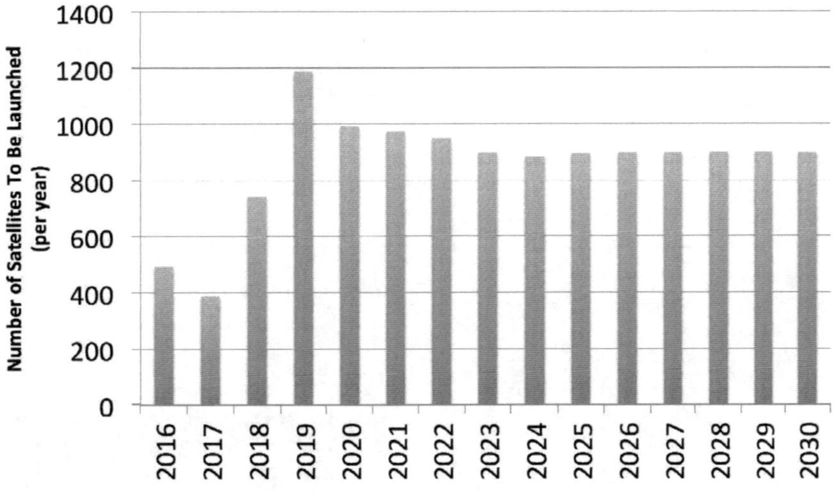

Figure 2-2 - Scenario for Satellites (2016-2030)

Tab. 2-1 shows a summary of our database for the period 2016-2020. Beyond 2020, most of our data has been calculated for replenishment launches, therefore are extrapolations of the first five years. The numbers indicated with underlined italic fonts in Tab. 2-1 are our estimations. Fig. 2-2 shows a summary graph generated from the database for the time interval 2016-2030.

As seen from Tab. 2-1 and Fig. 2-2, the main drivers of the sharp increase are the constellations for telecommunication (i.e. OneWeb and SpaceX). If the installations of these constellations are carried on as announced, these two alone will provide half of the annual launches starting from 2018.

3. LAUNCHED ROCKET BODIES

As aforementioned, satellites are not the sole source for space debris; the rocket bodies that carry them into orbit also contribute to the orbital debris problem. Fig. 3-1 shows the SpaceTrack data for the annually cataloged rocket bodies. This graph illustrates the number of rocket bodies injected into orbit per year peaked in 80s and decreased once satellites became smaller and shared launches popularized, stabilizing around 75 rocket body/year for the last 15 years. Around half of these bodies were positioned at LEO altitudes.

As a cross-check, Tab. 3-1 shows the number of LEO launches for the last five years [9]. Roughly half of the R-7 launches indicated in the table were missions that carried crew and cargo to the International Space Station. It is worth mentioning that the numbers given per year in Fig. 3-1 are not identical to the number of launches in that specific year in Tab. 3-1. This is due to the fact that for some of the launches, there are multiple upper stages remaining in orbit. Conversely, in some missions, the rocket bodies re-enter the atmosphere immediately after the deployment of their payloads; such objects may not be included in the SpaceTrack catalog.

Considering Fig. 3-1 and Tab. 3-1, it is reasonable to estimate that typically 40 rocket bodies have been injected into LEO every year since 2005. However, our analysis on the SpaceTrack catalog shows that some portion of these rocket bodies decays within a couple of days after their launch date and it is not meaningful to consider those in our long-term debris simulations. To find out this ratio, we divide the catalog into six 10-year periods and perform a histogram analysis within these blocks. Tab. 3-2 shows that, historically, 25% to 40% of the rocket bodies injected into orbit re-entered the atmosphere within 10 days. This temporal analysis reveals another interesting result: for the last 50 years, a shrinking percentage

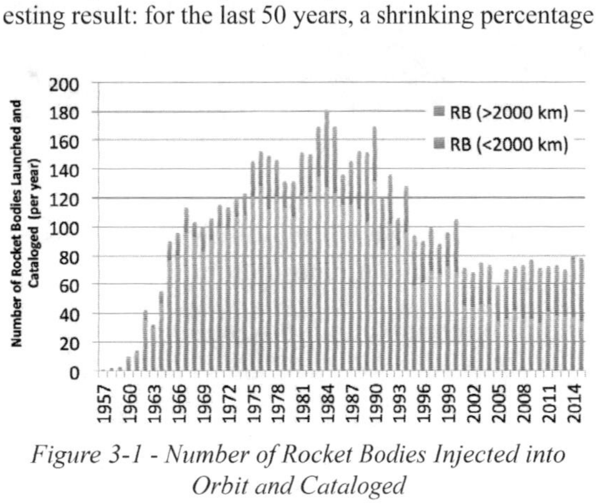

Figure 3-1 - Number of Rocket Bodies Injected into Orbit and Cataloged

Table 3-1 - Number of Successful Launches (LEO only)

Launch Vehicle	2011	2012	2013	2014	2015
R-7 (Soyuz/Molniya)	12	12	13	15	11
Long March)	9	10	10	13	10
Atlas 5	2	2	2	2	3
Ariane 5	1	1	1	1	0
Falcon 9	0	1	2	3	3
Delta 2	2	0	0	1	1
Delta 4	1	1	1	1	0
H-2A	2	1	1	2	2
H-2B	1	1	1	0	1
PSLV	2	2	1	1	3
Antares	0	0	2	2	0
Dnepr	1	0	2	2	1
Rokot	0	1	4	2	2
Vega	0	1	1	1	2
Strela	0	0	1	1	0
Kuaizhou	0	0	1	1	0
Minotaur 1	2	0	1	0	0
Uhna	0	1	0	0	0
Safir	1	1	0	0	1
Pegasus XL	0	1	1	0	0
Shavit	0	0	0	1	0
Epsilon	0	0	1	0	0
Angara	0	0	1	0	0
Zenit	1	0	0	0	0
Sum	**37**	**36**	**47**	**49**	**40**

of the rocket bodies decayed soon after their launch and hence contributed more to the orbital debris problem. However, with the expectation of stricter rules and a potential use of reusable launchers, we build our scenario around an assumption that 30% of the rocket bodies will be de-orbited in the future and 70% will be left for their natural decay.

Having all these historical data, it remains difficult to estimate future trends for the number of rocket body deployments into orbit. Since the satellites are getting smaller in size and weight, more of them fit into a launch vehicle. Therefore, the boom in the small satellite market is unlikely to trigger a sharp increase in demand on the launch sector.

Conversely, there is a widespread effort to enhance orbital injection capabilities and accuracy. A long list of companies such as Microcosm, Rocket Lab, Firefly Space Systems, Sierra Nevada Corporation and Arca Space Corporation are developing new launch vehicles dedicated for small satellites. There are other companies which intend to develop interstages with propulsive capabilities, which will allow the deployment of satellites into their desired orbits beyond the restrictions of the launch vehicle used.

Table 3-2 - Percentage of rocket bodies decayed in 10 days after their launch date, per decade

	1957 - 1966	1967 - 1976	1977 - 1986	1987 - 1996	1997 - 2006	2007 - 2006
# of RB deorbited in 10 days	94	417	495	357	154	110
Total # of RB cataloged	315	1062	1267	1012	583	445
Ratio of RB deorbited in 10 days	30%	39%	39%	35%	26%	25%

Table 3-3 - Average Number of Satellites Launched During Shared Missions

Launch Vehicle	Number of Secondary Payloads Launched	Number of Launches	Average Number of Payloads Launched per Shared Mission
Dnepr	122	12	10.2
PSLV	52	15	3.5
Atlas V	46	4	11.5
Minotaur 1	46	7	6.6
HII-A	27	6	4.5
Soyuz-2	23	5	4.6
Long March	22	5	4.4
Falcon 9	19	3	6.3
Delta II	11	4	2.8
Vega	11	2	5.5
TOTAL	**379**	**63**	**6.0**

Considering these aspects as a whole, we decided to correlate the rocket body deployment scenario with our satellite deployment scenario explained in Section 2. For this purpose, we analyze rideshare missions (i.e. missions containing at least one secondary payload with the primary payload) performed in the last 15 years to find out how many payloads were aboard in each of those launches. Tab. 3-3 lists the main launch vehicles used in these missions. As seen from the table, an average of six payloads was carried per mission. However, there had been launch campaigns which carried more than 30 payloads to space. In May 2016, a Falcon 9 rocket is expected to carry 88 satellites utilizing the Sherpa deployer.

Considering these advancements, we find it reasonable to assume that a LEO launch campaign, on average, will carry nine satellites into orbit in the near future. We also assume that only 70% of the rocket bodies will stay in orbit as explained above. Within this framework, Fig. 3-2 shows the scenario for rocket bodies to be included in our debris simulations. The apogee, perigee and inclination data for these objects were estimated in correlation

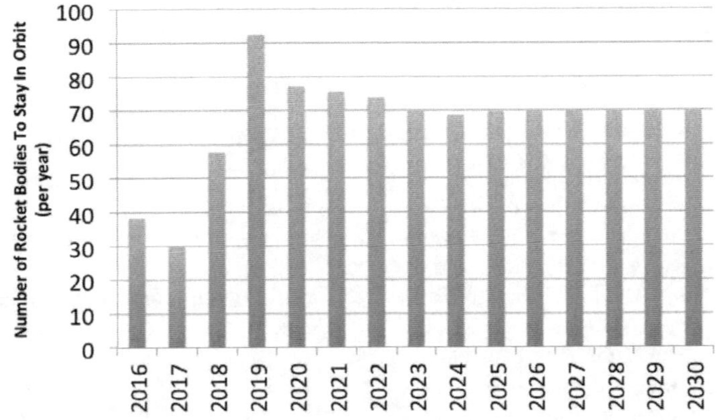

Figure 3-2 - Scenario for Rocket Bodies to Stay In Orbit (2016-2030)

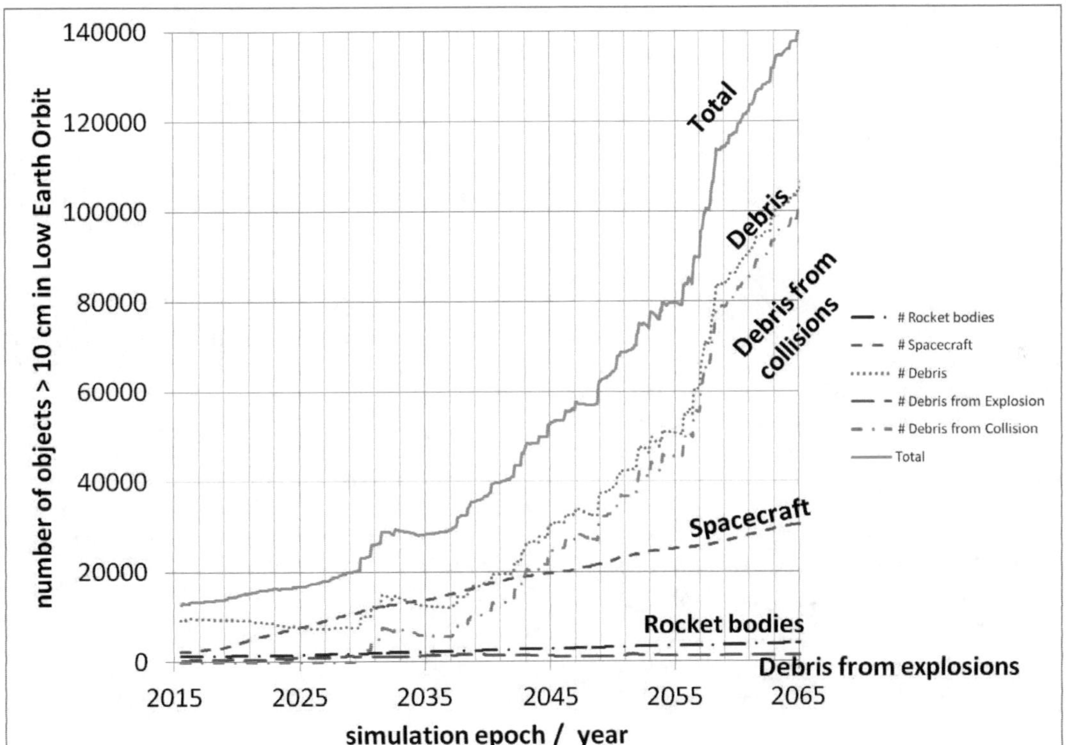

Figure 4-1 - Number of debris objects in LEO over time for a single simulation run using the deployment scenario.

with the information gathered on announced spacecraft launches and the historical trends.

4. SIMULATIONS

The developed deployment scenario feeds into a simulation tool that is capable of propagating the objects with variable-sized time-steps as small as one second. An automated script pulls the necessary parameters from the database and converts them into a suitable format to be fed into the simulation. Launch epoch dates were assigned randomly within the launch year for each constellation from the database. A maximum of 15 objects are allowed for a single launch. Additional parameters, i.e. area-to-mass ratios, drag coefficient and reflectivity, are assigned to each object according to their physical specifications. For modeling 2030 onward, we chose to build the scenario extrapolating the 2016-2030 data on yearly basis. Over the course of the run, the software also detects collisions; additional debris objects are then created according to the NASA breakup model and are then fed back into the simulation framework. More detailed information about the simulation tool can be found in [10] and [11].

5. RESULTS

Fig. 4-1 shows the results of a single simulation run for the number of objects greater than 10 cm in LEO. This run uses an initial population from the SpaceTrack catalogue as of June 2015; additional objects are introduced to the population over time according to our deployment scenarios. Full collision functionality of the code is enabled.

In this (singular) simulation run, the first collisions occur in the late 2029-2032 timeframe. A portion of that debris decays between 2032 and 2036, but from 2036 collisions start to occur regularly, dominating the increase in object population.

In addition to the number of objects, the tool also tracks conjunctions, which are close encounters between space objects. Fig. 5-1 shows all detected conjunctions with a probability of collision P_c larger than 10^{-4}. In the given scenario, there are only 1126 in 2016, rising to a maximum of 179 thousand in 2062. The pronounced spikes are follow-up conjunctions after breakup events, when objects in a debris cloud are still close to each other. Fig. 5-2 shows only conjunctions involving at least one spacecraft and hence omitting the direct after effects of a breakup event. In this case, the number rises from 415 conjunctions with $P_c>10^{-4}$ in 2016 to 44 thousand in 2064. The consequence is that operators might be overwhelmed by those numbers and do not perform collision avoidance maneuvers.

6. CONCLUSIONS AND OUTLOOK

While a single run already provides interesting information, it is necessary to obtain error bounds and average projections with a full Monte Carlo treatment. This task is to be implemented in the future.

Examining the simulation results, the total number of particles to accumulate in different orbits can be monitored and the number of conjunctions can be tracked to assess the collision risks.

The simulation makes it possible to follow the short- and long-term effects of a particular satellite or constellation in the space environment. Likewise, the effects of changes in the debris environment on a particular satellite or constellation can be evaluated.

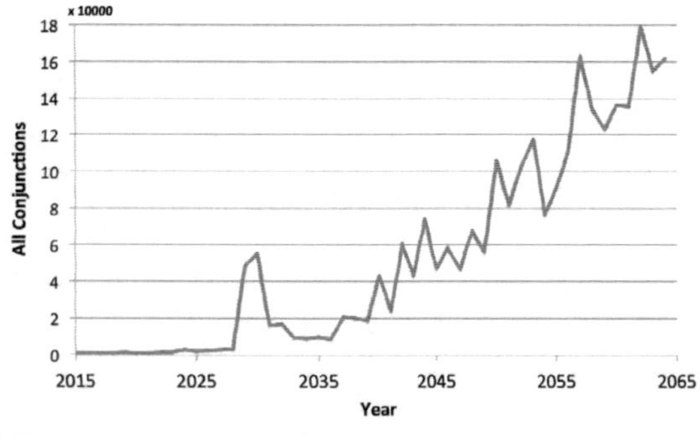

Figure 5-1 - Number of conjunctions with a probability of collision larger than 10^{-4} per year.

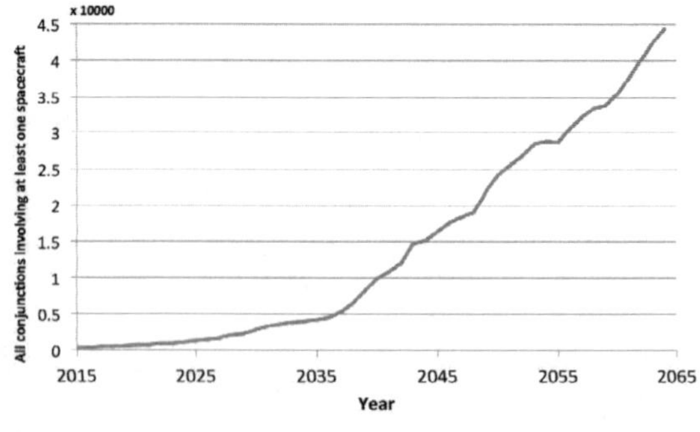

Figure 5-2 - Number of conjunctions involving at least one spacecraft , with a probability of collision larger than 10^{-4}, per per year

It is the authors' hope that the results of this paper and further utilization of the developed simulation tool will assist in the investigation of more accurate deorbiting metrics to replace the generic 25-year disposal requirement, as well as to guide future launches toward more sustainable and safe orbits.

7. ACKNOWLEDGMENTS

While this paper focuses on future launch activities, the data in Fig. 4-1, 5-1 and 5-2 was produced using the simulation tool described in [10] and [11]. We would like to thank Bron Nelson and Fan Yang Yang who are the primary authors of those papers and the tool. We also would like to thank Roberto Carlino, Andres Dono Perez, Nicolas Faber, Cyrus Foster, Chris Henze, Creon Levit, Conor O'Toole, Jason Swenson and Wang Ting for their contributions.

8. REFERENCES

1. Satellite Catalog, accessed on 15 April 2016, <www.space-track.org>
2. UCS Satellite Database, accessed on 15 April 2016, <http://www.ucsusa.org>
3. "Understanding Space Debris Causes, Mitigations, and Issues", Crosslink Fall 2015, The Aerospace Corporation
4. D.J. Kessler, B.G.Cour-Palais, "Collision frequency of artificial satellites: The creation of a debris belt", J. of Geophys. Res., 83(A6), 2637–2646, 1978.1978
5. Clean Space, ESA, accessed on 15 April 2016, <http://www.esa.int/Our_Activities/Space_Engineering/Clean_Space/Space_debris_mitigation>
6. Hildreth, S.A. & Arnold, A., 2014. Threats to U.S. National Security Interests in Space: Orbital Debris Mitigation and Removal.
7. J.-C.Liou,N.L.Johnson,N.M.Hill, "Controlling the growth of future LEO debris populations with active debris removal", Acta Astronautica 66 (2010).
8. Spaceworks Nano / Microsatellite Market Assessment, 2014.
9. Space Launch Report, accessed on 15 April 2016, <www.spacelaunchreport.com/log2015.html>
10. Nelson, B.C. et al., "Implementation of an Open-Scenario, Long-Term Space Debris Simulation Approach", AMOS 2015, September 15-18 2015, Wailea, HI
11. Yang Yang, F. et al., "Lightforce Photon-Pressure Collision Avoidance: Efficiency Analysis in the Current Debris Environment and Long-Term Simulation Perspective", 66th International Astronautical Congress, October 12-16 2015, Jerusalem, Israel.

ORBITAL DEBRIS: WHAT ARE THE BEST NEAR-TERM ACTIONS TO TAKE? A VIEW FROM THE FIELD[i]

Dr. Mark A. Skinner[(1)]

*[(1)] The Boeing Company, 4411 The 25 Way NE, Ste. 350, Albuquerque NM 87109 USA,
Email: Mark.A.Skinner@Boeing.com*

ABSTRACT

The geostationary Earth orbit (GEO) satellite belt* is a unique location above the earth affording a continuous line-of-sight to satellite uplink and downlink stations. The volume defined by this belt is large, but available slots are limited. During the last fifty years of the space age, this volume has become more crowded, as humankind has launched more and more satellites into this particular orbital regime, and satellites that suffered incapacitating anomalies and space debris have remained in the belt. The latter pose a hazard since they are uncontrolled, and the only way for satellite operators to avoid collisions with space objects is to maneuver. Knowing when and where to maneuver requires space situational awareness (SSA), but this is just one aspect needed to maintain safety of flight in this very valuable orbital regime.

This paper reports on, from the point of view of an SSA practitioner, what the key issues and dangers surrounding the current situation in the GEO belt are, and what the best possible set of near-term actions could be, involving international cooperation (through bodies such as the United Nations Committee On the Peaceful Uses of Outer Space (UN COPUOS)), data sharing between actors in the space arena, public and private sector SSA efforts, and nascent research efforts into active space debris removal. Where should limited available resources be applied to affect the best possible outcome?

1. KEY ISSUES AND DANGERS IN GEO

Operating in the geostationary Earth orbit (GEO) orbital regime entails operating in a harsh environment (including such factors as being in a vacuum, and having to deal with natural and artificial hazards such as increased radiation and charged particles, orbital debris, space weathering, *etc.*), at long distances that generally preclude making repairs to orbiting systems. But a slot at GEO is a valued, finite resource, and there is competition between countries and companies to utilize the GEO orbit. Below are described the major issues and hazards that are part of operating in the GEO orbit.

1.1 Space Debris Near GEO

The year 1963 saw the dawn of the commercial satellite era with the launch of the first GEO satellite, Syncom 2. Geosynchronous satellite communications have since provided many benefits to mankind. However, "as a

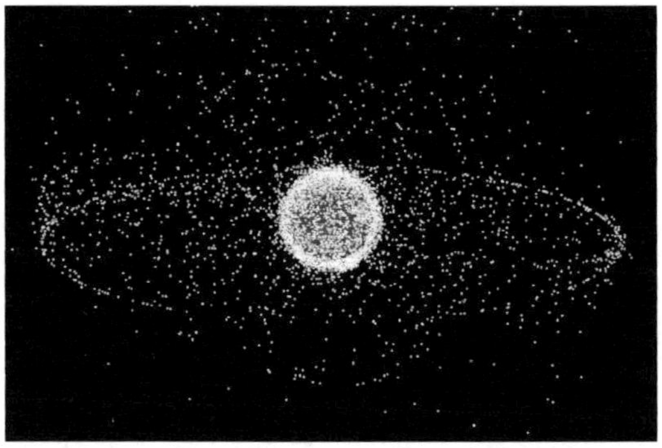

Figure 1. Computer generated images of objects in Earth orbit that are currently being tracked. Approximately 95% of the objects in this illustration are orbital debris, i.e., not functional satellites. The dots represent the current location of each item. The orbital debris dots are scaled according to the image size of the graphic to optimize their visibility and are not scaled to Earth. This image was generated from a distant oblique vantage point to provide a good view of the object population in the geosynchronous region (around 35,785 km altitude). Note that the larger population of objects over the northern hemisphere is due mostly to Russian objects in high-inclination, high-eccentricity orbits. Image and caption courtesy of the NASA Orbital Debris Program Office [2].

* In this work we will refer to the geostationary ring as the altitude at which the orbital period exactly matches the rotation of the earth, with zero eccentricity and zero inclination, such that the object appears exactly fixed in the sky from an earth-based observer. Similarly, the geosynchronous belt, as, similarly to the geostationary ring, that volume where the orbital period is almost synchronous with the rotation period of the earth, and the inclination and eccentricity are not constrained to being zero. Though this implies an unlimited region for geosynchronous objects, the limitation of launch geometries and perturbations constrain the geosynchronous objects to a "band" around the Earth.

result of past activities in space, a massive amount of space debris – non-functional and uncontrolled objects – has been left in Earth orbit which poses a serious challenge to the sustainability of outer space"[1]. Fig. 1 illustrates the concentrations of tracked, man-made objects in Earth orbit, with a concentration of objects evident in the neighborhood of the GEO ring.

One source of larger debris in the GEO ring is the population of non-functioning payloads and upper stages that were placed into the GEO ring, but were unable of being disposed into orbits higher than GEO (as is the current practice) upon their end of life [3]. These objects are trapped in one or both of the two gravitational wells ("pinch points") caused by the gravitational anomalies (more specifically, perturbations related to the tesseral components of the spherical harmonic expansion of the Earth gravitational model) of the Earth at the Equator [4]. Objects that are trapped in one (or both) of the wells oscillate back and forth, passing through the wells, the period and amplitude of the oscillations being dependent on the specific orbital geometry. The objects thus trapped are primarily defunct payloads, including the first commercial GEO communications satellite, Intelsat-1 F1 ("Early Bird") [5]. Tab. 1 summarizes the trapped object population. Fig. 2 shows, for the western pinch point (105° W), the cumulative growth of trapped objects.

Table 1. The oblateness of the earth causes the existence of two stable gravitation wells [6], which 'trap' non-station-kept objects in the geostationary ring. The trapped objects are mostly old payloads [7].

Characteristic	75° East Well	105° West Well	Trapped in Both Wells
Payload: Radugas (29), Gorizonts (9), Ekrans (8), etc	83	39	15
Rocket Body: Largely Proton-K Fourth Stages	17	0	3
Debris: 2006 Feng Yun and 1978 Ekran 2	2	0	0
Total	102	39	18

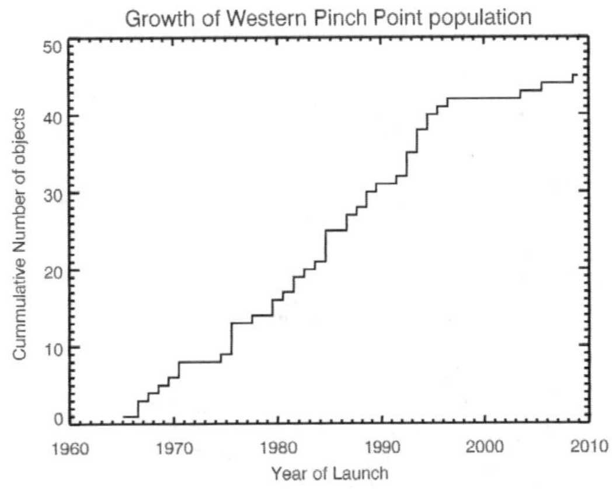

Figure 2 shows the cumulative growth of the number of defunct satellites that orbit about the western pinch point (the geo-potential well at 105° W).

The trapped objects represent a variety of ages (~45 years) and sizes (up to a factor of 30 in size difference), as well as a number of manufacturers and launching states. As may be seen in Fig. 2, the trend has been for fewer objects to become trapped over the years, as operators follow improved "best practices" [8]. However, the trapped objects will remain in GEO, and constitute a continuing threat to operational spacecraft.

Besides the larger (greater than a few square meters in cross sectional area) tracked objects found in the public domain catalog [9], observational campaigns estimate ~500 unknown, un-catalogued debris objects brighter than 18.5 visual magnitudes (about 29 cm in size assuming an albedo of 8%) in the geostationary ring [10]. This figure does not include high area-to-mass (HAMR) objects drifting through the GEO ring, nor does it include large station-kept satellites not included in the public catalog.

In addition to the objects enumerated above that exist in, or drift above or below the GEO belt, there is another class of objects shown in Fig. 3 that are also of concern. These objects have high area-to-mass ratios, and their orbital elements can change on time-scales of weeks to months by solar radiation pressure, which causes them to periodically pass through the GEO belt [11, 12, 13, 14], as may be seen in Fig. 4.

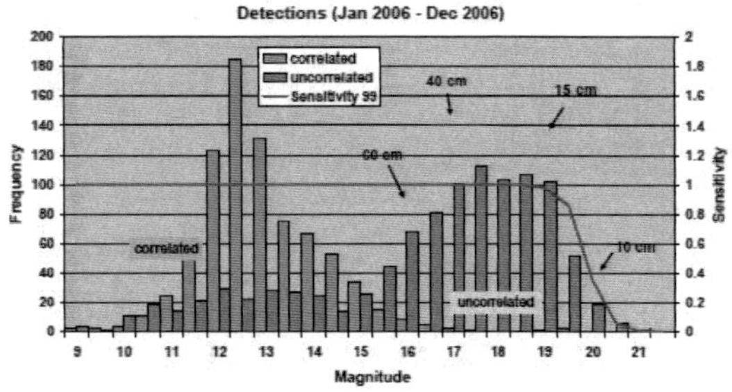

Figure 3 shows the distribution of objects that can be correlated (in blue) with the public catalog, as well as a new population of faint objects that are not found in the catalog (in red). The right-hand distribution of uncorrelated objects contains the high area-to-mass (HAMR) objects [15].

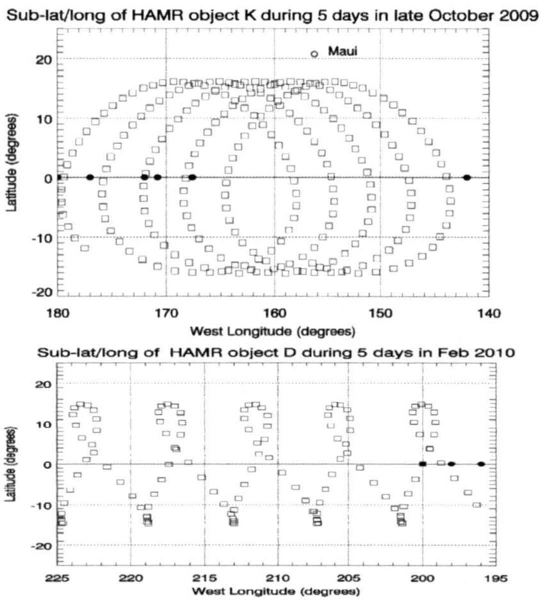

Figure 4 shows the calculated sub-latitude and sublongitude (ground tracks) for two HAMR objects observed from Maui, spanning five days. The filled circles at 0° latitude represent the locations of a subset of the geostationary satellite [16].

1.2 Orbital Congestion and Interference

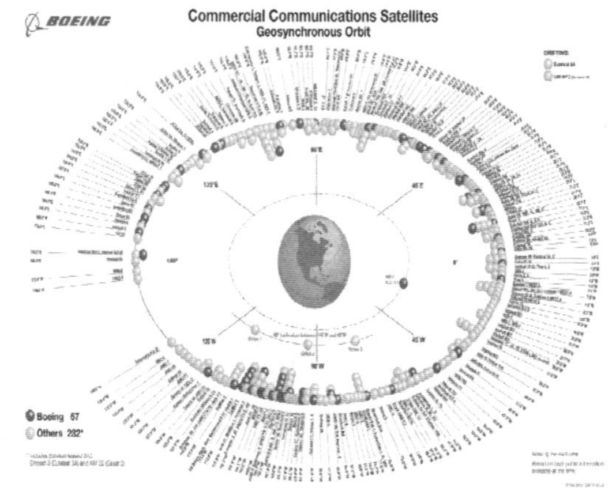

Figure 5, an historical illustration of the distribution of commercial communications satellites in the GEO ring [17] which is an important location for a wide variety of commercial and military satellite users – especially for communications satellites [see 18, 19, 20, and references therein].

1.3 Space Weathering and Satellite Aging

A modern communications satellite is a complex system involving numerous subsystems that have finite lifetimes due to mechanical wearing (*e.g.*, sun-tracking solar panel bearings, momentum wheels, *etc.*), limited propulsion resources (gas jets or similar), and carefully engineered thermal surfaces, which can be subject to degradation due to space "weathering" [21] or collisions with small debris.

Aging can be viewed as a continuous slight degradation of the performance of some surface (*e.g.*, weathering of thermal surfaces), or it can be a discrete event, punctuated by "anomalies" (*i.e.*, a critical component on the satellite fails). Even though satellite subsystems are instrumented with sensors that provide health and status ("housekeeping") data, not everything can be sensed, and sometimes the housekeeping data stream is unavailable. It is important that GEO satellites remain functional until the end of their useful life when they can be re-orbited into the GEO disposal ("graveyard") orbit (Fig. 6) [22], deplete their stored energy sources, and be decommissioned. A satellite that is not properly disposed of only becomes a new element of the debris picture, although it has been speculated that even satellites in the graveyard orbit may possibly shed debris that could drift into the protected GEO zone (or lower)[23].

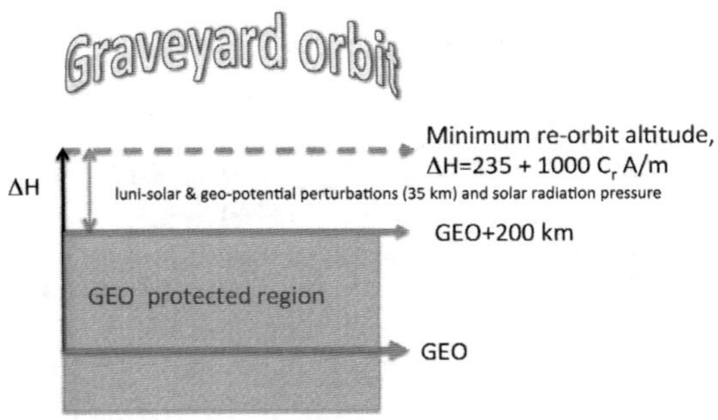

Figure 6 shows the geometry for the post-mission disposal region for geosynchronous satellites, a minimum height above the GEO altitude, which depends on the satellite's area-to-mass ratio.

2. NEAR-TERM ACTIONS

What courses of action are currently underway, or being actively studied for possible implementation to ameliorate some of the issues with operating at GEO? These include recognizing that, as a shared global resource, mankind for the long-term must preserve the GEO belt to allow its sustainable use, and thus efforts are underway at the international level (UN COPUOS and others) to allow lessons to be learned from past activities in space. Especially with recent satellite-debris collisions (e.g., the Iridium-Cosmos collision [24]), there is a better understanding of the need for timely and accurate orbital information on objects both active and inactive. Given that many of the assets currently in GEO or planned for GEO are not those of traditional space-faring national governments, but are owned or controlled by private sector entities, or international or non-governmental organizations, there is a growing need for tailored and timely information for these non-traditional satellite operators. Additionally, a cultural shift is beginning with regards to the proprietary, sensitive nature of some of the data on man-made space objects, and that such data must be shared broadly amongst actors in the space arena to ensure safety of flight in space. Finally, the last half-century of the space age has magnified the realization that, defunct space objects have formed a background of dangerous space trash that could negatively impact the GEO regime due to the possibility of collisions with other space debris or active satellites, and that something must be done in order to lessen the dangers posed by these objects.

2.1 International Cooperation

International efforts are underway to promulgate standards and best-practice guidelines for space actors [25], which will aid in decreasing the amount of orbital debris introduced into the GEO orbit. This will help to contain the debris growth problem. However, this will not prevent the possibility of an existing defunct payload shedding debris objects, nor will it prevent two larger pieces of debris from colliding and creating a cloud of smaller debris objects. Activities of nations acting outside of established international norms, as well as unexpected failures and/or anomalies of operational satellites, also present potential hazards that must be addressed.

However, these efforts do nothing to remediate the current problem, but do attempt to make things better for the future (or at least not make them any worse). In effect, it "buys time" so that other solutions addressing the present situation can be brought to bear.

2.2 Space Situational Awareness (SSA)

The current strategy for active payloads to avoid collisions is to obtain orbit information about objects in space, and then to propagate the various orbits, and their uncertainties, forward in time, so as to estimate the probability of collision between the two objects (a "conjunction assessment"), to determine if it might be necessary to maneuver the operational satellite out of harm's way. This orbital information generally is sourced by the United States Strategic Command (USSTRATCOM) and delivered on the Space-Track.org website, although a disclaimer does warn about the ineffectiveness of using the available two-line element sets on the public site for conjunction assessment. The Joint Space Operations Center (JSpOC) offers more "detailed" data to satellite operators under data sharing agreements, however there have been reports that such detailed data are unreliable [26, 27]. Additional factors adding to the unreliability of the current situation is that maneuvers (*e.g.*, for station-keeping) are not accounted for until after the fact [28], nor is the uncertainty of the prediction quantifiable in a statistical fashion.

Most satellite operators derive positions in space (orbital information) for their own geosynchronous satellites using a technique known as ranging. But, as the satellites appear fixed over a given location above the Earth, it is hard to determine an orbit using only data from a fixed ranging site, as the relative positions of the satellite and the ground site change very little over time. Usually satellite operators combine the ranging data with azimuth and elevation angle data from a high gain antenna that is 'locked-on' to the satellite's maximum signal strength. Orbital information gleaned in this fashion only has an accuracy of about 0°.01 (36 seconds of arc)[29]. This positional knowledge is adequate for a single satellite with no close neighbors for ordinary station-keeping purposes. However, for satellites on-orbit close to one another, it is possibly too crude a measurement for safety of flight. In these cases a second, geographically distant ranging station ("two station ranging) is utilized to provide more accurate trajectory knowledge.

There are some other sources of SSA information, from sources like the International Scientific Optical Network (ISON)[30], as well as nascent commercial SSA operations [31], utilizing electro-optical sensors ("telescopes"), that are helping satellite operators to better determine the orbits of their satellites.

2.3 Data sharing

Currently, data are "shared" (or perhaps "broadcast" would be more appropriate) by the US Space Command to the public via Space-Track.org, and amongst some commercial and civil satellite operators who are members of the Space Data Association (SDA)[32]. Both of these entities serve useful purposes, but have limitations. However, ideas exist for new types of sharing paradigms for data on man-made space objects.

a) Satellite operator-to-operator data sharing
To enhance the quality and timeliness of satellite orbit information, a number of operators have joined together to form the SDA [33]. Its vision is to create a clearing-house for members to exchange information that would overcome some of the issues surrounding the *ad hoc* exchange that had evolved over the years. The SDA serves a useful function, and net membership has grown since its inception.

How well the sharing model works is dependent on the accuracy and timeliness of the data to be shared, which is hindered by so-called "non-participating operators." In addition to the satellite operators that could potentially share orbital data but chose not to participate, there is also the entire population of space debris previously described. Information on these objects could be supplied by the publicly available two-line element sets (TLEs), however with the same caveats as described above, as many debris objects are not found in the public catalogue.

b) International Space Object Data Center
One idea under discussion would be to create an international center for the exchange of data on orbital debris. Space debris information is generally not available within the SDA (debris is not "active," so it is not ranged, nor does it maneuver in addition to the absence of telemetry). Also, much of the debris that is tracked by Space Command is not included in the public catalogue (Space-Track.org), as there is no trace-back to a known launch; thus obscuring the origins of the space debris. One possible model for such an organization (perhaps to be called the 'the International Space Object Data Exchange, or "ISODEX" [34]) could be the Minor Planet Center (MPC). "The MPC is responsible for the designation of minor bodies in the solar system: minor planets; comets [...] and natural satellites [...] The MPC is also responsible for the efficient collection, computation, checking and dissemination of astrometric observations and orbits for minor planets and comets" [35]. The MPC functions as a repository for information on asteroids and similar bodies that is sent by international observers. The ISODEX could function in a very similar fashion, except it would deal with man-made space objects. Setting up and running such an organization might be the job of interested and willing participants, not unlike the SDA.

Observers vetted for data quality and accuracy could send in rapid communications to the ISODEX (utilizing a "crowd-sourced" data-gathering paradigm, like the MPC) of orbital information on newly discovered, or follow-on observations of existing space debris. The ISODEX could maintain "custody" for objects so observed, and publish orbital ephemerides for these objects on a public website, which would be useful information for satellite operators to use in conjunction assessments, as well as for other SSA research.

c) Global SSA center
A related idea recently under discussion at the United Nations Committee on the Peaceful Uses of Outer Space (UNCOPUOS)[36] is that of a global SSA data center, either as a physical center run by the UN, or as a "framework" for a virtual center. While this would certainly be a tremendous help to space actors, a cultural shift in thinking about data on space objects (especially active space objects) would be needed by States, as these data have historically been regarded as either proprietary or sensitive for national defense needs.

2.4 Active Debris Removal

One method to reduce the probability of collision between satellites and debris is to physically remove the debris from the orbital planes of the active satellites. This concept, known as "active debris removal" (ADR), is being discussed and studied by the international community [37]. However, given the difficulty and expense of removing significant amounts of debris from the GEO ring, this is not a tenable solution for the near-term [38], but nonetheless warrants investigation.

As well, there are a number of issues that would need to be resolved before one could begin actively removing debris. Such issues include legal questions, *e.g.*, ownership and liability [39], and in many instances, a system capable of removing debris could also be used to remove an active satellite — creating the risk that such dual-use technology could be considered a "space weapon" [40].

The major short-coming of ADR (aside from stratospherically high cost estimates) is that it is probabilistic in nature; as astrodynamicists can only model resident space objects' interactions with their environment to a certain level of fidelity, estimating their future trajectories is limited to finite time spans, based on current observations. Even though an object might be considered "dangerous" (*i.e.*, it might be massive and have a high probability of collision with another/other object(s)), that is not a certainty, and it might not ever collide with anything. Meanwhile, something with a lower probability of collision (that was not removed) might actually hit something. To truly reduce the risk to the active fleet, a great many objects would need to be removed, at great cost.

Another technical challenge for ADR is that due to external torques (due to, amongst other things, solar radiation pressure, and the gravitational fields of the Sun, the Moon, and the Earth), most space debris is rotating or tumbling in an uncontrolled fashion. How the object is rotating will depend on the details of its construction, and its internal configuration at the time of abandonment (*e.g.*, the off-sets of the moments of inertia from the object's center of mass). Researchers have measured the tumble rates for a variety of defunct satellites and rocket bodies in the GEO belt, and find that their periods of rotation vary from seconds to tens of seconds, translating to rotation rates of 0.33-20 RPM [41, 42, 43]. For a "typical" satellite in GEO (Fig. 7), with ~15 m long extended solar arrays, this translates to the ends of the solar panels moving at speeds on the order of 9 m/s (~30 km/hour), with respect to the center of mass of the satellite. Any attempt to "dock with" the object, or otherwise "capture" it, would be challenging. Additionally, given the typical mass and rotation rates for rocket bodies and inactive satellites, a substantial amount of energy would need to be provided by the removal satellite (the rotational energy of the object = ½ x Moment of Inertia of the object x (its rotation rate)2) to "de-tumble" the debris object.

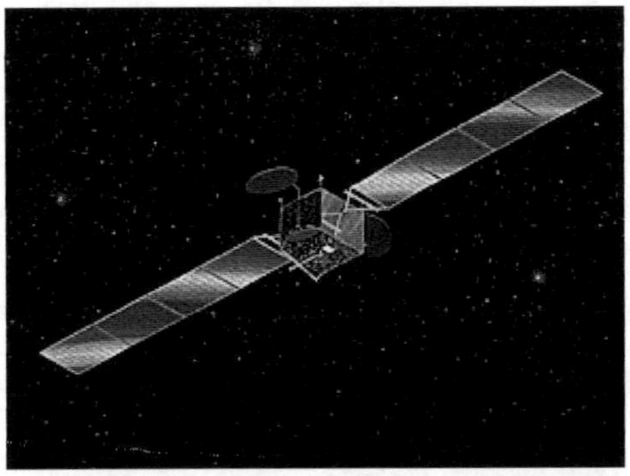

Figure 7 shows an artist's rendition of a Boeing/Hughes model 601 GEO communications satellite [44].

3. WHERE SHOULD LIMITED AVAILABLE RESOURCES BE APPLIED?

Given the various near-term actions (*e.g.*, international cooperation, data sharing, SSA, or ADR-type schemes) that might enhance the safety of flights for satellites at GEO, where should the "next dollar" be spent? What could have the biggest impact in the short term? Let us examine the various possibilities in this light.

International cooperation, in this regard, means multilateral discussions, conferences, and workshops (e.g., the UN COPUOS meetings). Certainly a large COPUOS-style meeting might cost about $4M for a 10-day meeting (estimated by labor, travel, lodging, and per diem costs), which is much less expensive than the $50M projected price for a Falcon 9 launch (excluding the cost of the satellites)[37]. As is said, "talk is cheap." These efforts are certainly needed to keep the existing problem from getting worse, as the ranks of actors in the space area continues to grow. Given the deliberative nature of these bodies, however, and the need for consensus, nothing happens very quickly. If the UN COPUOS were to increase its meeting schedule by a factor of two, could it increase its output by a similar factor, or decrease the time needed to get "results" (treaties, guidelines, reports on best practices, etc.)? Probably not, as time between meetings is utilized by delegations to confer with their Capitals, to read, discuss, and form positions on issues.

Data sharing is also relatively inexpensive, and can allow information to flow from the "best" sources to those who need it. All operators would benefit from exchanging information on their satellites with the owners & operators of other satellites, along the lines of the SDA, and it is important to have contact information readily available for emergency situations. Yet the likelihood of two operational satellites colliding with each other has been, and continues to be low. Of greater concern are the defunct satellites and populations of space debris, for which there are no operators providing information. Something of great utility that could be useful immediately, however, would be a center for the exchange of data collected on debris. The model for such could be the Minor Planet Center, which acts in a similar function for near-Earth objects, and utilizes a staff of only six full-time employees [45]. The MPC is currently funded via five-year grants from NASA. Such a center would fill a need, as currently there is no good location from which to exchange comprehensive orbital information on space debris (many debris objects do not appear in the SpaceTrack.org catalog). A global SSA data center, while worthy of discussion in an international context, is no doubt something that would be further removed in the future, given the timescales on which international discussions take place, but could happen much sooner, if interested participants join together and make it happen.

SSA. But while a space debris data center would be a convenient and useful tool to allow the exchange of information, we still need to develop additional sources for the information that is to be exchanged. While there may exist a market for commercial SSA services for GEO satellite operators [46], it is not clear if a similar situation exists on a large scale for space debris. So how are we to collect such data? Happily, ground-based optical telescopes provide a good, proven, inexpensive solution to this problem. As the astronomical community advanced the state-of-the-art observatory from 1-2 meter diameter telescopes to 4 m, and then to 8-10 m, and soon to 30 m or larger, there exist a number of functioning (or recently mothballed) smaller observatories that could be utilized, if funding were available to operate and maintain them. This seems another best use of the "next" dollar; funding additional sources of high-quality observations of space debris. A nation's "space administration" or "science foundation" might set up a series of operations & maintenance service contracts with science and engineering departments at colleges and universities that have such underutilized assets (perhaps even with motivated teams of experienced amateur astronomers). Industrial groups might also be willing to participate, perhaps motivated by profit, or for reasons of outreach to enhance STEM (science, technology, engineering and math) participation by students or the public [47].

Active Debris Removal will need to be accomplished at some point; even under best-case scenarios, we cannot expect 100% compliance with any set of disposal guidelines. With more objects being launched to GEO, there will always be a certain fraction of items that are not de-orbited prior to their end-of-life. Additionally, there exists the ever-present danger of debris-on-debris collisions for objects trapped in the GEO belt.

Yet ADR will be expensive; it will require multiple launches of many satellites that carry sufficient fuel to impart the necessary "delta-v" to the spinning, tumbling debris objects that need to be removed from the GEO belt. It will also necessitate the resolution of some thorny legal issues in an international setting. Thus, ADR is a high-cost, long lead-time activity. But are there any aspects of it that might be done for less funding and/or on shorter timescales?

Certainly characterization of space objects (an SSA task) is key to ADR. Before one might attempt to do anything with a candidate object, one would want to know as much as possible about that object; how big is it? At what rate and about what set of axes does it rotate? What is the area-to-mass ratio? What surface and thermal properties can be determined for the object [48]? These are all questions that can be answered by ground-based optical observations using existing technology.

Are there other schemes that might serve as well as full-blown removal, or might act as stopgaps until ADR is achievable? One idea is "just in time" debris trajectory adjustment [49]; this scheme involves expelling compressed cold gas into the orbit of a conjuncting debris object to change its course some ("short") time before it is predicted to collide with another object. It might be achievable with a few satellites, stationed near the two GEO "pinch points," that would be called upon as needed. It does not solve the long-term problem (it does not actually remove debris), but it solves the immediate problem of preventing collisions. It gets around the "probability" problem by only interacting with objects that are actually a threat, and it greatly reduces the energy needed, compared to actually de-orbiting the object.

Another possible first-step could be to attach small beacons on debris and defunct satellites, to aid in their tracking. It might have similar legal and political problems to actual ADR, but it could greatly aid in tracking and identifying such objects, and it would be technically much easier.

Therefore, given that ADR will require international cooperation, technology development, and multiple launches, it is neither cheap, quick, nor easy. Yet it is something that will need to be done in the long-term, if we want to maintain a sustainable use of the GEO belt. Thus, we should continue to investigate ADR, its derivatives, and similar alternatives, but it is probably unfeasible to begin constructing any deployable systems for the near future.

4. CONCLUSIONS

The GEO belt is a shared resource for use by the human race, but continued safe access to this resource is not assured, chiefly for reasons of our own making. Existing and future space debris threatens safe operations of satellites in this orbital regime. The space-faring nations whose actions contributed to the current situation will need to show leadership regarding actions needed to ameliorate the problem. Happily, there are a number of near-term steps that can be taken that do not entail the expenditure of vast resources. These include:

- Continue to enumerate, agree-upon, and promulgate space safety of flight "rules of the road" for GEO (and other orbital regimes as well)
- Detect, track, and characterize resident space objects; like rocks and shoals in shipping lanes, this information is fundamental. Unlike rocks and shoals, however, the space objects are in constant motion, and must be frequently monitored.
- Perform additional measurements of resident space objects to enhance the accuracy and precision of our knowledge of these objects. Share the orbital data as widely as possible among actors in the space arena. Incorporate operator knowledge (*e.g.*, past and future trajectory changes) into the data on active space objects.
- Conduct system architecture and legal studies to better understand the issues surrounding active debris removal and its alternatives.

These activities are moderately priced (compared to developing and launching space systems), and can have beneficial impacts on immediate and short-term timescales. They are all key components needed to ensure the long-term sustainable use of this important shared resource.

5. REFERENCES

1. UNCOPUOS, *Active Debris Removal - An Essential Mechanism for Ensuring the Safety and Sustainability of Outer Space*, UN Doc A/AC.105/C.1/2012/CRP.16 (2012).

2. NASA, Orbital Debris Graphics, online: <http://orbitaldebris.jsc.nasa.gov/photogallery/beehives.html>, accessed 4/12/12.

3. MA Skinner, *et al*, *"Hazards at GEO: The Space Debris population in the vicinity of the GEO belt and mitigation of its impact on commercial satellites"* (Address to the 30th AIAA International Communications Satellite Systems Conference (ICSSC), Ottawa, Canada, 24-17 September 2012).

4. EA Taylor, *Orbit Theory and Applications, General Electric training material* (1991).

5. R Choc, T Flohrer & B Bastida, *Classification of Geosynchronous Objects*, European Space Agency/European Space Operations Centre, Issue 13 (February 2011).

6. See Taylor, *supra* note [4].

7. D McKnight & F Di Pentino, *"New Insights on the Orbital Debris Collision Hazard at GEO,"* (Address to the IAC-11-A6.2.6, 62nd International Astronautical Congress, Capetown, South Africa, October 2011).

8. *ibid.*

9 See online: SpaceTrack.org <http://www.space-track.org>.

10. R Jehn, S Ariafar, T Schildknecht, R Musci, & M Oswald, "Estimating the number of debris in the geostationary ring" (2006) 59 Acta Astronautica 84.

11. Schildknecht, *et al*, *"Properties of the High Area-to-mass Ratio Space Debris Population in GEO"* (Address to the AMOS Technical Conference, Wailea, Hawaii, September, 2005).

12. MA Skinner, *et al*, *"Time-Resolved Infrared Spectrophotometric Observations of HAMR objects in GEO"* (2011) 69 Acta Astronomica 1007.

13. MA Skinner, *et al*, *"Further Analysis of Infrared Spectrophotometric Observations of High Area to Mass Ratio (HAMR) Objects in GEO,"* (2012) 80 Acta Astronomica 154.

14 J-C Liou & JK Weaver, "*Orbital Dynamics of High Area-To-Mass Ratio debris and their distribution in the geosynchronous region"* (Proceedings of the 4th European Conference on Space Debris (ESA SP-587), April 2005, ESA/ESOC, Darmstadt, Germany, April 2005) [unpublished], at 285.

15. See Schildknecht, *et al*, *supra* note [11]. Used with permission.

16. MA Skinner, *et al*, "Time-Resolved Infrared Spectrophotometric Observations of HAMR objects in GEO", *supra* note [12].

17. *See* Boeing, Commercial Communications satellites in geosynchronous orbit, online: Boeing <www.boeing.com/defense-space/space/bss/launch/980031_001.pdf>.

18. Satellite Industry Association, State of the Satellite Industry Report 2013, online: SIA <http://www.sia.org/wp-content/uploads/2013/06/2013_SSIR_Final.pdf>.

19. Satellite Industry Association, State of the Satellite Industry Report 2012, online: SIA <http://www.sia.org/wp-content/uploads/2012/06/Final-2012-State-of-Satellite-Industry-Report.pdf>.

20. AL Allison, T*he ITU and Managing Satellite Orbital and Spectrum Resources in the 21st Century*, SpringerBriefs in Space Development, DOI 10.1007/978-3-319-05314-1_1, 2014.

21. K Jorgensen, *et al*, *"Reflectance Spectra of Human-made Objects,"* (Address to the 2004 AMOS Technical Conference, Wailea, Maui, Hawaii, 8 – 12 September 2004).

22. *IADC Recommendation & Re-orbit Procedure for GEO Preservation*, Inter-Agency Space Debris Coordination Committee, IADC- 97-04, 1997.

23. M.A. Skinner, et al., *"Comparison of Thermal IR and Visible Signatures of Graveyard Orbit Objects,"* IAC-15.A6.1.4, 66th International Astronautical Congress, Jerusalem Israel October 2015.

24. B Iannotta & T Malik, *"U.S. Satellite Destroyed in Space Collision,"* online: Space.com (11 February 2009) <http://www.space.com/5542-satellite-destroyed-space-collision.html>, accessed 3/30/2016.

25. United Nations Committee on the Peaceful Uses of Outer Space, *Report of the Scientific and Technical Subcommittee on its forty-ninth session*, UN Doc A/AC.105/1001 (2012).

26. W Ferster, *"JSPOC Conjunction Alerts Could Be Improved, Group Says,"* Space News (9 March 2012), online: SpaceNews < http://spacenews.com/jspoc-conjunction-alerts-could-be-improved-group-says/>.

27. F Morring, *"USAF Notices Unreliable on Orbital Risks"*, Aviation Week (24 February 2012).

28. J Chan & R DalBello, *"Data Sharing to Improve Close Approach Monitoring and Safety of Flight"* (Address to the 3rd IAASS Conference, 21-23 October 2008).

29. M Rosengren, J De Vincente-Olmedo & F Pedersen, *"Keeping track of geostationary satellites,"* ESA Bulletin 119 (August 2004) 64.

30. See online: ISON <http://lfvn.astronomer.ru/report/0000029>.

31. MA Skinner, et al, *"Commercial Space Situational Awareness- An investigation of groundbased SSA concepts to support commercial GEO satellite operators"* (Address to the 2013 AMOS Technical Conference, Wailea, Hawaii, September 2013).

32. J Chan, et al, *"Data Sharing to Improve Close Approach Monitoring and Safety of Flight,"* (Address to the Fifth European Conference on Space Debris, Darmstadt, Germany, March 2009).

33. ibid.

34. MA Skinner, *"ISODEX: An entry point for developing countries into space activities,"* IAU General Assembly, Meeting #29, id. 2241978, Honolulu HI USA, August 2015.

35. See International Astronomical Union, Minor Planet Center, online: IAU Minor Planet Center <http://www.minorplanetcenter.net/iau/mpc.html>.

36. UNCOPUOS, *Long-term sustainability of outer space activities (basic elements of the concept of establishing a unified Centre for Information on Near-Earth Space Monitoring under the auspices of the United Nations and the most topical aspects of the subject matter)*, UN Doc A/AC.105/C.1/2014/CRP.17 (2014).

37. See *Active Debris Removal - An Essential Mechanism for Ensuring the Safety and Sustainability of Outer Space*, *supra* note [1].

38. D McKnight, *"Pay Me Now or Pay Me More Later: Start the Development of Active Orbital Debris Removal Now"* (Address to the AMOS Technical Conference, Wailea, Hawaii, September, 2010).

39. *Convention on International Liability for Damage Caused by Space Objects*, 29 March 1972, 961 UNTS 187, 24 UST 2389, 10 ILM 965 (1971) (entered into force 1 September 1972) [Liability Convention].

40. *Treaty on Principles Governing the Activities of States in the Exploration and Use of Outer Space, Including the Moon and Other Celestial Bodies*, 27 January 1967, 610 UNTS 205, 18 UST 2410, TIAS No 6347, 6 ILM 386 (entered into force on 10 October 1967) [Outer Space Treaty].

41. P Sommers, *"Cylindrical RSO Signatures, Spin Axis Orientation and Rotation Period Determination"* (Address to the AMOS Technical Conference, Wailea, Hawaii, September 2011).

42. P Papushev, Y Karavaev & M Mishina, *"Investigations of the evolution of optical characteristics and dynamics of proper rotation of uncontrolled geostationary artificial satellites"* (2009) 43:9 Advances in Space Research 1416.

43. Y Karavaev, *et al*, *"The Dynamic Porperties of Rotation and Optical Characteristics of Space Debris at Geostationary Orbit"* (2004) 119 Advances in the Astronautical Sciences, 1457.

44. Boeing 601 satellite bus, online: Boeing <http://www.boeing.com/boeing/defense-space/space/bss/factsheets/601/dbs/dbs.page>.

45. International Astronomical Union, Minor Planet Center, *"What is the Minor Planet Center?"*, online: Minor Planet Center <http://www.minorplanetcenter.net/about>.

46. MA Skinner, *et. al.*, *"Hazards at GEO"*, *supra* note [3]; and MA Skinner, *et al.*, *"Commercial Space Situational Awareness"*, supra note [31].

47. MA Skinner, *"ISODEX: An entry point"*, *supra* note [34].

48. MA Skinner, *et al*, *"Further Analysis of Infrared Spectrophotometric Observations of High Area to Mass Ratio (HAMR) Objects in GEO"*, *supra* note [13].

49. D McKnght, et al, *"System Engineering analysis of derelict collision prevention options"* (Address to the IAC-12-A6.5.2, 63rd International Astronautical Congress, Naples, Italy, October 2012).

[i] Portions of this paper were previously published by Mark Skinner, Orbital Debris: *What are the best nearterm actions to take? A view from the field*, 2nd Manfred Lachs Intl. Conference on Global Space Governance, May 29-31, 2014, Montreal, Canada. This is an update of that previous work.

ANALYSIS ON SPACECRAFT SAFETY ENHANCEMENT THROUGH ON-ORBIT SERVICING

Aureliano Rivolta(1), Yuval Porat(1), Nicolò Carletti(1), JeremyWang(1), Caroline Thro(1), Ali Nasseri(1), Joao Lousada(1), Matteo Emanuelli(1)

(1) Space Generation Advisory Council, Schwarzenbergplatz 6, 1030 Vienna, Austria,
Email: aureliano.rivolta@spacegeneration.org

ABSTRACT

The ever-growing space debris population has reached critical levels, rendering space safety a high priority issue for the space sector. On Orbit Servicing (OOS) could play a major role in enhancing space safety by extending the life of spacecraft. However, implementing OOS requires sacrifices on the part of all stockholders both in economic terms and in terms of the technical design of systems for serviceability. Long term benefits of OOS seem to outweigh these sacrifices. Moreover, this active method provides a deployable intervention which could complement passive safety measures. This paper presents an analysis of the impact of OOS on the safety of the space sector, underlying the contribution of OOS as an active safety measure to combatting the space debris problem and showcasing the complementary role of OOS in enhancing current safety policies.

1. INTRODUCTION

The term "safety" has been in use for centuries; yet, in the last century, the aerospace industry has redefined design and operational safety. Ever since the advent of flight, safety has been a major issue. Many pioneers of atmospheric flight lost their lives pursuing mankind's dream of flight and, thanks to their sacrifices and experiences, a large body of knowledge was developed to improve safety of aerospace vehicles. Nowadays, aircraft are probably the safest vehicles. It is no surprise then that spaceflight safety is a hot topic.

Although spaceflight was rooted in our capability to learn to fly in our atmosphere, there are historical differences in the evolution of safety in space and in air. While the first aircraft were controlled by an on-board crew, the first human satellites (even the Vostok 1) were automated systems controlled from the ground. Hence, the safety measures introduced for spaceflight were not the same as crewed air flight.

This different in evolution influenced the vision of safety for artificial satellites. While damaged aircraft had to reach ground, be repaired and reused, satellites were seen as disposable objects, and no life cycle analysis was performed to assess their long term impact. As a result, the impact of orbital debris was underestimated for many years until the Kessler syndrome [1, 2] changed our views, and highlighted debris population as a safety risk for both unmanned and manned space mission.

In atmospheric flight, safety is guaranteed by passive measures like redundancy of subsystems, and active measures like inspection and maintenance. This is caused by the repetitive but short missions a single aircraft has to carry out in its lifetime.

Spacecraft have (in most cases) a single mission and do not return to the ground for maintenance. Hence, only passive measures have been employed for their safety. Historically, many missions have exceeded their programmed lifetime by several years, thanks to the meticulous work of designers and operators; however, this is not the norm and it is not wrong to assume space reliability as a myth [3].

Active safety measures on satellites will not be similar to those in aircraft as satellites cannot be returned to ground facilities. The only alternative is to bring the maintenance facility to them, using another spacecraft. Technically speaking, it is possible to execute active tasks on satellites and it has been proven in orbit several times, with some experiments on the Space Shuttle [4], the demonstrative missions ETS-VII [5, 6] and Orbital Express [7-11]. These experiments demonstrate that this concept it is not impossible, but is complex.

In this paper, On Orbit Servicing (OOS) will be discussed, highlighting some of the hurdles to implementing it for current satellites. The economic aspects of such framework will be presented and discussed to showcase its feasibility. Finally, recommendations will be presented on how international guidelines could help bring us one step closer to implementing this solution in the future.

2. OOS ENHANCEMENT OF SAFETY

OOS can revolutionise space safety and our approach to alleviate the space debris problem. The ability to fix a problem directly on orbit can shift the whole approach from passive safety to active safety. Nevertheless, a proper combination of pre-emptive measures, scheduled maintenance, and in-situ repair is the best solution to maximize safety.

On orbit repair is the "substitution" of elements that have been damaged or have ceased to work on a customer satellite. Failed subsystems can be substituted in two possible ways: physical replacement or bus enhancement. In the former case, the parts are extracted using a robotic device and replaced with compatible units. In the latter case, the malfunctioning system is not removed from the spacecraft bus; instead another similar element is attached to the bus, resulting in an upgrade of the satellite, increasing its mass. The main difference lies in the adjustments needed in the serviced satellite design in order to make a subsystem removable or to connect an appendix.

On orbit refuelling is the most probable application of OOS in the upcoming years for a very simple reason: there is a market for it. Fuel depletion is one of the major elements restricting the lifetime of a satellite. In very lucrative areas of application, for example GEO communication satellites, the ability to increase the lifetime of a satellite translates into extended revenue. Since refuelling operations do not require complex part substitution, all that is needed is to have a fuel storage line designed to accept external mass transfer. This includes a proper design of the pipeline (for example avoiding elements that prevent backflow in the line or even doubling the line itself if needed), as well as a proper interface with the exterior part of the satellite where the servicing satellite can link.

On orbit monitoring is another interesting OOS scenarios that requires the servicing satellite to inspect for damages and could help with providing evidence for legal proceedings and insurance issues. Having an eye on the exterior of a satellite can increase the possibility of weaknesses detection and noticing problems before they can harm the satellite. The concept has already been used during Shuttle missions for ablative protection checks prior to re-entry manoeuvres.

On orbit towing is the generalization of orbit insertion manoeuvres. These manoeuvres are usually provided by launchers or upper stages until the satellite can operate on its own to change its orbit following its mission profile. Orbit insertion failures have happened in the past and most likely to occur in the future. A recent example is the failed insertion of Galileo positioning satellites on 22/07/14. Such events may cause a satellite to require extra manoeuvres reducing its lifetime or could cause the loss of the whole mission if the primary propulsive system is unable to perform the corrections. A wandering satellite, especially in crowded orbits like LEO, may jeopardize the safety of many other satellites as well as proving to be a huge economical loss. The ability to recover such satellites and put them in the right orbit can solve the issue as well as opening new OOS scenarios. A servicer satellite capable of such tasks can be used to transport satellites from the launcher injection orbit up to their target orbit, reducing substantially the weight of the customer satellite and the launch costs connected.

The OOS scenarios above can be seen as tools to increase the lifetime of satellites as well as reducing the chances of debris generation. The major drawback is that there are some important issues to be solved. First and foremost, of economical nature: OOS is business and it should be enforced in such a way that servicing companies get revenue and customers are willing to pay for the aforementioned services. An obstacle is that satellites currently in orbit were not designed to be serviceable; thus, the interfaces for most of OOS services should be designed in a case by case basis and in some cases servicing might prove impossible. Such approach might be a financial suicide. To foster a change in this situation, we propose the adoption of policies to modify the next generation of satellites to be serviceable. A satellite that is designed to accommodate servicing with a standardized approach would increase the chances of birth and survival of servicing companies [12].

3. TECHNICAL ANALYSIS

At this point in time, the most economically justifiable implementation of OOS is on-orbit refuelling, mainly of communication satellites in the GEO belt. A similar scenario has been studied in the past by NASA and this study will be based on the refuelling scenario, dubbed Notional Mission 2, as described in [13]. It is worth noting, that although this analysis considers only GEO refuelling, the results found in this study could also be extended to other OOS mission types involving a mass exchange (module repair, segment docking, etc.). In those cases, instead of having fuel as payload of the servicer to be delivered to the clients, specific hardware component (computer, batteries, antenna etc.) would be the payload.

In the NASA OOS report, conventional chemical propulsion was considered for the servicer segment [13], while in this work we explore low-thrust electric propulsion (EP) following the new trend of propulsion systems for

GEO. Due to their higher efficiency, it is obvious that EP would minimize propellant mass in comparison to high thrust chemical propulsion. However, due to their very low accelerations, EP would also mean much longer transfer times will have to be taken into account.

The study proposed here is conceptual; detailed analysis is out of the scope of this study and will be considered as part of future developments.

3.1 Mission Overview

The servicing mission will comprise of 2 segments: the servicing module ("Servicer") and the depot station ("Depot") which will hold the reservoir of fuel required for the entire mission. Both the depot and the servicer will be initially brought to the depot operational orbit which will be 100 km above GEO. This transfer can be achieved by several means.

After initially reaching the depot operational orbit (and after completing the commissioning phase), the servicer will embark on its first sortie to deliver fuel to the client satellites in GEO. During the course of a sortie, the servicer will transfer from one client to the next using EP, deliver the payload to the client and move on to the next client. After refuelling the last client in the sortie, the servicer will return to the depot station, where it will replenish both its own fuel supply as well as that intended for the client satellites.

The client satellites are communication satellites located at the GEO belt. The clients for the simulation are chosen randomly from a database of currently active satellites in GEO [14].

For the client selection, the complete GEO satellite database is refined to include only satellite whose inclination is smaller than 1.0° and who are positioned in a unique location, i.e. if two satellites are positioned in the same slot, they are considered as one satellite. These assumptions simplify the problem without losing relevance.

For every sortie, 6 unique clients are randomly sampled from the refined database, i.e. during the entire mission all 24 satellites will be different, such that no satellite would be refuelled twice during one mission. Servicing operations duration for each client is assumed to be 7 days (including automated rendezvous and docking manoeuvre, and the refuelling of the client) while the resupply at the Depot station is assumed to last 10 days. Tab. 1 and 2 outline the assumptions about the Servicer and Depot used for the simulations reported here.

Table 1. Servicer spacecraft data

Isp [sec]	3100
Thrust [mN]	214
Dry Mass [kg]	1840
Propellant Mass [kg]	30
plus 30% margin [kg]	40
Propellant gas	Xenon
Power [w]	6000

Table 2. Depot data

Dry Mass [kg]	1326
Fuel for 24 Clients [kg]	480
Fuel for Servicer [kg]	125

3.2 Mathematical Model

In order to select which of the Clients in GEO will be the first one to be serviced, the transfer time between the Depot orbit and GEO is calculated with eq. (1) [15]

$$t = \frac{m_0}{\dot{m}_p}\left[1 - \exp\left(\frac{v_{out} - v_{in}}{v_*}\right)\right] \qquad (1)$$

Where, m_0 is the servicer's mass, $\dot{m}_p$ is the propellant mass flow rate, v_{out} is the circular orbital velocity of the outer orbit, v_{in} is the circular orbital velocity of the inner orbit and v_* is the thruster's effective exhaust velocity. Then, the exact arrival longitude of the servicer in GEO can be computed by integrating the servicer's longitudinal drift rate (which is a function of altitude) throughout the transfer.

Once the servicer's longitudinal position in the GEO belt is established, selecting the first client to be served is achieved by looking for the client satellite that is closest in longitude to that of the servicer. **Error! Reference source not found.** shows this process schematically:

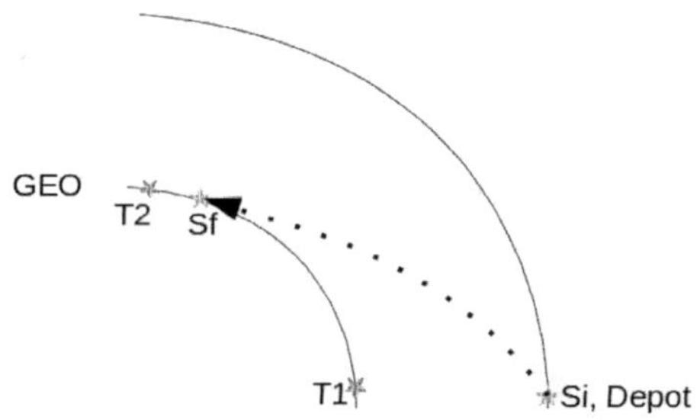

Figure 1. Depot-GEO Transfer

The servicer starts at the Depot station in point Si. During the transfer, as it lowers its altitude towards the GEO, both its longitude and its longitudinal drift-rate change continuously. Upon arriving at the GEO orbit, at point Sf, the servicer finds itself closer to target T2 than target T1, and therefore, in this case, target T2 will be the first target to be refuelled.

Next, the servicer's exact longitudinal position would have to be adjusted to match that of the first client satellite. Since the servicer is coming from a higher orbit, in case the client is behind the servicer w.r.t orbital motion, a proper drift time can be exploited in the depot orbit to reduce the number of manoeuvres. Otherwise, if the client is ahead of the servicer, the servicer would have to lower its altitude to gain orbital velocity and then increase its altitude again to rendezvous with the client satellite.

When transferring from client to client the initial and final orbits are both GEO. That means that saving up fuel by waiting in a higher orbit, which was possible in the previous discussion is no longer possible here. Therefore, depending on the relative location of the next client w.r.t the current client, a manoeuvre for either increasing the semi-major axis or decreasing it will have to be performed.

In the simulations clients are randomly sampled from the database: before each transfer, the next client would have to be selected by comparing the angular difference between the current client and both the next closest client ahead of the current client and the next closest client behind the current one. The client with the smallest longitudinal difference would be selected as the next one to be serviced. In real applications the selection of target satellite is more complex and contract dependent, however for the case in study such approach is legitimate.

After the last client has been serviced, the servicer would have to return to the depot to resupply. In principal, the process for going from GEO to Depot is analogous to the process of transferring from Depot to GEO (just in the opposite direction) without considering the selection step and taking into account the Depot longitudinal drift-rate due to its placement in a higher orbit.

3.3 Simulation Results

Each simulation run calculated the amount of time and propellant mass required for an entire mission consisting of visiting 24 clients in 4 sorties. To get an estimate for the average total mission time and required average total propellant mass, 150 simulation runs were carried out, where in each one different client satellites were randomly sampled from the refined list.

Figure 1 shows the results of the 150 simulations for the total propellant mass. The average value from the 150 runs was found to be $\bar{m}_p \cong 95$ kg. Adding a 30% margin results in a total propellant mass of 125 kg which will be loaded as payload into the Depot station. Figure 2 shows a histogram of the mission time calculated by the simulations. The average total mission time for servicing 24 clients was $t_{tot} = 615$ $days \cong 1.7$ $years$.

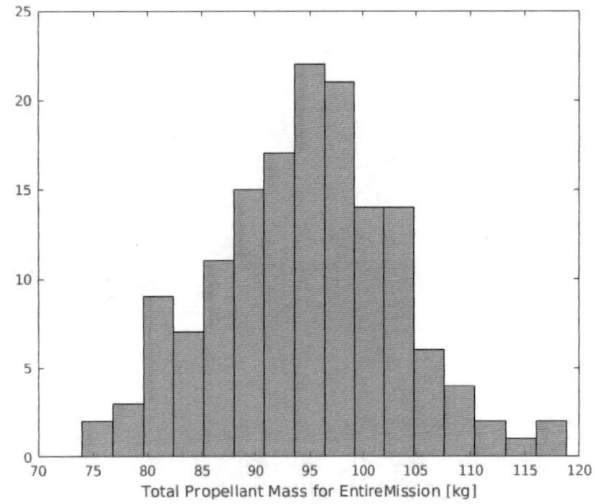

Figure 1: Histogram of servicer required propellant mass from 150 simulation runs

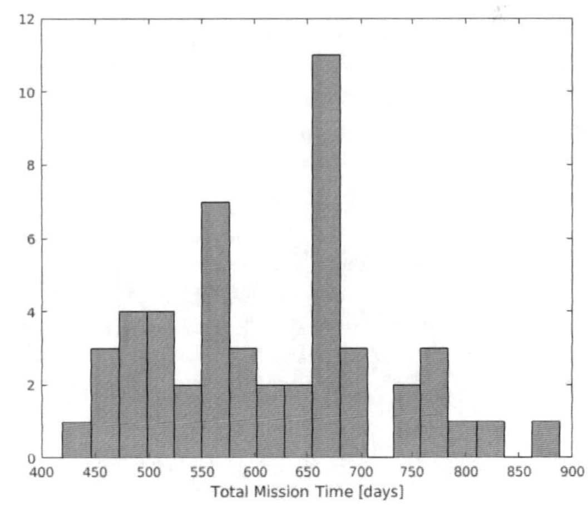

Figure 2: Histogram of total mission time

The NASA scenario envisioned a mission lifetime of 10 years, which includes the launching and commissioning phase of the satellite and depot station, the servicing of 26 GEO satellites and the eventual removal of both the servicer and the depot to a graveyard orbit after the end of the mission. They have also assumed the client satellites are evenly distributed around the GEO belt. In our study, we have mainly concentrated on the operational phase of the mission, i.e. from the time the servicer embarks on its first sortie until it reaches the depot after the very last (24th) client has been served. In both studies, the mission was assumed to be continuous, i.e. the servicer is always in operation – either refuelling, transferring or resupplying at the depot. There are no idle moments, in which the servicer awaits a mission.

The results acquired in this study in terms of propellant mass are obviously quite low, and it will be inappropriate to compare them to the NASA reference scenario which used chemical propulsion. What is interesting to note is the total mission life time. We have previously seen that the average total mission time is 615 days. By examination of Figure 2 we also see that in the worst ,and much less frequent, case the total mission time is approximately 930 days. These results are surprisingly low compared to the total 10-year mission time, considering that the client satellites were sampled randomly from a real list of operational satellites and were not distributed evenly around GEO as in the NASA report.

The explanation for the relatively low total mission time could be attributed to the fact, that due to the high efficiency of the EP system, relatively fuel-expensive manoeuvres could be performed for the transfers between clients, that could expedite the transfers. Such manoeuvres would be deemed impractical with a chemical propulsion because they would require an unreasonable amount of propellant. In comparison, in the NASA reference scenario, in order to save fuel, the servicer would always raise its altitude to a drift orbit (127 km above GEO) and would wait there for the longitudinal difference to the next client to diminish. These drifting arcs are quite time consuming and are avoided in this study.

In conclusion, the study carried out in this work showed that multiple target OOS using low-thrust propulsion could be feasible and even economical in terms of both propellant mass and mission time.

4. ECONOMIC ANALYSIS

An economic evaluation of the previous scenario is mandatory: a good service is not doable if it does not guarantee a profit. Two paths are followed for this study: first the evaluation of the costs related to the mission, second the verification of client profitability. The costs of the mission are obtained using the Unmanned Spacecraft Cost Model, eighth edition (USCM8) [16]. This model relies on cost estimation relationships (CERs), empirical equations based on statistics. These CERs are applicable for automated satellites, with masses between 500 and 5000 kilograms, fitting this scenario.

Both non-recurring and recurring costs are estimated by the model. Non-recurring costs are incurred during the development of the system and first model production; recurring costs are incurred during the production of the final product and ground operations.

The USCM8 model estimates costs for every section of the vehicle and mission based on the mass of the system. This analysis is usually performed several times during a feasibility study: at the beginning using statistical mass distributions, and later on using more refined numbers, until the final mass budget is reached. Recurring costs are increased by 20% to take into account the low TRL of the technologies being used, and provide estimates for the fiscal year 2020.

To verify the results, we compared our estimated to an existing study for a refuelling mission (that uses the USCM7 [17]), and with a rough rule-of-thumb (ROT) model [18], that gives the order of magnitude of the cost via a simple equation based on dry mass. In order to make this prior study comparable to the present case, a scaling factor for the mass, the 20% margin and the actualization to the fiscal year 2020 are added.

Table 1: Cost results for the depot and servicer across three studies. USCM8 and ROT were calculated specifically for this case whereas USCM7 is the reference study.

Cost	USCM8	USCM7	ROT
Depot [M$]	142.11	178.39	213.20
Servicer [M$]	202.26	180.54	216.50

Tab. 3 shows our estimates (USCM8) along with estimates from the other two methods. It is evident that the USCM8 seems to underestimate the depot cost. This is due to the assumed simplicity of the design – the depot has little to no fuel transfer robotics, and no propulsion system (only attitude control).

Next we can verify the recurring costs for the case of refuelling. Starting from assumptions, it is possible to evaluate the cost of the standard client satellite, then determine how many clients must be refuelled to make the servicing case profitable for the client, rather than sending a new satellite. Tab. 4 summarizes the assumptions used for the analysis, and defines the "standard" client. From the mission analysis in the previous section, it follows that each client is refuelled 20 kilograms. This fuel is enough to extend the lifetime (ΔLT) of the client satellite for more than 2 years (813 days).

Table 2: Assumed client profile

Ariane 5 launch cost (A)	22 $/kg
Client satellite cost (CSC)	100 K$/kg
Nominal lifetime of the client satellite (LT)	15 years
Mass of the client satellite (CSM)	2500 kg
Specific impulse of client's engine (Is)	1700 s
Δv needed for the SK/OCM	60 m/s/year

Eq. (2) from [17] gives the break-even point for the number of refuelling needed.

$$n = \frac{SC + A\,(DM + SM)}{CSM(CSC + A)\Delta LT} = 9.515 \qquad (2)$$

Where SC is total cost of the servicer and the depot, DM is the depot mass and SM is the servicer mass.

The USCM8 seems reasonable and comparable to the other models. It is important to notice that the break-even point is lower than 24 (number of satellites planned to be refuelled in this example) suggesting that the mission is cost-feasible and highly profitable.

Even if this analysis is preliminary, it shows that OOS can be profitable following the mission profile presented in the previous section where EP is considered. This statement is important because it is mandatory to have an economically viable solution to implement OOS in order to gain the aforementioned benefits for the space sector safety and sustainability.

5. LEGAL ISSUES

The very unique activity of OOS requires dedicated policies for spacecraft design. The idea first presented in [12] is to make all future satellites suitable for orbital servicing by recommending one specific design for a docking area to allow a standardized approach for all potential OOS companies. Concerning OOS, liability and Intellectual Property Rights are major legal issues to discuss.

5.1 Liability

The international Legal Framework for Outer Space Activities includes the so-called Liability Convention, which foresees the liability regime for damages caused by space objects. Under this Convention the Launching State is internationally liable for damages caused by its space objects even if the activity carried out is of private nature. According to the Liability Convention, the launching State is the State that: launches the satellite; procures the launch; from whose territory the space object is launched; or with whose facilities the space object is launched [19].

This is the reason why the Launching States start to implement national legislation to have an oversight procedure for any space activity before, during and after the Launch to minimise any risk of damage to another spacecraft. In fact, Article VI of the Outer Space Treaty [20] foresees that "State Parties to the Treaty shall bear international responsibility for national activities in outer space, [...] whether such activities are carried on by governmental agencies or by non-governmental entities". This means that the State is responsible for its public and private space activities.

The liability convention is applicable to OOS satellites as well. These missions carry higher risk of damages than other space activities due to the necessity to interact with another spacecraft.

According to Article 3 of the Liability Convention, "in the event of damage being caused elsewhere than on the surface of the earth to a space object of one launching State or to persons or property on board such a space object by a space object of another launching State, the latter shall be liable only if the damage is due to its fault or the fault of persons for whom it is responsible". In fact, the Convention foresees a fault based liability regime for any damages caused in Outer space. In principle, the fault is not easy to prove. However, in case of OOS, a fault presumption might exist as the servicing spacecraft is the spacecraft making contact. While negotiating its contracts, the service provider should keep this in mind.

5.2 Intellectual Property Rights

The technology used in the OOS satellite, and that this satellite mates and has access to the spacecraft being services, raises intellectual property rights [21]. Is the company, which is developing the OOS, obliged to give out a free licence to design the docking port for OOS? Or can a price be discussed within the frame of the contract? Should there be different stages of protection of intellectual property rights depending on the sensibility of the satellite refuelled, repaired or serviced?

Up to now, no specific international or national legislation has been issued regarding intellectual property rights for space activities. In fact, the existing rules are always applied as such, with some adaptations foreseen in the service contract or licence agreement. It is anyhow questionable if an international or at least national legislation would allow to unify and harmonise the system, even though this would deeply facilitate encouraging one docking port design worldwide. It is more certain that the service provider and the user will negotiate the specific terms and conditions in the contract, representing the clauses applicable to this specific activity.

6. THE NEED FOR SPACE POLICIES

In the authors opinion, the UN COPUOS should play in a first step an important role towards sustainability project in Outer space (where one of them is the OOS) by establishing and promoting a frame encouraging manufacturers to move towards maintainable spacecraft. Such intervention would encourage the birth and growth of OOS companies, since up to date no company has ever succeeded to carry out a commercial OOS mission. It is a recent news that Orbital ATK has announced a contract for OOS with Intelsat [22]. The effective operations are expected to be exerted approximately in 2019 and would be a milestone for OOS. The Orbital ATK approach presented there is to "enhance" the serviced bus through un-conventional docking. A similar approach has been suggested also in [12] but encouraging the adoption of standard docking interfaces to boost the market of logistics services. Such approach, if proven in flight, would require less modification in the serviced module, but no standardization is possible. Rather than establishing a single provider single servicer relationship it would be wise to support measures to ensure a competitive market.

It should be noted that Intelsat had an agreement with MDA prior to the one with Orbital ATK. This implies both interest in OOS and issues related to its implementation [23].

7. CONCLUSION

In this article OOS has been presented as a complementary solution to passive safety measures usually employed in space activities. OOS as an active maintenance and lifetime enhancement technique requires the right legal environment. An analysis on major technical and economic aspects has been performed using reliable models and encouraging results have been found. There are still open problems about some legal issues regarding liability and intellectual property and technical compromises that could help the development of OOS.

It has been found that with commonly agreeable guidelines and support from UN OOSA the OOS market could become a reality in the next decades. As presented beforehand in this article, this could have huge benefits for the safety of the space industry and environment.

8. REFERENCES

1. Kessler, D. J., & Cour-Palais, B. G. (1978). Collision frequency of artificial satellites: The creation of a debris belt. Journal of Geophysical Research: Space Physics, 83(A6), 2637-2646.

2. Kessler, D. J., Johnson, N. L., Liou, J. C., & Matney, M. (2010). The kessler syndrome: implications to future space operations. Advances in the Astronautical Sciences, 137(8).

3. Ellery, A. Kreisel, J. & Sommer, B. (2008). The case for robotic on-orbit servicing of spacecraft: spacecraft reliability is a myth. Acta Astronautica, 63(5), 632–648.

4. Chato, D.J. (2000). Technologies for refueling spacecraft on-orbit. AIAA Space. Vol. 200.

5. Yoshida, K. (2003). Engineering test satellite VII flight experiments for space robot dynamics and control: theories on laboratory test beds ten years ago, now in orbit. The International Journal of Robotics Research, 22(5), 321–335.

6. Kasai, T. Oda, M & Suzuki, T. (1999). Results of the ETS-VII mission-rendezvous docking and space robotics experiments. In Artificial Intelligence, Robotics and Automation in Space, volume 440, page 299.

7. Pinson, R.M., Howard, R.T. & Heaton A.F. (2008). Orbital express advanced video guidance sensor: ground testing, flight results and comparisons. In Proceedings of the AIAA Guidance, Navigation and Control Conference and Exhibit, volume 1, pp 1–9.

8. Howard, R.T., Heaton A.F., Pinson, R.M. & Carrington, C.K. (2008). Orbital express advanced video guidance sensor. In IEEE Aerospace Conference, pp 1–10.

9. Heaton A.F., Howard, R.T., & Pinson, R.M. (2008). Orbital express AVGS validation and calibration for automated rendezvous. In AIAA guidance, navigation and control conference and exhibit, Honolulu, USA, pp 18–21.

10. Dipprey, N.F. & Rotenberger, S. (2003). Orbital express propellant resupply servicing. In Proceedings of the 39th AIAA/ASME/SAE/ASEE Joint Propulsion Conference and Exhibit, pp 20–23.

11. Ogilvie, A. Allport, J. Hannah, M. & Lymer J. (2008). Autonomous satellite servicing using the orbital express demonstration manipulator system. In Proc. of the 9th International Symposium on Artificial Intelligence, Robotics and Automation in Space (i-SAIRAS'08), pp 25–29.

12. Rivolta, A. Ferrario, L. Borriello, C. Ramon Medel, J. Ghasemzadeh, L. Chan-Hao Wang, J. Nasseri, S.A. Emanuelli, M. (2015). International spacecraft design policies for orbital servicing, International Astronautical Congress 2015, Jerusalem.

13. National Aeronautics and Space Administration Goddard Space Flight Center (2010). On-orbit Satellite Servicing Study: Project Report

14. Satellite internet. (n.d.). Retrieved March 10, 2016, from http://www.satsig.net/

15. Walter, U. (2012). Astronautics: the physics of space flight. John Wiley & Sons.

16. Wertz, J. R., Everett, D. F., & Puschell, J. J. (Eds.). (2011). Space mission engineering: the new SMAD. Microcosm Press.

17. Hibbard, R. L. (1996). Satellite On-Orbit Refueling: A Cost Effectiveness Analysis. Naval Postgraduate School Monterey Ca.

18. Lamassoure, E. (2001). A framework to account for flexibility in modeling the value of on-orbit servicing for space systems (Doctoral dissertation, Massachusetts Institute of Technology).

19. Convention on the International Liability for Damage Caused by Space Objects, 29 March 1972, 961 UNTS 187.

20. Treaty on Principles Governing the Activities of States in the Exploration and Use of Outer Space, Including the Moon and Other Celestial Bodies, 27 January 1967, 610 UNTS 205.

21. Intellectual Property and Space Activities, WIPO, April 2004

22. Orbital ATK. Retrieved April 20, 2016, from http://www.orbitalatk.com/newsroom/release.asp?prid=137

23. MDA, Intelsat Scrap In-orbit Servicing Deal - SpaceNews.com. (2012). Retrieved April 20, 2016, from http://spacenews.com/mda-intelsat-scrap-orbit-servicing-deal/

PROCEEDINGS OF THE 8TH INTERNATIONAL SPACE SAFETY CONFERENCE

SESSION 19 DESIGNING SAFETY - II

A PROPOSED METHOD TO QUANTIFY ORBITAL DEBRIS IMPACT RISKS ON SPACE PRESSURE VESSELS

Michael S. Surratt[1], Michael Kezirian, PhD[2]

[1] *Graduate Student, Department of Astronautical Engineering, University of Southern California, 854B Downey Way, RRB 225, Los Angeles, California 90089-1192, Email: msurratt@usc.edu*
[2] *Adjunct Associate Professor, Department of Astronautical Engineering, University of Southern California, 854B Downey Way, RRB 225, Los Angeles, California 90089-1192, Email: kezirian@usc.edu*

ABSTRACT

Quantifying risk to pressure vessels from Hypervelocity Impact (HVI) by orbital debris in space is a multi-step process with many analysis components. Many of the current challenges surrounding this analysis are in the impact and immediate post-impact phase. Intended differences in selected vessel materials, vessel design, manufacturing techniques, and unintended unit to unit manufacturing variability even for supposedly identical vessels pose significant challenges to developing a risk quantification technique suitable for a priori prediction of pressure vessel post-impact behavior. This investigation attempts to characterize and organize those challenges into a comprehensive program of study.

1. OVERVIEW OF THREATS TO SPACE PRESSURE VESSELS

On-orbit threats to pressure vessels can come from different internal and external factors. Internal factors could be such short term issues as overpressure rupture while longer term factors could be time dependent issues such as stress rupture in the case of composite overwrapped pressure vessels (COPVs). External factors are more environmental such as on-ground or in-space pressure vessel mishandling as well as on-orbit micrometeoroid and orbital debris (MMOD) impact. MMOD impacts are more likely the longer a vessel is in use exposed to incident MMOD particle flux. Impact probability for any particular orbit and spacecraft arrangement/orientation scales linearly with exposure time. The purpose of this investigation is to propose a means of systematically quantifying risk of MMOD impact on pressure vessels.

2. MMOD IMPACT PROBABILITY VARIABLES

Risks for Micrometeoroid and Orbital Debris (MMOD) impacts are dependent on the flux of particular size of MMOD and the size/mass of the MMOD in question. Current models for the MMOD population can be consulted for the specific orbit of interest. Two popular tools for computing present-day MMOD flux estimates are ORDEM 3.0 (NASA, United States) and MASTER (European Space Agency). ORDEM will be the tool used in this investigation.

ORDEM inputs are orbital elements for the chosen orbit (primarily semi-major axis, eccentricity, and inclination) which completely specify the orbit around the Earth. One output type from ORDEM is flux distributions as a function of MMOD object size with +/- 1 standard deviation curves (Fig. 1). Another output is MMOD flux versus incident velocity for a particular debris size range (Fig. 2). Finally, another common output is debris flux as a function of MMOD approach direction for a particular debris object size range (Fig. 3).

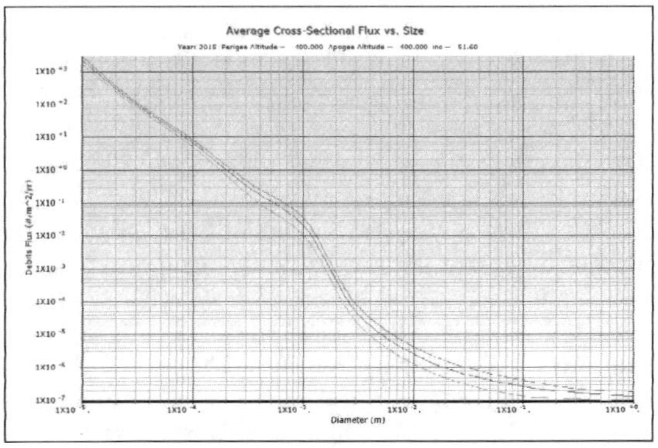

Figure 1. ORDEM flux output for ISS orbit

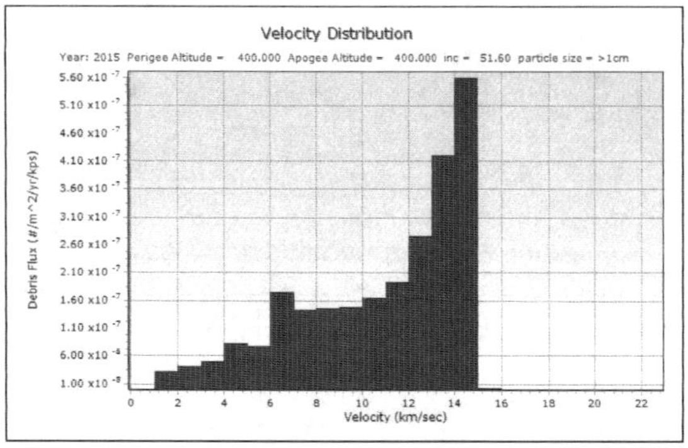

Figure 2. ORDEM velocity distribution for >1cm MMOD in ISS orbit

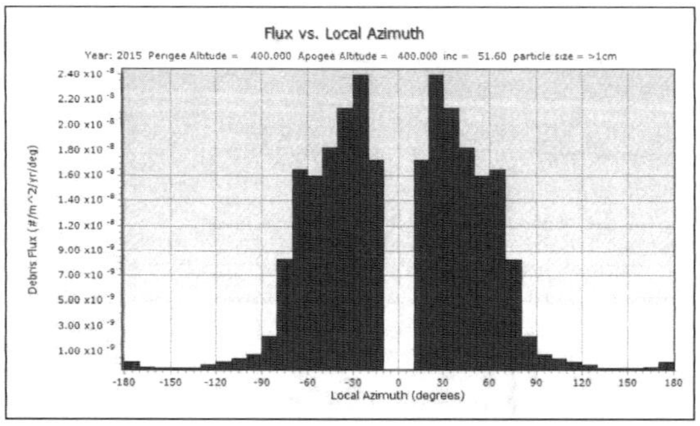

Figure 3. ORDEM incident direction plot for >1cm MMOD in ISS orbit

Impact probabilities (P_{impact}) are then computed using orbital target size such as a complete spacecraft, an at-risk spacecraft subsystem, or even a single spacecraft structural element such as an exposed pressure vessel. The area of interest (A_{target}) is multiplied by the size specific MMOD flux value (*Flux*, number of that MMOD size per unit area per unit time) and the time period of exposure ($t_{exposure}$) to create the number of impacts over the time period of concern. If the value is less than unity then it is interpreted as a probability of impact over the time period. Results equal to or greater than unity indicate the total number of impacts for that MMOD size over the time period.

One challenge of MMOD impact probability estimation is how the MMOD population is slowly changing over time, barring some sudden event. Events like the Chinese ASAT test in 2007 are very sudden MMOD population increases on the order of a few hundred to thousands of objects across the entire spectrum of trackable and smaller than trackable sizes. Tools like ORDEM and MASTER are nearly point in time systems that use current MMOD population estimates. Uncertainties on debris populations increase over long periods of time when using these models. However, for the purposes of this investigation, the time scale of our problem is modest on the order of days to weeks and period of interest is in the very near future. Therefore, ORDEM is a valid MMOD impact probability assessment tool for the current investigation.

3. MMOD IMPACT DAMAGE CHARACTERISTICS

Impact damage from MMOD is a high specific energy (energy per unit impactor mass) event. Damage modes in the target structure include ablation, spallation, and "cracking". Ablation and spallation are similar processes for either metal or composite exterior pressure vessels. "Cracking" is used broadly for material specific damage or microstructural changes. Metal pressure vessels can have cracks form inside the vessel wall thickness as well as on the inside wall of the pressure vessel as the impact shockwave exits the vessel wall into the interior volume. In similar fashion, "cracking" in a composite pressure vessel could be impact induced delaminations, matrix cracking, or inner metal liner surface cracking from a shock wave exiting the interior surface just as in the case of metallic pressure vessels.

4. IMPACT SEVERITY MITIGATION

Spacecraft structures can be shielded from MMOD fluxes in two general ways. One way is purely coincidental placement of the target in such a way to reduce or eliminate MMOD impact line-of-sight. The butterfly plots from ORDEM can help assess the benefit from this situation. Another way to get shielding is to intentionally place one or more layers of material at some standoff distance(s) from the sensitive structural element. This arrangement is commonly called a Whipple Shield. NASA Glenn's work with metallic pressure vessels was after the test impact object went through a single layer metal shield.

Shielding disperses the impact energy of MMOD strikes but also complicates the modeling of impact events since many secondary or tertiary particles are created and driven along the initial impact object's path toward the protected structure. Simply put, depth of penetration is typically reduced at the expense of a wider region being affected by the impact event.

5. CHOSEN EXAMPLE PRESSURE VESSEL

One object of particular interest to the authors is the Nitrogen-Oxygen Recharge System (NORS) pressure vessel. The NORS pressure vessel is a carbon fiber COPV with an Inconel metallic liner. The COPV is delivered to the International Space Station (ISS) on one of the Commercial Resupply Services vehicles fully charged with nitrogen or oxygen. The purpose of the NORS COPV is to replenish gas losses in the ISS Environmental Control and Life Support System (ECLSS). Dozens of NORS vessels will be manufactured and subsequently delivered to the ISS over the remaining years of planned ISS residency.

The NORS COPV has a 6000psi operating pressure. In case of catastrophic failure, a fully pressurized NORS tank would release the energetic equivalent of pounds of TNT. This poses a significant threat to the station structure and all residents if a catastrophic pressure vessel failure were to occur.

6. CHOSEN EXAMPLE OPERATIONAL CASE

Since the NORS vessel is intended for delivery to and use on the ISS, the orbital debris environment for our analysis will be for the ISS orbit. Recent orbital information for the ISS was taken from [http://spaceflight.nasa.gov/realdata/sightings/SSapplications/Post/JavaSSOP/orbit/ISS/SVPOST.html on 15 May 2016]. We find that the ISS orbit is a nearly circular ellipse with the following:

Semi-major axis, $a = 6787751.16$ meters
Eccentricity, $e = 0\ .0015021$
Orbital inclination, $i = 51.37026°$

MMOD flux distribution is not dependent on orientation of the target vehicle. However, we have to assume that the most strenuous case may exist for the MMOD incident velocity since the station's orientation with respect to its velocity vector in the future is not completely known. In such a situation, the prudent course of action is to assume the highest intersection velocity for that orbit. This will give us the upper bound for MMOD damage risk to pressure vessels external to the ISS.

Exposure of the NORS vessel to MMOD flux and the time of such exposure come from the chosen situation. In this case, we assume that a NORS recharge extravehicular activity (EVA) would be aborted while the NORS tank is near the ISS structure but otherwise exposed to MMOD incident flux from the oncoming hemisphere of possible MMOD paths. In our example case, no action would be able to be taken to shield the NORS vessel or move it to a shielded position with ISS Canadarm 2. Two weeks of this MMOD exposure is taken as the time period of interest.

7. MMOD SIZE AND DAMAGE

Physical size of a debris object is only part of the information needed for an impact analysis. The object's material composition will drive its mass and hence its total impact energy. Nylon and aluminum are two typical proxy materials for MMOD laboratory HVI tests. Considering that millimeter scale objects can cause significant damage through light shielding, a 1 cm MMOD object direct impact should be devastating and cause immediate pressure vessel rupture. Fortunately, such sized objects are expected to be rare. The challenge is then how to model impacts of more likely sizes in the 1-10 millimeter range. Then there is the final issue of post-impact pressure vessel use and modified lifetime.

8. IMPACT & IMMEDIATE POST-IMPACT ASSESSMENT

This part of the risk quantification is the most difficult for many reasons. One key reason is that there are no facilities to replicate the peak debris impact velocities (14+ km/s) in terrestrial laboratories. This makes modeling extremely problematical. Any computational analysis tool must be validated against test data to be ready for application to further cases. NASA Glenn Research Center has been able to perform HVI tests on metallic pressure vessels and COPVs up to approximately 8 km/s. Validating a post-impact pressure vessel computational model against this small but valuable data set is a critical next step in refining this risk quantification method proposed.

Data set size is expected to be a chronic problem for this type of work since pressure vessels in general and COPVs in particular are expensive to manufacture for destructive testing purposes. These complementary problems of low available testing energetics and small numbers of expected test articles are what drive the community to seek robust, flexible, and extensible computational models for simulating both HVI events as well as the post-impact degraded performance behavior of pressure vessels

9. CONCLUSIONS & FURTHER WORK

The objective of this paper was to highlight the multifaceted challenge in quantifying MMOD impact risks to pressure vessels used in space, especially Earth orbit missions. Early steps in the process have reasonable toolsets with wide industry acceptance and use. The significant technical challenges appear later in the process when first attempting to model HVI impact damage events accurately. Later challenges come when modeling pressure vessel behavior in a potentially degraded performance mode for the remainder of a space mission. In the case of larger manned or unmanned systems, there may be years remaining in the mission. The authors recommend that any experimental data collection be well-coordinated with members of the modeling community to ensure the most benefit from expensive destructive testing to all stakeholders.

10. ACKNOWLEDGEMENT

The authors would like to thank the Orbital Debris Program Office of NASA's Johnson Spaceflight Center for access to the ORDEM 3.0 application and its database.

EARLY ENGAGEMENT OF SAFETY & MISSION ASSURANCE EXPERTISE USING SYSTEMS ENGINEERING TOOLS: A RISK-BASED APPROACH TO EARLY IDENTIFICATION OF SAFETY AND ASSURANCE REQUIREMENTS

Scott Darpel, Sean Beckman

NASA John H. Glenn Research Center, 21000 Brookpark Rd, Cleveland, Oh 44135,
Email: scott.e.darpel@nasa.gov, sean.m.beckman@nasa.gov

ABSTRACT

Decades of systems engineering practice have demonstrated that the earlier the identification of requirements occurs, the lower the chance that costly redesigns will needed later in the project life cycle. A better understanding of all requirements can also improve the likelihood of a design's success. Significant effort has been put into developing tools and practices that facilitate requirements determination, including those that are part of the model-based systems engineering (MBSE) paradigm. These efforts have yielded improvements in requirements definition, but have thus far focused on a design's performance needs. The identification of safety & mission assurance (S&MA) related requirements, in comparison, can occur after preliminary designs are already established, yielding forced redesigns. Engaging S&MA expertise at an earlier stage, facilitated by the use of MBSE tools, and focused on actual project risk, can yield the same type of design life cycle improvements that have been realized in technical and performance requirements.

1. THE "HIDDEN FACTORY" OF REQUIREMENTS/DESIGN REWORK

Within the process improvement profession, a lot of time is dedicated to unearthing what is called the "hidden factory". This is a term to describe the myriad of steps in a process that evolve to deal with defects or other unplanned outcomes. These steps represent a drain on, or redirection of, valuable resources. They are often taken for granted, but do not add any value to the product. One example of a hidden factory would be the lost luggage department of an airline, with the defect addressed being a piece of lost luggage. No airline set out to design the need to deal with lost luggage into their original baggage handling process, as they did not intend to lose any pieces. And yet, all airlines have that counter at airports to do just that. Over time, these hidden drains on resources can creep into any process, even product development, where unplanned changes can be considered a defect.

Any product development team, whether in the aerospace, automotive, or consumer goods industries, can relate a simple fact: the later in the life cycle a change is made, the higher the impact to cost, schedule, and likely performance or success. Many such changes are the result of late identification of, or modification to, a product requirement. The more complex a system to be developed is, the higher the likelihood that such a miss or modification occurs. Just as with lost luggage, these unplanned changes are something that teams need to spend time addressing, pulling resources away from the true design effort, or increasing cost and schedule. Figure 1 illustrates the "hidden factory" steps that result from requirements "defects" in a the early steps of a typical NASA design life cycle.

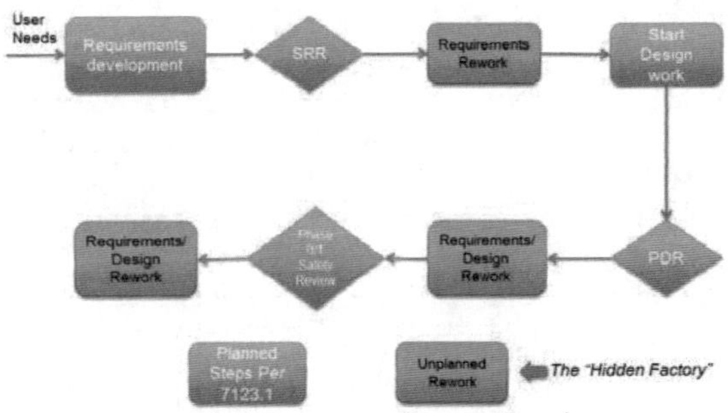

Figure 1: The "Hidden Factory" of requirements/design rework within the NASA design life cycle.

A significant amount of effort has gone into improving the process of capturing requirements within product development, such as that within systems engineering. A great deal of emphasis within systems engineering is focused on the identification, development, and management of project requirements. In his paper, "Systems engineering return on investment", Eric Honour cites the potential payback of 7 to 1 on project cost savings vs. investment in systems engineering efforts[1], much of which is due to this focus.

Systems engineering has evolved both as a discipline, and as a set of tools and practices that facilitate the kind of payback discussed by Honour. The "V" development cycle depicted in Figure 2 illustrates both the relative timing and interaction of all the key life cycle activities, including requirements development and management. This development of requirements begins with the translation of customer and stakeholder needs during concept development.

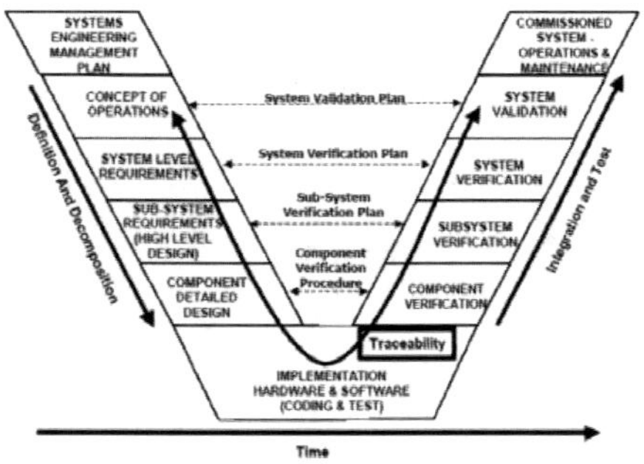

Figure 2: The Systems Engineering "V" approach to project life cycle [2].

It is during the concept development phase where projects begin to capture what is often referred to as the "voice of the customer" (VOC), which can be stated in terms of performance needs, goals, or objectives (NGOs). Teams translate these NGOs into requirements that are actionable, and are able to be verified. This VOC effort exists in product development in all industries, and has inspired the development of such practices as Design for Six Sigma, Concurrent Engineering, Concept Engineering, and Pugh Concept Selection. Within the aerospace and defense industry, this tends to be part of a stage-gate design philosophy capped with some sort of requirements review. Development projects within NASA most often follow the systems engineering practices and reviews specified in NPR 7123.1, "NASA Systems Engineering Practices and Requirements".

2. MODEL BASED SYSTEMS ENGINEERING

While the concepts of, and practices for systems engineering, including the "V" approach, have been implemented to some degree for decades, the recent efforts surrounding the Model Based Systems Engineering (MBSE) movement has provided a significant set of new tools which has greatly improved the effectiveness of the translation of VOC into requirements. These tools allow teams to model and simulate systems at early stages, before committing to any specific hardware architectures or choices.

MBSE makes use of the Systems Modeling Language (SysML) as a means of creating a representation of the system under development. SysML is an offshoot of the Universal Modeling Language (UML) that was developed in the 90's to help software engineers create a representation of the software system that was to be developed. UML is managed by the Object Management Group [3] and has been accepted by the International Organization for Standardization (ISO) as a modelling standard. The overlap and extensions of each set is shown in Figure 3. Both SysML and UML allow teams to represent how their systems should behave and interact with those who are using them.

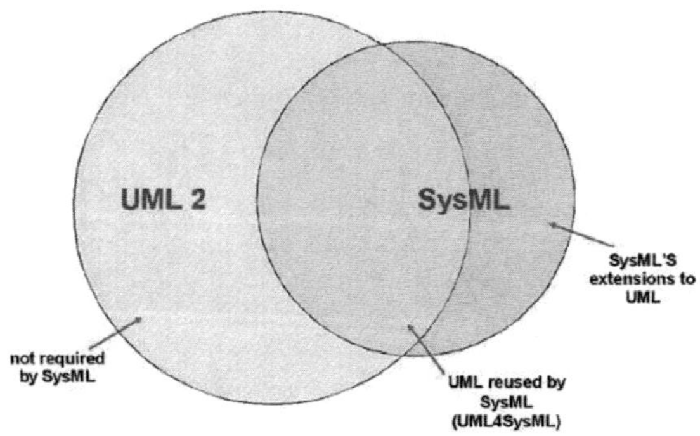

Figure 3: Overlap between UML & SysML tools

Within SysML, there are two main categories of diagrams that allow teams to represent the system: behavior diagrams and structure diagrams (see Figure 4). Early on, it is the set of behavior diagrams that have become valuable in the VOC to requirements translation process. Activity, sequence, state-machine, and use case diagrams give teams a way to represent what the system to be developed is supposed to do, per the NGOs of their customers and stakeholders. For the purposes of this paper, the Activity and Use Case diagrams will be explored further for their utility in improving the engagement of S&MA expertise.

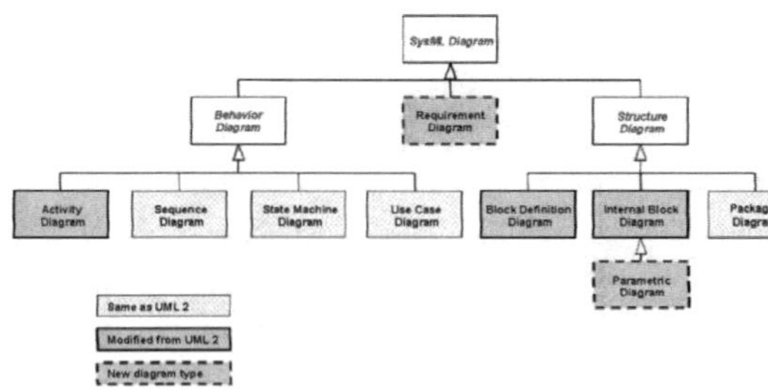

Figure 4: The 9 Diagrams of SysML

Activity diagrams help teams represent a process a given system is to perform. Any one system may be comprised of a number of activities, based on what the end-user needs are. These diagrams can be thought of as the modelling equivalent of a process map or operational concept. Lenny Delligatti, author of *SysML Distilled*, describes them as "a dynamic view of the system the expresses sequences of behaviors and event occurrences over time" [4]. They detail what steps are to be taken, and in what order.

Use Case diagrams represent how the system and its activities interact with users, also called actors. These diagrams allow teams to depict and model how these actors, including things like the environment the system is to be used in, "touch" the system. Use of activity and use case diagrams, along with MBSE in general, has improved the effectiveness of requirements identification and definition, in terms of the performance requirements for a desired system but has not eliminated the need for re-design on many projects

3. CURRENT PARADIGM FOR S&MA ENGAGEMENT & REQUIREMENTS IDENTIFICATION

Within projects following the stage-gate design process described in in NPR 7123.1, a project's requirements are reviewed at the System Requirements Review (SRR) stage, but this, again, focuses on the performance requirements. Figure 5 shows the design milestone reviews from NPR 7123.1 and typical safety reviews, superimposed onto the systems engineering "V" paradigm.

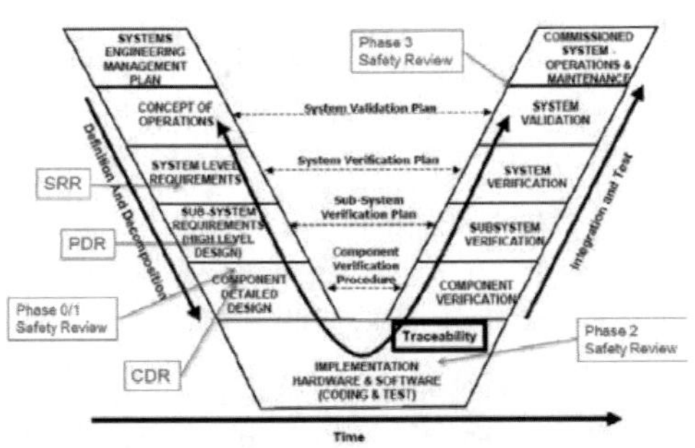

Figure 5: Systems engineering "V" paradigm with NPR 7123.1 and typical safety reviews superimposed. PDR is Preliminary Design Review; CDR is Critical Design Review

With the exception of some risk management focusing on performance NGOs, it is not until after a preliminary design is in hand that significant S&MA expertise is brought to bear on the project. Typically, the preliminary hazards assessment (PHA) is performed based on the preliminary design. Indeed, the first safety review is most often held after the preliminary design review (PDR), and can result in the need to redesign the system to address required safety features. The worst case scenario of waiting until this point in the design life cycle could be a complete abandonment of the preliminary design or concept if it is not possible to address or include required factors of safety or hazard controls. Figure 6 superimposes typical S&MA engagement on the same "V" paradigm.

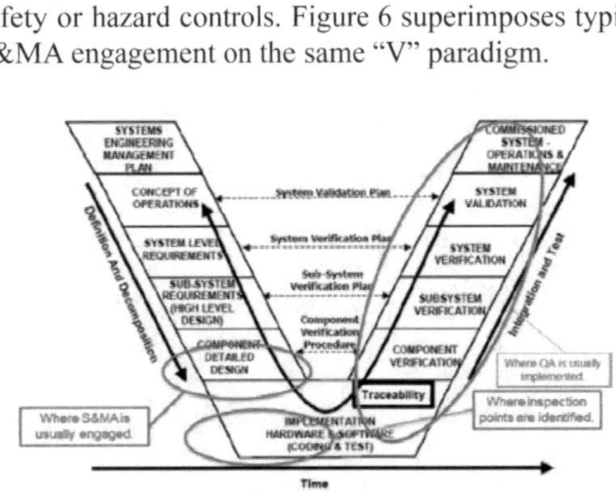

Figure 6: "V" paradigm with typical S&MA engagement

Although the S&MA community contains a rich mixture of expertise that can be thought of as another stakeholder in the system, it is often treated with skepticism in terms of adding value to a project. This is primarily due to a large set of requirements being added late in the life cycle that may or may not be associated with actual project risks. By delaying S&MA engagement, projects can arrive at a point that selection of an alternate concept is infeasible from a cost or budget standpoint, requiring concessions to function or performance to incorporate required safety or hazard controls.

A new paradigm, one that includes early engagement of S&MA expertise following a risk-based approach to the application of appropriate requirements and practices, can lead to a reduction in redesign efforts. The question considered was how to effectively implement such a paradigm shift, such that S&MA requirements can be both identified earlier, and provide real value to project success. Looking to the success of MBSE efforts can be used as an inspiration for such a shift.

4. INTEGRATING S&MA TOOLS INTO MODEL BASED SYSTEMS ENGINEERING, CURENT EFFORTS

The advantages of modeling behaviors and interactions using MBSE methods has been recognized by a growing number of S&MA leaders. As far back as 2004, there have been efforts to examine how the S&MA expertise might either benefit from, participate in, or enhance the model development. Initially, these efforts have focused on the application of such S&MA domain tools as PRA, FMEA, or reliability analysis within a MBSE-type concept model. Or, within a similarly focused, early-life cycle model that could be used to enhance requirements identification and definition.

In his paper *"Integrating Failure Modes and Effects with the System Requirements Analysis"*, Dr. Ronald Carson, introduces the concept of functional failures to allow analysts to identify potential system failures while the requirements are still being formulated and assigned to a piece of the functional architecture [5]. In essence, the proposed method has requirements analysts assign failure modes to functions that are modelled within the SE tool, thereby getting a jump on the failure modes and effects analysis that might yield insight into additional potential requirements. While this represents a significant step forward in early risk-informed requirements, some project and engineering leads may avoid the implied formality of a tool like FMEA. In 2014, Hecht, Dimpfl, and Pinchak examined the ability to auto-generate an FMEA from SysML models [6]. Again, a powerful step towards earlier S&MA consideration and potential requirements, but still focused on a potentially limited audience, current practitioners of FMEA.

In 2015, John Evans (NASA HQ), Steven Cornford (JPL), and Martin Feather (JPL) published a white paper and article through the NASA Office of Safety & Mission Assurance (OSMA) examining the concept of Model-Based Mission Assurance (MBMA) and its potential to enhance reliability and maintainability analysis during concept development [7]. The article surveys the landscape of efforts like Carson's, and determined the need to *"clearly and unambiguously establish the roles of uncertainty and risk in the system model"*. The goal has been to identify how to incorporate such tools as Probabilistic Risk Analysis (PRA) and Continuous Risk Management (CRM) earlier into the project life cycle.

5. CREATING A MORE INCLUSIVE FRAMEWORK FOR S&MA ENGAGEMENT DURING CONCEPT DEVELOPMENT

The efforts described, as well as many others, will go a long way to improving the quality of the models used to assist with requirements identification and definition. They have, in large part, focused on the specifics of tool interaction, such as in how one software tool used in the development of a reliability and maintainability (R&M) model can interact with those doing system functional modeling. What is also needed, however, is a way to actively engage the overall S&MA expertise and improve the value-proposition of using such tools, but without the specifics often involved in the software packages. The question under consideration is how to engage people with expertise in S&MA domains without the often complicating factor of software tool specifics. Referring back to Evans et al, what role can S&MA expertise play in concept and early stage model development?

When teams set out to create an Activity or Use Case diagram, one of the most effective methods may be considered one of the most old-fashioned: live and in person. Since the early days of their use, the creation of process maps is best done using a team approach, including those people impacted or served by the process. Developing an Activity Diagram, where a desired system process is going to be represented is no different. The more care given to having the correct expertise at the activity, including S&MA, will yield a better, more accurate depiction and model.

When constructing an activity diagram to depict a desired behavior for the system, the team starts by listing those steps it believes are the proper ones to execute, in their relative order. By taking the additional step of asking "is there anything hazardous about this step", "how can this step go wrong or fail", and "is there any kind of risk associated with performing this step" the team may identify unconsidered steps or conditions. This represents a risk-aware view of the concept under development. It forces the team to consider not just how they want a system to behave, but the ways by which it may fail to do so, or create hazards for the users. Not only is this a useful method by which to prevent redesigns later in the life cycle, it can also cause team members to think about and question the robustness of the activities. S&MA personnel often have a very deep understanding of typical system failure modes and can facilitate this type of discussion. Engaging S&MA expertise during the development of activity diagrams will help teams to consider whether certain safety or assurance requirements may result in a better concept.

Use case diagrams can be thought of in much the same way. As stated earlier, these diagrams depict how users, or actors, interact with the system. For example, a "crew member" may interact with the activity called "enter mode". In the early concept development stage, there will not be a specific manner by which this interaction occurs, only that it does. Still, the team can consider what the different ways this interaction can go wrong are. This could include "crew member enters the wrong mode", or "system executes the wrong mode". Both of these items represent a risk to the system not behaving as required. Teams might consider adding activities for the system to verify that the proper mode is being selected. While this simplified example may appear to be common sense, something addressed through good design, it may not be identified until after a first set of software or sample hardware is delivered.

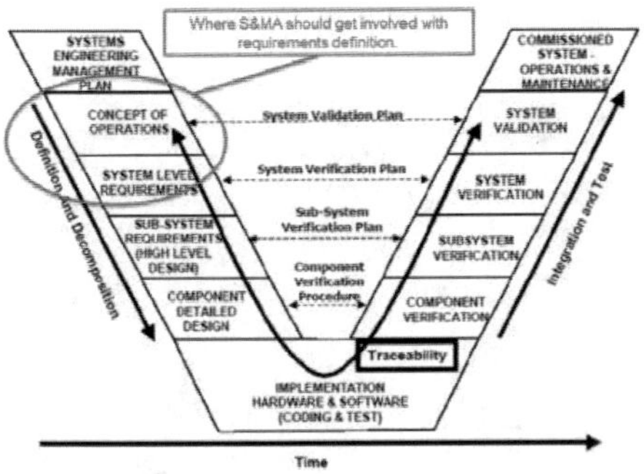

Figure 7: ""V" with Earlier S&MA Engagement

The objective is to draw S&MA expertise earlier into the concept development and requirements identification steps, as seen in Figure 7. Using it in the development of activity or use case diagrams, or similar early system behavior determination, allows the team to:

- Identify potential hazards and their requisite levels of control
- Identify a larger pool of potential risks that can help with concept trade studies
- Identify good practices early that could limit consequences or reduce the likelihood for performance risks

S&MA personnel, typically being those charged with the later development of safety data packages and reviews, are often keenly aware of what different carriers or programs may require for hazards of any type. For example, by understanding very early that a certain fluid is required for an experiment, teams can make an informed choice about controls that may be required due to toxicity levels.

For the purposes of concept exploration, the notional example of a human powered delivery system is considered. A team is tasked with developing a new delivery system that will increase the company's range, while adhering to a 30-minute cycle time. The following use case and activity diagrams can be considered an example of the typical output from such effort, following traditional flows without early S&MA engagement.

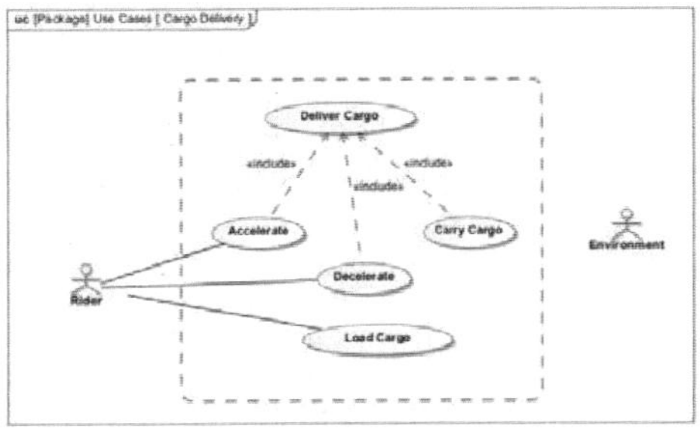

Figure 8: Typical use case diagram for example delivery system

From the use case depicted in Figure 8, the team has identified two potential actors: the "rider", being the human in the human-powered system, and the environment. The specific use case shows what activities the "rider" interacts with. In this case, the rider must load cargo, accelerate and decelerate.

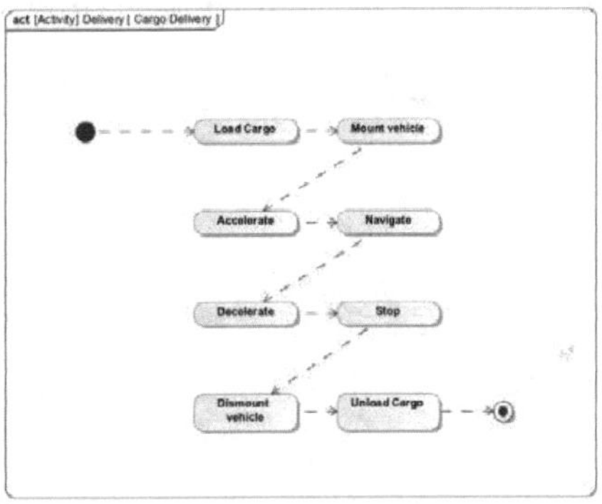

Figure 9: Typical activity diagram for example delivery system

In the activity diagram in Figure 9, the team has captured what was felt to be a primary process, starting with load cargo, and completing with unload cargo.

After going through the exercise of creating the use case and activity diagrams, the team could have come up with a set of performance requirements, such as those listed in Figure 10.

#	▲ Id	Name	Text
1	HPPDV1	Carry Capacity	The HPDV shall carry cargo of at least A m by B m by C m in dimension.
2	HPPDV2	Speed	The HPDV shall be capable of traveling at a speed of at least TBD kph.
3	HPPDV3	Acceleration	The HPDV shall be capable of accelerating at a rate of TBD kph/sec.
4	HPPDV4	Deceleration	The HPDV shall be capable of decelerating at a rate of TBD kph/sec.
5	HPPDV5	Maneuverability	The HPDV shall be maneuverable.
6	HPPDV6	Mass	The HPDV shall weigh no more than TBD kg.
7	HPPDV7	Size	The HPDV shall fit in an envelope no larger than X m by Y m by Z m.
8	HPPDV8	Cargo mass	The HPDB shall carry cargo of at least TBD kg in mass.

Figure 10: Example requirements derived from use case and activity diagrams.

User needs, as modelled through the diagrams, are translated into a requirements set. This initial set of requirements is reviewed at a System Requirements Review, or similar. Once this set of requirements passes review, work begins on the preliminary design, reviewed at the Preliminary Design Review (PDR). Typically, there are some concerns that require redesign as determined by reviewers. At times, this can require requirements re-work. After the PDR is passed, teams work on the initial safety reviews (Phase 0/1). This review can then lead to even more requirements rework.

If the team in the example, however, includes S&MA expertise during the development of the diagrams, expertise that can facilitate asking "what can go wrong with this step or interaction", it may determine that extra attention should be paid to the activity "load cargo". Additionally, the S&MA expertise involved with the development of the diagrams can help determine if there are any applicable regulations or laws that should considered.

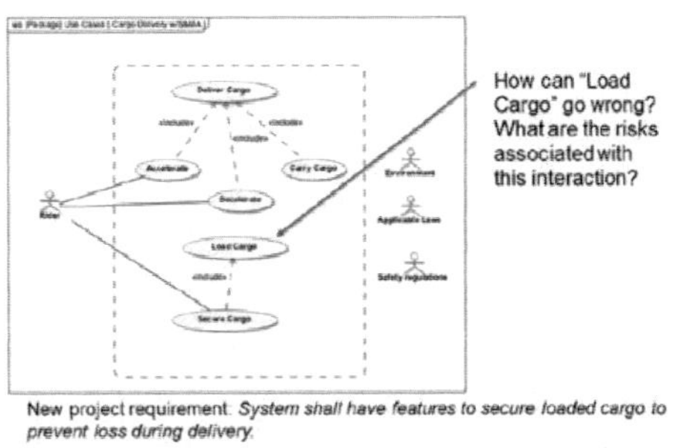

Figure 11: Enhanced use case with S&MA engagement

As an example, the city in which the company operates may have helmet ordinances which could prompt the team to include a requirement for helmet stowage. Per Figure 11, the team also determined that, as the cargo could be improperly loaded, a new requirement for a locking feature would be included. The team may even consider an operational requirement to ensure the cargo is locked in place. This, in turn, can lead to a requirement to ensure that the rider can inspect the condition.

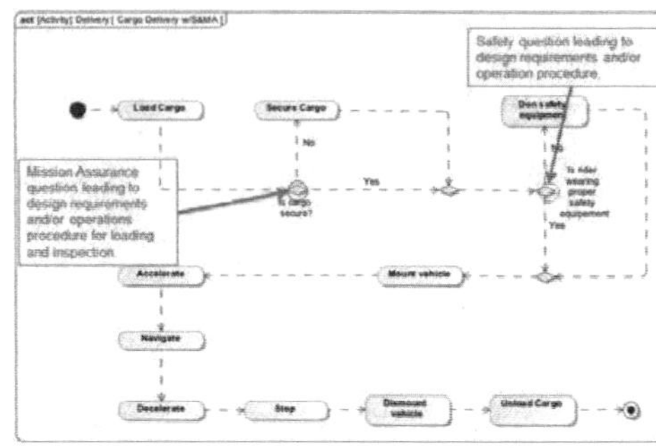

Figure 12: Enhanced activity diagram with S&MA engagement

Looking at the revised activity diagram in Figure 12, the team chose to depict the questions being asked as decision points in the activity flow. They could continue by examining each step on the activity diagram, determining if additional requirements can benefit the system, or reduce the risk associated with its success. This, in turn, may be just one of the activities they need to consider.

The proposed method focuses on personnel engagement in a software tool-neutral setting to bring more expertise into the early stages of determining what a system should do. There are a great many different tools on the market for creating MBSE models, as well as performing S&MA activities such as FMEA or PRA, with some organizations even creating in-house packages, as well. Teams should allow the process of development to determine the needs of software tool use and interaction, rather than having those tools dictate the process followed.

6. SUMMARY

While there are a great many possibilities for bringing specific S&MA models and analyses into direct interaction with those models created during early-phase concept development (within MBSE), the first step is engaging the S&MA personnel and expertise earlier in the requirements definition phase. This can be accomplished by adding the additional considerations and personnel during the development of such MBSE products as the activity and use case diagrams, exploring the risks inherent in the concepts under development. Doing so, engaging S&MA expertise to facilitate the questions "what can go wrong" or "what is hazardous about this", can significantly reduce the need to perform redesign later the in the life cycle due to missed requirements. A team may also find they have an advantage in creating initial sets of risks and hazards that can improve concept trades or design reviews. The ultimate evolution of Model-Based Mission Assurance will be the natural off shoot of this engagement, creating much more robust system models. Finally, the proposed paradigm reinforces a risk-aware or risk-informed process for the application of additional requirements, again, reducing the likelihood of redesigns later in the life cycle.

7. BIOGRAPHIES

Scott Darpel, MSIE
Program & Projects Assurance Division
Safety & Mission Assurance Directorate, NASA GRC

Scott Darpel earned a bachelor's and master's degrees in Industrial & Manufacturing Engineering from Cleveland State University. He has worked with organizations across many industries on process improvement, product development, systems engineering, and quality engineering. He has been a trainer for six sigma black belt, lean enterprise, set-up reduction and FMEA, and taught undergraduate and graduate level courses in statistical analysis, design of experiments, and quality management.

Mr. Darpel has served as the Chief S&MA Officer (CSO) for GRC's ISS Physical Sciences & Human Research Program, and as the Orion Multi-Purpose Crew Vehicle's external interface requirements manager. He is currently serving on the S&MA Team for NASA's Exploration Systems Development Division, focusing on cross-program risk and the secondary/co-manifested payloads safety review process. Scott is a member of the Cleveland-Northern Ohio chapter of INCOSE

Sean Beckman, CSEP
Quality Engineering and Assurance Branch
Program & Projects Assurance Division
Safety & Mission Assurance Directorate, NASA GRC

Sean Beckman received a Bachelor of Science degree in physics from the University of Akron and is pursuing a Master's of Science degree in Systems Engineering from Worcester Polytechnic Institute. He has spent over 15 years as a systems engineer working various projects for NASA, Boeing and DoD, including NASA GRC's Fluids and Combustion Facility, Orion and Altair spacecraft, the 747-8 aircraft, and the Army's Armored Multi-Purpose Vehicle. While specializing in requirements management he has taken on other systems engineering roles and tasks for these programs. Sean is now a quality assurance engineer at NASA's Glenn Research Center.

A member of the Cleveland-Northern Ohio chapter of INCOSE Sean is an advocate of Model Based Systems Engineering and continues studying SysML and ways it can be used for system development most recently looking at Model Based Mission Assurance (MBMA) in the NASA community. He currently holds an INCOSE Systems Engineering Professional certification.

Mr. Beckman and Mr. Darpel will be co-leading a NASA community of practice in Model-Based Mission Assurance.

8. REFERENCES

1. Honour, Eric, 2013, Systems engineering return on investment, PhD Thesis, University of South Australia. 12 p.

2. www.engineering.com/DesignSoftware/DesignSoftwareArticles/ArticleID/7352/Model-Based-System-Engineering--Beyond-Spreadsheets.aspx

3. http://www.omg.org

4. Delligatii, Lenny (2014). SysML Distilled: A Brief Guide to the Systems Modeling Language. Addison-Wesley, pp 89.

5. Carson, Ronald, Integrating Failure Modes and Effects with the System Requirements Analysis, 14th Annual International Symposium Proceedings, 2004, Boeing Integrated Defense Systems, Seattle, WA

6. M. Hecht, E.Dimpfl, J. Pinchak, "Automated Generation of Failure Modes and Effects analysis from SysML Models", 2014 IEEE International Symposium on Software Reliability Engineering Workshop (ISSREW). IEEE 2014.

7. https://sma.nasa.gov/news/articles/news-item/2015/11/09/is-model-based-mission-assurance-the-future-of-nasa-sma

"SAFE USE OF ELECTRICAL COTS HARDWARE IN HUMAN SPACE FLIGHT"

James A. Runnells

HX5, LLC, JSC Engineering and Technical Services Contract (JETS)
2224 Bay Area Blvd., Houston, TX 77058
E-mail: james.runnells@jacobs.com

ABSTRACT

The move to reduce the costs of space flight hardware development has led to the increased use of Commercial Off-The-Shelf (COTS) hardware on human space flights. Many of these commercially available devices are electrically powered, and typically use 120VAC (Volts Alternating Current) for the power source. The current human rated space vehicles were not designed for the use of commercial electrical items, and typical human space flight hazard controls (design features) may be difficult to verify. Specifically, the AC electrical COTS devices are designed and tested to meet commercial or consumer safety standards that differ significantly from manned space flight safety standards and requirements. This presents a unique challenge for the "safe" use of cost-effective 120VAC powered COTS hardware. To minimize any increase in risk to the crew or vehicle, an adaptation of the human space flight electrical safety requirements is necessary when utilizing AC COTS hardware.

1. ELECTRICAL COTS USE IN HUMAN RATED SPACE VEHICLES

The use of electrical COTS devices in Human Rated Space Vehicles is a challenge due to the differences in the design of the Space Vehicle versus the typical application envisioned for the commercially available devices. Virtually all space vehicles are fairly austere and constructed with metallic surfaces that are routinely mounted to vehicle structure, or vehicle electrical ground. While the internal metallic surfaces of today's space vehicles are usually anodized or chemically treated with similar processes for corrosion control among other factors, these surfaces do experience wear that exposes the bare metallic (conductible) surfaces. Also, distinctive areas such as seat tracks and other metallic surfaces are not anodized or otherwise treated, so they can be used as conductive surface grounding as necessary. In contrast, the typical home, workshop, or industrial location that electrical COTS devices are designed for are conspicuously different from today's space vehicles. The lack of metal or conductive surfaces, especially metal which is grounded to earth ground, is common in terrestrial applications and helps isolate a user from electrical shock hazards. Furthermore, modern applications of AC power use in high risk environments such as a garages and wet areas like patio/pool areas or bathrooms, require a Ground- Fault Circuit Interrupter (GFCI) be employed as an integral safety component of AC power circuits per building codes. The combination of hazard controls provided by today's homes and commercial buildings, combined with electrical product design features required by today's consumer and industrial safety standards, ensures relatively safe AC electrical products for terrestrial-based applications. However, even with these safety controls hundreds of people are electrocuted or shocked annually.

These same safety features (lack of a conductive ground path or use of a GFCI circuit) are not initially applicable for COTS AC electrical devices used on the International Space Station (ISS). The ISS is primarily a metallic based platform. The majority of the internal surfaces may be anodized, or otherwise chemically treated, which reduces the likelihood for contact with a grounded metallic surface. Treated metallic surfaces do not eliminate the potential to come in contact with a sufficiently grounded surfaces and, in fact, the opposite is more accurate given the high likelihood that a crewmember on the ISS will use a hand to stabilize or guide themselves when working in a specific area. It is postulated herein that there is a significantly higher likelihood that a crewmember will contact a metallic grounded surface on ISS than in terrestrial applications while handling COTS AC electrical devices or power connectors. If a crewmember is grasping a grounded metallic surface or metallic handhold, and grasps a failed 120VAC electrical device or energized connector with the other hand, an electrical shock is likely to occur across the crewmembers heart, which could be catastrophic. The hand-to-hand electrical path (across the heart) is a valid electrical shock scenario that can cause heart defibrillation, or lead to loss of a crewmember. This electrical shock hazard scenario caused by contact with exposed conductors during a mate/demate can be adequately controlled per MA2-99-170 [1]. Given the potential hazard severity (catastrophic) and the current designs for human space flight vehicles, Program guidelines meant to minimize the potential increase of an electrical shock (or molten metal generation) must be implemented for AC electrical COTS devices used on ISS or other human space vehicles.

2. BENEFITS FROM USE OF AC COTS DEVICES ON ISS

Safety challenges notwithstanding, the advantage of using COTS AC electrical devices on ISS, or other human rated space vehicles, stems from a cost and schedule reduction effort to use items available in the commercial market. These low-cost AC COTS devices can provide simple solutions to program needs, and for conducting experiments on orbit. The move to replace more costly items that are designed and built exclusively for use in space can be a significant savings for a program, as well as eliminate lengthy schedules for hardware/software design and development. Additional benefits include the time and costs saved by reducing extensive ISS power quality testing, and by minimizing other required electrical or mechanical safety testing of hardware for the space environment. It's a unique approach to provisioning the ISS, but it does require clear safety guidelines and additional precautions for the safe use of COTS AC electrical devices in the ISS or human spacecraft environment; one that recognizes the failure rates and reduced safety features of AC COTS devices in the ISS user environment. Most COTS AC devices are not designed for use in a space environment, or even in an equivalent wet, high risk, environment. This use in a space environment (i.e., increased internal operating temperatures, radiation impact to electronics, etc.) can lead to a diminished operational life with recurring replacement of these items, and also introduces high failure risk electrical items to the crewed vehicle. These same concerns for diminished life also drive the concern for failure modes in these high voltage (HV) devices that can impact the space vehicle or cause serious injury to a crewmember. Correspondingly, HV for NASA human spaceflight is a voltage greater than 30V (or 32V, depending upon the vehicle) due to the higher risk for contact with energized devices and station metallic surfaces.

3. GUIDELINES FOR USE OF COTS AC ELECTRICAL DEVICES ON ISS

In the case of the ISS, the Program has assessed the potential increase in risk from use of COTS AC electrical devices on ISS. This resulting hazard assessment has levied certain guidelines and hazard controls for the acceptable use of COTS AC devices on the ISS. The ISS Program instituted the use of AC Inverter power supplies onboard ISS that converts 28VDC (Volts Direct Current), or 120 VDC, from an ISS DC power outlet to 120VAC. Descriptions of each 120 VAC Power Inverter can be found in document JSC-66202 [2]. These AC Inverters have undergone a safety assessment by the ISS Safety Review Panel (iSRP) with design and procedural controls that strive to minimize the electrical shock or molten metal hazard potential for the ISS crew. ISS AC Power Inverter electrical hazard causes, controls, and verifications are documented in hazard report (HR) IPVR-017 [3]. Additionally, the Program was apprised of an increased risk due to use of high voltage AC Inverters on ISS, and the use of portable COTS AC electrical hardware on ISS that are typically not designed, built, or tested to operate in a human spacecraft environment. The areas of non- compliance with existing ISS electrical safety requirements were documented in non-compliance report (NCR) NCR-IPVR-001 [4], and the Program accepted the documented risk increase with the application of specific guidelines for various types of AC COTS devices. The ISS AC Power Inverter user safety guidelines help minimize the increase for potential electrical shock or molten metal hazards on ISS from the use of these COTS AC electrical devices.

The guidelines for use of COTS AC electrical devices utilized on-board ISS are captured in NASA Letter EP-GL-2014-001 [5], developed by the JSC Engineering Directorate, Energy Systems Division (JSC Code EP). The guidelines recognize areas of increased risk, including COTS AC connectors (NEMA 5-15 Receptacles (R) and Plugs (P)), COTS DC power connectors, and COTS electrical designs that are noncompliant with ISS Program fault tolerance requirements for electrical shock and molten metal generation hazards. Additional concerns for potential hazards not addressed in NCR-IPVR-001, which primarily documents COTS AC electrical shock hazards, include touch temperature exceedances from excessive heating (electrical items on the ISS lack convection cooling as a result of zero gravity), and possible materials concerns such as flammability and off-gassing.

4. COTS AC HARDWARE ELECRICAL DESIGN LIMITATIONS

COTS AC electrical devices are not designed, developed, or tested to NASA standard verification practices, nor are they designed for use in space environments. Industry Standards such as Underwriter Laboratory (UL) and Canadian Safety Agency (CSA) 60950 – Information Technology (IT) Equipment Safety, [7], or other user specific electrical safety standards (i.e., UL/CSA 60001, Medical Electrical Equipment [8]), document design and manufacturing standards that often result in single fault (1FT) tolerance to a shock hazard from an AC device, as specifically noted in the standards.

However, NASA and International Partner designed and developed HV electrical devices require two fault tolerance (2FT) to electrical shock hazards. ISS Program hardware typically requires 2FT from redundant ground paths and electrical isolation between primary power inputs and chassis ground, per MA2-99-142 [9], in addition to other design requirements captured in ISS power interface and control documents. Therefore, AC COTS electrical devices do not comply with the ISS safety requirements for 2FT to a catastrophic hazard, and require additional risk mitigation steps.

5. HAZARDS FROM EXPOSED ELECTRICAL CONNECTORS

The first area of concern would be the NEMA 5-15 (R and P) power connectors. These 120VAC power connectors do not meet ISS Safety Policy for power connector safety requirements for safe mate or demate, especially with regards to the lack of scoop- proof connector per MA2-99-170, and the lack of power connector positive locking features once mated. The lack of a scoop proof connector, a mechanical design feature that shrouds the connector contacts (sockets/pins) before pin/socket contact during the mate/demate process, exposes the crew to a potential molten metal hazard during such activity. Furthermore, electrical shock hazards are present during mate/demate of the AC power connector, as well as an electrical shock hazard potential from a partially demated power connector with exposed energized pins (possibly from crew contact or a lack of a locking connector). In terrestrial applications, the exposed NEMA 5-15 blades on an AC plug can generate a molten metal hazard. Molten metal generation during power connector mate/demate, or from partially demated NEMA 5-15 power plugs is considered a catastrophic hazard since molten metal ejected during arcing from opening/closing powered contacts can contact a crewmember in the ISS environment (zero-g). The primary concern is molten metal contact with the eye(s). However, the design of the 120VAC Power Inverter reduces the potential for molten metal to an unlikely event due to limiting the maximum arc time (8.3 milliseconds) with a reliable interrupt at the current zero-crossing, and by limiting the available AC current. This effectively minimizes the molten metal potential, as documented in program NCR-IPVR-001. However, the shock hazard from crew contact with energized and exposed NEMA 5-15 Plug blades still exists and is considered a significant risk for the crew.

6. AC COTS ELECTRICAL SHOCK HAZARDS

A significant hazard potential for the crew is electrical shock from unintentional or inadvertent crew contact with energized and exposed power connectors. This hazard can occur during crew activities such as mate/demate of the AC Power connectors or from inadvertent contact with partially demated NEMA-15 AC Plugs. Requirements for crew electrical interfaces are included in SSP 50005 [6], including Sections 6.4.3.3, 6.4.3.7 and 11.10.3.2. These requirements help provide risk mitigation features including protective connector covers, self-locking connectors, and other design features to protect the crew from inadvertent/partial disconnect of electrical power connectors. The AC Power Inverter provides spring loaded protective caps for the four NEMA-15 AC receptacles. Additionally, Interpretation Letter MA2-99-170, "Mate/Demate of Powered Connectors", provides design and procedural requirements for electrical device providers to follow when the crew connects or disconnects power connectors from the ISS or other power sources. The design features ensure a 2FT equivalence for electrical shock and molten metal hazards. The MA2-99-170 Letter requires power connectors with greater than 3A (but less than 65A) available upstream current to have; a verifiable upstream inhibit (i.e., a power switch or crew/ground controlled power relay with monitoring), the connector on the power side has to be terminated in sockets (versus pins), and the connector design must shroud the sockets and pins and preclude contact (or FOD) between them prior to mate/demate of the connector (i.e., a scoop-proof connector). Mil-Spec connectors used in systems designed for human space vehicles generally provide mechanical design features (terminated in sockets on power side and scoop-proof connector body), and power inhibits as standard design features. Unfortunately, AC COTS device connectors do not provide the required mechanical design features. This leaves the upstream power switch at each NEMA-15 power receptacle as the only design feature protecting the crew from electrical shock during mate/demate of the AC Power Cable to the AC Power Inverter.

To ensure a standardized approach to the use of AC COTS devices, the Program developed an AC Powered COTS Hardware Guidance Letter, EP-GL- 2014-001. These guidelines provide additional risk mitigation for crew AC COTS electrical mate/demate activities, as well as ensuring the AC COTS loads provide sufficient design features to minimize crew risk for use on orbit. The risk mitigation guidelines include securing power cables with strain relief measures at the AC Power Inverter AC switch guards (using zip-ties or other non- flammable methods) to minimize the potential for inadvertent disconnect in overnight or long term applications. It also provides for crew training on the proper methods for safe mate/demate of AC power plugs on orbit. Other mitigation measures include crew procedures and operational controls alerting the crew to observe an AC Power Inverter for partially demated AC connectors as they approach an inverter. If they observe any partially demated AC connectors, the crew removes power (inhibits power) for any partially demated plug and reconnects that power plug to remove the hazard potential prior to reapplying power for that device. These guidelines and procedures help reduce the risk of an electrical shock hazard during mate/demate of the AC power plug, and from partially demated AC power plugs

7. AC COTS DEVICES CATEGORIZED TO SIMPLIFY GUIDELINES FOR SAFE USE ON ISS

The ISS Program guidelines establish AC Load categories (*Category A, B, or C*) with specific hazard controls and verifications for each to ensure any risk increase from the use of AC COTS devices is minimized. *Category A and B* rely on electrical safety standards from organizations such as UL to ensure a safe design for AC COTS devices. Additionally, the Program requires UL Certification and Listing for the AC COTS device. This certification ensures independent UL (or an approved surrogate Test Lab) assessment and testing of the device design and manufactured product to the relevant safety standards. Also, the UL Listing ensures the device is manufactured and tested in the factory for specific safety features, with periodic site visits by UL.

The categories specified by the Program and documented in EP-GL-2014-001 are:

AC COTS Load Category A – *Double Insulated Loads*. These load types utilize either one layer of reinforced insulation, or two distinct layers of insulation materials (typically called basic and supplementary insulation) per electrical safety standards such as UL 60950. This design is designated by a double box symbol on the device label, and a double insulated AC cable is either included in this certification, or a suitable UL approved cable is used. This type of device should not have any penetrations or metallic surfaces (i.e., screws, etc.) exposed to the crew. Penetrations are a concern for floating (i.e., zero-g effect) conductive foreign object debris (FOD) that could enter device penetrations and cause molten metal or shock hazards.

AC COTS Load Category B – *Fully Insulated Loads*. This Load category allows AC COTS devices that are fully insulated from the crew (device fully enclosed in non-conducive insulating material), but not designated as double insulated by the safety assessment and test Lab. Once again, the connecting AC cable must be double insulated (or overwrapped with a non-flammable material, which is typically the case for most cables due to materials flammability concerns). Also, the same stipulation applies for no penetrations and metallic surfaces.

The typical hazard controls for both Category A and Category B devices include double or fully insulated AC devices, electrical isolation of AC power input from the secondary (downstream or converted) power by at least 1 MOhm, and a current UL Certification/Listing. The electrical circuit isolation is a significant feature when coupled with a double or fully insulated device. The UL Listing ensures the device is tested and certified to meet minimum electrical safety standards for its intended use. Another important hazard control is the confirmation that the device is not counterfeit, which would raise significant concerns with the devices safety design. Underwriter Laboratories (UL) has a Counterfeit Group that responsively answers requests to confirm a device is not counterfeit when queried by AC COTS device providers for the ISS.

The Program also has generic crew procedures in place for use with the AC COTS devices with all three AC COTS Load Categories. These generic procedures, or operational controls, ensure the AC COTS provider has ISS Program approved procedures that govern the safe application and use of the AC COTS device with the AC Power Inverter (including mate/demate procedures, securing the AC cable/plug against inadvertent demate, etc.). These generic procedures and controls are documented by the AC COTS device provider in their relevant safety documentation and Program deliverables (i.e. safety analysis and hazard reports).

AC COTS Load Category C – *Devices with Modifications or non-double insulated with conductive surfaces*. These devices have either been modified or do not provide sufficient design features to ensure crew safety when using the AC COTS device. Modifications can be as simple as replacing the cable connector with a non-standard or non- integral AC plug, modifications of the AC COTS device enclosure, or other modifications that would invalidate the electrical safety implemented by the manufacturer. AC Device modifications are a significant safety risk for the crew and Program since in most cases knowledge of the device design and safety features from the manufacturer are unknown or unverifiable by the ISS Program after modifications are made. Another area of concern is devices with penetrations larger than 0.125 sq. inches. Which as previously noted, could be a conductible FOD risk.

In the *Category C* case, the Program has implemented hazard control guidelines that requires at least three controls. These controls include; the use of a Program provided GFCI cable between the ISS AC Power Inverter and the AC COTS device, minimum electrical isolation of 1 MOhm between the AC power input and the device enclosure or chassis ground, and an acceptable AC COTS device grounded through the AC Power Inverter NEMA-15 receptacle ground socket, or electrically bonded and grounded to a suitable ground location inside ISS.

Other hazard controls are available for use by the AC COTS device to provide alternatives, or to augment the typical *Category A, B, or C* guidelines and recommended controls. The EP-GL-2014-001 guidelines outline several alternatives to help an AC COTS device provider apply clear, suitable hazard controls for their specific device. These guidelines and typical hazard controls help to minimize the increase in risk to the crew for use of commercial AC devices on ISS, or other human space vehicles, for which the AC COTS devices were not originally intended for use.

8. ELECTRICAL COTS DEVICE DC CONNECTOR HAZARDS

One additional area for consideration is the typical application of AC COTS AC/DC power supply devices on ISS. Many AC COTS devices are AC/DC converters (power supplies) such as a Laptop power supply (IT equipment), or a similar AC/DC power supply used as a battery charger or other COTS device power input (i.e., camera power supply or other commercial device power supply when low DC voltages are required by a COTS device). Many of these devices are as simple as an AC dongle used by personal electronic devices (tablets, etc.) for charging or operations. However, most are for AC/DC power supplies that can provide significant current (greater than 3A).

COTS AC/DC converters with DC outputs greater than three amps (A) require specific controls per MA2-99-170. A COTS AC electrical device with upstream available current greater than 3 amperes (A) and less than 32V can still be a molten metal hazard. The electrical shock hazard is abated with voltages below 32V. Depending upon the type of DC connector used between the AC/DC COTS device and the COTS DC Load device, there may be a need to remove upstream power. If the DC output can supply greater than 3A, and the connector is considered a high risk for potential molten metal generation, a suitable upstream power inhibit is generally required to reduce the risk of molten metal for the crew. A simple solution can be to have the crew, through approved procedures, connect the COTS load to the AC/DC power supply prior to plugging the AC/DC Power supply NEMA-15 plug into the AC Power Inverter and applying power at the AC Inverter. This is called dead-facing the load, and ensures no downstream electrical hazards exist for the crew until AC Power is applied at the AC Power Inverter.

There are other considerations for the COTS AC/DC power supplies that do increase the risk to a crewmember. Once particular area is the fact that a short or component failure in the power supply could allow high voltage on the DC power connector. These COTS DC connectors generally have exposed conductive surfaces and if the connector design is insufficient it could present a 0FT risk to the crew for electrical shock. Most COTS DC connectors are barrel type connectors, and these barrel connectors typically do not expose the crew to energized conductors, or both conductors simultaneously, to allow current flow. The risk of crew hand-to-hand current from a failed AC/DC power supply is still a concern. However, the risk of electrical shock for the crew is considered low likelihood for the barrel type COTS DC connectors. Other COTS DC connectors are assessed on a case-by-case basis for a crew electrical shock hazard from failed/shorted AC/DC power supplies.

9. APPLYING COTS ELECRICAL SAFETY GUIDELINES TO MINIMIZE RISK ON ISS

AC COTS devices are cost-effective hardware that allow prompt solutions for crew and ISS applications and experiments on orbit. AC COTS device (or other HV COTS devices) use on ISS, or any human space vehicle, can be accomplished while minimizing the risk increase for the crew, provided the application is properly characterized and the risks clearly defined. The ISS AC Power Inverter safety guidelines ensure a clear path to approval for AC COTS device use on ISS. Approved applications for safe COTS AC device use of the AC Power Inverter on ISS will ensure any risk increase to the crew on ISS is minimized and adequately controlled per Program approved and accepted guidelines.

10. REFERENCES

1. MA2-99-170, Crew Mating/Demating of Powered Connectors, Feb. 11, 2000, SSP 51700 Appendix E

2. JSC-66202, ISS Power Inverter to 120VAC 60Hz Loads Interface Definition Document (IDD), July 2013, Revision F

3. NCR-IPVR-001, 120V and 28V AC Power Inverter with AC loads non-compliance, Nov. 13, 2012

4. HR IPVR-017, ISS Power Inverter UNQ-IPVR- 17 Electrical Shock Hazard which includes the GFCI cable assembly, March 8, 2013

5. EP-GL-2014-001, AC Powered COTS Hardware Guidance Letter, April 2014

6. SSP 50005, International Space Station Flight Crew Integration Standard, June 30, 2006, Revision E

7. UL/CSA 60950, UL Standard on IT Information Technology Equipment Safety, Part-1, March 27, 2007, 2nd Edition

8. UL/CSA 66001, Medical Electrical Equipment, Part 1: General Requirements for Safety, April 25, 2003

9. MA2-99-142, On-Orbit Bonding and Grounding, Oct. 12, 1999, SSP 51700 – Appendix E

INTRODUCING IAASS-ISSB-S-1700-REV-B SPACE SAFETY STANDARD COMMERCIAL HUMAN-RATED SYSTEM

Tommaso Sgobba

International Association for the Advancement of Space Safety, The Netherlands
email: tommaso.sgobba@iaass.org

1. CONSENSUS STANDARDS

In February 2005 suborbital spaceflight industry leaders met to discuss establishing a voluntary industry consensus standards organization with the purpose of developing industry standards to ensure the safety of passengers [1]. After more than 11 years there is no news about any substantial progress.

Voluntary consensus standards are standards developed or adopted by voluntary consensus standards bodies. Even if nowadays the FAA/AST is forbidden to issue safety standard for the safety of those on board a commercial spaceflight, it is general policy of US Government to mandate the use of voluntary consensus standards in lieu of government-unique standards in procurement and regulatory activities, except where inconsistent with law or otherwise impractical, and to encourage US Government agencies to take part in voluntary consensus standards bodies [2].

Recently the US Government has opened the door to the flexibility to choose among standards developed in the private sector, particularly standards developed for emerging technologies that may or may not precisely follow the traditional voluntary consensus standards development process. Factors to consider include, among others, the effectiveness and suitability of the standard for meeting agency regulatory need [2].

The IAASS-ISSB-S-1700 "Safety Standard for Commercial Human-Rated System" is an heritage-based standard derived from NASA NSTS 1700.7b "Safety Policy and Requirements for Payloads Using the Space Transportation System" that responds perfectly to those criteria for effectiveness and suitability.

Initially issued in 2006, the IAASS standard was created by by taking into account some modifications made when NSTS 1700.7b was adapted for use on by the International Space Station Program, and considering NASA's post-Columbia accident safety requirements policy in NPR 8705 "Human-Rating Requirements for Space Systems". The only non-heritage requirements included in the IAASS standard are quantitative probabilistic safety risk goals (as guidance: "should"), a requirement on crashworthiness, and a requirement on data collection (black-box).

2. SHUTTLE PAYLOADS HERITAGE

2.1 Made to Fit Many Customers

The Shuttle program, officially known as National Space Transportation System (NSTS), was announced by President Nixon on January 5, 1972. The Shuttle was considered at the time the reusable launch system of the future, promising to revolutionize launch availability and costs. All kind of users were expected from government agencies, to commercial customers and other NASA programs. As Shuttle development was progressing, NASA started to work on payload accommodation, interfaces, and safety requirements.

A wide variety of payloads were expected including many that had never flown before in space. The Shuttle cargo bay was sized to accommodate huge payloads like Hubble telescope, entire satellites including rocket upper stages like Galileo and Ulysses, and habitable modules

Fig. 1 Hubble Telescope being released from the cargo bay of Shuttle Discovery

like SpaceLab and SpaceHab. Smaller payloads, research facilities and experiments would be accommodated in racks, lockers in the pressurized crew compartment or externally in dedicated canisters in the cargo bay called Get Away Special (GAS).

2.2 Shuttle payloads requirements

Shuttle payloads preparatory activities started already in 1972 with work on safety and interface requirements, payload integration process, and implementation guidelines. There was also the preparation of so-called interpretation letters (because of their format) meant to be advisory. Interpretation letters were needed because the Shuttle program in view of providing transportation services in particular to commercial customers did not want to dictate design solutions. On the other hand, being many safety requirements generic and goal oriented there was the need to help customers understand purpose and rationale of some requirements, and to advise on design solutions that could be acceptable.

Verification requirements (i.e. analyses and/or tests, inspections, and demonstrations) were also established as well as guidance for the preparation of payloads verification plans to be issued by users to prove compliance with requirements.

For customers benefit the Shuttle program created also documents describing accommodations in the way say of electrical power, cooling, or safety inhibits like switches and valves that could be made available (upon negotiation) on the Orbiter side to help meet some safety requirements levied on payloads. Whole series of technical standards support standards and handbooks were developed to help customer, design and verify their systems and to plan operations. Those standards were in areas of structures, mechanical systems, batteries, pyros, wiring, safe distances before allowing switch off of a safety inhibit for firing payloads propulsion system, etc.

In establishing payloads safety requirements, the Shuttle program decided to take some "comfort margin" or in other words to be more conservative than for Shuttle systems. This was not meant to be a penalty but just the price to be paid for allowing more design freedom to payloads developers. For example, for Shuttle systems rigorous requirements were applied on procurement, selection and testing of EEE components, on software development, and on quality control and configuration management. Instead payload developers were allowed to use whatever EEE components they considered suitable, and their own software development methods. In the case of small companies or universities it was even expected that some of them could lack familiarity with basic quality control and configuration management techniques. Overall it was a matter of trading reliability for cost. Payloads failures were acceptable as long as they did not lead to higher safety risk. (Years later during actual operations the reliability of some experiments was found to be so poor that it risked defeating the entire scientific purpose of their research program. Some payloads were failing 2-3 days into the mission!).

In general, payloads do not have 'must-work-function' (active functions vital for keeping the crew alive), but can often include 'must-not-work functions'. These are functions that if operated inadvertently or untimely (e.g. propulsion ignition, or access to a live high-power laser) can kill or injury the crew, or cause damages. Payloads may include some inherent hazards such as high voltages, high temperature, toxic compounds, etc.

The classification of 'hazard severity' was defined to be used to drive the amount of fault tolerance required. According to requirements, potential catastrophic hazards required two-fault-tolerance, while critical hazard required single-fault-tolerance. A critical hazard was defined as potentially resulting in injury to the crew or impairment to the vehicle or the mission. A catastrophic hazard could lead instead to loss of crew, loss of Orbiter, and loss of mission.

An important concept was to assume payload software as zero-fault-tolerant thus counting it as providing only one level of hazard control. All additional controls had to be hard-wired, unless payload developer opted to apply the same strict software development and independent verification and validation processes required for Shuttle systems. The possibility to achieve two-fault tolerance (i.e. 3 controls) by computer was basically discouraged. (The Shuttle system relied extensively on computer control of many 'must-work-functions'. For example, the flight control system had 4 redundant computers plus a back-up to be fail-operational/fail-safe).

The 'margin of comfort' consisted essentially in requiring up to two-fault-tolerance and larger safety factors for hazard controls, and applying some conservativism when classifying the severity of a hazard. For example, structural failures and pressurized systems failures where always classified as 'catastrophic'. Some conservativism would be also applied when classifying the toxicity level of chemical compounds.

2.3 How the safety standard was used

The selective application of safety and technical requirements was the source of the versatility that made the NSTS 1700.7b standard a great success. The performance of a hazard analysis was the prerequisite for using the standard because it allowed to identify the hazards that actually existed in the payload. Furthermore, not all requirements would usually apply because not all subsystems envisaged by the standard would be actually present in a payload. In other words, the NSTS 1700.7b requirements were not meant to be a blueprint on how to design a specific type of payload to be safe, but gave rules for controlling hazards determined to exist in the payload. The complexity of the payload and the hazards it actually included determined the selection of safety requirements, which in turn drove, without prescribing, the detailed design. Differently from a 'safety code' that prescribes in detail how to build a product to make it safe, the Shuttle payloads safety standard left up to the developer to decide how the detail design would meet the safety requirements.

2.4 The payload safety review panel

As work on safety and technical requirements progressed, consideration was given to the organization of the payload safety review panel (PSRP). This was to be a novel kind of organization.

Until Apollo 1 fire accident in 1967 there was no safety program at NASA and no safety organization. After the accident, hazards analyses were systematically performed, and safety divisions established starting with Johnson Space Center.

Safety divisions fulfilled their function through so-called 'concurrent engineering', consisting in participating with representatives to various project reviews, but without playing a role in the decisional process. In other words, the safety function had no independent voice and had to compete with engineering and operations representatives to have its voice heard during project review meetings.

In 1986, the Aeronautics and Space Engineering Board that evaluated the Shuttle risk assessment and management processes after the Challenger disaster stated in their report [6] that: *"The multi-layered system of boards and panels in every aspect of the NSTS may lead individuals to defer to the anonymity of the process and not focus closely enough on their individual responsibilities in the decision chain. The sheer number of NSTS related boards and panels seems to produce a mindset of 'collective responsibility.'"* They recommended that *"NASA should periodically remind all NASA personnel that boards and panels are advisory in nature. He should specify the individuals in NASA, by name and position, who are responsible for making final decisions while considering the advice of each panel and board. NASA management should also see to it that each individual involved in the NSTS Program is completely aware of his/her responsibilities".*

The Shuttle Challenger disaster investigation report branded the safety program as the 'silent program'. In 1977 the Shuttle Program decided to establish the PSRP (Payload Safety Review Panel). The approach was completely different from Shuttle systems project reviews, and gave to safety a leading and authoritative role, clear lines of responsibility, and ample technical support.

The safety review panel would be chaired by a safety staff and composed by representatives of different technical groups and organizations (engineering, operations, astronauts office, etc.). The chair had sole authority for signature of hazard reports, and had direct access to Shuttle Program Manager. A pillar of the overall approach was the concept that the payload organization (i.e. the customer) bore responsibility for the design and verification of their payload. The safety review panel had instead the responsibility to ascertain that customers understood the safety requirements and carried out correctly identified hazards, hazard causes, and relevant controls.

The customer had also to prove to the panel that agreed controls had been incorporated first in the design and then in the builds. In the process the payload organization had to obtain concurrence from NASA specialist groups on specific items such as construction materials usage, as meeting requirements for flammability, offgassing and stress corrosion, toxicological assessment of hazardous chemicals, etc.

At the end of the safety review process, the chair would issue a letter certifying that safety reviews had been successfully completed for that payload on a specific Shuttle mission. In case of re-flight on a subsequent mission any change had to be discussed with the panel. Any anomaly encountered in the previous mission had also to be discussed with the panel for safety relevance, and proposed corrective actions agreed.

Following the Challenger disaster it was decided to deemphasize the commercial aspects of the Shuttle Program and encourage commercial customers to fly on expendable rockets. Commercial customers had been a central consideration when establishing the Shuttle payloads safety requirements and the *modus- operandi* of the PSRP. Now the Shuttle was expected just to serve other NASA programs. Things may have been organized differently if this had been known from the very beginning. However, the goodness of the overall payloads safety approach was confirmed when an unexpected customer appeared: the Russians.

The PSRP had gained experience in dealing with foreign payloads since the early Spacelab missions with the European Space Agency. ESA had already adopted/adapted many of NASA technical standards as their own standards. Working with Europeans was not much different than working with a NASA center. In 1994 NASA agreed with Russians to use the Shuttle to deliver a Russian docking module to the Space Station Mir. So the Russian docking module became a Shuttle payload and PSRP had to take care of the safety review process. *"The Russians had no clue about what are these silly payload requirements, what's all this paper you guys are working, we don't need all that paper business."* [7]

What NASA learned was that the Russians had a robust and safety-minded design that had been thoroughly tested. It was not difficult to show that it met the Shuttle payloads safety requirements because they were goal oriented and not prescriptive. Eventually the NASA-JSC engineering Directorate ended up doing most of the paper work, and some analysis like fracture mechanics. Not a big deal!

The 'natural' separation between Shuttle program and payloads projects gave the rare benefits of independence, strong management support and outstanding technical support to safety reviews. The PSRP operated flawlessly since the beginning of Shuttle operations, and was reconfirmed after Challenger and Columbia accidents. But there was more come!

3. THE INTERNATIONAL SPACE STATION

3.1 Historical overview of the ISS

Until May 1960, when the U-2 spy aircraft flown by US Central Intelligence Agency pilot Francis Gary Powers, was shot down over Soviet Union, aerial reconnaissance was considered superior to satellite reconnaissance because of the efficiency engendered by the pilot on-board. The heritage of the International Space Station can be tracked back to the political decision to develop orbital reconnaissance capabilities with human on-board as a substitute for risky aerial reconnaissance. The USAF initiated in 1963 the development of the American military space station MOL (Manned Orbiting Laboratory), the Soviet Union responded with the Almaz program which was a series of military space stations under the cover of the Salyut civil station program.

The early US civil station developed by NASA was the Skylab. However, it was in 1984 that the US began designing Freedom, a permanent manned space station, and called upon the G7 Group of countries to join in the program. Two years later the Soviet Union launched Mir, the first modular space station.

In 1993 the Clinton Administration decided to broach the subject of joint cooperation to construct the ISS between US and Russia, together with the international partners of the previous Freedom station program, Japan, Europe, Canada, and Italy. In 1998 Russia launched a Proton rocket from Baikonur Cosmodrome in Kazakhstan, carrying the first element of the ISS, the Zarya module. Later before the end of the same year the US Space Shuttle was launched carrying the second element of the ISS, the Unity Module.

3.2 ISS agreements

Legally, ISS is not a single national spaceship, although the US has a leading program management role, and it is also not truly an international one.
The ISS partners had four possible legal options to organize the program: 1) a national space station, under the jurisdiction and control of one country; 2) a multinational space station, under the joint jurisdiction and control of several nations; 3) a multinational space station, the individual modules of which are under the jurisdiction and control of separate nations; or 4) an international space station, under the jurisdiction and control of an international governmental organisation similar to INTELSAT.

The option 3) above was the one eventually selected for ISS, thus recognising the *"fundamental objective" of each partner to retain "responsibility for design, development, exploitation and evolution of, ...identifiable elements of the space station together with the responsibility for their management..."* (ESA/C-M/LXVII/Res.2, January 31, 1985).

The ISS Inter-Government Agreement (IGA) states in fact in Article 5 that: *"Each Partner shall retain jurisdiction and control over the elements it registers...and over personnel in or on the Space Station who are its nationals".*

Safety is an exception to the multinational set-up above in the sense that it is defined in the ISS IGA, and in the MoUs between NASA and the international partners, as a shared responsibility which is discharged by each partner under the lead role and overall certification responsibility of NASA.

3.3 ISS safety requirements, organization and processes

When NASA started discussing possible requirements and safety organizations for the ISS certain similarities with the early considerations about Shuttle payloads emerged. A large number of ISS elements would be developed at remote locations and not under the direct control of NASA ISS Program Manager, and following to a large extend local technical standards. In addition, all modules and payloads except those from Russia would be transported to the station by Shuttle and had therefore to comply with Shuttle payload safety requirements.

It was in the best interest of everybody that the ISS safety requirements of ISS would not conflict but complement those of Shuttle payloads. The solution was very simple: adopt the Shuttle payloads requirements with some adaptations. There were two main kind of modifications. One was the use as applicable standards the space station technical standards, the other was to remove requirements about propulsion and pyros safety (the only module including propulsion was Russian and relevant safety requirements were left TBD).

For what pertains compliance assessment with safety requirements, the safety review process and panel organization was modelled after the Shuttle payloads PSRP. The ISS payload safety reviews became a responsibility of the Shuttle payloads PSRP that was renamed NSTS/ISS PSRP. A separate SRP (Safety Review Panel) was created for ISS systems safety review, which later was merged the PSRP in a single safety review panel when the Shuttle was retired from service in 2011.

In 2002, a new chapter was inaugurated with the decentralization of safety review panels. A "franchised" NSTS/ISS Payload Safety Review Panel was established at the European Space Agency in The Netherlands, to operate according to the same rules and practises and under the oversight of NASA. It was soon followed by a similar agreement with the Japanese Space Agency JAXA, while equivalent arrangements had been also made with Russians.

In the words of Bob Wren, engineering representative in NSTS/ISS and ISS SRP:
"We set it up, copied the exact same approach that we did for the PSRP, for the Shuttle payloads. It turned out that that was a good move. That was a smart move. I guess, and it worked fine. So that way we could come in and have reviews, phase reviews, and I guess I didn't talk about that, but we'd have a series in both payload process and in this SRP process where we keyed the reviews to where the customer was in their development cycle.... So we set up that same approach for the Station modules and it worked great" [7].

4. REFERENCES

1. Erik Seedhouse, *XCOR, Developing the Next Generation Spaceplane*. Springer, 2016.

2. OMB Circular A-119: *Federal Participation in the Development and Use of Voluntary Consensus Standards and in Conformity Assessment Activities*. January 27, 2016.

3. NSTS 1700.7b *Safety Policy and Requirements for Payloads Using the Space Transportation System.*

4. Hugh Blair-Smith, *Space Shuttle fault tolerance: Analog and digital teamwork*, 2009 IEEE/AIAA 28th Digital Avionics Systems Conference, 23-29 Oct. 2009, Orlando (FL).

5. NSTS/ISS 18798 Rev.B *Interpretations of NSTS/ISS Payload Safety Requirements*, September 1997.

6. Committee on Shuttle Criticality Review and Hazard Analysis Audit of the Aeronautics and Space Engineering Board *Post-Challenger Evaluation of Space Shuttle Risk Assessment and Management*, National Academy Press, January 1988.

7. NASA Johnson Space Center Oral History Project, *Edited Oral History Transcript of Robert J. Wren Interviewed by Jennifer Ross-Nazzal* Houston, Texas – 27 November 2007.

8. NPR 8705 *Human-Rating Requirements for Space Systems*, NASA.

9. SSP 50021 Appendix C, *Traceability of NSTS 1700.7B to SSP 50021*, NASA.

10. Sally K. Nash, Raymond B. Rehm, Darren M. Santiago, Teresa K. Wong, Scott L. Wolf, *Building on the Past – Looking to the Future: A Focus on Payload Safety*, 3rd IAASS Conference Rome, 2008.

PROCEEDINGS OF THE 8TH INTERNATIONAL SPACE SAFETY CONFERENCE

SESSION 20 HUMAN PERFORMANCE - I

HUMAN FACTOR IN FLIGHT SAFETY

H. Poussin[(1)], L. Rochas[(1)], T. Vallée[(1)], R. Bertrand[(2)]

[(1)] *Centre National d'Etudes Spatiales, Guiana Space Center, BP 726 - 97387 Kourou, French Guiana (France)*
Email: herve.poussin@cnes.fr, ludovic.rochas@cnes.fr; thierry.vallee@cnes.fr
[(2)] *Centre National d'Etudes Spatiales, Centre Spatial de Toulouse, 18 avenue Edouard Belin - 31401 Toulouse (France)*
Email: regis.bertrand@cnes.fr

ABSTRACT

A chief goal of the launcher design philosophy is to build launchers offering operational efficiency. Therefore, safety issues come in a 2nd step with a lower level of priority. Moreover, launch operator focuses the mission design on success criteria for the payload and then often mitigates launch risks.
These sole conditions clearly appear to be inadequate to ensure safety during a flight neither to be up to the safety challenges.

Flight Safety at CNES/CSG (Centre National d'Etudes Spatiales/Centre Spatial Guyanais) is considered to be a full part job, to be performed separately from the launcher mission. Dedicated ground operators, namely Safety Officers, who are independent from launcher teams, are ultimately responsible for ensuring the safety. During the flight of a launcher, they are in charge of interrupting actively the erroneous flight of a launcher before it endangers people or properties, making use of a flight termination telecommand from ground.

Human factor is therefore of fundamental importance in flight safety at CNES/CSG.

After a quick overview of CNES/CSG, this paper, based on the flight safety way of operation and on the Safety Officers recruitment, training and qualification, aims at declining how the human factor is handled throughout all flight safety activities.

1. A QUICK OVERVIEW OF THE GUIANA SPACE CENTRE (CNES/CSG)

CNES/CSG is located near the equator (5 deg north latitude) in French Guiana, north of Brazil, South America. It offers a wide (~100 deg) opening over the Atlantic ocean and is well adapted to a great variety of missions from northern to eastern trajectories.

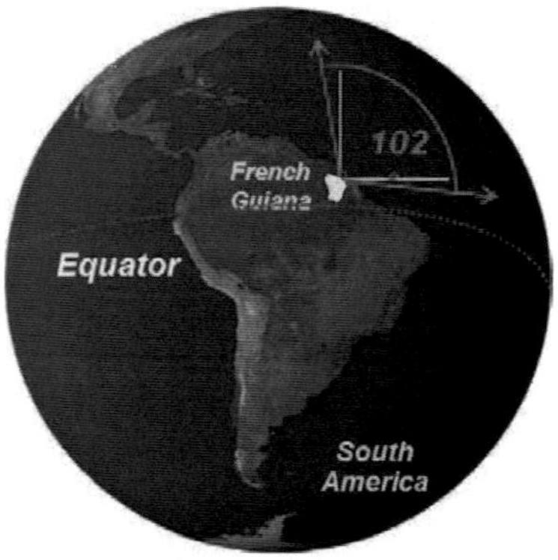

Figure 1: CNES/CSG location

Arianespace, as launch operator, operates 3 launchers from CNES/CSG: Ariane 5 (the historical European launcher) and, since 2011, Soyuz (Russia) and Vega (Italy) performing 10 to 12 launches per year

Figure 2: launchers operated from CNES/CSG

CNES/CSG, in support to Arianespace, operates the launch range facilities and, by delegation of the French government, ensures flight safety for the 3 rockets Ariane 5, Soyouz and Vega in compliance with the French Space Operations Act.

2. FLIGHT SAFETY WAY OF OPERATION

The flight safety mission starts at lift off and ends at confirmation that the upper stage is satellised in a stable orbit or at the end of upper stage controlled reentry, when applicable.
The flight safety operations are split into 2 phases run sequentially along time during a flight:

- The 1st phase starts at lift off and ends at the expected LOS (Los Of Signal) of telecommand stations (all stations being located in French Guiana) or as soon as the launcher ballistic impact point encounters lands.
  This phase, called MSI (Mission Sauvegarde et d'Intervention), lasts 5 to 8 minutes, depending on the launcher considered.

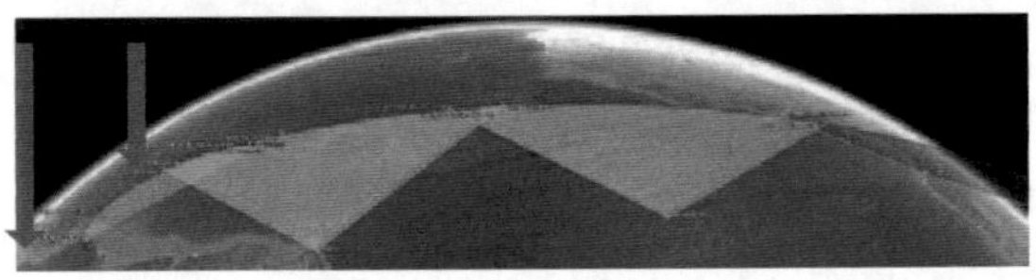

Figure 3: MSI illustration

  This 1st phase is both safety critical, as the launcher has a maximum of energy and public areas (at beginning of the flight) are close to the nominal trajectory, and safety active as the flight safety officer can interrupt an erroneous flight by sending a telecommand from ground.

- The 2nd phase, called MSA (Mission Surveillance et d'Alerte), comes immediately after the MSI and lasts until the end of the safety mission.

Figure 4: MSA illustration

  This 2nd phase is safety passive as flight safety performs a monitoring of the flight and a near real time assessment of the fall down areas of the various stages. It can last hours, depending on the mission. During this phase, flight termination from ground telecommand is forbidden (and not possible) and the associated launch risk is managed, before the launch, through a statistical approach based on launcher failure probability and population density on ground.

Obviously, the 1st phase (the MSI, safety active) of the safety mission is the most important in terms of human factor.

Before the flight, and based on the planned nominal trajectory, flight safety defines a safety qualified flight corridor for the launcher. This safety corridor is compatible with launcher mission and ensures that people and properties are safe (protected against thermal flow, overpressure, projection of debris and toxicity) whatever the launcher failure scenario occurs.

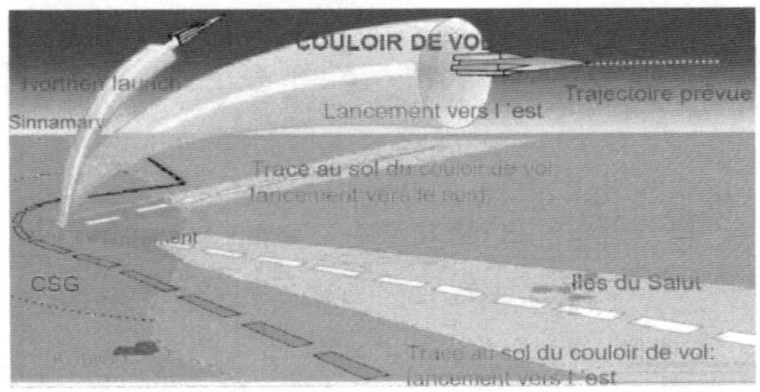

Figure 5: Safety corridor illustration

During the flight, should the launcher cross the limit of this safety corridor, the safety officer sends a flight termination command from ground.

The 3-Dimension trajectory is displayed on two screens; one for the impact point (horizontal plane) and the other one for the altitude against the distance to the launch pad (vertical plane). These displays show the nominal trajectory and the alert and safety limits, allowing detecting rapidly a launch vehicle that gets out of respectively its flight corridor and may become dangerous. Two safety officers have the same type of displays but they use two different localization means to protect against single point failure.

The flight termination decision follows precise operational criteria that must be identified and applied in real time during the flight. It takes into account not only the contingency itself but also any degraded performance or failure of the ground safety facilities, including humans. These criteria are based primarily on the trajectory of the vehicle. They differ for Ariane 5, Soyuz and Vega, due to the specific characteristics and neutralization mode of these launchers.

Safety decisions and actions are made in real time by a team of 4 safety officers, located in a dedicated safety room, fully independent from the Jupiter mission control center.

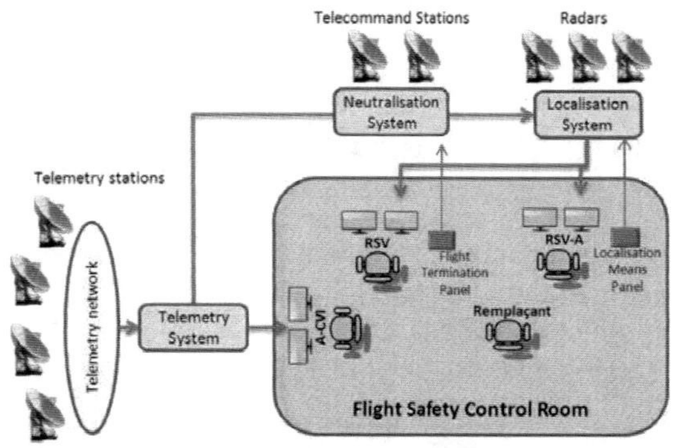

Figure 6: Flight Safety room

They exchange real time status at voice as follows:

- ACVI (telemetry officer) monitors the Telemetry received on ground to make early warnings of any launcher failure (propulsion, occurrence of separation events, on board computer and navigation status and status of the on board flight safety algorithm when applicable) and to monitor telecommand link quality
- RSVA (deputy safety officer) manages the localization facilities and monitors the trajectory of the launcher. He warns about any deviation of the launcher from the nominal trajectory and supports the RSV in his decisions.
- RSV (chief safety officer) manages the telecommand facilities, confirms the trajectory deviation, compares it to the flight corridor and sends the flight termination telecommand if needed.
- REMPLAÇANT is also present as back up operator and can substitute in real time any of the 3 other operators.

From this way of operation, securing as far as possible the safety team reactions and time to react, launcher and ground facilities failure diagnostics and decision making appears to be of critical importance to avoid wrong, early or late decision.

This is achieved through:

- Clear, simple and robust flight management procedures and flight termination criteria.
- Safety criteria are fully based on the launcher localization. This represents 1 macroscopic, physical and simple parameter, easy to manage in real time and not confusing as can be the numerous on board and on ground device failures and associated failure mode analysis.
- Fail Operational (FO) criterion applied to both safety facilities (on board and on ground) and operators.
- This requires to have full redundancy with no single point failure in safety facilities, a REMPLAÇANT operator that can substitute any member of the team, and confirmation of safety diagnostic between chief officer and deputy officer as a consequence of the "confirm or stop an iterate" mandatory principle
- The application of the nominality principle that protects launcher from safety device failures. This principle states that if the localization of the launcher is nominal with one localization facility, a deviation observed simultaneously with another localization facility is false and the associated facility is to be considered out of order
- Full independency between on board safety devices and launcher navigation facilities. In particular, launcher localization must be performed using 2 different sensors: one of them being a radar on ground
- Standard communication protocol and short, and precise dialogues shared among all the safety team members
- Clear, simple and understandable Man Machine Interface (MMI). MMI are to display decision making information only using a very limited set of colors (ex: alarms in red, nominal in green, etc.)

Based on the experience gained over last decades and few hundred flights, this way of operation with human deeply involved in the safety loop proved to be both well under control and efficient with human's strength over the machine lying in its ability to adapt to an unknown situation.

3. SAFETY OFFICERS SELECTION, INSTRUCTION, TRAINING AND CERTIFICATION

In the Flight Safety way of operation at CNES/CSG, human operators play a key role and therefore special attention is to be paid to selection, instruction, training and certification of safety officers.
This process is clearly defined and controlled and has been fine-tuned over many years.
It starts with the selection of a candidate: this is based on good technical background in engineering together with "human qualities", such as team spirit, decision-making ability, calmness, stress resistance, etc.
Then the process "instruction, training and certification" applies with the following characteristics:

- It is mandatory for all,
- It is sequentially ordered by increasing level of operational responsibility: starting from ACVI operational position then continuing with the RSVA, RSV, REMPLAÇANT positions and finally ending with the Instructor position.
- It is practical: as soon as 2 months after arrival, the safety officer is regularly put in operational situa-

tion during training sessions making use of a flight simulator.

- Training session typically consist in a succession of 10 simulated flights (or "simulations") built from a combination of pre-recorded malfunction scenarios for the launcher (trajectory or equipment malfunction), the telemetry, telecommand and the localization ground facilities. Each simulation re-creates as closely as possible the conditions of a launch, from typically one minute before lift-off up to the simulated neutralization of the launcher or up to the end of the MSI phase. The sessions are supervised by an Instructor, who assesses each participant and provides feedback during a short debrief taking place after each simulated flight.
- It is adapted to the 3 launchers operated from CNES/CSG: Ariane 5 instruction process is the blackbone for other launchers instruction process and additional launcher dedicated modules and training sessions are organized for Soyuz and Vega launchers.

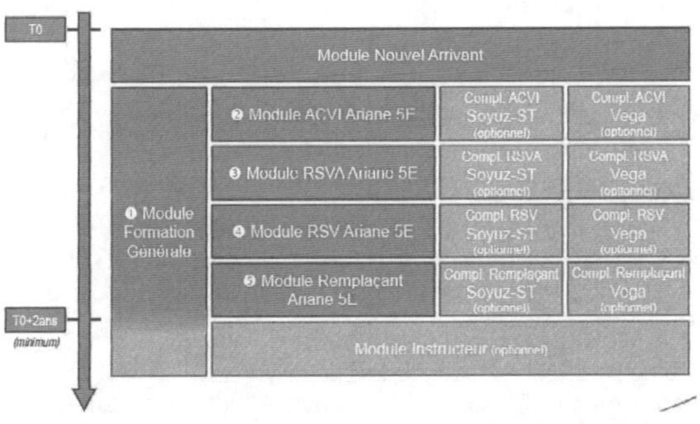

Figure 7: mapping of launchers instruction modules

It is monitored: certification to an operational position is acquired after the candidate has:

- passed an oral examination to validate the acquisition and understanding of the "theoretical corpus" (position specific operations, operational organization and interfaces, CSG means, intervention criteria, flight safety tools, etc...).
- passed 3 to 6 trainings (depending on the position). This step is required to assess the aptitude to hold the position and allows to take part to a first launch campaign
- participated to a launch campaign (final validation in real conditions)

- It is coached during all the process by a dedicated safety officer, a senior instructor and the head of the flight safety department
- It is continue along time: certification for an operational position is valid for 6 months and time validity extension requires participation to refresher training session

Figure 8: Flight Safety simulator (EMUL)

4. CONCLUSION

Based on the experience gained over last decades and few hundred flights, the safety way of operation in place at CNES/CSG, with human deeply involved in the safety loop, has proved to be both well under control and efficient.

It is nevertheless to recognize:

- the non-negligible effort for recruitment, learning and training of safety officers
- and, with humility, that human errors are inevitable and therefore it is unreasonable to expect error-free human performance.

Otherwise, considering rapid gains in technology and data processing science, a continuous and interesting debate is ongoing about how to best integrate human with machine, or even how to replace human by machine.
For the time being, the human's strength over the machine lies on human ability to adapt to an unknown situation. But for how long?

MANAGING COGNITIVE BIAS IN SAFETY DECISION MAKING: APPLICATION OF EMOTIONAL INTELLIGENCE COMPETENCIES

Walter S. Hersing

Kennedy Space Center, Integrated Mission Support Services
Mail Code: IMSS-005
Kennedy Space Center, FL 32899 USA
walter.s.hersing@nasa.gov

ABSTRACT

Safety leadership is often tasked with making critical decisions in the face of complex information and external pressures. Given situations with many variables and potential courses of action to be considered, the brain has evolved to attempt to simplify these judgmental operations. Consequently, at times the result may be flawed judgments concerning our assessments of future probabilities due to certain cognitive biases. These cognitive biases are systematic distortions of decision making arising from innate heuristics used to simplify large quantities of data (Krause, 2009) [1]. By developing a better understanding of the types of cognitive biases that influence decision making, steps can be taken to more effectively manage these bias. However, research has shown that technologies associated with training in statistical and probabilistic reasoning and cognitive repair techniques do not sufficiently address elements of intuitive belief and individual affect. Therefore, this paper suggests that certain Emotional Intelligence (EI) competencies, when applied to the individual decision making process, can be useful in mitigating the effects of cognitive bias. Emotional Intelligence can be defined as an array of interrelated emotional and social competencies, skills and facilitators that impact intelligent behavior (Bar-On, 1997) [2]. For the purposes of this paper competencies associated with the EQ-i $^{2.0}$, a revision of the original model by Reuven Bar-On, will serve as a template for discussion. The composite scales include Self-Perception, Self-Expression, Interpersonal, Decision Making and Stress Management. Certain scales are considered to be especially relevant to affect-based debiasing strategies. In addition to the potential benefits for the individual decision maker, leveraging EI competencies may positively influence team interactions in the process of safety decision making and promote an organizational culture more open to diversity of perspective.

1. INTRODUCTION

Safety and risk management (RM) protocols continue to evolve as evidenced by the extensive body of research in these areas and the applications of this evolving knowledge by many organizations, particularly those in the aerospace industry. Even though probabilistic risk assessment (PRA) continues to be refined, the complex environment of many safety decision scenarios makes clear the fact that decision makers will often still be influenced by the natural tendency to incorporate heuristics and their associated cognitive biases in the decision making process.

There is a wealth of important research on the broad topic of judgment and decision making (JDM). However, due to the limited scope of this paper, discussion of decision making theory will be confined to those elements directly related to heuristics and bias. According to the heuristics and bias paradigm of human judgment, people typically use cognitive short-cuts to make probability assessments simpler, but at the expense of being more prone to error. This is a factor not only in predictions but also in retrospective judgments of probability .

The effects of bias on JDM have been the subject of investigation in many disciplines including safety, psychology, economics and systems engineering. Cognitive bias is not due to any inherent defect in the decision maker, but rather is related to natural cognitive processes involving efforts to simplify complex decision making operations. Although some attempts have been made to uncouple the rational from the more intuitive decision making process, in practice both will continue to exert their effects in natural settings.

In recent decades the construct of Emotional Intelligence (EI) has developed in an attempt to describe and quantify those aspects of intelligent functioning not measured by standard Intelligence Quotient (IQ) tests. Given that EI address elements of self-awareness, emotional regulation, decision making and stress tolerance, it is believed to have the potential to contribute to debiasing strategies in naturalistic settings. Additionally EI incorporates elements of social intelligence which could serve to improve interactions among those involved in a comprehensive RM program. This may be especially beneficial when considering multiple stakeholders with diverse values and perspectives.

2. JUDGMENT AND DECISION MAKING

Many disciplines have contributed to the rich body of research in the area of JDM. For the purposes of this paper the focus is on the cognitive, affective and emotional components of JDM and their relationship to risk analysis. It is important to note that deterministic approaches to safety decision making have been supplemented with more comprehensive probabilistic risk assessment (PRA). This has led to a more integrated approach to safety decision making, which has been incorporated by agencies such as the National Aeronautics and Space Administration (NASA) and the United States Nuclear Regulatory Commission (USNRC) [15]. In addition, these agencies have developed a risk-informed decision making (RIDM) process which is integrated into their comprehensive RM processes. Zio and Pedroni (2012) [3] differentiate NASA's RIDM process from the USNRC in that NASA's focus seems to be on a combination of a better decision structure crossing internal organizational boundaries and support for multi-criteria decisions under uncertainty. NASA also describes RIDM as a fundamentally deliberative process that uses a diverse set of performance measures, along with other considerations, to inform decision making .The role of human judgment in decision making is also acknowledged [3]. Given this role in the deliberative process, NASA describes "decision traps" to be considered [5],[6]. These decision traps fall under the broader category of heuristics and bias. NASA describes the RIDM process as a mechanism to avoid these traps by establishing a rational basis for decision making, ensuring that the implications of each decision alternative have been thoroughly analyzed and by providing a structured environment for deliberation, where the merits and drawbacks of each alternative can be discussed in the context of the risk analysis [5]. Despite the best of technical processes to avoid or mitigate errors in JDM, heuristics and bias will continue to be an area requiring consideration in safety and risk management programs.

2.1 Heuristics and bias

Extensive research has been done on what influences decision making, especially under conditions of uncertainty. Traditional models of decision making are constructed on logic and rationality. However in reality, decision making very often is influenced by a complex set of cognitive and affective variables. This process was described by Tversky and Kahneman (1974) [7] in their work on judgment and uncertainty. They proposed that intuitive judgments under uncertainty are typically controlled by judgmental heuristics rather than laws of probability. Generally a heuristic can be described as a "rule-of-thumb" or simplified, intuitive approach to decision making that may offer expedience at the risk of cognitive error (bias). It occurs most often in situations where information or cognitive capacity is limited and other pressures are influencing the process. Tversky and Kahneman highlighted the availability, representativeness, and anchoring and adjustment heuristics, and their associate biases. The availability heuristic describes the perception that the probability of an event occurring is based on the ease with which examples or occurrences come to mind. The representativeness heuristic refers to the tendency to evaluate probabilities based on initial impressions about how much something resembles a class or type. The anchoring and adjustment heuristic refers to the tendency to remain tied to an initial estimation or judgment, even when provided with information supporting significant readjustment. As research on JDM has evolved, the number of identified biases has grown significantly.

Cognitive Biases

Anchoring – tendency to overweight the first information received

Confirmation Bias – placing more emphasis on information that supports currently held views

Framing – tendency to respond based on how information is presented

Hindsight Bias – tendency to see things as more predictable than they really are

Status Quo Bias – inclination to prefer the current state of affairs

Sunk-cost Bias – tendency to continue with a course of action in an attempt to recoup losses, despite rationale to do otherwise

Figure 1. Common cognitive Bias

It has been proposed by Das and Teng (1999) [8] that the multitude of identified cognitive biases could be grouped into four categories: 1) prior hypothesis and focusing on limited targets, 2) exposure to limited alternatives, 3) insensitivity to outcome probabilities and 4) illusion of manageability. Other researchers (Teovanovic, et al, 2015) [9] have devised classifications of cognitive bias based on individual differences in information obtained from multiple tasks and measures. According to Korte (2003) [10] research has shown that decision makers tend to (a) reduce problems into simple constructs and (b) use information selectively based on their beliefs (assumptions and mental models) and preferences (biases). Decision makers often create analyses and solutions that reflect their experience and beliefs and interpret their experience in ways that support and preserve their beliefs [10],[11]. Arkes (1991) states that a few general causes underlie a wide range of biases. He proposed three categories. The first two are attributed to an intuitive, non-conscious system (System 1) and the third attributed to an analytical, conscious system (System 2). The categories are: 1) psychophysically-based error, 2) association-based error and 3) strategy-based error [12]. The two system model will be described in more detail.

2.2 Two System Model

In an effort to further explain the role of heuristics and bias in decision making, theories have evolved citing two primary processing modes for a cognitive task. Based on this dual process theory (DPT), dual-system models have been proposed. System 1, is considered to be intuitive, automatic and non-conscious. The second, System 2 is considered to be analytical, deliberate and conscious. This model has been described extensively in the literature on dual-process and dual-system theory [13], [14]. It's of interest to note that many cognitive psychologists believe that we spend about 95% of our time in the intuitive mode [16]. Associated with this intuitive mode is affect, which can be described as a feeling state that reflects the positive or negative quality of a stimulus [17]. Results of Slovic's (2004) research reinforces the idea that people base their judgments of an activity or technology not only on what they think about it but also on how they fell about it.

2.3 Somatic Marker Hypothesis

To further elaborate on the complex role of affect and emotion in decision making it is important to look at the work done by the neurologist Damasio. He proposed the somatic marker hypothesis (SMH). In this hypothesis he describes the neural basis for the interaction between emotion and cognition in decision making [15]. In work done with individuals with damage to the ventromedial prefrontal cortex (VMPFC), it was demonstrated that emotional context (information) normally provided as a result of the somatic markers was essential for decision making [18].

As has been demonstrated in much of the research on JDM - cognition, affect and emotion have a complex relationship. Despite our best efforts to rely on analytical and Bayesian approaches to decision making, it's clear that intuitive factors will continue to exert a profound influence.

3. EMOTIONAL INTELLIGENCE

Although it has roots in the early 20th century in Thorndike's concept of "social intelligence", Emotional Intelligence (EI) is a relatively new construct. Gardner wrote about "multiple intelligences" in 1983 [19]. Salovey and Mayer coined the term "emotional intelligence" in 1990. The concept was popularized in 1995 in the book by Daniel Goleman [20]. Much research continues to be done on the theory and measurement of EI. There are currently several commonly used instruments to assess EI. For the purposes of this paper, the revised Bar-On EQ-i 2.0 (MHS) [2] will be used a template for discussion of those attributes and competencies that are most relevant to managing bias in the decision making process. The EQ-i 2.0 technical manual defines emotional intelligence as a set of emotional and social skills that establish how well we perceive and express ourselves, develop and maintain social relationships, cope with challenges and use emotional information in an effective and meaningful way. The five composite scales and the fifteen subscales are shown in Fig. 2 for reference. Rather than focusing on the intricacies of the EQ-i 2.0, we wish to consider how the emotional and social competencies associated with EI may be relevant to debiasing strategies in safety decision making. It has been reported by Stankov (2013) [21] that given the knowledge that probabilistic and scientific reasoning are known to be open to cognitive biases, which often do not correlate with intelligence, it is important to consider non-cognitive as well as cognitive aspects of intelligence. Others like LeDoux (1996) [22] provide compelling information on the complex interactions between the purely cognitive and the emotional aspects of the brain. A select number of EI competencies will be referenced to highlight the value of incorporating these elements into a debiasing protocol.

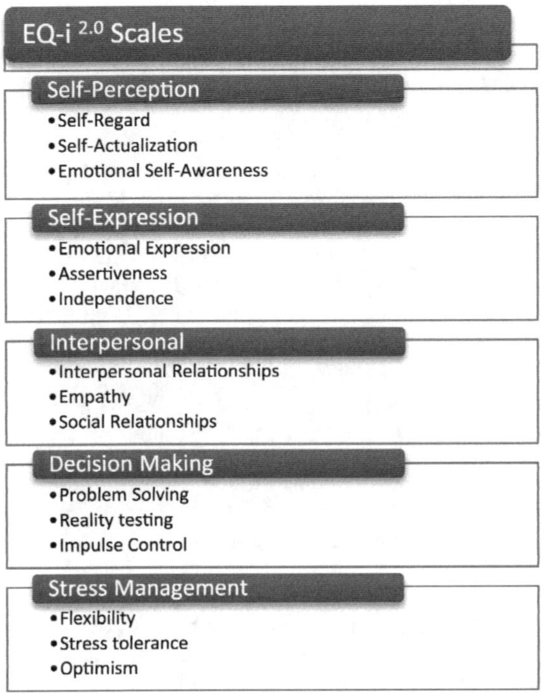

Figure 2. EI Composite and Subscales (©MHS)

In the realm of Self-Perception, Emotional Self-Awareness is crucial in recognizing the subtleties of one's own emotions, their genesis and the impact they have on cognition and behavior. Within the Decision Making composite, the three subscales all offer relevance to a debiasing strategy. The Problem Solving subscale reflects one's ability to understand and manage the impact of emotions on decision making. Reality Testing refers to one's capacity to recognize when emotion or bias is hindering objectivity. The subscale Impulse Control reflects the ability to manage frustration and impulsiveness in decision making. In the Stress Management composite, two subscales are considered. First, Flexibility, which demonstrates the ability to adapt to unpredictable and dynamic circumstances. Stress Tolerance involves coping with demands or difficult situations in an effective manner.

Two competencies in the Self-Expression composite are noted. Assertiveness reflects the ability to express beliefs and feelings in an open and respectful manner. Independence demonstrates the ability to be self-directed and free from emotional dependence. Finally, in the Interpersonal composite we highlight two competencies. Interpersonal Relationships, which reflects the ability to develop and maintain healthy relationships. Secondly, Social Responsibility which demonstrates the ability to function within a group in a responsible and effective manner.

The use of the EQ-I 2.0 as a frame of reference is not intended to suggest that this particular assessment is more valid or preferable than other assessments of EI. Its use is based on knowledge and familiarity with this particular instrument. Certainly there continues to be much discussion about how to best describe and measure the traits or abilities associated with EI [23]. As with any psychometrically valid instrument, it's important that the research is extensive and robust.

As has been noted previously, bias in decision making is generally operating at a level that may not be in the conscious awareness of the decision maker. Although training in decision making and risk management can serve to educate on the presence and impact of bias, real world scenarios will continue to elicit the more fundamental drivers of bias such as emotion and affect. The construct of EI may provide a means of bridging the gap between the analytical and the more intuitive cognitive processing systems. This may intern allow operationalization of a process wherein bias in decision making can be better mitigated in naturalistic settings.

4. DEBIASING

Organizations with robust RM programs have made great strides in attempting to address heuristics and bias in safety risk analysis. However, these approaches have primarily been descriptive and procedural. Although procedural errors are usually highly visible, cognitive errors can only be inferred from what people do or through accurate self-descriptions of thinking processes [24]. This highlights one of the primary challenges for any debiasing protocol. In order for debiasing to have a significant impact, the procedures need to be accessible or transferable to situations encountered in natural settings [25]. Incorporation of EI competencies in RM and safety decision making offers the potential to increase the effectiveness with which decision makers identify personal biases and mitigate the impact of these biases individually and in situations involving team collaboration and multiple stake holders.

The SMH proposed by Damasio (1994) appears to provide a good model for conceptualizing how affective and emotional information might be integrated with more analytical cognitive processes. The physiological signals (somatic markers) and related emotions are consciously or unconsciously associated with past outcomes, and bias decision-making for or against certain behaviors. Increased awareness of physiological and emotional states, which coincide with elements of the EI composites of Self-Perception, Decision Making and Stress Management may alert the decision maker to the increased probability of bias. Cognitive scientists have reported cognitive biases in a wide variety of experimental situations, which has led to predictions of the conditions under which biases are likely to occur. Their likelihood increases under conditions of uncertainty, especially when thinking is hurried or pressured [24]. Researchers have cited the role of ambient and contextual factors in creating high risk situations that predispose decision makers to particular biases. Factors identified include sleep deprivation, fatigue and cognitive overload [25].

In addition to benefits for the individual decision maker, success with debiasing would also have implications for teams and organizations. As has been noted, when risk analysis and decision making become more complex or increase in uncertainty, heuristics and bias are likely to play a role in the process. Although there are potentially increased pressures and demands with multiple internal and external stakeholders, potential benefits also exist. One of the challenges is to leverage the diversity of perspectives and preferences in order to mitigate the possible detrimental effects of cognitive bias in any decision making process. The EI composite scales of Self-Expression and Interpersonal relate well to dynamics associated with team interactions and conflict management. As reported in research by Barrick et al. (1998), emotion management, a component of EI, is a characteristic ability of cohesive, effective work teams [26]. Other studies have shown that emotionally intelligent team members are less likely to allow interpersonal conflict to interfere with team goals [27]. With improved interpersonal relationships, related to EI, it has been demonstrated that team members are more comfortable voicing opposing opinions (error-checking) believing they will be given honest consideration. This in turn promotes consideration of multiple alternatives and a more thorough decision making process [28].

Various approaches for managing bias in decision making have been proposed. Most focus on intervening in the conscious, more analytical cognitive system (System 2). Although some of the approaches are promising, the greatest challenge continues to be having influence on the intuitive, non-conscious system (System 1). In his work in the medical field, Croskerry (2003) incorporates metacognition to implement "cognitive forcing strategies" in an attempt to mitigate the errors associated with cognitive bias [24]. Bazerman believes effective debiasing requires some disequilibrium on the part of the decision maker, perhaps even an emotion-laden experience to effect cognitive change [29]. Based on significant findings in the research on debiasing it's evident that affect and emotion play a major role in the manifestation of bias in the decision making process. It is also clear that much work needs to be done in order to develop strategies to help decision makers more effectively manage affect and emotion endogenous to decision making.

5. DISCUSSION

Decision making is a process that is often taken for granted when the stakes don't seem high. However in the world of safety and RM, an ill-informed decision can be catastrophic. Fortunately, in high stakes endeavors such as aerospace and nuclear energy, the technologies associated with risk analysis continues to be refined. Agencies such as NASA and the USNRC have RM programs that are continually evolving. A manifestation of this evolution is the development of Risk-Informed Decision Making (RIDM) as a component of a more rigorous quantitative RM process [3]. Cognitive biases or "decision traps" are understood to be inherent in the decision making process and efforts are made to addressed them through the RIDM process. Despite integrated and robust approaches to RM, heuristics and bias will continue to provide challenges for risk analysts and risk managers.

The field of judgment and decision making (JDM) is rich in research and theories on heuristics and bias. Decision making models based on logic and rationality have been built upon to contend with the complexities of cognitive and affective variables. Early researchers such as Tversky and Kahnman (1974) have provided invaluable information on judgment under uncertainty. Heuristics provide us with a "rule of thumb" process to attempt to cope with uncertainty or complexity and make decisions more efficiently. However the trade-off may be cognitive biases that can lead to errors in judgment and decision making. Many cognitive biases have been identified and efforts have been made to catalogue them descriptively or based on functional effects.

A promising conceptualization for the role of heuristics and bias in decision making is the dual process theory. Based on this theory, a Two System model has been described by Kahneman and others [13],[14]. It consists of an intuitive, automatic and non-conscious system (1) and an analytical, deliberate and conscious system (2). In his somatic marker hypothesis (SMH), Demasio [18],[30] describes a neural basis for the interaction between emotion and cognition [11]. These theories provide intriguing insights into the neurophysiological basis for decision making.

The construct of emotional intelligence (EI) has received much attention since the popular book by Goleman in 1995. Subsequent to that publication, a great deal of research has been done on the theory and measurement of EI [2],[31]. Opinions differ regarding the best methods for describing and measuring EI. Subsequently, a variety of measures for EI have been developed. The EQ-I 2.0 based on the model from Bar-On was used in this report as a guide for discussing the competencies associated with EI[34]. Elements such as Emotional Self-Awareness, Problem Solving, Impulse Control and Stress Tolerance appear to relate well to cognitive and emotional aspects of the decision making processes described in this paper. Other EI competencies such as Emotional Expression and Interpersonal Relationships are especially relevant when making decisions as part of a team. Individual biases can interfere with our ability to be objective and open-minded, and hinder deliberation and collaboration.

Studies on how to mitigate the effects of heuristics and bias are relatively limited compared to the wealth of information in the domain of JDM [32],[33]. Certainly, just providing education on the prevalence and types of common biases can provide some benefit. However, as studies have shown, a significant challenge involves developing debiasing strategies that will operate in naturalistic settings. The strategies will also likely need to engage our more intuitive and automatic cognitive processes (System 1) as well as address elements of affect and emotion. This is likely an area where EI competencies could be effectively leveraged in the decision making process. Regardless of the robustness of a RM program, there will always be a human element to consider.

6. REFERENCES

1. Krause, T. (2005). Leading with Safety, John Wiley & Sons: New Jersey

2. Bar - On , R. (1997). Bar - On Emotional Quotient Inventory: Technical Manual. Toronto: Multi - Health Systems Inc.

3. Zio, E., and Pedroni, N. (2012). Risk-informed decision-making processes. FonCSI

4. NASA Risk-Informed Decision Making Handbook NASA/SP 2010-576, v1.0 (2010)

5. NASA Risk Management Handbook, NASA SP-2011-3422, v1.0 (2011)

6. Hammond,J. Keeney, R., and Raiffa, H. (1998). The Hidden Traps in Decision Making, Harvard Business Review.

7. Tversky, A., & Kahneman, D. (1974). Judgment Under Uncertainty: Heuristics and Biases. Science, 185, 1124 – 1131

8. Das, T.K., and Teng, B.S. (1999). Cognitive Bias and Strategic Decision Process: An Integrative Perspective. Journal of Management Studies, 36 757-778.

9. Teovanovic, P., Knezevic, G., and Stankov, L. (2015) Individual Differences in Cognitive Biases: Evidence Against One-Factor Theory of Rationality. Intelligence 50, 75-86.

10. Korte, R.F. (2003).Bias in Decision Making and Implications for Human Resource Development, Advances in Developing Human resources Vol.5, No.4, 440-457.

11. March J.G. (1999). The Pursuit of Organizational Intelligence. Malden, MA:Blackwell.

12. Arkes, H. R. (1991). Costs and Benefits of Judgment Errors: Implications for Debiasing. Psychological Bulletin, 110, 486–498

13. Kahneman, D. (2003) A Perspective on Judgment and Choice: Mapping Bounded Rationality, American Psychologist, 58, 697-720.

14. Stanovich, K. E. (1999). Who is Rational? Studies of Individual Differences in Reasoning. Mahway, NJ:Lawrence Elrbaum Associates.

15. Bechara & Damasio (2005) The Somatic Marker Hypothesis: A Neural Theory of Economic Decision Games and Economic Behavior 52, 336-372.

16. Lakoff G. Johnson,M. (1999). Philosophy in the Flesh; the Embodied Mind and its Challenge to Western Thought. New York: Basic Books.

17. Slovic, P., Finucane, M., Peters, E., MacGregor, D. (2004). Risk as Analysis and Risk as Feelings: Some Thoughts About Affect, Reason, Risk and Rationality. Risk Analysis, Vol.24, No.2.

18. Damasio, A.R. (1994). Descartes' error: Emotion, Reason and the Human Brain. New York: Grosset/ Putnam Book.

19. Gardner, H. (1983). Frames of Mind: the Theory of Multiple Intelligences. New York: Basic Books.

20. Goleman, D. (1995). Emotional Intelligence: Why it can matter more than IQ. London: Bloomsbury.

21. Stankov, L. (2013). Noncognitive Predictors of Intelligence and Academic Achievement: An Important Role of Confidence. Personality and Individual Differences, 55(7), 727-732.

22. LeDoux, J., 1996. The Emotional Brain: The Mysterious Underpinnings of Emotional Life. Simon and Schuster, New York.

23. Petrides, K. V. (2011) Ability and Trait Emotional Intelligence, in The Wiley-Blackwell Handbook of Individual Differences (eds T. Chamorro-Premuzic, S. von Stumm and A. Furnham), Wiley-Blackwell, Oxford, UK.

24. Croskerry, P. (2003) Cognitive Forcing Strategies in Clinical Decision Making. Annuals of Emergency Medicine 41:110-120

25. Croskerry, P., Singhal, G., Mamede, S. (2013). Cognitive Debiasing 1: Origins of Bias and Theory of Debiasing. BMJ Qual. Saf. 22:58-64

26. Barrick MR, Stewart GL, Neubert MJ, Mount MK. (1998). Relating Member Ability and Performance to Work-team Processes and Team Effectiveness. Journal of Applied Psychology, 83, 377–391.

27. Liu, Y., Prati, L. M., Perrewé, P. L. and Brymer, R. A. (2010), Individual Differences in Emotion Regulation, Emotional Experiences at Work, and Work-related Outcomes: A Two-Study Investigation. Journal of Applied Social Psychology, 40: 1515–1538.

28. Phillips, J.K., Klein, G., and Sieck, W.R. (2004) Expertise in Judgment and Decision Making: A Case for Training Intuitive Decision Making Skills. Blackwell Handbook of Judgment and Decision Making, pp. 297-315.

29. Bazerman, M. (2002) Judgment in Managerial Decision Making, 5th Ed. New York: Wiley & Sons.

30. Bechara, A., Tranel, D., Damasio, A.R. (2002). The Somatic Marker Hypothesis and Decision-Making. In: Boller, F., Grafman, J. (Eds.), Handbook of Neuropsychology: Frontal Lobes. Elsevier, Amsterdam, pp. 117–143.

31. Salovey, P., and Mayer, J.D. (1990). Emotional Intelligence. Imagination, Cognition and Personality, 9, 185-211 (1990).

32. Dunbar, N. E., Wilson, S., Adame, B., Elizondo, J., Jensen, M. L., Miller, C., et al. (2013). The Development of a Training Game for the Mitigation of Cognitive Bias: The Case Study of MACBETH. International Journal of Game-Based Learning, 3(4), 7–26.

33. Larrick, R. P. (2004). Debiasing. In D. J. Koehler & N. Harvey (Eds.), Blackwell Handbook of Judgment and Decision Making. Oxford, UK: Blackwell.

PROCEEDINGS OF THE 8TH INTERNATIONAL SPACE SAFETY CONFERENCE

SESSION 22 SPACE DEBRIS - II

GOOD PRACTICE FOR UPPER STAGES GOING TO LAGRANGIAN POINT APPLICATION TO THE ARIANE 5 JWST MISSION

D.-A. Handschuh[1], J. Campedelli[2], N. Lidon[3], E. Canalias[4]

[1] CNES Launcher Directorate, 52 rue J.Hillairet, 75012 Paris, France, david-alexis.handschuh@cnes.fr
[2] ALTEN, 221 bis bd Jean Jaurès, 92100 Boulogne-Billancourt, France, jean.campedelli@alten.com
[3] ESA Launcher directorate, 52 rue J.Hillairet, 75012 Paris, France, norbert.lidon@esa.int
[4] CNES, 18 Avenue Edouard Belin, 31401 Toulouse, France, elisabet.canalias@cnes.fr

ABSTRACT

More and more scientific space missions are sent to Lagrangian points of the Sun-Earth system, especially to L2. Recent years have witnessed Herschel, Planck and Gaïa missions to L2 and JWST, Euclid, or Athena will come in next few years. To achieve these missions, the launcher's upper stage is generally left on a very eccentric orbit but still geocentric. That means that the upper stage could remain in Earth vicinity for a long time. Due to Lagrangian point local instability, this could engender risks of collision with the Moon, with another artificial satellite on Earth orbit or, potentially, an Earth uncontrolled reentry. To limit these risks, an ideal solution would be to free the upper stage from the Earth attraction placing it on a heliocentric orbit beyond the L2 Lagrangian point. This article deals with the definition of a liberation criterion of the Ariane 5 ECA launcher's upper stage for the James Webb Space Telescope's specific mission. Once this criterion is defined, an optimized maneuver to be applied using the upper stage's propulsion system after the satellite release is proposed to ensure its liberation.

1. INTRODUCTION

In the last few years, a global awareness of the space debris problematic led on focusing on effective solutions for man-made space missions. Latest examples of in-orbit collisions as the one that compromised the operational mission of the Ecuadorian NEE-1 Pegaso satellite in 2013 or the one that interrupted the mission of the Russian BLITS satellite also in 2013 illustrate the emergency and the importance of this issue.

First of all, two protected in-orbit regions have been internationally identified close to the Earth. These areas have been defined to maintain the most useful operational regions as free of debris as possible. The first one is a sphere around the Earth, 2000km high and called the LEO region. The second one is a torus around the geostationary arc. These protected areas have particularly to be addressed when the space mission concerns a spacecraft to be injected in the Earth vicinity.

More and more scientific missions are targeting the Lagrangian points (especially L2 located 1.5 million km far from the Earth) and are frequently designed to abandon the launcher's upper stage used to reach the transfer orbit to L2 on a very elliptical orbit around the Earth. That means that this upper stage will continue turning around the Earth for many years proceeding to many revolutions further or closer to the Earth. However, it is well known that these orbits are not stable in time because the injected object will go through numerous perturbations (especially the combined attraction of Sun and Earth near to the Lagangian point and the Moon perturbation when orbiting close to its altitude). Thus, sooner or later, there is a non-negligible risk for the object to interfere with the protected areas and even to re-enter Earth atmosphere in an uncontrolled way.

In December 2010, the French Space Operations Act (FSOA) and its associated Technical Regulation (TR) entered into force in France. This act contains articles aiming at limiting the proliferation of space debris, stating quantitative objectives for human safety or officially asking to comply with the international "Planetary protection policy" standard published by the Committee on Space Research (COSPAR). This FSOA requests to study the risk of Earth return or space debris production in its vicinity for these missions targeting L2.

The James Webb Space Telescope (JWST) mission is one of these coming scientific missions. This spacecraft will be launched with an Ariane 5 ECA (A5ECA) from French Guyana. Questions were raised with regard to the upper stage's becoming.

This article, deals briefly with the history of launch operated from Kourou to Lagrangian points and specific studies led in their framework. A theoretical analysis of phenomena implied in transfer orbits to L2 will be presented and finally the sizing of an adapted maneuver to release the Ariane 5 upper stage of the JWST mission from the Earth attraction will be exposed.

2. ACRONYMS

A5ECA	Ariane 5 - "ECA" version
CNES	French Space Agency
COSPAR	Committee on Space Research
ESA	European Space Agency
GSC	Guyana Space Centre of Kourou
FSOA	French Space Operation Act
GEO	Geosynchronous Orbit
JWST	James Webb Space Telescope
LEO	Low Earth Orbit
wrt	with respect to

3. MISSIONS TO LAGRANGIAN POINTS

3.1 Past missions from the GSC

Many launches have been operated from the Guyana Space Centre of Kourou (GSC) to Lagrangian point L1 or L2 of the Sun-Earth system. Recently, Herschel and Planck have been launched to L2 on an A5ECA in 2009; GAIA has also been launched to L2 but on a SOYUZ rocket in 2013. These missions have generated specific analysis to minimize the risk of collision with the Moon or the risk of uncontrolled return to Earth.

A5ECA has been used to launch Herschel and Planck to L2. The launcher flexibility allowed establishing large launch windows increasing the operation availability and offering a better launch service to the payloads. CNES conducted statistical studies to determine launch date and time that induced collision risk with the Moon or unexpected return to Earth on a very short-term delay. To take into account the results of these studies, launch-window closures were prescribed, slightly decreasing the availability of the launch operation.

GAIA laboratory has been launched using a SOYUZ rocket. Arianespace proposed to use the reaction control system of the upper stage (called Fregat) to achieve an additional maneuver after the payload release. This maneuver has been designed and produced to put the Fregat on an heliocentric orbit beyond the L2 point. It has been implemented into the flight software, avoiding in this way the suppression of potential launch dates.

Many other scientific missions targeting Lagrangian point L2 are expected in the next few years such as Euclid or Athena for the European Space Agency. JWST will be the next payload sent from GSC to L2 by an A5ECA at earliest in October 2018. To be sure not to penalize the launcher's availability CNES, as technical directorate of the European Space Agency (ESA) for launchers, proposed the addition of a dedicated maneuver to allow the ESC-A upper stage to be released from the Earth attraction as it has been done for the Fregat stage that injected GAIA into orbit.

3.2 The JWST's launcher mission

JWST is the next generation space telescope supposed to replace Hubble for a part of its mission. It is dedicated to the outer-space observation from the L2 Lagrangian point. It will be launched by A5ECA on a transfer orbit with a perigee located at an altitude around 300 km and an apogee slightly above 1e6 km. JWST will then join the L2 point with its own propulsive system. If the A5ECA upper stage is abandoned on this transfer orbit, as this orbit is elliptical and geocentric, that means that it will come back in the vicinity of the Earth at every orbit. The stage is submitted to orbital perturbations close to the Earth or the Moon and therefore represents a risk for the orbital population and the Earth.

The most efficient way to limit the risk of an Earth uncontrolled re-entry is to make sure that this stage is freed from the Earth attraction and placed on a final heliocentric orbit. To realize the necessary orbital modification, Ariane 5 can use its attitude control system to produce an additional boost increasing its velocity and raising up its apogee altitude beyond L2. As discussed later, the best way to manage the problem is to use an energetic approach: it is then considered to add enough energy to the initial orbit of the upper stage to allow it to join a heliocentric orbit.

Once the quantity of energy to be added will be defined, specificity will have to be addressed: as it is not possible to re-ignite the main engine of the A5ECA's upper stage, the maneuver will have to be realized through the attitude control system. It has to be noted that this system is a cold gas one which implies that the maneuver will not be impulsive. A continuous and possibly long-duration boost will have to be designed to generate the required additional energy.

4. LIBERATION CRITERION

4.1 Trajectory extrapolation beyond the Earth and goal of the study

A trajectory extrapolation tool developed by CNES was used to compute the trajectory of the launcher and its Payload (JWST) from the injection point to the L2 Lagrangian point and beyond.

The simulator takes into account the gravitational forces of the Sun, the Earth and the Moon. For the last two ones, the gravitational field was modelled with a large number of harmonics as long as the launcher is inside their respective spheres of influence.

As previously mentioned, the osculating orbit at injection is highly elliptic with an apogee located at about 1e6 km from the Earth (L2 being at 1.5e6 km). In the L2 point neighborhood, the orbit becomes unstable because of gravitational potential perturbations caused by the combined attraction of the Sun and the Earth.

A liberation criterion has been defined using an approach based on energy, in order to evaluate the efficiency of the orbit modification maneuver that will be done to ensure that the launcher will reach a point beyond L2 where the gravitational potential can only make it go outside of the influence of the Earth and the L2, leaving it on a heliocentric orbit.

4.2 Trajectory analysis and energetic approach

The trajectory analysis is done a posteriori using the theory of the gravitational three body problem, making the assumption that one of the bodies has a negligible mass compared to the two others which are massive and both describing circular orbits around their common barycenter. The principle is to write down the Newtonian mechanics equations in a rotating frame of reference having the two massive body barycenter as origin.

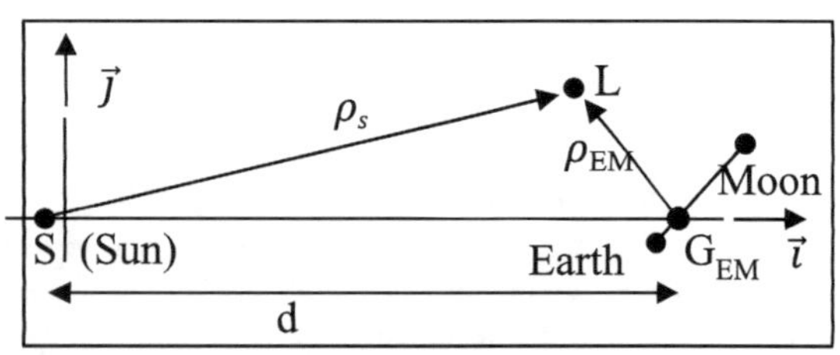

Figure 1. Launcher and massive bodies

In our case, the most massive body is the Sun and the other one is the Earth-Moon system, for which an approximation is done: the total mass of these two bodies is considered sitting the Earth-Moon barycenter as shown in Fig. 1. The origin of the rotating frame of reference is the Sun-Earth-Moon barycenter.

x, y and z are the notations for the launcher coordinates in the rotating frame of reference, normalized by the distance d between the Sun and the Earth-Moon barycenter. z is the out of plane component.

n is the notation for the angular speed of the rotating frame of reference with respect to an inertial frame of reference.

$\dot{x}$, $\dot{y}$ and $\dot{z}$ are the notations for the velocity coordinates in the rotating frame of reference normalized by $n \times d$.

μ is a constant which depends on the mass of the Sun (m_S), the Earth (m_E) and the Moon (m_M):

$$\mu = \frac{m_E+m_M}{m_S+m_E+m_M} \approx 3.04e-6 \quad (1)$$

The normalized distances between the launcher and the Sun or the Earth-Moon barycenter are:

$$\rho_S = \sqrt{(x+\mu)^2 + y^2 + z^2} \quad (2)$$

$$\rho_{EM} = \sqrt{(x-1+\mu)^2 + y^2 + z^2} \quad (3)$$

The Ω function only depends on the space variables and it is defined as:

$$\Omega = \frac{1}{2}(x^2 + y^2) + \frac{1-\mu}{\rho_S} + \frac{\mu}{\rho_{EM}} \quad (4)$$

The restricted circular three body problem assumes that d and n are constants. Then the dynamical equations of the launcher submitted to the gravitational attraction of the Sun and the Earth-Moon system (modelled as punctual masses) has a first integral called Jacobi constant. The Jacobi constant can be expressed as follows:

$$C = 2\,\Omega - (\dot{x}^2 + \dot{y}^2 + \dot{z}^2) \quad (5)$$

By definition, the value of a first integral of a dynamical system is constant when evaluated on any point belonging to a solution of the system.

Actually, the distance d fluctuates between 147e6 km and 152e6 km and the instantaneous angular speed n fluctuates accordingly.

However it is still possible to compute C using Eq. 5 even if it will not be strictly constant, especially for the regions in which the real physics differ from the simplified restricted three body problem modelisation.

For instance, in the Earth-Moon neighborhood the launcher dynamics does not comply with the approximation done when considering these two bodies as one punctual mass $m_E + m_M$ placed on G_{EM} and Eq. 5 is based on this approximation. Thus, the value of C strongly changes in this neighborhood but converges toward a stable value when the launcher moves away from it.

Moreover, the value of C in the restricted three body problem has an interpretation in terms of energy of the orbits. It is clear from equation (5) that the value of 2Ω-C cannot be negative. Thus, the Jacobi constant determines the regions in space accessible to the launcher.

When the launcher is heading toward the regions in space where Ω increases, the normalized velocity decreases and can possibly become equal to zero but it can never be negative. The value of C acts as a frontier that pushes the launcher back to regions where the velocity increases.

The L2 Lagrangian point is a local extremum of the Ω function. It can be demonstrated that its normalized positions are ($x = 1.01008, y = 0, z = 0$).

For the launcher to be able to reach this point, the stabilized Jacobi constant must be lower than twice the Ω value at L2: $\Omega(L2) = 1.500447$

For convenience, the following definitions, relatives to the L2 point will be used:

Kinetic energy: $E_k = \frac{1}{2}(\dot{x}^2 + \dot{y}^2 + \dot{z}^2)$ (6)

Potential energy: $E_p = \Omega(L2) - \Omega$ (7)

Total energy: $E_t = E_k + E_p = \Omega(L2) - C/2$ (8)

According to these definitions, the potential energy of the L2 point is 0 and the necessary, but not sufficient, condition for the launcher to be able to go beyond the L2 point is its total energy to be positive. The total energy equals the remaining kinetic energy when situated exactly at L2.

4.3. Analysis of liberation trajectories

The potential energy as a function of x and y is drawn using a color gradation in Fig. 2 with examples of liberated (in green) and unliberated (in red) launcher trajectories computed over 120 days.

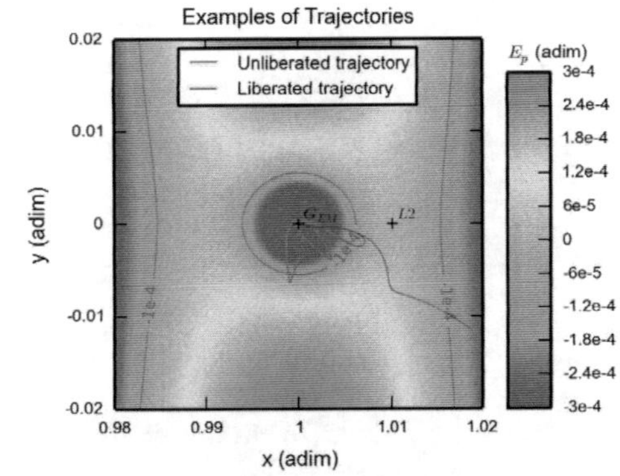

Figure 2. Example of liberated/unliberated trajectories

Energies E_p, E_k and E_t of the unliberated and liberated launcher are drawn in Fig. 3 and Fig. 4 over 120 days.

In Fig. 3, E_t converges to a stable negative value a few days after launch, which indicates that no liberation is possible. E_k decreases down to a minimum value after 27 days while E_p increases. Then the launcher moves back toward the Earth, while the gravitational potential of the Earth is causing the velocity to increase.

Every time the launcher passes close to the Earth-Moon neighborhood, E_t fluctuates greatly as expected, and then converges to a new different level each time. The total energy change is caused by gravity assist effects when passing close to the Earth and/or the Moon. Depending on the orientation of the velocity vector of theses massive bodies with respect to the launcher velocity vector, the total energy can either increase or decrease. These effects can sometimes make E_t become positive after several fly-bys and make the liberation possible through the L1 or L2 Lagrangian point.

When the launcher has already enough speed to be potentially released beyond the L2 point at injection time, E_t converges directly to a positive level as shown in Fig. 4

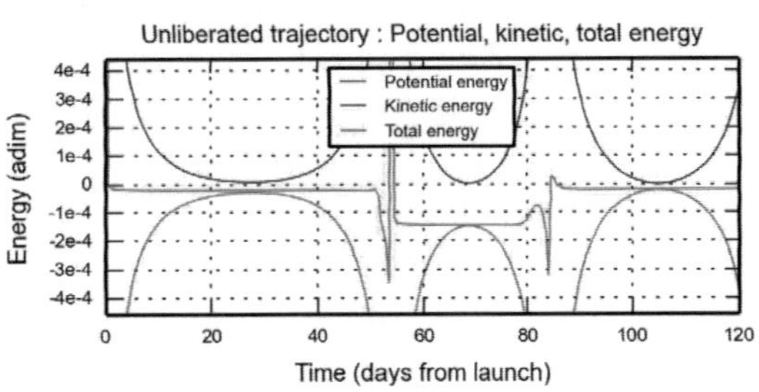

Figure 3. E_p, E_k and E_t for an unliberated trajectory

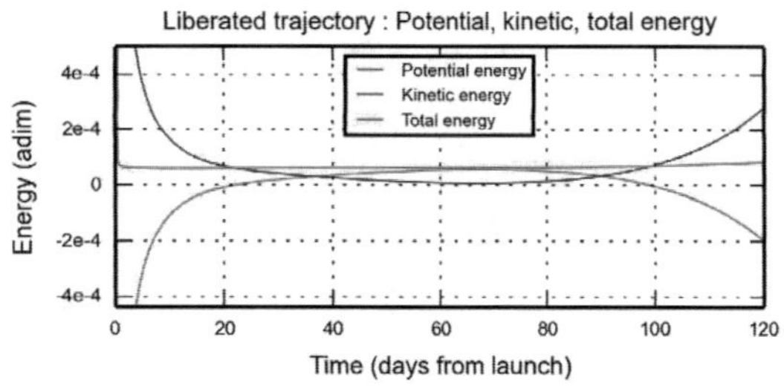

Figure 4. E_p, E_k and E_t for a liberated trajectory

In order for the launcher to be effectively liberated, its trajectory must lead it beyond the L2 through a region in space where the maximum potential energy, which is necessarily positive, is lower than the launcher total energy, as shown in Fig. 2 and 4.

4.4 Liberation criterion and liberation time

In order to determine if the launcher is liberated beyond the L2 point directly, without going back through the Earth-Moon neighborhood, the following method is used to analyze simulation results:

- Find the time t_1 corresponding to the first minimum of E_k of the launcher,
- Find the time $t_2 > t_1$ when E_k gets for the first time under -1e-4.

As shown in Fig. 2, at time t_2, if the launcher has come back in the Earth-Moon neighborhood, it is located on an almost spherical surface centered on G_{EM} and passing half way between G_{EM} and L2.

Otherwise, the launcher is necessarily located on a surface beyond the L2 point where the E_p gradient, which is strongly negative, will not allow the launcher to come back towards the Earth.

Thus, the launcher is considered liberated directly, without going back through the Earth-Moon neighborhood, if at time t_2, the launcher is located out of a sphere centered on G_{EM} and passing through L2:

$$\sqrt{(x-1)^2 + y^2 + z^2} > 0.010078 \tag{9}$$

If the launcher is liberated, t_2 is called liberation time.

5. SENSITIVITY ANALYSIS TO THE LAUNCH DATE AND TIME

5.1 Effect of the launch instant on the total energy

The nominal injection point of the launcher is a fixed location relatively to the Earth, whatever the launch instant (date and time).

As the launch instant modifies the Earth position and orientation with respect to the L2 point, the injection point relatively to the L2 point also changes.

The theory previously explained was applied to the JWST mission ensured by an A5ECA launcher. Trajectory extrapolations were done starting from the nominal injection point scanning every launch window with a 5 minutes step (1240 simulations over a year).

The total nominal energy $E_{t,nom}$ is drawn as a function of the launch instant in Fig. 5. The color indicates the time inside each launch window.

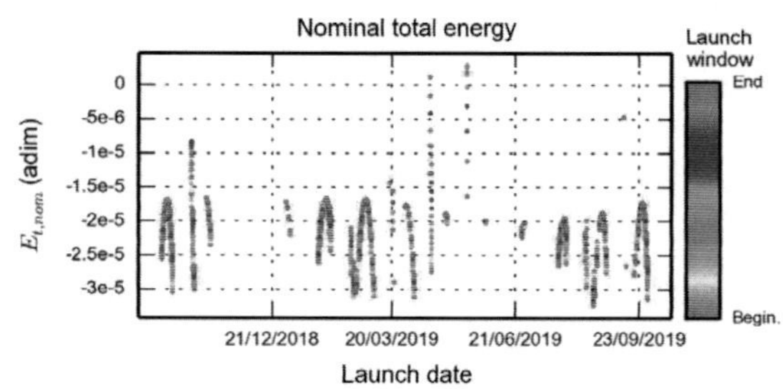

Figure 5. $E_{t,nom}$ wrt instant of launch

Three phenomena are identified:

a) $E_{t,nom}$ depends on the launch time of each day. The maximum value is either at the beginning, or at the end of each launch window, depending on the season.

b) $E_{t,nom}$ has cyclic variations over approximately a month

c) For a few particular dates, $E_{t,nom}$ is exceptionally high.

The reason of the cyclic variations is a gravity assist effect in the Earth neighborhood: as shown in Fig. 6, the Earth follows an orbit around the Earth-Moon barycenter G_{EM} lasting 28 days with a velocity $\vec{V_E}$, while the launcher velocity is $\vec{V_L}$, both in the rotating frame of reference.

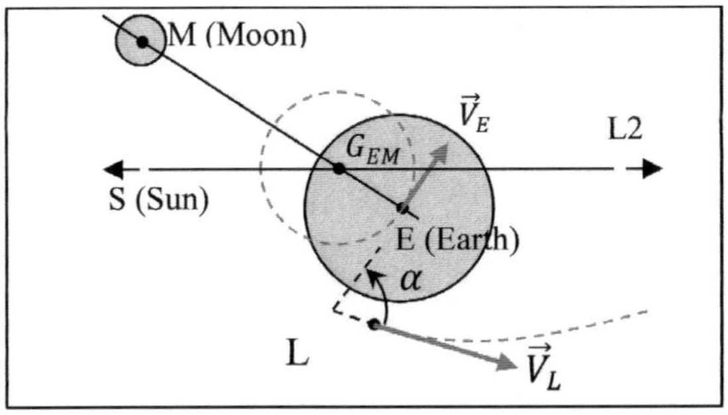

Figure 6. Earth and Launcher movements

At injection time, the angle between $\vec{V}_L$ and $\vec{V}_E$, in projection to the Earth-Moon orbital plane is called α.

The nominal total energy wrt α is drawn in Fig. 7 for every launch instant: a strong correlation appears between $E_{t,nom}$ and α, with seasonal variations due to the off- Earth-Moon orbital plane component of $\vec{V}_L$.

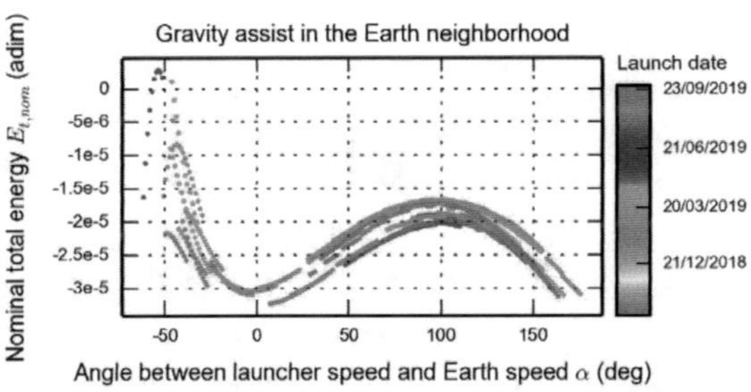

Figure 7. Gravity assist caused by the Earth

The minimum value of $E_{t,nom}$ is obtained when α is close to 0° or 180° and the maximum value is when α is close to 90° as the result of a gravity assist caused by the Earth.

This gravity assist effect is negative when α lies between -180° and 0° but it is compensated for α close to -50° by a very positive gravity assist of the Moon.

The Fig. 8, which draws the minimum Moon flyby distance as a function of α, shows that the launcher is only 1/3 of the Earth-Moon distance from the Moon when α = -50°.

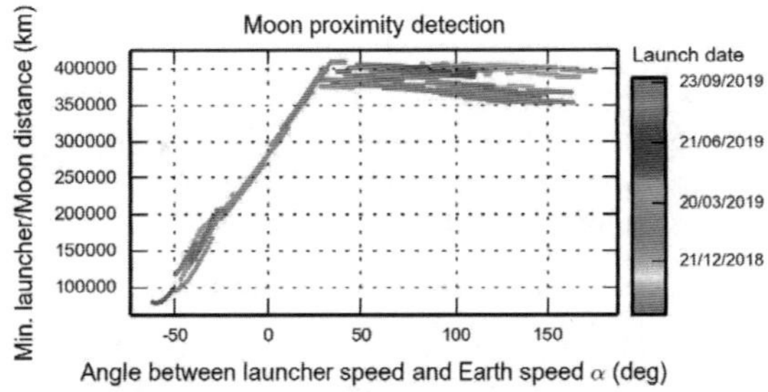

Figure 8. Minimum Moon flyby distance

5.2 Sensitivity of the total energy to small changes in the injection conditions

A sensitivity study has been conducted on the total energy and the liberation time when the injection conditions are submitted to small position and velocity changes. To do this, a random drawing of position and velocity scattering is added to the nominal injection point. Each position and velocity component is drawn independently with the same amplitude in all space directions. The statistical distribution allows covering the effect of:

- a realistic impulsive ΔV
- the injection accuracy of the launcher

For four different launch dates (the choice of these dates will be discussed later at §5.3) a random sample of 1000 values has been generated.

The liberation time wrt the total energy is drawn in Fig. 9 for liberated trajectories: whatever the launch date, the liberation time only depends on the total energy and the function between these two parameters depends on the launch date.

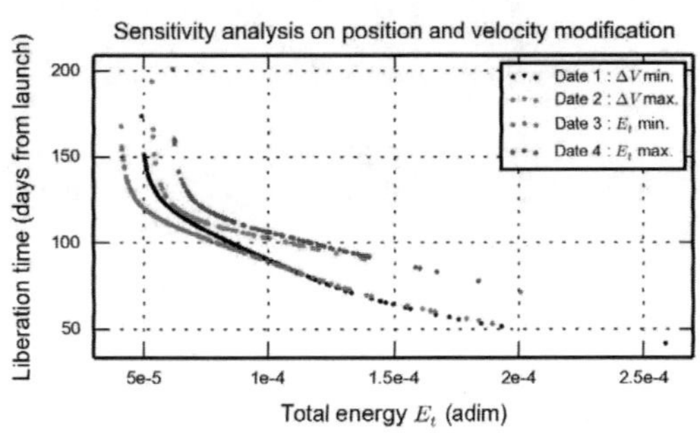

Figure 9. Total energy wrt liberation time

The liberation time rises to infinity when the total energy decreased towards a threshold limit which is function of the launch date.

The trajectories of all liberated cases are drawn for date 1 in Fig. 10 and for date 2 in Fig. 11.

The beginning of the trajectory, up to half-way between the Earth and the L2 point depends on the date but is poorly sensitive to the scattering.

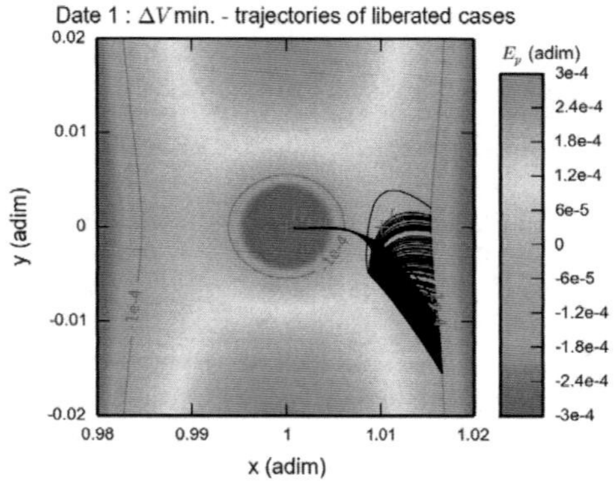

Figure 10. Liberated trajectories for launch date 1

However, the end of the trajectory depends a lot on the total energy changes caused by the position and velocity scattering: if the total energy is barely larger than the liberation energy, the launcher will describe a loop around the L2 point that will increase the liberation date.

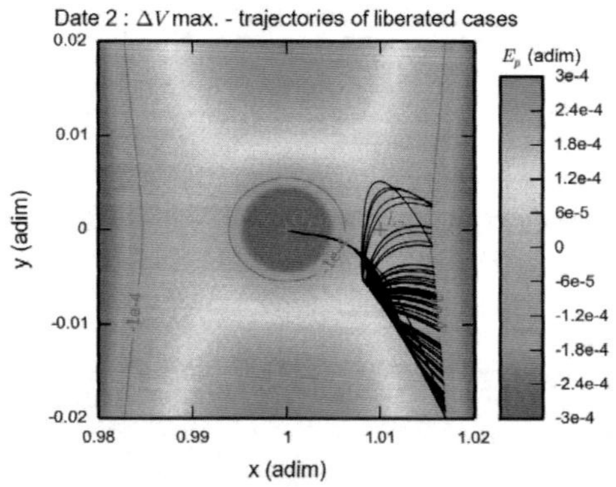

Figure 11. Liberated trajectories for launch date 2

Furthermore, regardless of whether the launcher is liberated or not, it has been empirically found that the total energy scattering obeys to Eq. 10. It is the sum of:

- the nominal total energy of the date: $E_{t,nom}$(t),
- a term which depends on the projection of the velocity scattering on the nominal velocity vector at injection, ΔV_x,
- a term which depends on the difference of altitude Δalt between the scattered and the nominal injection position,
- a small estimation error ε.

$$E_t(t) = E_{t,nom}(t) + c_V\,\Delta V_X + c_A\,\Delta alt + \varepsilon \qquad (10)$$

The kinetic energy is expected to depend on the square of ΔV_x. However, the linear approximation of Eq. 10 is valid because only small changes of the injection conditions are considered here.

Coefficients C_V and C_A were determined for each of the 4 dates using a linear regression from the 1000 total energy values as a function of ΔV_x and Δalt. The values of C_V and C_A slightly depend on the date. As explained later at §5.3, the most conservative date from the point of view of the impulsive ΔV to provide, is date 2, so this date is the best choice to determine C_V and C_A that will be used at all dates.

The estimation error ε obtained for the 1000 values of each of the 4 dates is drawn in Fig. 12.

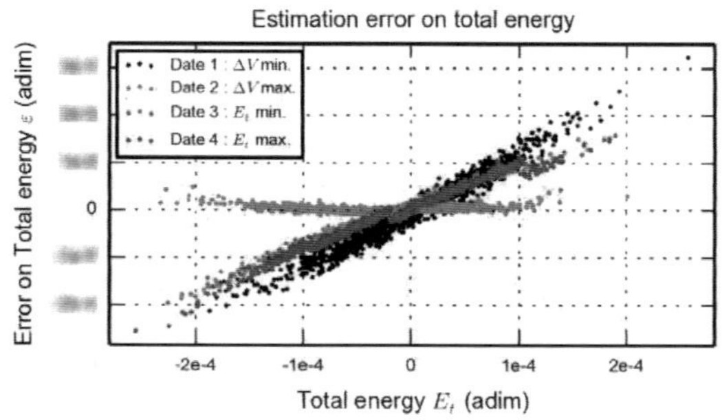

Figure 12. Estimation error on the total energy

The maximum estimation error observed at date 2 is smaller than those of all other dates because C_V and C_A were optimized for this date. The maximum value of the equivalent velocity error ε/C_V is very low: 0.04 m/s.

For all other three dates, the maximum value of the equivalent velocity error ε/C_V is 0.27 m/s.

Furthermore, Eq. 10 demonstrate that the most effective way to increase the launcher total energy with an impulsive ΔV is to apply it in the direction of the nominal velocity vector at injection time.

5.3 Determination of the total energy of liberation

For each of the 4 dates previously considered, an impulsive ΔV starting from 0 and increasing with a 0.1 m/s step was applied in the direction of the nominal velocity vector at injection time.

Among the 1240 launch dates considered at §5.1, dates 1 and 2 are the ones that allow to liberate the launcher with respectively the smallest and the biggest ΔV, as shown in Fig. 13, which draws the liberation time as a function of impulsive ΔV.

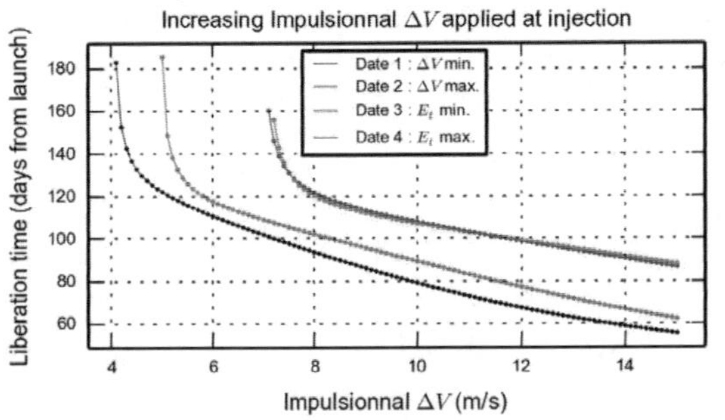

Figure 13. Impulsive ΔV of liberation

Fig. 14 draws the liberation time as a function of the total energy. Dates 3 and 4 are the ones that allow liberating the launcher with respectively the smallest and the largest total energy.

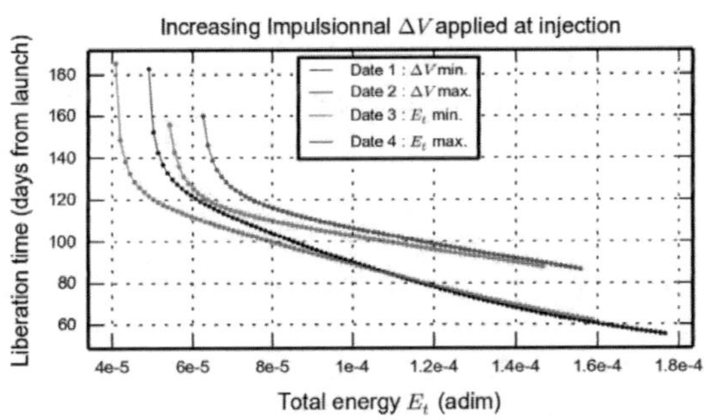

Figure 14. Total energy of liberation

Thus, the most conservative date from the point of view of the total energy (date 4) is different from the most conservative date from the point of view of the impulsive ΔV (date 2).

The impulsive ΔV necessary for the launcher liberation, ΔV_{lib}, provides an additional energy on top of already acquired nominal total energy, $E_{t,nom}$ (that depends on the date) to reach the total energy of liberation $E_{t,lib}$.

$$E_{t,lib}(t) \approx E_{t,nom}(t) + c_V \, \Delta V_{lib}(t) \tag{11}$$

Fig. 15 shows the liberation trajectories obtained applying the minimum impulsive ΔV that allows liberation, $\Delta V_{lib}(t)$, for the 4 launch dates .

For all of the 1240 launch instants, a scan in impulsive ΔV with a 0.1 m/s step was performed to determine the impulsive ΔV of liberation ΔV_{lib} .

Fig.16 draws ΔV_{lib} as a function of the launch date.

Fig.17 draws the corresponding total energy of liberation $E_{t,lib}$.

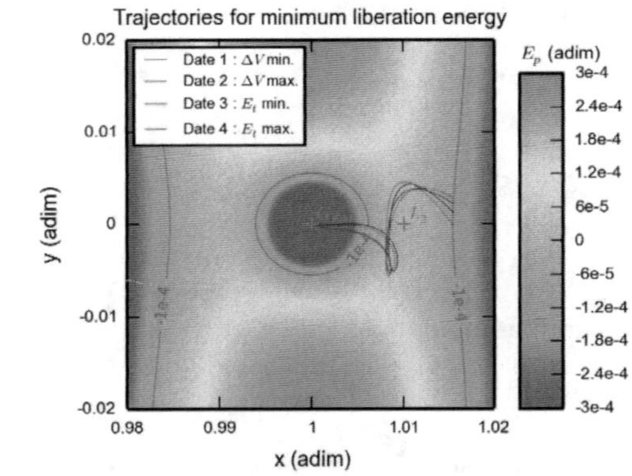

Figure 15. Liberation trajectories with minimum ΔV

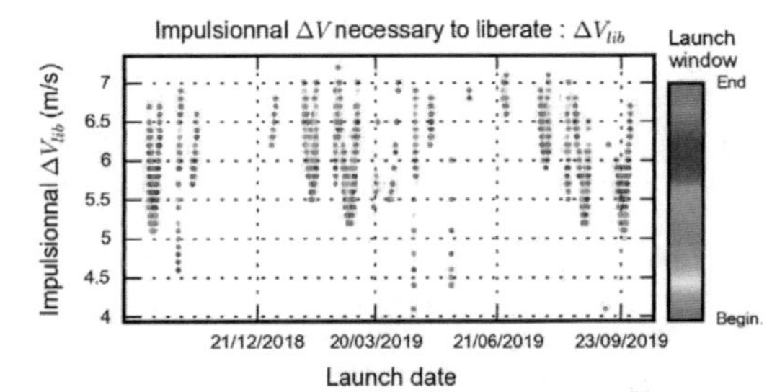

Figure 16. Impulsive ΔV of liberation by launch date

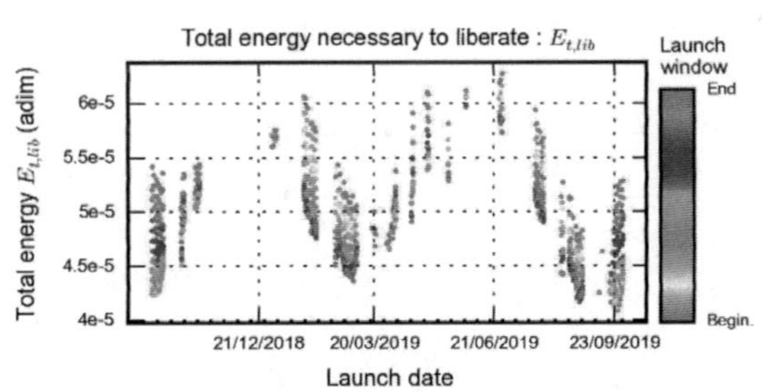

Figure 17. Total energy of liberation by launch date

5.4 Conclusion on the effect of the launch date

The nominal total energy and the total energy of liberation depend on the launch date.

The change of energy provided by applying a scattering in position or velocity or an impulsive ΔV does not depend on the launch date and is only function of the change of altitude and the change of speed along of the nominal velocity vector.

For a particular launch instant, the liberation time only depends on the launcher total energy.

The most conservative date for dimensioning a liberation manoeuver was determined: it is the date that has the greater gap between the nominal total energy and the total energy of liberation.

6. INJECTION ACCURACY

The injection accuracy of the A5ECA launcher in position and velocity is modelled with a normal distribution law defined by a covariance matrix on the 3 components of position and the 3 components of velocity, all 6 components being coupled.

The purpose of this study is to determine the impulsive ΔV to be applied to reach a certain level of probability to liberate the launcher at the most conservative date, taking into account the launcher accuracy.

The first step is to find the statistical distribution of the total energy corresponding to the covariance matrix. A random sample of 50000 values was generated and extrapolations were computed over 10 days to determine the scattered total energies which are considered to be stable outside of the Earth-Moon neighborhood.

The probability density function of the scattered total energy $E_{t,scat}(t)$ follows a normal law, as shown in Fig. 18. The mean value is the nominal total energy at the considered date, $E_{t,nom}(t)$ and the standard deviation is σ_{Et}.

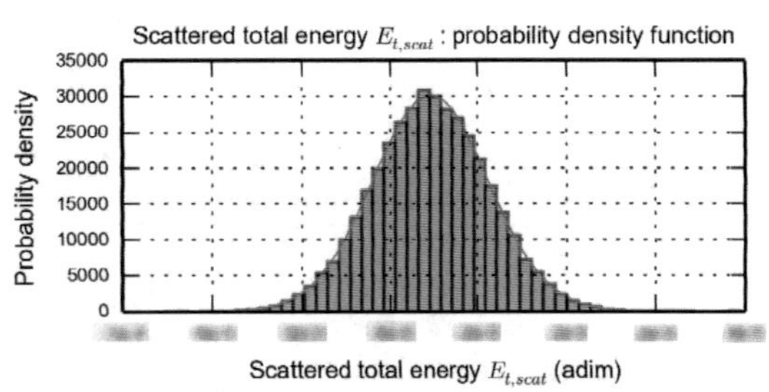

Figure 18. Probability density function of the scattered total energy

The standard deviation σ_{Et} depends on the covariance matrix but is independent of the launch date, as shown by Eq. 10, as each scattering value corresponds to values of ΔV_x and Δalt.

The probability for the scattered total energy of drawing i to be larger than a certain level defined by the value n is a property of the normal law:

$$P(E_{t,scat}(i) > E_{t,nom} - n\,\sigma_{Et}) = p(n) \qquad (12)$$

If, in addition of the scattering, an impulsive ΔV, designated as ΔV_{imp}, is applied at injection time along the nominal velocity vector, then the total energy of the drawing i is the one of Eq. 13.

$$E_t(i) = E_{t,scat}(i) + c_V\,\Delta V_{imp} \qquad (13)$$

Thus, the liberation probability equals $p(n)$ if Eq. 14 is fulfilled.

$$P(E_t(i) > E_{t,lib}) = p(n) \qquad (14)$$

Eqs 12-14 implies Eq. 15.

$$E_{t,lib} = E_{t,nom} - n\,\sigma_{Et} + c_V\,\Delta V_{imp} \qquad (15)$$

Thus, the impulsive ΔV required for the probability of liberation to be $p(n)$ is given by Eq. 16.

$$\Delta V_{imp} = (E_{t,lib} - E_{t,nom} + n\,\sigma_{Et})/c_V \qquad (16)$$

Eq. 16 can also be expressed as a function of the impulsive ΔV of the most conservative date using Eq. 11.

$$\Delta V_{imp} = \Delta V_{lib} + n\,\sigma_{Et}/c_V \qquad (17)$$

Furthermore, the pointing accuracy of A5ECA is to be taken into account when producing the impulsive ΔV: a loss of thrust due to a misalignment angle α_m both in pitch and yaw is considered. The new impulsive ΔV to compensate this effect is given by Eq. 18.

$$\Delta V_{cor} = \Delta V_{imp}/\cos^2(\alpha_m) \qquad (18)$$

7. SET UP OF A REALISTIC MANEUVER

7.1 Realistic maneuver sequential

An impulsive ΔV is not achievable by the launcher as it will use its propellant reserves to achieve tank depletion, producing non-impulsive maneuvers composed of the following phases, as illustrated by Fig. 19:

- Starting at time t_{inj}, ballistic phase to achieve the separation conditions needed by JWST,
- Separation of the JWST payload with a relative separation impulsive ΔV between bodies,
- Ballistic phase of the launcher with a pointing phase to prepare the thrust phase
- Series of boosts, all in the same inertial direction, and ballistic phases, ending not later than $t_{inj} + \Delta t_{man}$. (Δt_{man} is related to the launcher design limit)

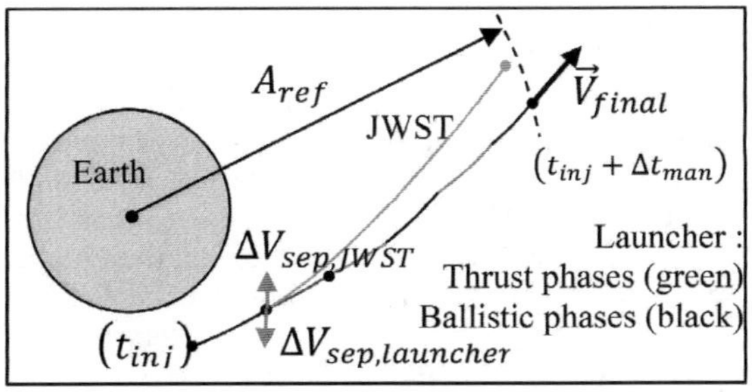

Figure 19. Realistic maneuver sequential

7.2 Equivalence between an impulsive ΔV at

injection time and a non-impulsive maneuver

The goal is to establish equivalence between a realistic maneuver and an impulsive ΔV applied at injection time from the point of view of its capacity to liberate the launcher beyond the L2 point, which depends mostly on the total energy when comparing two similar trajectories.

The later the impulsive ΔV is applied after injection, the smaller its effects are because the launcher's orbit is elliptic with a perigee located before the injection point. With this kind of orbit, the most effective way to increase the apogee, and then to liberate the launcher, is to apply a thrust at the perigee, or at least as soon as possible after t_{inj}.

The equivalent impulsive ΔV at injection is then always lower than the propulsive ΔV.

When the minimal impulsive ΔV necessary to liberate the launcher, ΔV_{ref}, is applied at injection time along the nominal velocity vector at injection, $\vec{V}_{inj}$, the launcher reaches the altitude A_{ref} at time t_{ref} with a final velocity vector $\vec{V}_{final,ref}$ as shawn in Fig. 20.

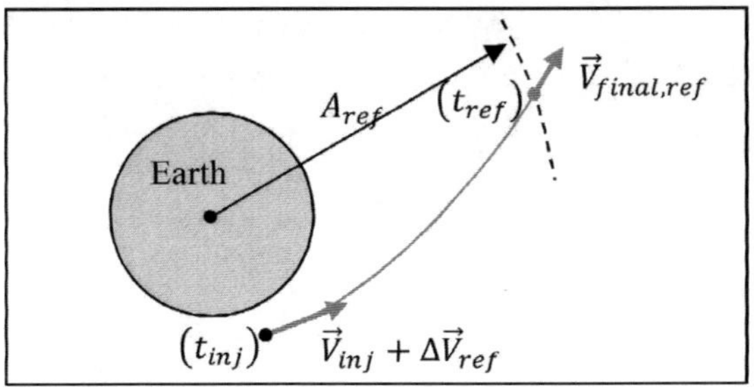

Figure 20. Reference impulsive maneuver

This trajectory, whose total energy is $E_{t,ref}$, will be used as a reference.

If the launcher is submitted to a different impulsive ΔV or to a realistic non-impulsive maneuver starting from the same injection point, it will reach the altitude A_{ref} at a location close to the one of the reference trajectory and a final velocity vector slightly different from $\vec{V}_{final,ref}$ in terms of norm and direction.

Applying Eq. 10 on the final reference state, the launcher total energy when arrived at altitude A_{ref} with a velocity $\vec{V}_{final}$ will be the one given by Eq. 19, where ΔV_x is defined by Eq. 20.

$$E_t \approx E_{t,ref} + c_V \, \Delta V_X \tag{19}$$

$$\Delta V_X = \left(\vec{V}_{final} - \vec{V}_{final,ref}\right) \cdot \frac{\vec{V}_{final,ref}}{\|\vec{V}_{final,ref}\|} \tag{20}$$

Thus, the final state obtained with an impulsive ΔV has the same total energy than another one obtained with a realistic maneuver if their final speed at altitude A_{ref} have the same projection along $\vec{V}_{final,ref}$.

A set of trajectories extrapolations have been achieved applying several impulsive ΔV with increasing amplitude at injection point as shown in Fig. 21.

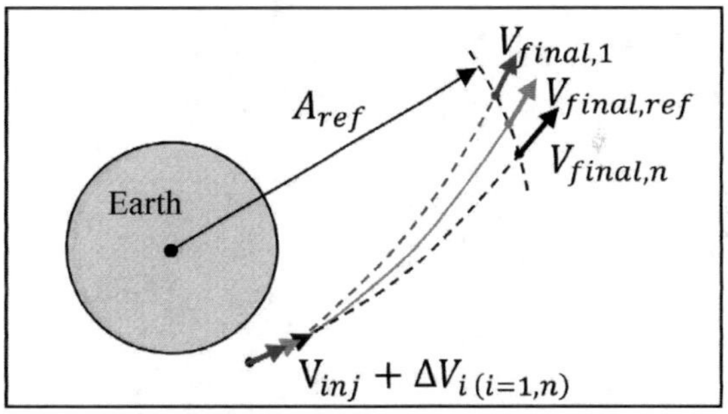

Figure 21. Increasing impulsive ΔV

The projection of the final velocity vector at the reference altitude along $\vec{V}_{final,ref}$ is drawn in Fig. 22 as a function of the applied impulsive ΔV. The relationship between these two parameters is quite linear.

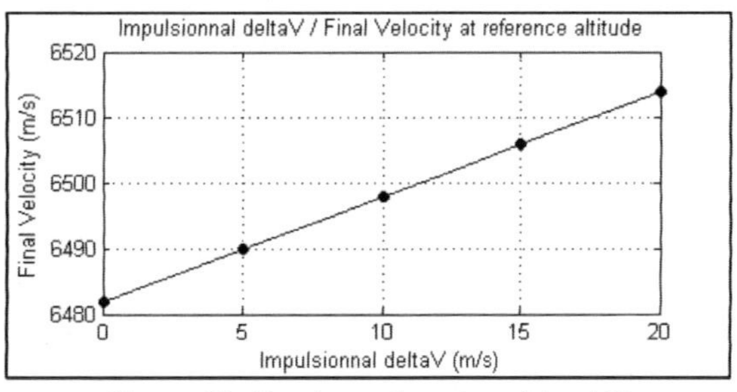

Figure 22. Final speed wrt impulsive ΔV

The equivalent impulsive ΔV at injection of a realistic maneuver is determined from the projection of its final velocity vector at the final reference altitude using a linear interpolation from the data of the Fig. 22.

7.3 Thrust direction optimization

The realistic maneuver is composed of several boosts, all applied along the same inertial direction, as illustrated by Fig. 23: the thrust is applied inside the orbital plane and forms an angle θ with the nominal velocity vector at injection time

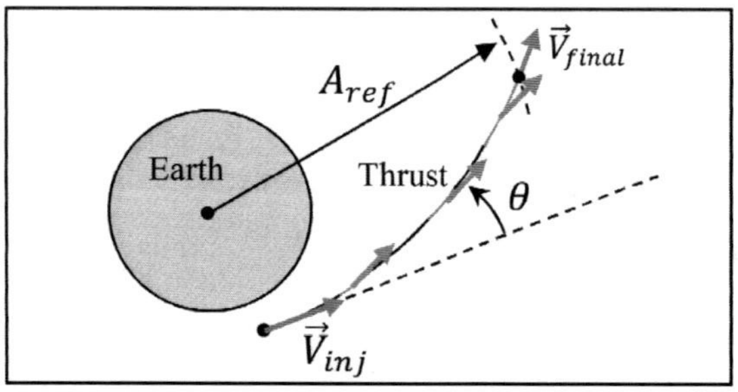

Figure 23. Thrust direction

An optimization algorithm has allowed determining the value of θ which maximizes the final speed and therefore the equivalent impulsive ΔV at injection as shown in Fig. 24.

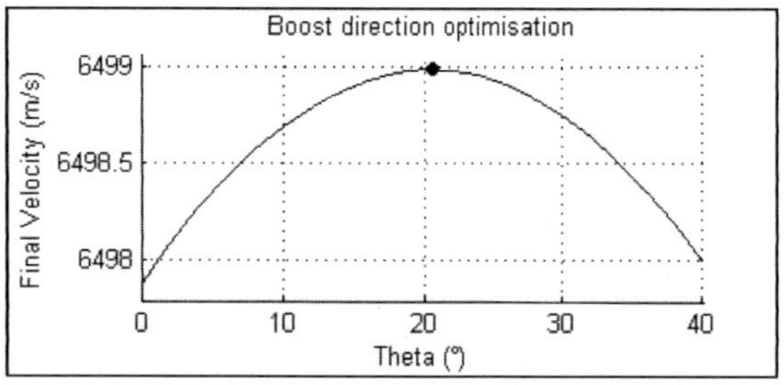

Figure 24. Optimum thrust direction

8. CONCLUSION

When a scientific space mission is launched targeting the Lagrangian point L2 of the Earth-Sun system, the launcher upper stage is generally abandoned on a transfer orbit close to the one of the spacecraft. This transfer orbit is often a very elliptical orbit but still geocentric. That induces a perturbation risk for operational orbital areas in the Earth vicinity and sooner or later, a risk for an uncontrolled reentry in the Earth atmosphere.

In the framework of the JWST mission to be launched on A5ECA at earliest in October 2018, ESA intends to minimize these risks by proposing an end-of-life maneuver for the upper stage which aim is to place it on an orbit that will never come-back close to the Earth. In this paper, physical phenomena to be taken into account for designing such a maneuver have been exposed (spatial three-body problem, influence of date and time of launch, influence of injection accuracy) and a methodology to size this maneuver has been introduced. Moreover, this paper proposes a methodology to take into account some A5ECA specific operational constraints to implement a realistic maneuver at the end of the launcher mission to free the upper stage. Proceeding like that, the mission safety would be increased.

9. REFERENCES

1. G. Gómez, W.S. Koon, M.W. Lo, J.E. Marsden, J. Masdemont, and S.D. Ross (2001). *Invariant Manifolds, The Spatial Three-Body Problem and Space Mission Design*. In AAS/AIAA Astrodynamics Specialist Conference, 2001. AAS 01-301.

2. B.T. Barden, K.C. Howell, ans M.W. Lo (1996). *Application of dynamical systems theory to trajectory design for a libration point mission*. In American Institute of Aeronautics and Astronautics, 1996. AIAA 96-3602

SATELLITE DESIGN FOR DEMISE: UPDATED STATE OF THE ART AND INNOVATIVE CONCEPTS

S.Heinrich [1], F.Leglise[1], F.Renard [2] L.Grassi, R.Destefanis [3]

[1] *ALTRAN France, 2 rue des Cormorans, F-06210 Cannes,*
Email: stephane.heinrich@altran.com florent.leglise@altran.com;
[2] *ALTRAN France, 1, place Verrazzano, F-69258 Lyon,*
Email : frederic.renard@altran.com
[3] *THALES ALENIA Space, Str. Antica di Collegno, 10146 Torino,*
Email: lilith.grassi@thalesaleniaspace.com, Roberto.Destefanis@thalesaleniaspace.com

ABSTRACT

The recent evolutions of Space Debris Mitigations policy all implemented the requirement of ground casualty risk for LEO satellites re-entry. Either a satellite has a Design for Demise (D4D) and can demise enough either it should go to Atmospheric Controlled Re-entry (ACR), and control the re-entry by end of life disposal manoeuvres to target safe areas like the SPOUA (South Pacific Ocean Unpopulated Area). The recent game-changing observed on new missions (SWOT, LOFT, METOP-SG...) impacted the entire space industry; ALTRAN Research initiated an internal project working on Orbital Debris Atmospheric Re-entry (ODAR) options since 2013. Team attended ESA Cleanspace CDF sessions on D4D before ESA initiation of activities on S/C D4D techniques. Recently ALTRAN was involved inside THALES ALENIA Space consortium in those ESA activities on D4D techniques and proposed successfully several D4D concepts for ESA Cleansat Building Blocks.

A previous paper described the trade-off between both options concluding that to mitigate system and mission impacts induced by controlled re-entry, all current LEO missions would have benefits from D4D techniques and innovations (mostly the satellites in the range of 750Kg – 2 tons). The intention of this paper is to update current state of art on D4D techniques and propose some innovative concepts:

1. INTRODUCTION

The scope of this paper is obviously restricted to satellite in LEO missions where satellites have still the ability to re-enter the earth atmosphere naturally (due to remaining atmospheric drag) or intentionally (when it is still economically viable). Those orbits were usually the place of following activities: manned space flight (i.e. Space Shuttle, Russian, Chinese and International Space Stations), earth science, observation and monitoring (ie ESA Living Planet & Earth Explorer programs) and communication constellations (Iridium, Orbcomm, Globalstar, and future Oneweb..).

Preliminary analyses at ESA have shown that space objects in LEO with masses above 500 kg might already imply an on-ground casualty risk higher than 1E-4 in case of uncontrolled re-entry. Compliance to 1E-4 casualty risk requirement may be achieved through controlled re-entry, but this solution has a major impact at system level, sometimes requiring the full re-design of the spacecraft and may involve switching to a completely different launcher performance (consequently a significant mission cost impact).

The need for a controlled re-entry may be avoided through design options aimed at achieving complete ablation (demise) of the spacecraft upon uncontrolled atmospheric re-entry.

Design-for-Demise is a highly multidisciplinary approach that can bring significant benefits in the future missions in the medium to long term.

2. D4D STATE OF THE ART

This section will introduce a dedicated focus on the demise capability and space agencies involvement.

ESA created in 2012 the "Cleanspace" initiative and team to promote actions on green aspects and debris remediation. This initiative was assigned with a top management priority and promotes researches on ADR concepts and Technos (E.Deorbit CDF, harpoon, solar sail, robotic arms, net, Debris attitude motion measurement) on D4D techniques (CDF Micra D4D which was open to industry, D4D Techniques, (characterisation of H/W ablation), and ODMG (orbital debris mitigation guidelines as passivation techniques on propulsion, pressurant and electrical systems). In 2014 ESA adopted the Ground Risk requirement as applicable on all its new programs and launched the CLEANSAT initiative throughout space industry which is promoted via industrial workshops on a yearly basis since 2015.

At NASA, [7] Global Precipitation Measurement (GPM) mission launched early 2015 was decided to be the first fully designed for demise LEO mission. The core spacecraft is similar to the highly successful Tropical Rainfall Measuring Mission (TRMM) spacecraft with a mass approximately of 3850 kg. [9]

The post- mission disposal objective is to meet Requirement 4.7-1 (56626) of NASA Standard [1] exclusively by hardware parts design practices. This requirement resulted in the need for a propellant tank that would also demise or ablate to an appropriate degree upon re-entry. Other elements such as reaction wheels and thermal hardware have been designed to demise better. The tank was designed and qualified with an Aluminium liner overwrapped with Carbon tensile fibres [7] [8] [9]. Reaction wheels where redesigned with a new demiseable material for the wheel rim replacing the original stainless steel. [9]

Demiseable at Earth's atmosphere re-entry

Fig. 1: GPM Satellite (NASA GSFC mission) [21]

Several space agencies such as ESA, CNES, DLR has initiated a set of DEMISE material characterisation test Main material characterisation at ESA [38] we based upon ITT7585 and focus on following materials

- Ti6Al4V
- Stainless Steel 316L
- Aluminium alloy 7075
- Silicon carbide
- Carbon fiber/epoxy CFRP

(Cycom 950-1/M55J UD fiber + M54J Hexcel resin) Unfortunately the selected facility "University of Stuttgart" broke its facility then up to now, very few results are available.

CNES [39] on its own mostly selected "Four Solaire d'Odeillo" since 2014 for materials characterisation Glass and Ceramics (Zerodur, SiC,Si3N4). Next year should be realised tests on CFRP.

NOTA: Carbon material represents today a main worry in D4D, as most S/W are predicting full demise of carbon overwrap and layers using graphite as a similarity. But recent re-entries of COPV tanks presented a recovered H/W with no ablation at all and very few damage on the overwrap.

Other works at CNES concerned activities up to TRL 3 with the outcomes of CNES patents:

- SELF FRAGMENTATION» TECHNOLOGY Linear shaped charges cords and thermal detonators
- BURNING ASSISTANCE» TECHNOLOGY Burning of thermite pellets on Ta6V plates (Tanks)

Fig. 2: CNES THERMITE patch on Tank[39]

3. CASUALTY RISK ANALYSIS

This section reminds the basics of casualty risk analysis of an atmospheric re-entry as required by debris mitigation guidelines and regulations [1] [3] [4].

Definition of the Acceptable Risk: [1] [3] [4]. The casualty risk limit according to current regulation is

Risk Acceptability: Ec < 1E-4

Definition of the Casualty Risk: [11] [17] [4].
The casualty risk is defined as the risk to hurt a standing man on ground:

Casualty Risk: Ec = Dca x Pd

Pd: Population Density of overflown surface:
Dca: Debris Casualty Area

Influence of Pd: Population Density
This parameter is function of mission or disposal orbit inclination (in fact maximum latitude overflown) and world population at re-entry date (assumed at 2050 with a world population of 9-10 billion people).

Table 1: DCA as a function of orbit inclination & time

	1994		2050	
Inclination	Population density	Allowed DCA	Population density	Allowed DCA
0°	6.6 /km^2	15m^2	15 /km^2	6,5m^2
50°	17 /km^2	6m^2	26 /km^2	3,8m^2
80-100°	10 /km^2	10m^2	20 /km^2	4,9m^2

Influence of DCA: Debris Casualty Area
This parameter is function of debris areas added with safety margins of about 30cm around each debris (representative of a standing man).
At least 30% margin is lost in case of one single debris Added 30% margin is lost in case of ten 10 debris 100% margin can be lost in case of lot of debris (20-30).

The D4D philosophy (main drivers, recommendations, techniques, technologies) can be summarised in detail as is:

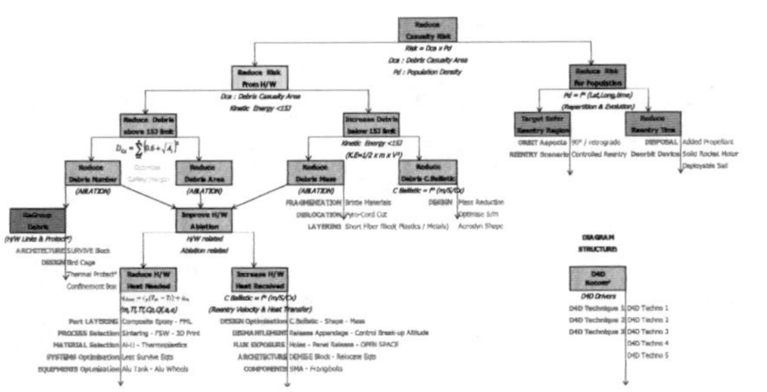

Fig. 3: D4D Philosophy Drivers Summary.

The four main families of techniques to mitigate the ground risk can be splinted in four main drivers:

- Reduce risk against population:
  => (orbit selection, earlier re-entry date, controlled / semi-controlled re-entry)
- Reduce risk via debris impact energy:
  => Promote solutions inducing less than 15J impact Kinetic Energy of the debris (brittle, fragmentation, (Then, debris are neglected in the DCA calculation)
- Reduce Risk via debris Ablation (Debris surface):
  => Improve heat received during re-entry
  => Reduce heat needed for ablation
  (material process selection, equipment selection)
- Reduce Risk via DCA optimisation
  => Reduce DCA via debris regroupment.
  (This technique has to be counter-balanced by the increased impact energy, in case of a new requirement of "sheltered man" is raised.

4. H/W STATUS

The main basic design driver is to identify Demise-critical H/W and to select and promote materials that are suitable for demise.
The decision parameter at first order is the melting temperature with a threshold around 1000°C.
In detail, the relevant parameter is the heat needed for melting (Qdemise) with could be set as good rule of "the thumb" at a decision limit around 1000KJ/Kg.

$$q_{dem} = c_p(T_m - T_i) + q_m$$

For instance, Aluminium and Cupper with clear different melting temperature have theoretical similar demise capability (even better for cupper).
Tungsten with a much higher melting temperature than Titanium has an equivalent Qdemise than Aluminium.
Beryllium alloys with melting temperature around Aluminium could in fact have worse Qdemise than Titanium.
Then following space H/W materials list can be raised:

- Material suitable: Al, Li, Mg, Pb, Sn, Cu..
- Material not suitable: Stainless Steel, Ti, Be, SiC, Ce/SiC, C/C, RCC, CMC, CFRP

This theoretical statement shall be adjusted with other material consideration occurring in plasma re-entry environment such as oxidation, etc...which only can be characterised by tests.
This statement can also be modified by the consideration of material thermos-optical parameters (radiative or conductive coupling) found in the real H/W architecture.
Another fundamental parameter can deeply impact this theoretical statement view factor and the ballistic coefficient provided by the H/W during re-entry to the aerodynamic flux.
If most of elements prefer to be exposed free and early to the flux to better demise, some hard-Demise material added with low ballistic coefficient released at higher altitude could then slow-down too quickly and then not achieve enough speed and heat to achieve their melting temperature.

Table 2: Release altitude effect comparison for marginal Ballistic Coefficients

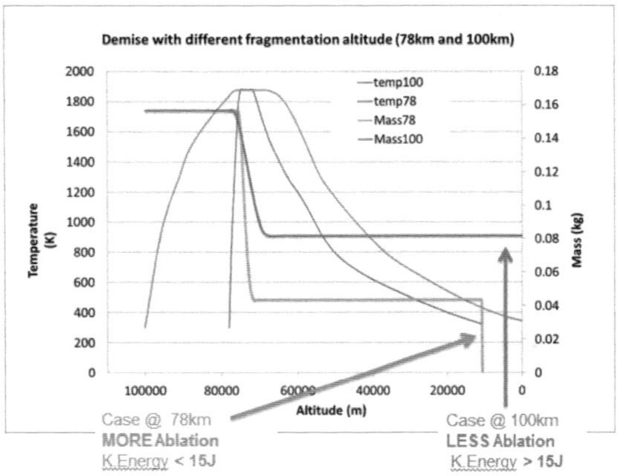

Consequently, the following LEO S/C components can be black-listed and may require alternative approaches:

- Propellant tanks (made of titanium or stainless steel with a bad area/mass ratio for demise),
- Reaction wheels (when inertia rims are in stainless and motor & core bearing and shafts),
- Magneto-Torquer Bars (stainless steels, cupper overwrapped on long cylindrical bars),
- Batteries (old generation Ni-Cd was an issue but now solved by Li-ion technology),
- Optical elements (SiC & Beryllium Mirrors, Titanium frames, C/C baffles).
- Mechanisms (motor, ball bearings and magnets, crown and gear wheels)
- Propulsion equipment (titanium, steel, nozzle materials but small size)

NOTA: The characteristics of brittle materials can be used intentionally for D4D purpose:
The sensitivity to Thermal stress (re-entry overheating followed by brutal atmospheric cooling at lower altitude) could be a useful technique, but Glass, Ceramics used on space application are usually intentionally "relaxed" with thermal post-treatments to release internal constraints. The capability to combine both functional stability and brittleness at re-entry has to be investigated. Nevertheless the challenge of this technique will be to demonstrate the break up into parts small enough

5. D4D TECHNIQUES & TECHNOLOGIES

Basic D4D technique will then focus mainly on replacing Demise-Critical materials.
In early space exploitation, these materials were not used much as aluminium (or magnesium); even if stiffness/density ratio is almost the same as titanium, the latter provided the same function at less volume and surface and better specific yield limit.

- ⇨ The swap to aluminium can be promoted when volume is not a constraint using innovative Alloys
- ⇨ The swap to aluminium can be promoted if reinforcement can be provided by an added composite
- ⇨ The swap to thermoplastic parts (such as PEEK, PEKK, PAEK etc...) can be achieved when structural constraint is not a driver (Propulsion Tubing studs) or with reinforcement by the addition of short fibres.

Table 3: TA6V and PEEK characteristics comparison

		VICTREX PEEK 90HMF40	TA6V
Density (Rho)	g/cm3	1.45	4.43
Young modulus	Gpa	43	113.8
E/Rho		29.7	25.7
Ultimate Tensile Strength	Mpa	330	1100
UTS/Rho		227.6	248.3
Thermal conductivity	W/(m.K)	2	6.7
Melting temperature	°C	343	1600
Electrical resistivity	Ohm.cm	100000	0.000178

The main drawback of this technology is the likely anisotropy provided by the material with long-fibre content (out of plane characteristic may be more driven by thermoplastic characteristics).
This technology can now be grouped with innovative process technologies (as 3D Printing).
The following Finite Element Study demonstrated the theoretical capability of high performance PEEK (with a high level of Carbon fibre content) to replace titanium in a design of a monolithic insert used on an optical payload. As this part was shaped with a lot of hollow cavities the plastic part was able to stay within the initial overall dimensions and obtained its strength through added internal thickness.
NOTA: This study was purely preliminary and no other consideration has been made for other physical functions required.

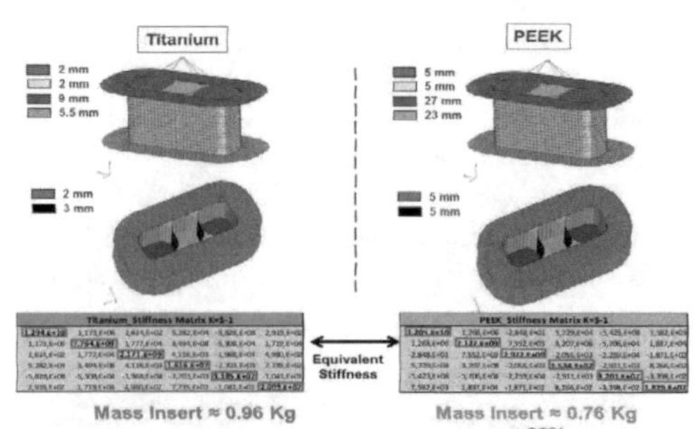

Figure 4: Sizing of a Titanium insert via PEEK

CONSTELLIUM, ALCOA, NOVELIS developments in Aluminum-lithium alloys are now so good they can still challenge recent implementation of carbon fiber structures.

Mass Specific stiffness Al/Ti is now in favor of Aluminium (due to Aluminium alloy density reduction down to 2.5).

Mass Specific Yield limit Al/Ti is now close to equivalence.

Table 4: Innovative Alu Alloys comparison

	Material	Density (g/cm3)	Ultimate Limit (Mpa)	Yield Limit (Mpa)	E (Gpa)	Criterion E/Rho	Criterion Yield/Rho
Al alloys	2024-T351	2.78	469	324	69	24.8	116.5
	7050-T7451	2.83	524	469	69	24.4	165.7
Al-Li	2050-T84	2.7	510	476	76.5	28.3	176.3
	2199-T8F80	2.64	428	380	77.3	29.3	143.9

OSPREY-SANDVIK now offers a range of materials with a controlled CTE.

Those materials are Al-Si alloys between 12-70% of Si allowing CTE between 5-20 ppm/°C

The following parameters appear to be very attractive:

- Attractive mechanical strength and density
- High thermal conductivity, low heat capacity
- Good electrical conductivity
- Good alternative for CTE around 7 to standard
- used Cu-W, Be-BeO, Ti Kovar

(lower cost, lower mass and easier safer manufacturing process)

- ⇨ Those innovative materials appear potentially attractive for a lot of applications as a replacement for Titanium for CTE advantages and where mechanical strength is not an important driver.
- ⇨ Nevertheless the melting temperature, Heat for demise has to be clarified and confirmed to assess the Qdemise and the asset in term of D4D.

Table 5: Al-Si characteristics

CE Alloy designation	Alloy composition	CTE, ppm/°C, 25-100°C	Density, g/cm^3	Thermal Conductivity at 25°C W/mK	Bend Strength, MPa	Yield Strength, MPa	Elastic Modulus, GPa
CE20	Al - 12%Si	20.0	2.70				
CE17	Al - 27%Si	16.0	2.60	177	210	183	92
CE17M	Al - 27%Si*	16.0	2.60	147			92
CE13	Al- 42%Si	12.8	2.55	160	213	155	107
CE11	Si - 50%Al	11.0	2.50	149	172	125	121
CE9	Si - 40%Al	9.0	2.45	129	140	134	124
CE7	Si - 30%Al	7.4	2.40	120	143	100	129

*CE17M also contains minor additions of Fe, Mg and Mn

At Process Level:

In this technique, the main driver is to exploit the parameters suitable for ablation:

- ⇨ All metallurgic process able to decrease the melting temperature of an original non-Demiseable material (Ti-AL,Ti2-Al,Ti-Al3) or increase the mass specific structural performance (Ti-Al Alloys, Al-Li, Al-Li-Sc)
- ⇨ All composite assemblies using demiseable material as a matrix (for demise) and a high tensile fibre (for structural purposes are good candidate to replace structural parts made of non-demiseable material (such as titanium). The use of short-fiber is less problematic in terms of demiseability and is implemented in metallic alloys (Al-C, Al-SiC, Al-CNT)
- ⇨ All sintering processes using different metallurgic powders: one for specific purpose and another one easy to demise as a binder material. Sintering process of Titanate, Alu, Li, Mg powders have been considered but were found unattractive. Some satellite used for instance INERMET – WOLFMET sintered alloys to allow heavy tungsten compound to melting at 3420°C be demised in powder.
- ⇨ All layering process such as used on Fibre Metal Laminate (FML families: GLARE, ARALL, and CARALL) will give the capability to dismantle large panel at resin evaporation instead of needing the melting a monolithic part.
- ⇨ Architectural assembly of several parts of smaller size can be easier to demise than a huge part (external satellite panels composed of several sub-panels sometimes present better demise results etc...)
- ⇨ Large parts with non-demisable materials can be processed via innovative 3D Print process to achieve a design with micro-mesh structure based on small lattice.

From this technique, there is potential for a reasonable mass benefit and a better wetted surface allowing a deeper exposure of aerothermal flux is foreseen, this could be understood as a demise benefit (mostly for parts with low melting temperature). However the overall volume or the cross section is likely to be increased. If the mass reduction leads to a worse ballistic coefficient the intensity of heat received during re-entry is reduced and could delay or prevent reaching the melting temperature. This technique must be used with care on high-temperature melting materials.

- ⇨ Glued structures also present good weakness points as they release at high temperature. (Most honeycomb panels used on satellites are in fact composed of glued metallic inserts forming the interface with structural elements. If the panel is shown to be able to demise alone, this technique can be promoted as should provide an early exposure of internal elements

At Component Level:
The main D4D-attractive applications of components devices are presented below:

- Application to appendages release
  - ⇨ To optimise S/C Ballistic Coefficient in early phase of re-entry (120-80km) and obtain better ablation conditions and demise for other S/C elements. (Higher re-entry speed and thermal flux)
- Application to P/F bus panels release
- Application to P/F bus dismantlement
  - ⇨ To obtain earlier and better exposure of internal elements to aerothermal flux during re-entry before spacecraft break-up

The dismantlement process of an entire S/C could be a long process as demonstrated by SCARAB simulations performed by HTG during ESA D4D S/C activities. This process could take minutes and release critical equipments in the altitude range of 80-65km.
(Very late to time enough or efficient demise)
It shall be noticed that SCARAB has few latitude to simulate I/Fs has very conservative assumptions:

- Release after complete ablation
- No consideration of mechanical failure
- No consideration of mechanical loads
- No consideration of loss of mechanical resistance in temperature

In any case, every device allowing to introduce flux inside a S/C of initiating internal heating and ablation would have a fruitful effect .

Main device that can be investigated are mostly related to release mechanism based on

- Shape Memory alloy (Low/ High T° SMA)
- Low Qdemise materials for washers (EZAC)
- Glued assemblies.

The mix of all those techniques should allow to have a complete set of temperature triggered devices.

ALTRAN proposed successfully to ESA Cleansat a Building Block with partner NIMESIS_F based on High-Temperature SMA intended to investigate:

- European purchasing source of standard Ti-Ni Frangibolt I/Fs
- High –Temperature SMA (aimeded for external applications around 100-200°C
- Very High Temperature SMA (aimed for applications above 150-250°C)
- Release mechanisms (inserts, washers, screws)

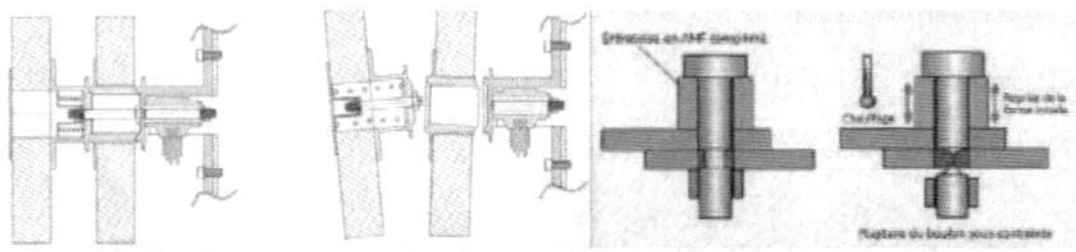

Propriétés	Unités	NiTi	CuZnAl	CuAlNi	CuAlBe
Taille de grain	µm	20-100	50-300	30-300	100-500
Domaine de transformation	°C	-100 +100	-100 +100	-100 +170	-200 +150
Hystérésis (As – Mf)	°C	20-40	10-20	20-25	20-25
Étalement (Af – As)	°C	30	10-20	20-30	15-20
Déformation maxi :	%				
- Mémoire simple effet		8	3-5	3-6	3-5
- Mémoire double effet		5	2	3	2
100 cycles		5	1	1.2	
10000 cycles		2	0.8	0.8	
1000000 cycles		0.5	0.5	0.5	
Température maxi d'utilisation	°C	400	160	300	400
Déformation maxi superélastique :	%				
- Polycristal		4	2	2	3
- Monocristal		10	10	10	10
Amortissement	SDC %	15	30	10	
Résistance à la corrosion		Excellente	Moyenne	Bonne	Moyenne
Biocompatibilité		Bonne	Mauvaise	Mauvaise	Mauvaise

(Source Nimesis)

Figure 6: Frangibolt- Like mechanism & NIMESIS synthesis on High Temperature SMA

Already implemented design on standard S/C panels should already have a better release capacity than predicted by simulations.
But this capability need to be further investigated and characterised in temperature

- Glued inserts with 5DOF (degrees of freedom) are assumed " easily releasable" (glue insert is assumed demising releasing insert form before panel demise TBC)
- Glued inserts with 3 DOF are assumed too trapped inside panel skin to release easily (before panel demise TBC)

This capability need to be further investigated and characterised in temperature

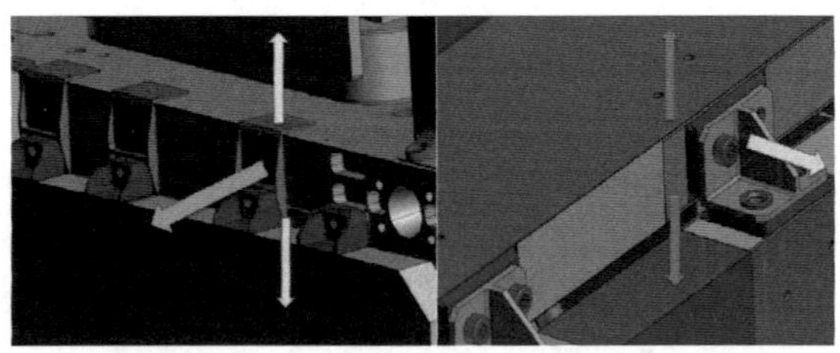

Figure 7: 5DoF and 3DoF Glued inserts

At Equipment Level:
Technologies for Demisable Tank
Below is a reminder of the demiseable tank solutions identified today in the space industry:

- Cobham Tank [] is about 45,226 cubic inches/ **1150L** Volume with an aluminium liner **6061-T62** of 0.05 inch/ 1.27 mm thin overwrapped with T1000 carbon fibre (Toray Carbon Fibers America Inc., Flower Mound, Texas) impregnated with an average thickness 0.111 inch/ 0.281 cm of EPON 862 bisphenol-F toughened epoxy resin

- IHI, Japan, [] is under development for JAXA of 350/430L Hydrazine COPV Tank designs (Al 6061 – CFRP).
- This development included some tests of Aluminium Demise under a Carbon overwrap.

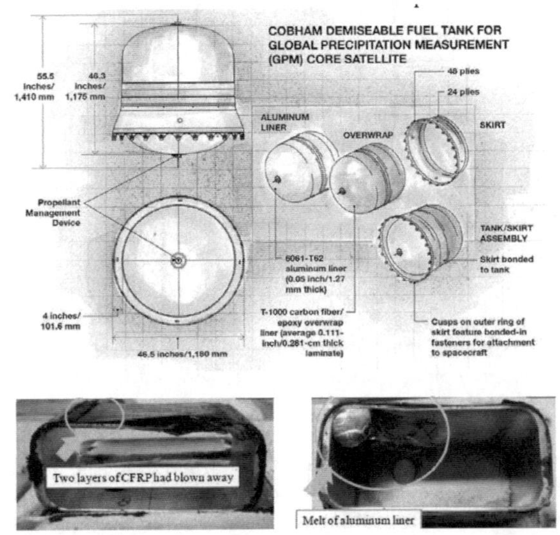

Figure 8: Ti -> Al Tank example NASA & JAXA

- MT-Aerospace_Germany developed a 154-180 litre, 35 bars MEOP Tank up to PDR for VEGA AVUM. (Tank is equatorially mounted with membrane and made of full Al2195 and no carbon overwrap)
- AVIO presented [] a preliminary COPV tank development based on Aluminium liner

In the following table, a simulation of Titanium to Aluminium monolithic Tank conversion is performed using the Sentinel-3 Tank case (130Kg-180L internal with membrane):

Table 9: Ti -> Al Tank Simulation

TANK Mass	TANK Standard	Material	Kg/dm3 Density	Kg Mass	Litre Volume	Mpa Stress	Mpa Yield	mm Min Thick	Mpa.m1/2 Klc	mm Min Flaw	Mpa.m1/2 Kiscc	mm Min Flaw
12,1	Tank TITANIUM	Ti	4,43	12,058	190,375	780	775	1,5	50	1,32491	20	0,21199
7,3	Sphere Standard	Ti	4,43	7,322	142		Worst case	2,55				
4,8	Cylinder Standard	Ti	4,43	4,736	48		at 60°C	Welds				
14,87%	Tank ALUMINIUM	Alu	2,7	13,851	187,225	390	400	3	25	1,2434	5	0,04974
21,43%	Sphere Alu	Alu	2,7	8,891	141							
4,74%	Cylinder Alu	Alu	2,7	4,960	47							
27,05%	Tank ALUMINIUM	Alu	2,7	15,329	186,681	334	350	3,5	25	1,62403	5	0,06496
41,48%	Sphere Alu	Alu	2,7	10,359	140							
4,74%	Cylinder Alu	Alu	2,7	4,960	47							
63,40%	Tank ALUMINIUM	Alu	2,7	19,702	185,058	234	250	5	25	3,1831	5	0,12732
101,34%	Sphere Alu	Alu	2,7	14,742	138		Min	Local				
4,74%	Cylinder Alu	Alu	2,7	4,960	47		Welds					

The existing design has been correlated with mass and average thickness and then a minimum thickness is assessed to fit (with some margins) proof pressure requirement (36Bar) considering yield limit assumptions of considered material and process.
The assumptions used the best new generation of Al-Li alloys give equivalent mass specific Stiffness to Titanium (same expansion in pressure) and almost same specific yield limit than Titanium (around 10- 15% less)

This lead to the following summary in Tank mass increase versus standard design:

- Realistic Option: +15%
  (High performance considered in skin & weld)
- Pessimistic Option: +30%
  (High performance in skin & Lower in weld)

In the present study, 2 major drivers have been identified for further investigation to achieve competitive Aluminium tanks

- The Yield limit of the selected material due to weld process (Driving the selection of best alloys and process to fit the tank assembly process). Usually Aluminum alloys experienced a severe yield limit decrease during TIG welding for instance as the Heat affected Zone is large. Only the innovative FSW can maintain high performance characteristics after welding

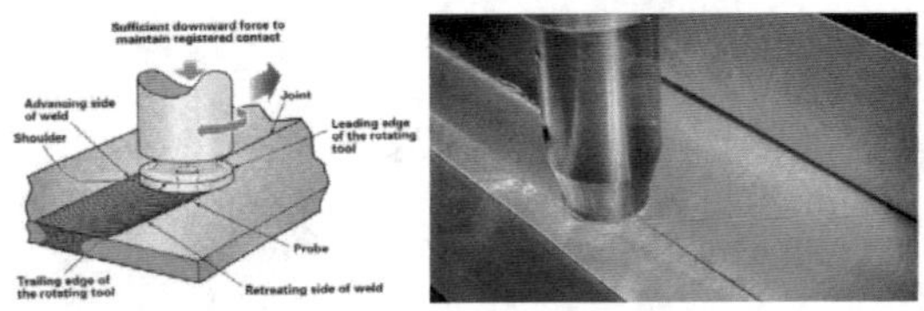

Figure 9: Friction Stir Welding

- The related fracture coefficients have to be considered. (Mostly the one related to pressure cycling and crack propagation in corrosive propellant environment). These parameters are already known as worst in Aluminium than Titanium by a factor of 4. They need to be carefully characterized with the correct alloys, weld process and associated propellant and could finally drive the choice of techniques. If it turns out not to be practical the overwrapped solution would remain a back-up as the Aluminium would then be less loaded.

A conservative approach would be to consider a (15-30%) mass increase in the preliminary definition phase considering that knowledge in innovative material & process could probably lead to more optimistic results.

All alternative solutions can be now compared:
(+[3-5]Kg for a 15Kg tank of 150Kg/200L capacity)
Intentional Fragmentation of this tank would lead to cut tank in titanium plates of 20x20cm (thickness 1mm). Pyrocord devices would weight probably much more.
A 10Kg Parasail device to slow down a tank below 15J limit would fit a 5-7 Kg Tank but would require 200m^2 of sail for a 15Kg tank and then very innovative materials (this fits the best performance parasail area/mass but without mass for any attachment !!)
A 10Kg parasail with current technology would glide

the tank at 50Joule impact energy (assumed non-lethal). Unfortunately this technique will need added mass to protect the device from reentry and to trigger the deployment of the sail.

Technologies for Demisable RW

First the problem of Reaction Wheels demise starts mainly above a certain type of high dimension wheel Considering mostly Rockwell Collins current H/W family via datasheet, this limit occurs at Wheel Diameter = 310mm /350 mm:

- RSI 68 (20N.m.s @ 1800 RPM) have Rim & Arms in Stainless Steel
- RSI45 (15N.m.s @ 1800 RPM) have Rim & Arms in Aluminium)
- Wheel Core demiscability remains TBC

ALTRAN proposed to ESA CLEANSAT a Building Block aimed to investigate mainly two following options in partnership with Rockwell Collins Deutschland (the major European supplier for Reation Wheels):

The proposed study will focus on how to achieve an Iso-Torque and Momentum with regard to RSI68 performances (upper class of LEO Applications) with a demisable unit:

1. Change of Mechanical Design:
   RSI 68 Mechanical Change with Aluminium Flywheel
   ⇨ -10% inertia expected at Iso-Diameter
   ⇨ Equivalent inertia with +10% volume
   Cumulating effect of rim Diameter, thickness and height increase.

Table 10: Reaction Wheel –Electrical Change

		Kg/dm3	mm	mm	mm	mm	Kg	Kg.mm²
RIM INERTIA	**Mat**	**Density**	**Dia Ext**	**Rad Ext**	**Thick**	**Height**	**Mass**	**Inertia**
Standard S/Steel	S/S	7,8	330	165	10	45	3,529	90 421
Standard Pb	Pb	11,3	330	165	10	31	3,522	90 241
Standard Pb	Pb	11,3	330	165	6,7	45	3,460	90 461
Standard Alu	Alu	2,7	330	165	10	45	1,221	31 300
Lumped Mass	Pb	11,3	320	160	5	43	2,404	59 655
Alu + Lumped Mass	Alu + Pb		330	165	15	45	3,626	90 955
Alu - Dia Increased	Alu	2,7	466	233	10	45	1,741	90 525
Alu - Thick increased	Alu	2,7	330	165	37	45	4,138	90 228
Alu - Height increased	Alu	2,7	330	165	10	130	3,529	90 421
Alu Optimised design	Alu	2,7	350	175	28	45	3,441	89 880

2. RSI 45 Electrical Change with upgraded of electronics
   ⇨ Mechanical Down –sizing at Higher speed
   ⇨ Increased Torque Power (X2 ?)

Following solutions have been investigated on are foreseen mostly oriented to Aluminium Rim and arms for the wheel.

Table 11: Reaction Wheel –Electrical Change

Kg.mm²	RPM	W	N.m.s	N.m	RCD
Inertia	**RPM**	**PWR**	**Momentum**	**Torque**	**RSI**
110 000	1 800	41	20,735	0,215	20-215/18
100 000	3 000	88	31,416	0,280	30-280/30
40 000	4 500	104	18,850	0,220	18-220/45
110 000	6 000	107	69,115	0,170	68-170/60
105 000	5 250	49	57,727	0,090	RDR57-0
109 000	6 000	47	68,487	0,075	RDR68-3
73 000	2 000	45	15,289	0,215	15-215/20
73 000	6 000	47	45,867	0,075	45-75/60
73 000	4 500	101	34,400	0,215	PWR increase
73 000	6 000	135	45,867	0,215	PWR increase

At Architecture Level:

The following D4D Techniques will focus mainly on capability to improve exposure, duration, and intensity of the aerothermal flux received by the element to be demised.

The **AltranSat v1** design concept presented the implementation of those technologies on a S/C platform:

A very dense architecture will provide a lot of shadowing on internal elements with regard to aerothermal flux and then resulting in less internal heating.

A better exposure of heat flux can be provided by an "OPEN SPACE" mechanical architecture such as the use of an internal structural frame made of strut design instead of a central tube, cone or shear webs.

This would allow the external panels to be without any equipment and the harness and could be made of "soft materials" easier to DEMISE in early re-entry phase at a lower temperature 200-400°C]

As a bonus a strut-based mechanical architecture, with tightening clamp band devices to join panels to the central tube could reduce the number of I/Fs to be released for an intentional controlled break-up managed with SMA mechanisms

This configuration will then include several D4D techniques to optimize the ablation and also speed-up the dismantlement with a less number of I/Fs to be released.

- Cylindrical shape for a better S/C ballistic coefficient (instead of rectangular shape)
- Soft easy demise panels to expose earlier internal equipments and quickly lowering S/C cross section during reentry (optimized Ballistic coefficient)
- Structural frame made of internal struts easing a controlled dismantlement (less I/Fs, less release devices)
- OPEN Space architecture avoiding shadowing on equipments and promoting aerothermal flux exposure deep inside S/C

Easily DEMISEABLE panels can be obtained as no equipment has been located on external panels. From a mechanical point of view, those panels can then be made of very soft and demiseable material as few loads are on external panels but essentially transferred on internal strut structure. From a technological point of view, those panels can be made of very thin aluminium panel of low density aluminium honeycomb or even high-performance plastic material (as far as RF EMC & faraday cage requirement are obtained by a metallic mesh grid or conductive fibres). This should theoretically provide flux exposure and external temperature deep inside the platform 200-400°C very early in the re-entry phase (above 100Km).

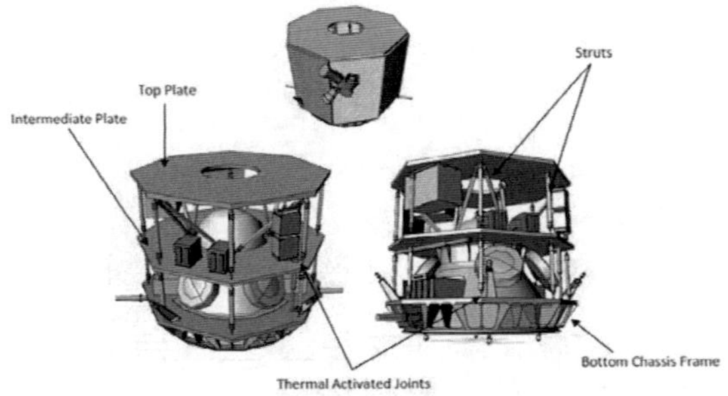

Figure 10: AltranSat Concept v1.0 (P/F only)

AltranSat v2 has been initiated to validate the concept of DEMISE Block(s) and SURVIVE Block(s). The intention is to force the design to regroup SURVIVE parts together with the objective of having only one or two blocks of each kind. The compliance is then easier to achieve through the limited surface of DEMISE block(s) as casualty area (no extra-safety margin lost due to number of loose parts). This design can be improved with the implementation of dismantling techniques with strut and panel release. In addition panels can be hinged & locked in controlled position

- To promote SURVIVE block protection and aerodynamic stability at initial re-entry.
- To promote DEMISE block in a controlled stabilized attitude and an increased mass to area ratio (higher ballistic coefficient) to optimse the ablation effect.

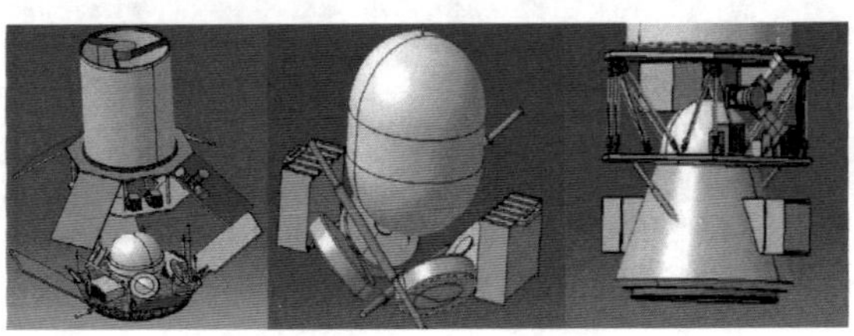

Fig. 11: AltranSat D4D techniques
Demise & Survive Block / Confinement Box

The **SURVIVE Block(s)** are designed to survive the re-entry environment and maintain all elements together without losing any lethal debris. This can only be achieved by a compact and optimized design reducing distance between lethal elements I/Fs and adding shadowing and protection from the aerodynamic and aerothermal flux. The Block shall integrate some thermal protection (backward heat shield) and maintain link structure between parts.

This Block has to demonstrate that the structure used to maintain the elements can withstand the mechanical loads (6-8g constant plus vibration) & thermal loads of re-entry:

- ⇨ Could be around 500°C inside the block if well protected, via thermal coupling or shadowing
- ⇨ Could be more than 1500°C if the block has an inappropriate ballistic coefficient or shape. If the concept is not viable for thermal or mechanical reasons a CONFINEMENT Box principle can be applied: a SURVIVE Block with overall box for thermo-mechanical protection. Even if the interfaces of internal parts degrade, the box confines the debris. For practical reasons, the box would have to be part of the platform mechanical structure withstanding launch loads. (SiC or C/C material could be then implemented Structural protective panels can be made of SURVIVE materials and specific innovative processes to provide a honeycomb"- like light-panel:
- ⇨ Titanium SPF –DB (Super Plastic Forming or Diffusion Bonding)
- ⇨ Titanium 3D Printed (SLS, SLM Selective Laser Melting or Sintering)

- THERMAL Protected Block solution:
  Intention is to protect thermally the NON DEMISE elements enough that the global block would survive. This local thermal environment has to be optimised to be thermally protected by insulating foam or ablative materials: NORCOAT-Liège (cork particles and phenolic resin 430g/L) or PICSIL (silicone and silica and micro balloons 410g/l, AG60 silica fibres in phenolic resin 300g/l). Then to ensure the assembly to survive, probably the part could to be enveloped inside a light and flexible thermal insulation (NEXTEL ceramic fibres, Carbon fibres, others?...) Structural protective panels can be made of SURVIVE Material and specific innovative processed to provide a honeycomb"- like light-panel:
  - ⇨ Titanium SPF –DB (Super Plastic Forming or Diffusion Bonding)
  - ⇨ Titanium 3D Printed (SLS, SLM Selective Laser Melting or Sintering)

- BIRD CAGE solution :
  - ⇨ To trap the elements in a light cage surviving the reentry conditions and still exposing survive parts to the flux to reduce the surviving mass (light structural frame to restraint mechanically NON DEMISE Elements not used as S/C struc-

tural elements).

- DEMISE Block solution :
  If SURVIVE parts are demonstrated to not create loss surviving debris or parts I/Fs do not detach from Block I.F, the same DEMISE Block concept (mostly aimed for Demise parts) can be applied to SURVIVE parts without the need of restraining elements from the BIRD Case solution. The DEMISE Block will then generate improved ablation by increased ballistic coefficient with always the same DCA reduction brought by the assembly of debris.

A preliminary simulation of a DEMISE Block made of a compact 4 Wheels configuration will represent a block of 4 RWA mass in the cross-section of around one. This configuration simulated by an equivalent steel sphere can probably achieve more than 70% ablated mass, reduce the DCA by a factor 3/4 and then represents the impact energy of a single RWA on the ground.

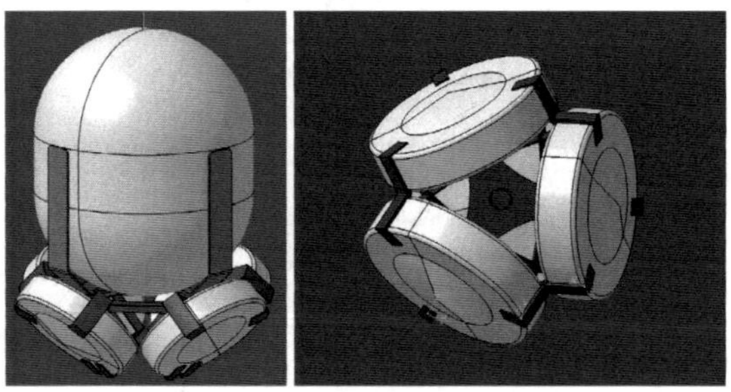

Figure 12: Bird Cage / Demise Block: Tank + RWA / 4 RWA

At MISSION LEVEL

This sections intents to detail all techniques that can be achieved at disposal phase of a mission that can impact ground casualty risk.

Orbit Inclination Selection

Considering the driver of the population repartition on the ground surface, the orbit inclination leads to a different calculation of acceptable Debris casualty area due to different population density of overflown regions.

The usual orbit inclinations lead to the following results:

- ISS-like (51.6°) can be considered as the worst case (-50%DCA versus nominal case)
- SSO (90-100°) being considered as the nominal case (reference risk 0%)
- Equatorial (0°) can be considered as the best case (+50% DCA versus nominal case)

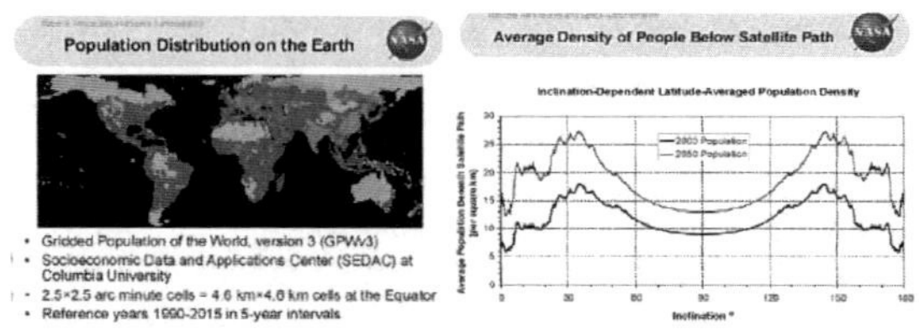

Figure 13: Population distribution

Considering the driver of ablation phenomenon impacting the capability to demise S/C hardware, a different orbit inclination provide a different re-entry air speed consider the ground velocity induced by earth rotation versus the orbital flight velocity. To the maximum extent, retrograde orbits can provide a re-entry air/atmospheric speed cumulating orbital velocity (7.8Km/s) and ground velocity (up to 0.5Km/s at low equatorial latitude)

This phenomenon is summarised in DEBRISK analysis on ablation performance [RD32] below:

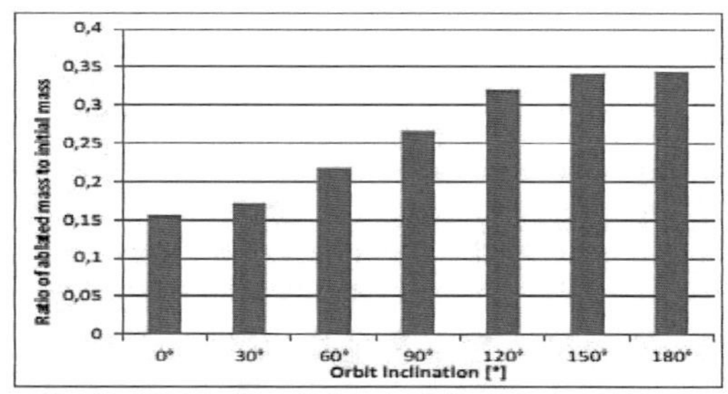

Figure 5 - Ratio of ablated mass to initial mass for orbit inclination variations

Figure 14: Orbit inclination influence on demise

Considering this driver, the theoretical orbit inclinations can be compared:

- Equatorial (0°) can be considered as the worst case (-40% ablation versus nominal)
- ISS-like (51,6°) can be considered as a worse case (-20% ablation versus nominal)
- SSO (90-100°) can be considered as the nominal case (nominal ablation rate)
- Retrograde (120°) can be considered as a better case (+20% ablation versus nominal)
- Retrograge (180°) can be considered as the best case (+40% ablation versus nominal)

Consequently, the previous recommendation of promoting equatorial orbital planes remains attractive (more DCA acceptable) but the best D4D benefits would be on a complete retrograde orbit (180°). Recent NRO

So, mostly equatorial retrograde orbits provide real D4D improvements, even if few missions can be performed on those inclinations but benefits are significant enough to be investigated.

Nevertheless, it is identified that a LEO constellation (today mostly implementing orbital planes around 60° with higher ground risk) could theoretically implement a combination of polar orbital planes added with equatorial orbital plane (at lower ground risk) in order to better cover the equatorial region which is harder to be covered correctly by polar orbital planes.
NOTA: Israeli military satellites have been already launched retrograde for political reasons and and some launch pads (as Vandenberg, Sea Launch) have the capability to launch on west bearings.
Recent USA 267 (NROL-45) assumed to be a military SAR reconnaissance satellite was launched in retrograde orbit from Vandenberg

Re-entry Date

The foreseen re-entry date is an important factor in natural re-entry scenario as casualty risk versus ground population is driven by the facts that:

- Current regulations allow space debris to rerenter in less than 25 years after their mission flight time. From this, current missions under project may potentially re-enter earth more than 50 years from now (sum of time periods of development, ground storage, flight lifetime and natural atmospheric decay).
- World population is still increasing in time (9-10 billion expected around 2050). Consequently, the following rough estimation on DCA reduction can be considered
- -50% in 50 years / -10% per decade /
- 1% per year

20% won if time re-entry is reduced from 25years to 5 years. This could be achieved by lowering perigee from 600-> 400km (based on a 7m²/1ton dry mass S/C like Sentinel-3).

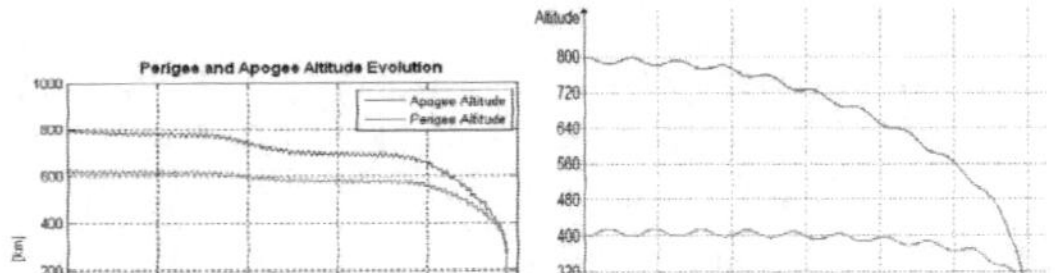

Figure 15: Natural Re-entry Time comparison (Initial Perigee 600km/25 yrs & 400km/3Yrs)

So considering all techniques to reduce earth re-entry date is a technique to mitigate ground risk while human population is still increasing. In addition, reducing the overall spaceflight lifetime will also decrease the collision risks for the space traffic.

Several techniques can be listed:

- Shorter development time, flight lifetime :
- Shorter natural atmospheric decay
  - o Lower altitude mission selection
  - o Lower altitude disposal
  - o Passive deorbit devices
  - o Active deorbit devices

The dedicated propellant budget to achieve compliance to 25 years re-entry requirement ist the one needed to lower a reach a perigee 800 -> below 600km
This dedicated delta-V is around 70m/s for the same study case (1ton dry mass S/C like S3). It can be noted that an equivalent added boost (70m/s) would set the same spacecraft perigee below 400km and involve a re-entry in less than 5 years (based on the same assumptions).

Mission Scenario	Debris Mitigation Requirement	Disposal Altitude	Propellant budget	Prop – S/C Mass Ratio	Constraints
Former Business as Usual	No 25 years Reentry	None	100 %	5 %	None
Today Business as Usual	25 years Reentry	< 600 km	200 %	10 %	D4D
Future Business as Usual?	5 years Reentry	< 400 km	300 %	15 %	D4D + Propellant Budget
FSOA Compliance	Controled Reentry	< 100 km	400 %	20 %	Propulsion & System

Figure 16: Rough Assessment of Debris Mitigation Requirements on S/C Mass & Propellant Budget

This intermediate solution of "short-term re-entry" seems a better trade-off between a 25 year re-entry and controlled re-entry, which is very demanding in term of propellant budget and system constraints.
Another solution could come from innovations such as GREEN propellant HPGP, which promises higher efficiency:
The European fuel is LM-103P using systems provided by ECAPS Sweden and the US has a fuel AF-M315E The following preliminary values compared to Hydrazine (N_2H_4, Isp=205s, density=1) can be considered:
[28] LM-103P: +6% Isp (iso-mass) with +24% density gives +25-30% added mass for iso-volume
[27] AF-M315E: Isp :+10-15% Isp (iso-mass) with +40-45% density gives +50% added mass for iso-volume

6. D4D TECHNIQUES OVERVIEW

The main D4D techniques are ranked from short term to long term approach (TRL) with a colour code:
High casualty risk mitigation, Low system impact
Low casualty risk mitigation, High system impact

- MATERIAL Change in Equipment:
  Tank, RWA, MTB, Optical payloads.
- FRAGMENTATION:
  Design debris below 15J aerodynamic shape
- COMPOSITE: DEMISE Matrix – Structural Fibres
- LAYERING: Design subdivision & Dismantlement under re-entry heat
- SINTERING: DEMISE Binder– Structural Element

(SURVIVE)

- SYSTEM architecture: Avoid NON –DEMISE Equipments
- EQUIPMENT D4D: Tank, RWA, MTB ,Mechanisms, Optical P/L
- EARLY DISMANTLEMENT: Only if released elements DEMISE
- EARLY internal EXPOSURE to flux: (Panels devices – Aperture holes)
- BALLISTIC Coefficient: Space (Cannonball) – Ground (Parachute)
- SURVIVE Block / CONFINEMENT Box (Regroup Debris)
- MISSION Parameters: Orbit inclination / Reentry Time
- SAT Mass: Promote SmallSat: 2 SmallSat (1 Ton) instead of 1 big (2 Tons)
- DISPOSAL Conditions: Re-entry Timeline, Passive deorbit device
- RE-ENTRY Scenario: Semi / Controlled / Quicker Re-entry

7. CONCLUSIONS

As mentioned, the variations of D4D limits are mostly driven by satellites with a dry mass of around 1 metric ton. This limit can be influenced by a magnitude of 2 depending on the nature of the payload and selected materials used in the design. Currently the D4D limits of standard design (business as usual, not specifically demise-oriented) can be summarised as follow today:

- Optical observation around 750 Kg
  (like Sentinel-5P)
- Scientific monitoring around 1000 Kg:
  (like Sentinel-3&2)
- Radio-Frequency payload around 2000 Kg
  (like Sentinel-1)

A lot of techniques and innovations can be investigated in order to improve the D4D capabilities but the time constraint to flight readiness has to be considered. If NASA claims to be able to Demise a satellite of almost 4metric tons today, it is only after a long-term development qualification process (around 10years for its main propellant tank); furthermore it only concerns a satellite with a RF payload with mainly aluminium components. The current ESA D4D study in which ALTRAN was involved within the TAS consortium had the objective to find D4D solutions for the Sentinel-1 study case (around 2 tonnes). This objective is about to be achieved. So
achieved. So

- **D4D Past Design:** S/C dry mass: 500 -1000 Kg
  (Business as Usual)
- **D4D Next Design:** S/C dry mass: 1t – 2t
  (easy modifications to be implemented)
- **D4D Future Design:** S/C dry mass: 2t-4t
  (major modifications needing requalification)

This paper has addressed an updated state of the art on European researched on Design for Demise

This paper has detailed a lot of recommendations and innovative guidelines for the improvement of D4D techniques.

8. REFERENCES

1. B. O'Connor, "Process for Limiting Orbital Debris", NASA-STD-8719-14ACHG-1, NASA 2012

2. EU, "European Code of Conduct for Space Debris Mitigation", 28.6.2004

3. ESA, "ESA/ADMIN/IPOL(2014)2 – Space Debris Mitigation Policy for Projects, ESA DG, 28/03/2014

4. CNES, "French Space Operations Act - Technical regulations", 31.03.2011

5. ESA Space Debris Office, "ESA and Space Debris", BR-309, March 2013

6. John C. Adams, Jr., "Atmospheric Re-Entry"

7. Peter M.B. Waswa, MIT, "S/C Design for Demise: Implementation strategy & decision-making methodology for low earth orbit missions", 2009

8. Robert.L.Kelley@nasa.gov ESCG/Jacobs Technology, "USING THE DESIGN FOR DEMISE PHILOSOPHY TO REDUCE CASUALTY RISK DUE TO REENTERING SPACECRAFT", IAC-12-A6.2.18, 2012

9. david.j.bundas@nasa.gov NASA GFSC, "Global Precipitation Measurement Mission - Architecture and Mission Concept", 2006

11. Heiner Klinkrad, Space Debris: Models and Risk Analysis,p241-287, 2006

12. William Ailor, Michael Weaver The Aerospace Corporation, REENTRY BREAKUP RECORDER: 5th IAASS Space Safety Conference, 2011

13. William Ailor, Wayne Hallman, Gary Steckel, Michael Weaver(4), The Aerospace Corporation" TEST CASES FOR REENTRY SURVIVABILITY MODELING", 5th IAASS Space Safety Conference, 2011

14. Pierre Omaly, Martin Spel, "CNES, DEBRISK, A TOOL FOR RE-ENTRY RISK ANALYSIS", 5th IAASS Space Safety Conference, 2011

15. Andrew S. Feistel, Michael A. Weaver, William H. Ailor Aerospace Corp°, "COMPARISON OF REENTRY BREAKUP MEASUREMENTS FOR THREE ATMOSPHERIC REENTRIES", 6th IAASS Space Safety Conference,

16. Pierre Omaly, Christophe MagninVella, Stephane Galera, "CNES, DEBRISK,: VALIDATION AND SENSITIVITY ANALYSIS", 6th IAASS Space Safety Conference, 2013

17. Carmen Pardini, Luciano Anselmo ISTI/CNR, "RE-ENTRY PREDICTIONS FOR UNCONTROLLED SATELLITES: RESULTS AND CHALLENGES", 6th IAASS Space Safety Conference, 2013

18. Ailor,W., Hallman,W., Steckel, G., and Weaver, "Analysis of Re-Entered Debris and Implications for Survivability", 4th Space Debris Conference, 2005

19. Koppenwallner, G., Fritsche, B., and Lips, T, "Survivability and Ground Risk Potential of Screws and Bolts of Disintegrating Spacecraft During Uncontrolled Re-Entry", 3rd Space Debris Conference, 2001 [20] CDF team ESA, "D4D MiCRA CDF study Webcast", ESA, 2013

21. G.Gardiner, AIAA, "Space Mission Maximized via Minimised Survivability", Aerospace America 2013-02,

22. Leonard David, IAASS, "Spacecraft Reentry: Safety by Design", Space Safety Magazine Winter 2012, 2012-03

23. Joel Spark, IAASS, "Coolant Tank Crashed in Brazil", Space Safety Magazine Winter 2012, 2012-03

24. Spacenews Editor, "ESA leadership in debris mitigation", http://spacenews.com/38113editorial-leadership-in-debris-mitigation/, 2012.11.03

25. C.Kjiellgaard AIAA, "Making SpaceFlight Greener", Aerospace America 2013-03, March 2013, p24

26. Franck Moorings, Jr, "Green Space", Aviation Week 2010, November 1st, p69

27. "IATA TECHNOLOGY RoadMap - Airframe materials", Georgia Institute of technology, 2014

28. "Applications of Osprey Lightweight Controlled Expansion (CE) Alloys",David M. Jacobson, Andrew J.W. Ogilvy and Alan G. Leatham Sandvik Osprey Ltd., Sandvik Materials Technology,

29. "Low cost production of nano-structured super-THERMITES", Mrs. Janet McGovern - navair.sbir@navy.mil, U.S.Navy, Navy SBIR 2008.1 - Topic N08-020

30. "Wolfmet Introduction to Wolfmet Tungsten Alloys", Wolfmet, Aug 2010

31. "Space mission maximized via minimized survivability", Ginger Gardiner, Composite World, Jan 2013

32. "Demise Characteristics Evaluation for Melting Promotion Type Tank", (JAXA) Japan Aerospace Exploration Agency, Tsukuba, Ibaraki, 302-0127, Japan and IHI Aerospace co. ltd, Tomioka, Gunma, 370-2398, Japan, AIAA Space Propulsion 2012,

33. RD6_PWT_Characterisation of materials Demisability through Plasma Wind Tunnel Test Experiments_86962_Pagan_Instut of Spacex System_ University of Stuttgart

32. ITT AO8413 D4D Optical Payloads, RD3_ESSB-HB-U-002-Issue1 (19 February 2015) ESA Space Debris Mitigation Compliance Verification Guidelines

33. ITT AO8287 Cleansat ITT, RD2-ESA-TEC-SC-TN-2015-001_Techno Overview of Technologies for the Evolution with the SDM Requirements_ESA Cleanspace team

34. NASA GSC RWA Patent, NASA GSC_14845-1_A Demisable Reaction Wheels_16/12/2005_demiseable-RWA@gsfc.nasa.gov

35. Demiseable Momentum Exchange System, Case Number: GSC 14845-1, Patent Number: 7,290,737

36. Analyticalgraphics Launch Notification: USA267 (NRO L-45) http://spaceflightnow.com/2016/02/10/delta-4-goes-against-the-grain-to-backwards-orbit-for-spy-bird

37.. DESIGN AND MANUFACTURING OF A COPV PROTOTYPE, Andrea Mataloni*, Giuseppe Pantanella* and Pierluigi Perugini* *Avio S.p.A Via Ariana, km 5.2, Colleferro (Rome) Italy 6th EUCASS

38. Characterization of Demisable Materials TEC-QTM/2013/100/BB issue 1.0

39. Arnaud VARINOIS, Denis DILHAN, Julien ANNALORO "SRL - CNES initiative for satellite compliant with French Space Operations Act" CNES Presentation Cleansat workshop ESA-ESTEC 19/03/2015

40. The ESA Policy on Space Debris Mitigation Prepared by: Independent Safety Office (TEC-QI) CleanSat Workshop ESTEC, Noordwijk, 18th March 2015

9. ABBREVIATIONS AND ACRONYMS

ACR	Atmospheric Controlled Reentry
AD	Applicable Document
AR5, 6	Ariane 5, 6
ATV	Automated Transfer Vehicle
BB	Building Block
CDF	Concurrent Design Faciliy
CFD	Computed Fluid Dynamics
CFRP	Carbon Fiber Reinforced Panel
D4D	Design For Demise
DAS	Debris Assessment Software
DCA	Debris Casualty Area
DEBRISK	Debris Risk Assessment (CNES S/W)
DRAMA	Debris Risk Assessment Mitigation & Analysis
DRA	Debris Reentry Area
EDT	Electro Dynamic Tether
EED	Electro Explosive Device
EOL	End of Life
ESA	European Space Agency
ESC-B	Etage Supérieur Cryogénique (Cryogenic Upper Stage)
ESTEC	European Space Research & Technology Centre
FSOA	French Space Operation Act
GTO	Geostationary Transfer Orbit
GEO	Geostationary Orbit
GFRP	Glass Fiber Reinforced Panel
GLARE	Glass Laminate Aluminium Reinforced Epoxy
H/W	Hardware
HC	Honeycomb
HPGP	High Performance Green Propellant
IAASS	International Association for the Advancement of Space Safety
IADC	Inter-Agency Debris Committee
ISON	Internation Space Observatory Network
Isp	Specific Impulse
L/V	Launch Vehicle
LEO	Low Earth Orbit
LVA	Launch Vehicle Adapter
MCC	Mission Control Centre
MiCRA	Mission Concept and Requirements Assessment,
MTB	Magneto Torquer Bar
NOTAM	Notice to Airmen
ODAR	Orbital Debris Atmopsheric Reentry
ODAR	Orbital Deris Assessment Report
ORSAT	Object Reentry Survival Analysis Tool
P/F	Platform
P/L	Payload
R/B	Rocket Body
RW	Reaction Wheel
RWA	Reaction Wheel Assembly
S/C	Spacecraft
SCARAB	Space Craft Atmospheric Re-entry and Aero-thermal
SESAM	Spacecraft Entry Survival Analysis Module
SMA	Shape Memory Alloy
SPOUA	South Pacific Ocean Unpopulated area
SSO	Sun Synchronous Orbit
TAS	THALES ALENIA Space
TBC	To Be Confirmed
TBD	To Be Defined
THR	Thruster
TRL	Technology Readiness Level

DEBRIS REMEDIATION EXAMINED VIA AN OPERATIONAL SUCCESS FRAMEWORK

Dr. Darren McKnight, IAI

ABSTRACT

The activity of debris remediation is essential to assure future safe access to space. By preventing the existing on-orbit debris population from proliferating via collision of objects the collision risk to future space systems can be moderated to permit everyone to operate reliably in space without the disruption of debris impacts. Debris remediation, in most discussions, is considered synonymous with active debris removal (ADR); ADR has been studied by many and determined to provide positive ramifications.

However, by examining debris risk management within an operational success[1] framework, other response options are identified and characterized. The "operational success" framework is reviewed and is organized around the temporal taxonomy of four phases of activities:

(1) PREPARE: identify, characterize, and predict;
(2) PLAN: dissuade, deny, and deter;
(3) ACT (During): interdict, suppress, and mitigate; and
(4) ACT (After): remediate, recover, and reconstitute.

An alternative definition of debris remediation as derelict collision prevention is reviewed and a complete list of debris remediation options are presented. Key linkages within the newly formulated operational success framework highlight the benefits and limitations of threat warning, situational awareness, and asset reconstitution. Lastly, the concept of risk transfer is also addressed; enhancing success in one area may simply result in the transfer of the risk to a different domain.

1. INTRODUCTION

Modeling has shown that debris mitigation will likely be insufficient to control the debris environment long-term; debris remediation will also be necessary to assure future safe access to space. [1] By preventing the existing on-orbit debris population from proliferating via collision of massive objects, the collision risk to future space systems can be moderated to permit everyone to operate reliably in space without the disruption of debris impacts.

[1] Success = 1 - Risk

Debris remediation, in most discussions, is considered synonymous with active debris removal (ADR). While ADR has been studied by many and determined to provide positive ramifications, this paper strives to identify other remediation options to complement ADR. First, an operational success framework is developed as a way to motivate new debris remediation options and to better describe how the entire spectrum of modeling, risk communication, warning/monitoring, and operational activities work together to enhance the two high-level global objectives of (1) space flight safety and (2) space sustainability.

2. OPERATIONAL SUCCESS FRAMEWORK

Risk management is often represented as a cyclical spiral of action items seemingly focused more on bookkeeping than keeping a system or domain safe. [2,3] Embedded in this approach are often generic management terms rather than mission-specific, people-oriented action words.

Prescriptive feedback on which techniques are the best to enhance success is limited and the circular process does not naturally integrate with a domain's safety which evolves over time but focuses on operational deployment.

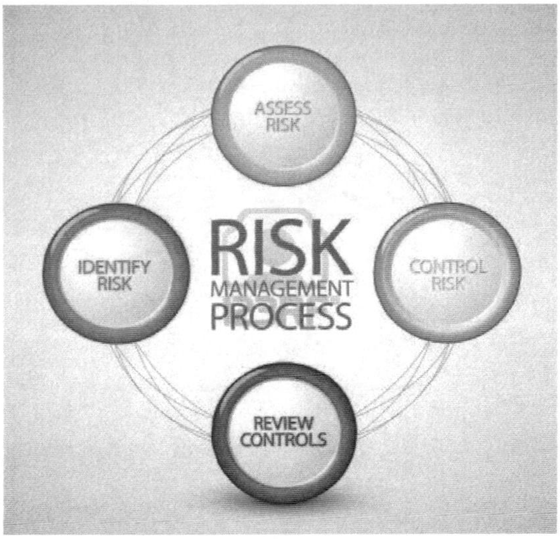

Figure 1. Risk management is often represented in a continuous circular

Applying positive (i.e., talk about success rather than failure) terminology in a linear spectrum may provide some benefits. The positive linear construct proposed is the operational success framework.

Early in the pursuit of operational success, as we PREPARE, we have many options available and their costs are very reasonable. However, as we progress through PLANNING for operational success, the options dwindle and the costs increase. Further, once the operations begin and barriers are introduced (e.g., threats activated) the options for how we ACT (During) are even more limited and the costs even higher. After the threats have acted on our operations and their full effects felt (i.e., ACT (After)), there are even fewer options and they will likely be increasingly more expensive.

If we use this construct, the activities earlier in the process (e.g., PREPARE and PLAN) are emphasized since they are generally more efficient and effective at assuring operational success. This framework reinforces a very important concept of any solution development: without a thorough problem statement any resulting solution will likely be unsatisfying. It is the upfront work (i.e., PREPARE and PLAN) that enhances operational success with a minimum of cost by greatly reducing the need to deploy implementation countermeasures once the threat (i.e., barrier to success) has actually been manifest (i.e., ACT (During) and ACT (After)). The best preparation and planning will prevent many barriers from ever being realized as threats.

The **PREPARE** phase includes three levels of analysis: identify, characterize, and predict (in that order).

We must first identify both the barriers to success (i.e., threats which may manifest as risks) and enablers for success. There are four families of barriers and enablers to success: cognitive, mechanical, natural phenomena, and manmade threats.[2]

1. **Cognitive** features include requirements definition, design, relevancy of testing conditions, operator error, deception (by an adversary), decisionmaking/analysis, and concept of operations.
2. **Mechanical** features contain material & component reliability & availability plus manufacturing & integration quality.
3. **Natural** phenomena comprise conditions imposed by the natural (i.e., non-manmade) environment that might affect the chance of a system's operational success (e.g., small aircraft are very sensitive to high crosswinds).
4. **Manmade** threats are those activities created by an adversary with the sole purpose of preventing your operational success. For a bomber, this might include jamming of navigation signals, general high altitude flack, or surface-to-air missiles.

Taking the time to identify each of these barriers and enablers at this early stage provides an opportunity to compare and contrast these aspects. This will be important as techniques to enhance operational success later are considered and prioritized; it allows "racking and stacking" (i.e., prioritizing) of options later.

Each barrier (and enabler) to success has the following dimensions that must be characterized:

- ***Effects:*** type, persistence, and rate (e.g., instantaneous, temporary, electronic system upset due to electrostatic discharge or permanent, delayed, software

[2] The comprehensive list permits this framework to be used for everything from physical systems in harsh environments to organizations in complex business ecosystems.

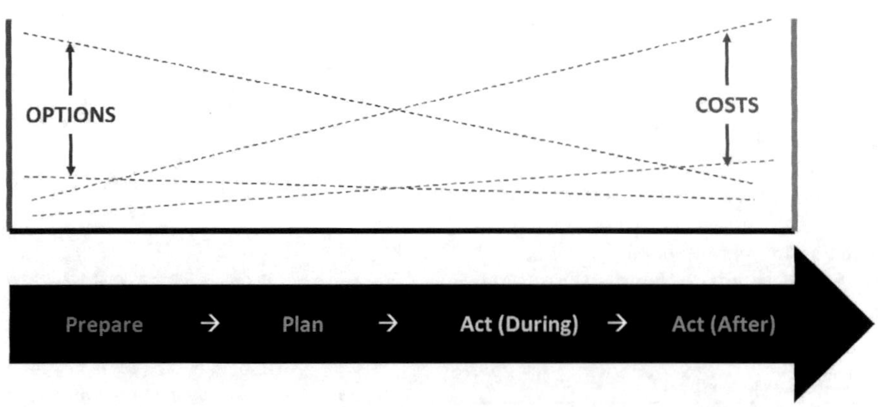

Figure 2. The operational success framework follows a linear process. continuous circular.

failure due to ingestion of malware).

- ***Uncertainty:*** variability in space and effect due to inherent behavior, lack of awareness, or limited understanding (e.g., average risk of antenna being struck by lightning is concentrated during afternoon hours in the summer especially if a freestanding deployment).
- ***Predictability:*** ability to project effects and uncertainty into the future (e.g., solar activity levels are fairly predictable to oscillate in 11-year cycles but UK scientists believe a significant reduction will occur in the 2030s).

It is important to examine each of these barriers in this order: effects, uncertainty, and predictability. It should be noted that uncertainty/variability may not be known so an average effect may be used. It is also optimal to consider how well we can predict effects and uncertainty in the future. These are problematic for collisional dynamics in LEO as we really have no way of identifying potential collisions more than a few days in advance and so we use the average (or statistical) value for most modeling scenarios.

and PLAN phases but become countermeasures and response options in the last two ACT (During and After) stages.

By including how each of these features interact in the PREPARE and PLAN stages, the possible synergistic effects can be caught early (e.g., if the aspects of the attitude and control system (ACS) keep being identified in the barriers analysis then it will become clear that more enablers will be needed there and that marginal enhancements to other subsystems may be less critical than improvements to the ACS). This helps to identify high priority components, subsystems, and systems.

While it is important to acknowledge a fairly complete suite of barriers, it is not necessary to consider all possible barriers to operational success. Planning for a diverse (though maybe not 100% complete) set of barriers will likely be sufficient to posit an effective range of planning options to enable operational success against a wide range of scenarios. The key is to consider a diverse (i.e., coming from different experts, using different physical modes, and different rates of execution) set of barriers.

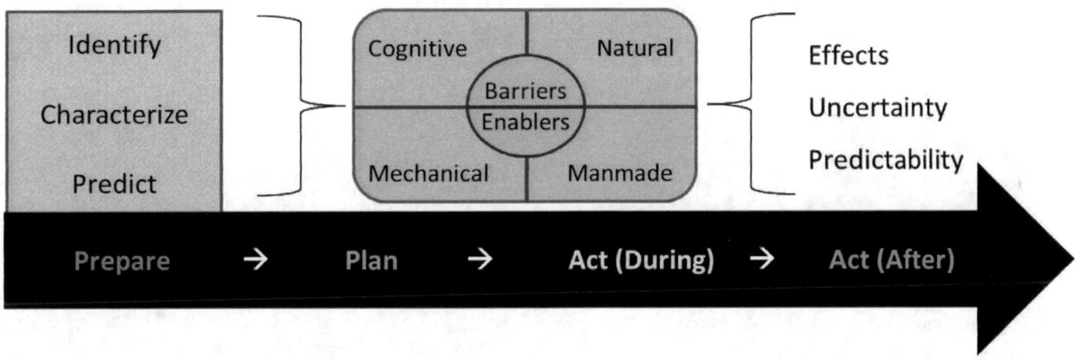

Figure 3. The PREPARE stage is the most important one in the operational success framework; do not try to develop a solution until you understand the problem.

Enablers are aspects of our efforts that enhance chances for operational success often in direct response to identified barriers. Enablers of success scrutinized during the PREPARE step include, but are not limited to:

(1) mirror the four families of barriers (i.e., cognitive, mechanical, natural, and manmade);
(2) a clear definition of success (or failure which is simply "1 minus success");
(3) requirements that are traceable to the definition of success; and
(4) all enabling features are done well or such that they increase likelihood of operational success (e.g., design and manufacturing standards).

Enablers are quality measures, modeling tools, and warning systems applied or developed during the PREPARE

It should be noted that the PREPARE and PLAN phases are very much related; there is a seamless transition between the two. The PREPARE stage is more about data collection and the PLAN phase is more about data reduction and data aggregation. The PREPARE and PLAN phases are difficult to separate in reality.

Cognitive Biases

Many of the cognitive barriers to operational success addressed during PREPARE activities depend on the basic way in which we make decisions on our own and as teams. As a result, it is worthy of attention before proceeding to the next steps of the operational success framework to examine cognitive biases that can lead to poor decisionmaking. This is also true in later stages of the operational success framework when barriers to success start acting on our assets; decisionmaking is even more important when under time and operational pressures (i.e., in the "heat of battle").

My observation is that there are four families of cognitive biases that are neatly divided into two analytical and two emotional, as depicted in the upper panel of the figure to the right. These four families were developed and validated by examining over 60 individual cognitive biases. [4]

The lower panel of the figure below provides a review of the families of cognitive biases and five ways to protect against these intellectual parasites.

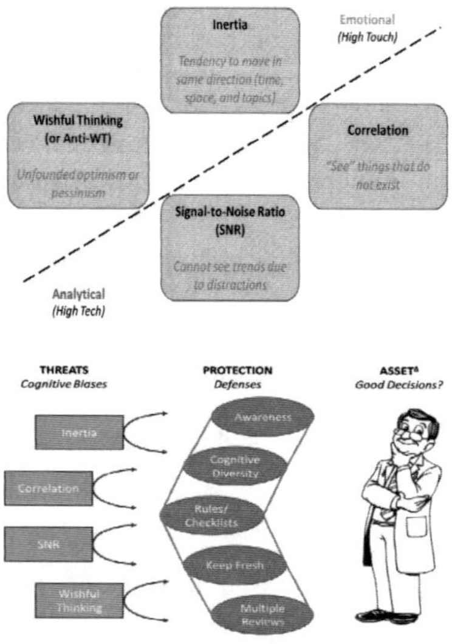

Any resemblance of the decision maker in the figure to the author is purely coincidental.

Figure 4. Identifying and combating cognitive biases is critical to the process

The defenses range from simple advice such as awareness and keeping fresh, all the way to specific calls for including more people (i.e., embrace cognitive diversity) and multiple reviews or even using documented rules and checklists.

PLAN and Beyond

The next three phases provide a tiered response to barriers to operational success that depend on a complete, coherent, and compelling representation of the factors identified in the PREPARE stage.

PLAN: Before barriers to success (i.e., threats) become realized as risks, there are opportunities to increase operational success by combating them by:

Dissuade/Defend: create your asset to be able to absorb the effects of the barriers to success (e.g., have redundant subsystems/components for your system). Do not try to avoid the threat, just build your system to deal with it. We dissuade with defense. [5]

Deny: change the operations of the system to avoid the threat (e.g., move your satellite to a higher altitude to avoid atomic oxygen prevalent below 650km that erodes exposed polymers) or even move a subsystem to a different location on a space platform to avoid a threat (e.g., move a processor for attitude and control system to behind a battery pack to use battery's mass as expedient shielding from potentially disruptive high energy protons).

Deter: This dimension usually implies the involvement of a human who you can convince to not mobilize their threat against you if they feel that they will not be able to achieve their desired results because of your preparation and planning; deter by fear of reprisal and response (i.e., offense). [6] In a military sense, the effectiveness of deterrence is based on perceived response being (1) certain, (2) timely, and (3) overwhelming. [7]

ACT (During): There are three key activities that encompass means to respond to barriers to operational success that are realized (i.e., threats become risks). These opportunities to ensure operational success at this stage include:

Interdict: Once the potential barrier to success has been put into play (e.g., counterfeit microprocessors found in existing weapon systems) you can actively try to thwart the effect of the threat. In a classical force protection situation, this includes your security forces attempting to repel or kill individuals that are attacking your site. Just mobilizing forces (e.g., cars approaching with sirens wailing or releasing your dog into the backyard after the motion sensor has triggered) may be sufficient to equalize the barrier to your operational success. Consideration of the speed of interdiction relative to the speed of threat execution (i.e., "response physics") is paramount; the faster the threat executes relative to your response options, the more important warning is to the overall calculus of efficient operational success.

Suppress/Mitigate: However, you may only be able to minimize the effects of the event (e.g., shielding does not prevent anti-tank rounds from being fired but it does reduce their effectiveness). The result from this option may be instantaneous or take hours to days to act. Suppress and mitigate are included in the same activity as they have similar effects but are traditionally used in specific domains (e.g., mitigation for orbital debris and suppression for firefighting).

The enablers to operational success (i.e., countermeasures to barriers or threats) may transition from mitigation/suppression into remediation-type activities once the barrier or threat has executed and is not being continually fueled or powered in any way (e.g., once burglars have fled your home you have transitioned into ACT (After) activities such as remediation, recovery, and reconstitution).

ACT (After): Again, the three terms for potential ACT (After) activities may not represent three distinct responses; it is more about terminology that spans many domains or scenarios comprehensively.

Remediate: When you prevent the deployed threat from affecting your mission by acting on it (i.e., the threat) is considered remediation. Remediation activities are instituted which may include encapsulation, prevention, or removal. This is any activity to alleviate the effects from an already executed barrier to success (i.e., a threat); debris remediation via active debris removal is reacting to the abandonment of massive spacecraft in orbit by deorbiting them.

Recover: If the barrier to success (e.g., a burglar broke into your home by breaking a window and stealing your refrigerator) has made you susceptible to future incursions or unable to safely operate then making your home secure again is recovery (e.g., replacing your window).

Reconstitution: In this last analogy, replacing your refrigerator is considered reconstitution (i.e., you have put your asset back to its original state). For orbital concerns this issue is much more problematic. The reconstitution of a constellation of satellites may take a long time due to launch delays and lack of inventory of satellites to replace destroyed ones. Alternatively, on-orbit spares provide a potentially responsive, though possibly expensive, means to reconstitute. Lastly, you may just reconstitute the capability in a consistent way but not exactly. For example, a ground network of warning sensors may be mobilized to replace a previous space-based capability or communications that previously used dedicated military satellites might be transferred to a commercial satellite network.

Deception as a Barrier to Operational Success

In this world that seems to be getting more technical every day, a fairly low tech technique, deception, requires continued attention. This is especially true as many of our adversaries are refining their deception skills while generally in the U.S. we do not actually consider it as a key contributor to operational success.

Deception is the intent to purposely distort one's perception of reality (and possibly their beliefs). There are two main types of deception. The first is when we distort our own perception of the world through either delusion or illusion. Delusion is considered an idiosyncratic belief or impression that is firmly maintained despite being contradicted by what is generally true. Illusion is simply something that looks or seems different from what it is. While these are interesting, barriers to operational success primarily come from the second type, the classic deception of others where one attempts to distort the perception of someone else.

The left branch in the figure to the right contains the terms and techniques relevant to this development. Misinformation is the less active (i.e., more passive) form of deception (though it still requires some action from the perpetrator): providing facts that confuse the observer as to reality. This term is equally applicable to social situations (interactions between people) and highly complex environments (such as the battlefield).

The misinformation techniques (via hiding the truth) often used are masking, repackaging, and dazzling,:

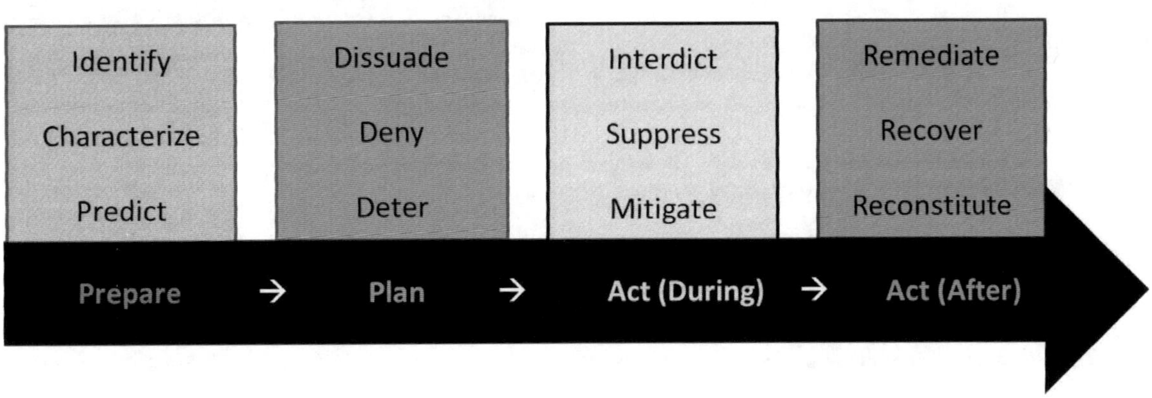

Figure 5. The full operational success framework has four phases with clear options in each.

Mask: Hide the actual object or state with something that is not real. The chameleon is expert at this mode as they change their color to blend in with their surroundings. ***Blends an old pattern with the background.***

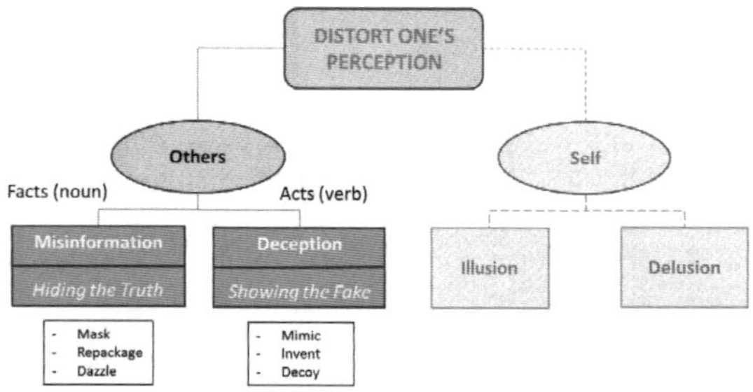

Figure 6. The distorting of our perception by others provides potential means by adversaries to inde gthrae delef to burr acnhachn coef s thfoer fiogpuerrea.t ional success, seen

Repackage: When the actual object is built to look like its surroundings so its presence is not sensed is repackaging. For example, a stick insect looks like a dry twig to blend into the background and look unappealing to birds. Modifies an old pattern by matching another.

Dazzle: One may produce some obscurant to confuse an observer as to their exact location or state. (This is often used when both masking and repackaging fail). A biological dazzling example is when an octopus releases an ink cloud to provide an opportunity for escape. ***Blurs an old pattern.***

On the deception (i.e., acts) side, there are three families of techniques that collectively "show the fake" [9,10]:

Mimic: Act like the situation is different than it really is (e.g., putting high security on one location implying important people or things are present). ***Imitates an old pattern.***

Invent: Create a new appearance, do not just try to look like something else. For example, one moth actually has the image of an eye on the end of its long tail so birds attack the easily detachable tail, thinking it is the eye of a large insect. The moth may lose its tail but it has bought time so that it can then go hide. ***Creates a new pattern.***

Decoy: Create a distraction that lures an attacker or observer away. For example, a mother bird will pretend to be crippled and flap around uselessly, as it wanders away from her nest. The attacker follows the flailing bird so is in essence lured away from the nest. ***Provides an alternative pattern.***

Monitoring and Warning

During the linear operational success timeline there are two overarching, and inter-related, activities that provide valuable contributions: (1) success monitoring/predicting & risk/response interactions; and (2) interplay between warning systems & "response physics."

First, there is a need to continually monitor and predict operational success as a measure of effectiveness of the team's/industry's efforts (from PLAN through ACT phases). This is facilitated by an ongoing set of interactions which include communications (e.g., sharing information on precursors and indicators of barriers to operational success becoming realized and messaging to adversaries and stakeholders about countermeasures in place) and transfer of assets from one activity to the next (i.e., resources that transition from planning to interdiction or remediation). In the mission assurance capacity this is represented by a series of reviews and audits. Risk communications is an art of its own, with the U.S. Environmental Protection Agency (EPA) having developed best practices that can be leveraged. [11]

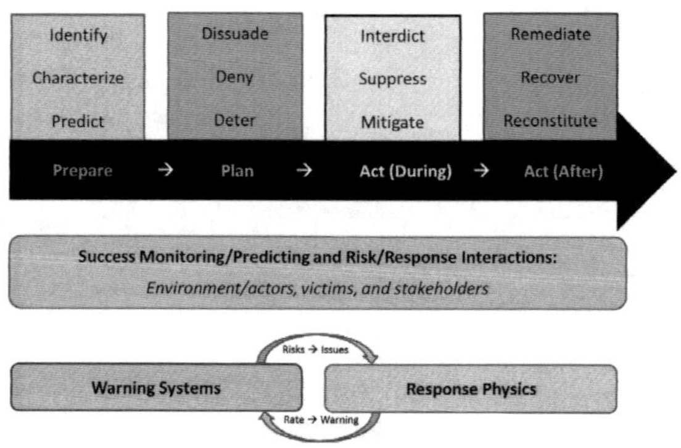

Figure 7. The operational success framework leverages success monitoring & predicting and risk & response interactions plus the interplay between warning systems and response

Second, the relationship between warning systems and response physics is important to consider. The rate at which barriers/enablers to success act and react relative to each other (e.g., such as how quickly must computer networks be "protected" after a malware intrusion is detected to remain viable) will drive the relative investments between PREPARE, PLAN, and ACT phases.

This interplay is important to understand and track. Warning systems defined and established during the PREPARE and PLAN phases may improve the responses to enhance operational success once the risk has been manifested (e.g., ACT (During and After)) by providing information on the existence, location, movement, and extent of the risk.

The ability to leverage warning depends on "response physics": how quickly do cognitive, natural, mechanical, and manmade threats act relative to warning system capabilities and response options (i.e., countermeasures or enablers you have available). For example, once a forest fire has reached a certain intensity and size, only certain countermeasures will be effective or efficient. Proper warning will not only provide insight into where a fire is located but what sort of response is required (e.g., do not want to send a truck with four firefighters and shovels to fight a raging 200 acre fire.)

Some catastrophes are easily handled if caught early – forest fires and anthrax. Other events, exemplified by influenza, are more cost-effective to respond to after they have been realized with simple response options (e.g., flu shots and hand washing) rather than costly warning capabilities.

Historically, orbital debris risk management generally has focused more on enhanced response than enhanced warning. However, current active collision avoidance warnings/maneuvers and the proposed massive collision warning monitoring activity (proposed later in this paper) lean more towards the warning end of the spectrum.

The operational success framework has similar characteristics to the SCRUM or Agile software product development methodology to include cross-functional teams working in overlapping phases to enhance flexibility and effectiveness of final solution with a focus on monitoring and communications with stakeholders to produce the desired results efficiently and effectively. [12]

3. OPERATIONAL SUCCESS FRAMEWORK APPLIED TO ORBITAL DEBRIS

Now that we have established a new framework for ensuring operational success, let's apply it to the orbital debris problem.

PREPARATION started in the 1970's as Don Kessler and Burton Cour-Palais first identified the fact that Delta second stages were exploding in space. Debris can be considered an unintentional manmade barrier to operational success.

As shown in the table below, two separate but related success domains are considered: space flight safety and space sustainability. The identification, characterization, and prediction activities have been gaining momentum ever since orbital debris was "discovered" in the 1970's. However, predictability is still problematic. There are sound statistical algorithms for calculating average collision probability (for an individual object) and collision rates (between sets of objects), however, there have not been sufficient events to be able to accurately describe the distribution of collision events about the likely mean values.

More specifically, on average, there is a one in five chance each year that any two cataloged objects will collide but we do not know which ones. This ambiguity makes it impossible to discern the consequence of such a collision and assuming an "average" event is even more distressing and limiting as we have had only one catastrophic collision in space between two cataloged objects; what is average?

Table 1. Classes of problems to handle with debris remediation have significant overlap.

DOMAIN	RISKS	SOURCE
Space Flight Safety *- Operational spacecraft disrupted or destroyed*	Lethal nontrackable (LNT) debris	- **Fragmentation events** - Surface sloughing - Operational releases
	Cataloged objects	- **Fragmentation events** - Abandoned payloads and rocket bodies - Mission-related debris
Space Sustainability *- Long-term viability of satellite operations in Earth orbit*	Cascading sequence of fragmentation events	**Vast numbers of large debris** (> 1/1000 of larger object in LEO, assumes 10km/s impact velocity)

In our analysis, we propose that it is best to examine the collision events that are the most impactful (i.e., have the greatest consequence) and then try to determine their likelihood more accurately (i.e., enhance predictability) to better combat deterioration of our operational success domains. The table below provides a high level examination of the barriers to success for both space flight safety and space sustainability.

Debris mitigation guidelines developed in the 1990's and refined/applied since represent the PREPARE and PLAN phases. However, when we turn our efforts to dealing with derelict hardware already left in orbit (i.e., debris remediation) it involves the ACT phases. ACT (During) is for intact derelicts removed via active debris removal while ACT (After) represents options available to manage the collision risk once the derelict has fragmented. As stated in the operational success framework, as we move further along the continuum, the options are usually less appealing (i.e., expensive, inconvenient, etc.) and less effective. This is indeed true in the case of orbital debris as the concept of sweeping up hundreds or thousands of lethal nontrackable (LNT) fragments has proven very difficult to even conceptualize an effective method. Sweeping up many LNT fragments has been shown to be much more difficult than dealing with the intact derelict such as through active debris removal. Indeed, the prevention of massive-on-massive collisions is the most important single type of event to prevent.

The common theme up to this point in dealing with preventing future collisions as an enabler to space sustainability (i.e., prevent a cascading of collisions in Earth orbit) has been to mainly consider only active debris removal. However, further investigation has identified the potential for a variety of other remediation approaches to include:

- Just-in-Time Collision Avoidance (JCA): nudge one of two derelicts out of harm's way just before (hours to days) a collision is predicted to occur providing a potentially cost-effective, responsive means to prevent massive derelicts from colliding [13,14];[3]

- Salvage: collect derelict and repurpose them as an alternative spacecraft or as spare parts [15,16];

- Consume: convert derelict hardware into fuel by a device that in turn powers the move of the device to the next derelict object (and so on) [17]; and

- Disintegrate: apply some external chemical/physical catalyst to derelict hardware to disintegrate it quickly into sizes smaller than will pose a hazard to other spacecraft. This new concept was inspired by the intellectual exercise provided by the operational success framework and dialogue with a diverse set of people.[4]

These remediation options along with debris mitigation and active collision avoidance, are included in Figure 9 to summarize all debris management options to enhance the operational success for both space flight safety and space sustainability. The lower half of the figure represents debris remediation options which are most accurately depicted as derelict collision prevention. The new options to eliminate derelicts in situ clearly are just concepts at this point but as a community we must continue

[3] JCA will require enhanced positional awareness of derelict objects than we do now and to establish go/no-go criteria similar as to currently exercised with Conjunction Data messages for operational go/no-go criteria similar as to currently exercised.

[4] Neil Arora, IAI, suggested making the hardware biodegradable just like trash bags, as the catalyst for this new approach.

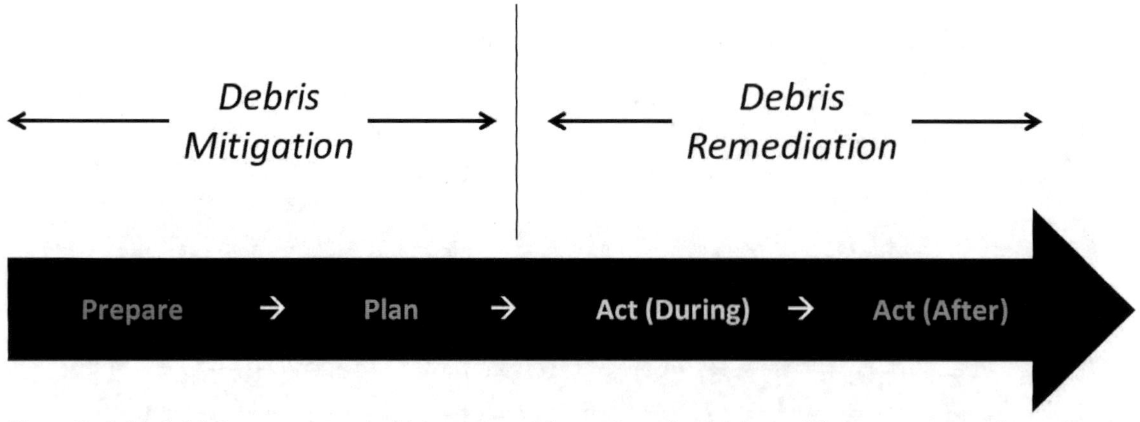

Figure 8. Orbital debris management is fairly mature with a variety of activities in debris mitigation and remediation spanning the entire operational success continuum.

to examine new ways to improve the likelihood that we can assure space flight safety and space sustainability.

Risk Transfer

It should be noted that while we are striving to enhance operational success in space (i.e., space flight safety and space sustainability) some of our options may actually transfer the hazard to another domain. For example, ADR will permanently eliminate the potential of a derelict from colliding with another object in space but its reentry poses a hazard to assets on the ground. In essence, by enhancing the operational success of space systems we may transfer the barrier to operational success (i.e., threat) to ground assets.

4. CONCLUSIONS

In closing, this paper introduced three concepts: (1) examine debris management options through development of the operational success framework; and (2) consider debris remediation as derelict collision prevention, not just ADR, and (3) contribute new potential remediation options.

The operational success framework is linear, reinforcing, action-oriented, and traceable to the key objectives (e.g., space flight safety and space sustainability). Looking at the area of debris remediation in a more inclusive way reinforced the concept of JCA that has the potential of assuring space flight safety and space sustainability as an adjunct to ADR. This observation is not new but may be more coherent from the basis of our operational success framework. In addition, some new ideas have been added to the overall suite of derelict collision prevention options as ACT (During) activities that should be examined more closely.

Similarly, Swiss Re (a worldwide reinsurer) summarizes approaches to maintain safety in any domain as (1) avoid, (2) manage, and (3) embrace (i.e., operate-through). [18] Further, this treatise summarizes means to create resiliency through the following activities:

- Learn better and more;
- Refine ability to judge difficulty of a task;
- Leverage experience; and
- Garner relevant knowledge to apply to extreme situations.

This source and perspective provide a complimentary call to arms to the operational success framework.

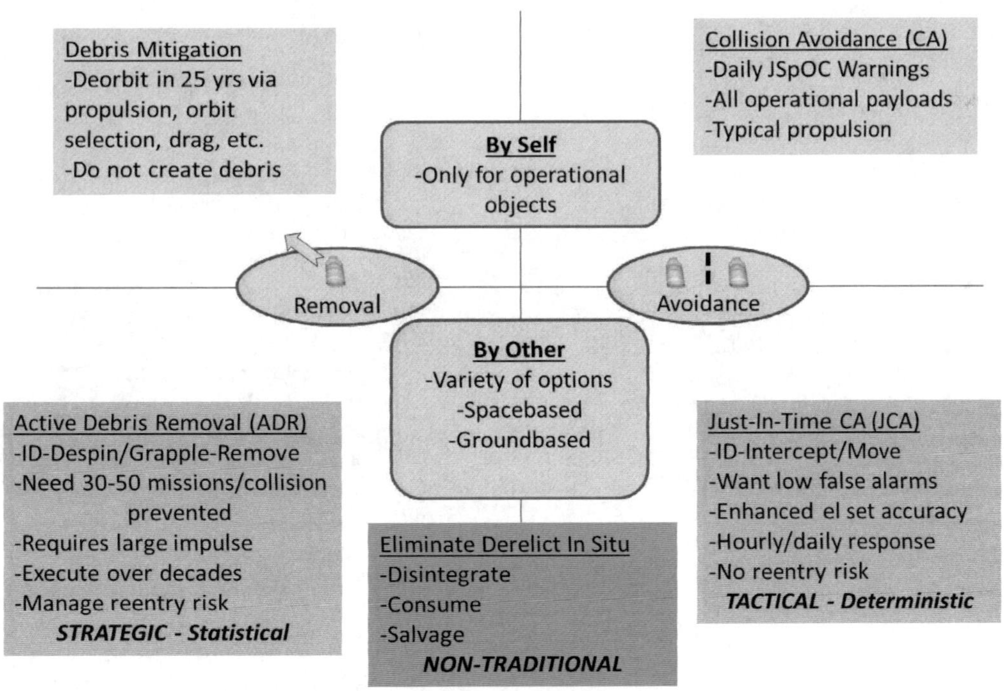

Figure 9. Derelict collision prevention as a remediation framework contains many more options than just active debris removal. [Note: If collisions of massive derelicts are not prevented then the sweeping up of thousands of mm-to-cm fragments may have to be considered as yet another "last ditch" debris remediation option, a very difficult and expensive recovery option.]

Lastly, key linkages within the operational success process highlight the benefits and limitations of threat warning, situational awareness, and asset reconstitution. The difference between JCA and ADR is largely a realization of (1) our inability to predict the most likely "next collision" between all cataloged objects and (2) the hypothesis that the most impactful collision events may actually be more likely and more predictable (i.e., derelicts in clusters).

Further Research

As a result, a major area of research being pursued is to develop a more deterministic and predictive model for cluster collisions.[5] This can only be developed in tandem with a Massive Collision Monitoring Activity (MCMA) where large groups of massive objects in very similar orbits have their encounter dynamics with each other monitored constantly. This experimental tasking will serve to support/refute hypotheses about these potentially more likely (i.e., greater probability than predicted by the kinetic theory of gases-inspired algorithm) and more consequential (i.e., many more fragments produced) events. The MCMA results will be detailed in an upcoming paper. It should be noted that another benefit of MCMA will be the opportunity to establish and refine go/no-go thresholds for JCA operations.

5. REFERENCES

1. Liou, J.-C., "An Active Debris Removal Parametric Study for LEO Environmental Remediation," Advances in Space Research, 47 (2011) 1865-1876.

2. Hubbard, Douglas. The Failure of Risk Management: Why It's Broken and How to Fix It. John Wiley and Sons. 2009.

3. Dorfman, Mark. Introduction to Risk Management and Insurance. 9th Ed. Englewood Cliffs, NJ. Prentice Hall. ISBN 0-13-224227-3. 2007.

4. McKnight, D.,Cognovation Dialogues, August 2014.

5. Fox, Scott, "Deterring and Dissuading in Space: A Systems Approach," USAWC, 2008.

6. MG Crosier, STRATCOM, Dec 2015: "Juice is not worth the squeeze."

7. Doug Leverro, OSD (Space Policy), Dec 2015.

[5] This has been partially motivated by awareness during the PREPARE phase of the operational success framework of the current inability to reliably predict future collisions.

8. Caddell, J. W., "Deception 101 – Primer on Deception," Strategic Studies Institute, December 2004.

9. Bell, J.B. and Whaley, B. Cheating and Deception. Transactions Publishers, NJ, 2010.

10. Caddell, J. W., "Deception 101 – Primer on Deception," Strategic Studies Institute, December 2004.

11. McKnight, D. and Di Pentino, F., "Massive Collisions in LEO – A Catalyst to Initiate ADR," IAC-14-A6,2.1x21048, IAC Toronto, CN, September 2014.

12. McKnight, D., "Engineering and Operational Issues Related to Just-In-Time Collision Avoidance (JCA), 6th European Conference For Aeronautics and Space Sciences (Eucass), Krokaw, Poland, June 2015.

13. Ramihalli, K., Mitchell, D., and Taft, B., "Autonomous Space Processor for Orbital Debris (ASPOD)," University of Arizona, Under the Smithsonian/NASA Astrophysics Data System, 2011.

14. DARPA Phoenix Program established to examine the efficacy of enabling GEO robotics servicing and asset life extension, while developing new satellite architectures to reduce the cost of space-based systems, http://www.darpa.mil/program/phoenix.

15. "Junk-eating rocket engine could clear space debris," MIT Technology Review, 3DEC2015.

16. The EPA's Seven Cardinal Rules of Risk Communication Adapted by Dr. Vincent T. Covello from the 1988 EPA Seven Cardinal Rules of Risk. Accept and involve the public/other consumers as legitimate partners (e.g. stakeholders).1. Plan carefully and evaluate your efforts with a focus on your strengths, weaknesses, opportunities, and threats (SWOT). 2. Listen to the stakeholders' specific concerns.3. Be honest, frank, and open. 4. Coordinate and collaborate with other credible sources. 5. Meet the needs of the media. 6. Speak clearly and with compassion. http://www.orau.gov/cdcynergy/erc/Content/activeinformation/resources/EPA_Seven_Cardinal_Rules.pdf

17. Brian Cheng, IAI, identified this similarity between the operational success framework and Agile/SCRUM techniques during an early reading of the document. A coherent explanation of SCRUM and Agile can be found at http://scrummethodology.com/.

18. Amalberti, R. and Vincent, C., "Continuum of Safety Models," Swiss Re Report, 19 Sept 2014.

FAST AND FLEXIBLE SPACE DEBRIS RISK ASSESSMENT FOR SATELLITES

Max Gulde[1*], Scott Kempf[1], Frank Schäfer[1]

[1] *Fraunhofer Ernst-Mach-Institut, Department of Systems Solutions, Eckerstr. 4, 79104 Freiburg, Germany*
**Email: gulde@emi.fhg.de, Phone: +49 761 2714 414*

ABSTRACT

The rapid increase in uncontrolled high velocity objects orbiting earth poses a continuously growing failure potential for ongoing and future space missions [1-2]. Despite increasing efforts to mitigate collisional risks, e.g. via active debris removal, potential solutions are currently only emerging at the horizon, emphasizing the importance of space debris related risk assessment [3].

However, traditional impact shielding requirements for manned mission cannot directly be transferred to unmanned missions. Instead of restricting the analysis to the outer hull, it is necessary to directly assess the failure risk on the level of individual, internal components.

Here, we outline a newly developed deterministic methodology to enable the evaluation of system level effects on the spacecraft [4]. In particular, we employ a computational model based on available probabilistic debris models and component-specific ballistic limit equations [5]. The methodology has been implemented into the easily accessible software tool PIRAT (Particle Impact Risk and Vulnerability Assessment Tool) [6].

1. INTRODUCTION

The continuing accumulation of space debris in heavily used orbits such as the LEO presents an increasing threat for civil and military satellite infrastructure. Until effective debris removal strategies are designed and set into motion, space missions will have to rely on a combination of collision avoidance maneuvers for large, traceable objects and damage mitigation by appropriate shielding of the spacecraft for smaller, untraceable debris [3].

Several computational tools, including NASA's BUMPER-II and ESA's ESABASE2/Debris have been developed to assess the risk of mission failure caused by impacts of high-velocity micrometeoroids and manmade space debris [7-8]. Traditionally, the assessment is based on the probability of no penetration (PNP) of the outer pressure wall of the spacecraft [6, 8]. While this approach is well suited for the manned space missions it has been originally developed for, its application to unmanned satellites results in often unnecessary steep requirements stemming from unjustified high failure probabilities. Furthermore, for the majority of unmanned missions, a thorough risk assessment is out of the question due to the connected prohibitive costs.

Here, we present a newly developed software tool, PIRAT, which has been specifically designed for the risk assessment of unmanned missions. It provides an intuitive, reliable, fast, thorough, and resource-effective analysis of a spacecraft for a given set of mission parameters. In particular, we improve on the traditional risk assessment methods by looking beyond structural penetration and additionally investigate impact effects at the level of internal components.

The approach is in large parts based on the Schäfer-Ryan-Lambert (SRL) ballistic limit equations (BLE), which is capable of making damage predictions for objects behind multi-plate structure walls such as Whipple shields.

In combination with knowledge about the functional architecture of the spacecraft, this allows for a detailed component-level, functionality-based failure analysis.

Furthermore, PIRAT is continually developed, maintained, and tested against available models following the guidelines of the Inter-Agency Space Debris Coordination Committee (IADC) [12]. The performance-oriented approach of PIRAT delivers results within minutes, making it an ideal tool for in- session support, e.g. during the early design phase of European and American space missions [9].

2. METHODOLOGY

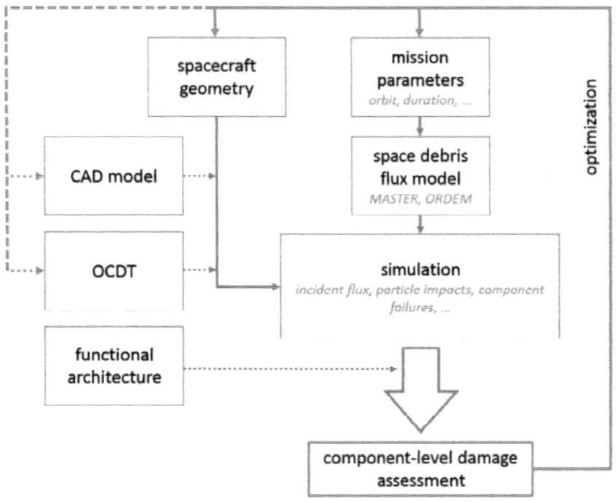

Figure 1: PIRAT operation scheme. 3D geometry and space debris flux data (green) for given mission parameters provided to compute component-level damage on spacecraft. Output (orange) fed back into refinement loop for optimization.

The PIRAT methodology is schematically depicted in Fig. 1 (for a detailed description, see [6, 9]): it requires a 3D satellite geometry as well as space debris flux data for a given set of mission parameters. PIRAT combines this data to compute and display impact-related information, including the total impact flux, the cratering area, as well as the failure probability for every component integrated over the mission duration. The results can be used, e.g., in conjunction with knowledge about the functional architecture of the vessel, to systematically modify the satellite geometry as well as the mission parameters. This optimization loop is repeated until the overall mission supports a sufficiently small failure probability [9].

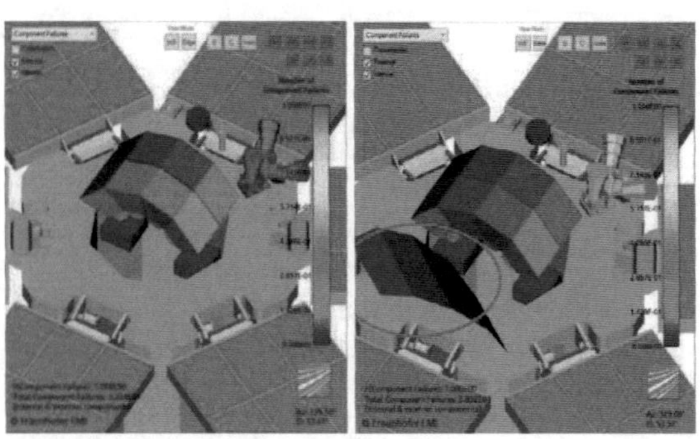

Figure 2: Exemplary component vulnerability reduction by addition of a sun shield (red circle) to the spacecraft geometry. Image adapted from [9].

Catastrophic failure is modelled via energy coupling coefficients of each component and is based on a predefined energy-to-mass ratio. Fig. 2 depicts an exemplary optimization of the failure probability for several satellite components via a slight modification of the spacecraft geometry. The graphical user interface (GUI) is illustrated in Fig. 3.

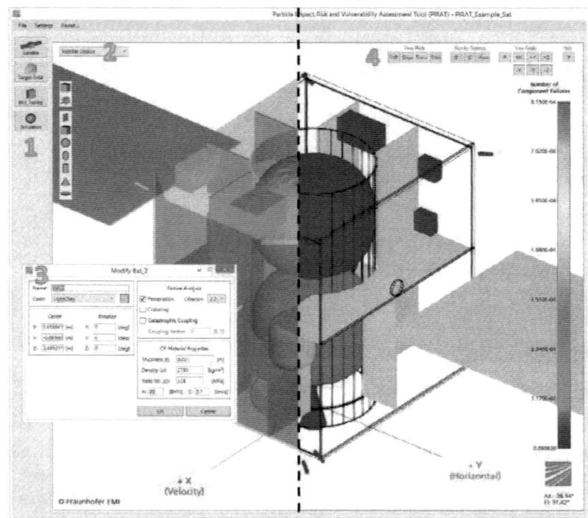

Figure 3: PIRAT GUI in visualization mode. Left half: geometry and material editing view. Right half: results visualization. Numbers: (1) Main menu, (2) viewing mode selection, (3) component material editor, (4) viewport options.

PIRAT provides an interface to ESA's MASTER and NASA's ORDEM space debris models up to the most recent versions, allowing to assess American as well as European missions [10-11].

Internally, PIRAT uses a deterministic approach based on multi-wall BLEs, in particular the SRL BLE [5], to account for failures of internal as well as external components. The SRL is a three-wall equation, which inherently considers failure mechanisms of different component types, including pressure vessels, electronic boxes, cable harnesses, and batteries. It supports opaque impacts, different shielding materials as well as configurations and is founded on a large and continuously growing experimental basis.

The SRL requires a set of material as well as geometrical parameters. Geometrical information are directly derived from the 3D satellite model, which can be constructed within PIRAT using geometric primitives. However, in order to provide greater design flexibility and tap the full potential of current computer-aided design (CAD) software, PIRAT supports direct model import via the Standard for the Exchange of Product model data (STEP) format. Hence, previously constructed models can be directly evaluated within the computational framework of PIRAT.

Material properties, such as physical density or hardness, can also be individually entered in PIRAT for every component and structure. Additionally, PIRAT provides an interface to existing models and parameter sets via ESA's concurrent design facility (CDF). This interface allows to import necessary design as well as mission characteristics via ESA's new mission data sharing system, Open Concurrent Design Tool (OCDT).

3. CONCLUSION

Until effective measures are taken to mitigate collisional risks, impact shielding against untraceable space debris will stay indispensable for spacecrafts. While traditional risk assessment is well suited for manned missions, unmanned missions exhibit inherently different requirements. However, prohibitive costs and limited time do often not allow for thorough risk assessment. Here, we provide a flexible, performance-oriented methodology particularly designed for the risk assessment of unmanned missions in the early design phase. With support for CAD geometry data and different space debris models, interfaces to ESA's CDF as well as component level threat evaluation, PIRAT provides the opportunity to quickly perform optimizations on the geometrical model and mission layout. We believe that our methodology will substantially facilitate early mission design by providing a fast, easy-to-use, and resource effective way for satellite optimization.

4. REFERENCES

1. Liou, J-C., and Nicholas L. Johnson. "Instability of the present LEO satellite populations." *Advances in Space Research* 41.7 (2008): 1046-1053.
2. Liou, J. C. "USA Space Debris Environment, Operations, and Measurement Updates." (2015).
3. Shan, Minghe, Jian Guo, and Eberhard Gill. "Review and comparison of active space debris capturing and removal methods." *Progress in Aerospace Sciences* 80 (2016): 18-32.
4. Welty, N. et al. "Computational methodology to predict satellite system-level effects from impacts of untrackable space debris." *Acta Astronautica* 88 (2013): 35-43.
5. Schäfer, F. K. et al. "Ballistic limit equation for equipment placed behind satellite structure walls." *International Journal of Impact Engineering* 35.12 (2008): 1784-1791.
6. Lang, T., et al. "Short and long term efficiencies of debris risk reduction measures: Application to a European LEO mission." *Advances in Space Research* 55.1 (2015): 282-296.
7. Bjorkman M. D., E. L. Christiansen, D. M. Lear, "Bumper 3 Software User Manual", NASA TM-2014-218559, (2014).
8. Grassi, L. et al. "Satellite vulnerability to space debris–an improved 3D risk assessment methodology." *Acta Astronautica* 99 (2014): 283- 291.
9. Kempf, S. et al. "Simplified S/C vulnerability assessment at component level in early design phase at ESA's CDF." *66th International Astronautical Congress*, Jerusalem, Israel (2015).
10. Flegel, S. et al. "The MASTER-2009 space debris environment model." *Proc. 5th European Conference on Space Debris* (ESA SP-672, 2009).
11. Krisko, P. H. "The New NASA Orbital Debris Engineering Model ORDEM 3.0." In *Proceedings of the AIAA/AAS Astrodynamics Specialist Conference*. 2014.
12. IADC space debris mitigation guidelines. Inter- Agency Space Debris Coordination Committee, 2002.

PROCEEDINGS OF THE 8TH INTERNATIONAL SPACE SAFETY CONFERENCE

SESSION 24 MATERIALS

INVESTIGATION OF TI-6AL-4V ALLOY RESPONSE TO ATMOSPHERIC RE-ENTRY EXPOSURE

Jessica L. Buckner[(1)], Stephen W. Stafford[(1)], Darren M. Cone[(1)], John D. Olivas[(1)]

[(1)] *Center for Advancement of Space Safety and Mission Assurance Research, The University of Texas at El Paso, 500 W. University Ave, El Paso, TX 79968*

ABSTRACT

Ti-6Al-4V is a widely used aerospace alloy for it's high strength-to-weight ratio and high operating temperature properties. Despite widespread use, titanium and its alloys have been shown to ignite in oxygen and nitrogen rich test streams. The reactivity of titanium is attributed to the high solubility for oxygen that increases with temperature, accelerating the oxidation rate and resulting in a combustion reaction. When introduced to the monatomic oxygen rich and high enthalpy re- entry environment, Ti-6Al-4V X-link components from the space shuttle *Columbia* exhibited accelerated oxidation and combustion behavior. Ti-6Al-4V metal plates tested in the simulated re-entry environment of an arc-jet facility have also exhibited ignition and combustion responses. A forensic material characterization study of the X-links and arc-jet samples is necessary to gain a better understanding of the material response in the dynamic and extreme environment of re-entry. Data from X-link components is compared to the Ti-6Al-4V samples previously tested in the simulated re-entry arc-jet facility conditions. The observation of similar microstructural and solidification features in conjunction with similar oxide compound detection suggest that arc-jet testing is representative of the re-entry environment effect on titanium and its alloys. A thorough materials characterization of these materials will yield insight into the discrimination between the oxidation, ignition, and combustion response of titanium alloys. Widespread use of titanium alloys in space vehicles warrants a thorough characterization of their potential failure modes to help ensure the safe and reliable operation of future spaceflight vehicles.

1. INTRODUCTION

Ti-6Al-4V is a common aerospace alloy utilized for its high strength-to-weight ratio, corrosion resistance, and high operating temperature capabilities [1]. Yet, research has shown that practically every commercial titanium alloy can ignite [2]. Despite a relatively mature understanding of combustion involving gaseous and liquid fuels, knowledge regarding combustion of bulk metals is not as extensive. This is likely due to the many compounding factors affecting oxidation and combustion that complicate analysis, to include purity, gas composition, pressure, velocity past the surface, state of subdivision, and previous processing [3]. Titanium is recognized as having a high heat of combustion, a high flame temperature, and a high solubility for oxygen [4-7]. The high heat of combustion of titanium is related to its relatively modest thermal conduction properties when compared to other structural aerospace alloys.

Conduction of heat within a material as it affects surface temperature is an important factor in ignition [8]. With a lower thermal conduction and thus lower thermal inertia in the material, the surface temperature will rise more quickly and lead to the creation of a "hot spot", in which a localized ignition reaction can occur. The reactivity of titanium is also attributed to the high solubility for oxygen within titanium [9]. At temperatures above the β transus (approximately 1950°F), the solubility of oxygen in titanium and the oxidation rate increase. Similarly, an increase in pressure has proven to lower the critical temperature necessary for ignition [10]. Historical review of titanium testing at NASA, Department of Energy (DOE), and Department of Defense (DOD) facilities [7] revealed that freshly exposed titanium surfaces are known to react violently with gaseous oxygen, while oxygen deficient test streams did not result in ignition. Titanium was also found to combust vigorously in ambient pressure gaseous nitrogen, with higher pressures contributing to decreasing ignition temperatures. Ti-6Al- 4V has exhibited an exothermic oxidation reaction on fresh surfaces that leads to the formation of oxides with the evolution of a large amount of heat and ignition [2, 11-13]. At high oxidation rates, as would be expected in a re-entry environment, the heat of reaction will not dissipate quickly enough and the material will begin to burn. These studies explored varying geometries, angles of attack, and methods to achieve ignition, illustrating the complexity of compounding factors that must be considered.

On February 1st, 2003, NASA's Space Shuttle *Columbia* endured a catastrophic failure during re-entry into Earth's atmosphere. Ultimately, all seven crew members lost their lives in this tragic event. Root cause analysis determined that a breach of the reinforced carbon-carbon (RCC) leading edge of the port wing was caused by a piece of insulating foam that detached from the external fuel tank during lift-off. This breach allowed superheated gas ingestion into the vehicle during re-entry, which ultimately degraded the substructural components and integrity of the space shuttle [14].

The X-links, a substructural component composed of Ti- 6Al-4V that connected the crew module, forward fuselage, and mid-body structures of the vehicle in the longitudinal direction, were identified as a component of interest due to the physical features observed. The X- links were secured to the shuttle structures with fasteners made of MP35N, a nickel-cobalt-chromium- molybdenum alloy trademarked by SPS technologies.

The as-received port and starboard X-link are shown in Fig. 1 as they were configured within the space vehicle. Upon recovery of the X-links after the accident, the starboard X-link exhibited thermal degradation patterns that could not be explained by re-entry alone. Relative to the port X-link, the starboard X-link exhibited more mass loss, discoloration, and is missing the aft-most fastener. The fastener heads exhibit evidence of exposure to forward-to-aft flow conditions, while the webbing of both X-links contain evidence of exposure to aft-to- forward flow conditions. Structural debris analysis suggests that initial failure occurred immediately aft of the starboard X-link from a combination of structural loads and thermal degradation, and the vehicle proceeded to separate from starboard to port side [15]. Despite similar flow patterns, the starboard X-link was recovered in the forebody debris field and the port X-link in the midbody debris field; this separation is theorized to have occurred after the combustion event.

In terms of vertical temperature, there are four atmospheric layers. In order of increasing altitude is the troposphere, stratosphere, mesosphere, and thermosphere. Modeling results published in NASA's *Columbia* Crew Survival Report indicate that breakup of *Columbia* occurred in the mesosphere, between 180-210 thousand feet altitude [15]. In the mesosphere, low temperatures prevail and molecular oxygen and nitrogen still compose 99% of the air, with virtually the same mixings as air at sea level (78% N_2, 20.9% O2) [16]. The high velocities produced by vehicle deceleration can result in the dissociation of O_2 and N_2 gases, converting the species to their atomic forms. The presence of atomic nitrogen and oxygen create a more aggressive environment in which oxidation and combustion are accelerated.

During accident reconstruction, NASA's Object Re-entry Survival Analysis Tool (ORSAT) predicted entry heating alone was not sufficient to explain the observed damage, most notably the burn through in the flange area [15]. In order to cause the observed damage, analytical predictions from ORSAT estimated a required heat rate an order of magnitude higher than the maximum peak aero heating the vehicle could have experienced. These findings led to the consideration of other possible material interactions: shock-shock interaction and metal combustion. The focus of this research is on the latter.

During *Columbia* accident reconstruction, overhead windows from the forward fuselage were observed to contain a metallic deposit. This 'char' layer was found to be largely composed of metal oxides, and an oxidized layer of Ti-6Al-4V was deposited prior to that of an aluminum alloy [17]. Simply from the standpoint of relative melting temperatures between aluminum and titanium alloys, and given the larger overall weight percentage of aluminum alloys comprising the shuttle construction, the observations from the overhead window char layer were puzzling. To determine parameters that could account for these observations, an arc-jet test facility was utilized to simulate the oxygen rich and high enthalpy re-entry environment to characterize the ignition and/or combustion response of Ti-6Al-4V [18]. Arc-jet testing represents the best ground-based simulation of a re-entry environment by providing the ability to explore the oxidation behavior of materials under extreme conditions [19]. The arc-jet is a high altitude, hypersonic wind tunnel facility that heats and accelerates O_2 and N_2 gases through a nozzle onto the surface of a sample [20]. Three different classifications of material response were observed from arc-jet testing (Fig. 2). Class I represents material passivation, Class II represents ignition and initiation of melting with no combustion, and Class III represents ignition with sustained combustion. It should be noted that ignition and combustion only occurred when there was a protuberance of the sample from the mount, thus causing the test coupon to have a higher effective ballistic number. In cases where the test coupon was mounted without protuberance, only the Class I oxidation response was observed. Because the X-links have a complicated geometry and thus a high effective ballistic number, it is likely that many separate oxidation and combustion states were occurring simultaneously.

The results of the arc-jet testing provide evidence that ignition and combustion are plausible phenomena for titanium alloys in certain re-entry conditions

A forensic materials characterization study of two separate Ti-6Al-4V components that have been exposed to re-entry conditions, both actual and simulated, will further the understanding of bulk metal combustion in the extreme environment. Qualitative and quantitative correlations will be made between the X-links and the arc-jet samples in an effort to relate the known parameters of testing to the unknown parameters of the X-links re-entry into the atmosphere. This comparison will help to determine if ground-based testing methods are sufficiently representative of the actual environment to encompass conditions supportive of metal combustion. These results will also aid in differentiating separate material reactions in an effort to identify the mechanism of Ti-6Al-4V combustion in a re-entry environment.

Given the extensive use of titanium alloys as structural components in modern aerospace vehicles, consideration for the accelerated thermal degradation due to exposure of these alloys to the re-entry environment should not be neglected. The aerospace community must be aware of a well-predicted and graceful degradation of materials being sent into space. Detailed materials characterization of the *Columbia* X-links and arc-jet samples will advance the understanding of titanium reactivity in the extreme environment of atmospheric re-entry.

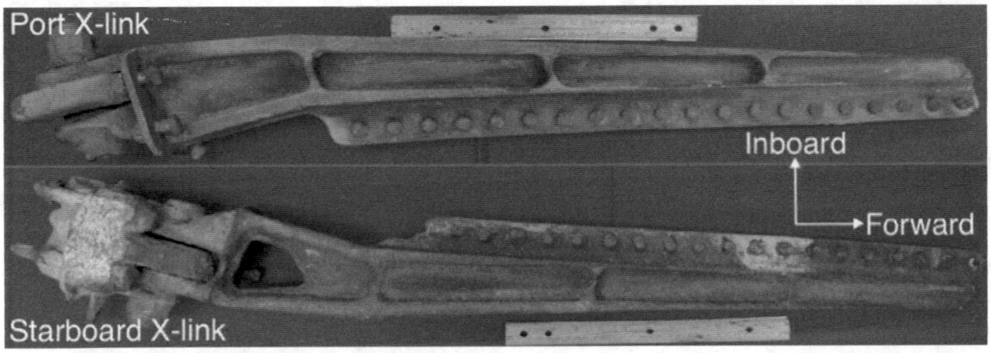

Figure 1. As-received photographic documentation of the port and starboard X-links.

Figure 2. The three different classifications of material response observed from arc-jet testing.

2. METHODS

Metallography and Light Optical Microscopy (LOM), Scanning Electron Microscopy (SEM), Energy Dispersive X-ray Spectroscopy (EDS), X-ray Diffraction (XRD), and microhardness testing were used to characterize the material. A significantly

degraded region of the starboard X-link, located on the aft side, immediately next to the aft-most burned off fastener, was chosen for analysis; two metallographic sections were taken from this region (Fig. 3). For comparison, the metallic portions of two arc-jet samples, one Class II and one Class III, were analyzed with complementary methods (Fig. 4). Dotted lines indicate secondary sections, and arrows indicate observation orientation of the metallographic samples. For ease of explanation, the X-link and arc-jet samples will be referred to as sample 1-4, as described in Tab. 1.

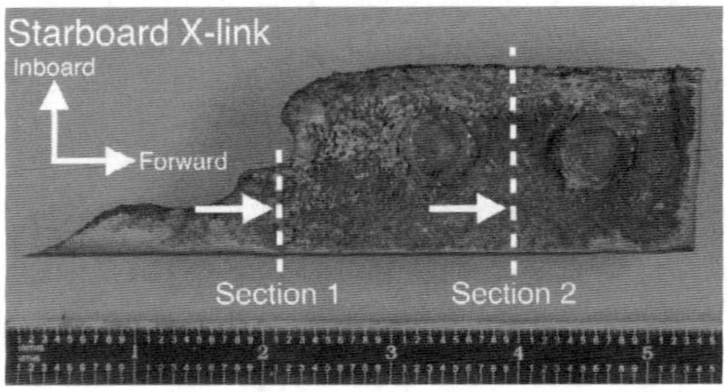

Figure 3. Starboard X-link region of interest with numbered sections.

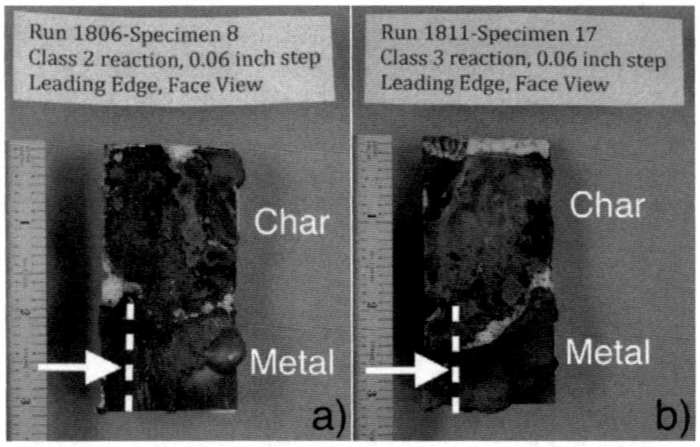

Figure 4. Class II (a) and Class III (b) arc-jet samples.

Table 1. Nomenclature of samples discussed in paper.

Sample No.	Source: X-link or Arc-jet	Additional Information
1	X-link	Section 1 (Fig. 3)
2	X-link	Section 2 (Fig. 3)
3	Arc-jet	Class II (Fig. 4a)
4	Arc-jet	Class III (Fig. 4b)

3. RESULTS

While the purpose of this work is to correlate material responses from the simulated and actual re-entry environments, it is necessary to make some distinctions between intrinsic differences that exist. The history of the X-links, to include temperature exposure and initial microstructural state, are unknown. Therefore, assumptions about the history are made based upon observations on the macro and micro scale. As the position of the X-links decreased in altitude it is likely that they experienced a variable cooling rate as it transitioned through the atmospheric levels. Ballistic heating, exothermic reactions occurring near the X-links, and other heat sources could have also influenced temperature and cooling rate. The arc-jet samples essentially experienced an air quench upon removal from the arc-jet flow stream at the end of the test. Cooling rate directly affects the microstructure. Geometrical variations between the X-links and arc-jet samples could have affected observations. The arc-jet samples originate from 1/8 inch, uniform thickness plate metal while the X- links were geometrically complex parts with varying thicknesses up to 3 inches. The thinnest, ¼ inch, sections of the X-links were examined and compared to the arc-jet samples. The X-link samples examined within this paper transitioned to a thicker geometry nearest the outboard side. The conductive heating effects from the thicker material may have caused the material nearest the outboard side to remain hot for longer periods of time, converse to the relatively uniform cooling rate of the arc- jet samples. Contamination from surrounding material systems for both arc-jet and X-link samples also exists. The X-links were surrounded by several material systems comprising the shuttle, and the arc-jet samples were fixed to a SiC ceramic plate. As illustrated, several different factors are present which could influence the characterization results.

References to the relative flow directionality observed on the X-links will be attributed to the similar characteristics observed on the arc-jet samples. Physical features on the X-links suggest that flow originated from the inboard side. The configuration of the arc-jet creates a leading and trailing edge on tested specimens. Throughout this work, inboard and leading edge features will be referred to in terms of forward facing flow, while outboard and trialing edge features will be referred to in terms of back facing flow. Discrete changes in features are observed with arc-jet samples, while more pronounced transitions are described for the X-link samples.

Microstructural Features

In order to evaluate the effects of the combustion reaction on the microstructure, it is helpful to know the initial microstructural state. The initial microstructure of the arc-jet samples, taken from untested metal plate, was a fine-grained bimodal structure of slightly elongated α in a lamellar α/β matrix (Fig. 5), a sub-β transus structure. Assigning an initial microstructural state to the X-links without any processing history was a more difficult task, for which the following method was used. To mitigate heat effects in the material, metallography was performed on the thickest cross-sections (~1 inch) on both the port and starboard X-link in the same relative location away from any extreme thermal effects (Fig. 6). The port X-link, possessing minimal thermal degradation and mass loss, contained a bimodal structure, with elongated α in a lamellar α/β matrix. The starboard X-link contained transformed β with platelike α and more clearly defined prior β grain boundaries. The port X-link suggests a sub β transus temperature exposure while the starboard X-link suggests exposure above the β transus. The initial state of the arc-jet samples is similar to the microstructure observed on the port X-link, though arc- jet has a finer grain size. The port microstructure is believed to adequately represent the original microstructural state of the X-links, and the starboard section appears to be moderately heat affected. The microstructures in both port and starboard X-link were uniform and no gradients were observed.

Forward facing flow microstructural features were similar for all samples (Fig. 7), consisting of martensitic α prime within large prior β grains. Martensitic α formation occurs with a quick cooling rate, as in quenching.

Grain sizes were comparable though not identical in all samples. Smaller grains were observed in X-link sample 2 than X-link sample 1, possibly due to the closer proximity of sample 1 to more aggravated flow, evidenced by the burned off bolt immediately aft of the parent piece. X-link section 2 also exhibited martensitic and serrated α; serrated α is typically observed in weld zones due to overheating [21] and when there is a rapid cool from the β phase field [22]. Oxygen enriched, α stabilized phases, referred to as α cases, were observed in all forward facing flow material, due to the high temperature diffusion of oxygen into the surface.

Back facing flow microstructural features differed between the samples (Fig. 8). X-link samples 1 and 2 possessed a microstructural gradient, exhibiting a basketweave structure of acicular α colonies in an aged, transformed β matrix with acicular α with some α islands dispersed in the structure. The acicular α colonies are thinner, with a more defined β in sample 2 than sample 1. Cooling rate typically determines the thickness of α lamellae, suggesting that sample 2 cooled more quickly than sample 1. Sample 2 appears less thermally degraded, so it is plausible that the material in sample 1 was at a higher temperature prior to cooling, resulting in a slower cooling rate than sample 2. The acicular α microstructure appears to be an intermediate structure resultant from the transition from a martensitic α to the initial globular α observed in Fig. 5. Arc-jet sample 3 exhibited a less extreme microstructural gradient, with larger grains at the forward facing flow side, transitioning to slightly smaller grains at the back facing flow side. Serrated α grains were also observed. Arc-jet sample 4 did not exhibit a microstructural gradient and had a relatively constant grain size across the length of the sample consisting of acicular α within large prior β grains. The back facing flow side of arc-jet samples 3 and 4 did not transition to the intermediate acicular α structure as X-link samples 1 and 2 did. This is likely due to the geometry of the X-links versus the arc-jet samples. As discussed previously, it is plausible that conductive heating effects from thicker material in X-links sample 1 and 2, causing the back facing flow material to remain hot for longer, producing a microstructure corresponding to a lower cooling rate than the forward facing flow material.

The absence of a microstructural gradient in sample 4 (a Class III reaction) may rely several factors. One possibility relates to the large consumption of titanium metal during testing for Class III combustion reactions. The Class III samples often had molten material encompassing the entirety of the remaining metal, causing the metal to remain at a higher temperature longer and produce even cooling within the metal. It is also possible that arc-jet sample 4 reached some critical threshold where even heating was established, resulting in the absence of a microstructural gradient. Arc-jet sample 3, a Class II reaction, has a slight microstructural gradient, thought to be negligible due to the martensitic α and large grained structure retained throughout. X-link samples 1 and 2 were not able to reach this even heating state due to the close proximity of thicker Ti-6Al-4V material connected to the back facing flow material.

Microstructural features are dependent on many processing parameters. A higher cooling rate from the β phase field results in smaller α colony size, α plate width, and thickness of α at grain boundaries for the lamellar microstructures observed. [6]. Longer times at temperatures above the β transus result in a higher percentage of β in the microstructure. Larger β grain sizes result from high solutionizing temperatures. Because exact parameters are not known for the temperature exposure of the X-links, it is difficult to ascertain what parameter caused each microstructural characteristic.

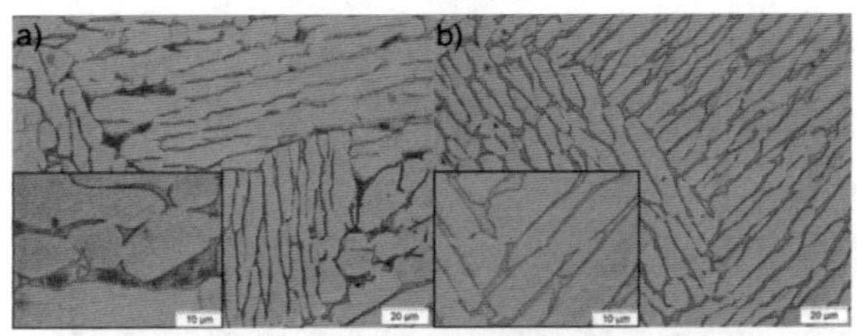

Figure 5. Complementary microstructures through thickest section of port (a) and starboard (b) X-links.

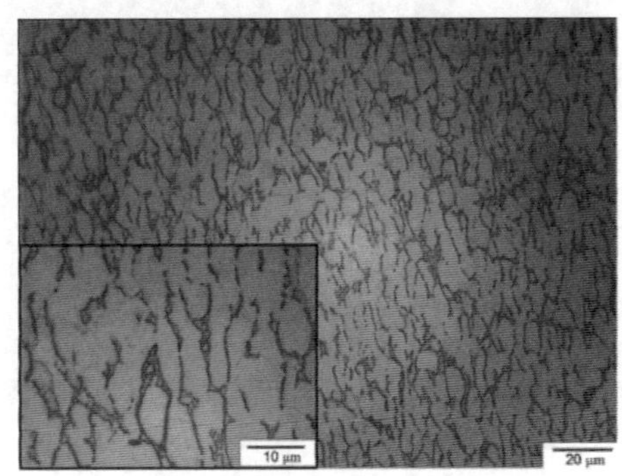

Figure 6. Initial microstructural state of the arc-jet samples.

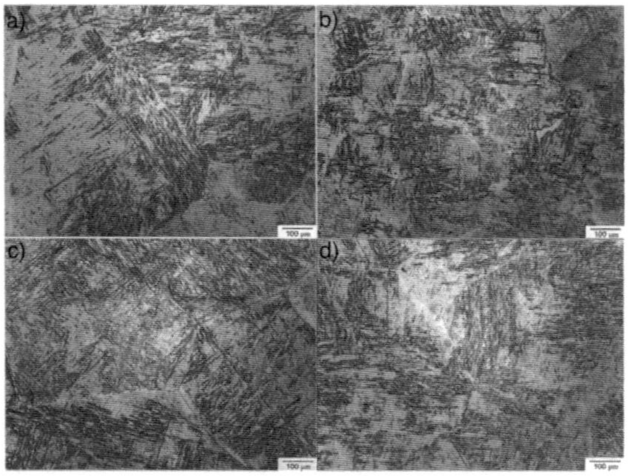

Figure 7. Forward facing flow microstructures for starboard X-link section 1 (a), starboard X-link section 2 (b), arc-jet sample Class II (c), and arc-jet sample Class III (d).

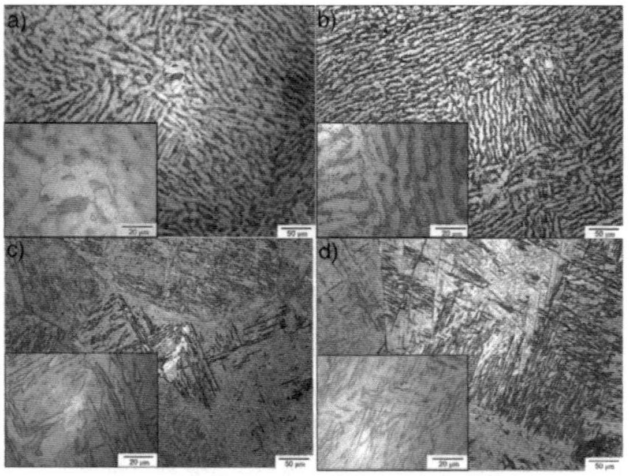

Figure 8. Back facing flow microstructures for starboard X-link section 1 (a), starboard X-link section 2 (b), arc-jet sample Class II (c), and arc-jet sample Class III (d)

Scanning Electron Microscopy

Oxide cracking, gas porosity, and shrinkage porosity were observed within the deposits. Several different oxide layers exist due to the dynamic nature of the event and the tendency of oxide layers to flake at higher temperatures. All samples contained small amounts of chemical contamination. Contamination of X-links samples 1 and 2 was consistent with MP35N fastener constituents and other constituents present in shuttle structural materials. Arc-jet samples 3 and 4 were contaminated by the SiC ceramic plate fixture used for testing. Solidification features observed were similar between samples. Granular features coalescing to form a layer at the outermost surface were observed in X-link sample 2 and arc-jet sample 3 (Fig. 9). EDS scans of all samples were titanium and oxygen rich, with arc-jet scans for samples 3 and 4 containing trace amounts of nitrogen (Fig. 10). The ambient air content used in the arc-jet facility is similar to the air content in the mesosphere where breakup occurred. It is possible that detection limits of the instrumentation prevented small concentrations of nitrogen from resolution in the spectra. EDS also lacks the sensitivity necessary to accurately discriminate between titanium, vanadium, oxygen, and nitrogen energies. More sophisticated equipment will be utilized in further research to determine with certainty the elemental composition of the deposits.

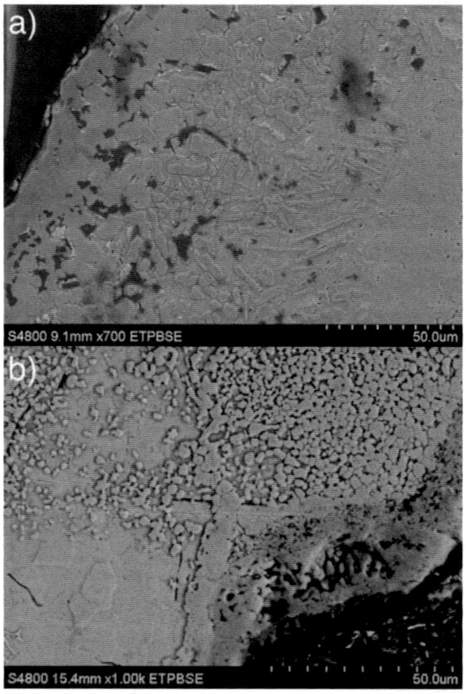

Figure 9. Backscattered electron images of granular features coalescing to form a solid layer in X-link sample 1 (a) and arc-jet sample 3 (b).

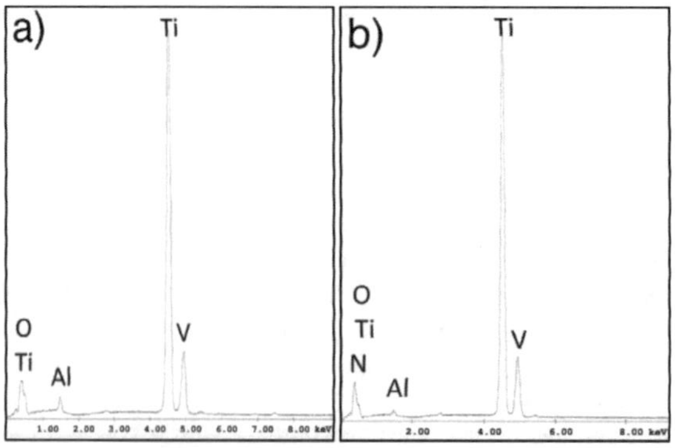

Figure 10. EDS spectrum of granular features in X-link sample 1 (a) and arc-jet sample 3 (b).

X-Ray Diffraction

XRD was performed at the forward facing flow side for all samples. Fig. 11 shows XRD patterns, grouping the more thermally degraded X-link sample 2 and arc-jet sample 3 and the less thermally degraded X-link sample 1 and arc-jet sample 4. The most common titanium oxide form, TiO_2 (rutile), was not detected because this oxide typically forms between 300°C to 1000°C [23] [24], indicating that these oxides formed at higher temperatures. High temperature oxides of the form Ti_xO and Ti_xO_{2x-1} in addition to hexagonal α titanium were detected in all samples. The less degraded sample scans exhibit titanium nitrides, consistent with EDS scans, and Ti_2O_3, a lower titanium oxide formed by reacting TiO_2 (rutile) with titanium metal at roughly 3000°F [25]. The more degraded sample scans show titanium rich Ti_2O. Oxygen deficient hexagonal oxides of the form Ti_xO, detected in all samples, are formed when oxygen orders in hexagonal α Ti; these oxides were likely detected at the metal/scale interface and were shielded from physical removal early in the deposition event. The oxide Ti_6O is a lower temperature oxide formed from the solid solution, and Ti_3O is a higher temperature oxide that crystallizes from the melt [26]. The higher oxide, Magneli-type phase Ti_4O_7, found in all samples, forms when TiO_2 (rutile) is reduced [27].

The results indicate that more extreme degradation of samples formed oxygen deficient oxides and the less degraded samples formed nitrides and Ti_2O_3. The non-uniformity in the color of the scales suggests different oxides in different areas of the samples.

Similar oxides were detected, but a more diverse set of oxide species were detected in the less thermally degraded samples. Lower stabilities of titanium nitrides versus titanium oxides resulting in reduction of the nitrides [28], physical processes to include ablation and dislodging of oxide layers, and instrumentation limitations may account for this finding. The scans suggest mixing of the oxides and the possibility of non- stoichiometric titanium oxides, as indicated by peak shifting and broadening. It is possible that several oxides and nitrides coexist, considering the similar atomic radii and tendency of titanium to combust in ambient pressure gaseous nitrogen in addition to oxygen. More sensitive methods are necessary to discriminate separate oxide and nitride species.

Microhardness Testing

Microhardness readings were taken traversing the sample from forward to back facing flow sides and averages lie within the typical range of commercial Ti-6Al-4V. Scatter in readings is due to the intrinsic nature of the two phase Ti-6Al-4V structure. A spike in hardness is observed at the forward facing flow side due to the presence of an α case. Arc-jet sample 4 exhibited an α rich phase at forward and back facing flow surfaces. X- link samples 1 and 2 have relatively constant microhardness values after the initial α case despite the drastic transition in microstructural features. This may be explained by the acicular α present throughout the entire microstructure. Arc-jet sample 3 shows a slight increasing trend in hardness values as the grain size decreases from forward to back facing flow material while arc-jet sample 4 remains relatively constant, agreeing with the microstructures. It is noted that the bulk microhardness values of X-links samples 1 and 2 are lower than arc-jet samples 3 and 4, possibly due to the air quench or initial fine grained microstructure, resulting a harder material.

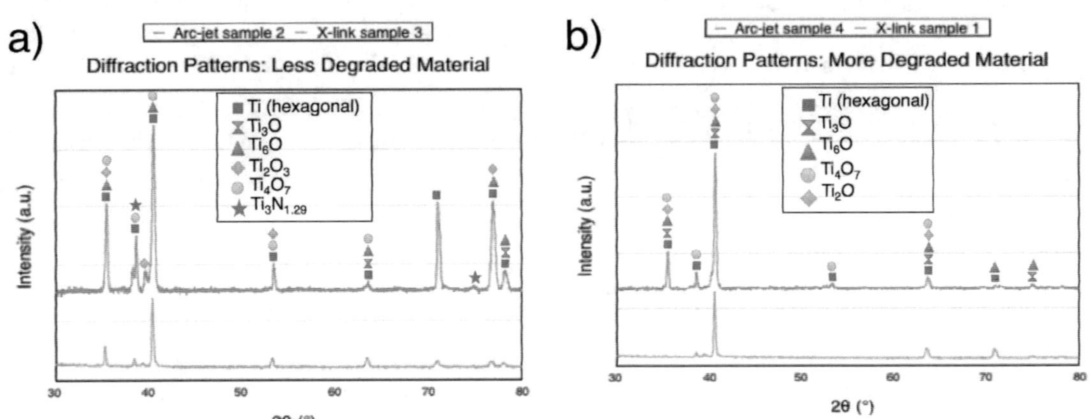

Figure 11. X-ray diffraction patterns for the less thermally degraded X-link sample 2 and arc-jet sample 3 (a) and the more thermally degraded X-link sample 1 and arc-jet sample 4 (b).

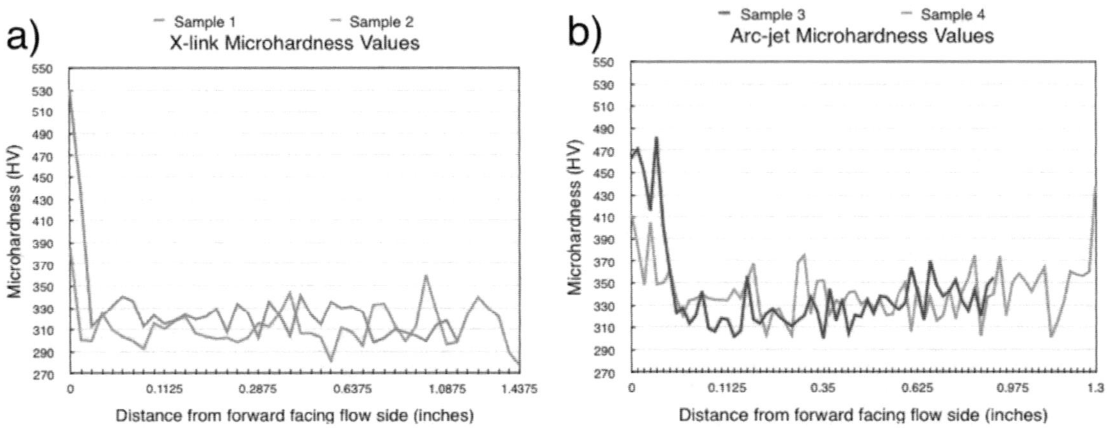

Figure 12. Microhardness values traversing from forward to back facing flow side for samples 1 and 2 (a) and samples 3 and 4 (b).

Table 2. Average microhardness and population standard deviation for samples 1-4

	X-link Sample 1	X-link Sample 2	Arc-jet Sample 3	Arc-jet Sample 4
Average Microhardness (HV)	328	313	341	342
Pop. Standard Deviation (%)	40.1	19.4	40.4	24.3

4. CONCLUSION

The widely used aerospace titanium alloy, Ti-6Al-4V, was forensically characterized utilizing multiple techniques to better understand the material response in the monoatomic oxygen rich and high enthalpy re-entry environment. Material previously exposed to the actual and simulated re-entry environment were analyzed and compared by examining artifacts from the Space Shuttle *Columbia* and arc-jet tested samples, respectively. While the purpose of this work is to relate similarities, there are intrinsic differences affecting the results, including interaction with surrounding material systems, cooling rate, geometry, thickness, conductive heating effects, air content, length of combustion reaction, and possible encapsulation by surrounding materials. Similarities and differences were discussed relative to these factors.

The initial microstructural state of the X-links is suggested to be a sub β transus structure of elongated and equiaxed α in a lamellar α/β matrix, based upon observations of a physically unaffected portion of the port X-link. The initial microstructural state of the arc-jet samples was established to be slightly elongated α in a lamellar α/β matrix, similar except in grain size to the port X-link. Forward facing flow microstructural features are similar in all four samples, consisting of martensitic α within large prior β grains. The strongly microstructural similarities suggest conditions of the arc-jet are relatable to the environmental conditions of the X-links at the forward facing flow side. Back facing flow microstructural features differed between samples, yet similarities were distinguished between X-link samples 1 and 2 and arc-jet samples 3 and 4. The X-links exhibited a drastic microstructural transition to a basketweave structure of acicular alpha colonies and aged, transformed beta at the back facing flow side that is likely due to conductive heating effects from a transition to thicker material. Arc-jet sample 3 had a less extreme transition to smaller grains at the back facing flow side. Microstructural gradients were only observed in material that physically appeared to have combusted, with the exception of sample 4, and uniform microstructures were observed in physically unaffected material. This implies that a combustion reaction may create a non-uniform heating condition and manifest through microstructural gradients, but it is possible for the combusted material to reach some steady state condition, as evidenced in sample 4.

SEM and XRD revealed qualities of the deposits, but these methods lacked the sensitivity to accurately discriminate between the oxide and nitride species present. XRD patterns suggest the more physically degraded material of X-links and arc-jet differed from the less physically degraded material, with a more diverse oxide species in the less thermally degraded material.

Reduction of the nitrides, physical oxide removal processes, and instrumentation sensitivity may account for this observation. Microhardness testing revealed a higher average hardness in the arc-jet samples and a large degree of scatter due to the intrinsic nature of the material.

The similarities between features in samples 2 and 3 and samples 1 and 4 evidenced through XRD and SEM characterization indicate that different areas of the X- links exhibited different material reactions. It is possible that X-link sample 2 underwent only an ignition and melting event, and not a combustion event, as was documented through testing for arc-jet sample 3. X-link sample 1 appears to have ignited and combusted, like arc-jet sample 4. Recognition of these patterns will allow for information to be gleaned on the differing material evidence present in oxidation, ignition, and combustion events. Material with a higher effective ballistic number, i.e. regions with an out-of-plane protrusion (arrow in Fig. 1, shown in Fig. 3) on the X-links and Class III arc-jet specimens, result in an uncontrolled combustion reactions limited by consumable material and removal from air stream. Lower effective ballistic numbers, i.e. planar material and Class II arc-jet specimens, appear to only ignite but not sustain combustion. Considering the complicated design of many structural components, the possibility of destructive material behavior resulting in the compromise of integrity of the space vehicle should not be ignored.

Several questions remain regarding the nature of the deposits and oxide/nitride species observed using the current techniques. A quantitative and more sensitive technique, Wavelength Dispersive Spectroscopy (WDS), will be used to further distinguish between the deposit oxide/nitride species. The more precise energy resolution achievable with WDS will allow for better differentiation between the titanium, vanadium, oxygen, and nitrogen energies and provide an improved understanding of the oxide species present using quantitative data. WDS will assist in understanding the complex deposit characteristics and reactions.

Given the extensive use of the Ti-6Al-4V alloy in past and future spacecraft, it is necessary to establish a known material reaction if exposed to the re-entry environment. Understanding the combustion response by defining thresholds and parameters necessary for a combustion reaction to occur will improve safety in future space travel. This and future work contributes to the sparse knowledge regarding bulk metal combustion and material reactions in the reentry environment. Thermal degradation of titanium components can compromise the structural integrity of spacecraft; therefore, it is of interest to establish an expected combustion mechanism and graceful degradation pattern to ensure safety and reliability of future space travel.

5. ACKNOWLEDGEMENTS

The authors would like to acknowledge the Herrera, Stafford, and Associates employees for assistance with laboratory services. The authors would also like to thank Mike Ciannilli of the *Columbia* Research and Preservation Office at NASA-KSC for facilitating access to Columbia artifacts and Rick Russell of the Spacecraft and Launch Vehicle M&P Engineering group at NASA- KSC for his support.

6. REFERENCES

1. R.R. Boyer, "An overview on the use of titanium in the space industry," *Materials Science and Engineering A*, vol. 213, no. 1-2, pp. 103-114, 1996.
2. E.A. Borisova and Bardanov, K.V., "Ignition of titanium alloys in media containing oxygen," *Metal Science and Heat Treatment*, pp. 37-40, 1963.
3. A.V. Grosse and J.B. Conway, "Combustion of metals in oxygen," *Ind. Eng. Chem.*, vol. 50, pp. 663-672, 1958.
4. A. Abbud-Madrid, C. McKnight, M.C. Branch, and J.W. Daily, "36th AIAA Aerospace Sciences Meeting and Exhibit," in *Buoyancy and Pressure effects of bulk metal-oxygen reactions*, Reno, NV, 1998, pp. 1-7.
5. G.H. Markstein, "Combustion of metals," *AIAA*, vol. 1, pp. 550-563, 1963.
6. G. Lutjering and J.C. Williams, *Titanium*. Berlin, Germany: Springer, 2003, vol. 1.
7. W.C. Rochelle, "Suvery of titanium testing at NASA, DOE, and DOD test facilities," National Aeronautics and Space Administration, ESCG- 4380-06-AFD-MEMO-0011 , 2006.
8. National Fire Protection Association, "Guide for fire and explosion investigations," NFPA 921, 2014.
9. I.I. Kornilov, "Effect of oxygen on titanium and its alloys," *Metal Science and Heat Treatment*, vol. 15, no. 10, pp. 826-829, 1973.
10. V.I. Bolobov, "Mechanism of Self-Ignition of Titanium Alloys in Oxygen," *Combustion, Explosion, and Shock Waves*, vol. 38, pp. 639-645, 2002.
11. V.I. Bolobov, "Possible mechanism of autoignition in titanium alloys in oxygen," *Combustion, Explosion, and Shock Waves*, vol. 39, pp. 677-680, 2003.
12. N.N. Bakhman, G.P. Kuznetsov, and V.M. Puchkov, "Combustion of titanium plates," *Combustion, Explosion, and Shock Waves*, vol. 36, pp. 470-475, 2000.

13. B.G. Efimov and P.N. Kuzyaev, "Influence of angle of attack on the combustion characteristics of titanium plates in an airstream," *Combustion, Explosion, and Shock Waves*, vol. 31, pp. 653-654, 1995.

14. "Columbia Accident Investigation Report," 2003.

15. "Columbia Crew Survival Investigation Report," 2008.

16. T.W. Schlatter, "Atmospheric Composition and Vertical Structure," *Environmental Impact and Manufacturing*, vol. 6, 2009.

17. J.D. Olivas, M.C. Wright, R. Christoffersen, D.M. Cone, and S.J. McDanels, "Crystallographic Oxide Phase Identification of char deposits obtained from the space shuttle Columbia," *Acta Astronautica*, vol. 67, pp. 553-560, 2010.

18. J.D. Olivas, B. Mayeaux, P. Melroy, and D.M. Cone, "AIAA Space 2009 Conference and Symposium," in *Study of Ti alloy combustion susceptibility in simulated entry environments*, Pasadena, CA, 2009.

19. R. Savino, M. De Stefano Fumo, D. Paterna, A. Di Maso, and F. Monteverde, "Arc-jet testing of ultra-high-temperature-ceramics," *Aerospace Science and Technology*, vol. 14, pp. 178-187, 2010.

20. National Aeronautics and Space Administration. Arc-jet Testing Facility. [Online]. http://www.nasa.gov/centers/johnson/engineering/human_space_vehicle_systems/atmospheric_re-entry_materials/index.html

21. American Society for Metals, *Atlas of microstructures of industrial alloys*, 8th ed. Metals Park, OH, USA: American Society for Metals, 1972, vol. 7.

22. Frauke Hogue, "Take a closer look at titanium," *The International Metallographic Society Newsletter*, vol. 31, no. 2, pp. 8-9, June 2003.

23. A.S. Khanna, *Introduction to High Temperature Oxidation and Corrosion*, 1st ed. Materials Park, OH, USA: ASM International, 2002.

24. O. Kubaschewski and B.E. Hopkins, *Oxidation of Metals and Alloys*, 1st ed. London, Great Britain: Butterworths Scientific Publications, 1953.

25. N.N. Greenwood and A. Earnshaw, *Chemistry of the Elements*, 2nd ed. Boston: Butterworth- Heineman, 1997.

26. I.I. Kornilov, "Interaction of titanium with elements of the periodic system," in *The Science, Technology, and Application of Titanium*, London, 1968, pp. 407-418.

27. The Electrochemical Society, "Effect of embedded titanium containing particles on the corrosion of particulate alumina reinforced aluminum-matrix composite,“ *ECS Transactions-Light Alloys 3*, vol. 11, no. 15, pp. 109-120, 2008.

28. J. Hong, Ed., *Preparation and Characterization of materials*. New York, USA: Elsiever, 2012

THE SPACECRAFT FIRE EXPERIMENT (SAFFIRE) - OBJECTIVES, DEVELOPMENT AND STATUS

William Schoren(1), Gary A. Ruff, David L. Urban

(1) NASA John H. Glenn Research Center, 21000 Brookpark Road, Cleveland, OH 44135, Email: william.r.schoren@nasa.gov

ABSTRACT

Since 2012, the Spacecraft Fire Experiment (*Saffire*) has been under development by the Spacecraft Fire Safety Demonstration (SFS Demo) project that is funded by NASA's Advanced Exploration Systems Division in the Human Exploration and Operations Mission Directorate. The overall objective of this project is to reduce the uncertainty and risk associated with the design of spacecraft fire safety systems for NASA's exploration missions. This is accomplished by defining, developing, and conducting experiments that address gaps in spacecraft fire safety knowledge and capabilities identified by NASA's Fire Safety System Maturation Team. This paper describes the three Spacecraft Fire Experiments (Saffire-I, -II, and -III) that were developed at NASA-GRC and that will conduct a series of material flammability tests in low-gravity and at length scales that are realistic for a spacecraft fire. The experiments will be conducted in Orbital ATK's Cygnus vehicle after it has unberthed from the International Space Station. The tests will be fully automated with the data downlinked at the conclusion of the test and before the Cygnus vehicle reenters the atmosphere. The objectives of these experiments are to (1) determine how rapidly a large scale fire grows in low-gravity and (2) investigate the low-g flammability limits compared to those obtained in NASA's normal gravity material flammability screening test. The hardware for these experiments has been completed and is awaiting their respective launches, all planned for 2016. This paper will review the objectives of these experiments and how they address several of the knowledge gaps for NASA's exploration missions. The hardware development will be discussed including several novel approaches that were taken for testing and evaluation of these series payloads. The status of the missions and operational status will also be presented.

1. INTRODUCTION

Tests on full-scale transportation vehicles, buildings, homes, and habitats have been common on earth to more fully understand the fire hazards associated with each application and how best to protect the passengers and inhabitants from a potential fire. Fire safety on spacecraft has been a significant concern for NASA during the decades of crewed spaceflight and will remain so as NASA plans future exploration missions. Many microgravity combustion experiments have been performed on the Space Shuttle and continue to be performed on the International Space States (ISS) and the data from these experiments have contributed to our understanding of how to prevent and protect crewed spacecraft from fire. However, none of these experiments have studied sample and environment sizes typical of those expected in a spacecraft fire [1]. Prior experiments have been limited to samples no larger than 10 cm in length and width whereas a serious spacecraft fire would likely be much larger and consume much more material. Because of the large differences between fire behavior in normal and reduced gravity, there is a significant lack of data available for spacecraft designers on which to base their fire safety designs and procedures. The use of terrestrial fires and fire standards to design spacecraft fire safety systems presents an inherent risk to the vehicle because of the significant level of uncertainty. While this approach has been successful thus far, the uncertainty and risk will increase as exploration missions venture further from earth with considerably longer transit times for a safe return. Despite their obvious importance, full scale spacecraft fire experiments have not been possible because of the inherent hazards involved with conducting a large fire test in a crewed vehicle. To address this knowledge gap, a project was proposed to conduct large-scale fire safety experiments in an expendable spacecraft without risk to a crew.

In October 2011, the NASA Advanced Exploration Systems (AES) Division of NASA's Human Exploration and Operations Mission Directorate funded this project to develop and demonstrate spacecraft fire safety technologies in relevant environments. The keystone demonstration to be performed was a large-scale fire safety experiment to be conducted on an ISS re-supply vehicle after it has left the ISS and before it enters the earth's atmosphere. The project team led from NASA's John H. Glenn Research Center (GRC) was identified and began formulating such an experiment. The NASA team was augmented by an international topical team assembled by the European Space Agency (ESA) [2]. The participation of members from other countries and space agencies not only brings additional skills to the science team but also facilitates international cooperation in the development of an approach to spacecraft fire prevention and response for future exploration vehicles. The current spaceflight experiment funded and being developed by NASA addresses two objectives. The first objective is to understand the flame spread and growth of a fire over a large piece of flammable material, consistent with clothing or other fabric likely to be in a spacecraft cabin. This sample material is approximately 0.94 meters long and 0.4 meters wide. This will be at least an order of magnitude larger in both dimensions than any prior low-g flame spread experiment. The second objective is to examine the flammability limits of materials in low gravity for comparison with results of NASA's terrestrial normal gravity evaluation tests. The status of the development of the flight experiment and the individual contributions of the topical team will be discussed in subsequent sections.

2. EXPERIMENT DESCRIPTION

The premise for the design of the flight experiment was that practically all of the hardware would be identical in the three units except possibly for the sample material(s) to be burned. The concept for this experiment focuses on conducting two types of material combustion tests to be performed on different flights of the *Cygnus* vehicle. The experiment package consists of a flow duct and an adjacent avionics bay. A schematic of the flow duct is shown in Fig. 1. The flow duct forms the primary chamber of the experiment while the avionics bay is connected to the side of the flow duct as shown in the figure. A Lexan™ panel forms the wall between the flow duct and the avionics bay. Air is drawn through the flow duct by fans located at the top of the duct with flow straighteners at the bottom of the experiment module. The flow duct/avionics bay assembly is a rigid structure and will be secured with the standard stowage straps used in the *Cygnus* vehicle. This duct will provide a uniform flow across the samples, maintain a clear flow path within the experiment module, and prevent burning debris from interacting with the rest of the cargo. The sample card is positioned in the center of the flow duct as shown in Fig. 1.

2.1 Sample Materials

The sample material for Saffire-I and III is a composite fabric consisting of cotton on a fiberglass substrate (75% cotton, 25% fiberglass by weight). It is called SIBAL cloth because it was developed for the Solid Inflammability Boundary at Low-Speeds (SIBAL) spaceflight experiment and has been used in many space experiments because it burns like cotton but maintains structural in-

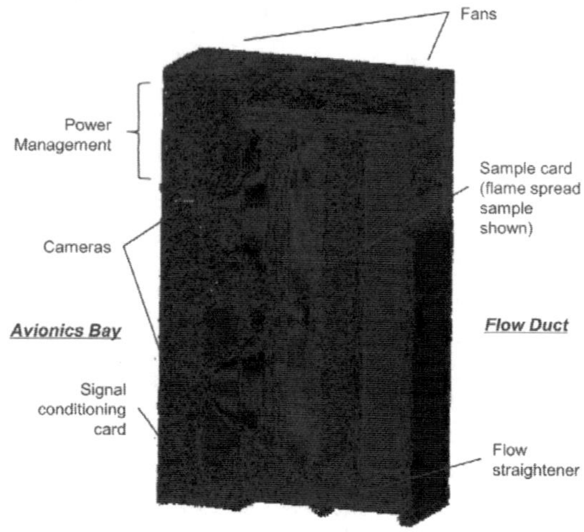

Figure 1: Schematic of the Spacecraft Fire Safety Demonstration Experiment. The experiment module consists of a flow duct containing the sample card and an avionics bay. All power, computer, and data acquisition modules are contained in the bay. The experiment module is approximately 53- by 90- by 133-cm.

tegrity after the cotton is consumed. The objective of the Saffire-I and III tests will be to determine how rapidly a large-scale fire grows in low-gravity and whether it reaches a steady-state spread rate. Saffire-I will be conducted at an air flow velocity of 20 cm/s while Saffire-III will be run at a flow velocity of 30 cm/s .

Saffire-II has a sample card that contains 9 samples, each 5 cm x 29 cm. Table 1 lists the samples contained on Saffire-II while Fig. 2 shows their locations on the sample card. These materials were selected because they

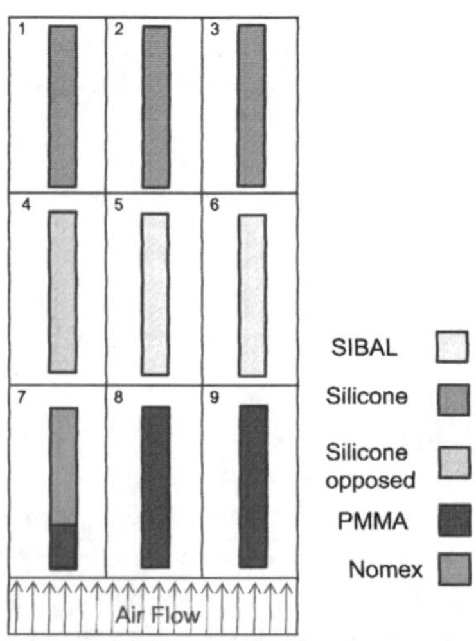

Figure 2: Sample layout for Saffire-II

Table 1. Saffire-II Samples

Sample Number	Material	Sample Thickness	Air Flow (cm/s)	Igniter Position
S1	Silicone	0.25 mm	20	Upstream
S2	Silicone	0.61 mm	20	Upstream
S3	Silicone	1.02 mm	20	Upstream
S4	Silicone	0.36 mm	20	Downstream
S5	SIBAL fabric	0.33 mm	20	Upstream
S6	SIBAL fabric	0.33 mm	30	Upstream
S7	PMMA to Nomex	0.33 mm	20	Upstream
S8	Structured PMMA	10 mm (with tapered edge for ignition)	20	Upstream
S9	Flat PMMA	10 mm (with tapered edge for ignition)	30	Upstream

have normal gravity Maximum Oxygen Concentration (MOC) flammability limits that indicate they will burn in normal gravity just above and below 21% O_2 by volume, *i.e.*, the expected concentration of O_2 when Cygnus unberths from ISS. Drop tower testing has demonstrated the MOC flammability limits of a material can be different in low-g than in normal gravity [3-5]. Therefore, the objective of the Saffire-II tests is to quantify that affect in long-duration low-gravity using samples of the same size as those used in the in NASA-STD-6001 Test 1 normal-gravity tests [3]. Some of these samples are expected to burn while others may not.

Quantifying the difference in flammability limits for large samples is the objective of the Saffire-II experiments.

Table 2. Saffire-I, II, and III Diagnostics

Measurement	Method	Location
Flow temperature	thermocouples	inlet and outlet of flow duct
Flame temperature	thermocouples	embedded in sample
Plume temperature	thermocouple	
Flame radiation	radiometers	two viewing front of sample; two viewing back of sample
CO_2 mole fraction	solid-state sensor	in avionics bay
O_2 mole fraction	solid-state sensor	inlet to avionics bay
Pressure	solid-state sensor	in avionics bay
Flow velocity	anemometers	one on each side of sample
Flame image	cameras	two viewing front of sample
Flow uniformity	smoke wire	leading edge of sample

2.2 Diagnostics

Camera images of the flame are the primary data measurement obtained from the Saffire experiment but there are other diagnostics to measure other flow, flame, and atmospheric parameters. The diagnostic suite used in Saffire are shown in Table 2. The primary experiment data will be provided by two cameras that view one side of the sample card.

2.3 Experiment Development

The Saffire flight systems (hardware and software) underwent the NASA flight experiment development reviews and testing although these processes were heavily tailored to realize savings in cost and schedule. Because all three units were being developed in parallel, some components were ready to progress to the next development phase before the entire system for all three units had completed that phase. The logic behind the tailoring was to tailor the review criteria to allow components that were ready to progress onto the next development phase while the remainder of the system was completed. Given the design of the system and the risk tolerance advocated by the Advanced Exploration Systems Division, this approach was warranted. A Mission Concept Review/System Requirements Review was held in November 2012 and followed approximately 7 months later by a Periodic Technical Review-1 (PTR-1) that combined NASA's typical Preliminary Design Review/Critical Design Review.

Fabrication of the mechanical components for Saffire-I was performed at NASA-GRC to ensure the accuracy of the drawings. NASA's Fabrication Alliance, a coordination of the fabrication shops at all NASA centers, was used to manufacture some of the components for Saffire-II and III again to help maintain project schedule without being limited by the staffing demands of other projects at NASA-GRC. Components for the second and third flight units was performed by the Fabrication Shops at NASA Johnson Space Center (JSC) and White Sands Test Facility (WSTF). All avionics manufacturing was conducted at NASA-GRC.

Periodic Technical Review-2, a tailored System Integration Review marking the start of system assembly was held in February 2014 with the assembly of the flight system being completed in October 2014. Because of the similarity of the flight systems and the risk posture of AES projects, the project tailored the approach to environmental testing and functional testing. Table 3 shows the environmental tests performed on each flight unit along with the date this testing began.

Various levels of functional testing was performed on all three flight units before and after all environmental testing to verify the health of the flight unit being tested and to collect performance data. Additional functional testing was performed throughout 2015 for validation and verification of software and other systems. These test results were compared across all three flight units to verify and demonstrate consistency between the flight units. The System Acceptance Review marking the completion of the flight hardware was held in September 2015.

Table 3. Summary of Saffire Environmental

Tests	Saffire Unit/Date		
	I	II	III
Safety-critical Relay Component Vibe Test	Nov-13	Nov-13	Nov-13
Camera Component Vibe Test	Dec-13	Dec-13	Dec-13
Power Management System Vibe Test	--	Mar-14	Mar-14
System Vibe Test	Oct-14	--	--
System Thermal Test	Nov-14	Oct-14	Nov-14
Off-gas System Test	--	Nov-14 (samples only)	Nov-14
EMI/EMC System Test	Dec-14	Jan-15	Feb-15

2.4 Launch and Operation

The Saffire-I flight unit was successfully launched within the Cygnus PCM on an Atlas V launch vehicle on March 22, 2016 as part of the OA-6 mission. The Cygnus vehicle was successfully berthed to the ISS on March 26, 2016. As of the writing of this paper, the un-berthing of the Cygnus is scheduled for June 2, 2016 and Saffire-I is scheduled to operate a few hours afterwards. Data from Saffire-I will be downlinked over the next several days prior to destructive re-entry of Cygnus about June 10. Saffire-II and III are planned to launch from the Wallops Flight Facility on Orbital ATK's Antares rocket with the tentative dates shown in Table 4.

Table 4. Launch and Operation Dates for Saffire-I-III (as of 4/30/2016)

	Mission	Launch	Operations
Saffire-I	OA-6	22-Mar-2016	2-Jun-2016
Saffire-II	OA-5	NET 6-July-2016	19-Sep-2016
Saffire-III	OA-7	30-Dec-2016	3-Mar-2017

NET - no earlier than

3. SAFFIRE EXPERIMENT SAFETY

Even though this experiment will be conducted during the un-crewed portion of the flight, the Cygnus vehicle berths with the ISS and, therefore, the experiment is subject to many of the same safety considerations as other ISS payloads. The major safety hazard is inadvertent igniter activation prior to the Cygnus de-berthing from the ISS. This would result in the exposure of the flight crew to toxic gases that are produced from the combustion event. It should be noted that this hazard is a concern prior to Cygnus berthing to the ISS since the toxic gases would be present when the Cygnus hatch is opened to the ISS. The power and communication interfaces between Cygnus and Saffire are utilized to control this hazard. Cygnus provides two power feed circuits to Saffire due to power requirements. One power feed circuit provides power to the Saffire avionics, the other provides power to the igniter circuitry. There are two latching relays in each of these two power feed circuits, one in the power feed side, one in the return. These four latching relays are contained within the Saffire flight unit but are controlled and monitored by Cygnus avionics. A separate ground command to the Cygnus avionics is required to close each of these four latching relays. Igniter activation requires that all four relays be closed. This is because Saffire avionics provides an enable command to a DC/DC converter in the igniter circuitry and also provides commanding to solid state relay switches on the igniter control card within the igniter power feed circuitry. In addition, a ground command to the Saffire avionics via Cygnus avionics is required to initiate an igniter activation for an experiment run. Cygnus avionics and the associated ground commanding provide the required failure tolerance for control of the Saffire hazard of inadvertent igniter activation.

Significant focus was placed on these four latching relays due to their safety criticality. This attention is reflected in the environmental testing performed as shown in Table 3. The relays for all three flight units were vibration tested at the component level. Tailoring of the environmental testing resulted in only Saffire I being vibration tested at a system level. Therefore, vibration testing was performed on the Saffire II and III Power Management Systems that includes the relays to address the criticality of these relays and the associated risk of this tailored system-level vibration testing. Safety engineering input was critical to developing a design solution that was acceptable from an ISS payload safety perspective due to the interface limitations between Saffire and Cygnus.

This hazard along with the other typical ISS payloads hazards are documented in the Saffire Unit I Flight Safety Data Package. The ISS Payload Safety Review Panel

Table 5. Spacecraft Fire Safety Demonstration Experiment Objectives

Area	Objective	Comment	Saffire-I, II, III	Saffire-IV, V, VI	Ground
Fire behavior/ modeling:	Quantify growth and end state of realistic fires in spacecraft and their influence on vehicle habitability	Require to validate computational models	x	x	
Fire growth/ dynamics	Flame behavior in complex geometries	More realistic configurations than Saffire-I, II, and III		x	
	Flame behavior for planar and complex geometries in exploration atmospheres	Elevated O_2, lower P; compare with Saffire-I, II, III; supplement small-scale tests in Combustion Integrated Rack		x	
	Measure flame behavior over large-scale planar surfaces	Continues Saffire-I and III investigations	x	x	
Post-fire monitoring	Demonstrate performance of prototype Orion and ISS combustion product monitor	Demonstration of prototype flight hardware		x	x
Fire Detection	Obtain data to validate transport and detection models	Required for model development		x	x
	Demonstrate fire detection with multi-moment sensors	Demonstrate capability to reject nuisance alarms			x
	Evaluate performance of hybrid fire detection (smoke and gaseous products)	Combustion product detection by prototype combustion product monitor			x
Post-fire monitoring	Quantify rate of decay of gas species after a spacecraft fire	Required for model development		x	x
Post-fire cleanup	Quantify atmosphere cleanup rate with prototype smoke-eater	Demo of prototype flight hardware		x	x
Fire Suppression	Performance of low-momentum water mist suppression	Effectiveness of fire ports using water mist fire suppression			x

(PSRP) reviewed and approved this Package at the Phase I/II level in August of 2013 and again at the Phase III level in March 2015. These Payload Safety Reviews were augmented by a handful of Technical Interchange Meetings (TIMs) with the PSRP including TIMs to address the inadvertent igniter activation hazard. These TIMs were of tremendous value in the overall safety approval process by providing a focused review of specific safety issues.

4. SPACECRAFT FIRE SAFETY KNOWLEDGE GAPS AND FUTURE EXPERIMENTS

The experiments being developed for Saffire-I, -II, and –III will be the first of their kind to evaluate low-g material flammability with direct implications for fire safety on future exploration vehicles. One of the drawbacks of these experiments is that a relatively small number of tests will be conducted (two large samples and nine smaller samples). A thorough evaluation of these phenomena would require many more samples and range of materials. Also, if they are typical of most low-g combustion experiments, the findings will raise additional significant questions for material flammability in spacecraft. The Saffire hardware and the Cygnus vehicle provides a unique opportunity to demonstrate other fire safety technologies including fire detection, fire suppression, post-fire cleanup and monitoring. Needs in these areas have been identified by NASA's Spacecraft Fire Safety Systems Maturation Team and are summarized in Table 5. Because of the importance of addressing these remaining needs to crew safety in exploration missions, AES management authorized the SFS Demo project to plan for three more Saffire experiments, Saffire-IV, V, and VI, that will address objectives not achieved in Saffire-I-III. Of particular note is that many of these objectives require supporting ground-based experimentation to perform the parametric investigations that are required to fully understand the phenomena in question. The objective of the flight experiment is to anchor the ground-based experiments with flight data.

While the planning for the follow-on experiments is underway, the development process is intended to mirror the processes executed with Saffire-I-III. The Mission Concept Review/System Requirements Review is planned for June 2016 with a design review planned for December 2016.

5. CONCLUSIONS

The Spacecraft Fire Safety Demonstration Project has designed, manufactured, assembled and tested three Saffire flight units. The decision to develop flight units of nearly identical design allowed for the tailoring of environmentally testing that resulted in cost and schedule savings without introducing significant risk. This design similarity also allowed for the comparison of test results to verify consistency when testing was performed across all three flight units. Safety engineering input was critical to developing an acceptable design solution to control the inadvertent igniter activation hazard due to the interface limitations between Saffire and Cygnus. Technical Interchange Meetings (TIMs) with the ISS Payload Safety Review Panel proved to be invaluable to the overall safety approval process by providing a focused review of specific safety issues.

Conducting these three experiments will be a milestone for spacecraft fire safety. The data and follow on analysis will provide verification of spacecraft fire safety design and operations for future exploration vehicles and their crews. Development of the Saffire-IV – VI flight units will extend spacecraft fire safety knowledge related to the combustion by-products and fire detection.

6. ACKNOWLEDGEMENTS

The members of the Fire Safety in Space International Topical Team are acknowledged for their continued contributions to the development of the Saffire series of experiments. The members are: Gary A. Ruff , David L. Urban, and Sandra Olson (NASA Glenn Research Center), Carlos Fernandez-Pello (UC Berkeley), James S. T'ien and Ya-Ting Liao (Case Western Reserve University), Jose L. Torero (University of Queensland, Brisbane, Australia, Guillaume Legros (Université Pierre et Marie Curie, Paris, France), Christian Eigenbrod (University of Bremen (ZARM), Bremen, Germany, Nickolay Smirnov (Moscow Lomonosov State University, Moscow, Russia), Osamu Fujita (Hokkaido University, Sapporo, Japan), Adam J. Cowlard (University of Edinburgh, Edinburgh, UK, Sebastien Rouvreau (Belisama R&D, Toulouse, France), Balazs Toth (ESA ESTEC, Noordwijk, Netherlands), and Grunde Jomaas (Technical University of Denmark, Kgs. Lyngby, Denmark).

The Spacecraft Fire Safety Demonstration project is funded by the Advanced Exploration Systems Division in NASA's Human Exploration and Operations Mission Directorate.

7. REFERENCES

1. Ruff, G.A., Urban, D.L., Pedley, M.D., Johnson, P.T., Safety Design for Space Systems edited by Gary Musgrave et al. Chapter 27 "Fire Safety," Elsevier, 2009.

2. Ruff, G.A., Urban, D.L., Fernandezs-Pello A.C., T'ien, J.S., Torero, K. L., Legros, G., Eigenbrod, C., Smirnov, N., Fujita, O., Cowlard, A.J., Rouvreau, S., Minster, O., Toth, B., and Jomaas, G. (2013) Development of Large-Scale Spacecraft Fire Safety Experiments, Paper No. AIAA-2013-3410, 2013 International Conference on Environmental Systems, Vail, CO.

3. Hirsch, D.B., Williams, J.H., Haas, J.P., Beeson, H.D., and Pedley, M.D., "Oxygen Concentration Flammability Thresholds of Selected Aerospace Materials Considered for the Constellation Program," Proceedings of the 2nd IAASS Conference, Chicago, May 14-16, 2007; (also ESA SP-645, July 2007).

4. Olson, S.L., Ruff, G.A., Miller, F.J., Microgravity Flame Spread in Exploration Atmospheres: Pressure, Oxygen, and Velocity Effects on Opposed and Concurrent Flame Spread," Paper No. 2008-01-2005, 38th International Conference on Environmental Systems, San Francisco, CA, June 29-July 2, 2008.

5. Olson, S.L., Hirsch, D.B., and Ferkul, P.V., "Improving Materials Flammability Screening For Spacecraft Fire Safety," Proceedings of the 2010 Technical Meeting of the Central States Section of the Combustion Institute, Champaign, Illinois, March 2010.

ABOUT THE DEMISABILITY OF PROPELLANT TANKS DURING ATMOSPHERIC RE-ENTRY FROM LEO

T. Lips[(1)], B. Fritsche[(1)], R. Kanzler[(1)], T. Schleutker[(2)], A. Gülhan[(2)], B. Bonvoisin[(3)], T. Soares[(3)], and G. Sinnema[(3)]

[(1)] *HTG GmbH, Albert-Einstein-Str. 11, 37191 Katlenburg-Lindau, Germany, Email: t.lips@htg-hst.de*
[(2)] *DLR, Linder H¨ohe, 51147 K¨oln, Germany*
[(3)] *ESA/ESTEC, Keplerlaan 1, 2200 AG Noordwijk, Netherlands*

ABSTRACT

The concern about the on-ground risk caused by spacecraft fragments surviving atmospheric re-entry has significantly grown during the past decade, resulting in numerous activities of the space community like re-entry simulation tool development, improvement, and validation, as well as research for design-for-demise (D4D). The purpose of the first group is to increase the confidence in numerical re-entry risk predictions, while the second activity aims to develop new spacecraft design techniques which are likely to improve the demise behavior significantly.

Standard (Titanium) propellant tanks for LEO (Low Earth Orbit) missions appear regularly within the group of spacecraft (satellite) components causing the highest on-ground risk, i.e. in the order of 20% for a 1-ton class Earth observation satellite. Thus, tanks are one of the primary targets for any D4D activity. Various ideas for demisable propellant tank concepts have been published recently, including aluminum tanks (e.g. based on aluminum-lithium alloys) or composite tanks (e.g. Carbon-fiber overwrapped metallic liners).

The demonstration of the actual demisability of such new tank concepts has turned out to be difficult, sometimes even questionable. Intensive laboratory tests in arc-jet heated or plasma wind tunnels and laser facilities have provided extensive results. Numerical simulations have been conducted to reproduce these experimental findings.

This paper will contain a critical review of the latest experimental and numerical results about the demisability of satellite propellant tanks during atmospheric re-entry from LEO.

NOMENCLATURE

D	Tank diameter
L	Tank length
r	Radius of toroidal tank section
V	Tank volume
m	Tank mass
A_s	Tank surface area
$\dot{q}$	Heat flux
p	Pressure

1. INTRODUCTION

In 2013, at the 49th AIAA/ASME/SAE/ASEE Joint Propulsion Conference and Exhibit, Masuoka et al. have presented a paper on the demise characteristics of CFRP (carbon fiber reinforced plastics) over-wrapped aluminum tanks [1]. In 2015, this paper was also published by the AIAA in the *Journal of Propulsion and Power* [2].

Although not stated explicitly in the conclusions, this paper basically implies that CFRP over-wrapped aluminum tanks for spacecraft re-entering into the Earth atmosphere from LEO will demise completely. At least, this is how the paper was understood by several people of the European D4D community.

Recent discoveries of surviving CFRP over-wrapped tanks in Spain (see Fig. 1), resulting from an Altas Centaur upper stage re-entry in November 2015, have raised some questions concerning this D4D approach. The CFRP wrapping of the tanks seem to show only some minor destruction (fiber delamination/oxidation, resin pyrolosis).

As a consequence, the authors of this paper tried to reproduce the Japanese study using ESA's re-entry simulation tool SCARAB and the results of a recent material demisability study funded by ESA.

Figure 1. Surviving CFRP over-wrapped tank (Credits: ladepeche.fr)

1.1 SCARAB

SCARAB (Spacecraft Atmospheric Re-Entry and Aero thermal Break-Up) is a spacecraft-oriented soft- ware tool allowing the analysis of mechanical and thermal destruction of spacecraft and other objects during re-entry (controlled or uncontrolled). It is an integrated software package (six degrees-of-freedom flight dynamics, aerodynamics, aerothermodynamics, thermal- and structural analysis) used to perform re-entry risk assessments (quantification, characterization and monitoring of surviving fragments during re-entry). The software application has been validated with in-flight measurements, re-entry observations and wind tunnel experiments, and it has been compared to other re-entry prediction tools of the international community.

SCARAB has been developed under ESA/ESOC contracts since 1995 under the lead of HTG (Hypersonic Technology Göttingen) and with support from other European and international partners. It is considered as operational software. The software development has evolved over time, based on lessons learned from preceding software versions, upgrades and specific re-entry analyses performed for various satellites (e.g. ROSAT, BeppoSAX, TerraSAR-X, GOCE, Sentinel-1/2/3/5P/6, SWARM, Cluster-II, Integral, EnMap, EarthCARE), and for the ATV and the ESA launcher programs. Typical launch vehicle (or similar) re-entry applications have been: Ariane-5 stages (EPC, EPS/VEB, ESC-A), Vega stages (Zefiro-9, AVUM), and ATV.

SCARAB version 3.1L [3] has been used for this paper.

1.2 ORSAT/ORSAT-J

ORSAT (Object Reentry Survival Analysis Tool) is an object-oriented re-entry analysis tool. The original version of ORSAT (version 4.0, released in 1993) was developed by the NASA Johnson Space Center. Major updates, version 5.0 (1999) and version 5.8 (2002), were developed by Lockheed Martin for NASA. The current version 6.0 (2005) was developed by ESCG/Jacobs Sverdrup [4, 5, 6, 7].

ORSAT-J was developed by the JAXA Safety and Mission Assurance Department, with capabilities similar to NASA's ORSAT 5.0.

2. JAXA TANKS

Based on the information published in [1], we tried to reproduce the physical tank parameters used by JAXA. Fig. 2 shows the basic sketch of the outer shape of the tank design. All tank parameters – for tank volumes from 230 to 1,030 liters – are shown in Tab. 1. All five tanks – drawn to scale – are shown in Fig. 3. The SCARAB model of the 430 liter tank is shown in Fig. 4.

Table 1. Tank parameters

D [m]	L [m]	r [m]	V [m^3]	m [kg]	A_s [m^2]
0.760	0.662	0.201	0.230	14.937	1.534
0.820	0.980	0.217	0.430	22.519	2.470
0.900	1.174	0.238	0.630	29.247	3.254
1.000	1.261	0.264	0.830	35.164	3.880
1.250	1.094	0.330	1.030	41.050	4.170

The wall of all tank versions consists of an inner aluminum liner (1 mm) over-wrapped with CFRP (3.1 mm). The latter value has been taken from a D4D study conducted for ESA, and its application provides overall mass properties which are consistent with the mass of the 830 l tank stated in [1].

It has to be noted that these are not the exact tank parameters of the JAXA tanks. Some parameters had to be estimated because they were not included in [1].

3. ORSAT-LIKE SIMULATIONS

In the next step, we tried to confirm the demise altitudes of the JAXA tanks as published in [1]. In order to achieve this, ORSAT-like simulations have been conducted with SCARAB. As material properties (Tab. 2) we used those presented in [8]. The study behind this paper was a comparison between SCARAB and NASA's ORSAT. It is assumed that JAXA's ORSAT-J uses the same or at least similar material parameters as NASA's ORSAT.

The initial conditions used for the re-entry simulations are shown in Tab. 3. These are the same initial conditions as in [1]. The initial azimuth angle was not stated explicitly in this paper, but the initial velocity of 7.14 km/s at 78 km altitude is consistent with a re-entry from an equatorial orbit, i.e. towards the East direction. For a re-entry from a polar orbit the corresponding initial velocity should have been higher, in the order of 7.6–7.7 km/s.

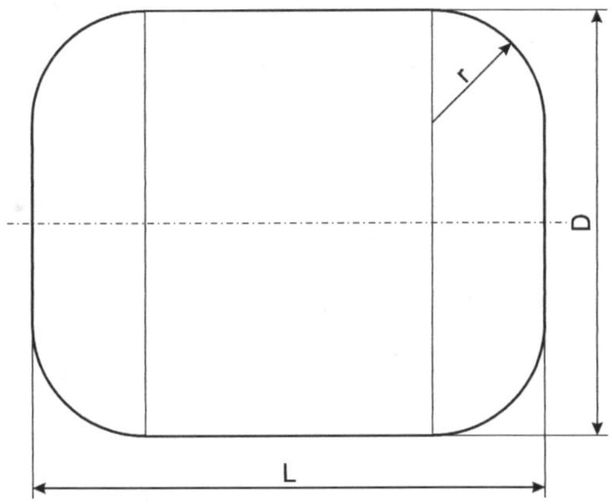

Figure 2. JAXA tank sketch

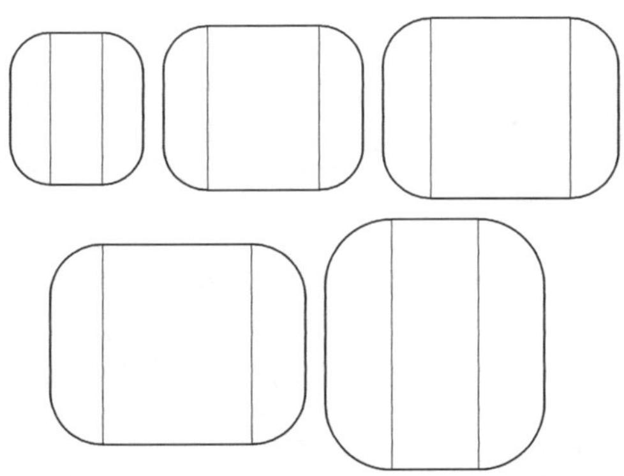
Figure 3. All JAXA tanks (drawn to scale)

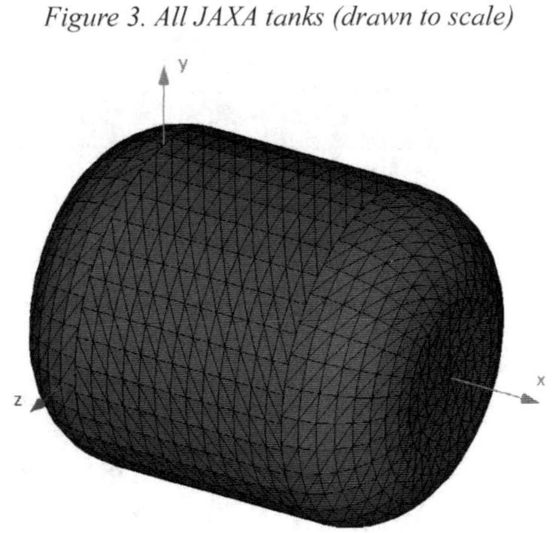

Figure 4. SCARAB model of the JAXA tank (430 l)

Table 2. Material properties

	Aluminum	CFRP
Density [kg/m³]	2700	1550.5
Melting Temperature [K]	850	700
Heat Capacity [J/kg-K]	1100	879
Heat of Melting [kJ/kg]	390	0.236
Thermal Cond. [W/m-K]	140	0.9-6.4
Emissivity [-]	0.3	0.9

In addition to the initial position/velocity state vector and the initial temperature, also an initial attitude state – a pitch rate of 10 deg/s – had to be assumed since SCARAB is a six degrees-of-freedom simulation tool.

In principle, we were able to confirm the demise altitudes from ORSAT-J with SCARAB, as shown in Fig.5, although the SCARAB demise altitudes are generally up to 1 km lower. This appears acceptable to us, considering the amount of assumptions which had to be made.

Table 3. Initial conditions

Altitude [km]	78
Longitude [deg]	272
Latitude [deg]	0
Velocity [km/s]	7.14
Flight Path Angle [deg]	-0.805
Azimuth Angle []	East
Temperature [K]	300
Pitch Rate [deg/s]	10

The small range of demise altitudes (74.3–72.4 km) also corresponds to a short time window (32.5–46.8 s). All tanks demise within less than 50 s after their release to the re-entry flow. The tank trajectories are shown in Fig. 6.

Figs. 7 and 8 show the maximum (stagnation point) and average heat fluxes on the tanks during re-entry. The average heat fluxes have been determined by dividing the integral heat input with the nominal tank surface area

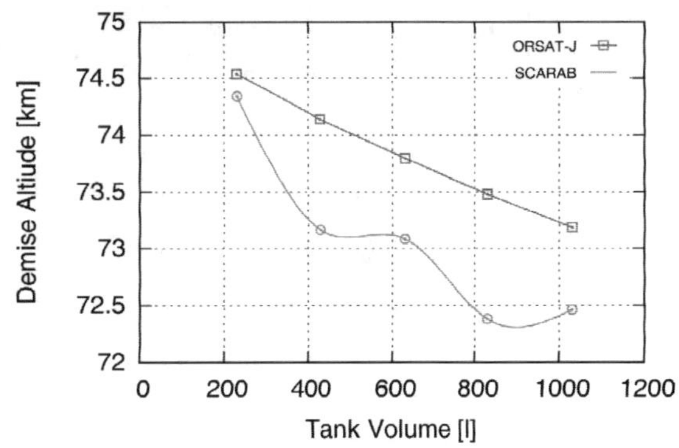

Figure 5. Demise altitudes

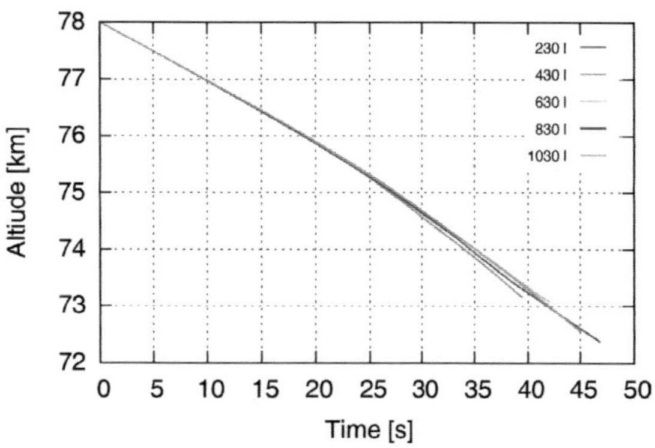

Figure 6. Tank trajectories (SCARAB simulations)

(see Tab. 1). The local (maximum) heat load on the tanks is in the order of 250–500 kW/m², while the average heat load on the whole tank surface is around a factor of four lower (before the melting starts and the actual surface area is reduced compared to the nominal tank surface area).

Fig. 9 shows the dynamic pressure history during reentry. The dynamic pressure is in the order of 650–850 Pa.

Fig. 10 shows the evolution of the maximum temperature during re-entry. Aluminum liner melting starts already after less than 10 s for the smallest and biggest tank. These tanks are actually flying stably because of their smaller L/D ratio. The initial pitch rate is compensated by the aerodynamic stabilization. That is why the temperature of the front increases much more rapidly. For the tumbling medium size tanks, the aluminum liner starts to melt after 15–20 s.

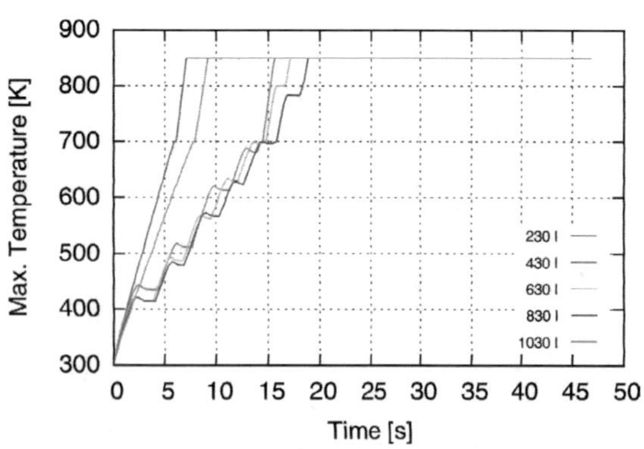

Figure 7. Maximum (stagnation point) heat flux (SCARAB simulations)

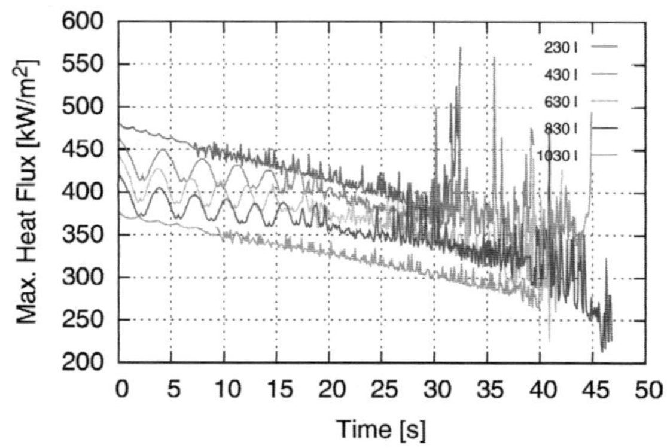

Figure 8. Average heat flux (SCARAB simulations)

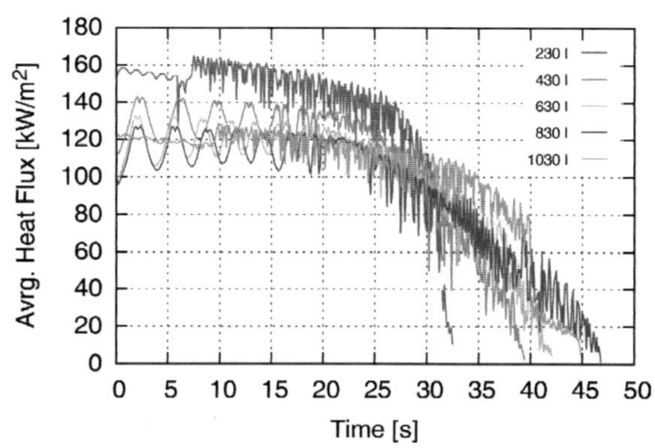

Figure 9. Dynamic pressure (SCARAB simulations)

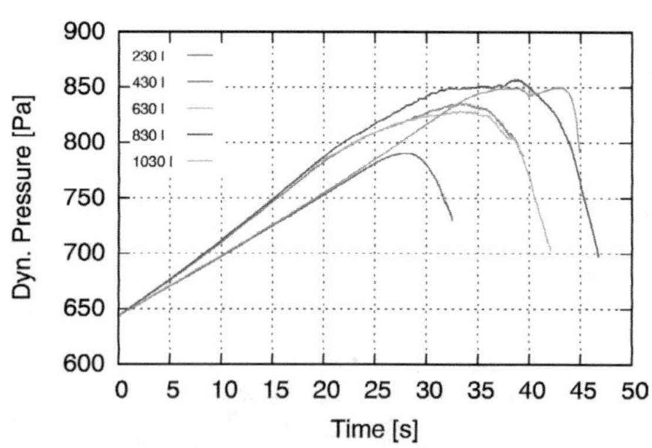

Figure 10. Maximum temperature (SCARAB simulations)

4. JAXA LABORATORY TESTS

After these promising results on the demisability of carbon-fiber overwrapped aluminum liner tanks, achieved by ORSAT-J simulations, Masuoka et al. con- ducted laboratory tests for CFRP/aluminum material samples. Laser heating and arc-jet heated wind tunnel tests were performed.

The laser heating tests were conducted with three different test conditions, as shown in Tab. 4. The CFRP wall thickness was 1.6 mm, and the aluminum wall thickness was 1 mm.

Table 4. Laser heating test conditions

$\dot{q}$ [kW/m^2]	p [Pa]
200	5
500	5
500	9000

The arc-jet heated wind tunnel tests were conducted with a heat flux of 200 kW/m^2. Unfortunately, [1] does not give any information about the dynamic pressure for these tests. The CFRP wall thickness was 0.5 mm, and the aluminum wall thickness was 1 mm.

The used heat loads apparently tried to reproduce either the local maximum heat flux (500 kW/m^2), or the average heat flux (200 kW/m^2) during re-entry. The used pressure conditions for the laser heating tests were either two or- ders of magnitude too low, or one order of magnitude too high. This might have an influence on the chemical destruction processes (resin pyrolysis, fiber oxidation) observed during the tests.

The general results of the tests were the following: For the laser heating tests with low ambient pressure (5 Pa), the resin of the CFRP samples was destroyed, but no aluminum liner melting occurred after testing times of 60.8 s with 200 kW/m^2, and 180.7 s with 500 kW/m^2. Only the laser heating test with high ambient pressure (9000 Pa) resulted into aluminum liner melting after 535 s with 500 kW/m^2. For the arc-jet heated wind tunnel tests, aluminum liner melting was observed after 85 s with 200 kW/m^2.

These test results do remarkably disprove the previously shown simulation results. In the simulations, the aluminum liner fails already after less than 10 s under heating conditions comparable to the 500 kW/m^2 test scenarios, and after less than 20 s under heating conditions comparable to the 200 kW/m^2 test scenarios. This is even more remarkable since the CFRP wall thicknesses used for the tests were smaller (0.5–1.6 mm) compared to the wall thickness used for the simulations (3.1 mm).

5. CFRP MATERIAL PROPERTIES AND ABLATION MODEL

The only reason for the disagreement between simulations and tests identified in the previous section can be the CFRP material properties or the inherent ablation model used for the simulations.

A critical review of the CFRP material properties shown in Tab. 2 reveals that CFRP is actually treated as a quasi-metal with a metal-like demise process. In the ablation model, the material is heated until a "melting temperature" upon which the "melting" process consumes a certain amount of heat ("heat of melting") before it is considered as completely demised. This "heat of melting" is assumed to be very low (three orders of magnitude lower than for an actual metal). The resulting effect is that, by using this kind of ablation model and material properties for CFRP, the material does disappear almost instantaneously upon reaching the "melting temperature". With only 700 K, this is even lower than the melting temperature of aluminum. This can be interpreted as an instantaneous "burn-up" upon reaching a critical temperature limit.

In reality, the ablation process for CFRP is much more complex. The process is more chemical and not a simple physical phase change from solid to liquid. The resin is decomposed by the heat (pyrolysis), and the fibers react with the oxygen reaching the surface. Endothermal and exothermal reactions result in additional heat sinks and sources. Outgassing reaction products cause additional cooling and blocking effects (i.e. effects reducing the effective external heat fluxes). Extremely high surface temperatures can make radiative cooling very relevant, combined with very high temperature gradients perpendicular to the wall due to the low heat conductivity of the CFRP in this direction. The material properties are very variable with time and position within the material.

HTG has developed a CFRP ablation model for SCARAB which takes into account all these effects [9]. This ablation model has recently been validated with arc-jet heated wind tunnel results conducted in the LBK facility of DLR in Cologne ("Characterization of Demisable Materials", CHARDEM, ESA Contract No. 40000110952/14/NL/PA). Fig. 11 shows the results for a real CFRP overwrapped tank sample with a 1 mm titanium liner and 8 mm CFRP. The sample was taken from a high pressure gas tank. That is why the liner is made from titanium, and the wrapping is relatively thick. Thermocouples have been placed at various depths relating to the front surface of the sample. The surface temperatures were measured with two pyrometers. This test was conducted with a cold wall stagnation point heat flux of 1.2 MW/m^2, corresponding to the peak heat flux typical for an uncontrolled re-entry from LEO. The dynamic pressure for these tests was 6,100 Pa.

Although simulated temperature results still show some deviations compared to the measurements, the overall observed ablative mass loss of 22.4 grams could be reproduced numerically quite well with a value of 24.2 grams.

Most striking in these wind-tunnel results is that the ablation of the first 3 mm of CFRP takes 112 s, and the ablation of the following 3 mm of CFRP takes another 120 s. This gives a mean ablation rate of only 0.026 mm/s – for a quite high heat load of 1.2 MW/m^2. Considering the fact that the "hot" phase of re-entry (during which ablation can occur) after release at 78 km altitude is not longer than 150 s, and that the average heat load on a tank is most likely much lower (100–200 kW/m^2), the maximum demisable wall thickness of a CFRP tank wrapping is probably less than 1 mm.

It has to be noted that these results were achieved for a certain type of CFRP: Toray T1000 fibers and UF3325 TCR resin. For another type of CFRP (XN-60-60S fibers and CP0031 resin), the results agreement was not as good. In fact, for the latter type of CFRP, the SCARAB CFRP ablation model underpredicted the mass loss by a factor of five. The reason for this disagreement was the delamination of undemised CFRP layers during the wind tunnel tests. Such an effect is not covered by the ablation model. It is also not clear if such an effect would occur during a real re-entry. The CFRP samples have been cut from larger pieces of CFRP. Therefore, they have sort of "open" edges. In reality, CFRP structures like in tank wrappings would be "closed" due to the winding manufacture process.

However, since the Toray/TCR CFRP is actually being used for tank over-wrapping, and since this type of CFRP does not show undemised layer detachements during the wind tunnel tests, the current SCARAB CFRP model appears to be valid for this application case. For other CFRP types and application cases, variations of the CFRP ablation model have to be developed and qualified.

6. CFRP ABLATION SIMULATIONS

This section summarizes the results of the SCARAB re-entry simulations for the JAXA tanks, applying the full CFRP ablation model to the CFRP over-wrapping layer. The re-entry trajectories are shown in Fig. 12. All tanks survive down to the ground.

Fig. 13 shows the total mass loss for each tank. The mass loss is between 1.5–4.5 kg. Please note that the ablation model stops at Mach number 3 in these simulations in order to keep the hypersonic approximations of SCARAB valid. The trend of the curves indicate that the mass loss will to some extent continue even in the trans- and subsonic flow regime, but this cannot be simulated with the current version of SCARAB.

The overall mass loss of the tanks is primarily driven by the pyrolysis of the resin. Figs. 14 and 15 show the relative mass loss for the carbon fibers and the resin, respectively. While the resin mass loss can reach orders of magnitude between 20–35 %, the carbon fiber mass loss stays between 2–4 %. Aluminum liner melting occurs only for the 230 l tank (only 1 gram; not shown in figures).

It is also interesting to note that the maximum temperature (Fig. 16) for all tanks stays below 1000 K, and even below 850 K (melting temperature of aluminum) for the medium size tanks (430–830 l).

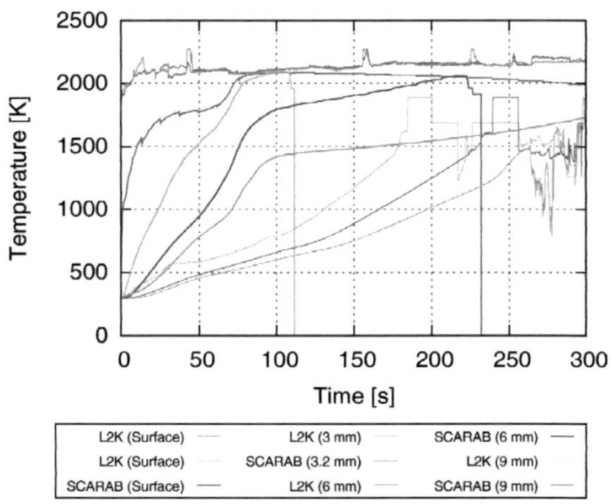

Figure 11. CHARDEM wind tunnel test

7. SUMMARY, CONCLUSIONS, OUTLOOK

In this paper we have evaluated the re-entry simulations conducted by Masuoka et al. [1] for CFRP over- wrapped aluminum liner tanks. The ESA SCARAB soft- ware was used for the numerical simulations. By using "metal-like" material properties for the CFRP wrap- ping, the ORSAT-J results could be reproduced (complete demise). A critical review of the laser heating and arc-jet heated wind tunnel results also conducted by Masuoka et al. revealed that these laboratory experiments actually disproved the "metal-like" ablation model approach.

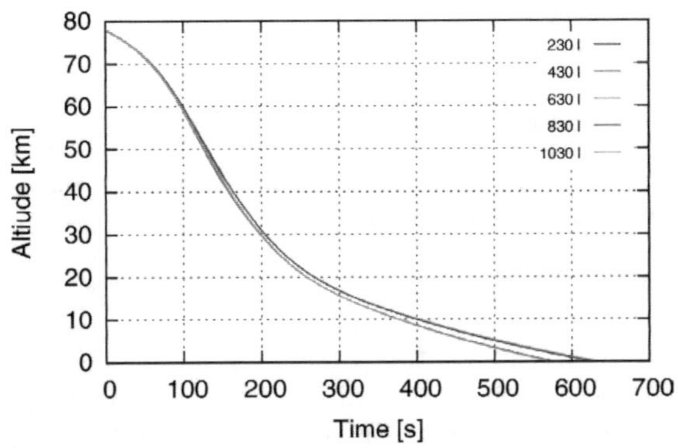

Figure 12. Tank trajectories (SCARAB simulations)

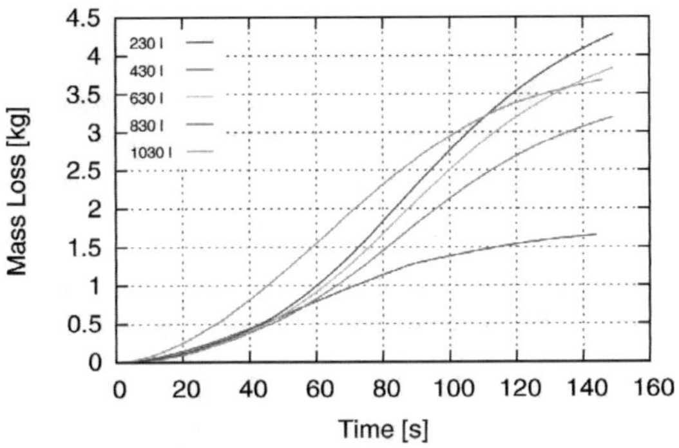

Figure 13. Mass loss (SCARAB simulations)

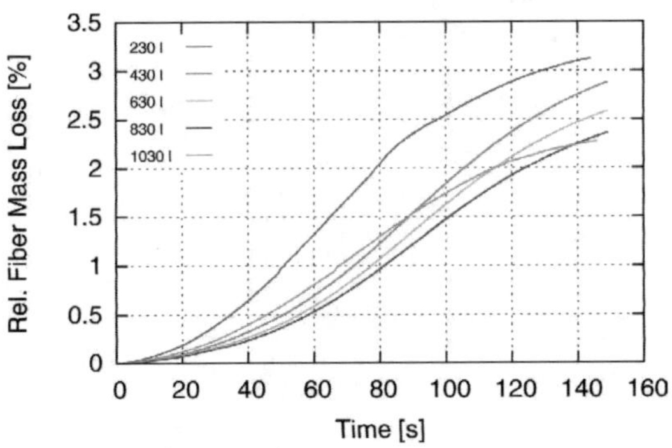

Figure 14. Relative fiber mass loss (SCARAB simulations)

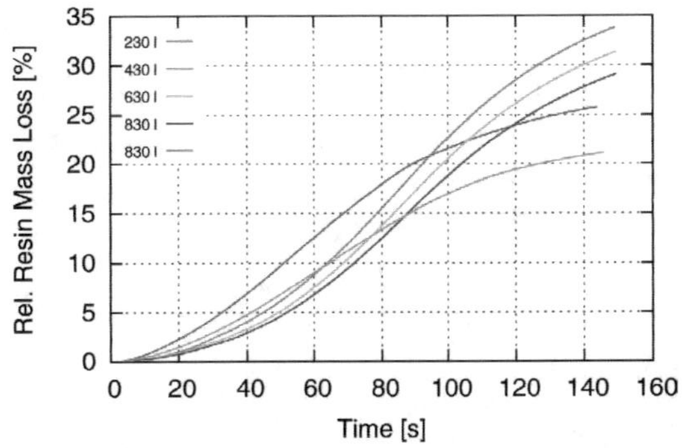

Figure 15. Relative resin mass loss (SCARAB simulations)

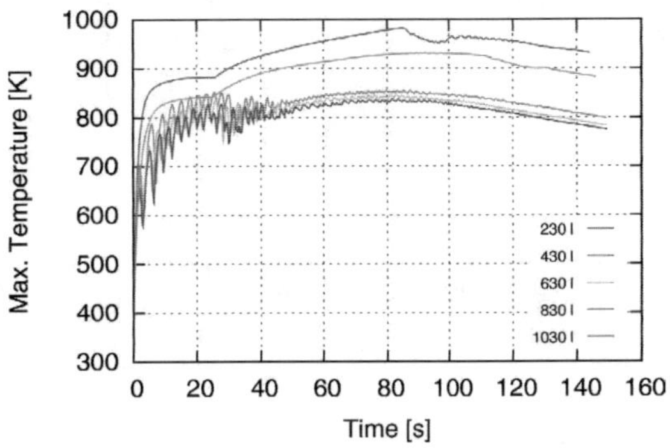

Figure 16. Maximum temperature (SCARAB simulations)

The implicitly assumed instantaneous burn-up of CFRP at only 700 K with almost no heat consumption is by far too optimistic with regard to realistic ablation simulation.

A full CFRP ablation model developed for SCARAB, taking into account resin pyrolysis and fiber oxidation, and validated with arc-jet heated wind tunnel experiments, has been applied to the JAXA tanks. These simulations provided results showing the complete survival of the tanks. The low overall mass loss (less than 5 kg for all tanks) is primarily caused by resin pyrolysis. How- ever, the relative resin mass loss is always less than 35 %. The relative carbon fiber mass loss stays below 4 %. The aluminum liner does not reach melting conditions in most of the analyzed cases.

In conclusion, it has to be stated that a positive effect of the D4D approach of substituting titanium tanks by ones made from CFRP over-wrapped metallic liners is questionable for re-entries from LEO if a CFRP behaving like Toray T1000 fibers and UF3325 TCR resin is used. Although the CFRP ablation model of SCARAB is already quite elaborated, further development work and experimental testing/ validation has to be done in order to achieve a generic ablation model, applicable to various types of CFRP.

The "metal-like" ablation model approach could still be an interesting (less computation intensive) alternative, but more realistic (equivalent) material properties for the "melting temperature" and the "heat of melting" have to be found, i.e. derived from material demise tests, as shown in [8].

Of important relevance will also be the question of the structural integrity of such tanks during re-entry. Although the aluminum liner might not reach melting conditions, its reduced strength at temperatures close to melting can result in deformations or even rupture of the tank. Especially the latter could increase heat loads and demisability. However, if no complete demise is achieved, and if more than one tank fragment survive the re-entry, the on-ground casualty risk will probably be higher than for standard titanium tanks.

8. REFERENCES

1. Masuoka T., et al., (2013). *Demise Characteristics Evaluation for Melting-Promotion Type Tank*, Presented at the 49th AIAA/ASME/SAE/ASEE Joint Propulsion Conference and Exhibit, AIAA Paper 2013-3984

2. Masuoka T., et al., (2015). Demise Characteristics Evaluation for Melting-Promotion Type Tank, AIAA *Journal of Propulsion and Power*, Vol. 31, No. 3, pp. 981–985

3. Lips T., Fritsche B., Homeister M., et al., (2007). Re- entry Risk Assessment for Launchers - Development of the New SCARAB 3.1L, *Proceedings of the Second IAASS Conference*, ESA SP-645, ESA Communication Production Office, Noordwijk, The Netherlands

4. Rochelle W.C., Kirk B.S. Ting, B.C., (1999). *User's Guide for Object Reentry Analysis Tool (ORSAT) - Version 5.0*, Vol. 1, JSC-28742, NASA Johnson Space Center, Houston, Texas, United States of America

5. Dobarco-Otero J., Smith R.N., et al., (2003). *Upgrades of Object Reentry Survival Analysis Tool (ORSAT) for Spacecraft and Lauch Vehicle Upper Stage Applications*, presented at 54th International Astronautical Congress, Bremen, Germany

6. Anonymous (2005). *The Orbital Debris Quarterly News*, NASA Orbital Debris Program Office, NASA Johnson Space Center, Houston, Texas, USA, Vol. 9, No. 4

7. Dobarco-Otero J., Smith R.N., et al., (2005). *The Object Reentry Survival Analysis Tool (ORSAT) - Version 6.0 and its Application to Spacecraft Entry*, presented at the 56th International Astronautical Congress, Fukuoka, Japan

8. Lips T., Dobarco-Otero J., et al., (2005). Comparison of ORSAT and SCARAB Reentry Survival Results, *Proceedings of the Fourth European Conference on Space Debris*, ESA SP-587, ESA Publications Division, Noordwijk, The Netherlands

9. Fritsche B., (2013). Modelling the Thermal Decomposition of Carbon Fibre Materials During Re-Entry, *Proceedings of the Sixth European Conference on Space Debris*, ESA SP-723, ESA Communications, Noordwijk, The Netherlands

DEMISE AND SURVIVABILITY CRITERIA FOR SPACECRAFT DESIGN OPTIMIZATION

Mirko Trisolini[(1)], Hugh G. Lewis[(2)], Camilla Colombo[(3)]

[(1)] *University of Southampton, University road, Southampton, UK, Email: m.trisolini@soton.ac.uk*
[(2)] *University of Southampton, University road, Southampton, UK, Email: H.G.Lewis@soton.ac.uk*
[(3)] *University of Southampton, University road, Southampton, UK, Email: C.Colombo@soton.ac.uk*

ABSTRACT

In a period where the evolution of the space environment is causing increasing concerns for the future of space exploitation and sustainability, the design for demise philosophy has gained an increased interest. However, satellites designed for demise still have to survive the space environment polluted by space debris. Within this context we are developing a model to evaluate the effect of preliminary design choices on the survivability and on the demise of a spacecraft configuration. Considering common spacecraft components such as tanks and batteries, a set of maps are presented, which shows the variation of survivability and demise criteria as function of the component geometry and material. Furthermore, a preliminary multi-objective optimization is performed to evaluate a simple spacecraft configuration and define an optimal design according to the demise and the survivability criteria.

1. INTRODUCTION

In the context of a sustainable use of the outer space, the major space faring nations and international committees [1-3] have proposed a series of debris mitigation measures, including the de-orbiting of spacecraft at the end of their operational life. The consequent growth of the ground casualty risk associated with the increased frequency of re-entering objects can be limited by designing a spacecraft through a design for demise philosophy, where most of the spacecraft will not survive the re-entry process. Such strategies may favor uncontrolled re-entry disposal options over the controlled ones, because a spacecraft will be more likely to meet the casualty risk requirements. This could make design for demise a simpler and cheaper alternative for the disposal of satellites at the end of their operational lives [4, 5]. However, a spacecraft designed to demise still has to survive for several years the space environment, which is populated by a large number of space debris and meteoroids. The impact with particles even of millimeter in size can be extremely dangerous for satellites and cause the loss of the mission [6-8]. This means that the spacecraft design has also to fulfill the requirement of survivability against on-orbit debris impacts. The demise and the survivability are both influenced by a set of common design choices such as the outer structure material, the geometry and the shape of the structure [4, 9]. Within this context, we developed two models to assess the demise and the survivability of a preliminary mission design concept. Two criteria are thus presented to evaluate the degree of demise and survivability of a spacecraft configuration against the different spacecraft and components design parameters. The link between the design parameters and the specific spacecraft components such as tanks and battery cells is also taken into account. In addition, mission characteristics such as the mission duration and operational orbit are taken into account, with a particular attention towards Earth observation missions and remote sensing missions. Results shows how spacecraft design parameters and mission constraints affect the design choices when considering the effects on the demise and survivability of a spacecraft. In a preliminary attempt to perform an optimization for the selection of the design parameters, the two developed models are used to construct the fitness functions in a multi-objective optimization framework. In this way, trade-off solutions that consider both the demise and survivability design drivers can be found. As the problem is nonlinear and involves the combination of continuous variable such as the thickness and the size of the objects together with discrete variables like the material type, classical derivative based procedures becomes unsuited. Thus a heuristic optimization approach based on genetic algorithms was selected. The optimization algorithm uses the demise and survivability models and criteria to evaluate possible design choices as function of the characteristics of the object. The solutions are presented as a Pareto front showing the best non-dominated individuals obtained from the evolved initial population.

It is important to consider demisability and survivability requirements since the early stage of the mission design process [5]. A late analysis of such requirements may cause an inadequate integration of these design practice, leading to late changes to the mission, which of course are more expensive and may cause delay in the mission timeline. On the other end, an early consideration of demise and survivability requirements can favor cheaper options such as the uncontrolled re-entry of the satellite, without losing survivability performance and, thus, mission reliability.

2. SURVIVABILITY AND DEMISE MODELS

To analyze spacecraft configurations and components against the demisability and the survivability requirements it is necessary to develop two separate models. The first model allows the user to perform a re- entry analysis of a simplified representation of a spacecraft and to evaluate its demise performance, the second model carries out a debris impact analysis and returns the penetration probability of the analyzed structure as a measure of its survivability during the mission lifetime. The two models are also used to compute the terms that composes the fitness function in a multi-objective optimization process. For this reason, throughout their development, much effort is made to maintain a comparable level of detail and computational time between them. A more detailed description of the model is presented in [10].

2.1 Demise Model

The demise model consists of an object-oriented code [11-14]. The main feature of object-oriented codes resides in the fast simulation of the spacecraft re-entry through the use of a simplified model of the spacecraft structure; this is achieved by reducing the satellite design and its components into primitive shapes (e.g. spheres, cubes). The re-entry trajectory simulation uses a three-degree of freedom dynamics where only drag forces are considered (i.e. ballistic motion) whereas lift and thrust are neglected (i.e. uncontrolled re-entry). In doing so, the computation of the attitude motion of the object is not required, but assumed as random tumbling (in the case of the algorithm developed for this work). The adoption of motion and shape averaged drag coefficients allow determining the pressure forces on each component [12, 15-18]. The computation of the thermal load on the components uses the Detra-Kemp-Riddel correlation [19, 20] to obtain the hypersonic heat rate at each instant of the re-entry and a set of motion and shape averaged shape factor to adjust the heat load to the specific shape considered [15, 21-24].

The re-entry analysis also requires the knowledge of the gravity acceleration and of the Earth's atmosphere characteristics in order to simulate correctly the descent trajectory. The model of the Earth's gravitational potential is zonal harmonic gravity model up to degree 4 [25]. The atmospheric model employed is the 1976 U.S. Standard Atmosphere [26], as implemented by the Public Domain Aeronautical Software (PDAS) [27].

The demise of an object, i.e. its mass loss during the re-entry, is analyzed with a lumped mass model where the temperature of the object remains uniform over the entire volume. After reaching the melting temperature, the object starts losing mass at a rate that is proportional to the net heat flux on the object and to the heat of fusion. The result of the demise simulation is here expressed in terms of an index called Residual Mass Fraction (RMF), which is defined as the ratio between the final mass at the end of the re-entry and the initial mass of the object as Eq. (1) shows.

$$RMF = \frac{m_{fin}}{m_{in}} \tag{1}$$

2.2 Survivability Model

The survivability model analyzes instead the satellite resistance against the impacts of untraceable space debris and meteoroids. The procedure involves representing the spacecraft structure with a panelized schematization of its shape. To each panel we assign a material selected from a predefined library and geometrical properties such as the type of shielding, the wall thickness and the failure modality (i.e. penetration, detached spall and incipient spall are possible options). The survivability model uses the same geometrical elementary shapes of the demise model to represent satellite structures, in order to keep the two models comparable. Beside the geometrical schematization of the satellite, a representation of the space environment is also needed. This is obtained using the European Space Agency (ESA) state of the art software MASTER-2009 [28] that provides a description of the debris environment via flux predictions on user defined target orbit. MASTER-2009 provides the impact fluxes with an impactor diameter between 0.0001 m and 0.1 m, and a set of 2D and 3D flux distributions as a function of the impact azimuth, impact elevation, impact velocity, and particle diameter. Then, we subdivide the space around the satellite in a set of angular sectors and associate to each sector a vector element containing the average of the impact flux, impact direction and impact velocity (see Figure 1).

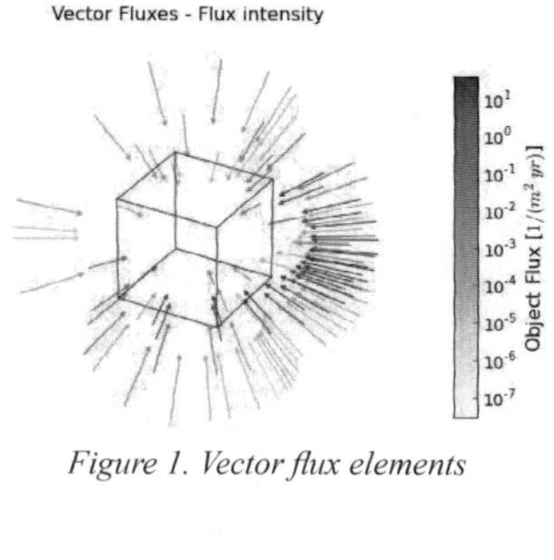

Figure 1. Vector flux elements

The code then uses the vector characteristics of impact flux, direction and velocity previously described to compute the critical diameter corresponding to each panel of the structure using Ballistic Limit Equations (BLEs).

Once obtained the critical diameter, the corresponding critical flux allows computing the penetration probability using Poisson statistics (see Eq. (2)).

$$P^{P}_{j,i} = 1 - e^{-\phi_{c,i} A^{\perp}_{j} t} \tag{2}$$

where the index j identifies the panel and the index i the vector flux. $\phi_{C,i}$ is the critical particle flux in 1/m2yr that is the flux of particles having a diameter bigger than the computed critical diameter, $A^{\perp}_{j}$ is the projected area of the j-th face in the direction of the i-th vector flux and t is the mission duration in years.

The penetration probability ($P^{P}_{j,i}$) is computed in this way for each vector flux on every panel of the structure and then we compute the overall penetration probability (P^{P}) with Eq. (3) [29].

$$P^{P} = 1 - \prod_{j=1}^{N_{panels}} \left(\prod_{i=1}^{N_{fluxes}} \left(1 - P^{P}_{j,i}\right) \right) \tag{3}$$

where N_{panels} and N_{fluxes} are the number of panels in which the structure is schematized and the number of vector flux elements, respectively.

3. SURVIVABILITY AND DEMISE MAPS FOR SATELLITE COMPONENTS

To have a better understanding of a combined demisability and survivability analysis, we present a set of contour maps showing the variation of the RMF (see Eq. (1)) and of the penetration probability (see Eq. (3)) as a function of the thickness, dimension and material of the object. In addition, the orbit characteristics for the survivability and the initial re-entry conditions for the demise are taken into account. Figure 2 shows an example of such contour maps for a cubic shaped object re-entering from an altitude of 120 km at a velocity of 7.3 km/s, with an operational orbit of 800 km of altitude and 98 degrees of inclination, for a mission lifetime of 3 years.

The x-axis and y-axis represent the thickness and the side length of the cube, respectively. The blue contour lines identify the RMF percentage of 1, 50 and 99 percent, meaning that an object with dimension under the 1 percent curve will have less than 1 percent of its mass remaining after the re-entry. The red contour lines, on the other hand, represent the 0.1, 1, and 10 percent penetration probability of a structure for the considered mission orbit and lifetime. The bottom right corner of the map (grey area) identifies a region of non-physical combination of thickness and side length that is when the thickness is greater than half the side length. The arrows indicate in which direction of the contour map the survivability and the demisability are improving. It is possible to observe that, as expected, an increased thickness results in a better survivability performance, whereas a thinner object will, in general, be more demisable. Both criteria instead favor smaller objects: for the survivability, a smaller cross-sectional area means a lower probability of getting impacted by space debris, and for the demisability, a smaller object is less massive and thus in general easier to demise. Of course also physical constraints need to be taken into account as, for example, the overall size of a spacecraft can be strongly influenced by the dimensions of the payload.

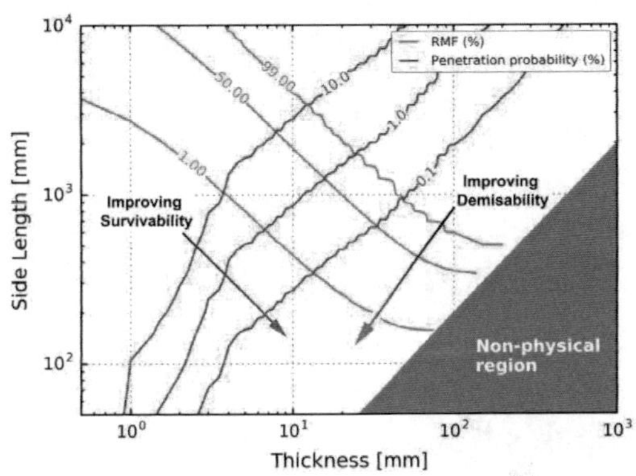

Figure 2. Combined survivability and demisability map

A set of results are here presented taking Earth observation and remote sensing missions as an example. Many of these missions exploit sun-synchronous orbits due to their favorable characteristics that allow the spacecraft to pass over any given point of the Earth's surface at the same local time. Examples of this kind of missions are Landsat 8 [30], MetOp [31] SPOT 6/7 [32] and many others. As already pointed out, sun- synchronous orbits have highly appealing features for Earth observation mission, and for this reason, they are much exploited. An indication about the current population of satellites populating the sun-synchronous region is represented in Figure 3. Figure 3 shows the current population of satellite between an altitude of 400 and 800 km and an inclination between 0 and 150 degrees (data from [33]). Each point represents a satellite, and it is possible to observe a very high concentration of satellites around the 98 degrees inclination. The number of satellites is even higher than it appears from the figure since the points superimpose each other because of the very similar inclinations of the satellites. The plot also includes a color-map showing the penetration probability on an aluminum alloy (7075-T6) cube of 1-meter side length and 4 mm thickness as a function of the orbital inclination and altitude. As it can be observed, the sun-synchronous region is not only very populated, but it is also one if not the most exposed to the debris environment with the highest penetration probability. Following these considerations, a set of three circular orbits with altitude of 600, 700, and 800 kilometers respectively and an inclination of 98 degrees were selected (see Figure 4). In fact, these orbital altitudes and inclination enclose many of the operational orbits of the previously cited sun-synchronous missions. Earth observation and remote sensing missions usually have a lifetime of at least 3 years, and up to 10 years.

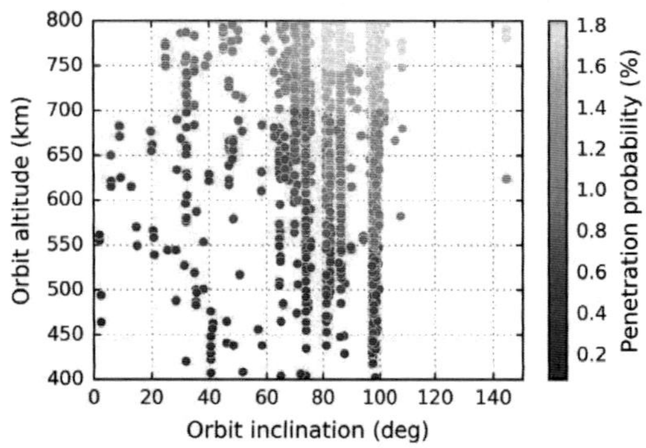

Figure 3. Satellite distribution for orbit between 0 and 150 degrees of inclination and 400 and 800 km of altitude, with associated esteem of their penetration probability

This consideration led to the decision of taking into account 4 different mission durations of 3, 5, 7, and 10 years respectively (see Figure 5).

The difference among the different orbits considered is observable in Figure 4 and it is caused by the different amount of space debris populating the selected orbits.

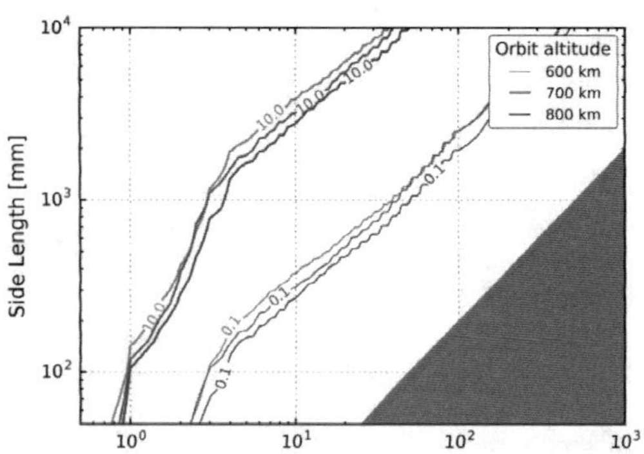

Figure 4. Penetration probability as function of orbit altitude (orbit inclination of 98 degrees)

The 800 km orbit has the highest particle density thus producing greater impactor fluxes, which in turn translate into a higher penetration probability on a structure.

On the other end, as expected, an extended mission lifetime results in a higher penetration probability. From Figure 5 the trend appears to be linear with correspondent penetration probability lines almost equally spaced as the mission duration varies.

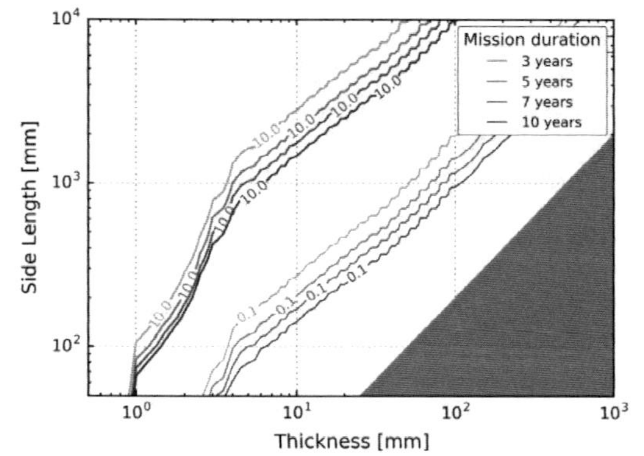

Figure 5. Penetration probability as function of mission duration

A consideration can be done on where the main spacecraft structures and components are located inside these maps; in other words, which are the usual ranges of thickness and size of common spacecraft components. We represents some these components, in particular at five categories which are summarized in Table 1 together with their ranges of thickness and size.

Table 1. Size ranges used in the article for typical spacecraft structures and components

S/C Component	*ts* range (mm)	Size range (mm)
Tanks	0.5 – 15	150 – 500 (radius)
EOS payloads	2 – 20	500 – 1200 (side length)
Typical structures	1 – 10	100 - 2000 (side length)
Reaction wheels	0.5 – 3	70 - 150 (radius)
Battery cells	0.5 - 1	55 – 85 (radius)

The maps presented from Figure 6 through Figure 11 have all a similar structure: two materials for each map, distinguished by the dashing of the lines; two color shades, with cold colors (blues) representing the survivability and warm colors (reds) representing the demisability. The colors gradient is such that darker colors represent better solutions for both the demise and the survivability so, for example, dark blue is better than light blue for the survivability.

The Figure 6 and the Figure 7 present the maps previously described, highlighting the region where most satellite tanks resides. The survivability part of the plots is representative of an 800 km altitude and 98 degrees inclination orbit and for a mission lifetime of 3 years, whereas the demisability analysis has been performed for an initial altitude of 120 km and an initial velocity of 7.3 km/s with an initial flight path angle of 0 degrees. In order to describe the behavior of a tank, a cylindrical shape with the diameter equal to its height was selected. Figure 6 shows the results for aluminum and stainless steel tanks. As it is possible to observe, there is a substantial difference between the two materials for both the criteria considered. The stainless steel is extremely resistant to debris impacts, indeed the 0.1% penetration probability line (dark blue dashed line) is very close to the 10% (light blue solid line) penetration probability line of the aluminum case. This means that there is a difference of two orders of magnitude in the vulnerability of two cylindrical tanks with the same dimensions but different material, i.e. aluminum or stainless steel. On the demise side, the exact opposite happens, as the aluminum configuration is more favorable with respect to the stainless steel one. It is possible to observe how the line representing the 99% RMF (orange dashed line) of the stainless steel crosses the line of the 1% RMF (red solid line) of the aluminum. This, in turn, indicates that when a cylinder made of aluminum has dimensions under the 1% line it will completely demise, whether, if it is made of stainless steel, almost all of its mass will survive.

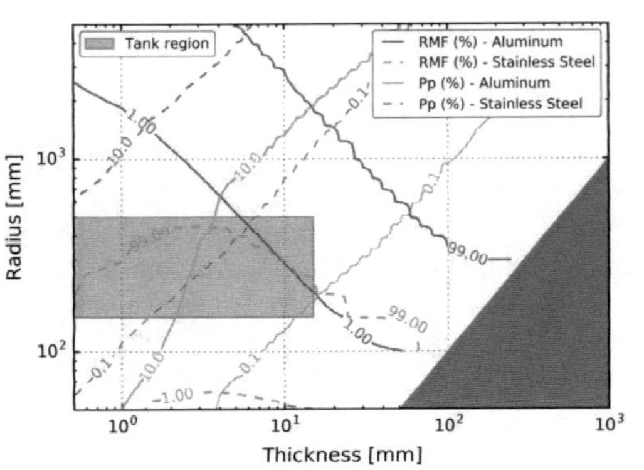

Figure 6. Survivability and demisability map for aluminum alloy and stainless steel with satellite tank region highlight

The Figure 7 shows the same map but for graphite epoxy and titanium material selection. In this case, a consistent difference between the demise behaviors of the two materials can be seen. The titanium tanks are extremely difficult to demise, due to the high melting point and heat of fusion of the titanium alloy. For this reason none of the titanium tanks analyzed fully demise, and only a small fraction (the region under the orange dashed line) suffer a partial demise. On the other extreme, the graphite epoxy components fully demise in almost all the cases analyzed. This is due to the way the graphite epoxy is modelled in DAS, that is as an equivalent material with very low heat of fusion. Such a model is used to render the charring behavior of the graphite epoxy. In fact, as the material reaches the melting temperature, it becomes very fragile and starts to char, thus making it much more demisable. This behavior is also the reason why only the 99% RMF line is represented on the graph: all the other lines almost superimpose the 99% line because the very law heat of fusion causes a very rapid demise once the melting temperature is reached.

On the survivability side, the difference is less evident but still important. As expected, the graphite epoxy is more vulnerable than the titanium alloy.

Looking at the shaded region representing common tanks dimensions, it is evident that tanks made of stainless steel will most probably survive the re-entry but they will be, at the same time, very resistant to particle impacts, at least one order of magnitude more than the other materials. On the other end, an aluminum tank will almost certainly demise but will also be much more vulnerable to debris impact. We can observe similar results for the graphite epoxy, which is more demisable with respect to aluminum but also has a higher penetration probability than the aluminum. For what concerns titanium, its resistance is higher than the one provided by aluminum and graphite epoxy but lower than the one stainless steel can provide.

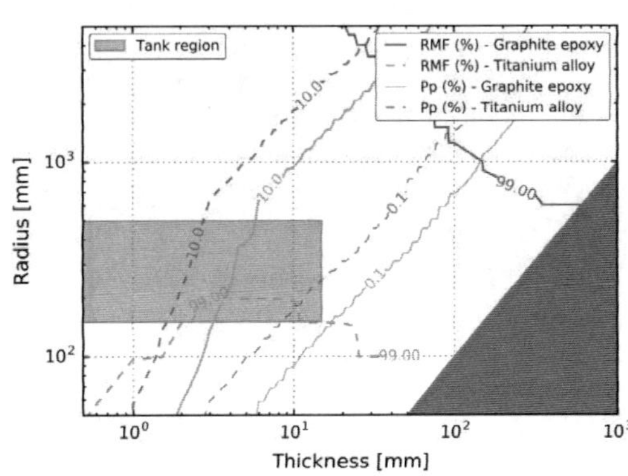

Figure 7. Survivability and demisability map for graphite epoxy and titanium alloy with satellite tank region highlight

Combining this with the fact that titanium is the least demisable among the four materials analyzed makes it the worst trade off solution.

At this point, it is important to discuss the region selected to represent cylindrical tank dimensions. The shaded area presented in Figure 6 and Figure 7 is quite extended since it covers many different options for tank designs. Now, most of spacecraft tanks are manufactured with stainless steel and titanium; usually theses tanks have thicknesses in the order of millimeters [4, 34, 35]. However, also a sub millimeter portion is represented in the graph in order to take into account for tank liners. Liners constitute the inner part of Composite Overwrapped Pressure Vessels (COPV) and are usually under the order of millimeter in size [4, 34]. The upper limit selection results from the composite part of the COPV tanks, which is made of graphite epoxy or similar composite materials.

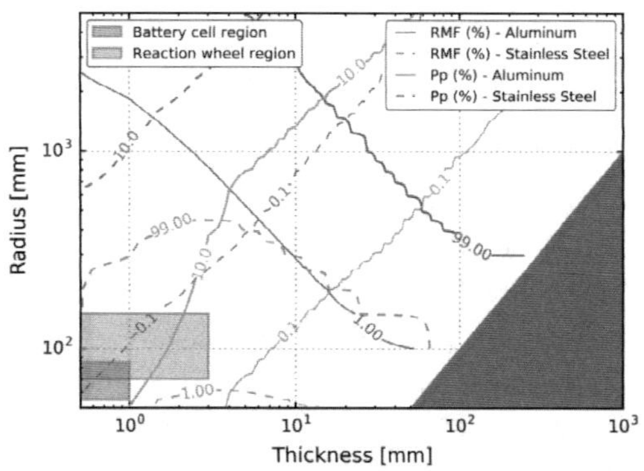

Figure 8. Survivability and demisability map for aluminium alloy and stainless steel with battery cell and reaction wheel region highlight

We present an analogous set of maps highlighting regions with a combination of radius and thickness typical of other spacecraft components. The components considered are battery cells and reaction wheels. Both objects are schematized with right cylinders (i.e. the height equals the diameter) and their regions are presented together in Figure 8 and Figure 10. The battery cell region and the reaction wheel region are represented with a green and pink area respectively.

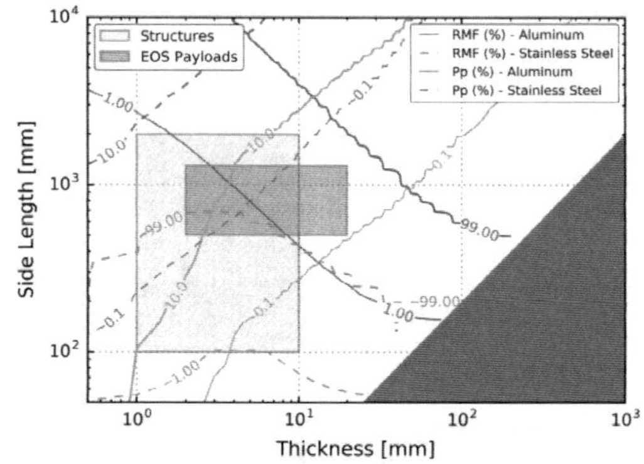

Figure 9. Survivability and demisability map for aluminium alloy and stainless steel with typical S/C structure and EOS payload region highlight

These regions were estimated looking at catalogues of battery cells [36] and reaction wheels [37, 38] manufacturers and using preliminary design relationship taken from [39]. We decided to represent these two components since they can usually survive the re-entry, thus posing risk for people on the ground. Both reaction wheels and battery cells are usually made of titanium and stainless steel; from the orange contours in Figure 8 and Figure 10 it is clear that these two materials produces non-demisable solutions.

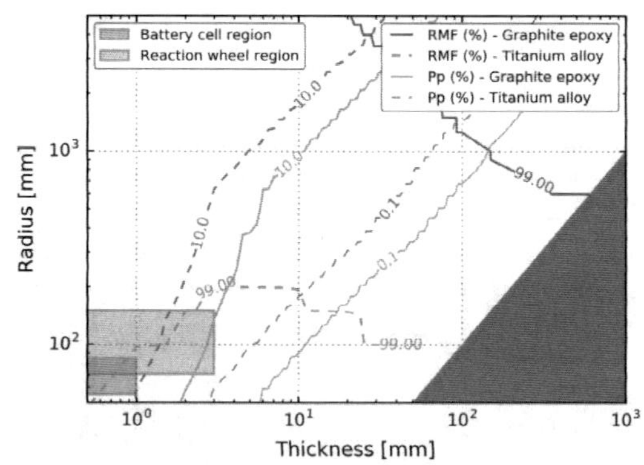

Figure 10. Survivability and demisability map for graphite epoxy and titanium alloy with battery cell and reaction wheel region highlight

In order to have demisable solutions one of the most effective solutions is to change the component material. Changing the material from steel to aluminum would make both this component demise upon re-entry. For battery cells, it is possible to consider Li-ion batteries instead of NiH and NiCd. The latter, in fact, are manufactured with stainless steel vessels whether the former are usually made of aluminum [4].

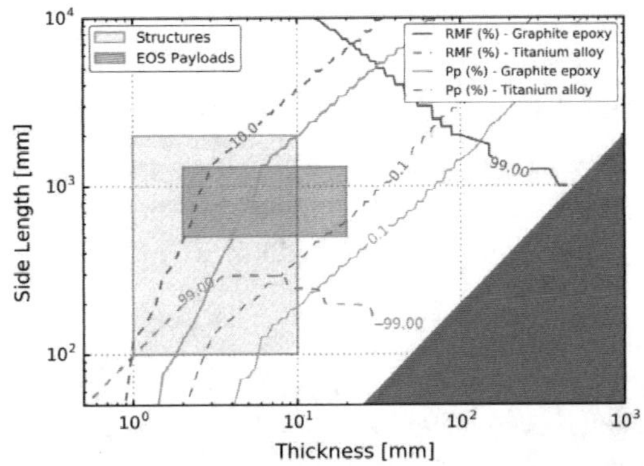

Figure 11. Survivability and demisability map for graphite epoxy and titanium alloy with typical S/C structure and EOS payload region highlight

This change of course has to come to a price in term of survivability: a more demisable aluminum battery cell will be certainly more vulnerable to debris impacts. Of course changing the material of the object is not the only option towards a more demisable configuration: considerations about the aspect ratio of the component, the position inside or outside the spacecraft are also extremely important. In addition, it is not important only for the demise but for the survivability.

It is beyond the scope of this article to investigate such possibility but they will constitute the natural development of the current work for a more detailed trade off design of spacecraft configuration for demise and survivability.

A final set of maps presented in Figure 9 and Figure 11 show the regions of dimensions for common spacecraft component and for Earth Observing System (EOS) payloads. In this case, we generated the maps for a cubic shaped structure since it better resembles the shape of common satellite structures and components such as electronic boxes and power units. The same orbital conditions have been used for the survivability analysis (800 km of altitude, 98 degrees of inclination and 3 years mission duration), as well as the same initial conditions for the demise (120 km of initial altitude, 7.3 km/s initial velocity and 0 degrees initial flight path angle). The limits for the typical spacecraft structure were taken form [35], whether the limits for the EOS payloads have been deduced from the data available for actual payloads used into NASA Earth Observing Systems Missions [40]. Their dimension were traced back to a cubic shape with equivalent volume and the range of thickness obtained assuming they were made with the lightest and heaviest material considered, i.e. graphite epoxy and stainless steel respectively. Most spacecraft structures and component casings are manufactured with aluminum. It is possible to observe that such option produces configurations that usually demise. Considering this, in most situations a configuration could be optimized by changing the geometry and the thickness of a components rather than the material itself unless specific needs for demise or survivability arises. Possibly, the most convenient option would be to switch towards a graphite epoxy configuration that provides a much better demisability, sacrificing some protection from the debris environment. This could be the case in missions where the satellites needs to have a very high rigidity such as the case of the GOCE spacecraft [41], whose primary structure is in fact made entirely from graphite epoxy. On the payload side, the situation is much more variable since there is usually a wide variety of instruments and sensors. Their requirements and integration with the satellite are also less flexible and a change of material or geometry may be possible but not in every situation. Other options could be exploited in case of payloads.

4. MULTIOBJECTIVE OPTIMIZATION

The ultimate goal of the survivability and demise models will be their implementation into an optimization framework. The aim of this framework is to find optimized trade-off preliminary spacecraft configurations, which satisfies both the survivability and the demise requirements. In this way a more integrated approach to both this requirements can be achieved since the early stages of the mission design, thus saving late changes in the project that can lead to increased costs and delays.

A multi-objective optimization problem was formulated, whose fitness functions are represented by the criteria described in section 2.1 and 2.2. In its general formulation, a multi-objective optimization problem can be described in mathematical terms as:

$$\begin{aligned} &\text{Min / Max} \quad f_m(x), && m = 1,2,\ldots,M; \\ &\text{Subject to} \quad g_j(x) \le 0, && j = 1,2,\ldots,J; \\ &\qquad h_k(x) = 0, && k = 1,2,\ldots,K; \\ &\qquad x_i^{(L)} \le x_i \le x_i^{(U)}, && i = 1,2,\ldots,n. \end{aligned} \tag{4}$$

where x is a solution vector. The first line in Eq. (4) represents the set of m objective functions, and lines two to four represent constraints to the problem in form of inequalities, equalities and boundaries respectively. In the case in exam the constraints can be represented by dimensions limitations, mass and volume upper limits, structural resistance of the component, etc.

In multi-objective optimization, there is not usually a solution that minimizes all the objective functions simultaneously. Therefore, the concept of Pareto optimality has to be introduced. A Pareto optimal solution is a solution that cannot be improved in any of the objectives functions without producing degradation in at least another objective [42]. Expressed in mathematical terms, a solution x^1 is said to Pareto dominate a second solution x^2 if:

$$f_i\left(x^1\right) \le f_i\left(x^2\right) \ \forall \ i \in \{1,2,...,M\} \quad \text{and}$$
$$f_i\left(x^1\right) < f_i\left(x^2\right) \ | \ \exists \ j \in \{1,2,...,M\} \tag{5}$$

A Pareto optimal solution is a solution that is not dominated by any other solution. The set of Pareto optimal solution is referred as to the Pareto front.

In general, there is a large variety of optimization strategies. Many of these methods rely on the knowledge of the derivative of the functions to optimize. This requires the functions to have specific mathematical properties, i.e continuous functions and derivatives. For this reason they are not always applicable, especially when the problem is very complex. Other optimization strategies, such as search algorithms, instead do not require the knowledge of the derivative of the functions. They compare the target function at points in a distance defined by a step function and move until no further improvement is found. A final set of algorithms is represented by heuristic algorithm like genetic algorithm or simulated annealing, which applies mechanisms found in nature to the optimization of a problem. In particular, genetic algorithms use the principles of natural evolution: starting with a certain population of possible solutions, they evolve it using operators like mutation, selection and crossover. The selection of the individuals in the population that carry on in the evolution depends on their quality as well as on chance. These kinds of algorithms of course rely on a certain amount of randomness both in the population generation and in the genetic operators that evolve the population itself [43, 44].

In the case in exam, the problem, in its most complex form, has to take into consideration very diversified parameters, such as the shape of the object (sphere, box, and cylinder), its dimensions and material. As it possible to observe, these parameters are a mix of discrete variables such as the material and the shape, and continuous variables like the size and thickness of the component. Considering also a future development of the project, other parameters to take into account will be the position of the component inside and outside the spacecraft and the different type of shielding options adoptable.

With all these consideration in mind, the decision to adopt a genetic algorithm was taken. Genetic algorithms were selected because of their extended documentation and their relative simplicity of implementation. Moreover they are suitable for complex problems with a combination of continuous and discrete variables such as the preliminary design of a spacecraft configuration [45]. The development of such an optimization framework is at this point of the work at its early stages, but it is intended to provide a tool for a preliminary optimization, which will take into account survivability and demise requirements since the first stages of the mission development.

The implementation of the genetic algorithm was carried out using the Python framework Distributed Evolutionary Algorithms in Python (DEAP) [46]. DEAP provides the user with a series of pre- implemented multi-objective optimization strategies such as the Non-dominated Sorting Genetic Algorithm II (NSGA II) [44] and the Strength Pareto Evolutionary Algorithm 2 (SPEA2) [47] together with a set of standard mutation and crossover operators readily available.

For the case in exam, we set up a simple optimization using the NSGAII selection mechanism provided in DEAP, selecting the Simulated Binary Bounded crossover mechanism and the Polynomial Bounded mutation operator. As is commonly done in genetic algorithms, the crossover probability was set to a high value (0.9) and the mutation probability to a low value (0.05). The initial population has 30 individuals and the evolution is carried out for just 10 generations; every generation a number of offspring equal to the initial population is generated and evaluated by the NSGAII algorithm. The individuals coming from the selection mechanism carry on further in the evolution. The number of retained individual is again equal to the size of the initial population. As mentioned before, the objective functions selected are the two criteria previously described, and expressed by Eq. (1) and Eq. (3). The search space for the optimization is bounded with the limits showed for the satellite tanks (see Table 1) with hard boundaries, and the only variables considered in the optimization were the radius and the thickness of the cylinder.

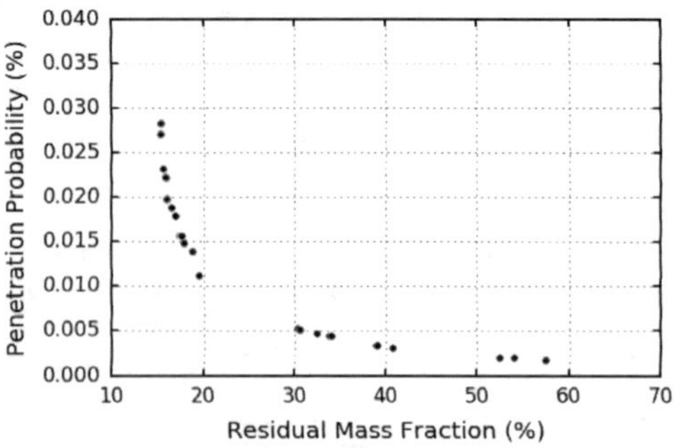

Figure 12. Pareto front

Figure 12 shows the Pareto front for the described simulation, for stainless steel tanks. Figure 13 shows the corresponding individual plotted against the correspondent survivability and demisability map. The result is what we were expecting: the optimizer identifies the solutions with the smaller diameter because, without any further constraints, it is the configuration that favors both the demise and the survivability. A smaller cylinder has less mass and it is thus more demisable, and at the same time, has a lower external surface, which in turn means a lower probability of getting hit by space debris.

The thickness on the other hand, has a maximum values that correspond to the biggest allowable thickness, which assures the highest possible survivability index. The lower value instead, in this case, does not correspond to the lower possible thickness since the highest RMF index is achieved at around 3 mm thickness. We can note that this is the value where we have the peak in the RMF contour lines for the stainless steel (see Figure 13).

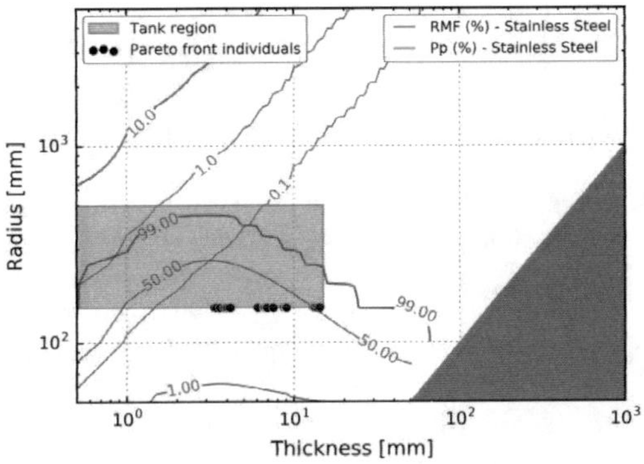

Figure 13. Pareto front individuals' survivability and demise map

5. CONCLUSIONS

A novel approach towards considering in a more integrated way the design requirements arising from the survivability and the design for demise was presented. Two models have been developed and two criteria to evaluate the level of demise and survivability have been presented. A set of maps showing simultaneously the survivability and demisability index as function of size, thickness, and material were produced. The maps shows the competing behavior of the two design requirements and have been contextualized adding satellite structures and components boundaries and considering orbit specific to Earth observation missions.

As the overall aim is to consider the two competing requirements and how they affect the preliminary design of a spacecraft, a multi-objective optimization framework looks like the natural step forward for the project. With this in mind, a preliminary optimization was performed, using as objective functions the developed criteria.

Future effort will be devoted towards the further development of the two models for the survivability and demise with the complementary objective of merging them in the outlined multi-objective optimization framework.

6. ACKNOWLEDGMENTS

Part of this work was funded by EPSRC DTP/CDT through the grant EP/K503150/1

7. REFERENCES

1. B. O'Connor 2008. Handbook for Limiting Orbital Debris. NASA Handbook 8719.14. *National Aeronautics and Space Administration*, Washington, DC.

2. F. Alby, D. Alwes, L. Anselmo *et al.* 2004. The European Space Debris Safety and Mitigation Standard. Advances in Space Research, 34, 1260-1263.

3. F. Schäfer, M. Lambert, E. Christiansen et al. The inter-agency space debris coordination committee (IADC) protection manual. 4th European Conference on Space Debris, 2005.

4. P. M. Waswa, and J. A. Hoffman 2012. Illustrative NASA Low Earth Orbit spacecraft subsystems design-for-demise trade-offs, analyses and limitations. *International Journal of Design Engineering*, 5, 21-40.

5. P. M. B. Waswa, M. Elliot, and J. A. Hoffman 2013. Spacecraft Design-for-Demise implementation strategy & decision-making methodology for low earth orbit missions. *Advances in Space Research*, 51, 1627-1637.

6. R. Putzar, and F. Schäfer, *Vulnerability of spacecraft equipment to space debris and meteoroids impacts*, Final Report, vol. 1, European Space Agency, 2006.

7. L. Grassi, F. Tiboldo, R. Destefanis *et al.* 2014. Satellite vulnerability to space debris – an improved 3D risk assessment methodology. *Acta Astronautica*, 99, 283-291.

8. E. Christiansen 2009. Handbook for Designing MMOD Protection. Huston, Texas, USA: NASA.

9. R. L. Kelley 2012. Using the Design for Demise Philosophy to Reduce Casualty Risk Due to Reentering Spacecraft.

10. M. Trisolini, H. G. Lewis, and C. Colombo. Survivability and Demise Criteria for Sustainable Spacecraft Design. International Astronautical Conference, 2015 Jerusalem.

11. NASA. 2015. *Debris Assessment Software* [Online]. NASA Orbital Debris Program Office Available: http://orbitaldebris.jsc.nasa.gov/mitigate/das.html [Accessed July 14 2015].

12. NASA. 2009. *ORSAT* [Online]. NASA Orbital Debris Program Office. Available: http://orbitaldebris.jsc.nasa.gov/reentry/orsat.html [Accessed July 14 2015].

13. P. Omaly, and M. Spel. DEBRISK, a tool for re-entry risk analysis. ESA Special Publication, 2012. 70.

14. T. Lips, and B. Fritsche 2005. A comparison of commonly used re-entry analysis tools. *Acta Astronautica*, 57, 312-323.

15. R. D. Klett, *Drag Coefficients and Heating Ratios for Right Circular Cylinder in Free- Molecular and Continuum Flow from Mach 10 to 30*, Technical Report SC-RR-64-2141, Sandia Report, SC-RR-64-2141, Albuquerque, 1964.

16. W. P. Hallman, and D. M. Moody, *Trajectory Reconstruction and Heating Analysis of Columbia Composite Debris Pieces*, The Aerospace Corporation, 2005.

17. D. J. Masson, D. N. Morris, and D. E. Bloxsom, *Measurements of sphere drag from hypersonic continuum to free-molecule flow*, U.S Air Force, 1960.

18. R. Kelley, N. Hill, W. Rochelle *et al.* Comparison of ORSAT and SCARAB reentry analysis tools for a generic satellite test case. 38th COSPAR Scientific Assembly, 2010. 3931.

19. T. M. Owens. 2014. *Aero-Thermal Demise of Reentry Debris: A Computational Model*. Master of Science, Florida Insititute of Technology.

20. J. Beck. May 5 2015. *RE: Personal Communication.*

21. J. Merrifield, J. Beck, G. Markelov *et al.* Aerothermal Heating Methodology in the Spacecraft Aerothermal Model (SAM). 7th IAASS Conference, 2014.

22. J. Beck, J. Merrifield, I. Holbrough *et al.* 2015. Application of the SAM Destructive Re-Entry Code to the Spacecraft Demise Integration Test Cases. 8th European Symposium on Aerothermodynamics of Space Vehicles, March 2015 2015 Lisbon.

23. E. Minisci 2015. Space Debris and Asteroids (Re) Entry Analysis Methods and Tools. Glasgow, UK: University of Strathclyde.

24. T. Lips, V. Wartemann, G. Koppenwallner *et al.* Comparison of ORSAT and SCARAB Reentry Survival Results. 4th European Conference on Space Debris, 18-20 April 2005 Darmstadt, Germany. 533.

25. A. Tewari 2007. *Atmospheric and Space Flight Dynamics: Modeling and Simulation with MATLAB® and Simulink®*, Birkhäuser.

26. NASA 1976. U.S. Standard Atmosphere 1976. Washington, D.C.

27. R. Carmichael. Public Domain Aeronautical Software. Available: http://www.pdas.com/atmosdownload.html [Accessed March 10 2015].

28. S. Fiegel 2011. MASTER-2009 Software User Manual. Institute of Aerospace Systems (ILR).

29. N. Welty, M. Rudolph, F. Schäfer *et al.* 2013. Computational methodology to predict satellite system-level effects from impacts of untrackable space debris. *Acta Astronautica*, 88, 35-43.

30. J. R. Irons. 2016. *Landsat Science* [Online]. National Aeronautics and Space Administration. Available: http://landsat.gsfc.nasa.gov/ [Accessed January 2016].

31. European Space Agency. 2012. *Metop - meteorological missions* [Online]. European Space Agency. Available: http://www.esa.int/Our_Activities/Observing_the_Earth/The_Living_Planet_Programme/Meteorological_missions/MetOp/Overview14 [Accessed January 2016].

32. European Space Agency. 2014. *SPOT-6 and SPOT-7* [Online]. European Space Agency. Available: https://directory.eoportal.org/web/eoportal/satellite-missions/s/spot-6-7 [Accessed January 2016].

33. Scitor. 2016. *Space-Track.org* [Online]. Space-Track.org. Available: https://www.space-track.org/auth/login [Accessed January 2016].

34. R. H. Estes, and N. Moore 2007. An Overview of Demise Calculations, Conceptual Design Studies, and Hydrazine Compatibility Testing for the GPM Core Spacecraft Propellant Tank.

35. B. Fritsche, T. Lips, and G. Koppenwallner 2007. Analytical and numerical re-entry analysis of simple-shaped objects. *Acta Astronautica*, 60, 737-751.

36. EaglePicher Technologies. 2016. *Nickel- Hydrogen Batteries and Cells* [Online]. EaglePicher Technologies LLC. Available: http://www.eaglepicher.com/technologies/battery-power/nickel-hydrogen-ni-h2 [Accessed January 2016].

37. VECTRONIC Aerospace. 2014. *Space Applications* [Online]. VECTRONIC Aerospace GmbH. Available: http://www.vectronic-aerospace.com/space.php [Accessed January 2016].

38. Rockwell Collins. 2016. *Satellite Stabilization Wheels* [Online]. Rockwell Collins. Available: https://www.rockwellcollins.com/Products_and_Systems/Controls/Satellite_Stabilization_Wheels.aspx [Accessed January 2016].

39. J. R. Wertz, and W. J. Larson 1999. *Space Mission Analysis and Design*, Springer Netherlands.

40. S. Platnick. 2016. NASA's *Earth Observing System* [Online]. National Aeronautics and Space Administration. Available: http://eospso.nasa.gov/mission-category/3 [Accessed January 2016].

41. European Space Agency. 2013. *GOCE* [Online]. European Space Agency. [Accessed January 2016].

42. K. Deb 2001. *Multi-objective optimization using evolutionary algorithms*, John Wiley & Sons.

43. H.-G. Reimerdes, and W. Wohlers. Optimization of micrometeoroid and space debris protection systems. Space Debris, 2001. 655-660.

44. K. Deb, A. Pratap, S. Agarwal *et al.* 2002. A fast and elitist multiobjective genetic algorithm: NSGA-II. *Evolutionary Computation, IEEE Transactions on*, 6, 182- 197.

45. T. Mosher 1999. Conceptual spacecraft design using a genetic algorithm trade selection process. *Journal of Aircraft*, 36, 200-208.

46. F. A. Fortin, F. M. De Rainville, M. A. Gardner *et al.* 2012. DEAP: Evolutionary Algorithms Made Easy. *Journal of Machine Learning Research*, 13, 2171-2175.

47. E. Zitzler, M. Laumanns, L. Thiele *et al.* 2001. SPEA2: Improving the strength Pareto evolutionary algorithm. Eidgenössische Technische Hochschule Zürich (ETH), Institut für Technische Informatik und Kommunikationsnetze (TIK).

PROCEEDINGS OF THE 8TH INTERNATIONAL SPACE SAFETY CONFERENCE

SESSION 26
SPACE TRAFFIC CONTROL - II

EXPLORING NECESSARY ALTITUDE AWARENESS AND RESPONSE TIMES FOR AIR TRAFFIC CONTROL DURING SPACE LAUNCH AND RE-ENTRY OPERATIONS

Zheng Tao(1), Paul D. Wilde, Ph.D., P.E(2), Jonathan L. Schwartz(3), Jon L. Semanek(4), Dr. Ganghuai Wang(5), Ashley G. Williams(6)

(1, 3, 4, 5, 6) The MITRE Corporation, 7515 Colshire Drive, McLean, VA 22030, USA, ztao@mitre.org, jschwart@mitre.org, jsemanek@mitre.org, gwang@mitre.org, ashley@mitre.org

(2) Federal Aviation Administration, 800 Independence Ave., Washington, DC 20591 USA, paul.wilde@faa.gov

ABSTRACT

The Federal Aviation Administration (FAA) and The MITRE Corporation (MITRE) are working together to more efficiently integrate growing space launch and re-entry operations while minimizing effects on all National Airspace System (NAS) users. Current operations entail segregation of large amounts of airspace using aircraft hazard areas (AHAs) to maintain safety. This requires other NAS users to reroute their operations around them or schedule flights around the AHAs' timeframes.

In order to enable more dynamic AHAs and be more responsive to unforeseen events (e.g., vehicle breakup) the FAA must know how much time (and altitude awareness) is necessary for Air Traffic Control (ATC) to separate aircraft from dynamic AHAs. The necessary time can be assessed through a parametric fast-time modeling and simulation capability built by MITRE. This paper describes the dynamic AHA separation concept, MITRE's modeling capability, and initial assessments of ATC response times in a variety of operational scenarios.

1. INTRODUCTION

Space transportation and services are experiencing a new phase of expansion, commercialization, and development. There are new operators that are envisioning significantly more operations that could occur daily with innovative launch and re-entry vehicle (LRV) designs that have not been seen before. As the industry has modernized, so too must the Federal Aviation Administration's (FAA's) standards, regulations, and operations for handling air and space transportation in the National Airspace System (NAS). The FAA is faced with the challenge of integrating growing space launch and re-entry operations while minimizing effects on other NAS users. Currently, these operations utilize AHAs to segregate large amounts of airspace in order to maintain safety. This practice requires other NAS, users including military, general aviation, and commercial flight operations to reroute their operations around the AHAs or schedule their operations to avoid the timeframes of the AHAs. As these operations grow, this practice will result in LRV operations having more noticeable and frequent effects on the NAS.

In order to make these operations more efficient and dynamic, the FAA and The MITRE Corporation (MITRE) are working together in a variety of areas to reduce the impact of AHAs and to enhance ATC responses to them. Thus far, there have been few quantitative studies on response time or advanced notice (and hence altitude awareness[1]) sufficient for ATC to safely separate aircraft from unplanned re-entries and dynamic AHAs.

2. DYNAMIC AIRCRAFT AHAS FOR LAUNCH AND RE-ENTRY OPERATIONS

For LRVs with operational histories that demonstrate low probability of an off-nominal event (e.g., vehicle breakup, loss of control, loss of surveillance, etc.), the FAA is exploring the use of a future separation concept during launch that employs much smaller AHAs during nominal operations ("the concept"), but a larger contingent AHA would not be activated during nominal operations. Hence, that contingent airspace would be available for NAS users to utilize. Currently for launches, the FAA has been able to release airspace after a launch vehicle has passed through it, but the future concept would allow more airspace to be open to NAS users during a lunch operation even before the vehicle transits the airspace. Only when an off-nominal event happens would portions of that contingent AHA be active, which would require aircraft in it to be evacuated. The FAA is envisioning a capability that would quickly generate AHAs in real time and display them to controllers so that they can evacuate

[1] In this paper, altitude awareness refers to how high in altitude the FAA needs to have awareness of a status (vehicle health and position) in order to provide safe and efficient operations.

aircraft and reroute traffic. The FAA conducted a Human-In-The-Loop (HITL) simulation on this concept in October 2014 [1] to assess the acceptability to controllers of dynamic AHAs. The feedback from controllers during those tests was positive and paves the way for additional research and maturation activities on the concept, such as assessing the required time necessary for ATC to safely evacuate and avoid AHAs.

This new separation concept and associated standards call for new and increased controller involvement during these operations. A key factor for assessing the operational viability and risk of the concept and associated standards would be the amount of time ATC needs to safely separate aircraft from dynamic AHAs and other aircraft during these scenarios.

3. HIGH LEVEL APPROACH

To determine the amount of time necessary to respond to off-nominal events requires a statistical analysis of numerous scenarios under many operational conditions. However, testing the use of dynamic AHAs with live traffic in numerous test flights is not a realistic approach for evaluating response times due to the costs and safety implications involved. Thus, fast-time simulations of a number of space launch and re-entry scenarios are necessary to help quantify risks.

MITRE is building a flexible, fast-time modeling and simulation capability that can help evaluate the amount of time necessary for ATC to safely separate aircraft from dynamic AHAs. The capability provides operational measures of safety that can be used to examine the operational risks and effects from different factors on safety during space launch and re-entry operations. The capability takes into account the characteristics of each operation, the level of available surveillance, the time required for ATC to respond to situations and execute commands, and historical air traffic.

The MITRE modeling capability builds upon MITRE's aviation domain expertise and capabilities by leveraging MITRE's aircraft, separation, conflict detection, and surveillance models. The analyst or researcher can vary the separation concepts and standards, LRV performance characteristics, ATC and pilot responses, and surveillance performance for each scenario.

4. BRIEF DESCRIPTION OF THE MODEL

The basis of the modeling framework is a Monte Carlo simulation tool that has been used for several tasks in the past, including evaluating Unmanned Aircraft System Sense and Avoid algorithms [2], En Route Automation and Modernization surveillance performance, and evaluating surveillance requirements for Automatic Dependent Surveillance – Broadcast (ADS-B) [3]. To evaluate space launch and re-entry operations, the simulation model was enhanced through new ATC control algorithms, an updated surveillance model for space launch and re-entry, safety metric calculations, and the incorporation of probabilistic ATC and pilot response times. Additionally, models for LRV trajectories and debris have been developed that leverage prior debris modeling by Stanford University as part of the FAA's Center of Excellence for Commercial Space Transportation. MITRE acknowledges that there are more robust and validated capabilities for debris modeling and producing AHAs than the Stanford University one. If a more robust analysis of the scenario is needed by the FAA, it is expected that validated models would generate AHAs as inputs into the MITRE model.

The Monte Carlo simulation capability comprises the following models and algorithms (Fig. 1): aircraft model, ATC control algorithms (including a AHA evacuation algorithm, a hazard-volume avoidance algorithm, and an aircraft-LRV separation algorithm), wind model, navigation sensor error model, and ADS-B model. A more detailed description about the model and its components is described in [4]. The capability is still in development and requires further validation, thus all quantitative metrics generated should be deemed preliminary, pending further validation.

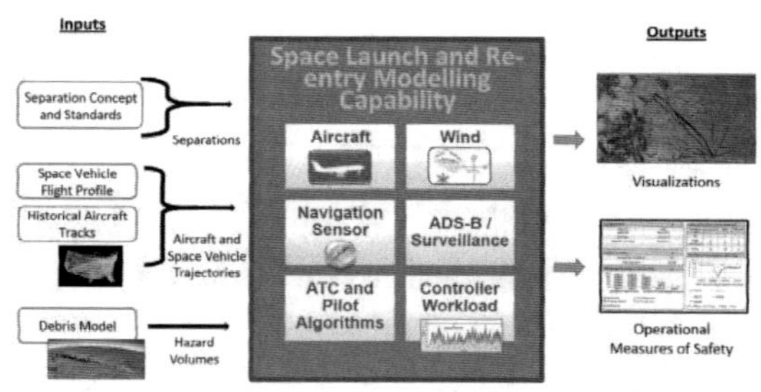

Figure 1. High-Level Diagram of Model

The simulation inputs include nominal and off-nominal AHA boundaries, LRV trajectories (including trajectories for the stages), stochastic response times for air traffic controllers and pilots (based on prior MITRE work [5]), aircraft traffic density, and surveillance performance. Based on these inputs, Monte Carlo simulations are run to generate various operational measures of safety that can be used to examine the operational risks for each space launch and re-entry operation. The results can be analyzed to determine the required warning time to evacuate and avoid dynamic AHAs. It also creates Google Earth files so that each scenario can be visualized.

Current key assumptions, limitations, and behaviors of the model include:

Assumptions:

- AHAs are convex polygons.
- ATC knows the shape of the AHAs
- Position and velocity of aircraft are known.

Limitations and specific behaviors:

- When avoiding AHAs, ATC directs aircraft to the vertices and edges of a buffered AHA, and then to the next waypoint on the other side of the AHA in a path that avoids intersecting the AHA.
- When evacuating AHAs, aircraft look for the shortest path to exit that does not put it in conflict with another aircraft.
- The algorithms currently only evacuate and avoid AHAs; the model and algorithms are not intended to ensure separation and spacing between aircraft pairs that are not directly impacted by the AHA or LRV.

5. MEASURES OF SAFETY

The FAA's Office of Commercial Space Transportation (AST) uses a figure called the mean number of casualties predicted (EC) to result from a given hazard to measure risk and help define an AHA [6]. However, calculating the EC requires risk sophisticated debris modeling and risk contours, which MITRE did not have access to at the time of this study. If MITRE had access to risk contours, MITRE's modeling capability could have assessed the EC risk per aircraft per unit of time.

In lieu of that, MITRE examined other metrics to assess the safety of space launch and re-entry operations with dynamic AHAs. Note that these are not necessarily the same measures of risk that AST would use when given a EC. However, these metrics are related to safety during launch and re-entry operations and can complement a risk assessment in addition to using EC. There are two metrics that directly relate to the necessary time to safely evacuate and avoid AHAs: the time it takes to exit the AHA and the number of aircraft inside one when it is activated.

The number of aircraft inside a dynamically activated AHA is an indication of the traffic density. It affects the associated ATC workload for communicating and executing reroutes for these flights out of the AHA; as a result, it may be used as a metric to show the risk to these aircraft. All other things being equal, the longer it takes to evacuate aircraft from an active AHA, the greater the safety risk is to aircraft; therefore, the time required to evacuate aircraft from an active AHA helps to quantify the safety risk to aircraft. Tab. 1 lists the metrics considered. Aggregation of all metrics totals is on a per scenario basis.

Table 1. Metrics Definitions

Metric	Units of Measure
1. Time in AHA	Minutes
2. Number of aircraft in AHA when activated	Aircraft Count
3. Time to exit hazard a volume	Minutes

Sensitivity analyses can be performed by adjusting scenario parameters (such as ATC notification of AHA activation, surveillance performance, orientation of traffic flows, traffic volume, and ATC and pilot responses) and comparing the metrics after each adjustment. While absolute values are potentially problematic until further validation work is complete, risk ratios can be used to compare the scenarios with different conditions. For example, the risk ratio between scenarios with different ATC delays can be determined by comparing the relative times in AHAs between them.

6. SCENARIO DESCRIPTIONS

Operational scenarios were used to demonstrate the modeling capability. Scenarios were derived from FAA recommendations and developed from historical missions. The scenarios facilitate the measurement of metrics associated with ATC response times regarding clearing known non-participant[2] air traffic away from AHAs associated with nominal and off-nominal events. Non-participant peak traffic times were selected in order to simulate stress on modeling outputs. The amount of notification time ATC receives of the status of AHAs were varied (e.g. two minutes, five minutes, ten minutes) to assess its impact on response times. Scenario descriptions are included below. Scenarios included hypothetical altitudes for a vehicle breakup and the estimated time for the first hazardous debris to reach 60,000 ft (the upper edge of FAA-controlled airspace). This time is based on findings from [7], which indicate that objects "could reach the airspace anywhere from 13 to 19 minutes" after a breakup observed at 255,906 feet (42 NM). Assuming a conservative 13 minutes, assuming that the fragments quickly reach terminal

[2] Aircraft or operations that are not authorized to operate inside the hazard volumes.

velocity, and assuming that the pre- and post-breakup ballistic coefficients are similar to the ranges detailed in [7], then the vertical velocity is about 15,000 feet/minute. This factor is used to calculate the time debris would fall to 60,000 feet altitude in the scenarios.

6.1 Scenario 1: Winged Re-entry

This scenario is based on a winged re-entry vehicle from space, similar to past Space Shuttle missions. The nominal re-entry trajectory that would cross over the coast of the western United States and land at the Kennedy Space Center (KSC) in Florida. It would have associated AHAs for various break-up times during the re-entry such as those shown in Fig. 2.

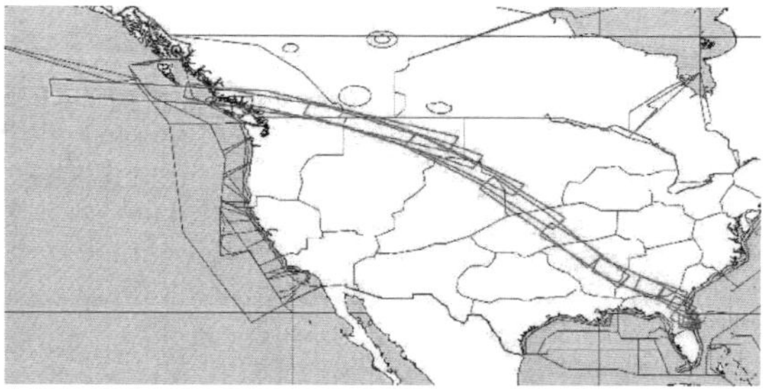

Figure 2. Example of AHAs for a Shuttle Re-entry (Source: FAA)

Debris AHAs were generated by the FAA during the planning phase of re-entry. The AHAs encompass airspace that would be affected by debris in the event of a re-entry breakup.

For this scenario, three AHAs from a past NASA Space Shuttle re-entry mission were selected due to the potentially high volumes of non-participant aircraft during re-entry. For the purpose of separating the results, the volumes mentioned above will be referred to as Scenario 1A (yellow aircraft and AHA in Fig. 3), 1B (green in Fig. 3), and 1C (blue in Fig. 3). The AHAs generally overlaid Atlanta and Jacksonville ARTCCs (1A), Kansas City and Memphis ARTCCs (1B), and Denver, Minneapolis, and Salt Lake City ARTCCs (1C). Upon the re-entry vehicle's breakup, ATC would need to be notified of the breakup (either by the vehicle operator or through ATC determining that a breakup occurred) and the FAA would generate more refined AHAs based on the last known state vector of the vehicle and forecasted meteorological conditions. This would have produced a significantly smaller AHA that would become active. However, in lieu of having refined AHA's, MITRE utilized the planned AHAs. Thus, the number of affected aircraft identified in these scenarios are likely to be overestimations.

Once the AHA's are activated, aircraft inside the AHAs would be rerouted by ATC to evacuate the airspace. Aircraft outside the volume will be rerouted to avoid the AHA. The date and timeframe selected for this scenario are March 19, 2015, 1500-1800 UTC in order to pull examples of peak non-participant traffic. The amount of time between the actual breakup and the start of ATC reroutes (delay) is varied – from one, three, five, and ten minutes. This delay represents the time needed for ATC to be made aware of, and confirm, the breakup event (either from the operator or from FAA monitoring capabilities) and the amount of time to communicate it to controllers.

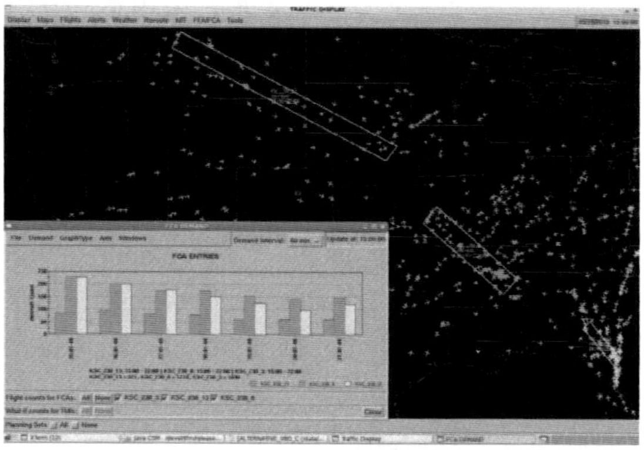

Figure 3. AHAs and Traffic for Winged Re-entry Scenarios (1A-1C)

Scenario 1A. This AHA (yellow) is nearest to the re-entry vehicle's landing at KSC and is the smallest of the Scenario 1 volumes. Non-participant traffic (yellow) controlled by ATC for this scenario mainly consists of north-south traffic volume operating in the south-eastern United States. The breakup point is at 149,611 feet MSL with the vehicle travelling at 7,838 feet/second. The size of the volume is about 150 NM by 65 NM. For simplicity, it is assumed that the AHA extends from 60,000 feet to the ground. Future iterations and refinements of the model can examine AHAs for multiple altitude bands. The first piece of debris reaches the top of the NAS about six minutes after breakup.

Scenario 1B. This AHA (green) is located in a very dense portion of the NAS near the central United States. This AHA is larger than Scenario 1A due to the altitude (198,564 feet MSL) and speed (17,656 feet/second) at the time of breakup. This scenario includes complex and dense non-participant traffic arriving and departing from several major U.S. hubs. The size of the volume is about 250 NM by 100 NM. The first piece of debris reaches the top of the NAS about 9.3 minutes after breakup.

Scenario 1C. This AHA is located in the upper central region of the United States and is significantly larger than the others. The altitude and speed at the point of breakup are 225,182 feet MSL and 22,578 feet/second. Non-participant traffic controlled by ATC consists of mainly east-west traffic from hubs at Chicago, Minneapolis, Denver, San Francisco, and Seattle. The traffic has peak periods and can be complex and dense at many times throughout the day. The size of the volume is about 885 NM by 65 NM. The first piece of debris reaches the top of the NAS about 11 minutes after breakup.

6.2 Scenario 2: Future West Coast Orbital Launch

This scenario is based on an orbital launch from Vandenberg Air Force Base (KVBG) in California that would put a satellite into a polar orbit. The AHAs from FAA Notices to Airmen (NOTAMs) for this particular launch included an ascent area extending approximately 72 nautical miles southwest of KVBG (green box in Fig. 4). Additionally, a large section of Warning Area 291, extending 670 nautical miles southwest of KVBG, was also active (magenta in Fig. 4). AHAs for an off-nominal event during this launch were not provided in NOTAMs. If they were, they would have covered a portion of airspace between the green and magenta AHAs. Routes are shown as yellow lines.

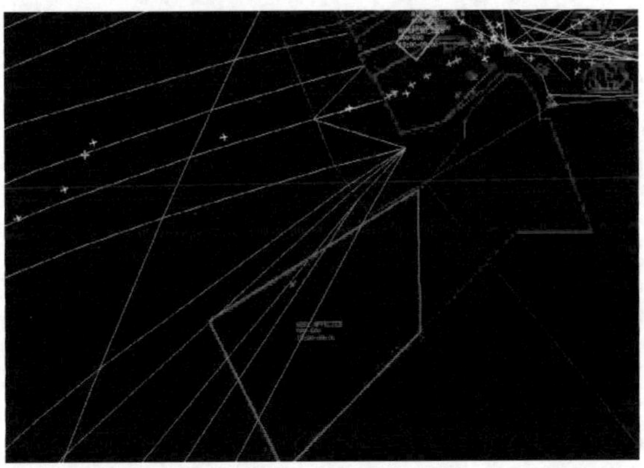

Figure 4. Historical AHAs for a Polar Launch from Vandenberg Air Force Base

This scenario explores whether the launch operation could be managed safely with smaller and dynamic AHAs. A potential future concept envisioned by the FAA and described in [1] is that only some portions of an AHA would be active in advance of a launch while the remaining portion would only be dynamically activated only during an off-nominal event such as a breakup. A notional dynamic AHA is shown in red in Fig. 5. It is envisioned that in the future there will be better reliability of LRVs, improved FAA awareness of their position and health status, new capabilities for ATC to be alerted of, and respond to, dynamic AHAs, and new procedures to respond to off-nominal events. These enablers would allow for smaller nominal AHAs.

Fig. 5 depicts notional AHAs for the launch based on MITRE's in-house debris model, which was adapted and enhanced from Stanford University's Range Safety Assessment Tool [8]. That model was based on debris modeled from the Space Shuttle Columbia accident [9][10]. The generated notional AHAs are not necessarily representative of what AST would be using and could be much smaller than what the actual AHAs might be. However, the purpose of this work is not to assess the accuracy or performance of MITRE's generated AHAs. MITRE acknowledges that there are more robust and validated capabilities for producing AHAs. If a more robust analysis of the scenario is needed by the FAA, it is expected that validated models would generate AHAs as inputs into the MITRE model. The volume closest to KVBG (light blue) is designed for protection of non-participants during launch. The AHA approximately 600 miles off shore is for first-stage drop protection, but is likely an underestimation of the actual size.

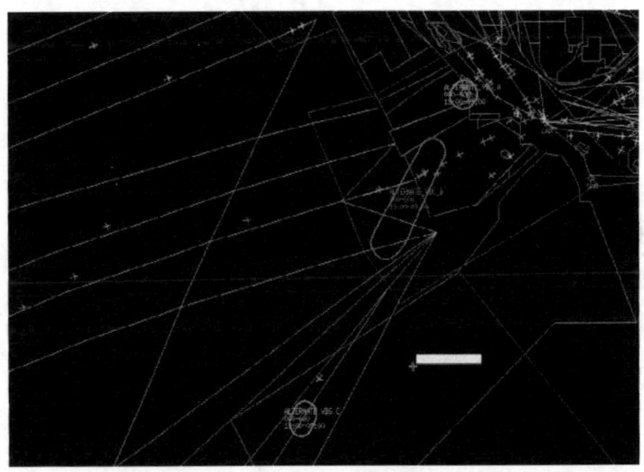

Figure 5. Hypothetical AHAs for Future West Coast Orbital Launch

In this scenario, oceanic traffic passes through the gap between the two light blue AHAs during a planned launch. ATC is aware of the nominal volumes (light blue) becoming active 30 minutes prior to launch. Non-participant air traffic is routed around the blue AHAs as part of normal NAS operations. Shortly after launch (about two minutes), an off-nominal event occurs on the launch vehicle at about 200,000 feet and the dynamic debris AHA (red) activates and the nominal first-stage drop AHA deactivates. The first piece of debris reaches the top of the NAS about 9.3 minutes later. ATC evacuates flights from, and reroutes the flights around, the off-nominal AHA (red). Note that half of the dynamic debris AHA is outside of radar coverage (dark blue line). The size of the volume is about 185 NM by 44 NM.

Historical non-participant traffic from March 13, 2015, 1100-1400 UTC is used in the scenario to simulate peak traffic for the areas. The amount of time between the actual breakup and the start of ATC actions (delay) is varied from one, three, five, and ten minutes.

7. EVALUATION OF SCENARIOS

MITRE's modeling capability was used to analyze times in a AHA and the number of aircraft when the volumes were active for each scenario. Each scenario is run multiple times with different samples of air traffic. For example, a single scenario would be run many times where the time of the launch or re-entry event shifting by 10 minutes. Varying the traffic sample helps identify outlier cases. ATC instructions are given to aircraft to avoid or evacuate flights from AHAs. The amount of time for ATC and pilots to issue, respond to, and execute commands (e.g., due to pilot read-backs and delay until maneuver initiation) are modeled as random variables and computed stochastically during the simulation run using parameters based on prior MITRE research on controller workload [5]. It was assumed that controllers had near-real-time updates of the aircraft position (similar to ADS-B performance). Tab. 2 summarizes the number of aircraft that were inside a AHA once it became active.

Table 2. Number of Aircraft in AHAs at Activation

Scenario	Average Number in AHA at Activation	Maximum Number in AHA at Activation
1A	10.4	13
1B	19.4	28
1C	8.8	21
2	0.6	2

Not surprisingly, scenarios 1A-1C had much higher numbers than scenario 2, which was focused mostly on oceanic traffic. The traffic in scenario 1B had the densest traffic due to the traffic arriving and departing from several major U.S. hubs, even though its size was roughly half of the size of the 1C AHA. Thus, the number of aircraft in a AHA at activation depends on both the size of the AHA and traffic density of the airspace. These results make it clear that AHAs in the continental United States (CONUS) can pose a challenge for ATC during peak traffic times.

Fig. 6 summarizes the average times in AHAs for each of the scenarios.

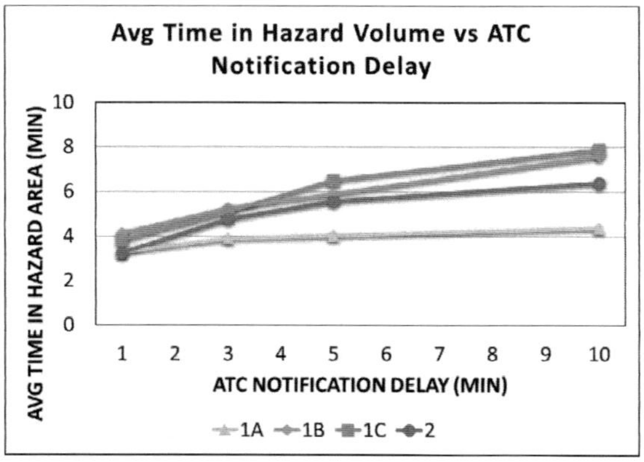

Figure 6. Average Hazard Times for Scenarios

The average times are much larger in scenarios 1B and 1C due to the larger hazard areas. The average times range from 3.8 to 7.9 minutes for scenarios 1B and 1C. For scenario 1A, the average times in the AHA range from 3.3 to 4.4 minutes. For scenario 2, the average times range from 3.2 to 6.4 minutes. The results show that the length of time in a AHA is largely dependent on the AHA's size and on the flow of traffic relative to the orientation and shape of the AHA. The 1A AHA is an example of both a small AHA and one where traffic flows in a direction that would lower time spent in a AHA. ATC notification delay also has a significant effect, but the effect plateaus at the time it takes to originally traverse the AHA airspace. For example, suppose an aircraft's original flight plan would have it fly seven minutes in an inactive AHA's airspace. If the AHA turns on, then suppose the aircraft could reroute and exit in three minutes or less. If there was a three minute ATC notification delay while the aircraft is in the AHA, then the aircraft would spend no more than six minutes in the AHA after the volume turns on (three due to the delay and three to exit). Regardless of how high the delay is, the aircraft would spend no more than seven minutes in the AHA, assuming the aircraft takes the shortest path out.

Rather than perform more simulations to attempt to arrive at a situation that would represent the maximum time to exit a AHA, analyses were performed to calculate the theoretical maximum time for the slowest (400 knots) aircraft to exit a AHA. The analyses also served as a check on the simulation results. The worst case time would occur when an aircraft is moving lengthwise through the middle of the AHA and at a heading parallel with the AHA's heading. The worst case time comprises two components:

- The time to exit a AHA from the center of a AHA
- The minimum of either the ATC notification delay or the time it takes to traverse the AHA's longer side.

The calculated results for the time it takes to exit from the center of a AHA and the time it takes to traverse a AHA's longer side are shown in Tab. 3.

Table. 3 Maximum Time to Exit AHA from Center of AHA

Scenario	Theoretical Maximum Time to Exit from Center of AHA (minutes)	Theoretical Maximum Time to Traverse AHA (minutes)
1A	5.2	23
1B	7.8	38
1C	5.2	133
2	3.5	28

The exit time from the center of the AHA is dependent on, although not proportional to, the minimum aircraft speed and the smallest dimension (e.g., the width) of the AHA. This assumes that the aircraft would take the fastest path to exit a AHA.

The theoretical maximum times to traverse a AHA are much higher than the maximum ATC notification delay of 10 minutes. Thus, the ATC notification delays are used to calculate the theoretical maximum time. The theoretical maximum times to exit the AHAs for different ATC delays is shown in Fig 7. The remaining warning time after breakup is the time for debris to reach 60,000 ft minus the ATC notification delay. It represents the amount of time ATC has remaining, after being notified, to move aircraft before the debris reaches the top of the NAS. It is shown as dashed lines in Fig. 7.

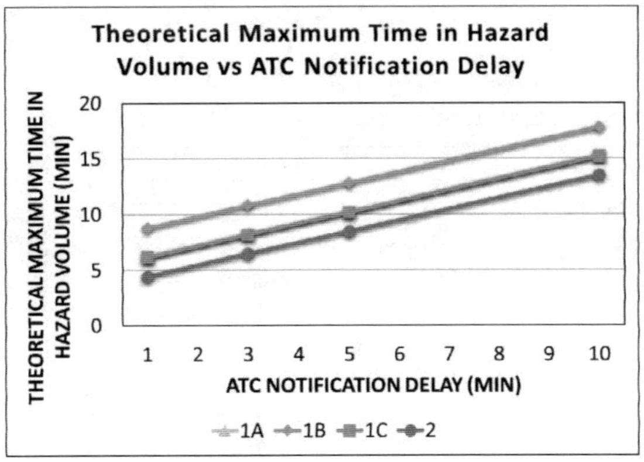

Figure 7. Theoretical Maximum Times to Exit a AHA vs ATC Notification Delay

The maximum ATC notification delay to accommodate the maximum time in a AHA is when the warning time is equal to the theoretical maximum time to exit. This occurs when the dashed line intersects the solid line in Fig 7. The maximum safe ATC delays are 0.4, 0.7, 2.9, and 2.9 minutes for scenarios 1A, 1B, 1C, and 2 respectively. The theoretical maximums for scenarios 1A and 1C are identical because their AHAs share the same width. They range from 6.2 to 15.2 minutes. The theoretical maximums for 1B and 2 range from 8.8 to 17.8 and 4.5 to 13.5 minutes respectively.

The theoretical minimum time needed for an aircraft to avoid entering a AHA when an aircraft flies perpendicular towards a boundary of a rectangular AHA was also calculated. The minimum time is equal to the reciprocal of the turn rate in radians per second. Notice that the minimum time is independent of the aircraft speed and is only dependent on the turn rate of the aircraft. At the standard turn rate of three degrees per second, the minimum time for an aircraft to avoid entering a AHA is 19.1 seconds.

7.1 Altitude Awareness

Altitude awareness needs defined here are based on the maximum safe ATC notification delay to clear a AHA (Fig. 7). If we assume that 15,000 feet/minute is still applicable for the downward velocity of debris after breakup, then Eq. 1 can be used to determine the minimum altitude awareness needed by FAA to safely clear a AHA.

$$\text{Minimal Altitude Awareness (feet)} = 60{,}000 + 15{,}000 * \text{minutes to clear} + \text{Buffer} \quad (1)$$

The minimum altitude awareness for scenarios 1 and 2 are shown in Fig. 8, and do not include any buffer. The buffer would incorporate a variety of considerations and uncertainties such as the delay of having to reissue commands to aircraft, or an additional safety margin. The altitudes of the debris when ATC receives the notification are shown as dashed lines in Fig. 8.

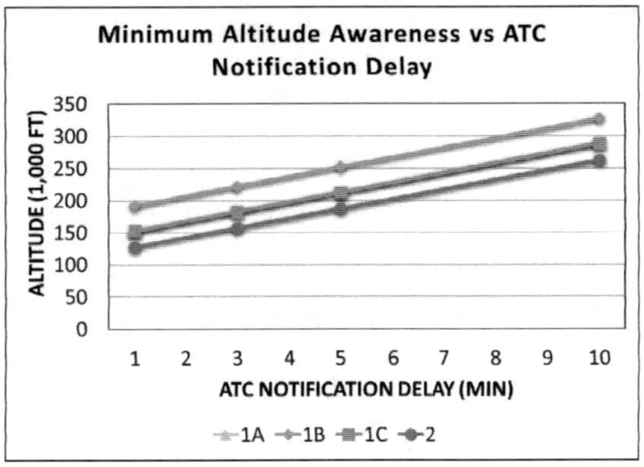

Figure 8. Needed Altitude Awareness for Scenarios

The minimum altitude awareness occurs at the maximum safe ATC notification delays, and are 144,000, 188,000, 182,000, and 156,000 ft for scenarios 1A, 1B, 1C, and 2 respectively. This means that ATC needs to have awareness of the health status of the vehicle at those altitudes or higher. If an additional buffer is incorporated, then the maximum safe ATC notification would be shorter, and the altitude awareness would be higher.

8. IMPLICATIONS FOR ATC

The results show that the response times to AHAs are primarily dependent on four factors:

- Size of the AHA
- Orientation of the flow of traffic to the AHA
- Density of traffic in and around the AHA
- ATC notification delay

The FAA can use these findings to help establish launch or re-entry trajectories and times with LRV operators in order to ensure safety. If the FAA finds that the average or maximum times to evacuate a potential AHA, (with an assumed ATC notification delay based on current FAA technology and capabilities) is longer than the time it takes for debris to reach 60,000 feet, then the FAA could ask the LRV operator if they have flexibility in any of the following:

- Launch/re-entry time and location, in order to reduce traffic density
- Launch/re-entry azimuth, in order to change the orientation of the AHA with the traffic flows

The FAA could also ask the LRV operator if they have any data or information that could justify a reduction in the size of the AHA or the buffers around it, such as data supporting the reliability, safety, and maturity of the LRV. The LRV operator could also potentially provide more detailed data about the LRV to more accurately represent a debris catalogue.

On the ATC side, the FAA could develop playbook routes that would be oriented in a manner that minimizes the time to fly through a AHA should the AHA become active. The FAA could also increase the separation of aircraft flying through the potential AHA thereby limiting nearby traffic density during the mission.

MITRE's modeling capability can be used to reevaluate the maximum and average flight times in a AHA after these adjustments and considerations have been made in order to determine if they sufficiently reduce the times in a AHA to below the time for debris to reach 60,000 ft.

8.1 Potential Technologies to Reduce Response Times

ATC notification delay plays a significant role in determining the time in a AHA. For scenario 1A, the delay needs to be less than one minute for ATC to safely evacuate the AHA. This delay could be reduced if the FAA had real-time access to the LRV's position and health status. Real-time position data can also be used to calculate more accurate (and potentially smaller) AHAs in real-time and reduce the buffer in Eq. 1.

In situations where the traffic density is high and complex, ATC should prioritize the reroutes of flights that are about to approach the riskiest portions of a AHA rather than trying to reroute every flight or focusing on flights that are in the AHA the longest. This is not an easy thing to mentally calculate when under pressure. Thus, an automation capability is needed to identify which flights are most at risk and to provide reroute advisories to the controllers on those flights. Equipping aircraft with data communications technology could potentially help speed up evacuations as well. MITRE's modeling capability is able to incorporate these considerations and re-evaluate the scenarios based on these updated conditions.

9. SUMMARY

This paper explored the ATC response times to and altitude awareness necessary to safely implement dynamic AHAs for a few winged re-entry and west coast orbital launch scenarios. The AHAs for the winged re-entry spanned large portions of the CONUS, and the AHA for the orbital launch was over the Pacific Ocean. The evaluation of these scenarios leveraged MITRE's modeling and analysis capabilities for evaluating launch and re-entry operations. The scenario evaluations examined the length of time aircraft would spend in a AHA, the number of aircraft in a AHA when it is active, and the maximum time it takes to exit an active AHA. The response times can be compared to the expected debris fall time to determine if dynamic AHAs for a proposed mission can be evacuated in time to mitigate the risk from a potential debris impact. The response times and other prior work [7] were used to help determine an equation for the minimum altitude awareness needed by the FAA to safely evacuate a dynamic AHA. The paper also identifies options, including collaborating with the LRV operator, that the FAA can examine to minimize time needed to evacuate a AHA.

The results showed that the average time to clear a AHA ranged from 3.2-7.9 minutes, and the theoretical maximum time ranged from 4.5-17.8 minutes in the scenarios. The maximum safe ATC delays are 0.4, 0.7, 2.9, and

2.9 minutes for scenarios 1A, 1B, 1C, and 2 respectively. The minimum altitude awareness (without a buffer) occurs at the maximum safe ATC notification delays, and are 144,000, 188,000, 182,000, and 156,000 ft for scenarios 1A, 1B, 1C, and 2 respectively. The altitude at breakup in the scenarios ranged from 149,611-225,182 feet. The response times were largely dependent upon four factors:

- Size of the AHA,
- Orientation of the flow of traffic to the AHA,
- Density of traffic in and around the AHA, and
- ATC notification delay.

Capabilities needed to reduce ATC response times were also identified such as real-time vehicle position and health status, and a reroute tool for ATC to use to identify, prioritize, and reroute aircraft most at risk in a AHA.

The results presented are only preliminary, but encouraging. Further refinement and assessment of the MITRE modeling capability are needed. Assessments of additional vehicle types and trajectories, ATC strategies for responding, and varying levels of surveillance performance are also needed, all of which can be performed with the MITRE modeling capability. A sensitivity analysis of how the different factors can affect the response times and safety of an operation can help the FAA better understand the risks of these operations.

10. REFERENCES

1. Hatton, K. (2014). Space Vehicle Operations Human-in-the-Loop. FAA. Washington, D.C., USA.

2. Wehner, P.J., Schwartz, J., Hashemi, D., et al. (2013). Evaluating Prototype Sense and Avoid Alternatives in Simulation and Flight. 13th Annual Association for Unmanned Vehicle Systems International Conference. Webster Field, MD, USA.

3. RTCA Special Committee-186 (2003). Minimum Aviation System Performance Standards for Aircraft Surveillance Applications (ASA) (RTCA DO-289). RTCA, Inc. Washington, DC, USA.

4. Wang, G., Tao, Z., Masek, T., Schwartz, J. (2016). A Monte Carlo Simulation Tool for Evaluating Space Launch and Re-entry Operations. Integrated Communications Navigation and Surveillance (ICNS) Conference. Herndon, VA, USA.

5. Marr, B., Lindsay, K. (2015). Controller Workload-based Calculation of Monitor Alert Parameters for En Route Sectors. AIAA Aviation and Aeronautics Forum and Exposition. Dallas, TX, USA.

6. Wilde, P.D. (2011). Public Risk Criteria and Rationale for Commercial Launch and Reentry. International Association for the Advancement of Space Safety (IAASS). Versailles, France.

7. Ailor, W., Wilde, P. (2008). Requirements for Warning Aircraft of Reentering Debris. IAASS. Rome, Italy.

8. Capristan, F.M., and Alonso, J.J. (2014). Range Safety Assessment Tool (RSAT): An analysis environment for safety assessment of launch and reentry vehicles. 52nd Aerospace Sciences Meeting. National Harbor, MD, USA.

9. Collins, J.D., Nyman, R., and Lottati, I. (2005). Estimation of Space Shuttle Orbiter Reentry Debris Casualty Area. AIAA Atmospheric Flight Mechanics Conference and Exhibit. San Francisco, CA, USA.

10. Graybeal, S.R., Mendeck, G.F., Mrozinski, R.B. (2006). Debris Field Estimation Methods for the STS-107 Columbia Investigation. AIAA Atmospheric Flight Mechanics Conference and Exhibit. Keystone, CO, USA.

11. ACKNOWLEDGEMENTS

The authors from MITRE wish to thank the invaluable contributions of team members Tom St. Clair, Tudor Masek, and Mark Banyai. The authors also wish to thank Win Heagy, Dean Fulmer, Emily Stelzer, Glenn Roberts, Marshall Koch, Mike Klinker, and David Kuperman for their review, support, feedback, and management oversight.

12. DISCLAIMER

This work was produced for the U.S. Government under Contract DTFAWA-10-C-00080 and is subject to Federal Aviation Administration Acquisition Management Clause 3.5-13, Rights in Data-General, Alt. III and Alt. IV (Oct. 1996). The contents of this material reflect the views of the authors and/or Director of the Center for Advanced Aviation System Development. Neither the Federal Aviation Administration nor the Department of Transportation makes any warranty or guarantee, or promise, expressed or implied, concerning the content or the accuracy of the views expressed herein. Approved for Public Release; Distribution Unlimited. 16-1761.

APPLICATION ANALYSIS FOR AEROASSISTED ORBIT TRANSFER IN NONCOPLANAR ORBIT DEBRIS REMOVE

YAN Ruidong, WANG Ronglan, LIU Siqing, SHI Liqin, GONG Jiancun

(National Space Science Center, Chinese Academy of Sciences, Beijing 100190, China)

ABSTRACT

Ncoplanar orbit transfer of debiris remove vehicle need lots of fuel, this paper presents optimal impulsive design for aeroassisted orbital transfer from Low Earth Orbit (LEO) or Geostationary Earth Orbit (GEO) to low-Earth orbit with an inclination change. Through comparison of the double-impusive hohmann orbit transfer and aeroassisted orbit Transfer in noncoplanar orbit, The fuel saving of different altitude or inclination from two noncoplanar orbits are concluded. The result show that minimumof orbit transfer between GEO and 300km LEO is 1.55 km/s. With the increase of orbital altitude difference, aeroassisted orbital transfer efficiency gradually reduce. Implementation of aeroassisted orbit transfer with the same orbital altitude, low orbit aeroassisted orbit transfer optimization efficiency is higher than high orbit. when the value of orbital inclination between two noncoplanar orbits is less than 5°, adopting aeroassisted orbit transfer and double impulse orbit transfer is equal with fuel consumption. Aeroassisted orbit transfer optimization between noncoplanar orbits with different altitude but different orbital inclination difference, with the increase of orbit inclination difference, fuel saving first increases then decreases, when the angle is about 20° fuel savings rate is the maximum.

1. INTRODUCTION

With the increasing of spacecraft launch and anti-satellite test carried out, and the earth orbit, especially some special sun-synchronous orbit is becoming to be limited natural resource. The increasing of space objects to some extent, increase the collision probability between the various objects. In February 2009, there was an collision happen between satellite iridium 33 and cosmos 2251. Cascading Kessler Effect suggests that collisions would produce new debris, thus increasing the collision probability of space objects. Even if the human no longer launch new spacecraft into space, debris number will keep rising[1].

NASA research showed that the collision is the main reason of the future debris growth. debris mitigation measures and spacecraft maneuvers cannot make debris environment stable, after debris mitigation there will be 8-9 times catastrophic collision still occur over the next 40 years[2]. In order to control the increase of cataloged objects and reduce in-orbit spacecraft space collision events occur, clearly active debris removal strategy must be made.

The in-orbit spacecraft's main problem is the fuel consumption, especially debris removal vehicle. Traverse the debris to remove need to optimize series of orbit maneuver sequence. Because the space debris in space distribution is not concentrated, so it needs to traverse debris orbits in the same plane or noncoplanar. It will consume a large amount of fuel. In the 2015-03-30, china launched a new generation of the upper stage expedition-1. Besides autonomous orbit maneuver ability, in-orbit flight time of the upper stage is long. It can generally start ignition for many times, meet the demand of different mission to send one or more of the payloads into orbit[3]. Spacecraft Like the expedition-1 is also called orbital Transfer Vehicle (Orbit Transfer Vehicle, OTV). OTV work range is mainly in the earth's atmosphere, so it can put forward to use aerodynamic to fulfil its mission, which is called aeroassisted orbit transfer vehicle (Aeroassisted Orbit Transfer Vehicle, AOTV)[4]. Although AOTV have to increase the quality of the part of the aeroassisted wing and the quality of the heat protective material, compared with horman orbit transfer, the propellant consumption is greatly reduced. Since the time of early aeroassisted orbit transfer research, a number of numerical optimization studies have been conducted on the problem of aeroassisted orbit transfer. But the real applications of aeroassisted orbit transfer are rare, and the optimization analysis of orbit altitude difference and inclination difference factors impact on the efficiency of aeroassisted orbit transfer are rare.

In the design of today's debris removal aircraft, aircraft design mission is more of big debris removal, orbit inclination between removal debris is inevitably exist. Under the background of active debris removal research needs, in order to save fuel, this paper proposed aeroassisted orbit transfer technology is applied to noncoplanar orbit debris removal tasks. Considering the application of aeroassisted orbit transfer during the different inclination of debris removal also need to evaluate the inclination or altitude such factors between the debris influence on aeroassisted orbit transfer fuel economical efficiency, It is need to select the most suitable debris for implementation of aeroassisted orbit transfer to remove in order to achieve maximum fuel economical efficiency. Additionally under the condition that optimization trajectory form initial orbit to target orbit is determination, the mean anomaly aircraft reach the target orbit is only related to the initial orbit maneuver position. It can reach target orbit at different mean anomaly from changing the initial orbit maneuvering position. And because time from initial orbit to the target orbit is short, the target orbit change of the ascending node right ascension here can be neglected. Therefore, in this paper, the short-period terms such as the ascending node right ascension of target orbit and the mean anomaly etc. do not set as limited value.

2. AEROASSISTED ORBITAL TRANSFER BETWEEN NONCOPLANAR ORBITS

2.1 Aeroassisted Orbit Transfer

Assume that the initial orbit and target orbit are circular orbits. Radius are r_1 and r_2 respectively. Orbit radius of the atmosphere is r_b = 649 8 km. De-orbit impulse applied at Initial orbit and vehicle flies into the atmosphere. Using the aerodynamic force vehicle gliding with orbit inclination changed, and away from the atmosphere into the target orbit. As shown in figure 1.

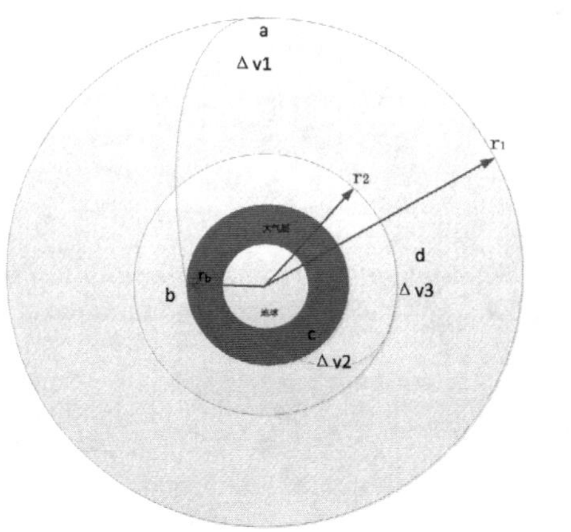

Fig. 1 Process of aeroassisted orbit transfer

As shown in figure 1, an impulse Δv_1 applied atin the opposite direction from the vehicle, the vehicle then enter an elliptical orbit that apogee is r_1 and perigee is r_{p1}. The vehicle moves along the elliptical orbit and at point *b* enters the atmosphere. In the atmosphere using aerodynamic force vehicle orbital inclination is changed. At point *c* vehicle leaves the atmosphere, at the same time an impulse Δv_2 is applied, which makes the orbital apogee *d* located in the target orbit. An impulse Δv_3 is applied at point *d* again to recircularize the orbit, then vehicle can stay in the target orbit. The equation of motion of the following describes the whole process.

1) A de-orbit impulse applied at point a, spacecraft descend from point a to point b

$$\left.\begin{aligned} \Delta v_1 &= \sqrt{\frac{\mu}{r_1}}\left(1-\sqrt{\frac{2r_{p1}}{r_1+r_{p1}}}\right) \\ v_b &= \sqrt{2\mu\left(\frac{1}{r_b}-\frac{1}{r_1+r_{p1}}\right)} \\ \gamma_b &= -\arccos\left[\frac{r_1}{r_b v_b}\sqrt{\frac{2\mu r_{p1}}{r_1(r_1+r_{p1})}}\right] \end{aligned}\right\} \tag{1}$$

Where v_b and γ_b are speed and inertial flight path angle respectively when spacecraft enter the atmosphere.

2) The equation of motion in the atmosphere

$$\frac{d\theta}{dt}=\frac{v\cos\gamma\cos\psi}{r\cos\phi} \tag{2}$$

$$\frac{d\phi}{dt}=\frac{v\cos\gamma\sin\psi}{r} \tag{3}$$

$$\frac{dr}{dt}=v\sin\gamma \tag{4}$$

$$\frac{dv}{dt}=-\frac{C_D\rho v^2 s}{2m}-\frac{\bar{\mu}\sin\gamma}{r^2} \tag{5}$$

$$\frac{d\gamma}{dt}=\frac{C_L\rho v^2 s\cos\mu}{2mv}-\left(\frac{v}{r}-\frac{\bar{\mu}}{r^2 v}\right)\cos\gamma \tag{6}$$

$$\frac{d\psi}{dt}=\frac{C_L\rho v^2 s\sin\mu}{2mv\cos\gamma}-\frac{v\cos\gamma\cos\psi\tan\phi}{r} \tag{7}$$

Where $\bar{\mu}$ is gravitational parameter, m is spacecraft mass, θ and ϕ are inertial longitude and latitude, r is geocentric radius, γ is inertial flight path angle, ψ is the heading angle, μ is bank angle. s is the vehicle reference area, ρ is atmosphere density, the density is modeled using msise2000 atmosphere model. C_D and C_L are the coefficient of drag and the coefficient of lift respectively. The lift and drag coefficients are modeled using a standard drag polar as

$$C_L = C_{L,\alpha}\alpha$$

$$C_D = C_{D0} + KC_L^2$$

Where C_{D0} the coefficient of zero drag; K is induced factor; $C_{L,\alpha}$ is lift slope; α is the angle of attack.

3) The rising after out of the atmosphere:

$$\Delta v_2 = \sqrt{\frac{2\bar{\mu}(\frac{1}{r_b}-\frac{1}{r_2})}{1-\left(\frac{r_b}{r_2}\right)^2\cos^2\gamma_f}} - v_f \tag{8}$$

$$v_D = \frac{r_b}{r_2}\sqrt{\frac{2\bar{\mu}(\frac{1}{r_b}-\frac{1}{r_2})}{1-\left(\frac{r_b}{r_2}\right)^2\cos^2\gamma_f}}\cos\gamma_f \tag{9}$$

$$\Delta v_3 = \sqrt{\frac{\bar{\mu}}{r_2}} - v_D \tag{10}$$

Where f denote the moment of vehicle leaving the atmosphere. At this stage the design objective is *c-d* elliptical orbital apogee is in the target orbit. The design of vehicle entering target orbit regardless of the short-period terms such as the ascending node right ascension of target orbit and the mean anomaly etc. mainly because under the condition that optimization trajectory form initial orbit to target orbit is determination, the mean anomaly aircraft reach the target orbit is only related to the initial orbit maneuver position. It can reach target orbit at different mean anomaly from changing the initial orbit maneuvering position. That is to say, we can design the initial orbit maneuver position allow vehicle to reach the target orbit when just a rendezvous with space debris. Therefore, this article does not set the mean anomaly of target orbit.

The difference in mass before and after an impulse can then be computed using Goddard rocket equation as

$$\frac{m_p}{m_0} = 1 - \exp\left(\frac{-\Delta v}{g_0 I_{sp}}\right)^{[8]} \tag{11}$$

Where Δv is speed increment, I_{sp} is the engine specific impulse, and is expressed as *s* in engineering units. Corresponding to the different types of fuel I_{sp} is different. m_0 is the initial vehicle quality, and m_p is the fuel consumption.

3. THE OPTIMAL CONTROL MODEL AND OPTIMIZATION

The indirect method and direct method is commonly used in aeroassisted orbit transfer optimization of 2 class methods. Direct method USES the parametric method will be continuous optimal control problem solving of the space is transformed into a Nonlinear Programming (Nonlinear Programming, NLP) problem, through the numerical solution of Nonlinear Programming problem to obtain the optimal trajectory [6].Direct method relatively indirect method applies more widely, and there are many different types of direct method. According to different parameterized method, the direct method is divided into two basic types: (1) the discrete control variables, the continuous space of parameterized control variables, and state variables are obtained by numerical integration;(2) the discrete control variables and state variables at the same time, the node USES the polynomial representation between state variables over time, the changes of continuous dynamic differential equation is converted to algebraic constraints [7].In this paper, the Radau pseudo-spectral method is adopted in the configuration points LGR (Legendre-Gauss-Radau) structure though laser polynomial approximation state variables and control variables, the differential equation constraints into algebraic constraint, dynamic control is transformed into static parameter optimization. Finally a method of nonlinear programming is adopted to optimize the aeroassisted orbit transfer paths in different planes.

According to the above for vehicle in and out of the atmosphere and aircraft motion parameters described in the atmosphere, aeroassisted orbit transfer vehicle nonlinear programming model is established.

$$J = \min f(X(t),U(t),t)$$

$$s.t\begin{cases} X(t)\in R^n \\ g_i(X(t_f),t_f) = 0 \\ g_j(X(t_f),t_f) \le 0 \\ i = 1,2,3,\mathrm{L}\ m;\ j = 1,2,3,\mathrm{L}\ n \end{cases} \tag{12}$$

$X(t)$ is the state variables, $g_i(X(t_f),t_f)$ is equality constraint at t_f, $g_i(X(t_f),t_f)$ is inequality constraint at t_f.

1) Optimization parameters

Select minimum pulse velocity increment as the optimization performance index, optimization performance indicators is the sum of three velocity increment.

$$J = \min(\Delta v_1 + \Delta v_2 + \Delta v_3)$$

2) Admissible control

The optimization design in this paper is aimed at vehicle movement in the atmosphere, and the vehicle control variables are lift coefficient and angle of attack, which meet: $|C_L| < C_{Lmax}$,

$$\alpha\in[0,\pi/2]$$

3) Target constraint

Target orbit constraint is orbital altitude and orbital inclination, but the short-period terms such as the ascending node right ascension of target orbit and the mean anomaly etc. are not limited. The effect of heat flow constraint in the process of optimization is not considered. The focus of this article is the design of the optimal orbit and orbit parameters influence on the optimization efficiency, rather than optimization efficiency of a specific debris orbit. Target constraints as follows:

$$\begin{cases} X(t)\in R^n \\ r_f = r_b \\ arc(\cos(\psi_f)\cos(\phi_f)) = i_f = i_{dev} \\ \gamma_b \le 0 \\ \gamma_f \ge 0 \end{cases} \tag{13}$$

i_{dev} is orbital inclination changed in the atmosphere, r_b is the radius of the atmosphere.

4. AEROASSISTED ORBIT TRANSFER OPTIMIZATION PERFORMANCE IN NONCOPLANAR ORBIT WITH DIFFERENT ORBITAL ALTITUDE

4.1 GEO-LEO Aeroassisted orbital transfer

4.1.1 16 ° angle optimization

Vehicle transfer from Geostationary Earth Orbit (GEO) to 300 km altitude Low Earth Orbit (LEO) as described in section 1. In the process aeroassisted orbit transfer is carried out in the atmosphere with orbital inclination changed 16°. Finally make an analysis of aeroassisted orbit transfer fuel saving rate compared to Homan double impulse orbit transfer method between noncoplanar orbits. The GEO initial six orbital elements are:

$a(t_0) = R_e + h_0$, $e(t_0) = 0$, $i(t_0) = 0°$,
$\Omega(t_0) = 0°$. $w(t_0) = 0°$, $M(t_0) = 0°$

1) Kinetic parameters and mass-area ratio, specific impulse parameters are shown in table 1.

Table 1 Simulink parameters

$C_{L,\alpha}$	m/s/kg·m^{-2}	K	$\rho 0$/g·ml^{-1}	C_{d0}	σ_0/°	Θ_0/°
0.569 9	172	1.4	1.225	0.032	0	0
r_1/km	r_2/km	r_{p1}/km	i_0/°	i_f/°	Ψ_0/°	I_{sp}/s
42 164	6678	6 428	0	16	0	310

The optimization result showed that the optimization velocity increment is 1.551 km/s, the result is in accordance with the result of reference [5]. Liquid rocket fuel consumption is about 687.7 kg. Horman double impulse velocity increment is 4.018 km/s, liquid rocket fuel consumption is about 1261.7 kg. In the atmosphere (within 120km altitude areas), the change of the orbital parameters are shown in figure 2.

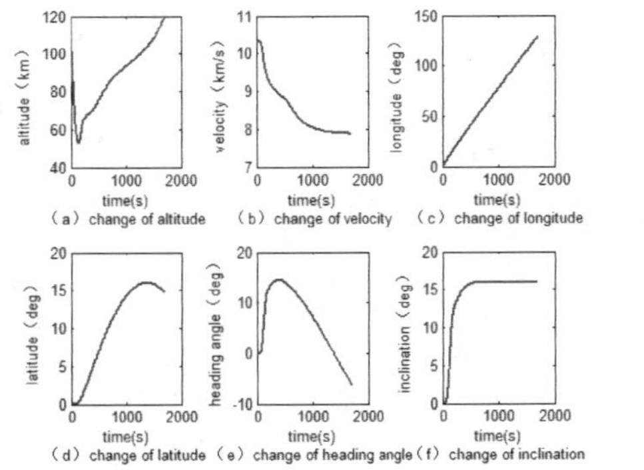

Fig. 2 Orbit parameters in atmosphere(16° inclination, mass 1720 kg)

2) Mass-area ratio is 300 kg·m^{-2}, other kinetic parameters and specific impulse parameters are same as shown in table 1.

The optimization result showed that the optimization velocity increment is 1.556 km/s. Horman double impulse velocity increment is 4.02 km/s., fuel consumption are 1202.4 kg and 2201.2 kg respectively. In the atmosphere, the change of the orbital parameters are shown in figure 3.

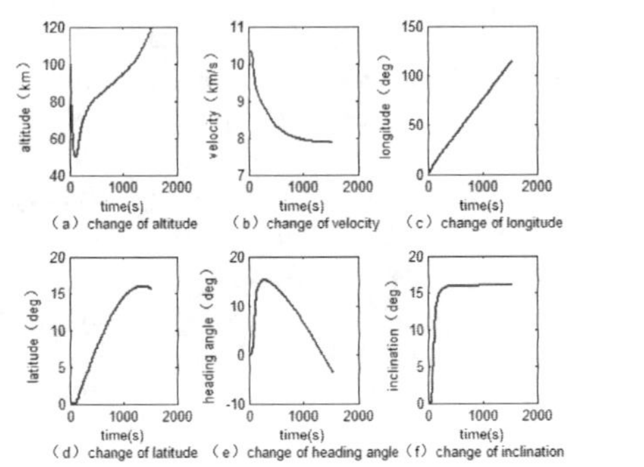

Fig. 3 Orbit parameters in atmosphere(16° inclination, mass 3000 kg)

4.1.2 Angle of 30° optimization

Vehicle transfer from Geostationary Earth Orbit (GEO) to 300 km altitude Low Earth Orbit (LEO) as described in section 1. In the process aeroassisted orbit transfer is carried out in the atmosphere with orbital inclination changed 30°. The GEO initial six orbital elements are same as section 3.1.1

1) kinetic parameters and quality area ratio, specific impulse parameters as shown in table 2.

Table 2 Simulink parameters

$C_{L,\alpha}$	m/s/kg·m^{-2}	K	$\rho0$/g·ml^{-1}	C_{d0}	γ_0/°	Θ_0/°
0.569 9	172	1.4	1.225	0.032	0	0
r_1/km	r_2/km	r_{p1}/km	i_0/°	i_f/°	Ψ_0/°	I_{sp}/s
42 164	6678	6 428	0	30	0	310

The optimization result showed that the optimization velocity increment is 1.552 km/s. Liquid rocket fuel consumption is about 688 kg. Horman double impulse velocity increment is 4.29 km/s, fuel consumption of liquid rocket is about 1301 kg. In the atmosphere, the change of the orbital parameters are shown in figure 2

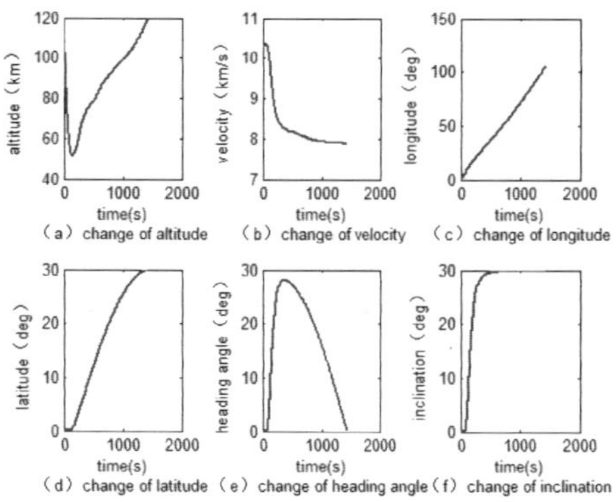

Fig. 4 Orbit parameters in atmosphere(30° inclination, mass 1 720 kg)

2) Mass-area ratio is 300 kg·m^{-2}, other kinetic parameters and specific impulse parameters are same as shown in table 2.

The optimization result showed that the optimization velocity increment is 1.553 km/s. Fuel consumption of liquid rocket is about 1 201 kg. Horman double impulse velocity increment is 4.29 km/s, fuel consumption of liquid rocket is about 2269.1 kg. In the atmosphere, the change of the orbital parameters are shown in figure

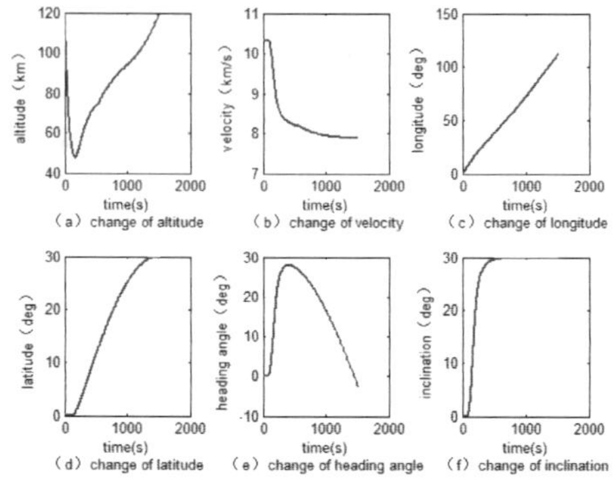

Fig. 5 Orbit parameters in atmosphere(30° inclination, mass 3000 kg)

4.2 LEO-LEO orbit transfer

4.2.1 Optimization aeroassisted orbit transfer between two different orbital altitude orbits

In order to analyze orbital altitude difference influence on fuel optimization efficiency of orbit aeroassisted transfer, assumes that the vehicle weight 1720kg, specific impulse parameters is 310 s. The optimization index is the sum of three speed increment same as section 2. Seven of the 900~1500 km altitude circular orbit are selected. Adopt aeroassisted orbit transfer optimization method to transfer 7 orbits to 300km orbital altitude with orbital inclination changed 16°. The velocity pulse and fuel saving is as shown in table 4 and figure 6.

Table 3 Optimization aeroassisted orbit transfer with different altitude difference

Orbital altitude/km	Velocity increment Optimization /(km·s^{-1})	Optimization fuel (kg)	Impulse velocity increment /(km/s)	Impulse fuel consumption (kg)	Fuel saving /%
900	1.955	816.24	2.208	888.45	8.13
1 000	1.896	798.52	2.218	891.19	10.40
1 100	1.843	782.30	2.229	894.18	12.51
1 200	1.794	767.05	2.240	897.17	14.50
1 300	1.749	752.83	2.251	900.14	16.37
1 400	1.707	739.37	2.262	903.10	18.13
1 500	1.694	735.16	2.273	906.06	18.86

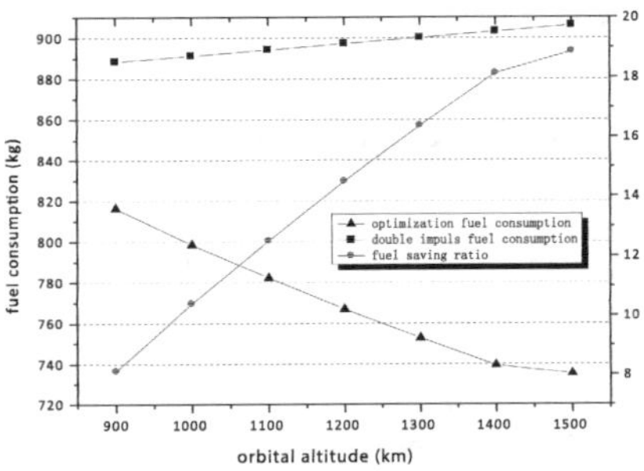

Fig. 6 Optimization aeroassisted orbit transfer with different altitude difference

It can be seen from table 5 and figure 6, the optimization efficiency of aeroassisted orbit transfer increase with the increase of the orbital altitude difference, but when the orbital altitude difference is greater than 1400 km, orbital optimization velocity pulse increases with its orbital altitude difference gradually tend to be stable. Calculated in the section 3.1 shows that when the altitude difference is 36 000 km, the optimized pulse value is 1.55 km/s. From 1400~36000 km orbital altitude, the variation of optimized velocity pulse is 0.15 km/s. Because orbit transfer fuel consumption increase with the orbital altitude difference increase, the overall aeroassisted orbit transfer fuel saving percentage increase along with the increase of orbital altitude difference.

4.2.2 Optimization aeroassisted orbit transfer between two same orbital altitude orbits

In section 3.2.1 it is analyzed Optimization aeroassisted orbit transfer between two different orbital altitude orbits. This section will analyze Optimization aeroassisted orbit transfer between two same orbital altitude orbits with inclination changed 30°. It is also assumed that the vehicle weight is 1720 kg, specific impulse parameters is 310 s. The optimization index is the sum of three speed increment same as section 2. 7 orbits from altitude 180~450 km are selected. It is analyzed of double pulse orbit transfer and aeroassisted orbit transfer between two same orbital altitude orbits. Optimization efficiency and results is as shown in table 5 and 7.

Table 4 Optimization aeroassisted orbit transfer when orbital altitude difference is zero

Orbital altitude/km	Velocity increment Optimization $/(km \cdot s^{-1})$	Optimization fuel (kg)	Impulse velocity increment /(km/s)	Impulse fuel consumption (kg)	Fuel saving /%
185.2	2.019	835.08	4.034	1264.12	33.94
230.0	2.224	892.83	4.020	1262.01	29.25
280.0	2.424	945.52	4.005	1259.75	24.94
300.0	2.528	971.59	3.999	1258.84	22.82
350.0	2.947	1068.01	3.984	1256.55	15.01
400.0	3.213	1122.67	3.970	1254.41	10.50
450.0	3.513	1178.83	3.955	1252.11	5.85

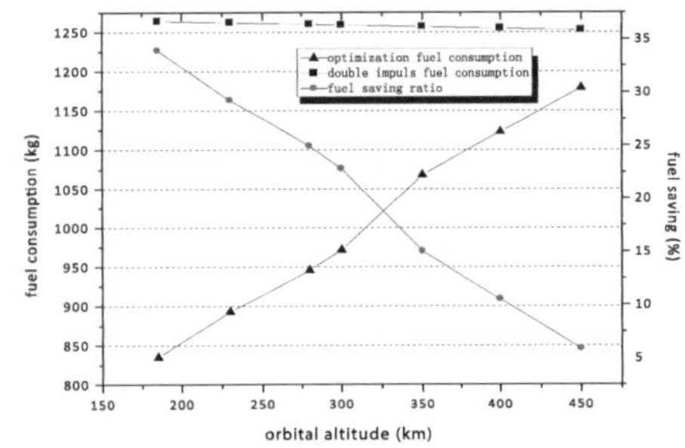

Fig. 7 Optimization aeroassisted orbit transfer when orbital altitude difference is zero

Figure 7 shows aeroassisted orbit transfer optimization in 7 selected orbits of different orbital altitude. From the fuel optimization results it can be seen that with the increase of orbital altitude, aeroassisted orbit transfer optimization efficiency gradually reduce. It also shows fuel saving percentage of aeroassisted orbit transfer optimization between two same orbital altitude orbits is reduced with the increase of orbital altitude.

4.3 Optimization aeroassisted orbit transfer with different orbital inclination changed

There is critical value of orbital inclination in optimization aeroassisted orbit transfer between two noncoplanar orbits of the same orbital altitude. When the inclination between two noncoplanar orbits is less than critical value, it is appropriate to use double impulse Hohmann orbit transfer between two noncoplanar orbits. When the orbital inclination between two noncoplanar orbits is greater than the critical value, it is better to use aeroassisted orbit transfer. In the section fuel consumption is analyzed using aeroassisted orbit transfer between the circular orbits of 185 km altitude with different orbital inclination changed. It is assumed that the vehicle weight 1720 kg, specific impulse parameters 310 s. The optimization index is the sum of three speed increment same as section 2. As shown in table 6 and figure 8.

Table 5 Optimization aeroassisted orbit transfer with different orbital inclination changed

Inclination difference/ (°)	Velocity increment Optimization /(km/s)	Optimization fuel(kg)	Impulse velocity increment /(km/s)	Impulse fuel consumption (kg)	Fuel saving /%
5	0.680	344.95	0.690	349.46	1.30
10	1.005	484.45	1.358	619.99	21.99
15	1.232	573.41	2.034	839.44	31.69
20	1.498	669.53	2.706	1014.18	33.98
25	1.790	765.80	3.373	1153.31	33.60
30	2.019	835.08	4.034	1264.12	33.94
35	2.278	907.40	4.687	1352.29	32.90
40	2.492	962.67	5.331	1422.53	32.32
45	2.742	1022.50	5.965	1478.56	30.85
50	2.973	1073.56	6.587	1523.26	29.52
55	3.232	1126.39	7.197	1559.05	27.75
60	3.412	1160.54	7.793	1587.72	26.90

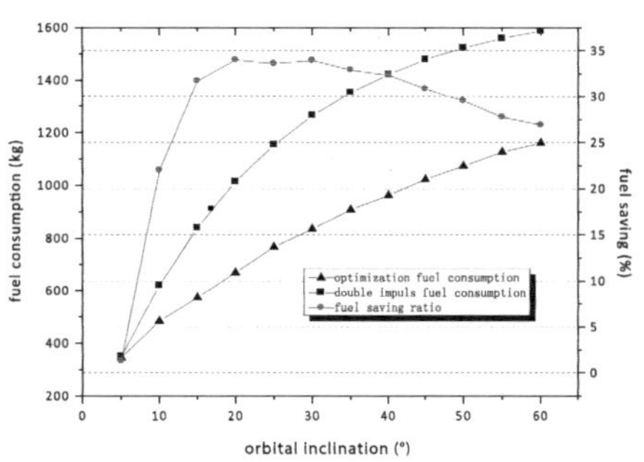

Fig. 8 Optimization aeroassisted orbit transfer with different orbital inclination changed

As is shown in figure 8 aeroassisted orbit transfer is implemented between two noncoplanar orbits with the same orbital altitude. With the increase of orbital inclination difference between two noncoplanar orbits, both fuel consumption of double impulse orbit transfer and aeroassisted orbit transfer are increased. But the rate of fuel consumption change of aeroassisted orbit transfer is less than the rate of fuel consumption change of double impulse Hohmann orbit transfer. With the increase of orbit transfer angle, its fuel saving of aeroassisted orbit transfer firstly increases then decreases. When the transfer angle is about 20° fuel savings rate is maximum. In addition, when the value of orbital inclination between two noncoplanar orbits is less than 5°, adopting aeroassisted orbit transfer and double impulse orbit transfer is equal with fuel consumption.

5. PROBLEMS AND PROSPECT

The technical problem aeroassisted orbit transfer facing is hypersonic flying technology in upper atmosphere. On the 2015-05-20 the united states fourth launch X–37B test vehicle - -OTV–4. One of the main task X–37B is space hypersonic flying in the near space with high Mach number. Near space altitude is between 30 and 100 km. The speed of X- 7B within the range of the altitude is $Ma = 25$. In a flight test in the atmosphere X-51A only reach 6 times the speed of sound. Time lasts 200s and there arise communication problems. While to X-37B the data is as an important part of the flight test index.

If speed $Ma = 25$ of X-37B in 30~100 km of near space can be achieved, and it has the ability to repeat flying. It may change the way the United States entering space from the ground. Namely, it can use ground plane send X–37B (OTV) carried payload to an altitude of 20 km (30 km) and release X-37B, then X-37B restart propulsion system to low-earth orbit. It is no longer to rely on disposable rocket launch, making the ground launch reusable truly. And it can greatly reduce the cost of entering the near-earth space.

6. CONCLUSION

In this paper, there are three aspects analysis of the aeroassisted orbit transfer between noncoplanar orbits are made:

(1) Aeroassisted orbit transfer optimization efficiency between noncoplanar orbits with different altitude;
(2) Aeroassisted orbit transfer optimization efficiency between noncoplanar orbits with the same altitude;
(3) Aeroassisted orbit transfer optimization between noncoplanar orbits with different altitude but different orbital inclination difference. As is known from the analysis:

The bigger is the orbital altitude difference, the better for aeroassisted orbit transfer between noncoplanar orbits, the fuel savings rate is higher; Implementation of aeroassisted orbit transfer with the same orbital altitude, low orbit aeroassisted orbit transfer optimization efficiency is higher than high orbit; Aeroassisted orbit transfer optimization between noncoplanar orbits with different altitude but different orbital inclination difference, with the increase of orbit inclination difference, fuel saving first increases then decreases, when the angle is about 20° fuel savings rate is the maximum. But the aeroassisted orbit transfer is still faced with many problems, with the continuous development of space technology, it will gradually solve these problems.

7. REFERENCES

1. 都亨, 张文祥, 庞宝军, 等. 空间碎片 [M]. 北京: 中国宇航出版社, 2007.

2. Liou J C, Johnson N L, Hill N M. Controlling the Growth of Future LEO Debris Populations with Active Debris Removal [J]. Acta Astronautica, 2010, 66 (5/6): 648-653. DOI: 10.1016/j.actaastro.2009.08.005.

3. Niels van der Pas, Lousda J, Terhes C, et al. Target Selection and Comparison of Mission Design for Space Debris Removal by DLR's Advanced Study Group [J]. Acta Astronautica. 2014, 102(9/10): 241-248. DOI: 10.1016/j.actaastro.2014.06.020.

4. 崔恩慧. 长三丙/远征一号火箭首飞成功 [N/OL]. 中国航天报, (2015-04-01) http://zhuanti.spacechina.com/n881117/content.html.

5. 吴德隆 王小军. 航天器气动力辅助变轨动力学与最优控制[M]. 北京：中国宇航出版社，2006

6. 胡松启, 陈雨. 伪谱法在飞行器轨迹优化中应用分析 [J]. 火箭推进, 2014, 40(5): 61-68.

7. 黄国强, 路宇平, 南英. 飞行器轨迹优化数值算法综述 [J]. 中国科学: 技术科学, 2012, 42(9): 1016-1036.

8. Portland State Aerospace Society. "Simple Rocket Science." Simplerockets 1d. Portland State University, 02 Mar. 2010. Web. 29 July 2013.

WHY A FUTURE COMMERCIAL SPACECRAFT MUST BE ABLE TO SWIM

Frank Morlang[1], Jorge Ferrand[2], Remzi Seker[2]

[1] *German Aerospace Center DLR, Lilienthalplatz 7, 38108 Braunschweig, Germany, frank.morlang@dlr.de*
[2] *Embry-Riddle Aeronautical University ERAU, Daytona Beach, FL 32114, USA, ferrandj@erau.edu*

ABSTRACT

Regarding current as well as future commercial space transportation (CST) projects, new re-entry trajectory profiles, differing from capsule and shuttle-like approaches, have to be taken into account. These characteristics are mainly ruled by the principle of "landing like an aircraft at an airport". This directly generates the need to be taken in a system wide information management (SWIM) consideration because of the fact, that all the future air traffic participants are requested to act as SWIM communicating sub-systems by the future Single European Sky Air Traffic Management Research (SESAR) SWIM "Intranet for ATM" concept. Against the background of the global character of future CST operations and the associated SWIM harmonization need referring the U.S. Next Generation Air Transportation System (NextGen) and SESAR, the presented solution bases on the already harmonized data format standards AIXM and FIXM.

1. INTRODUCTION

The concept of SWIM can roughly be condensed to the ability of delivering commonly understood quality information at the right time to the right target(s), resulting in a continuous data flow of the right sources to the right sinks. Strictly following this elemental request, generates the need that future commercial spacecraft will have to obey to it in order to build a fundament for seamless integration into normal air traffic [6, 7].

2. THE VALUE PROPOSITION

Preparing for the advent of Commercial Space and its impact to standard Air travel requires a Global, Enterprise-level approach that maximizes non-negotiable Safety and Cyber Security needs while addressing business and economic pragmatics. The approach presented within this Use Case leverages existing international data standards and infrastructures i.e., SWIM. This Use Case clearly demonstrates International Harmonization of Global Airspace to ensure key, critical benefits:

- All data exchanged is from a trusted, secure and authenticated source.
- All airspace stakeholders have shared situational awareness for real-time nominal and emergency situations – this is a globally available solution.
- Policies and Processes to be developed will use data from this approach for coordinated Collaborative Decision Making (CDM).
- Stakeholder notification, dynamic CDM and reconfiguration of the airspace back to a nominal state is achieved using this approach for safe and cost-effective airspace management.

3. THE CHALLENGE

Looking at the risk of falling debris from a spacecraft's in-flight explosion or breakup event, a large area can be covered by resulting fragments, depending on the position and velocity of the vehicle in the moment of disintegration. On the one hand, it must be assured that such a hazard area can be cleared of civilian/commercial air traffic in time, on the other hand, airspace segregation needs to be minimal in time and size, enabling an economic and efficient joint air- and space traffic operation. Facing this, the distribution of potential hazard area position information at the right time, to the right people is essential, for efficient flight planning purposes as well as for rapid and correct decisions in case of a fatal event. The optimization of the size and machine-aided decision-making aspects of this project are beyond the scope of this paper and will be handled as future studies as our research progresses.

4. THE SOLUTION

4.1 General Architecture

The solution's general architecture is shown in Figure 1. It consists of a data server implemented as a Hypertext Transfer Protocol Secure (HTTPS) web server and a provisional hazard area model assuming a Space Shuttle like trajectory of the hypothetical spacecraft. The data server is represented by an HTTPS web server, doing Simple Object Access Protocol (SOAP) message parsing for incoming requests and SOAP message creation for its responses. The provisional hazard area model is realized as an inter-/extrapolation model based on the United States National Aeronautics and Space Administration's (NASA) Columbia space shuttle accident debris data [1].

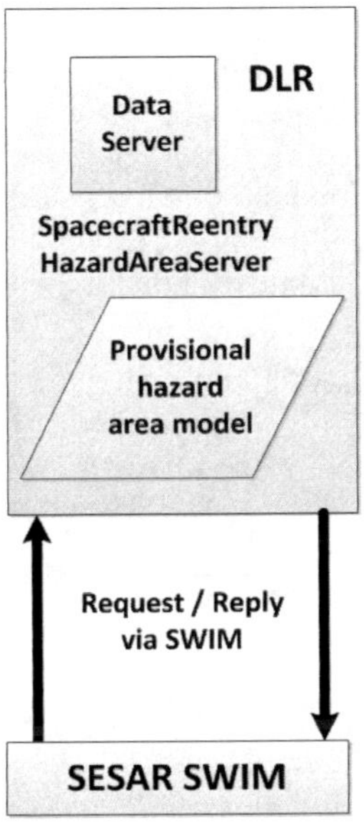

Figure 1. General architecture of the SpacecraftReentryHazardAreaServer

4.2 SESAR SWIM / FAA SWIM Use Case

The principle of SESAR SWIM / FAA SWIM demonstration use case is presented in Figure 2. It shows the connection to the United States Federal Aviation Administration's (FAA) concept of Global Enterprise Messaging Service (GEMS), where Service by NEAR (SBN) is one representational implementation of a GEMS provided by the Next Generation ERAU Applied Research (NEAR) lab. The demonstration use case includes Flight Object data exchange using Flight Information Exchange Model (FIXM) version 3.0.1 between an FAA research and development Flight Object Exchange Service, an airline EFB, and the SpacecraftReentryHazardAreaService, consuming the aircraft state data within the Flight Object FIXM message and publishes its output using AIXM version 5.1 for consumption and display by other interested parties.

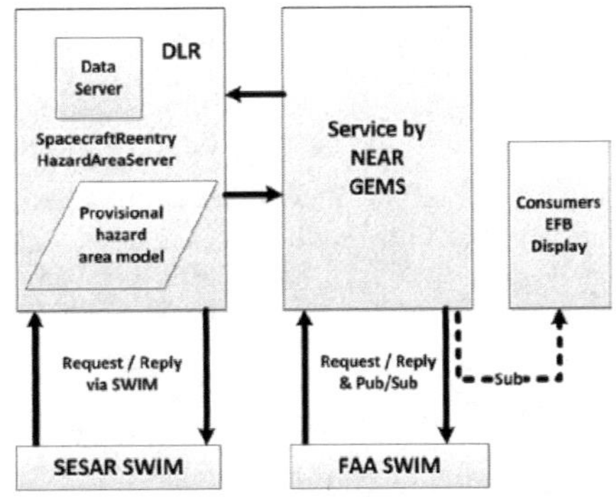

Figure 2. SESAR SWIM / FAA SWIM architecture

4.3 Data Server

The data server consists of an HTTPS web server, doing Simple Object Access Protocol (SOAP) message parsing for incoming requests and SOAP message creation for its responses. It benefits from using the following modules/extensions:

- Tcl httpd realizing the server side of the HTTP protocol handling and using Tcl's I/O channelling,
- Tls providing Secure Sockets Layer (SSL)-enabled TCP sockets in Tcl, supporting password protected SSL keys to improve security,
- WS::Sever exporting SOAP interfaces where HyperText Markup Language (HTML) page autogeneration, describing all methods and data types, is part of the service creation functionality,
- Rpcvar enabling complex data type definitions and their associated result packaging in the Remote Procedure Call (RPC) handler.

The service's reaction performance (server side time between receiving the request and sending out the response) is shown in Figure 3.

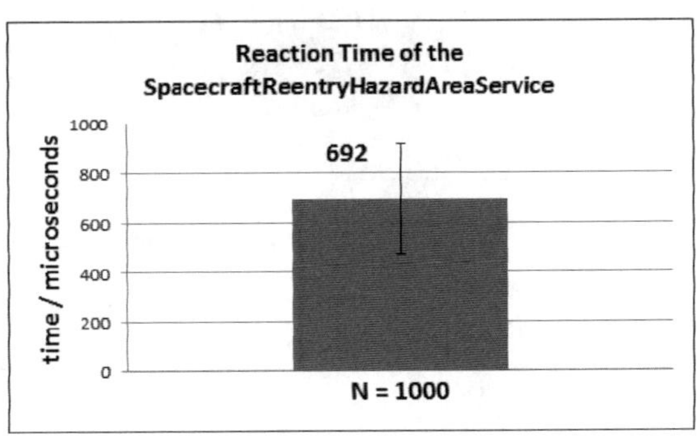

Figure 3. Reaction time of the SpacecraftReentryHazardAreaService

4.4 Provisional Hazard Area Model

The provisional hazard area model is currently characterized by a hard coded implementation of tabled debris data (derived from Columbia accident measurements), only representative for close to Space Shuttle like trajectory behaviour [1]. The relation in the tabled data can be summarized by:

$$\text{Debris Area length}_{\text{in km}} = \frac{Alt_{\text{in feet}}}{1000} \qquad (1)$$

$$\text{Debris Area width} = \frac{\text{Debris Area length}}{8} \qquad (2)$$

It must be emphasized, that these formula are not used, but one-dimensional table interpolation (first column is independent variable, all others dependent) is taken , facing the preparation for potential future tabled data of other vehicles, not reflecting a simple altitude-area relationship as in (1) and (2). The performance of the associated processing function is shown in Figure 4. Further, the hazard area calculation is prepared for:

- Lagrange interpolation for finding polynomial of degree n for a table of n sets of data,
- Neville algorithm usage for computing successively higher-degree approximations to the fit function by using the recurrence and take the straightest path possible,
- cubic spline interpolation for interpolating with different stepwise polynomials.

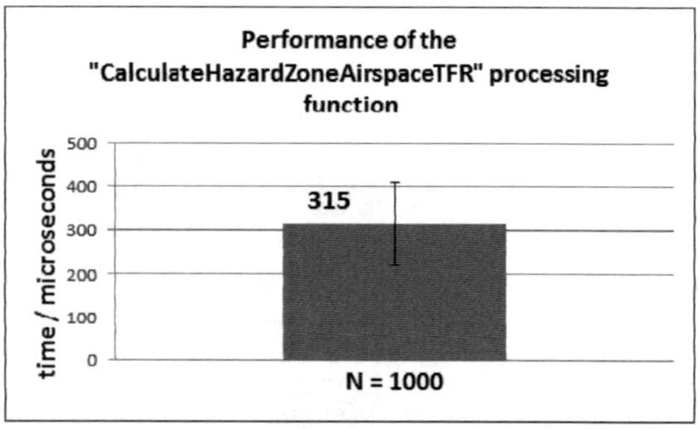

Figure 4. Performance of the "CalculateHazardZoneAirspaceTFR" processing function using one-dimensional table interpolation

5. OUTLOOK

5.1 Specific Hazard Area Model

There is different research work going on referring the development of debris catalogues and deriving vehicle specific debris area data of finer granularity [3][4][5]. With a further development of the hazard area calculation model, a solution directly benefiting from newest data sets can be delivered in future, where a debris catalogues database connection will be established as shown in Figure 5.

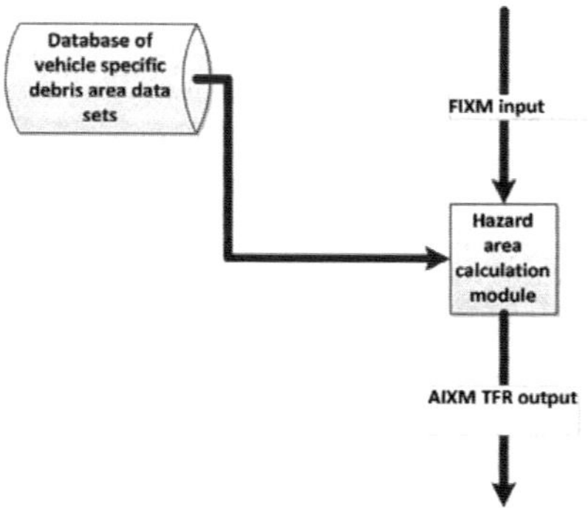

Figure 5. Vehicle specific hazard area calculation model

5.2 Improving Scalability And Performance

Tcl's underlying sophisticated I/O system multiplexes multiple clients without the complexities of threading. The service's reaction time performances for up to ten connected clients are shown in Figure 6. Each test ran for 1000 requests with a frequency of one client request per second.

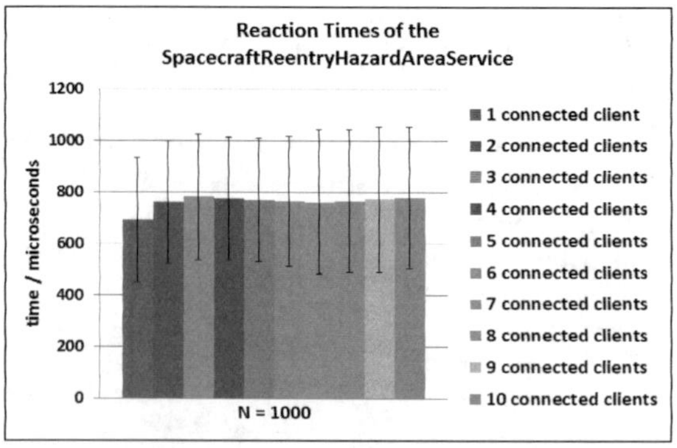

Figure 6. Reaction times of the SpacecraftReentryHazardAreaService

Because of the server's event driven I/O handling in a non-blocking fashion, the maximum number of simultaneously served clients expecting response according to the request frequency is anticipated as:

$$\text{clients}_{\max} = \frac{1 / ClientAverageFreq._{Request}}{\text{AverageReactionTime}_{\text{Service}}} \quad (3)$$

Taking an average reaction time of 800 microseconds from Figure 6 and putting it in (3) with the tests' client request frequency of 1 Hz reveals a maximum number of 1250 simultaneous clients. Addressing the intension of the implementation of a vehicle specific hazard area model () in the future, the $\text{clients}_{\max}$ number is estimated to be reduced by 30 to 50 percent, caused by needed database requests.

The following steps are planned to improve performance and scalability:

- Realization of a multi-server system with an on top application that directs clients to the appropriate servers,
- accompanying database split: a server only stores data related its subset of clients,
- full exploitation of Tcl 8.6's new Non-Recursive Engine (NRE) with its feature of supporting co-routines.

5.3 Cyber-Security for SWIM

Although the current prototype solution contains Transport Layer Security (TLS), it must be emphasized that this only secures point-to-point connections without intermediaries. Near future solution updates' developments will target to the implementation of message (end-to-end) security by using WS-Security technology throughout the whole chain of confidentiality, integrity, and availability as well as non-repudiation. Leveraging technologies such as Security Assertion Markup Language (SAML), a service provider can authenticate appropriate users so that they can gain access to secure content. There remains the issue of availability in terms of system's resilience in presence of distributed denial of service (DDoS) attacks. DDoS attacks have been increasing in intensity and frequency lately. These attacks can knock of services provided by large service providers with ample resources. In order to address this threat, we suggest the necessary DDoS prevention techniques (filtering, load balancing, input sanitation, as well as traffic redirection) and technologies should be utilized to protect this system.

The measures for preventing spoofing will also be implemented. The necessary policies for issuing certificates and management (including revocation) need to be included as part of the WS-Security technology will be clearly defined and adapted according to re-entry needs. WS-Security adaptation will include use of Public Key Infrastructure (PKI) with appropriate certificate authority (CA) in order to assure the integrity of the system. The adapted WS-Security system will be up to date enough to avoid some salient vulnerabilities such as the Weak Diffie-Hellman and the Logjam Attack [2].

The solution is already prepared with server side "intransform" and "outtransform" options for further SOAP header injections' handling and processing, building the fundament for the future WS-Security as well as on top WS Trust implementation development.

6. SUMMARY

The paper presents the need of future commercial spacecraft being able to act as communicating sub-systems in the system wide information management environment (SWIM). The air traffic integration challenge with the associated spacecraft specific properties is addressed as well as its global character's implications on ATM specific SWIM. A prototype solution with first test data results against an European / U.S. harmonization use case background is introduced with an outlook to future developments, covering precision, performance and security aspects.

7. REFERENCES

1. Robledo, L. F. (2004). Analysis and integration of a debris model in the VIRTUAL RANGE PROJECT. A thesis submitted in partial fulfillment of the requirements for the degree of Master of Science in Modeling and Simulation. Orlando, Florida: College of Engineering and Computer Science at the University of Central Florida.

2. Adrian, D., Bhargavan, K., Durumeric, Z., Gaudry, P., Green, M., Halderman, J.A., Heninger, N., Springall, D., Thomé, E., Valenta, L., VanderSloot, B., Wustrow, E., Zanella-Béguelin, S., Zimmermann, P. (2015). Imperfect Forward Secrecy: How Diffie-Hellman Fails in Practice. In Proc. 22nd ACM Conference on Computer and Communications Security, Denver, CO, USA.

3. Ailor, W., Wilde, P. (2008). Requirements for warning aircraft of reentering debris. In Proc. 3rd IAASS 'Building a Safer Space Together', European Space Agency, Rome, Italy.

4. Sarconi, M. (2013). A prototype system for simulating the risks of sub-orbital space flight for commercial aviation, Level 4 Project, School of Computing Science University of Glasgow, Glasgow, UK.

5. Falsone, A., Noce, F. & Prandini, M. (2014). A randomized approach to space debris footprint characterization. In Proc. 19th World Congress The International Federation of Automatic Control, Cape Town, South Africa.

6. Morlang, F., Hampe, J., Kaltenhaeuser, S., Jakobi, J., Schmitt, D.-R.. (2014). Commercial Space Transportation and Air Traffic Insertion – SESAR Requirements and the European Perspective. In Proc. 1st Annual Space Traffic Management Conference, Daytona Beach FL, USA.

7. Seker, R. Moallemi, M., Yapp, J., Towhidnejad, M., and Klein, R. (2014) "Performance Issues in Aircraft Access to the National Airspace SWIM Program" Integrated Communications Navigation and Surveillance Conference, Herndon, VA.

PROCEEDINGS OF THE 8TH INTERNATIONAL SPACE SAFETY CONFERENCE

SESSION 27 RE-ENTRY SAFETY - II

UPDATE OF AERODYNAMICS AND HEAT FLUX MODEL FOR ORSAT-J

Keiichiro Fujimoto, Hiroumi Tani, Hideyo Negishi, Yasuhiro Saito, Nobuyuki Iizuka, Koichi Okita

Japan Aerospace Exploration Agency, 2-1-1, Sengen, Tsukuba, Ibaraki, Japan, fujimoto.keiichiro@jaxa.jp

ABSTRACT

An improvement in the accuracy of re-entry safety analysis is getting increased interest in recent years. Important uncertainty factors are identified, and its quantification strategy is discussed. Core approaches to achieve update of current analysis method such as the ground and flight experiment and the high-fidelity numerical simulations are employed. Some of the findings related to the identified uncertainty factors are shown. Advantages to employ high fidelity numerical simulation are also discussed.

1. INTRODUCTION

1.1 Motivation and Objective

The destructive re-entry of satellites and rockets and the related ground risk due to fragments reaching the ground are getting increased interest in recent years. In order to minimize ground risk to be smaller than the acceptable level, the expected casualty is evaluated based on the estimated fragment dispersion area and the population density. By designing re-entry trajectory and control system of the vehicle, the expected casualty value is minimized to meet the safety requirement. Every space agency has their own software system to evaluate the re-entry expected casualty. Those software have been widely used for various types of re-entry safety analysis for satellites and rocket upper stages, and missions have resulted in successful destructive re-entry without any injury and fatal accident. It is essential to minimize re-entry fragmentation risk, however, it is also important to increase the accuracy of re-entry safety analysis to achieve further higher level of safety.

Current primary research scope of this study is on the improvement of re-entry safety analysis for the rocket upper stages. Objectives of this research campaign is to re-consider of the validity of the assumptions of current safety analysis method and to reduce the uncertainties of the re-entry risk analysis. It would contribute to improve space safety as well as the capability of the launch vehicles. In order achieve this ultimate goal, two new approaches are introduced, 1) ground and flight experiment and 2) high-fidelity numerical simulations. Research strategy will be discussed in the following sections with considering the uncertainty factors and uncertainty quantification process. Advantages to employ high-fidelity numerical simulations are also shown with some discussions on the computed result. Finally, further research topics are identified and concluded.

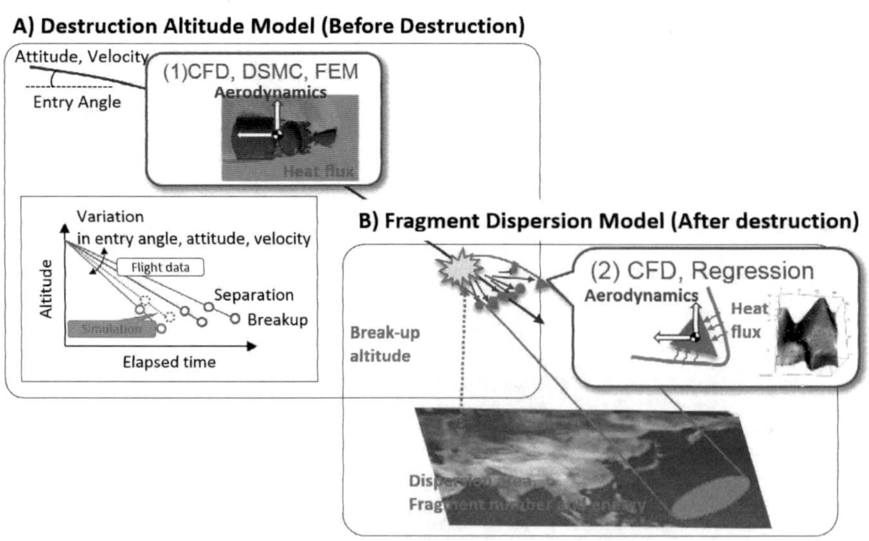

Figure 1 Overview of Re-entry Safety Analysis.

1.2 Re-entry Risk Analysis Tools

In order to evaluate the destructive re-entry safety, the complicated off-nominal physics such as the trajectory, aerodynamics, aerothermal issues, and thermal conductance issues should be considered for various types of geometries under the wide range of environmental conditions such as flow speed. As shown in Figure 1, both model consists of two part, A) destruction altitude model and B) fragment dispersion model. Fragment dispersion area depends on break-up conditions such as position, altitude, and velocity vector, and on the aerodynamic characteristics and heat rate for each fragment object. Break-up altitude is evaluated by A) destruction altitude model, and the fragment dispersion area, number and motion energy are evaluated by B) fragment dispersion model.

There are two types of models to evaluate destructive re-entry risk, one is "object-oriented" model and the other is "spacecraft-oriented" model. Comparison of "object-oriented" approach and "spacecraft-oriented" approach are compared in Table 1. Main difference of these two approaches is how to take into account the object shape changing in time. Object shape can be changed by the break-up, the material fusion, and structural failures.

Table 1 Comparison of re-entry analysis tools.

	Object-oriented - ORSAT (NASA) [1] - ORSAT-J (JAXA) - DEBRISK (CNES)[2]	Spacecraft-oriented - SCARAB (ESA)[3] - PAMPERO (CNES)[4]
Object shape	- Every object is treated as base shape - Shape change is not considered, except for that by break-up	- Real shape is considered - Shape changes in time including the change by material fusion
Breakup altitude	Decided based on past flight data or modelled based on thermal and/or stress analysis	
Equation of motion	3 DOF - Stable attitude - No lift considered	6 DOF - Integration of attitude motion
Thermal analysis	1D or 2D heat conduction model	
Aerodynamic model	Function based on base shape, motion, Knudsen number, and Mach number.	Evaluated based on local panel method
Heat-flux model		

Orbital re-entry survivability analysis tool of Japan (ORSAT-J) is re-entry risk analysis tool used in Japan Aerospace Exploration Agency (JAXA), which is derived from NASA-ORSAT ver. 4.0, and modified. It is one of the object-oriented methods.

In ORSAT, the drag coefficient C_D is determined based on the type of object (sphere, cylinder, box or flat plate), the flow regime, the orientation (spinning, tumbling, and so on), and the Mach number. Convective heating rate for objects in ORSAT are mainly computed based on object geometry, Stanton number, and Knudsen number. The stagnation point cold wall heating rate to a sphere in continuum flow is obtained using the Detra-Kemp-Riddell correlation [5]. To obtain averaging heating rates over the object, the sphere stagnation point heating rates are multiplied by factors that account for the average heating over the surface due to rotation and geometric dimensions of the object shape.

1.3 Uncertainty Factors

In the re-entry safety analysis there are various types of uncertainties. Major uncertainty factors are identified in this study. Those are shown in right of Figure 2 and as listed below. Brief description on for each factors and the difficulties to quantify those uncertainty is discussed.

(1) Model accuracy
In order to evaluate the destructive re-entry safety, the complicated off-nominal physics such as the trajectory, aerodynamics, aerothermal issues, and thermal conductance issues should be considered for various types of geometries under the wide range of environmental conditions such as flow speed. The re-entry safety analysis consists of various types of models. Main models are to evaluate break-up altitude, the aerodynamic characteristics, and the heating rate.

In terms of the numerical approaches, there are still big challenges to fill gap to achieve high accurate simulation of the destructive re-entry. Achievement of the aerodynamic and heating rate estimation by high fidelity simulation and the local panel method for complicated geometry is one of the challenges.

In terms of the experimental approaches, the comprehensive ground experiment campaign to obtain the mechanical and thermal properties of the various types of material, and the aerodynamic/aerothermal data for various types of shape is one of the challenges.

(2) Attitude stability mode
Especially in the non-controlled re-entry, there is large uncertainties in the attitude mode of the whole spacecraft before the break-up, and the fragments after the break-up.

In terms of the numerical approaches, validation of the model to estimate attitude stability is one of the challenges. In the ORSAT-J, although it is safer assumption, however, random tumbling is assumed. As will be discussed in the section 3.3, there is possibility the re-entry object to have aerodynamic stability depending on its shape.

In terms of the experimental approaches, uncertainty quantification for the initial condition of the objects such as attitude, angular rate, and velocities is one of the challenge. Therefore, as will be discussed in the section 2.1, the re-entry flight data acquisition is planned to obtain pressure, temperature, and acceleration data in this study.

(3) Shape complexity
Aerodynamic characteristics and the amount of the heating is significantly depending on the shape. In the ORSAT-J, complicated shape is modelled by using base geometry such as sphere, cylinder, and boxes. Since utilization of simpler shape is resulting in the smaller amount of the aerodynamic heat due to the smaller surface area, this safer assumption for the amount the heating. While, the aerodynamic characteristics has effects on the trajectory which is resulting in the change in the severity of the aerodynamic heating environment. Therefore, there is still discussions on the difference of "object-oriented" approach and "spacecraft-oriented" approach.

Therefore, the high fidelity numerical simulation method for complicated geometries has been developed and is employed in this study. As will be discussed in section 3.3, based on the results of high fidelity numerical simulation, the real shape effect is investigated in this study.

(4) Shape change
Shape of the spacecraft and fragments is changed in time due to the aerodynamic load and the structural fusion by aerodynamic heating. Change in the shape has effects on the aerodynamic characteristics thus trajectory, the heating rate, and the heat capacity.

In order to quantify uncertainties in treatment of the shape change, comparison of the results of "object-oriented" approach and "spacecraft-oriented" approach has been carried out.

(5) Initial conditions
In the re-entry safety analysis, various types of initial conditions are considered and the initial values are decided based on the knowledge of design and the flight data, and on the consideration of the worst case scenario. Major uncertain initial conditions which have dominant effect on the expected casualty are listed below.

1) Object's motion parameters such as the attitude, angular and translational velocity
2) Breakup altitude
3) Initial temperature distributions at the break-up

One of the challenges on this issue is difficulty to obtain flight data during destructive re-entry. It is easier to obtain the data during the orbital operations, data acquisition during the re-entry before the breakup is possible. Controlled re-entry is one of the approach to reduce the uncertainties related to the vehicle's motion. However, the data acquisition during the re-entry after the breakup is challenging.

As will be discussed in the section 2.1, development of re-entry flight data acquisition system is under the way in this study. By using this system, the breakup altitude, the temperature and pressure data can be obtained. This would greatly contribute to the uncertainty reduction.

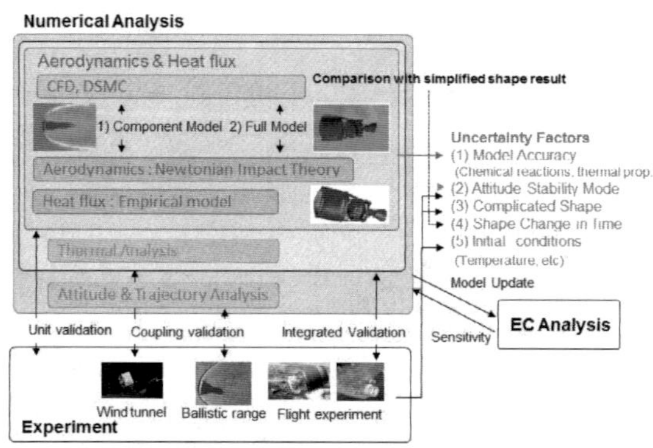

Figure 2 Uncertainty factors and its quantification strategy.

1.4 Uncertainty Quantification Strategy

It is essential to employ high-fidelity numerical simulations, and the comprehensive experimental data acquisition to reduce uncertainties in the re-entry safety analysis. In terms of the high-fidelity numerical simulations, the improvements in the accuracy and to realize complicated geometry handling are key issues. While, in terms of the experimental approaches, careful consideration on the parameter sensitivity analysis on the expected casualty is essential, and the acquisition of the major parameter data is key issue.

Detailed uncertainty quantification strategy is shown in Table 2, please not that this is still under the discussions. In Phase1, update of current model in ORSAT-J is carried out. Then, further improvement of the ORSAT-J is pursued in Phase2.

<u>Phase1</u>
In terms of the validation of high fidelity numerical simulations, simpler geometry such as base geometry such as sphere, cylinder, and boxes are considered as the scope of validation activity. Unit validation is carried out to validate each physical model such as aerodynamic and heating rate analysis. Coupling validation is then carried out to validate coupling analysis model such as aerodynamics and 6DOF trajectory analysis, and that of heating rate analysis and thermal conductance analysis. Based on the computed result obtained by the validated model, the current aerodynamic and heating rate model would be updated. It is one of the research activity candidate to introduce new base geometry to ORSAT-J.

In terms of experimental approaches, the flight data acquisition and the ground experiment is under the way to reduce major uncertainties, such as vehicle's attitude stability mode and the temperature distributions.

<u>Phase2</u>
It is still under the discussion, however, if the valuable advantages is found to be expected, further improvement in the current safety analysis model will be carried out. Candidates of modifications are introduction of the 6DOF trajectory analysis and the consideration of the real shape effect and the shape change effect. It is similar approach to the "spacecraft-oriented" approaches.

2. OVERVIEW OF RESEARCH ACTIVITIES

2.1 Re-entry Flight Data Acquisition

In order to reduce uncertainties on the attitude stability mode, break-up altitude, and an initial condition after the break-up such as temperatures on system components, re-entry flight data acquisition system is under the development and planned to use re-entry flight data acquisition of H-IIB launch vehicle in JAXA. By using this system, pressure and temperature data of various part can be obtained from entry to just before the break-up. Currently measurement system is connected sensors by wired connection, wireless sensing is one of the option in the future missions. Sensor positions are selected not only to reduce uncertainty in pressure and temperature distributions but to know break-up timing. As will be discussed in the section 3.3, results of high fidelity numerical simulation is used to design the sensor positions.

Table 2 Current uncertainty quantification strategy (still under the discussion).

	Phase1 (Update of current model) - Validation of high fidelity simulations - Update of assumptions and models	Phase2 (Modification of current model) - Further research to improve methodology
(1) Model accuracy	- Unit validation (Heat flux) - Coupling validation (Heat flux + Thermal conduction)	
	- Unit validation (Aerodynamics)	- Unit validation (Aerodynamics) - Coupling validation (Aerodynamics + Attitude and Trajectory)
(2) Attitude stability mode	A) Destruction altitude model - Quantified by flight data acquisition B) Fragment dispersion model - Same as current model (3DOF, no-lift), would be modified based on the validated numerical simulations	
		- 6DOF, all aerodynamic coefficients are considered
(3) Shape complexity	- Simple base shape	- Real shape effect is considered
(4)Shape change	- Not considered	- Considered
(5)Initial conditions	- Quantified by flight data acquisition	

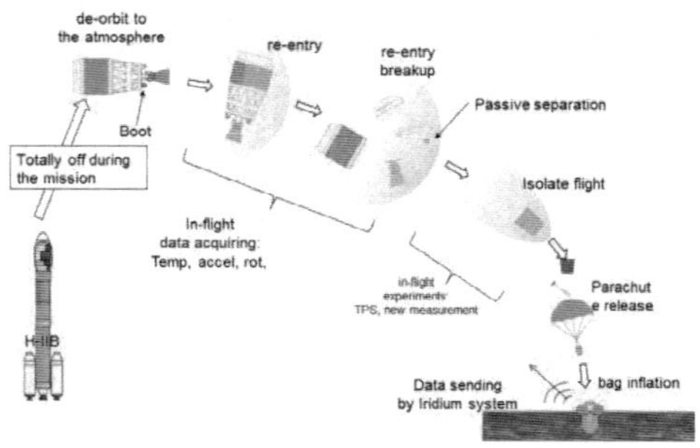

Figure 3 Sequence of re-entry flight data acquisition system.

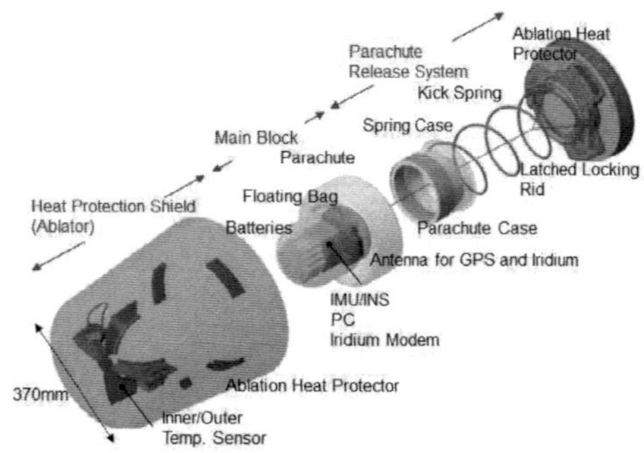

Figure 4 Schematic view of re-entry flight data acquisition system

2.2 High-fidelity Numerical Simulation

High fidelity numerical simulation is key technology to achieve higher-level of accuracy in re-entry safety analysis method. Once physical model used in the high-fidelity numerical simulation is validated by comparing with experimental data under the enough extent of the environmental conditions, mechanism of the complicated physics during the destructive re-entry can be understood, and current aerodynamic and heat flux model can be improved based on the computed results. Due to the significant improvements in the numerical methodology and the computational capability, an aerodynamic characteristics and heat flux estimation for complicated geometry is now available. Based on the computed results, re-considerations of assumption and the base geometry selection can be performed. Technical details on the computational fluid dynamics are described as follows.

Rarefied Flow Solver (UNITED)
At the high altitudes that are generally over 90 km, the effects of the rarefied gas dynamics, i.e. non-equilibrium of thermodynamics and chemistry, are significant. To evaluate heat flux and pressure on the spacecraft at such high altitude, the Boltzmann equation of kinetic theory was solved by Direct Simulation Monte Carlo (DSMC) [6]. In the present study, University of Tokyo Evolution DSMC (UNITED) code, originally generated by university of Tokyo and have been developed for years by JAXA, was used for DSMC calculations. The structure grid system is employed and it has the multi-level grid refinement and coarsening capability. By using cutting cell approach, complicated shapes and multi bodies of the practical spacecraft can be handled easily. The variable soft sphere [7] (VSS) and variable hard sphere (VHS) models[8] can be chosen as the collision model. UNITED code also has the capability of the sub-cells [9] and chemical reactions of air [10].

Continuous Flow Solver (LS-FLOW)
At the moderate altitude the flow can be treated as continuous flow. To evaluate heat flux and pressure on the spacecraft at such conditions, three-dimensional compressible Navier-Stokes equation solver LS-FLOW [11, 12] is used. The system of equations is discretized and solved by cell-centered finite volume method for unstructured grids with arbitrary polyhedral. It has also capability of overset grid method. Capability to handle arbitrary polyhedral meshes and over-set grid method is suitable to carry out analysis for complicated geometry such as full configuration of rocket upper stage. In the re-entry safety analysis, aerodynamic characteristics and heat flux analysis under the wide range of flow condition should be carried out including flow speed and angles. Since flow structure such as shock waves can be changed depending on the flow conditions. By using solution adaptive mesh refinement (AMR) technique, pre-processing time for the grid re-generation can be minimized. Both overset grid method based on prism layer grid and Cartesian volume grid and body-fitted Cartesian grid generated by LS-GRID [11, 13] is used. Utilization of Cartesian volume grid is key to utilize AMR technique to reduce computational turnaround time.

Table 3 Numerical methods for rarefied flow solver.

Governing equations	Boltzmann equation of kinetic theory
Chemical reaction	No chemical reaction is considered
Collision model	Variable Soft Sphere model [7]
Internal energy relaxation model	LB model for internal energy relaxation [14] Constant (Zr=5, Zv=50) model for relaxation coefficient of rotational and vibrational energy
Wall reflection model	Diffusion reflection boundary at 300K (wall)

Table 4 Numerical methods for continuous flow solver.

Governing equations	Three-dimensional, compressible Navier–Stokes eqs.
Chemical reaction	No chemical reaction is considered
Spatial discretization	Cell-centered FVM
Spatial reconstruction	1st order: No gradient is considered
Inviscid Term	SLAU2[16]
Viscous Term	Shima's Method
Turbulence Model	Laminar computation
Time Integration	2nd order Lower–upper symmetric Gauss–Seidel [17] with 8 inner iterations, and local time stepping
Grid	- Body-fitted Cartesian grid - Cell Reynolds number is less than 1 for all grids Section 3.1: 1.11 million cells Section 3.2: 2.64 million cells

3. RESULT AND DISCUSSIONS

In this section, some of the computed results obtained by using high fidelity numerical simulation are shown. In each following sections, advantages to use high fidelity numerical simulations as well as findings to reduce analysis uncertainty are shown. In the sections from 3.1 to 3.2, computed results of continuous flow solver LS-FLOW is discussed. In the section 3.3, computed results of rarefied flow solver UNITED is discussed. Numerical approaches used in this study is shown in Table 3 and Table 4.

3.1 Mach Number Effect on Aerodynamics and Heat flux

In the re-entry safety analysis, wide range of free-stream Mach number should be considered from hypersonic to supersonic. Typically, maximum stagnation heat flux condition corresponds to the Mach number range 10 to 20. As is described in the Detra-Kemp-Riddell correlation [5], stagnation heating is proportional to $\rho_\infty V^3_\infty$ where ρ_∞ is atmospheric density V_∞ s free-stream velocity. Therefore if aerodynamic characteristics has Mach number dependency, stagnation heating becomes different whether Mach number dependency is considered or not. Heat flux evaluation was carried out for the sphere which is one of the base geometry of the ORSAT-J under the flow condition of trajectory data for helium gas reservoir of rocket upper stage. Computed surface heat flux and temperature distribution in the pitch symmetry plane is shown in Figure 5. As shown in this figure, bow shock wave inclination angle near shoulder becomes steeper with increase in Mach number. This is the main reason of Mach number dependency on the aerodynamic drag and the heat flux as shown in Figure 6 and Figure 7 respectively. Stagnation point cold wall heating rate is basically evaluated by the Detra-Kemp-Riddell correlation in ORSAT-J. It was also clarified that difference between stagnation heating evaluated by Detra-Kemp-Riddell correlation and that evaluated by CFD is small. This is mainly due that this correlation is derived from the experimental data for hemi-sphere cylinders [5]. It was also found that ratio of averaged heating rate to the stagnation heating rate C has slight Mach number dependency. Validation of the CFD model and the sensitivity analysis to EC result is the next research topic.

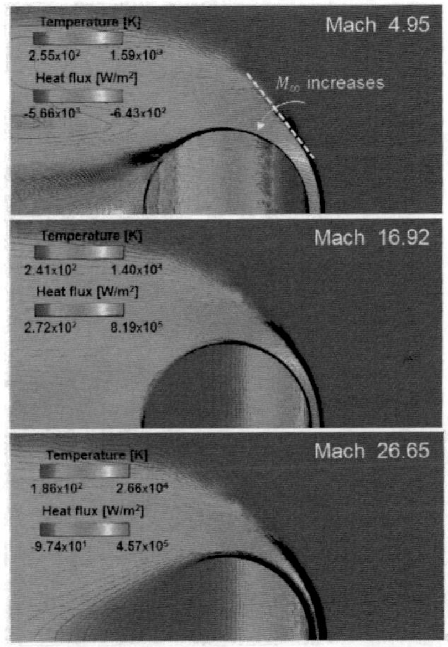

Figure 5 Surface heat flux and temperature distribution in the symmetry plane for sphere.

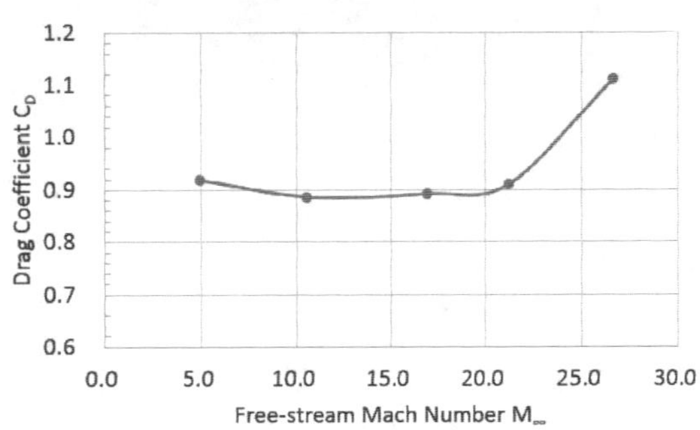

Figure 6 Mach number effect on drag coefficient of sphere.

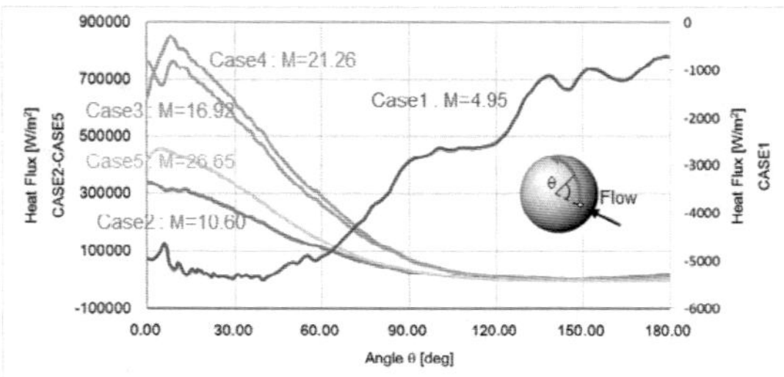

Figure 7 Mach number dependency on heat flux of sphere.

3.2 Local Curvature Effect on Heat flux

In the heating rate distributions for the sphere, position of the maximum heating rate is corresponding to the stagnation point. While in that for the cylinders and boxes, maximum heating rate appears near the shape edge of shoulder as shown in Figure 8. In the sharp corner of the object shock distance becomes smaller as comparing with that of bluff bodies such as spheres. This is the one of the main reason of the peak heating rate existence near the sharp corners. In ORSAT-J and also original ORSAT, such a heat flux distribution is considered to evaluated spatial averaged heating rate, however in most of cases 1D thermal conductance analysis under the assumption of material temperature distribution is uniform. In some of the spacecraft-oriented approaches such as PAMPERO of CNES [4], 3D thermal conductance analysis is used with considering of heat flux distributions including peak near the sharp corners. This is one of the key uncertainties which should be quantified in the further researches.

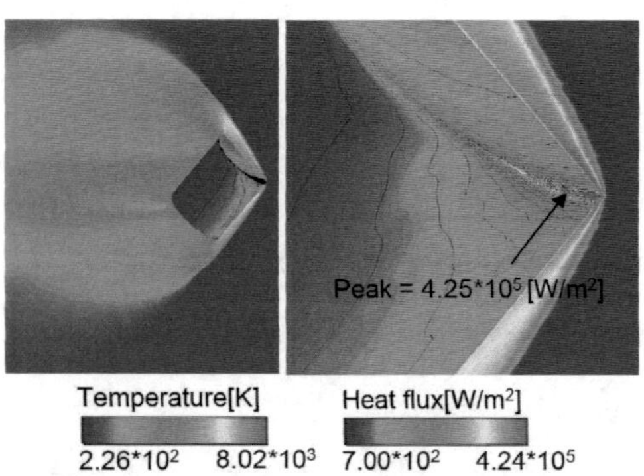

Figure 8 Surface heat flux distribution and temperature distribution in the symmetric plane for Mach number 14.35.

3.3 Real Shape Effect on Aerodynamic Stability

In ORSAT-J, object's motion is typically assumed to be random tumbling wherein it is assumed that object surface is heated uniformly. In case of the sphere-type objects such helium gas reservoir, this assumption is reasonable, if shape is perfectly sphere and its centre of gravity is not shifted.

While it was found that aerodynamic stability can be achieved depending on the object's shape such as entire rocket upper stages and liquid rocket engines. One of the computed result of pressure and heat flux distribution for rocket upper stage in the rarefied flow regime is shown in Figure 9. As shown in this figure, due to the larger projection area in the vehicle's upper part which is resulting in the stable pitching moment as shown in Figure 9. As far as investigated, at the static stable pitch angle the flow is coming from liquid rocket engine side. Especially in the case of un-controlled re-entry, there should be comprehensive consideration on the various uncertainties such as initial condition of the vehicles's attitude just before entering high density altitude, this results in the higher heat flux rate for liquid rocket engine. There is possibility to re-consider the assumptions on the object's attitude stability to improve current re-entry safety analysis method. This is the one of the advantages to use state-of-the-art simulation technology which can handle complicated real geometry. Further investigation will be continued including the aerodynamic stability analysis and the validation of the models.

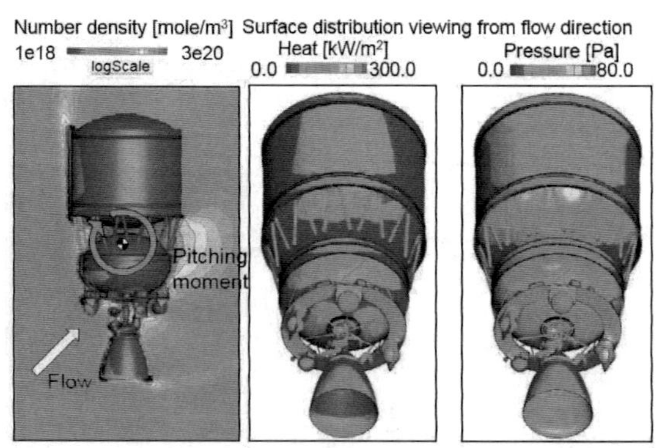

Figure 9 Surface heat flux and pressure distribution for rocket upper stage.

4. CONCLUSIONS

An improvement in the accuracy of re-entry safety analysis is getting increased interest in recent years. Important uncertainty factors were identified, and its quantification strategy was discussed. Core approaches to achieve update of current analysis method such as the ground and flight experiment and the high-fidelity numerical simulations are employed. Some of the findings related to the identified uncertainty factors were shown. It was shown that there is Mach number dependency on the aerodynamic characteristics and the heat flux distribution. It was also shown that there is local peak in the heat flux distributions especially near the sharp edges, which is mainly due to the smaller shock wave distance. Finally it was shown that the aerodynamic and heat flux evaluation for realistic complicated geometry is available. By using high fidelity numerical analysis for complicated geometry, it is possible to re-consider to the assumptions on the object's stability mode and the surface area related to the amount of the aerodynamic heat.

5. REFERENCES

1. Dobarco-Otero, J., et al., "Upgrades to Object Reentry Survival Analysis Tool (ORSAT) for spacecraft and Launch Vehicle Upper Stage Applications", AAS Science and Technology Se 4, pp ries, vol. 109, Space Debris and Space Traffic Management Symposium 2003, Univelt, 200 . 273-289.

2. Omaly P., Magnin Vella C., Galera S., DEBRISK, CNES tool for re-entry survivability prediction: validation and sensitivity analysis, Proc. 6th IAASS Conf. "Safety is not an option", Montréal (Ca), 21-23 May 2013.

3. Koppenwallner, G. et al. (2005). SCARAB – a Multidisciplinary Tool for the Destruction Analysis of Spacecraft during Re-entry, 5th European Symposium for the Aerothermodynamics of Space Vehicles, Cologne.

4. Annaloro, J. (2015) "Elaboration of New Spacecraft-orientated Tool: PAMPERO", 8th European Symposium on Aerothermodynamics for Space Vehicles, Lisbon, 2-6 March 2015.

5. Detra, R. W., Kemp, N. H. and Riddell, F. R., "Addendum to Heat Transfer to Satellite Vehicles Re-entering the Atmosphere," Jet Propulsion, vol. 27, no. 12, Dec. 1957, pp. 1256-1257.

6. G.Bird, "Molecular gas dynamics and the direct simulation of gas flows," Oxford Engineering Science Series, 42, 1994.

7. Koura, K. and Matsumoto, H., "Variable soft sphere molecular model for inverse-power-law or Lennard-Jones potential," Phys. Fluids, A3 (1991), pp.2459-2465.

8. Bird, G. A., "Monte-Carlo simulation in an engineering context, Proceedings of the 12th International Symposium on Rarefied Gas Dynamics," Prog. Astronaut. Aeronaut., 74, 239, 1981.

9. Boyles, K.A., et al., "The Use of Virtual Sub-Cells In DSMC Analysis of Orbiter Aerodynamics at High Altitudes Upon Reentry", AIAA-2003-1030, 41st AIAA Aerospace Sciences Meeting and Exhibit, Reno, NV, Jan. 2003.

10. G.A. Bird, "Chemical Reactions in DSMC," 27th International Symposium on Rarefied Gas Dynamics, Pacific Grove, CA, July 10-15, 2010.

11. Fujimoto, K., "Study on the Automated CFD Analysis Tools for Conceptual Design of Space Transportation Vehicles," Ph.D. Dissertation, Univ. of Tokyo, Tokyo, Japan, 2006.

12. Kitamura, K., Fujimoto, K., Shima, E., Kuzuu, K., and Wang, Z. J., "Validation of Arbitrary Unstructured CFD Code for Aerodynamic Analyses," Transactions of the Japan Society for Aeronautical and Space Sciences, Vol. 53, No. 182, 2011, pp. 311–319.

13. Fujimoto, K., Fujii, K., and Wang, Z. J., "Improvements in the Reliability and Efficiency of Body-Fitted Cartesian Grid Method," AIAA Paper 2009-1173, 2009.

14. Larsen, P. S., and Borgnakke, C., "Statistical Collision Model for Simulating Polyatomic Gas with Restricted Energy Exchange" in Rarefied Gas Dynamics I, edited by M. Becker and M. Fiebig, DFVLR Press, Porz-Wahn, 1974, paper A7.

15. Venkatakrishnan, V., "Convergence to Steady State Solutions of the Euler Equations on Unstructured Grids with Limiters," Journal of Computational Physics, Vol. 118, No. 1, 1995, pp. 120–130.

16. Kitamura, K. and Shima, E.,"Towards shock-stable and accurate hypersonic heating computations: A new pressure flux for AUSM-family schemes," J. Comput. Phys., Vol.245, 2013, pp.62-83.

17. Jameson, A., and Turkel, E., "Implicit Schemes and LU Decompositions," Mathematics of Computation, Vol. 37, No. 156, 1981, pp. 385–397.

ATMOSPHERIC ENTRY OF SPACE DEBRIS: OXIDATION AND EMISSIVITY DATA FOR MODEL IMPLEMENTATION

M. Balat-Pichelin (1), L. Barka (1), J. Annaloro (2), P. Omaly (2)

(1) *PROMES-CNRS laboratory, 7 Rue du four solaire, 66120 Font-Romeu Odeillo, France*
Marianne.balat@promes.cnrs.fr
(2) *CNES, Centre Spatial de Toulouse, 18 avenue Edouard Belin, 31401 Toulouse Cedex 9, France*

ABSTRACT

In order to mitigate debris in orbit and to avoid dramatic collisions on Earth after atmospheric re-entry at its end-of-life, the spacecraft missions have to take into account the influence of atmospheric re-entry conditions on the spacecraft survivability. In this way, oxidation in air plasma conditions was studied on alloys such as TA6V, Inconel 718 and SS 316L in order to obtain degradation kinetic laws at high temperature and for short time duration (few minutes). Moreover, measurements of the emissivity at high temperature were carried out on the same virgin alloys – in high vacuum to avoid oxidation – and on samples oxidized in air plasma conditions. An increase of the total emissivity for oxidized samples by a factor 3 to 5 is obtained compared to virgin samples. This effect can strongly influence the ablation during a trajectory and consequently the casualty area calculation.

1. INTRODUCTION

The materials to be used in the spacecraft must take into account the different new rules, LOS (Loi relative aux Opérations Spatiales, France), ISO 24113, ECSS of ESA… in order to be mainly or totally destroyed due to the severe aerothermodynamic and aerothermochemical conditions encountered during the re-entry phase.

The non-equilibrium air plasma flow impinging the material enhances oxidation phenomena and also heat flux by species recombination. Oxidation, ablation, catalycity and melting coupled phenomena can occur on the material leading to an important increase of heating and damage of the materials.

In this way, it is necessary to perform on-ground experimentation in conditions representative of the re-entry and to know the material degradation in such plasma environment. Moreover, the thermal radiative properties of the material under such conditions must be evaluated for implementation in the simulation codes.

The LOS approved by the French Parliament is applicable since December 2010, a main objective of this law is the security of space operations. To assess the risk of re-entry, CNES has decided to develop its own engineering tool, called DEBRISK to study the survival of debris during the atmospheric re-entry [1, 2]. DEBRISK is an engineering tool giving the 3D trajectory of the satellite components reentry phase and the surface heat flux and debris disappearance altitudes. Simulation results are very dependent on the value of the properties of materials such as fixed input conditions for each modeled object. The properties of the materials currently available in DEBRISK are limited. Most of the time, the bibliographic origin and validity of these values remain unknown. In some cases, a property as the emissivity is set equal to 1, which leads to an overestimation of the mass reaching the ground object. To increase the accuracy of the calculation with DEBRISK, CNES has decided to conduct a high-temperature characterization of the main materials used in spacecraft.

In the literature, there are few works on oxidation of alloys such as TA6V, Inconel 718 and SS 316L and the oxidation is always performed in standard air and at temperatures lower than the ones we are interesting in [3-7]. Emissivity data are also scarce and often obtained on partially oxidized surfaces [8-11] this is why oxidation in air plasma conditions and emissivity evaluation at high temperature were done in order to obtain representative data.

2. METHOD FOR EMISSIVITY EVALUATION AND EXPERIMENTAL DEVICE

The direct method developed at PROMES-CNRS for the evaluation of the emissivity is based on the measurement of the directional radiance of the material in a given wavelength range and of the true temperature [12-14]. For the evaluation of the total hemispherical emissivity the radiance measurement is made in the range of 0.6-40 μm for different incidence angles by 10° step (from 0° to 80°), 0° being normal incidence. The measurement can be performed in high vacuum (10^{-4} Pa) in the MEDI-

ASE facility installed at the focus of the 1 MW solar furnace of Odeillo (Fig. 1) that also allows the study of the physico-chemical behavior of materials under extreme conditions at very high temperatures (up to 2500 K). The experimental MEDIASE (Fig. 2) consists of a stainless steel vessel with a capacity of about 60 l, equipped with a turbomolecular pumping system. A hemispherical silica window, 35 cm in diameter, is placed in front of this vessel and allows concentrated solar radiation to heat the sample (up to 10 MW/m^2). The sample (40 mm in diameter, 2 mm thickness) is exposed to concentrated solar radiation by the gradual opening of the doors of the solar furnace. The sample temperature is measured on the rear face using the two-color pyro-reflectometer developed in our laboratory [15, 16]. The pyro-reflectometer (accuracy 1%) is calibrated on a blackbody before and after each series of measurements. The temperature of the blackbody that can reach 3000 K is measured with a calibrated pyrometer. The radiance is measured using a bolometer with an optical Cassegrain (KT4 Heimann). According to the required wavelength range, the radiometer is used without filter (0.6-40 μm) or with narrow spectral filters (2.7; 5 or 5.5 μm) and band filter (0.6-2.8; 3-5; 7-10 or 8-14 μm). The measurements are performed using a motorized three-mirrors system (goniometer) so the sample and the radiometer stay fixed. The quality of the window is adapted to the wavelength range and for these measurements the window used is made of iodo-bromide thallium (KRS5) to carry out measurements in the range from 0.6 to 40 μm. The entire optical radiometer window-goniometer system is calibrated before and after each series of measurements on a blackbody as for the optical pyro-reflectometer.

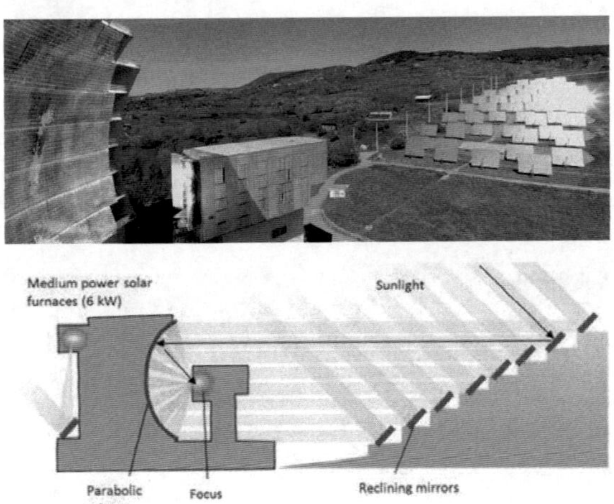

Figure 1. Picture and scheme of the 1 MW Odeillo solar furnace

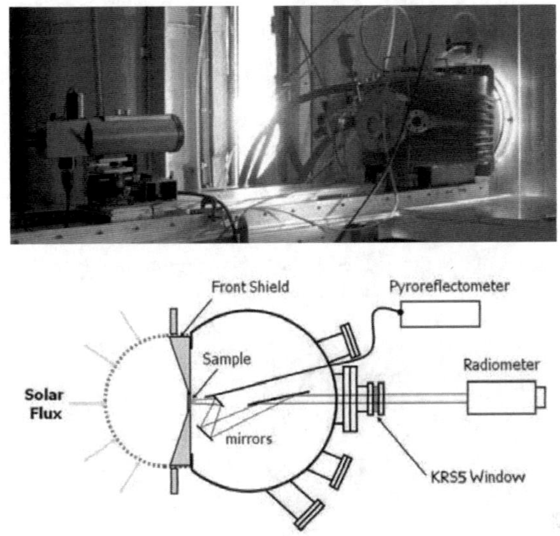

Figure 2. Picture and scheme of the MEDIASE facility for emissivity measurement

The sample is inserted into the sample holder and the vacuum is created in the chamber until it is around 10^{-4} Pa. Then, the sample is subjected to concentrated solar radiation by the gradual opening of the doors of the solar furnace to obtain the desired temperature on the rear face of the sample. It only takes a few seconds. Following the temperature stabilization, the rotation of the three-mirrors system of the goniometer is performed between 0° and 80° and the data are stored (at 1 kHz, 16-bit resolution). Radiance measurements are followed by a temperature measurement. This procedure is repeated for each temperature level. Directional values are then extrapolated to zero for the angle of incidence of 90° by cubic spline functions which are integrated to obtain the hemispherical emissivity. The uncertainty is around 3% between 1500 and 2300 K for the spectral range of 0.6 to 40 μm.

3. OXIDATION STUDY IN AIR PLASMA CONDITIONS

The MESOX facility installed at the focus of the 6 kW solar furnace (Fig. 3) is used to study the oxidation of materials in atmospheric reentry conditions [17, 18]. It is equipped with a microwave generator and a waveguide (2450 MHz, 1200 W max) for creating non-equilibrium air plasma with an oxygen dissociation rate of the order of 80% [19]. For these tests, the air plasma is generated by a 300W discharge surrounding the sample. A regulator, a pressure gauge and a vacuum pump allow the precise control of the total pressure during the experiments. The airflow is set at 4 l/h for maximum dissociation.

The high temperature is obtained by the concentrated solar radiation (5 MW/m^2, 2500 K max). Monitoring the oxidation in the MESOX is carried out also by in situ video and by measuring the mass variation before and after the test. Microstructural characterization of the samples before and after the test is carried out by techniques such as SEM (surface and cross-section), EDX, XRD to perfectly know the composition of the material surface and to correlate the results obtained on emissivity including the nature of the oxides that can be different in plasma or in standard atmospheric conditions.

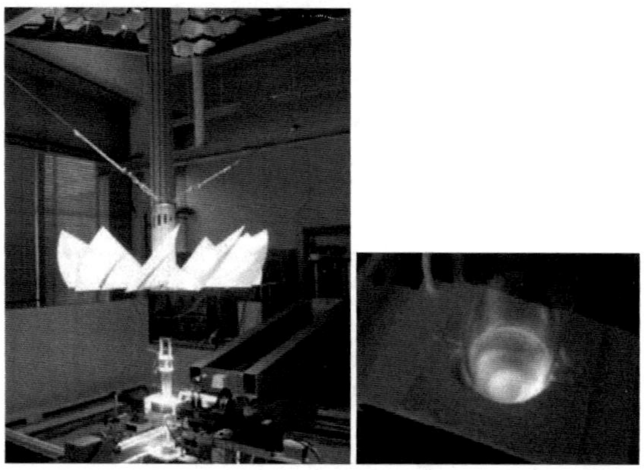

Figure 3. MESOX facility (on the left) and air plasma image (on the right)

4. EXPERIMENTAL RESULTS

In this section, emissivity measurements in high vacuum and at low pressures on oxidized, or in situ oxidized samples are presented together with oxidation study in air plasma conditions in order to have kinetic laws.

4.1. Materials

The materials used in this study are provided by different manufacturers: TA6V from Timet and Aludis, inconel 718 from Carpenter and Goodfellow, SS 316L from Ugine, Thyssen and Goodfellow.
The compositions for each material are close from one to another manufacturer and are the following for the main components:
- TA6V: Ti 90%; Al 6% and V 4%
- inconel 718: Ni 53%; Cr 19%; Fe 18%; Nb+Ta 5%; Mo 3% and Ti 1%
- SS 316L: Fe 70%; Cr 17%; Ni 10%; Mo 2% and Mn 1%.

4.2. Emissivity in high vacuum

Measurements of the total directional emissivity and calculation of the total hemispherical emissivity - obtained by integration of the directional values - were performed on the materials SS 316L, inconel 718 and TA6V alloys in high vacuum (10^{-4} Pa).

Figure 4 reports an example of the directional emissivity for the TA6V alloy (from Aludis) for several temperatures. The order of the temperatures in the figure caption represents the chosen protocol. The temperature is increased from around 1100 K up to the maximum one (here 1629 K) and then measurements are performed decreasing the temperature as the surface of the sample changed due to its smoothing after the high temperature exposure that leads to a grain growth as it can be seen on Fig. 4. Grains are clearly visible on the image of Figure 5 (on the right) after exposure at high temperature.

Figure 6 reports the evolution of the total hemispherical emissivity for the TA6V alloy (Aludis and Timet) in high vacuum. It can be seen that the total hemispherical of non-oxidized TA6V alloy is nearly constant along the temperature range and its value is around 0.15-0.18.

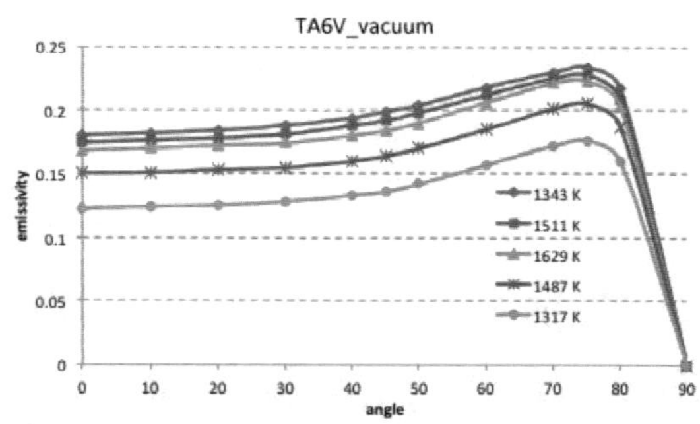

Figure 4. Emissivity versus angle for a TA6V alloy in high vacuum for several temperature levels

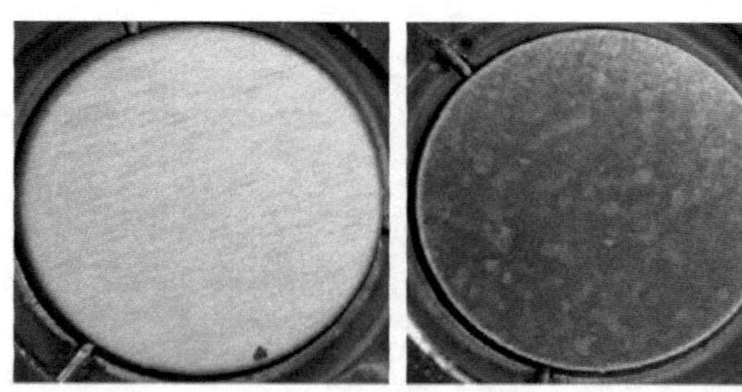

Figure 5. Images of the TA6V sample before (left) and after emissivity at high temperature (right)

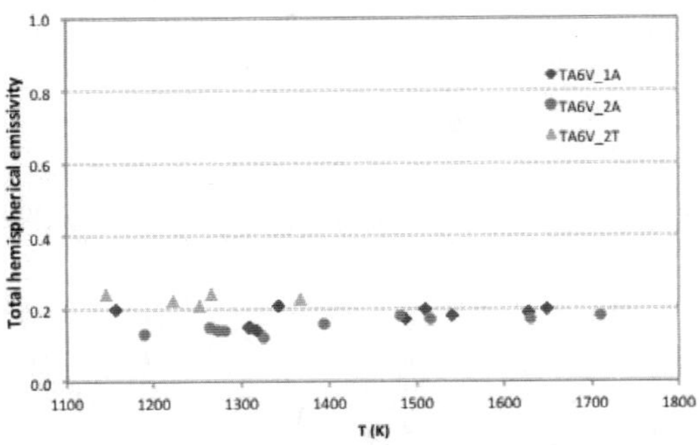

Figure 6. Total hemispherical emissivity versus temperature for two samples of TA6V(Aludis & Timet)

4.3 Oxidation in air plasma conditions

The three materials (TA6V from Timet, Inconel 718 from Carpenter and SS 316L from Ugine) were oxidized in air plasma in the MESOX for two total air pressures: 500 and 2000 Pa. The duration of the tests was comprised between 40 and 270 s. These preliminary results have to be taken as a trend, the samples being not homogeneous in temperature over the entire surface as it can be seen on the pictures of oxidized samples of Figures 7 and 8.

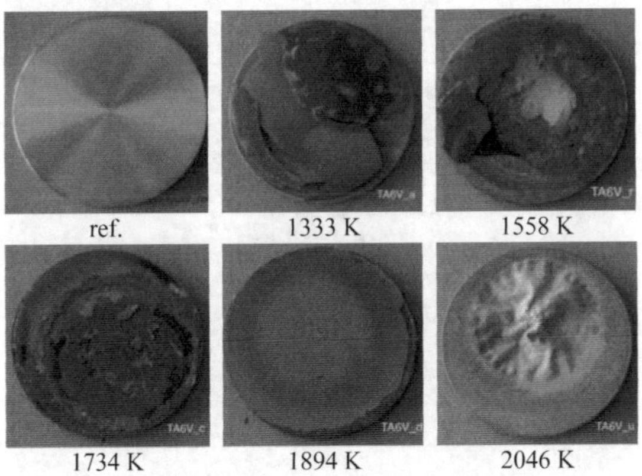

Figure 7. Pictures of TA6V samples oxidized at different temperatures for 500 Pa air plasma

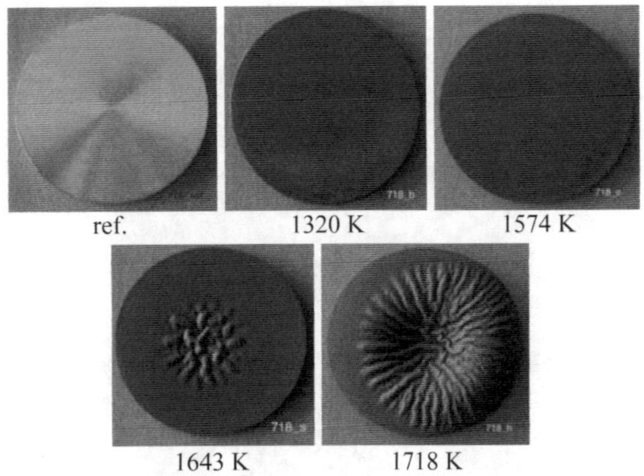

Figure 8. Pictures of inconel 718 samples oxidized at different temperatures for 500 Pa air plasma

This preliminary study was completed with more systematic testing on smaller surfaces (15 mm x 15 mm) to have a uniform temperature throughout the sample surface in order to obtain more precise degradation laws.

A strong mass gain is noted for the TA6V material but this is misleading because the oxides formed are not adherent and leave the surface in form of fine sheets (Fig. 7) at temperatures lower than the melting one that will be evacuated during the reentry phase as a result of the speed of the debris. Figure 9 presents the mass variation law versus time obtained for TA6V (from Aludis) around 1820-1840 K and the laws are linear as the oxides are not adherent and atomic oxygen can easily arrive at the metal interface. According to the total air pressure, the following laws were obtained with $\Delta m/S$ in g m^{-2} and t in s:

- 300 Pa: $\Delta m/S = 0.7775\ t$
- 2000 Pa: $\Delta m/S = 0.8724\ t$

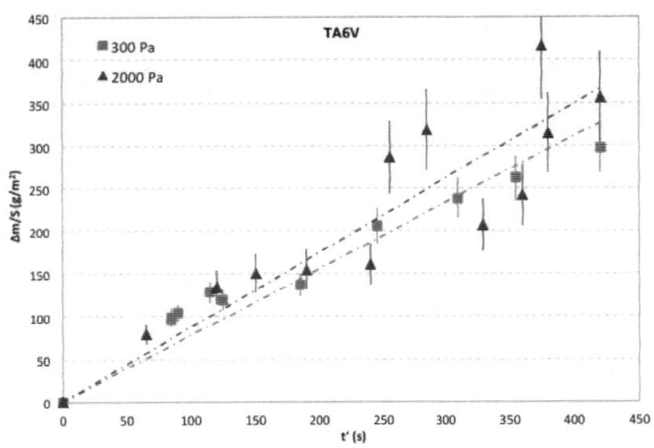

Figure 9. Mass loss variation versus time for TA6V

In the case of Inconel 718, the oxides formed (Fig. 10) are adherent and this irrespective of the temperature or the oxidation time. Around 1600 K, a slight mass gain is observed with a maximum value of 14 g m^{-2} very different from the maximum one measured for TA6V of nearly 400 g m^{-2}. As the oxide layers are adherent, the mass variation law is parabolic – instead of linear for TA6V – and the mean law obtained is, with Δm/S in g m-2 and t in s: Δm/S = 0.64 t.

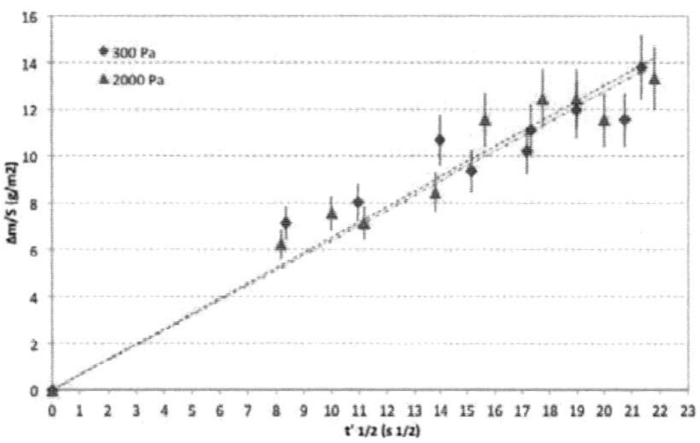

Figure 10. Mass loss variation versus time for Inconel 718

For both samples, it can be seen on Figures 7 and 8 that around the melting point of the pure metal, the oxides are still adherent and for the demise of pieces in TA6V and Inconel 718 alloys it could be better to do not reach the melting temperature as the mass will increase due to the stable and adherent oxide formed on the surface.

The surfaces of some oxidized samples were observed by Scanning Electron Microscopy (SEM). Figures 11 to 13 present the SEM images of the surfaces and cross-sections of oxidized TA6V, Inconel 718 and SS 316L with different crystal sizes according to temperature. For TA6V (Fig. 11), the growing of the oxide layer is visible with grains of increasing size with temperature and also of the thickness. From around 1500 to 1900 K, the thickness of the still adherent oxide layer is going from 10 to 30 μm with a more porous structure at the highest temperature.

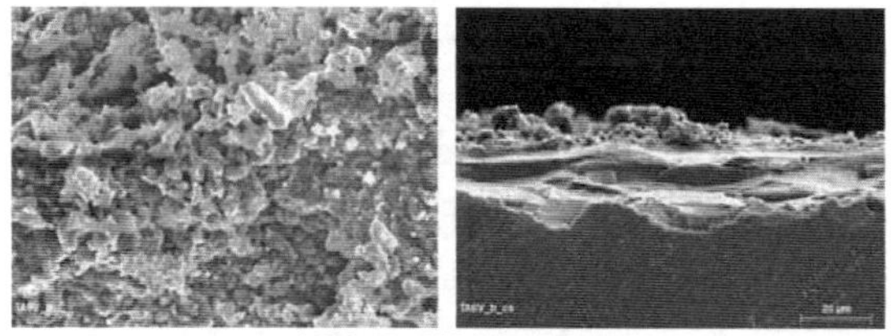

1528 K, 500 Pa, 240 s

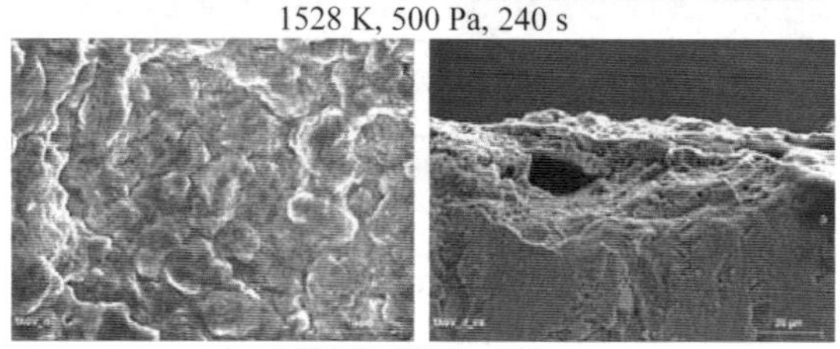

1894 K, 500 Pa, 240 s

Figure 11. SEM images of the surfaces and cross-sections of oxidized samples for TA6V. The scale bar is 10 μm for surface and 20 μm for cross-section

For Inconel 718 (Fig. 12), the oxide layers are well crystallized and with grain growth at higher temperature and are very adherent as it can be seen on the cross-sections with also more porous layers when the temperature increases. Around 1700 K, some cracks are visible on the surfaces.

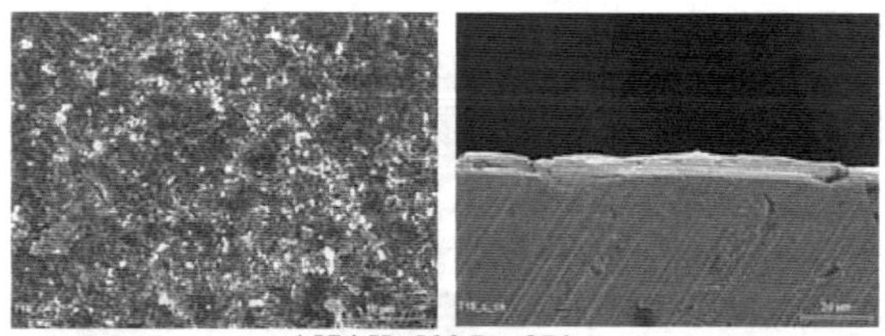

1574 K, 500 Pa, 270 s

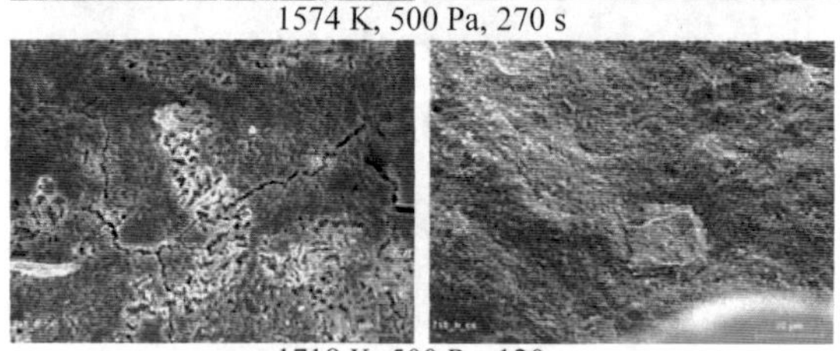

1718 K, 500 Pa, 120 s

Figure 12. SEM images of the surfaces and cross-sections of oxidized samples for Inconel 718. The scale bar is 10 μm for surface and 20 μm for cross-section

The surfaces of SS 316L (Fig. 13) present great differences in morphology from 1500 to 1900 K with an important grain growth with very well crystallized species. The oxide thickness also increases with temperature from around 5 μm at 1500 K up to 20 μm at around 1900K.

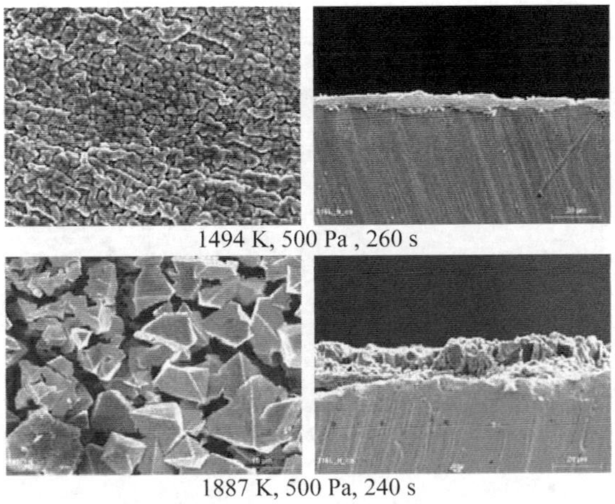

Figure 13. SEM images of the surfaces and cross-sections of oxidized samples for SS 316L. The scale bar is 10 μm for surface and 20 μm for cross-section

Taking into account that the oxides have a molar mass higher than each component of the metallic phase, the sample gains mass with adherent oxide layers. In such case, the use of Inconel 718 and SS 316L must be limited to small pieces.

4.4 Emissivity on oxidized materials

Several types of measurements were carried out. Measurements of the normal total emissivity were performed on the samples oxidized under 500 or 2000 Pa in air plasma in the MESOX: 718-p (1627 K, 500 Pa), 718-t (1643 K, 500 Pa), 718-k (1587 K, 2000 Pa), 718-n (1643 K, 2000 Pa), TA6V-e (1531 K, 500 Pa), TA6V-c (1734 K, 500 Pa), TA6V-p (1563 K, 2000 Pa) and TA6V-m (1707 K, 2000 Pa) from about 1050 to 1700 K depending on the materials. Directional measurements were performed up to an angle of 60° because the sample diameter is only 25 mm, this is why

Figures 14 and 15 report respectively only the data for the normal total emissivity for Inconel 718 and TA6V

For Inconel 718, some discrepancies can be seen in Figure 14 for emissivity measurements on samples oxidized in comparable conditions such as (p and t) and (k and n). A global increase of the normal total emissivity is observed when the temperature increases. Figure 16 reports the images of two Inconel 718 samples before and after emissivity measurement. The measurement for sample 718-p was stopped before the beginning of the metal melting and for the 718-t, measurements were carried out up to higher temperature leading to the melting of the metal substrate as it can be seen in Figure 16 with a great deformation of the sample.

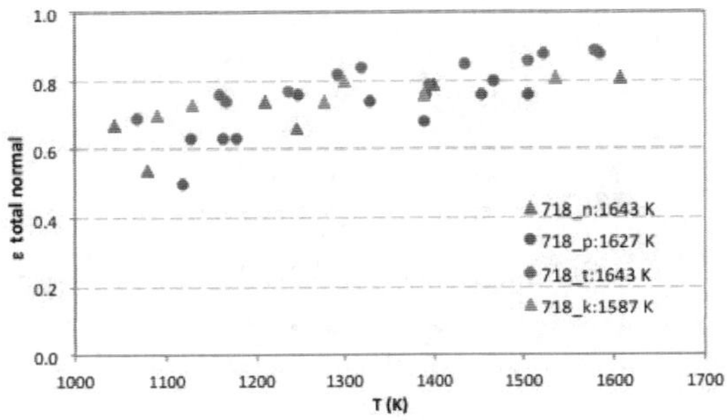

Figure 14. Normal total (0.6-40 μm) emissivity versus temperature for some samples of Inconel 718 pre-oxidized in air plasma conditions.

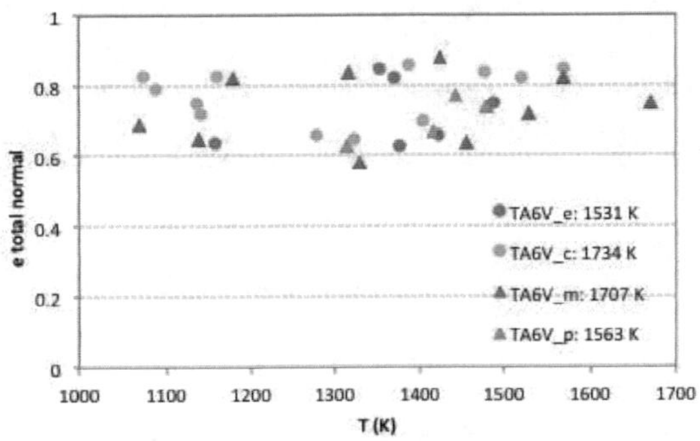

Figure 15. Normal total (0.6-40 μm) emissivity versus temperature for some samples of TA6V alloy pre-oxidized in air plasma conditions.

For TA6V alloy, even if some discrepancies exist among the data due to the partial detachment and peeling-off of the oxide layers, a mean value for the normal total emissivity along the temperature range can be envisaged around 0.75.

Nevertheless the data scattering, there is a huge increase, by a factor of 3 to 5 depending on the materials, of the total emissivity in the presence of oxides.

These first results will be completed by other measurements in different conditions of pressure, temperature and duration of the oxidation.

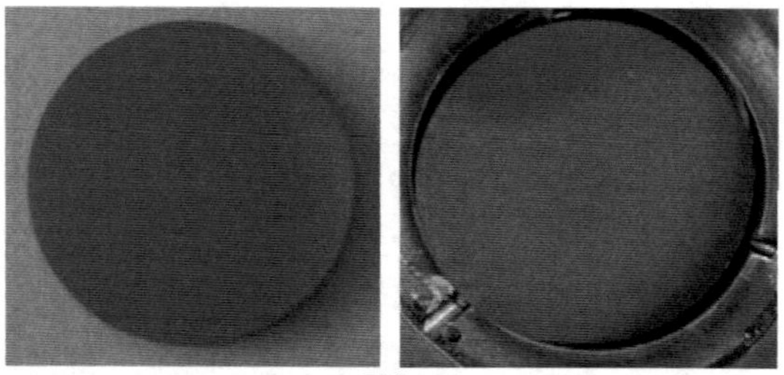

718-p

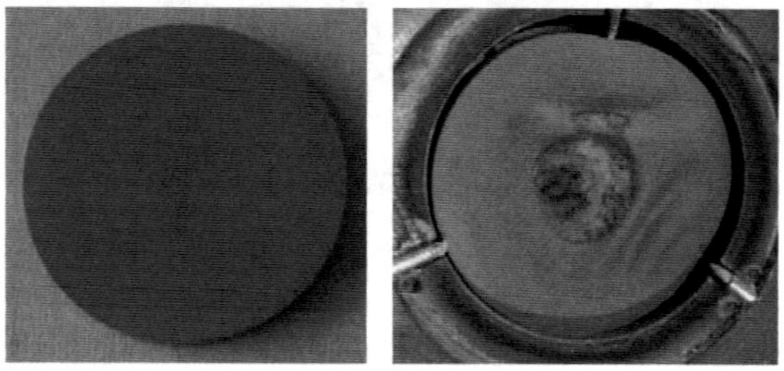

718-t

Figure 16. Pictures of some inconel 718 samples before (but after a first oxidation in air plasma) and after emissivity measurement

In a second time, emissivity measurements were performed during in situ oxidation in standard air in the MEDIASE facility.

Figures 17 and 18 report respectively the total hemispherical (as the sample were 40 mm in diameter) emissivity and some pictures of the samples after the emissivity measurement where the delamination of the oxide layers is clearly visible.

In Figure 17, the beginning of the growing of the oxide layer is visible with lower values of emissivity around 1100 K in comparison to the pre-oxidized samples (Fig. 15) where an emissivity of 0.75 is already obtained around 1100 K. This is confirmed by the evolution of the reflectivities at 1.3 and 1.55 µm in situ measured during emissivity measurement (Fig. 19). In high vacuum, as the sample becomes more and more reflective, the two reflectivities increase with temperature, but differently according to the thermal history of the measurement. In oxidizing atmosphere, after a very short duration, both the reflectivities are decreasing and become nearly equal to 0 as the oxides are very efficient emissive coating. Figure 18 confirms the presence of delaminated oxide sheets on TA6V alloys that will be flown away during the high speed atmospheric reentry.

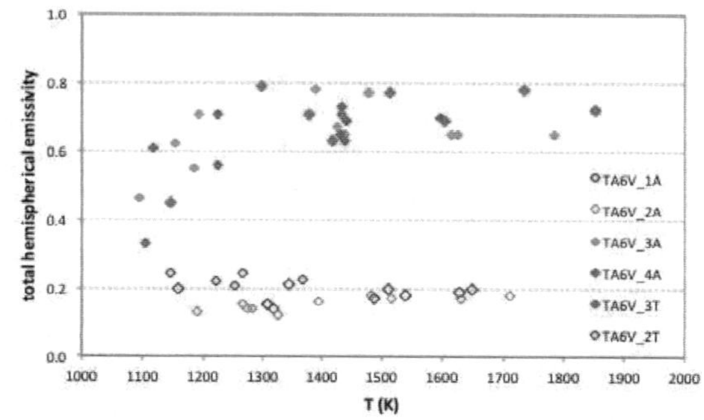

Figure 17. Hemispherical total (0.6-40 µm) emissivity versus temperature for some samples of TA6V alloy oxidized in situ (3A, 4A and 3T) compared with the non oxidized samples (1A, 2A and 2T)

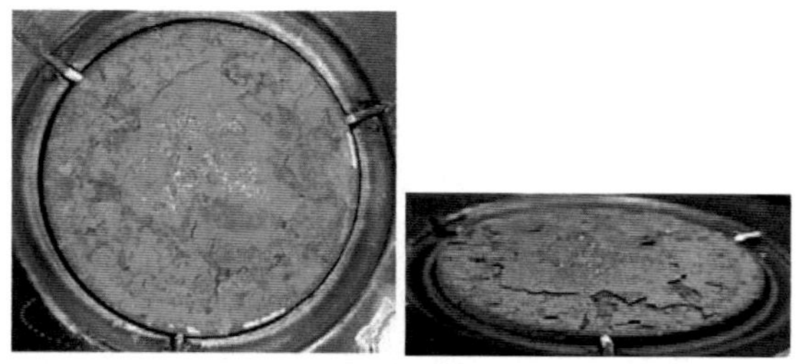

TA6V_3A

TA6V_4A

Figure 18. Pictures of some TA6V samples after emissivity measurement during in situ oxidation

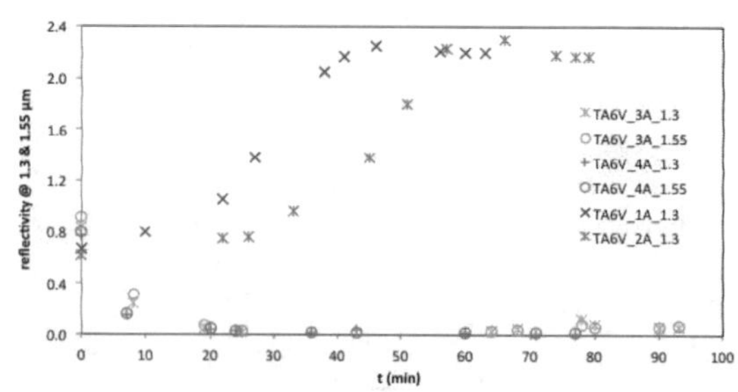

Figure 19. Evolution of the reflectivities at 1.3 and 1.55 µm for samples in high vacuum or during in situ oxidation for TA6V alloy

5. CONCLUSION

Emissivity measurements on Inconel 718 and TA6V alloys were performed on non-oxidized surfaces in high vacuum and after oxidation in air plasma or in standard air conditions. Additional measurements will be carried out in order to improve the accuracy - differences were observed between samples TA6V due to detachment and peeling-off of the oxide layers. Oxidation kinetics was studied on square (15 x15 mm^2) samples in order to obtain mass variation laws versus time for TA6V and Inconel 718 respectively around 1820 K and 1600 K.

This first study seems to favor the use of TA6V among the three metals as the oxides formed are separating from the surface (delamination), leading to a decrease in mass during the re-entry phase. Against by the oxides formed on Inconel 718 are always adherent even if the weight gain is low for the different temperatures studied, this material should be used sparingly. Moreover, it was shown that no peel-off exists when the temperature of the material is close to its melting one, even for the TA6V alloy. The oxides being more refractory than the metal alloys, the material keeps its integrity and its mass increases.

6. REFERENCES

1. Omaly P., Spel M., DEBRISK, a tool for re-entry risk analysis, Proc. 5th IAASS Conf. "A Safer Space for a Safer World", Versailles (Fr), 17-19 oct. 2011, ESA SP-699, Jan. 2012.

2. Omaly P., Magnin Vella C., Galera S., DEBRISK, CNES tool for re-entry survivability prediction: validation and sensitivity analysis, Proc. 6th IAASS Conf. "Safety is not an option", Montréal (Ca), 21-23 May 2013.

3. Greene G.A., Finfrock C.C., Oxidation of inconel 718 in air at high temperatures, Oxid. Metals 55, 2001, 505-521.

4. Kumar S., Sankara Narayanan T.S.N., Ganesh Sundara Raman S., Seshadri S.K., Thermal oxidation of Ti6Al4V alloy: microstructural and electrochemical characterization, Mater. Chem. Phys. 119, 2010, 337-346.

5. Biswas A., Dutta Majumdar J., Surface characterization and mechanical property evaluation of thermally oxidized Ti-6Al-4V, Mater. Charact. 60 , 2009, 513-518.

6. Wang X., Wang L., Zhu M., Zhang J., Lei M., Influence of high-intensity pulsed ion beams irradiation on oxidation behavior of 316 L stainless steel at 700°C, Trans. Nonferrous Met. Soc. China 16, 2006, 676-680.

7. Geng L., Na Y., Park N., Oxidation behavior of alloy 718 at a high temperature, Mater. Design 28, 2007, 978-981.

8. Gonzalez-Fernandez L., Del Campo L., Perez-Saez R.B., Tello M.J., Normal spectral emittance of inconel 718 aeronautical alloy coated with yttria stabilized zirconia films, J. Alloys Comp. 513, 2012, 101-106.

9. Gonzalez-Fernandez L., Risueno E., Perez-Saez R.B., Tello M.J., Infrared normal spectral emissivity of Ti-6Al-4V alloy in the 500-1150 K temperature range, J. Alloys Comp. 541, 2012, 144-149.

10. Wang Z.W., Wang Y.M., Liu Y., Xu J.L., Guo L.X., Zhou Y., Ouyang J.H., Dai J.M., Microstructure and infrared emissivity property of coating containing TiO_2 formed on titanium alloy by microarc oxidation, Curr. Appl. Phys. 11, 2011, 1405-1409.

11. Cao G., Weber S.J., Martin S.O., Sridharan K., Anderson M.H., Allen T.R., Spectral emissivity of candidate alloys for very high temperature reactors in high temperature air environment, J. Nucl. Mater. 441, 2013, 667-673.

12. Balat-Pichelin M., Robert J.F., Sans J.L., Emissivity measurements on carbon-carbon composites at high temperature under high vacuum, Appl. Surf. Sci. 253 (2), 2006, 778-783.

13. Balat-Pichelin M., Eck J., Sans J.L., Thermal radiative properties of carbon materials under high temperature and vacuum ultra-violet (VUV) radiation for the heat shield of the Solar Probe Plus mission, Appl. Surf. Sci., 258, 2012, 2829-2835.

14. Brodu E., Balat-Pichelin M., Sans J.L., Kasper J.C., Influence of roughness and composition on the total emissivity of tungsten, rhenium and tungsten-25%rhenium alloy at high temperature, J. Alloys & Compounds, 585, 2014, 510-517.

15. Hernandez D., Olalde G., Beck A., Milcent E., Bicolor pyro-reflectometer using an optical fiber probe, Rev. Sci. Instrum. 66, 1995, 5548-5551

16. Hernandez D., Sans J.L., Netchaieff A., Ridoux P., Le Sant V., Experimental validation of a pyroreflectometric method to determine the true temperature on opaque surface without hampering reflections, Measurement 42, 2009, 836-843.

17. Balat M.J.H., Determination of the active-to-passive transition in the oxidation of silicon carbide in standard and microwave-excited air, J. Eur. Ceram. Soc. 16, 1996, 55-62.

18. Balat-Pichelin M., Charpentier L., Panerai F., Chazot O., Helber B., Nickel K., Passive/active oxidation transition for CMC structural materials designed for the IXV vehicle re-entry phase, J. Eur. Ceram. Soc. 35 (2), 2015, 487-502.

19. Balat-Pichelin M., Vesel A., Neutral oxygen atom density in the MESOX air plasma solar furnace facility, Chem. Phys. 327 (1), 2006, 112-118.

STATISTICAL ISSUES FOR CALCULATING REENTRY HAZARDS

John B. Bacon[(1)] and Mark Matney [(2)]

[(1)] *Orbital Debris Program Office, NASA Johnson Space Center, Houston, TX, USA, john.bacon-1@nasa.gov*
[(2)] *Orbital Debris Program Office, NASA Johnson Space Center, Houston, TX, USA, mark.matney-1@nasa.gov*

ABSTRACT

A number of statistical tools have been developed over the years for assessing the risk of reentering object to human populations. These tools make use of the characteristics (e.g., mass, shape, size) of debris that are predicted by aerothermal models to survive reentry. This information, combined with information on the expected ground path of the reentry, is used to compute the probability that one or more of the surviving debris might hit a person on the ground and cause one or more casualties.

The statistical portion of this analysis relies on a number of assumptions about how the debris footprint and the human population are distributed in latitude and longitude, and how to use that information to arrive at realistic risk numbers. This inevitably involves assumptions that simplify the problem and make it tractable, but it is often difficult to test the accuracy and applicability of these assumptions.

This paper builds on previous IAASS work to re-examine one of these theoretical assumptions.. This study employs empirical and theoretical information to test the assumption of a fully random decay along the argument of latitude of the final orbit, and makes recommendations how to improve the accuracy of this calculation in the future.

1. INTRODUCTION

One of the hazards of the space age is that many objects in orbit eventually reenter the Earth's atmosphere. For many large objects, components can survive reentry and pose a hazard to persons on the ground. A whole science has developed around predicting what portions of a satellite survive the violent forces and heating of reentry.

For controlled reentries, the reentry target zone can be chosen to avoid populated areas on the Earth. For uncontrolled reentries, however, statistical tools must be used to map human distributions on the Earth under the satellite orbit. The exact time and location for uncontrolled reentries are notoriously difficult to predict with any accuracy. In the hours immediately before a reentry, it may be possible to narrow the possible ground tracks of the reentering object. But when the risk calculation is needed weeks or months before reentry, essentially any location on the Earth over which the satellite flies is a potential reentry landing site. Nevertheless, it is possible to use even this vague information to compute meaningful risk statistics.

2. REENTRY MODEL

Under the assumption that a satellite is reentering randomly, computations to date have assumed that there is no preferred position along the orbit that the reentry occurred. Using the simplification of a circular orbit, this assumption is that the argument of latitude (equal to the argument of perigee plus the true anomaly) is randomly distributed. This is a very good assumption if the Earth were a perfect sphere.

Using the satellite inclination, i, and the argument of latitude θ, the latitude λ can be computed

$$\lambda = ArcSin(Sin(i)\ Sin(\theta))$$

Assuming that any possible value of the argument of latitude is equally likely, the distribution in argument of latitude will be

$$P(\theta)\, d\theta = \frac{d\theta}{2\pi}$$

which corresponds to a normalized distribution in latitude of

$$P(\lambda)\, d\lambda = \frac{1}{\pi}\frac{Cos(\lambda)}{\sqrt{Sin(i)^2 - Sin(\lambda)^2}}\, d\lambda$$

where

$$-Sin(i) \le Sin(\lambda) \le Sin(i)$$

This distribution in randomized latitude is integrated over the latitude distribution of population on the Earth to get an average number of human beings underneath a particular orbit. This can be combined with assumptions on the size of the surviving debris pieces and the average size of human beings to compute a casualty expectation, E_c. This value is then used to determine the risk to humans on the ground [1],[2].

This set of assumptions assumes the Earth's gravitational field is that of a perfect sphere. In actuality, the Earth's gravitational field can be described by spherical harmonic terms that become more important to an orbit as it gets closer to the Earth's surface, which is what actually happens in the final stages of orbital decay.

In addition, the Earth itself is not a perfect sphere, but an oblate spheroid, with the equatorial radius slightly larger than the polar radius. Because the atmospheric density is primarily a function of altitude above the planet's surface, that means that a satellite crosses an enhancement of atmospheric density, and corresponding atmospheric drag, as it crosses the equator – an atmospheric "speed bump" that may affect the location of the final reentry.

In the case of a near-polar circular orbit, the difference is 21.4 km in local altitude from the pole to the equator, in a final decay region where the density scale height is only 5.4 km. This would result in a theoretical 50-fold difference in the local deceleration as the satellite crosses the equator. In the simplifying assumption of a delta function of density at the equator, one sees that the resulting decay would behave like a series of Hohmann transfers that would tend to force the final perigee of the orbit to the equator.

3. REENTRY MODEL CALCULATIONS

In an effort to quantify the effects of these effects, we have simulated the decay of satellites in NASA's Generalized Mission Analysis Tool (GMAT): a high-fidelity mission-certified software environment using the 4th order Earth spherical harmonics, the MSIS-E 90 atmosphere, and a 9th order Runge-Kutta orbit propagator. Decaying orbits were simulated from 200 km altitude with initial conditions set to minimize J_2 spherical harmonic effects on the radius vector: i.e., as nearly circular as the gravitational perturbations would allow. For moderate ballistic number, this resulted in approximately 40 orbits before final decay. A typical relative density along the decaying orbit is shown in Figure 1, with the density plotted as a function of latitude in the gradually decaying orbit, and Figure 2, with relative density plotted as a function of time, showing a significant pulsing near the equatorial regions of the drag rate that a satellite will experience as it decays.

The initial mass of the satellite was slowly varied over a total range of less than ±1.1% from the median value in uniform small steps of 0.025% to create a full 360 degree spread along the orbit of final decay locations ("decay" defined as 85 km). With such small steps over a limited range, a uniform step in decay location around the orbit would be expected, and any deviation from that spread should be the contribution of non-uniformities in the spherical-Earth model, such as the atmospheric bulge and gravitational perturbations.

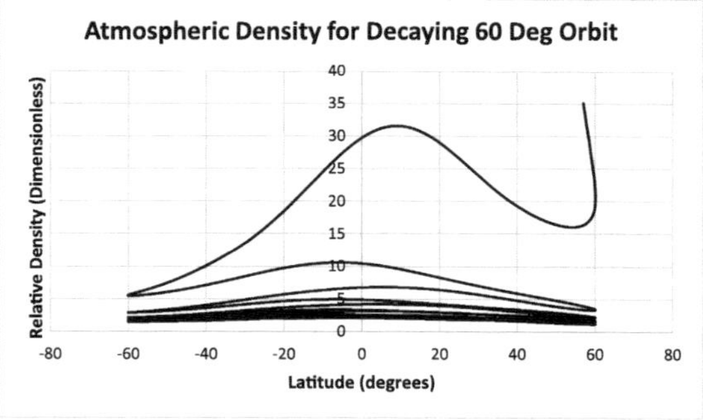

Figure 1 – As a satellite decays, it encounters varying atmospheric density. Here the density is plotted as a function of satellite latitude. Note how the atmospheric density is higher near 0° degrees latitude, near the equator.

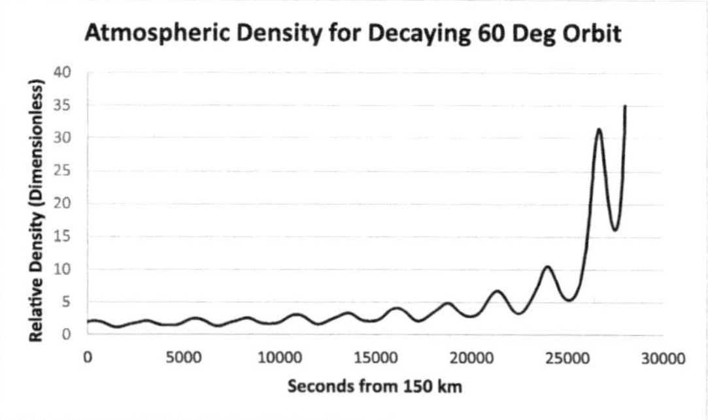

Figure 2 – This plot shows the changing atmospheric density as in Figure 1, but here it is plotted as a function of time, showing the repeating "pulses" in atmospheric density.

The integrated reentry location results show a bias of decay latitude towards the equator that increases with orbit inclination. The bias is symmetric such that there is no apparent tendency to reenter on the approach to the equator compared to departure from it. The difference between the simple spherical Earth model and the numerical integration is shown in Figure 3. The difference between the two models is approximated very well by a pure sine wave ±90° of arc around a nodal crossing. The perturbation is sinusoidal at twice the orbit rate, such that the perturbation is zero at the latitude extremes and at the equator.

As shown in Figure 4, the amplitude of the sinusoidal perturbation geometrically increases with the square of the orbit inclination, and reaches a peak amplitude of 11° bias in the case of a polar orbit, but only 3.5° for a 51.6° inclination orbit typical of Russian spacecraft. These calculations are all for the same 4050 kg, 14.6 m^2 hypothetical object decaying from low altitude with free molecular Cd of 2.07 (Ballistic Number = 134 kg/m^2). Initial studies show that the amplitude of the perturbation is less when the object is exceptionally low ballistic number, executing its final plunge from an altitude where the density scale height is larger, and the relative equatorial density perturbation consequently lower. The case of decay latitude for a 134 kg/m^2 object is shown for a polar orbit in Figure 3.

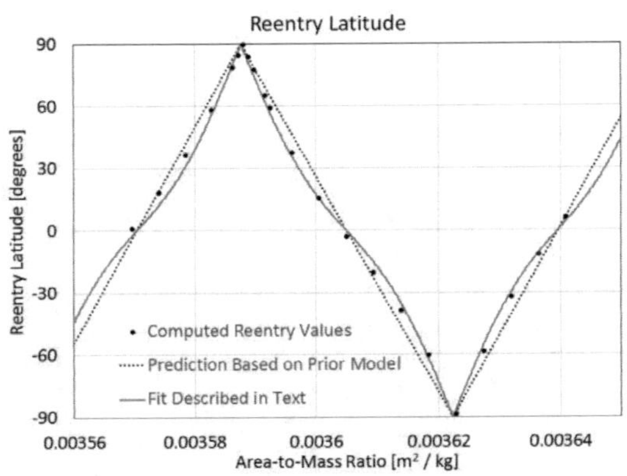

Figure 3 – Comparison of the calculated reentry locations for a polar orbit compared to the "prior model" based on a spherical Earth.

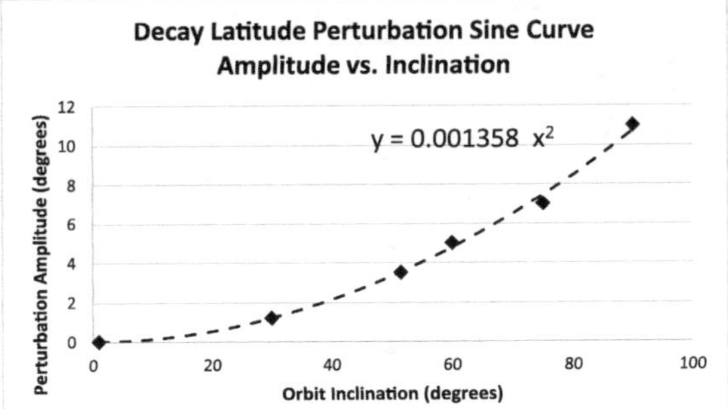

Figure 4 – The amplitude of the difference between the calculated reentry locations and the "prior model" based on a spherical Earth.

An extensive statistical study over different values of argument of latitude and inclinations gives the following empirical equation for the realistic decay latitude λ for a 134 kg/m^2 object.

$$\lambda = ArcSin(Sin(i)\ Sin(\theta)) - 0.001358^\circ\ (i^\circ)^2\ Sin(2\theta)$$

Our future work is to more fully evaluate the bias as a function of object ballistic number, and to evaluate a broader range of initial eccentricities and arguments of perigee.

4. EMPIRICAL DATA

In order to test the calculations against actual data, information on actual reentries were gathered from the final TIP decay messages for satellites in groups of common inclination orbits available from the Space-Track public database of the US Joint Space Operations Command's (JSpOC) [3]. Only TIP messages that posted after the declared decay time were gathered. There is reason to believe that not all the final reentry locations are reported accurately, and therefore may contain noise. However, the overall distribution of the reentry locations should provide useful data.

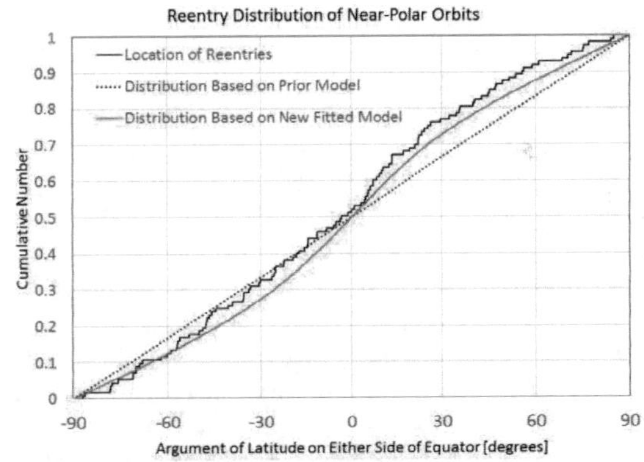

Figure 5 – The ordered cumulative distribution of 113 historical reentry locations for a set of near-polar orbiting satellites are shown here. To remove the inclination differences between the objects, the reentry latitudes have been converted to argument of latitude, and the range converted to between -90° (the southernmost point in the orbit) and +90° (the northernmost point in the orbit). Also shown are the straight-line curve expected under the spherical-Earth approximation and the equator-biased distribution described in the text. Statistical tests on the distribution indicate the data is outside the 2-sigma standard deviation for the prior model, but well within the 1-sigma range for the new fitted model, indicating that the distribution of actual reentries are consistent with the model described in the text.

Reentry data on 113 historical near-polar reentries are plotted in Figure 5, with the latitude converted into the argument of latitude in order to remove the effects of the different inclinations, and limited to ±90° to be analogous to latitude. Also shown are the prior model predictions assuming a spherical Earth, and the distribution computed using the formula described above, with reentries biased toward the equator. Under the prior model, the data should lie along a straight line with minor statistical scatter.

Using the Kuiper statistical test (described in [2]), the data curve lies outside the 2-sigma standard deviation for the prior model expected from sampling error, indicating that there is a statistically significant bias away from the prior model. When the same statistical test is applied to the comparison of the data to the new model described in the text, the data is now within 1-sigma. It appears that a bias similar to what we predict is being observed in actual reentry latitudes.

5. REENTRY GEOMETRY

Although the latitude bias varies geometrically, the actual perturbation along track varies more linearly as orbit inclination increases. This is because from simple geometry the flight path length to achieve a given latitude bias varies like the size of the bias divided by the sine of the inclination. The peak "along track" perturbation ranges from about 800 km to 1220 km as inclination increases. To see how significant this effect really is, one can ask what sort of ΔV would be required to achieve the same result. Under Clohessy-Wiltshire relations in zero atmosphere, one can see that downrange deflection of 500 km (500,000 meters) in a 51.6° inclination orbit, initiated one-half orbit before (approximately 2600 seconds downrange at low decay altitude) would result from an impulsive ΔV of about 64 meters/second. This is 42% of the ΔV accumulated during the final 2600 seconds of decay before the orbit reaches 85 km altitude.

The authors propose that this integral effect is significant enough that spacecraft with finite but inadequate ΔV for a direct plunge to a targeted entry may be able to strongly influence the final target location through minor orbit shaping that would take advantage of the driving density "bump" during the final orbit. In general, this would make it easier to target regions closer to the equator than farther from it, given natural biasing of the decay location towards there.

6. POPULATION RISK

The next obvious question is how this biasing in the reentry location affects the average population density beneath the orbit. Figure 6 shows the population density as a function of latitude once the longitude dependence has been averaged out. Note the preponderance of population in the northern hemisphere and the lack of population near the poles [4],[5].

To find the average density of humans beneath an orbit, this population density is integrated over the probability that the satellite will reenter at each latitude. Figures 7 and 8 show the result of this calculation using both the spherical-Earth random-angle approximation and the new biased equation derived from the GMAT simulations. The two curves are plotted for 2020 population data.

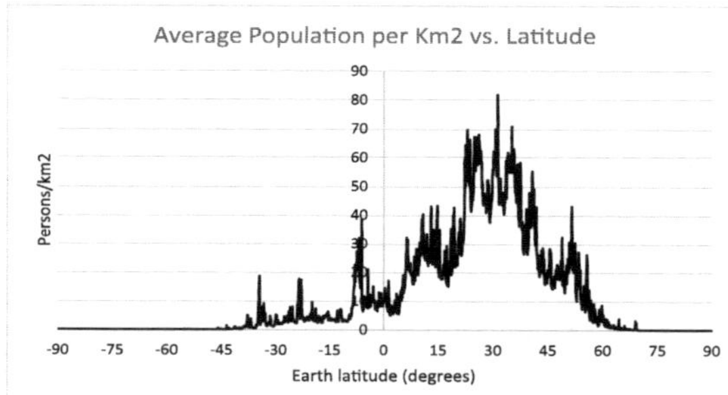

Figure 6 – the average population density as a function of latitude on the Earth (longitude dependence removed) based on the SEDAC World Population database. (2020 estimates are shown.)

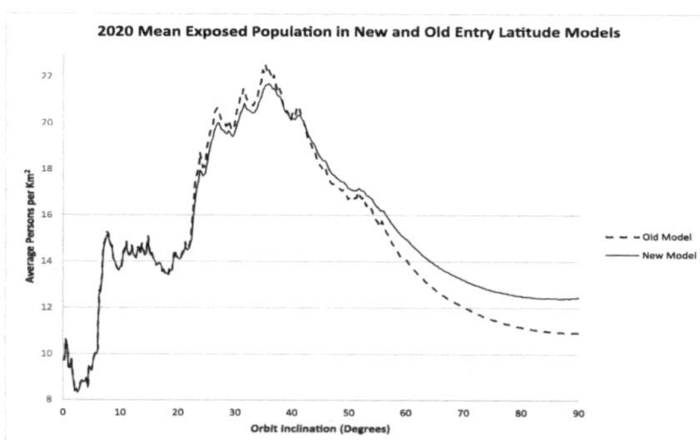

Figure 7 – Two curves that show the result of integrating orbits of different inclinations over the year 2020 population distribution in Figure 6, weighting the calculation by the probability that a satellite will reenter at a particular latitude. "Old model" refers to the spherical-Earth approximation, and "New model" refers to the new equations described in this paper. Note that for orbits with inclinations above 90°, the result will be the same as those with inclination equal to 180° minus the inclination.

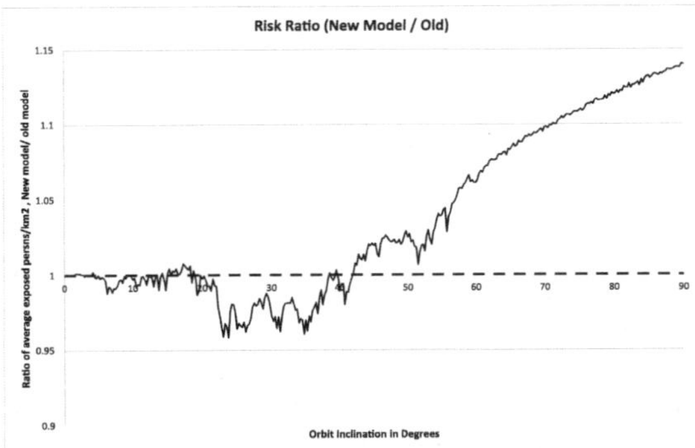

Figure 8 – This curve shows the ratio between the two curves in Figure 7.

There are two primary changes in the computed population densities. Because of the biases away from the poles and toward the equator, the new model gives a higher population density for high-inclination orbits > 42°. On the other hand, the regions between about 20° and 40° inclination, where the population density is a maximum, the new equations give a slight reduction in population density below an orbit. Lower inclination orbits do not show major differences between the two models.

Note that the updated calculations show changes in population densities at mid-latitudes of no larger than 4.5%, but the high-latitude changes reach nearly 14%.

7. CONCLUSIONS

The calculations presented in this paper have shown that calculations using a realistic non-spherical Earth and atmosphere result in a bias toward reentries near the equator. This results in modifications in the computed average density of humans that would be under a reentering satellite. At high latitudes, this results in a noticeable increase in risk, but a small reduction at mid-latitudes. While further statistical studies are needed to confirm these results and to evaluate for a range of ballistic numbers and initial orbits, this study points to a need to include these effects in computing future ground risks.

This equatorial "bulge" in the atmosphere could also be used by satellites undergoing controlled reentry to enhance their reliability of reentering over uninhabited regions, especially in cases where limited ΔV is available for the disposal operation.

8. REFERENCES

1. Matney, M. J., "Statistical Issues for Uncontrolled Reentry Hazards," 3rd IAASS Conference, Rome, Italy, 2008.

2. Matney, M. J., "Stastistical Issues for Uncontrolled Reentry Hazards: Empirical Tests of the Predicted Footprint for Uncontrolled Satellite Reentry Hazards", 4th IAASS Conference, Versailles-Paris, France, 2011.

3. Historical reentry data (TIPS) can be obtained from the space-track web site at www.space-track.org

4. The SEDAC World Population Databases and their documentation can be found on the internet at http://sedac.ciesin.columbia.edu/

5. Opiela, J. N., & Matney, M. J. (2003). Improvements to NASA's Estimation of Ground Casualties from Reentering Space Objects. Space Debris and Space Traffic Management Symposium 2003, Bremen, Germany, In *Proceedings of the International Academy of Astronautics Space Debris and Space Traffic Management Symposium held in conjunction with the 54th International Astronautical Congress* (Ed. J. Bendisch), IAA 03-5.4.03, Volume 109, Science and Technology Series, Univelt Publishers, San Diego, California, pp385-392.

THE UNCONTROLLED REENTRY OF PROGRESS-M 27M

Carmen Pardini[(1)], Luciano Anselmo[(2)]

[(1)] *Space Flight Dynamics Laboratory, ISTI/CNR, Via G. Moruzzi 1, 56126 Pisa, Italy (Carmen.Pardini@isti.cnr.it)*
[(2)] *Space Flight Dynamics Laboratory, ISTI/CNR, Via G. Moruzzi 1, 56126 Pisa, Italy (Luciano.Anselmo@isti.cnr.it)*

ABSTRACT

Uncontrolled reentries of sizable space objects are becoming of growing concern due to the increase of space activities around the Earth and population on the ground. After providing an updated review of various aspects of the problem, this paper presents the exemplar case of Progress-M 27M, whose control was lost immediately after launch, on April 28, 2015, and which reentered on May 8. As in similar previous occurrences, the Space Flight Dynamics Laboratory of ISTI/CNR, in Pisa, was in charge of reentry predictions for the Italian civil protection authorities and space agency. The first prediction was issued in the morning of April 30, in the morning of May 7 the only potentially risky reentry trajectory over central Italy was identified, and in the afternoon of May 7, about 12 hours before the actual reentry, any residual risk for Europe and Italy was finally excluded.

1. CATEGORIES OF UNCONTROLLED REENTRIES

The space objects subjected to uncontrolled reentries can be roughly grouped in five orbital categories:

1. Nearly circular low Earth orbits (LEO), with mean eccentricity $e \leq 0.015$, whose evolution is dominated by atmospheric drag;
2. Low or moderate mean eccentricity ($0.015 < e \leq 0.5$) orbits in LEO, or crossing both the LEO and the medium Earth orbits (MEO) regions, whose evolution is mainly driven by atmospheric drag around the perigee;
3. High mean eccentricity orbits ($e > 0.5$), whose evolution is driven by luni-solar third body attraction, but for which the atmospheric drag at the perigee can play an important role at the end of the lifetime;
4. High mean eccentricity orbits ($e > 0.5$), whose evolution is dominated by luni-solar third body attraction, with minor influences from solar radiation pressure;
5. Orbits initially above LEO, whose evolution is dominated by solar radiation pressure [1] [2].

The relatively massive intact objects for which reentry predictions are relevant are typically characterized by an area-to-mass ratio $A/M < 0.1$ m^2/kg, so the fifth category does not generally apply. Among the others, only for the objects belonging to the fourth category it is often possible to predict accurately in advance the location and time of atmospheric reentry. In fact, due to the Equivalence Principle and to the absolute predominance of gravitational forces, the orbit evolution does not (significantly) depend on the object physical characteristics and attitude, often unknown or known with considerable uncertainty, so it can be accurately modeled well in advance, if a reliable orbit determination is available. For example, the unidentified artificial object WT1190F, found in trans-lunar orbit and probably a remnant of the lunar exploration program, was predicted to impact a specific location of the Earth, 40 days before reentry, occurred on November 13, 2015, at 6:20 UTC, with a time error of less than one minute.

Unfortunately, this situation does not apply to the first three categories, where the complex interaction between the decaying space object and the Earth's thermosphere prevents the accurate prediction of a reentry location and time even a few hours before the event. For this reason, in most cases the reentry predictions of uncontrolled satellites are characterized by considerable uncertainties [3] [4].

The large majority of the potentially risky uncontrolled objects belong to the first two categories. Concerning the first category, thermospheric drag is effective along the entire trajectory, whose semi-major axis progressively contracts, leading ultimately to reentry. Conventionally, the orbital motion ends when the object reaches the atmospheric interface at the altitude of 120 km, with a flight path angle slightly negative, but close to zero (tangential reentry). At this point, a final plunge in the densest layers of the atmosphere is unavoidable, even though, in certain cases, if some lift force builds up, a skip reentry may occur, with the object briefly bouncing back above 120 km, describing a ballistic arc, and falling again towards the Earth.

The orbits of the objects belonging to the second category are principally affected by thermospheric drag around the perigee, where most of the mechanical energy is dissipated. This mainly produces a progressive decrease of the apogee height, i.e. a progressive circularization of the orbit. When the orbit becomes nearly circular, what already said for the first category applies. However, it should be pointed out that a small residual eccentricity, of the order of 0.001, cannot be eliminated in practice (this applies also to category one objects), and that in certain cases the reentry occurs when the circularization process is still underway and not yet completed, with $e > 0.015$.

The evolution of the apogee and perigee height of the objects belonging to the third category is driven by luni-solar perturbations, but the perigee altitude decrease is not always enough to cause a direct collision course with the Earth, as happens for category four objects. The evolution is therefore much more complicated and difficult to predict. The dive of the perigee in the thermosphere contributes, of course, to the orbit evolution, in terms of apogee and eccentricity decrease. In many cases a transition to the second category may occur. However, the perigee may also plunge several times well below the conventional atmospheric interface at 120 km, avoiding immediate reentry and regaining space each time to complete a further elliptical orbit. This situation is relatively common for satellites in Molniya orbits, which can experience aerodynamic fragmentations during these extremely low endo-atmospheric perigee passes. In these cases, the final decay of the main object occurs only when the combined effects of gravitational and air drag perturbations shift the perigee height below 50 km.

2. PHASES OF UNCONTROLLED REENTRIES

The objects belonging to the first two categories described in the previous section are subjected to a progressive mechanical energy loss due to thermospheric drag, with the consequent decrease of the semi-major axis. The atmospheric capture occurs when the objects, moving at that point in nearly circular orbit and with a slightly negative flight path angle, with a relative velocity of more than 7 km/s, descend at a height in between 120 and 110 km, initiating the reentry, i.e. the endo-atmospheric phase of their flight.

Even though some appendages and loose parts (like antennae, booms and solar panels) might already break away in such earlier phase, in general the main body of intact spacecraft and rocket stages fragments below the altitude of 80 km, due to a combination of thermal and aerodynamic loads. Therefore, the motion of a reentering object can be consistently modeled with the methods of orbital dynamics down to an altitude of 80-90 km. For this reason, very often reentry predictions epochs refer not to the crossing of the atmospheric interface, at 120 km, but to the altitude of 80 (or 90) km. For instance, current USSTRATCOM reentry assessments refer to 80 km and this is also roughly the altitude at which a reentry event becomes visible to eyewitnesses on the ground (if any). Last, but not less important, this is the lowest height for which a prediction is possible for any generic object, without a detailed knowledge of its design and construction details, coupled with a sophisticated endo-atmospheric fragmentation analysis.

Objects not specifically designed to survive the reentry, which is generally the norm in case of uncontrolled orbital decay, cannot avoid a catastrophic breakup below the altitude of 80 km. For intact objects with an initial dry mass of 800 kg or more, a mass fraction comprised between 60% and 95% is usually completely melted at high altitude [3] [4]. The surviving fragments, accounting for the remaining 5% to 40% of the total mass, will instead reach the ground, distributing them in a "footprint" which can be as long as 2000 km, aligned with the sub-satellite trajectory [5]. The cross-track dispersion, due to the fragmentation processes, lift forces and winds (during the fall through the stratosphere and troposphere) is typically less than 50 km with respect to the sub-satellite ground track [5]. It should also be emphasized that in the last 15 km the surviving debris falls nearly vertically, i.e. has a flight path angle close to −90° [3] [4].

The trajectories of the surviving fragments depend on their release time, melting history and physical characteristics. Those with a lower area-to-mass ratio tend to impact the ground earlier towards the tip of the footprint, while those with a higher area-to-mass ratio tend to fall later towards the heel [5]. Most of the macroscopic fragments impact the ground between 5 and 25 minutes after reentry at 80 km, but in certain cases debris may hit the Earth's surface after 45 minutes [3] [6] [7].

Concerning the reentries belonging to the third category, aerodynamic fragmentations are possible, before atmospheric capture of the main object, at each perigee pass below 110−120 km. At the atmospheric interface (120 km) the relative velocity is around 10 km/s and the flight path angle may be lower than −5°. Moreover, the main object can survive several perigee passes at 80-90 km without further breakup. When the atmospheric capture occurs, the surviving debris is scattered over a significantly shorter footprint and the dispersion of the impact times is quite smaller. However, the details depend on the specific characteristics and final evolution of the parent object trajectory.

In the case of high eccentricity orbits dominated by luni-solar third body perturbations and intersecting the Earth's surface (the fourth category of the previous section), the possible impact geometries may be extremely varied, as well as the relative velocities and flight path angles at the atmospheric interface. In general the reentries are faster and steeper, the surviving mass lesser, due to considerably higher heat peak loads and cumulative absorption, and the ground dispersion of the fragments smaller, both in space and in time. As an example, for WT1190F the relative impact velocity, at the atmospheric interface, was around 11 km/s and the flight path angle was approximately −20°.

3. ORBIT DETERMINATION DATABASES

The most complete catalog of artificial objects in orbit around the Earth is maintained by the US Strategic Command (USSTRATCOM). Currently it includes more than 17,600 objects, of which more than 6200 are intact payloads and upper stages, mostly abandoned, while the rest is just debris. The catalog is officially available to registered users through the Space-Track Organization (www.space-track.org). However, orbits are available only for unclassified objects, making approximately 88% of the total. Sensitive military and intelligence spacecraft, together with the related upper stages, mission released objects and, often, debris, are listed, but their orbits are not provided. Moreover, it could take one or two days to include in the catalog reliable orbits and identifications of newly launched space objects. This means that for new space objects, with orbital lifetimes of less than a couple of days, reliable reentry predictions are often impossible.

Even with these important limitations, nothing else currently available may be compared to the USSTRATCOM catalog in terms of completeness, rate of updating and quality of the information provided. Orbits, when available, are supplied, for historical reasons, in the so-called Two-Line Elements (TLE) format [8] [9].

Over the years, a global network of very motivated and competent amateur astronomers, specialized in satellite observations, has proved itself very effective in identifying and tracking classified spacecraft with optical telescopes. The orbits obtained are generally quite accurate [10] and can be found in specialized websites, as Heavens-Above (www.heavens-above.com), developed and maintained by Chris Peat and hosted by DLR at the German Space Operation Centre (GSOC).

Additional tracking information obtained on request by European sensors can be very useful, as demonstrated by the reentry test campaigns promoted yearly by the Inter-Agency Space Debris Coordination Committee (IADC), but can only be supplementary and not substitutive of US space surveillance data. Only those collected by Russian sensors may offer, in certain reentry campaigns, a practicable and effective alternative to USSTRATCOM orbits.

4. REENTRY PREDICTION UNCERTAINTY

Most of the uncontrolled reentries, including those with the highest casualty expectancy, are caused by thermospheric drag, so the accurate modeling of such perturbation is of paramount importance for reentry predictions as accurate as possible. To a large extent, a large fraction of the underlying uncertainties can be condensed into a single coefficient, the so-called ballistic parameter B, defined as follows:

$$B = \frac{C_D A}{M}, \tag{1}$$

where C_D is the drag coefficient, M the mass and A the average cross-section. In many cases the mass, the shape, the size and the attitude of the reentering object are not known, or approximately known, so the area-to-mass ratio may be affected by a significant uncertainty. The drag coefficient is uncertain as well, being a complex function of the atmosphere composition, density and temperature, and of the shape, size, surface composition and temperature of the object. Moreover, for a given atmospheric density, C_D also incorporates the errors of the model itself [11] [12] [13] [14] [15] [16].

Therefore, for the reentering objects belonging to the first two categories of Section 1, by far the most frequent, the prediction uncertainty is dominated by the modeling of thermospheric drag, which is characterized by inherent inaccuracies and biases related to the satellite physical properties and attitude, to the thermospheric density model itself, and to the prediction of solar and geomagnetic activity affecting the atmospheric temperature, and then the density, as a consequence. This means that even when the best practices are adopted and the best data available are used, the a priory uncertainty of the reentry predictions can be quite significant.

Focusing the attention on the last USSTRATCOM reentry predictions (www.space-track.org), they are currently based on orbits with a mean epoch of 5.0 ± 4.3 hours (1σ) before decay, with a cumulative non-Gaussian distribution detailed in Tab. 1. Comparing the last reentry predictions with actual observations in the 28 cases in which eyewitnesses on the ground were able to observe the reentry, from June 2010 to March 2016 [17], the results presented in Tab. 2 were found for the statistical distribution of the prediction uncertainty.

As an example, a reentry time uncertainty of ± 30%, able to include 90% of the events in the uncertainty window, applied five hours before reentry, i.e. at the mean epoch of the last available orbit, corresponds to ± 1.5 hours, i.e. to a couple of orbit tracks around the Earth. Consequently, for uncontrolled reentries driven by thermospheric drag, it is not possible to predict a reentry location, which remains quite undetermined until the end, but it is only possible to identify the areas of the planet where the reentry may no longer occur within a given confidence level [18].

Table 1. Epoch of last USSTRATCOM orbit determinations before reentry, from December 22, 2014, to January 30, 2016. The mean value was 5.0 ± 4.3 hours (1σ) before decay.

Cumulative probability	Reentry time since last orbit epoch (hours)
2.0%	< 1
19.6%	< 2
33.3%	< 3
50.0%	< 4
72.5%	< 5
90.2%	< 10
96.1%	< 15
98.0%	< 20

Table 2. Observed uncertainty of last reentry predictions at 80 km obtained by comparing USSTRATCOM data with eyewitnesses sightings (from June 19, 2010, to March 10, 2016).

Cumulative probability	Percentage reentry time absolute error (±) of last prediction at 80 km
25.0%	< 5%
57.1%	< 10%
64.3%	< 15%
71.4%	< 20%
89.3%	< 30%
92.9%	< 50%
96.4%	< 60%
100.0%	< 70%

5. UPDATED REENTRY STATISTICS

Since 1957, on average, reentered in the atmosphere 54 payloads, 64 upper stages and 289 debris per year, i.e. 2-3 intact objects per week. During the last decade, reentered on average 42 payloads, 40 rocket bodies and 361 debris per year, i.e. 1-2 intact objects per week [19].

Considering the uncontrolled reentries of intact objects occurred in 2015, 62% (64) were payloads and 38% (40) were upper stages, but 79% of the mass (i.e. 82 metric tons) was concentrated in the latter and the remaining 21% (i.e. 22 metric tons) in the former ones, consisting of small satellites with a mass lower than 50 kg in 83% of the cases [19].

Analyzing the uncontrolled reentries of intact objects occurred from 2004 to 2015 and for which a USSTRATCOM post-reentry time assessment with a claimed error of ± 1 minute was available (for 287 rocket bodies and 59 payloads/platforms), it was found a slight prevalence, 53.5% vs. 46.5%, of reentries occurring in the southern hemisphere. This bias was probably a real effect, reflecting the pattern of launches and mission orbits during the considered time interval. In fact, the probability that it was just the product of chance was 1.9%. The southern hemisphere bias was smaller for upper stages (51.6% vs. 48.4%), with a probability of chance occurrence of 4.1%, and higher for payloads (62.7% vs. 37.3%), with a probability of chance occurrence of 1.6%.

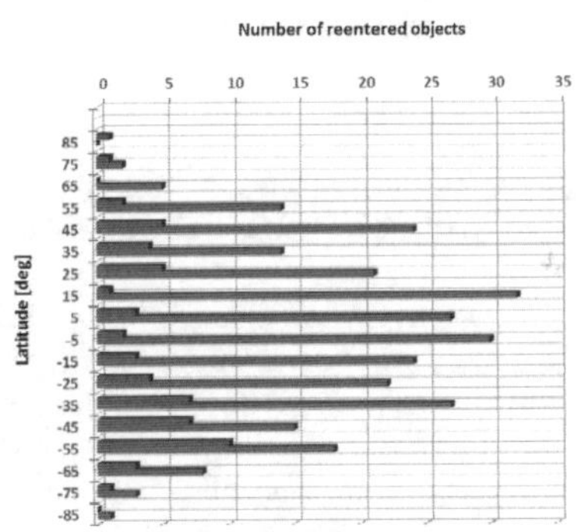

Figure 1. Distribution over 10° latitude bands of uncontrolled reentries occurred from 2004 to 2015 and for which a USSTRATCOM post-reentry time assessment with an error of ± 1 minute was available (payloads in red, rocket bodies in blue)

The reentry distribution over 10° latitude bands is presented in Fig. 1. 50.0% of the reentries occurred around the equator ± 30°, 42.5% at intermediate latitudes, between 30° and 60°, north and south, and 7.5% in the polar regions, above 60° of latitude north and south. The uncontrolled reentry flux for surface unit, compared with the average value over the total Earth surface, is detailed in Tab. 3.

Table 3. Uncontrolled reentry flux for surface unit, compared with the average value over the total Earth surface (2004-2015).

Latitude band	Reentry flux compared with the average
from 60° to 90°	38.8%
from 30° to 60°	99.5%
from 0° to 30°	102.9%
from −30° to 0°	98.3%
from −60° to −30°	132.7%
from −90° to −60	69.0%

It was lower than average, approximately one half, in the polar regions, relatively uniform and close to the average from 30° south to 60° north, and maximum, about 30% higher than average, from −60° to −30°.

Considering all the uncontrolled reentries of intact objects occurred from 2004 to 2015 (722, of which 447 were rocket bodies and 275 payloads/platforms), the mean flux of decaying intact objects over the Earth was 1.05×10^{-7} km^{-2} per year. This implies, on average, an uncontrolled reentry over Italy every 28 years and over Europe every 10 months.

6. THE FAILURE OF PROGRESS-M 27M

Progress-M 27M (Fig. 2), launched from the Baikonur Cosmodrome aboard a Soyuz 2-1A rocket at 7:09 UTC on April 28, 2015, was the 150th Progress mission to fly since 1978 and the 2nd Progress craft to use the modern 2-1A version of the Soyuz launcher. After nearly nine minutes of powered flight, the cargo ship was separated in orbit, but it was quickly evident that communications with the ground were deteriorating.

Figure 2. A Progress-M cargo vessel, with a length of 7.23 m, a maximum diameter of 2.72 m and a maximum solar panels span of 10.6 m (courtesy of NASA)

The loss of telemetry from the Soyuz Block I third stage and the Progress spacecraft in the crucial seconds around the shutdown of the upper stage and the separation of the spacecraft was a first sign of malfunction, indicating that a significant event had taken place at, or around, separation. In an attempt to gain time, the mission profile was changed from the rapid 4-orbit rendezvous to a 34-orbit flight plan, with docking with the International Space Station expected two days after launch.

Progress-M 27M (2015-024A) was later tracked in an orbit with an apogee 40 km higher than planned, while the third stage ended up into a lower orbit with a 20 km deviation against the planned apogee. At least 44 pieces of debris were detected close to the spacecraft by the US Space Surveillance Network and later on the cargo ship was found to be in a rapid tumble around its long axis, with body rates in excess of 60° per second, confirming a loss of control. After repeated vain attempts to regain control of the vehicle, the mission was officially declared lost by the Russian Space Agency on April 29, 2015. That same day, at 7:32 UTC, the Soyuz third stage (2015-024B), with a mass of 2300 kg, reentered the atmosphere after 17 orbits.

With no ability to command Progress-M 27M for a safe return, the out of control supply vessel was irremediably doomed to an uncontrolled reentry. Starting from an orbit of around 189 × 269 km, inclined by nearly 52°, the reentry was expected within two weeks from the failure. Considering the launch mass of 7289 kg, a dry mass in excess of 5 metric tons and the presence on board of 1373 kg of highly toxic propellants (UDMH and N_2O_4), the uncontrolled reentry was expected to violate the worldwide alert casualty expectancy threshold of 1:10,000, even though a detailed fragmentation and demise analysis was not available.

A number of parts were presumed to survive reentry. The components with the highest probability of reaching the ground were the docking mechanism and the spherical pressurant tanks, compact in size and consisting of titanium alloy. An additional unknown were the tanks holding the very toxic UDMH and N_2O_4 propellants. However, due to the very short permanence in orbit of the vehicle, it was unlikely that the propellants had time to completely freeze. Consequent-ly, their dispersion at high altitude during the reentry was considered the most probable scenario.

7. REENTRY PREDICTIONS

Taking into account the orbit inclination of Progress-M 27M, a reentering debris impact was a priori possible in the latitude belt between 53° south and north, therefore including Italy and a large fraction of Europe. Assuming a debris footprint of 800 km and the actual distribution of seas and land, the a priori probability of no impact on the ground was approximately 62%, reduced below 50% in case of a debris footprint 2000 km long.

As in previous similar occurrences since 1979, the Space Flight Dynamics Laboratory of ISTI/CNR, in Pisa, was in charge of the reentry predictions for the Italian civil protection authorities and space agency. Fig. 3 shows the time evolution of the mean observed semi-major axis from the TLEs issued by USSTRATCOM. In the final decay phase, orbital data was only sparsely available, with the final data set epoch at 19:30 UTC, when Progress-M 27M was still more than six hours away from reentry. Also, the data sets issued in between the last propagated by ISTI/CNR, with epoch at 12:12 UTC, and the last available, with epoch at 19:30 UTC, showed irregular slowing or rushing in the spacecraft decay rate, leading to questions on the quality and reliability of the data.

The ballistic parameter estimates and the reentry predictions were carried out, respectively, with the CDFIT and SATRAP software tools [15] [20]. The thermospheric density model adopted thorough the campaign was NLRMSISE-00 [21], with the appropriate observed and predicted solar flux and geomagnetic activity indices. Concerning the estimation of the ballistic parameter B, the results are summarized in Fig. 4. After remaining substantially stable during the first 5 days of the campaign ($B \approx 5.3 \times 10{-}3$ m^2/kg), from 2.91 to 0.82 days before reentry B increased steadily by nearly 27%, then decreasing by 10% during the last day. Such kind of variations is not unusual for uncontrolled objects in extremely low orbits. Moreover, the time interval of interest was characterized by a sharp increase of solar activity, with the flux at 10.7 cm passing from 100 standard flux units at the beginning to 155 standard flux units at the end, and by a small geomagnetic storm on May 6.

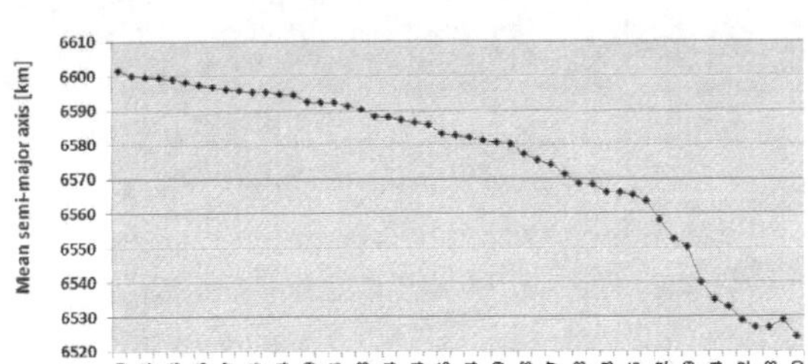

Figure 3. Time evolution of the mean observed semi-major axis of Progress-M 27M from the USSTRATCOM TLEs

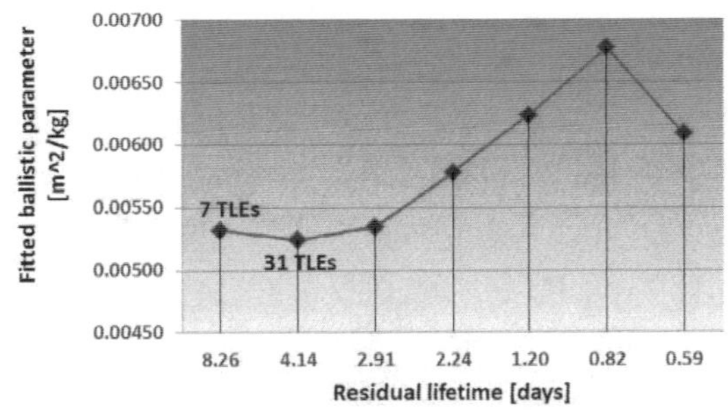

Figure 4. Evolution of the ballistic parameter of Progress-M 27M estimated during the reentry campaign

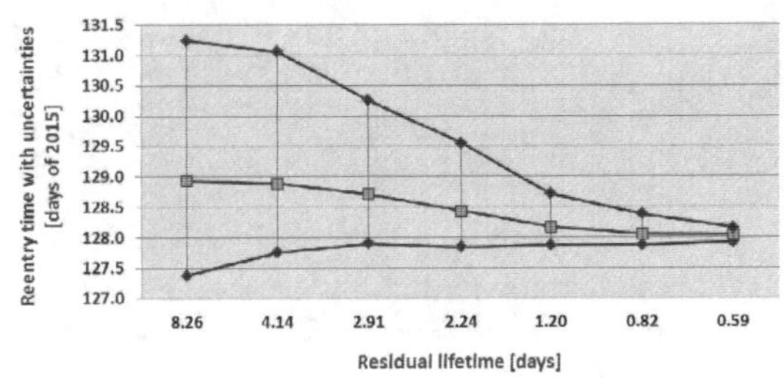

Figure 5. Evolution of Progress-M 27M reentry predictions and uncertainty windows (UTC)

In total, 7 reentry predictions at 80 km were issued to the Italian authorities (Fig. 5 and Tab. 4). The first uncertainty window was computed by varying by ± 20% the estimated ballistic parameter, while the following five windows were obtained with a ballistic parameter variation of ± 30%. For the last one, the start of the window was computed by increasing by 30% the last B estimated, while the end of the window was obtained by increasing by 30% the final nominal residual lifetime.

Table 4. Progress-M 27M reentry predictions and uncertainty windows issued by ISTI/CNR (UTC).

Propagation start epoch	Start of reentry uncertainty window	Nominal reentry prediction at 80 km	End of reentry uncertainty window
20150429 20:00	20150507 09:07	20150508 22:18	20150511 05:51
20150503 23:01	20150507 18:10	20150508 21:16	20150511 01:16
20150505 04:49	20150507 21:39	20150508 17:11	20150510 06:21
20150505 20:39	20150507 20:12	20150508 10:22	20150509 13:04
20150506 21:35	20150507 20:59	20150508 04:02	20150508 17:10
20150507 06:39	20150507 20:58	20150508 01:12	20150508 09:06
20150507 12:12	20150507 22:05	20150508 00:58	20150508 03:48

8. REENTRY GROUND TRACKS

With an orbit determination corresponding to an epoch 1.2 days before reentry, the reentry ground track still included in the uncertainty window is shown in Fig. 6. For the way in which the figure was obtained, not for inherent computational inaccuracies, the plotted sub-reentry ground track was reasonably accurate around the "nominal" reentry epoch, but, due to the size of the uncertainty window, was affected by a westward cross-track error of 210 km at the window start and by an eastward cross-track error of 300 km at the window end.

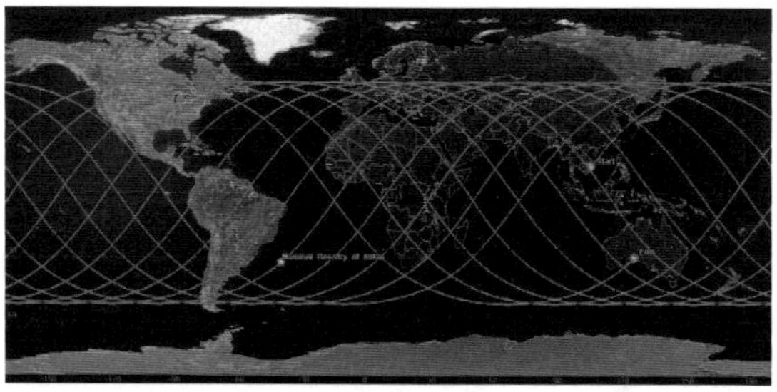

Figure 6. Progress-M 27M sub-reentry (at 80 km) ground track included in the uncertainty window corresponding to the 5th ISTI/CNR reentry prediction

The following prediction, the 6th issued by ISTI/CNR, was based on an orbit determination corresponding to an epoch 0.82 days before reentry. The reentry ground track still included in the uncertainty window was considerably reduced, as shown in Fig. 7, and comprised only one potential reentry opportunity at 80 km over Italy, around 5:53 UTC of May 8. Concerning Europe, the reentry opportunities still included in the uncertainty window are detailed in Fig. 8.

Finally, at 14:00 UTC of May 7, i.e. 12 hours and 20 minutes before the actual reentry, ISTI/CNR issued the 7th reentry prediction, based on an orbit with an epoch corresponding to 12:12 UTC. The new, reduced, uncertainty window finally allowed the exclusion of Europe from the reentry risk, as shown in Fig. 9, but no further prediction was issued later on because the new orbit determination data available until reentry was not considered reliable enough.

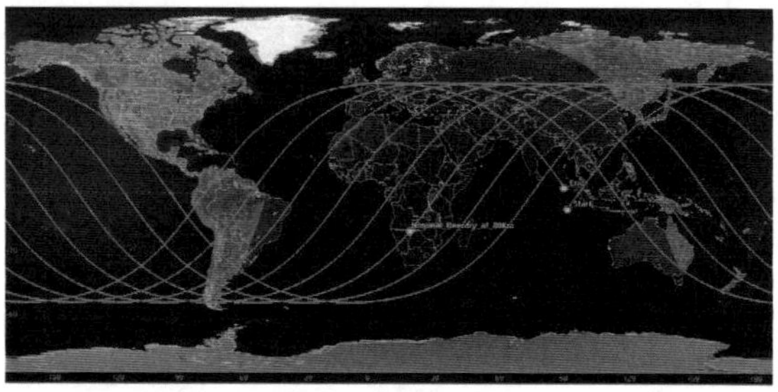

Figure 7. Progress-M 27M sub-reentry (at 80 km) ground track included in the uncertainty window corresponding to the 6th ISTI/CNR reentry prediction

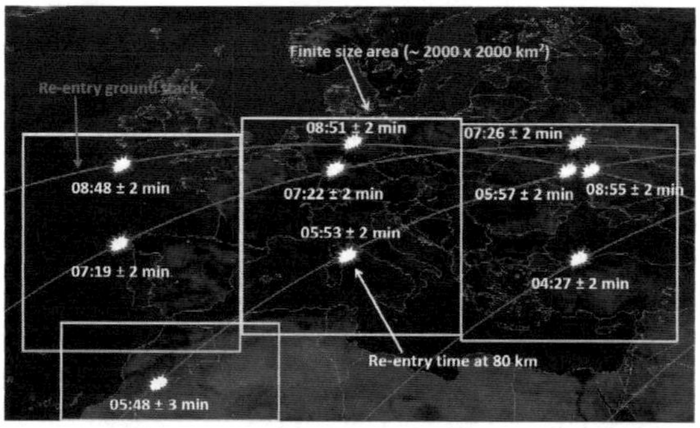

Figure 8. Progress-M 27M sub-reentry (at 80 km) ground track over Europe, and corresponding reentry times in each geographic area, included in the uncertainty window of the 6th ISTI/CNR reentry prediction

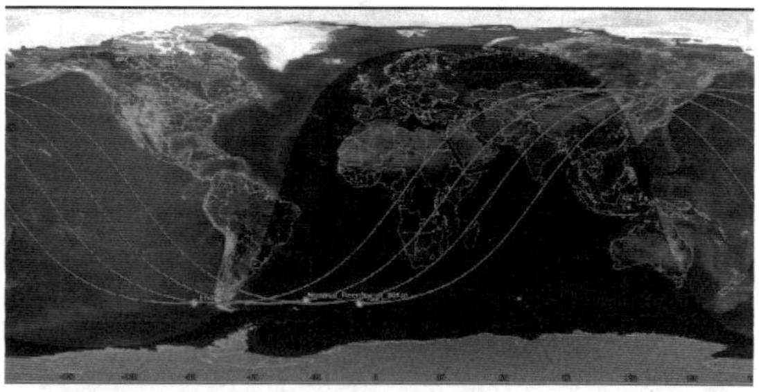

Figure 9. Progress-M 27M sub-reentry (at 80 km) ground track included in the uncertainty window corresponding to the 7th ISTI/CNR reentry prediction

The reentry at 80 km of Progress-M 27M at last occurred on May 8, 2015, at 2:20 UTC ± 1 minute, over the Southern Pacific Ocean, based on the post-reentry assessment carried out by the Joint Space Operations Center (JSpOC) of USSTRATCOM with classified satellite infrared observations (Fig. 10). This estimate placed the orbital decay of the unlucky cargo ship anywhere from 1300 to 350 km off the southern Chilean coast. Nevertheless, no sighting was reported and no debris was recovered.

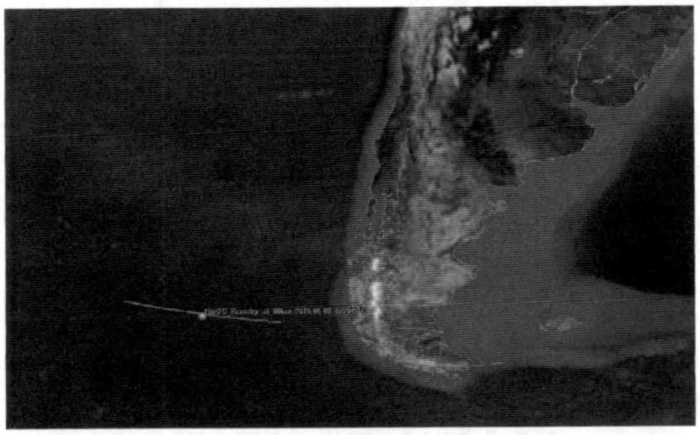

Figure 10. Progress-M 27M JSpOC reentry assessment at 80 km with ± 1 minute uncertainty

9. REENTRY PREDICTIONS ACCURACY

The Progress-M 27M reentry predictions campaign was set up in a few hours for an object having less than 10 days of residual lifetime. Moreover, the rotational dynamics was completely unusual, quite complex and rapidly evolving, and both solar and geomagnetic activity were on the rise. However, the results obtained where consistent and satisfactory, allowing the timely issuing of useful data and products for civil protection applications [18].

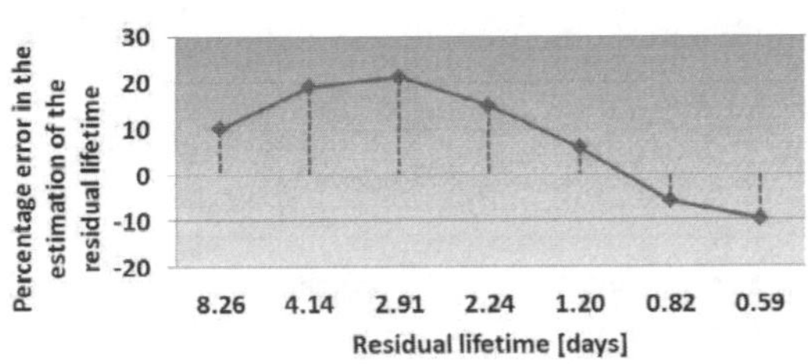

Figure 11. Percentage error of each reentry prediction issued by ISTI/CNR

The errors affecting the reentry predictions issued by ISTI/CNR are summarized in Figs. 11 and 12. The worse prediction was the third one, around three days before reentry, characterized by an over-estimation of the residual lifetime of little more than 20%. This was probably due to the occurrence, the following day, of a small geomagnetic storm, whose main effect was an increase of thermospheric density and a consequent acceleration of the decay. Nevertheless, a progressive acceleration of the Progress-M 27M decay was a constant of the campaign (see Fig. 5), due to a significant growth of solar activity and extreme ultraviolet emissions, leading to higher thermospheric density as well.

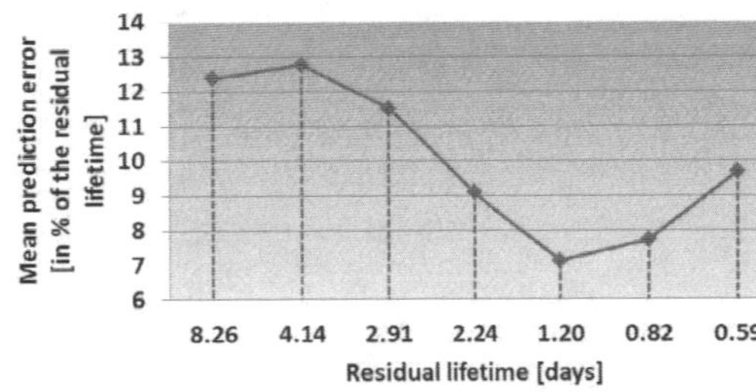

Figure 12. Running mean error of the reentry predictions issued by ISTI/CNR

The mean error of the predictions was approximately 12%, the residual lifetime was over-estimated in the first five and under-estimated in the last two, the most accurate estimates (with an error around 5%) were obtained around one day before reentry and the last prediction was affected by an error of approximately 10%. Therefore, the uncertainty windows adopted resulted fully appropriate, consistent and conservative through all the campaign.
The only significant problem encountered was the lack of reliable and consistent orbit determination data during the last 12 hours. Fortunately, a reentry over Europe and Italy had already been excluded at that time, but it was impossible to further reduce the uncertainty window on Africa, Asia and South America before the actual plunge of the spacecraft in the atmosphere.

10. ACKNOWLEDGMENTS

Part of the work described in this paper was carried out in the framework of the ASI-INAF agreement on "Detriti Spaziali: supporto alle attività IADC e validazione pre-operativa per SST" (F82I15000440005). We would like also to thank the USSTRATCOM and the Space-Track Organization for making available the two-line elements of Progress-M 27M and the data concerning historical reentries, the Space Debris Office of the European Space Agency for providing a couple of TLEs determined with the German TIRA radar and sharing relevant information, and the Mission Control Center of the Russian Space Agency for information exchange. Finally, some of the figures were generated with the AGI's Satellite Tool Kit (STK), Version 10.1.3.

11. REFERENCES

1. Pardini, C. & Anselmo, L. (2008). Long-Term Evolution of Geosynchronous Orbital Debris with High Area-to-Mass Ratios. Transactions of The Japan Society for Aeronautical and Space Sciences 51(171), 22–27.

2. Anselmo, L. & Pardini, C. (2009). Dynamical Evolution of High Area-to-Mass Ratio Debris Released into GPS Orbits. Advances in Space Research 43(10), 1491–1508.

3. Anselmo, L. & Pardini, C. (2005). Computational Methods for Reentry Trajectories and Risk As-sessment. Advances in Space Research 35(7), 1343–1352.

4. Pardini, C. & Anselmo, L. (2013). Re-entry Predictions for Uncontrolled Satellites: Results and Challenges. In Proc. 6th IAASS Conference 'Safety is Not an Option' (Ouwehand, L., Ed.), ESA SP-715 (CD-ROM), European Space Agency, Noordwijk, The Netherlands.

5. Wilde, P.D. & Ailor W. (Eds.) (2013). Re-Entry Operations Safety. In Safety Design for Space Operations (Sgobba, T., Allahdadi, F.A., Rongier, I. & Wilde, P.D., Eds.), Chapter 9, Elsevier & Butterworth-Heinemann, Oxford, UK, pp. 603–775.

6. Anselmo, L. & Pardini, C. (2013). Satellite Reentry Predictions for the Italian Civil Protection Authorities. Acta Astronautica 87(1), 163–181.

7. Pardini, C. & Anselmo, L. (2015). GOCE Reentry Predictions for the Italian Civil Protection Authorities. In Proceedings of 5th International GOCE User Workshop (Ouwehand, L. Ed.), ESA SP-728, European Space Agency, Noordwijk, The Netherlands.

8. Hoots, F.R. & Roehrich, R.L. (1980). Models for Propagation of NORAD Elements Sets. Spacetrack Report No. 3, Project Spacetrack, Aerospace Defense Command, United States Air Force, Colorado Springs, Colorado, USA.

9. Vallado, D.A., Crawford, P., Hujsak, R. & Kelso, T.S. (2006). Revisiting Spacetrack Report #3. Paper AIAA 2006-6753. Presented at the AIAA/AAS Astrodynamics Specialist Conference, Keystone, Colorado, USA.

10. Pardini, C. & Anselmo, L. (2009). USA-193 Decay Predictions with Public Domain Trajectory Data and Assessment of the Post-Intercept Orbital Debris Cloud. Acta Astronautica 64(7-8), 787–795.

11. Pardini, C. & Anselmo, L. (2001). Comparison and Accuracy Assessment of Semi-Empirical Atmosphere Models Through the Orbital Decay of Spherical Satellites. The Journal of the Astronautical Sciences 49(2), 255–268.

12. Pardini, C. & Anselmo, L. (2004). On the Accuracy of Satellite Reentry Predictions. Advances in Space Research 34(5), 1038–1043.

13. Pardini, C., Tobiska, W.K. & Anselmo, L. (2006). Analysis of the Orbital Decay of Spherical Satellites Using Different Solar Flux Proxies and Atmospheric Density Models. Advances in Space Research 37(2), 392–400.

14. Pardini, C. & Anselmo, L. (2008). Impact of the Time Span Selected to Calibrate the Ballistic Parameter on Spacecraft Re-entry Predictions. Advances in Space Research 41(7), 1100–1114.

15. Pardini, C., Anselmo, L., Moe, K. & Moe, M.M. (2010). Drag and Energy Accomodation Coefficients During Sunspot Maximum. Advances in Space Research 45(5), 638–650.

16. Pardini, C., Moe, K. & Anselmo, L. (2012). Thermospheric Density Model Biases at the 23rd Sunspot Maximum. Planetary and Space Science 67, 130–146.

17. The Aerospace Corporation (2016). Reentry Predictions. Center for Orbital and Reentry Debris Studies webpage (www.aerospace.org/cords/reentry-predictions/)

18. Pardini, C. & Anselmo, L. (2015). Satellite Re-entry Prediction Products for Civil Protection Applications. In Proc. 7th IAASS Conference 'Space Safety is No Accident' (Sgobba, T. & Rongier, I., Eds.), Springer International Publishing, Switzerland, pp. 453–462.).

19. Pardini, C. & Anselmo, L. (2016). ISTI-CNR Re-entry Campaigns in 2015 – Progress-M 27M & IADC test object CZ-2D second stage (2014-051C). In Proceedings of the 34th Inter-Agency Space Debris Coordination Committee Plenary Meeting, Harwell Campus, Oxford, UK.

20. Pardini, C. & Anselmo, L. (1994). SATRAP: Satellite Re-entry Analysis Program. Internal Report C94-17, CNUCE Institute, CNR, Pisa, Italy.

21. Picone, J.M., Hedin, A.E., Drob, D.P. & Lean, J. (2002). NRLMSISE-00 Empirical Model: Comparisons to Data and Standard Models. In Astro-dynamics 2001, Advances in the Astronautical Sciences Series, Vol. 109, Univelt Inc., San Diego, California, USA, pp. 1385–1398.

PROCEEDINGS OF THE 8TH INTERNATIONAL SPACE SAFETY CONFERENCE

SESSION 28 ORGANIZATION CULTURE - II

THE EVOLUTION OF CONTINUING EDUCATION & TRAINING FOR SAFETY & MISSION ASSURANCE PROFESSIONALS

Megan K. Stroud[(1)], Tom Pfitzer[(2)]

[(1)] *A-P-T Research, Inc, 4950 Research Drive Huntsville, AL 35805, United States Email:mstroud@apt-research.com*
[(2)] *A-P-T Research, Inc, 4950 Research Drive Huntsville, AL 35805, United States Email:tpfitzer@apt-research.com*

ABSTRACT

The international need for a safety and mission assurance (S&MA) training is narrow, but deep. Although some universities have ventured into this area and offered degree programs in the past, they quickly found that the narrowness made it unprofitable. The question arises as to how we as S&MA professionals train those in our field with the needed knowledge to prevent the potentially catastrophic events that may occur.

One organization who has accomplished this mission is the NASA Safety Center (NSC). NSC is tasked with the responsibility to provide S&MA professionals throughout NASA with the knowledge, skills, and tools they need to accomplish the mission of the organization. They have done so in large part through the development of their SM&A Technical Excellence Program (STEP). STEP provides training in several technical disciplines including system safety, quality engineering, reliability and maintainability, operational safety, software assurance, and Aviation Safety. Currently only available to NASA SMA civil servants and contractors, the key to the success of this training program is offering this via a virtual classroom, making the content available 24/7.

A-P-T Research, Inc. is currently in the initial stages of developing a web-based training program for S&MA professionals that will pool resources from multiple professional organizations and Government agencies. The S&MA virtual classroom will host online training classes in a variety of the S&MA disciplines, capture guest lectures, and offer professional certifications that are missing from our industry.

1. APT SEAC

In 2004, the safety Engineering and Analysis Center (SEAC) was formed as a division of A-P-T Research, Inc. As an APT Division, the SEAC supports all aspects of safety engineering with emphasis on three overlapping disciplines: System Safety, Range Safety, and Explosives Safety. Each of these three disciplines contains elements that are unique to it alone. Each also has elements that are common among the others, including risk management and risk assessment.

The mission of the SEAC focuses on the proper practice of safety engineering. In support of the mission, SEAC personnel provide analytical services, prepare special analyses including quantitative risk assessments, develop new methods, draft safety engineering standards, build software models, conduct safety training, serve as expert witnesses, participate in special studies, support accident investigations, perform third-party reviews, and conduct testing.

The SEAC is a collaboration of experienced practitioners and has attracted distinguished and experienced members of the S&MA community who have a desire to mentor the next generation of safety professionals. One of these members is the late Pat Clemons. Pat was widely recognized as a mentor of system safety and brought the insight of over 50 years of experience. Prior to joining APT Pat had trained over 2000 safety practitioners and assembled the first training classes offered by APT.

1.1 APT Continuing Education and Training Division

As a division of the SEAC the APT Continuing Education & Training Division (CE/T) supports the mission of the SEAC by providing both internal and external training to the S&MA community.

Each year APT trains a wide variety of S&MA professionals on topics such as system safety, software safety, explosive safety, and site planning software. Conducting training for more than 250 safety professionals annually. Safety practitioners from the Army, Air Force, Navy, National Aeronautics and Space Administration (NASA) as well as others from industry and internationally regularly take APT CE/T courses. The current offerings are shown in Figure 1.

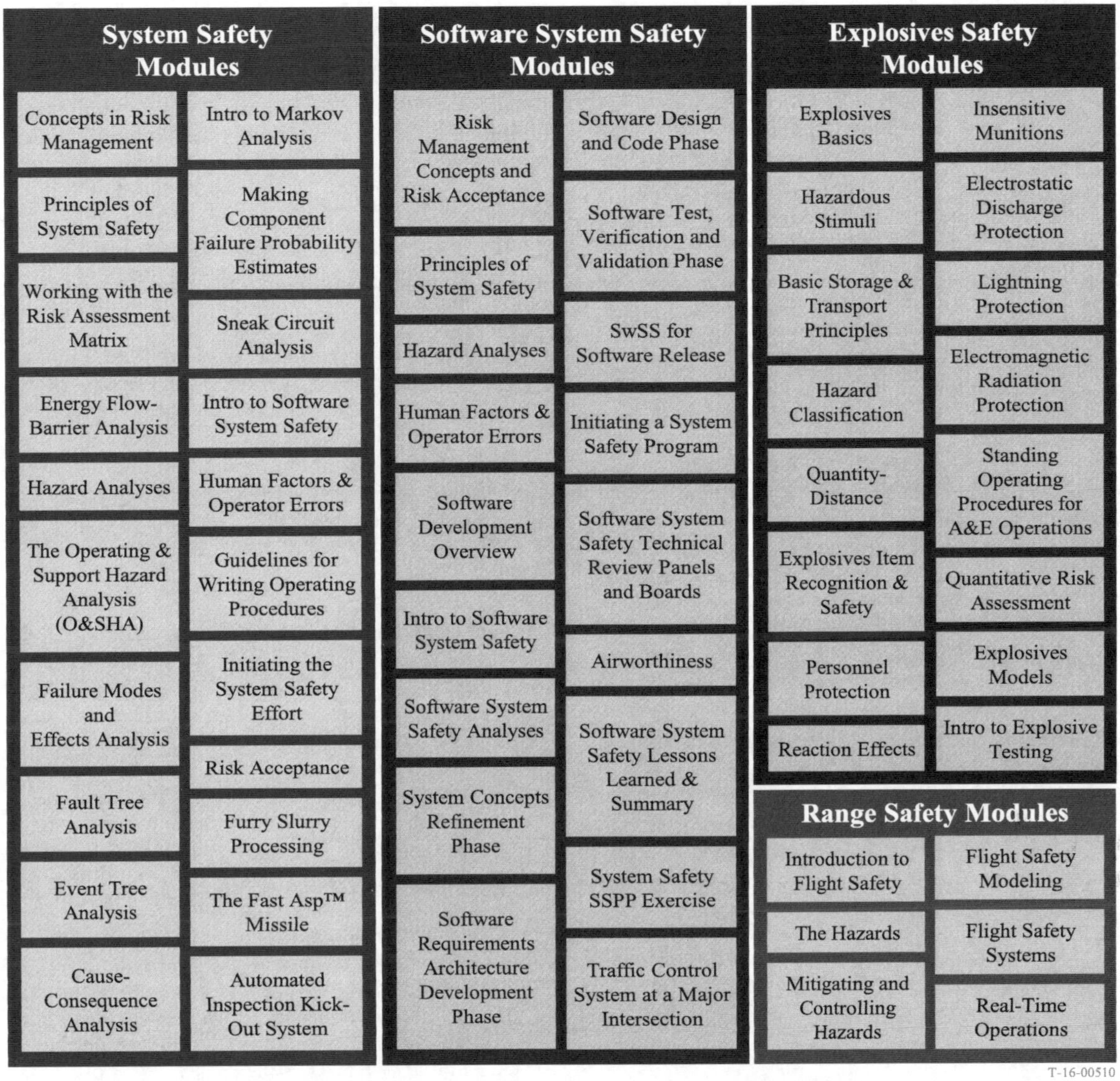

Figure 1. APT Training

2. THE CHANGING OF THE TIMES

In the past government organizations have developed training programs. In the 70's and 80's over 200 people went to the US Army Intern training program, many of which went on to careers in DoD and NASA. As time went on these types of formal training programs slowly evaporated from industry. Although, higher education attempted to duplicate these types of training programs for industry in the form of professional and continuing education it soon found that the audience for these types of programs was so vast that they soon found these types of programs unprofitable and ultimately these programs would also disappear from the list of options.

3. THE ADVANTAGE

Professional and Continuing Education has always been a pursuit of both individuals and corporate initiatives. The approach to Professional and Continuing Education should not be taken as a singular experience but rather as a career long practice. The benefits have been studied at length and result in increased cognitive abilities, spark interest and new methodology, and a curiosity to continue to learn more. There is also evidence that suggest that t here is a positive correlation between effective training and an increase in effective teamwork [1].

4. DEVELOPING PEOPLE, NOT COURSES

One of the most important aspects that should be considered is that our mission should be to develop people, not courses. The systems that S&MA professionals work on in this day and age can change rapidly. Looking closely at the evolution of NASA in recent years this is extremely evident. During the 80's and 90's NASA space travel was conducted by the shuttle. NASA's S&MA professionals only had a need to be concerned with a single system and the familiarity of the system was an advantage to the S&MA community. The day-to-day operations were familiar and it wasn't uncommon to find engineers or safety professionals that may have up to 20 years of experience on the same system.

This is no longer the case. Today, these systems are designed by multiple commercial companies including, but not limited to, Blue Origin, Sierra Nevada Corporation, Space Exploration Technologies (Space-X), & The Boeing Company. Even the systems within a commercial company are not fixed. Space-X for example utilizes several orbital launch vehicles including Falcon 1, Falcon 9, and the Falcon Heavy, some of which come in a variety of configurations.

In a report [1] published by the National Transportation Safety Board (NTSB) it was determined that failing to consider the possibility that the test pilot had the ability to unlock the feather early, a single point human error, ultimately resulted in the break-up of the vehicle. NTSB also determined that the test pilots were poorly trained and did not fully comprehend the risks of unlocking the feather early. Although, the issue was well known by designers the operators (pilots) were lacking the proper training and knowledge needed to safely operate the system, no doubt attributed to the unfamiliarity with the system itself.

Developing the S&MA community to be adaptable to multiple systems means giving them the professional training that they need and making this training accessible to them on as as-needed basis. By taking this approach it is more likely that our S&MA professionals will have the correct solution at the critical time. Resulting in an overall reduction to the risk imposed to both life and mission that our community works to improve on a daily basis.

5. THE WAY WE LEARN

It is undisputed that our universities and colleges have developed students utilizing a structure of curriculum based learning since the inception of formal education. But with the evolution of technology society has also changed the way learning and information transfer occurs. Understanding that our employees and our professional community has also changed is of upmost importance as we seek to design a new learning environment for professional and continuing education that will provide our employees and ultimately our discipline with the maximum benefit.

As a society we have become more focused on information seeking than retaining information and this really is an evolution of the way we operate in our work environments as well. There is constantly an influx of tools and information. Whether they are software tools, hardware tools, or even instructional manuals and knowing how to use these can provide a variety of benefits. However, we must educate the community on the proper way to use them. Software tools can provide advantages by giving quick analysis answers, but if the community using them doesn't understand the way they work they can also provide detrimental outcomes when used inadequately.

6. CONTENT

The APT S&MA classroom will focus on developing online training courses, capturing guest lectures from S&MA subject matter experts, and providing discussion boards for collaboration. The online classroom embraces the spirit of an information seeking user and will enable students to develop their own curriculum specific to the issues and topics that are most applicable to both their professional and personnel goals. The online classroom will be available 24/7 and have a fully functional searchable catalogue enabling students to quickly find the topics that are of need/interest.

Although the content may be expanded in the future, the initial focus of the classroom will be to provide content on the following subject matters.

- System Safety
- Quality Engineering
- Reliability and Maintainability
- Operational Safety
- Software Assurance
- Aviation/Range Safety
- S&MA Managerial and Leadership

The technical subject matter closely resembles the structure of NASA's Safety and Mission Assurance Technical Excellence Program (STEP) developed by the NASA Safety Center (NSC). STEP is one of the first comprehensive training programs that has been developed to address the needs of the S&MA community. The success of this program within the NASA community lays a foundation for the future success of the APT S&MA classroom that will be accessible to an international audience.

In addition to the topics presented above the online classroom intends to have other resources such as:

- Case Studies
- Guest Lectures
- Discussion Forum
- Resource Library

7. GETTING INVOLVED

The APT S&MA classroom will not be a success without the support of the S&MA community. There is already a vast database of material available to those in our community and APT would like to invite professional societies, organizations or individuals who have content to collaborate on this effort. Those organizations who have a desire to participate in the advancement of the discipline and the development of the APT S&MA classroom should provide contact information to training@apt-research.com.

8. REFERENCES

1. Take time for team training: the benefits of continuing education are numerous--a more qualified team and motivated employees are among them. Here's how to get the most out of these training sessions. (2010, August). Veterinary Economics, 51(8), 39+. Retrieved from https://portal.lib.fit.edu/login?url=http://go.gale group.com.portal.lib.fit.edu/ps/i.do?id=GALE%7CA238980808&sid=summon&v=2.1&u=m elb26933&it=r&p=AONE&sw=w&asid=d77d 27495a1391e9920e925ca9c83615

2. Lack of Consideration for Human Factors Led to In-Flight Breakup of SpaceShipTwo. (2015, Jul 29). US Fed News Service, Including US State News Retrieved from http://search.proquest.com.portal.lib.fit.edu/docview/1699512213?accountid=27313

PROCEEDINGS OF THE 8TH INTERNATIONAL SPACE SAFETY CONFERENCE

SESSION 29
DESIGNING SAFETY - III

HOW TO ENSURE MEDICAL SAFETY AND RESCUE IN HUMAN SPACEFLIGHT FOR THE FUTURE

Yacine Benyoucef (1)

(1) SPACEMEDEX R.&D., 74 Av. Michel Jourdan, 06150 Cannes (France)
Email: yacine.benyoucef@gmail.com

ABSTRACT

Spaceflight-related accidents and non-fatal incidents showed that there is a real need in developing rescue safety and capabilities. Principles of rescue were eluded within treaties and technical papers, including concepts, cooperation and planning. Despite the initiative to build in 1975 a movement toward an international rescue force, it's difficult now to evaluate the real improvement of such important project. The interest in international rescue studies was renewed, meetings were planned and conducted during the 1990's decade, but nothing seem to move because of difficulties and limitations such like launch readiness, orbital mechanics and also, implications of national agencies. New projects, bigger than ever, are now in charts to expand our frontiers to the Moon and beyond: to Mars. Because a medical issue or a loss of life include the suspension of flight operations and the associated project, because the development of commercial spaceflight and space-tourism will need a strong support in medical rescue and critical care capabilities, because it is essential for the future of space exploration to have an effective and strong medical rescue force able to take action as soon as the incident happened, we will discuss here how we could manage the inherent medical risk and address a proposal for a rescue plan as a continuation of the previous initiative and a start for the future.

1. INTRODUCTION

Human spaceflight began with Yuri Gargarin on 1961. Then, more than 400 astronauts have flown, twenty-one fatalities occurred, five catastrophic events and so many other mishaps that could have resulted in death [1,2]. Spaceflight events like fatalities, near-fatalities, flight contingencies and medical events are well documented and illustrate how hazardous this environment is and how complex it could be to build a safety mission [3]. To remember, several evacuation and emergent medical events have occurred in spaceflight between 1971 and 2005 [4] but most of fatal accidents were during ascent or reentry. Conscious efforts have been made to develop a rescue vehicle for rapid return if a life-threatening event occurs into the International Space Station (ISS) [5,6,7], but this seems to be the single option if something happens. Paradoxically, despite a serious lack in rescue capabilities, agencies and companies are targeting other planets and commercial spaceflight.

Developed to face war conditions, medical support and rescue is supposed to bring aid and to evacuate injured civilians and soldiers from the battlefield. Airplanes ameliorations and engineering allowed the development of news procedures: tactical or strategic air medical evacuation, whether the patient is treated onsite or evacuated to the closest hospital. Briefly, the main idea was to bring a strong medical support as far as possible, as fast as possible, and to shorten the delays with the highest quality of critical care capabilities. Since 1975, interesting efforts were done to promote space safety through an international cooperation but technical difficulties and political limitations restrained the action to theoretical considerations: more than 40 years after the international agreement on the rescue and recovery of international space explorers, there is still no international space crew rescue system available to respond to urgent situation in space [8,9]. However, things are moving, and it is urgent to have something effective to prove that we are also able to keep humans in space like in any other safety place. As a continuation of the previous initiative and a start for the future, we will discuss, after a short risks review and analysis, how we could manage medical emergencies to support long-duration spaceflight out of Low Earth Orbit (LEO). This to address a rescue proposal based on space-adapted medical evacuation methods.

2. RAPID RISKS ANALYSIS

LEO spaceflight risks were mainly evaluated by using ground-analog stations [10,11,12], and data from the Russian Space Program and the NASA Medical Operations since 1959 [13]: events were classified regarding their medical significance and the requirement of an evacuation or a management onsite, but the use of analogs for such epidemiological analysis provides only an approximation because of the absence of spaceflight operational and occupational risk, even if characteristics and medical events seem to be similar [14]. The largest ground-analog is the Antarctic McMurdo station with 1200 occupants during the austral summer months and 125 during the winter-over months. The evacuation rate was analyzed from 1992 to 1995 [10]. To resume, 71 medical evacuations were required with an incidence of 3.55 per month. With no evacuation capabilities during the winter, McMurdo station appears to be a good approximation of events onboard the ISS, leading to the extrapolation of 0.25 evacuations per year for a crew of seven, so one evacuation every 4 years. To add, three cosmonauts from the Russian Space Program were returned for medical reasons in 41.5 person-years of spaceflight, so an evacuation rate of one per 14 person-years [4]. Also, retrospective studies and data review from the NASA Longitudinal Studies of Astronaut Health (LSAH) reported 88 hospitalizations among active astronauts between 1959 and 1999 with a total of 2 715 person-years over this 40 years period [12]. From these, potential evacuation events applicable to the ISS were subtracted to generate an incidence about 0.02 person-years. The importance of an onboard medical system was well described and supported [4]. Further, if we consider a human mission to the Moon or Mars, despite the evident lack of data we can understand that the risk will be increased because of the distance and the duration of the mission: this could contribute to restrain the medical capabilities and the access to emergencies. Based on the 2012 NASA Space Cancer Risk model developed to estimate cancer risks and uncertainties for space missions [15,16], it has been predicted that a wide range of cancer would be observed in crews after a mission to Mars, with a probability of causation for longer post-mission times for most solid cancers and circulatory diseases [17]. It is tempting to say that duration and distance of a deep space mission multiply all the risks taken during a LEO spaceflight. Regarding space-tourism, as long as it follows conventional human spaceflight in the LEO and its medical requirements, risks would be the same as previously defined for the ISS, including launch, reentry and transportation [18].

3. EMERGENCIES MANAGEMENT IN SPACE

Providing an acute medical care solution is a fact, but the point is that for a long-duration orbital spaceflight or in deep space, patients will be deconditioned.

3.1 Deconditioned Patients

Neurovestibular disturbances and thoracic fluid shift are the two immediate responses to spaceflight [19]. The upper blood engorgement is visible through apparent facial edema and erythema with a diminution of the lower extremities circulation. This could last from a few hours to a few days. The sensitivity to motion is increased with headache, loss of appetite and nausea. The motion sickness could start at 15 min or 3 days after reaching orbit, lasting 2 to 3 days with a possible persistence up to 7 to 10 days. The cardiac output is increased due to increased filling and stroke volume. The mean arterial pressure is maintained giving a decrease of peripheral vascular resistance [20]. Also, there is around 10-15% volume loss in plasma being extravased from the vascular space to the intracellular space by increased capillary permeability [21]. Thus, the hematocrit is increased leading to the inhibition of erythropoietin [22], but normalization by a reduction in red blood cells number is seen over several days to stabilize plasma volume [22]. To add, a 10% decrease in erythrocyte mass was observed, supposed to involve a selective neocytolysis to facilitate a rapid circulatory adaptation to microgravity [23]. Regarding bones, loss of calcium and phosphate and a decreased calcium absorption lead to a loss of bone density. Minerals are lost first from the weight-bearing bones, lower extremities, lower pelvis and lumbar spine [24]. Concerning the gastrointestinal apparatus, a reduced absorption and increased urinary excretion were observed [25]. Results for parathyroid hormones are mitigated: globally increased [26,27,29] or unchanged during and after long-duration flight [28,30], whereas levels of active vitamin D are reduced during long-duration flights [30] but found increased during shorter missions [28]. Further, muscle atrophy was observed associated with losses of urinary nitrogen and phosphorus with a 10-20% reduction of strength in lower extremities [24,31]. The glomerular filtration rate increases early and remain elevated for at least a week with a decrease in creatinine clearance, an increase in ADH levels, a decrease in plasma renin activity and aldosterone levels until a normalization, and a decrease of atrial natriuretic peptide [21,32,33]. Further, evidence have highlighted a dysregulation of the immune system with lymphopenia, reduced NK cells cytotoxicity and decreased interferon production [34,35,36]. Those are a brief example of what is the response of the body to weightlessness.

3.2 Acute Care and Emergencies

Emergencies management in space goes through the use of minimal equipment. So, each crewmember receives medical care training for the mission (kits, references, computer-based training). Consultations with a ground-based medical team can be set-up, but they are most likely on their own to treat emergencies. Thus, motion sickness, cardiac dysrhythmias leading to heart failure, toxic chemicals and gases leaks, bone injuries and fractures, muscles and ligamentous injuries, blunt and penetrating injuries with/without hemorrhage, infectious disease and genitourinary sepsis, immune suppression and viral reactivation, long-duration flights psychiatric diseases and mental illness, radiation and cancer are a broad range of what could happen as long as the mission persists. Studies and database collected from previous spaceflights were exposed and detailed to compare the on-orbit medical resources with the potential clinical cases [37]. It seems that capabilities, including surgical, are limited, with a need for specialized equipment other than diagnosis and imaging. As long as we consider a limited area for emergencies, medical and surgical hardware will be limited too. Moreover, physiological effects of weightlessness have an impact on surgical care in terms of preoperative preparation, intraoperative response and postoperative recovery, limiting the medical act [38]. To conclude, we need a larger place: a site with a strong medical staff, able to provide medical and surgical support and able to enrich the panel of countermeasures in space.

4. TOWARD A PERMANENT SOLUTION

The best manner to develop a safety process is to have a deep understanding of the risks. Once acquired, we can build countermeasures and models to predict the danger. Including intervention and evacuation. To precise, a medical evacuation is an act in itself, because critical cares are provided during transport. However, ISS just have a crew rescue vehicle to use in an urgent situation for rapid return so that we cannot really consider this as a medical evacuation.

4.1 Tactical Considerations

Two months of spaceflight and the patient shows signs of tachycardia and hypotension illustrating a decreased tolerance for lower-body negative pressure, and thus for blood loss or shock [39]. If we consider the design of a tactical intervention, we should keep in mind that it has to be as fast as possible within the *golden hour*, which is the critical period immediately after a significant trauma to have the most favorable outcome. That period might be shorter than usually due to the deconditioned state of the patient in space. With this information correlated to those clinical data, we can conclude that an evacuation back to Earth constitutes a great danger for a patient in shock: a return for medical reasons or trauma would be on a middeck floor at 1.2G in a chest-to-back axis for several minutes facing reentry forces, then orthostatic hypotension and near-syncope state in addition of his trauma. Apparently, there is no adverse effect unless the hemorrhage is severe (30-50% loss of blood volume) or the forces are excessive (up to 8G) in primates with a controlled hemorrhage following centrifugation to mimic atmospheric reentry forces [6]. Nevertheless, uninjured but deconditioned patients use to display the hypovolemic profile of a class I hemorrhage (15% loss of blood volume) with small tachycardia and orthostatic hypotension; but to compare, a true class I hemorrhage in space correspond to a class II hemorrhage on Earth (30% loss of blood volume) with tachycardia, tachypnea and increased pulse pressure: the tolerance to the return is affected. Moreover, it seems that there is an increased risk of complicating factors for patients in a situation of medical return (neurovestibular disturbance, cardiac and muscular atrophy, immunosuppression, fractures, etc.) [4]. Also, in the case of an early in-flight resuscitation and stabilization, it would be difficult to evacuate immediately considering both the risk and the danger. It is tempting to say that it would be better to treat on site with a strong medical facility or to evacuate to a close medical platform avoiding the immediate reentry.

4.2 Ensuring Medical Safety and Rescue

The area available in the shuttle is limited and restricts any emergency care. So, an area for prolonged critical care and emergencies management is required. In-flight medical kits are restricted because of storage constraints and most of what we can propose are for symptomatic relief and first-aid supplies. At the light of this, it becomes necessary to set-up a medical checkpoint as a permanent space platform to design a pertinent medical evacuation and support for long-duration spaceflights. This has to be considered as a clinic in space with critical care and monitoring equipment, with a crew of surgeons, nurses and/or physicians. Such a medical base would avoid launch readiness, costs, orbital mechanics and delays of intervention. Spacecraft constraints would be respected until better technologies and shuttles will be designed. Moreover, this clinic would be an initiative to have a deeper understanding of the human physiology in space through research activities beside the medical support. With a platform like this, we would be able to develop appropriate evacuation methods after trauma. This means the development of a special medical and paramedical corps with operational techniques and technologies made in space for space. This checkpoint has to be at a place easily reachable from the LEO, or on the return from the Moon or Mars. The best would be a place with no gravitational influences to minimize readjustments and vibrations. This means also the design of appropriate vehicles for intervention and evacuation once the patient has recovered. This medical checkpoint would have to integrate the knowledge we have obtained from spaceflights studies and to continue the development of medical integrative countermeasures to facilitate therapies and physiological recovering. It would help to the development of space medicine and surgery; it would help to the adequate understanding of anesthesia and pharmacology in space [40]; it would help to create a strong medical support and rescue system for spaceflights in the future.

5. CONCLUSION

Human spaceflights have reached a point where it needs the development of infrastructures to support its future goals. There are a lot of opportunities and perspectives from this permanent solution: this could not just provide a strong medical support for the human side of space exploration, it would also serve as a clinical research institution in space, to bring down to Earth important and relevant data. Moreover, technologies we need to face such an important project would of course benefit the humankind on Earth in many ways. Obviously, we have the know-how on hands thanks to the ISS, we shall go further and enlarge our presence on orbit and facilitate human safety and rescue for space exploration and tourism.

6. ACKNOWLEGEMENTS

I am highly thankful to Mr. T. Zouari for his financial support for the venue to this congress, making it easier and possible. Also, thanks to Mrs. L. Bouzelifa and Mrs. H. H. Benyoucef for their mutual help and presence.

7. REFERENCES

1. Qian Z, Hao X. Autonomous Rescue System. Space Safety and Rescue 1995; Proceedings of the IAA Symposium, Oslo, Norway, Oct. 2–6, 1995, San Diego, CA, Univelt, Inc. (Science and Technology Series. Vol. 93), 1997, p. 77– 83. (IAA 95-6.1.08).

2. Fabian J. An historical perspective on crew rescue and the role of the association of space explorers, IAA 89-618, 22nd IAA International Space Safety and Rescue Symposium, Space Safety and Rescue 1988–89 Vol. 77. p. 227– 238.

3. Myers H. Assured Crew Return Capability Crew Emergency Return Vehicle (CERV) Avionics. In NASA, Washington, DC, Space Transportation Avionics Technol. Symposium. Nov. 7–9 1989, JSC-IA131, Vol.2: Conference Proceedings p 163–177 (N91-17025).

4. Barratt M.R., Pool S.L. Principles of Clinical Medicine for Space Flight. Johnston S.L., Arenare B.A., Smart K.T. Chapter 7: Medical Evacuation and Vehicles for Transport. 2008 Springer Science p. 139-162.

5. Griswold HR, Trusch RB. Emergency and rescue considerations for manned space missions. Acta Astronautica 1981; 8(9): 1123–1133.

6. Hamilton GC, Stepaniak PC, Stizza D, et al. Considerations for medical transport from space station via assured crew return vehicle (ACRV). Unpublished final report, NASA Grant NAG-9-263, 1989.

7. Assured Crew Return Vehicle (ACRV) Project System Engineering Data Book, JSC 34015, 1992, NASA Johnson Space Center, Houston, TX.

8. United Nations. Office for Outer Space Affairs. Agreement on the Rescue of Astronauts, the Return of Astronauts and the Return of Objects Launched into Outer Space (1967) RES 2345 (XXII).

9. Shayler DJ. Space Rescue. Ensuring the Safety of Manned Space Flight. Praxis Publishing Ltd, Chichester, UK, 2009.

10. Harrison AA, Clearwater YA, McKay CP. The Human Experience in Antarctica: Applications to life in space. Behav Sci 1989 Oct; 34(4): 253–271. Reprint in Harrison, AA, Clearwater, YA, McKay, CP, (eds.), From Antarctica to Outer Space: Life in Isolation & Confinement, NY, NY: Springer-Verlag. 1991.

11. Billica RD, Simmons SC, Mathes KL, et al. Perception of the medical risk of spaceflight. Aviat Space Environ Med. May 1996; 67(5): 467–473.

12. Johnston SL, Marshburn TH, Lindgren K. Predicted Incidence of Evacuation-Level Illness/Injury During Space Station Operation. 71st Annual Scientific Meeting of the Aerospace Medical Association, Houston, TX, May 2000.

13. Berry CA. Descent and landing of space crews and survival in an unpopulated area. In NASA, Washington Found. Of Space Biol. And Med., Volume 3, pp. 372–394, 1975, NASA I.D. 19760019754N (76N26842).

14. Johnston, SL. Medical Care at the South Pole, Presented at the 1st "Pushing the Envelope" Conference, Univ. of Texas Medical Branch, Department of Preventive, Occupational and Environmental Medicine, Nassau Bay Hilton, Clear Lake, TX 1998.

15. Cucinotta FA, Kim MY, Chappell L (2013) Space radiation cancer risk projections and uncertainties-2012. NASA TP 2013-217375.

16. Committee for Evaluation of Space Radiation Cancer Risk Model, National Research Council (2012) Technical evaluation of the NASA model for cancer risk to astronauts due to space radiation. The National Academies Press, Washington DC.

17. Cucinotta FA, Kim MHY, Chappell L, Huff JL (2013) How safe is safe enough? Radiation risk for a human mission to Mars. PLoS ONE 8(10): e74988.

18. Antuñano MJ, Gerzer R, Russomano T et al. Medical safety considerations for passengers on short-duration commercial orbital space flights. International Academy of Astronautics Study Group, 2010.

19. Barratt M.R., Pool S.L. Principles of Clinical Medicine for Space Flight. Baker E.S., Baratt M.R.,Wear M.L. Chapter 2: Human Response to Space Flight 2008 Springer Science p. 139-162.

20. Buckey JC, Gaffney FA, Lane LD, et al. Central venous pressure in space. J Appl Physiol 1996; 81:19–25.

21. Leach CS, Alfrey CP, Suki WN, et al. Regulation of body fluid compartments during short-term spaceflight. J Appl Physiol 1996; 81(1):105–116.

22. Alfrey CP, Udden MM, Leach-Huntoon C, Driscoll T, Pickett MH. Control of red blood cell mass in spaceflight. J Appl Physiol 1996; 81(1):98–104.

23. Rice L, Alfrey CP. Modulation of red cell mass by neocytolysis in space and on Earth. Pflugers Arch 2000; 441(2–3 Suppl):R91–R94.

24. Whedon GD, Lutwak L, Rambaut PC, et al. Mineral and nitrogen metabolic studies, experiment M071. In: Johnston R, Dietlein L (eds.), Biomedical Results from Skylab. Washington, DC: Scientific and Technical Information Office, NASA; 1977:pp. 164–174.

25. Smith SM, Wastney ME, Morukov BV, et al. Calcium metabolism before, during, and after a 3 months spaceflight: Kinetic and biochemical changes. Am J Physiol Heart Circ Physiol Regulatory Integrative Comp Physiol 1999; 277:R1–R10.

26. Caillot-Augusseau A, Lafage-Proust MH, Soler C, Pernod J, Dubois F, Alexandre C. Bone formation and resorption biological markers in cosmonauts during and after a 180-day space flight (Euromir 95). Clin Chem 1998; 44(3): 578–585.

27. Grigoryev AI, Dorokhova BR, Semenov VY, et al. Fluid-electrolyte metabolism and renal function in cosmonauts following 185-day spaceflight [In Russian]. Kosmicheskaya Biol I Aviakosmicheskaya Meditsina 1985; 19(3): 21–27

28. Morey-Holton ER, Schnoes HK, DeLuca HF, et al. Vitamin D metabolites and bioactive parathyroid hormone levels during spacelab 2. Aviat Space Environmental Medicine 1988; 59:1038–1041.

29. Tipton CM, Greenlead JE, Jackson CG. Neuroendocrine and immune system responses with spaceflights. Med Sci Sports Exerc 1996; 28:988–998.

30. Smith SM, Wastney ME, O'Brien KO, et al. Bone markers, calcium metabolism, and calcium kinetics during extended duration space flight on the Mir space station. J Bone Miner Res 2005; 20(2):208–218.

31. Thornton W, Hoffler G, Rummel J. Muscular deconditioning and its prevention in space flight. In: Johnston R, Dietlein L (eds.), Biomedical Results of Skylab. Washington, DC: Scientific and Technical Information Office, NASA; 1977:191–197.

32. Schrier RW, Berl T, Anderson RJ. Osmotic and nonosmotic control of vasopressin release. Am J Physiol 1979; 236(4):F321–F332.

33. Eversmann T, Gottsmann M, Uhlich E, Ulbrecht G, Von Werder K, Scriba PC. Increased secretion of growth hormone, prolactin, antidiuretic hormone, and cortisol induced by the stress of motion sickness. Aviat Space Environ Med 1978; 49(1 Pt 1):53–57.

34. Konstantinova IV. Problems of space biology. In The Immune System Under Extreme Conditions, Space Immunology Volume 59. Translated from Sistema V Eksytremai 'Nykh Usloviyakh, Problemy Kosmicheskoy Biologiya Vol. 56. Washington, DC: Natl. Aero. Space Admin.; 1990.

35. Konstantinova IV, Antropova EN, Legen'kov VI, Zazhirey VD. Study of reactivity of blood lymphoid cells in crewmembers of the Soyuz- 6, Soyuz-7, and Soyuz-8 spaceships before and after flight. Kosmi Biol Avikosmi Med 1973; 7:35.

36. Manie S, Konstantinova I, Breittmayer JP, Ferrua B, Schaffar L. Effects of long duration spaceflight on human T lymphocyte and monocyte activity. Aviat Space Environ Med 1991;62(12):1153–1158.

37. Barratt M.R., Pool S.L. Principles of Clinical Medicine for Space Flight. Mashburn T.H. Chapter 5: Acute Care. 2008 Springer Science p. 139-162.

38. Kirkpatrick AW, Campbell MR, Novinkov OL, et al. Blunt trauma and operative care in microgravity: A review of microgravity physiology and surgical investigations with implications for critical care and operative treatment in space. J Am Coll Surg 1997; 184:441–453.

39. Johnson RL, Hoffler GW, Nicogossian AE, et al. Lower body negative pressure: Third manned Skylab mission. In: Johnston RS, Dietlein LF (eds.), Biomedical Results from Skylab. Washington, DC: US Government Printing Office; 1977:284–312. NASA SP-377.

40. Stingl JC, Welker S, Hartmann G, Damann V, Gerzer R (2015) Where Failure Is Not an Option – Personalized Medicine in Astronauts. PLoS ONE 10 (10): e0140764.

PROCEEDINGS OF THE 8TH INTERNATIONAL SPACE SAFETY CONFERENCE

SESSION 31
NEO & COSMIC HAZARDS

SPACE SITUATIONAL AWARENESS PROGRAMME: ENHANCED NEO PROPAGATOR (NEOPROP2)

Valentino Zuccarelli (1), Sven Weikert (2), Celia Yabar Valles (3), Detlef Koschny (4)

(1) *Astos Solutions GmbH, Grund 1, 78089 Unterkirnach (Germany), Email: Valentino.Zuccarelli@astos.de*
(2) *Astos Solutions GmbH, Grund 1, 78089 Unterkirnach (Germany), Email: Sven.Weikert@astos.de*
(3) *ESA-ESTEC, Keplerlaan 1, 2201 Noordwijk, (The Netherlands), Email: Celia.Yabar.Valles@esa.int*
(4) *ESA-ESTEC, Keplerlaan 1, 2201 Noordwijk, (The Netherlands), Email: Detlef.Koschny@esa.int*

ABSTRACT

The overall aim of the Space Situational Awareness (SSA) Preparatory Programme is to support the European independent utilization of, and access to, space for research or services, through providing timely and quality data, information, services and knowledge regarding the environment, the threats and the sustainable exploitation of the outer space surrounding our planet Earth.

Currently there are over 600.000 asteroids known in our Solar System, where more than 13.000 of these are NEOs. They could potentially hit the Earth and depending on their size could produce considerable damage. For this reason, NEOs deserve active detection and tracking efforts. The role of the SSA programme is to provide warning services against potential asteroid impact hazards including discovery, identification, orbit prediction and civil alert capabilities.

In 2012, Astos Solutions developed for ESA a new orbital propagator algorithm in order to assess the potential risk of impact of a NEO. The objective of that activity was to come up with a different trajectory prediction algorithm, which allows an independent validation of the current algorithms within the SSA-NEO segment (e.g. NEODyS, JPL Sentry System). Analytical and numerical algorithms were developed in order to assess minimum orbital intersection distances and close approaches with other celestial bodies.

In 2015, a new version of the tool was developed to improve and extend its functionalities. The following new algorithms were added:

- Estimation of impact probabilities as Torino and Palermo scale values, after propagating virtual asteroids computed along the line of variation. These figures are important characteristics for Governments to identify potentially catastrophic scenarios and to prepare counter-measures or evacuations.
- New orbital perturbations (e.g. Poynting-Robertson effect, solar radiation pressure, outgassing) to improve the propagator accuracy and to allow the identification and propagation of any celestial body (not only NEOs but also moons, comets, planets, etc).

The pre-existing algorithms have been further improved in order to increase the performance and reduce the need for human intervention. Robust and redundant preliminary orbit determination techniques have been added in order to deal with very long and disrupted observational arcs, which usually would require a manual split of the observations.

Moreover, a graphical user interface has been developed in order to ease the setup of a new scenario, collecting observations and physical properties of the object to be analyzed and allowing the comparison of the results with those provided by other systems (e.g. NEODyS, JPL Sentry).

1. ARCHITECTURE

NEOProp2 improves and extends the functionalities of the NEOProp software. In this paper, only the new features are presented and discussed. Detailed information about existing functionalities of the tool can be found in [1].

As its predecessor, NEOProp2 has two separate modules:

1. The Analytical Module makes use of analytical algorithms in order to rapidly assess the impact risk of a NEO. Orbit determination algorithms determine the initial state, along with its uncertainty, and the minimum orbital intersection distances (MOID) of the NEO (computed analytically). If the MOID is under a certain threshold, a numerical analysis is recommended.
2. The Numerical Module makes use of numerical algorithms in order to refine and to better assess the impact probabilities of a NEO. The initial state provided by the Analytical Module is used for the numerical propagation of the trajectory, which can

be run in two modes: one faster, in order to get a quick estimate of the trajectory (fast analysis) and one more precise, taking into consideration more detailed and complex models (complete analysis). Along with the numerical propagation of the nominal state, Virtual Asteroids (VAs) can be numerically propagated in order to determine the (numerical) MOID and assess the impact risk of the asteroid, in case at least one virtual impactor is found. This MOID computation differs from the one performed by the Analytical Module since it takes into consideration the full dynamics of the problem (all the relevant perturbations).

An aspect common to both the modules is the usage of multi-threading. The tool is able to detect how many threads are available and to assign them in order to parallelize some computations.

Moreover, in order to improve the usability of the tool a Graphical User Interface (GUI) has been developed. It allows the user to quickly create a new scenario and to easily retrieve from internet observations and physical properties for the NEO to be analyzed. With the GUI, the user can run the modules without having to use the command-line interface and can rapidly visualize and export the most important results in an html file. If requested, the GUI can retrieve the orbital elements, MOID and risk figures from the NEODyS and Sentry systems and compare them with the results obtained by NEOProp2.

2. ANALYTICAL MODULE

This module allows determining the initial state (with uncertainty) of a NEO from a set of observations. First a preliminary Gauss algorithm is run in order to get a first solution of the orbit determination problem using only optical observations.

The Gauss method needs only three observations (triplet) in order to compute a preliminary initial state. Then a Linear Least Square (LLS) technique is used to refine the solution and minimize the root mean square (RMS). As any other least square technique, an initial guess is required in order to start the computation. The weak point of such technique is the initial guess itself. If it is too poor (too far from the real solution), the least square method may fail to converge.

In the previous version of the tool, the LLS technique did not always converge, especially when there were many observations, discontinuities or close approaches within the observation set. A first solution to this problem consisted in splitting the observations into shorter arcs, which would start to constraint the solution. In this way the Gauss method can work with consistent observations and find more easily a good solution. The set of observations had to be manually split into intervals, whereas first the orbit determination was performed just with the oldest interval and then further arcs were gradually added, always using the results of the previous run as initial guess.

Since this approach was time consuming and depended mostly on the ability of the user, a different algorithm able to 'help' the Gauss method, by weighting and selecting more carefully the initial observations to be used, has been implemented. In this way the output of the Gauss method may be more robust and good enough to allow the LLS technique to converge.

The new weighting algorithm identifies and weights all triplets matching the following criteria:

- the time difference between two consecutive observations must be at least 0.03 days;
- the time difference between two consecutive observations must be maximum 150 days;
- the weight (computed considering an ideal time difference between two consecutive observations equal to 20 days) assigned to a triplet must be lower than the weight assigned to the previously tested triplets;
- the statistical RMS assigned to the observatory related to each observation must be lower than 4.0 arcsec.

As many threads as possible are used to evaluate the triplets. The search is stopped once the specified minimum RMS is reached. If not reached, the triplet with the lowest RMS is used. Then the preliminary solution is improved by mean of a LLS technique, which eventually considers also radiometric observations.

Since it is not possible to guarantee the LLS convergence, an outer loop, able to restart the whole orbit determination process every time it fails, has been implemented. If the LLS technique does not converge, the Gauss method is rerun and a new preliminary solution is found (without using observations used for previous preliminary solutions). After 30 unsuccessful attempts the software stops, saying that the maximum number of iterations has been reached without obtaining convergence. This can happen if observations are not good enough, or if an abrupt event (e.g. close approach) modified the orbit within the observational arc.

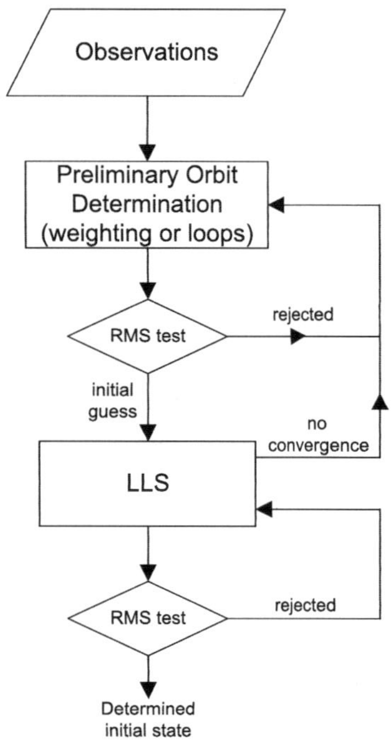

Figure 1. Orbit determination algorithm diagram

If requested, the initial states of a set of VAs are computed along the Line Of Variation (LOV). Finally, the determined initial state of the NEO is used in order to analytically compute its MOID by using the Sitarski method. The vector, identifying the LOV, can be computed from the covariance matrix, corresponding to the LLS solution, by using the first or the second eigenvalue (depending on the user settings). Each VA is computed from the previous one by moving along the LOV. An iterative technique is used to assure that the new VA belongs to the LOV, whose direction vector is indeed recomputed constantly to improve the convergence.

In this new version of the tool, it is possible also to perform orbit determination and trajectory propagation of comets. For this purpose solar radiation pressure and outgassing perturbation have been added and taken into account in the orbital propagators (analytical and numerical).

3. NUMERICAL MODULE

3.1 Propagator

The Numerical Module performs numerical integrations. When the MOID computed by the Analytical Module is small enough to be potentially dangerous for the Earth (e.g. smaller than 0.005 AU), a numerical integration is recommended.

This module can be run for a 'fast analysis', which uses a reduced set of models (and already pre-customized), and for a 'complete analysis', which allows the user to further customize the propagator using more complex models. For both analyses the user can select one of the following integrators: Runge-Kutta 4, Runge-Kutta 4 adapted, Runge-Kutta 45, Runge-Kutta 853, Dormand-Prince 8, Gauss-Jackson 8, Gauss-Jackson 8 adapted and Gauss-Jackson 8 self-adapted. The term "adapted" has been used to distinguish the original integrator scheme from a new algorithm implemented in NEOProp to improve the performance. The "adapted" integrators are fixed step-size integrators which are able to reduce the step-size whenever a certain condition applies: the NEO exits the sphere of influence of the Sun (adapted) or the ratio between the acceleration due to perturbations and the total acceleration is larger than 0.0001 (self-adapted).

The available perturbation models for each type of analysis are listed in Table 1.

Table 1. Available perturbation models

"Fast" Analysis	"Complete" Analysis
Third-Body (planets)	Third-Body (planets + 4 main asteroids)
-	Relativistic Effects
Solar Radiation Pressure	Solar Radiation Pressure
-	Yarkovsky
Earth Spherical Harmonics	Earth Spherical Harmonics
-	*Poynting Robertson drag
-	*Sun J2
*Outgassing	*Outgassing

*new in NEOProp2

3.2 Risk Assessment

Orbit determination algorithms are limited by finite precision in positional measurements and finite lengths of measured orbital arcs, there are natural uncertainties associated with close approach predictions. When a NEO is recognized to have a future close approach to Earth and has an orbital position uncertainty that intersects the known location of the Earth on a specific date, a collision probability can be calculated.

An object's orbit is never known exactly because the orbit calculation is based on measurements which always contain small errors. The level of uncertainty in an orbit is quantified by the orbital covariance matrix, which depends on several factors:

- time span of the observational arc;
- number, type and accuracy of the observations;
- object's distance from the Earth when observed.

In the prediction of close approaches of NEOs it is important to consider their uncertainties, which can be quite large. In particular, the impact probability P is a function of the position and velocity uncertainties during the encounter.

A straightforward method for computing the impact probability is simply to count the number of virtual asteroid with a minimum close approach distance within one planetary radius. This is computationally expensive because many samples are required in order to identify very low impact probabilities: if the impact probability is of the order of 1E-x, more than 1E+x VA are required to find a virtual Impactor (1 million VA to assess 1E-6 impact probability).

To reduce the necessary number of simulations, a statistical approximation can be adopted. If the PDF is Gaussian and if the expression is converted into spherical coordinates, the following relation provides the impact probability P [2]:

$$P = \int_0^{r_E} PDF(r)dr = CDF(r_E) - CDF(0) = \frac{\left[erf\left(\frac{r_E - \bar{d}}{\sigma\sqrt{2}}\right) - erf\left(\frac{-d}{\sigma\sqrt{2}}\right)\right]}{2} \quad (1)$$

where CDF denotes the cumulative density function, d is the mean distance from the Earth, σ is the standard deviation of the perigee distances from the Earth, r is the heliocentric position vector and r_E is the radius of the Earth.

Once the impact probability has been computed, the typical hazard scales used by the scientific community can be used to assess the impact risk:

- Torino scale, which is a 0-10 scale that takes into consideration multiple aspects as the impact probability and the kinetic energy of the object, the potential consequences should an impact occur, the time horizon for forecasting all potential impacts, and conveyance as to whether the specific threat posed is significantly higher or lower than that posed by the multitude of similar-sized objects that remain undiscovered [3]2. Unfortunately it tends to deemphasize the threat posed by very large impacts, while it raises smaller impacts to a level of concern that is perhaps more than necessary based on the potential threat alone.
- Palermo scale, which tries to correct some limitations of the Torino scale, taking into consideration the time remaining until the foreseen impact, which is also a way to express how pressing and sure is a certain event [4].

Another important new feature of the Numerical Module is the backwards propagation. In the previous version of the tool only forwards propagation was supported. This new functionality could be used for several purposes as recovery of lost objects, identification of NEOs, etc.

4. GRAPHICAL USER INTERFACE

The GUI developed for NEOProp2 does not add any new functionality to the software, but it tries to simplify and improve the usability of the tool. In this chapter its main features are presented.

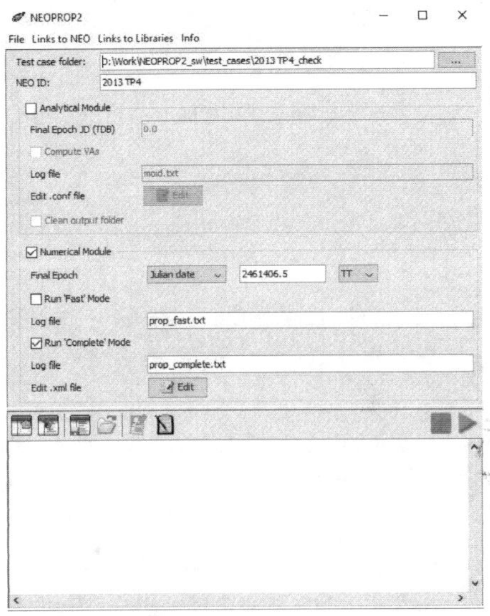

Figure 2. NEOProp2 GUI screenshot

The 'File' menu allows the user to create a new scenario, load an existing one, reload, save or create a backup copy of the current scenario.

The creation and execution of a new scenario are always delicate tasks if performed with software which does not provide a GUI. Especially for new users, retrieving the correct list of input files and data and understanding how to correctly setup the tool can be quite challenging. For this purpose this GUI provides a dedicated panel (see Figure 3) where a new scenario can be created, allowing the user to collect with few clicks all the inputs required to set up a new test case. All the requested input files are automatically created. Not all the input settings stored in the configuration files can be modified from the GUI, but the most common applications should be able to run successfully with the default values.

Once the 'Create' button in the 'New Scenario' dialog panel is pressed, a new scenario is created: a new folder with the provided name is generated and placed at the specified location. If the 'Download observations from MPC' checkbox is selected, the tool tries to download the observation file from the MPC website, using the provided identifier, which therefore needs to be consistent with the one used by the MPC website. The diameter of the NEO can be also retrieved from internet and used to estimate its mass. Alternatively, the user can provide numerical values.

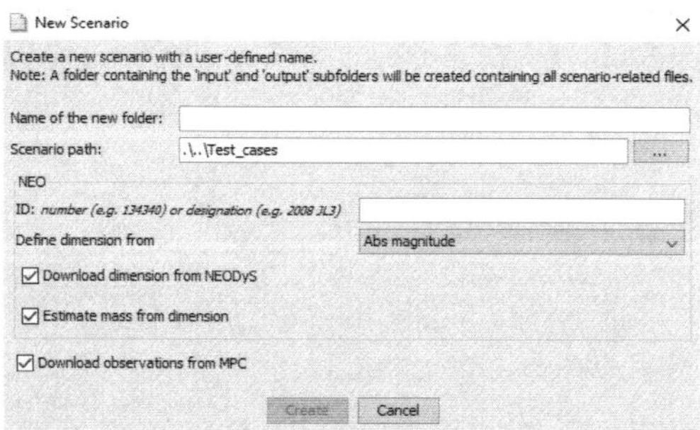

Figure 3. GUI dialog panel for the creation of a new scenario

Once a scenario has been created or loaded, the user can access some useful internet links to the MPC, JPL, ESA SSA, NEODyS and Sentry pages through the 'Links to NEO' menu. Moreover, links for the selected NEO to the NEODyS and Sentry pages are available. If the NEO is inserted in the risk list, links to the NEODyS and Sentry risk pages are accessible too, instead of being greyed out. Some useful internet links are provided also in the 'Links to Libraries' menu. From there, the user can open internet pages where most of the NEOProp2 libraries are taken from.

The configuration panel allows the user to modify some of the settings of the current scenario. In this panel the user can select also which modules have to be run.

The toolbar allows the user to run (and stop) any NEOProp2 module, to create a scenario summary or results comparison and also to save or load the content of the log pane, which visualizes in real time the output stream of a running executable. In this way the user is informed about the current status of a certain action/analysis. The log pane is also used to visualize (in html format) the scenario report and results comparison.

Users can generate a scenario summary and/or compare the results obtained by NEOProp2 with the values reported by NEODyS and Sentry. In both cases results are separated in three sections, one per task (Analytical Module, Numerical Module fast and complete).

The scenario summary functionality collects all the information from the output files stored in the output folder and writes them into the log pane. In this way the user gets a rapid overview of the results without the need of accessing manually all the files. Orbital elements, MOID, close approaches, Earth virtual impactors and risk figures are reported.

5. RESULTS

In order to verify the software requirements and to validate the tool, the original functionalities implemented in NEOProp had to be retested. Therefore, all the test cases already used to test NEOProp (and presented in [1]) have been re-run and the results successfully crosschecked. Additionally, new test cases were defined in order to verify whether the new functionalities introduced in NEOProp2 work correctly. Only these new test cases are presented in this chapter.

5.1 Analytical Module Improved Convergence

In the previous version of the tool, sometimes, especially when dealing with long observational arcs, the set of observations had to be split in order to make the LLS converge. The new algorithms implemented in NEOProp2 should be able to overcome this issue. To verify the new implementation, all the NEOs, which required splitting the observational arc [1], were again analysed and tested with the whole observational arc at once. The identified NEOs are: Apophis, 2004DC and 2012DA14. This last NEO is of particular interest since it required a three-step orbital fit, due to a close approach that this NEO had with the Earth. NEOProp2 successfully managed to perform orbit determination in a unique shot and the obtained orbital elements were in line with the reference (NEODyS) and reported below.

Table 2. NEODyS-NEOProp2 results comparison (99942, 2009FJ, 2011AG5)

	99942	2009FJ	2011AG5
Epoch [JD]	2456400.5	2456200.5	2456400.5
Δ-a [AU]	3.5950E-09	-0.00046	1.19E-08
%-a	0.000%	-0.021%	0.000%
Δ-e [-]	-2.8690E-09	-9.2E-05	2.54E-08
%-e	0.000%	-0.016%	0.000%
Δ-i [°]	1.7410E-08	-0.00182	0.001224
%-i	0.000%	-0.206%	0.033%
Δ-Ω [°]	-1.9900E-07	-0.01126	0.018876
%-Ω	0.000%	-0.003%	0.014%
Δ-ω [°]	-4.2500E-07	0.011993	-0.01893
%-ω	0.000%	0.008%	-0.035%
Δ-M [°]	6.2300E-07	0.123461	2.1E-06
%-M	0.000%	0.339%	0.000%
Δ-MOID [AU]	0.000%	3.8E-07	-3.8E-06
%-MOID	0.000%	0.030%	-1.042%

5.2 Comet Orbit Determination and Propagation

NEOProp2 is now able to identify and propagate orbits of comets, while NEOProp could analyse only NEOs. In order to validate this functionality, the new algorithms for orbit determination and trajectory propagation were tested with two comets: 319P/Catalina-McNaught and 67P/Churyumov-Gerasimenko. These comets have been selected because they provide not only the possibility to test the orbit determination module, but also the numerical one, since they both present close approaches to at least one of the planets. Their reference values have been taken from the Sentry system since NEODyS does not consider comets at the moment.

Comet 67P is of particular interest since it is the comet targeted by the Rosetta mission. It also gives the possibility to test the implementation of the outgassing perturbation, since the required input coefficients for the outgassing model are available for this comet and have been published by Sentry.

The analytical and numerical modules have been both run, once including the outgassing perturbation and once without it. The orbit determination produced almost the same values in both cases. Therefore, only the results obtained considering the outgassing perturbation are presented in Table 3. The determined orbital elements are very close to the reference ones, except for the mean anomaly, since the reference orbital elements were available at a different epoch (7 years later) from the one computed by NEOProp2. Nevertheless, the computed analytical MOID is almost identical to its reference value.

Table 3. NEOProp2 results and comparison with Sentry for comet 67P

	NEOProp2	Difference
Epoch [JD]	2454642.574744	-2619.92526
Δ-a [AU]	3.466402	0.13%
Δ-e [-]	0.640288	-0.09%
Δ-i [°]	7.041595	0.02%
Δ-Ω [°]	50.210012	0.15%
Δ-ω [°]	12.652102	-1.13%
Δ-M [°]	322.087716	-1761.42%
Δ-MOID [AU]	0.2610731909	1.11%

On the other hand, the effect of the outgassing perturbation plays a big role in the long term trajectory propagation. In the next table the close approach to Mars, that will take place in 2067, has been compared once with the results obtained without considering the outgassing perturbation (first row) and once with the results computed including that effect. From the numerical values, the benefit due to the use of the outgassing force is obvious.

Table 4. 67P Close approach comparison with Sentry with and without considering outgassing

#	Δ-Date	Distance %
1 (no outgassing)	-0.71693	-6.752%
1	0.03307	-0.001%

The 319P comet could not be tested with the outgassing perturbation, since it was not possible to find the numerical values for its coefficients. Nevertheless, the computed orbital elements and MOID are very close to the reference values (considering again the different reference epoch).

Table 5. NEOProp2 results and comparison with Sentry for comet 319P

	NEOProp2	Difference
Epoch [JD]	3.568053	-0.10%
Δ-a [AU]	0.666418	0.18%
Δ-e [-]	15.101114	0.15%
Δ-i [°]	111.386396	0.01%
Δ-Ω [°]	203.635114	-0.04%
Δ-ω [°]	358.049192	77.39%
Δ-M [°]	2454727.989414	1394.51059 [d]
Δ-MOID [AU]	0.1943621846	-2.40%

This comet is also very interesting due to the two close approaches to the Earth. Both of them have been successfully identified by NEOProp2.

Table 6. 319P Close approach comparison with Sentry

#	Δ-Date	Distance %
1	-0.01059	-0.021%
2	0.03108	0.09%

5.3 NEOs Risk Assessment

The last test was intended to validate the accuracy of the new impact risk assessment computation. NEOProp2 is now able to estimate the impact probability, which can be used then to compute the Torino and Palermo scale values. Three NEOs (2007 SN6, 2012 VE77 and 2013 TP4) having an impact probability non-zero were fully analyzed and their observations used to feed the orbit determination module. The obtained nominal initial states and VAs were used to estimate the impact probability and assess the impact risk. For each NEO, orbital elements, analytical MOID, virtual Earth impactors and Palermo scale values were compared. Torino scale values could not be compared since no NEO having a value above 0 has been discovered yet. Table 7 shows that the orbital elements found by NEOProp2 are perfectly in line with the NEODyS and Sentry values. MOIDs computed by NEOProp2 are very close to the values provided by Sentry, which quite differ from those provided by NEODyS.

Table 7. NEO orbits and MOIDs comparison between NEOProp2 and NEODyS-Sentry

	2007 SN6 NEODyS	2007 SN6 Sentry	2012 VE77 NEODyS	2012 VE77 Sentry	2013 TP4 NEODyS	2013 TP4 Sentry
Epoch [JD]	0.35	-0.07	-0.11	-0.13	-2.10	-4.56
Δ-a [AU]	0.00%	0.00%	-0.03%	0.04%	0.09%	0.02%
Δ-e [-]	0.00%	0.00%	-0.03%	0.04%	0.04%	0.00%
Δ-i [°]	0.00%	0.02%	-0.02%	0.00%	0.03%	-0.02%
Δ-Ω [°]	0.00%	0.00%	-0.02%	0.00%	0.04%	0.00%
Δ-ω [°]	0.02%	0.00%	-0.06%	-0.01%	0.01%	0.00%
Δ-M [°]	0.03%	-0.16%	-0.06%	-0.01%	-0.29%	-8.58%
Δ-MOID [AU]	94.598%	0.00%	28.915%	0.06%	-12.143%	0.275%

In Table 8 the Earth virtual impactors reported by NEODyS and Sentry are compared with those computed by NEOProp2. The comparison shows that NEOProp2 results are closer to Sentry's values for the 2007 SN6 NEO, closer to NEODyS for the 2012 VE77 asteroid and close to both of them (since they are in agreement) for the last asteroid. This result shows that NEOProp2 is independent from both these systems, which was one of the fundamental software requirements in order to make of NEOProp2 an independent benchmarking tool for these existing systems.

The computed impact probabilities are not always very close to the reference values, but as mentioned in [5] differences up to one order of magnitude are considered fully acceptable.

Table 8. Earth virtual impactors' comparison between NEOProp2 and NEODyS-Sentry

NEO	NEODyS			Sentry		
	Δ-Date [d]	Δ- Impact prob. [-]	Δ-Palermo [-]	Δ-Date [d]	Δ- Impact prob. [-]	Δ-Palermo [-]
2007 SN6	0.00715	5.376E- 06	1.552	-	-	-
	-	-	-	-	-	-
2012 VE77	0.03477	-2.9E-05	-0.53	0.03546	-7.03E- 05	-0.9
	-	-	-	0.03378	7.9E-06	0.684
2013 TP4	0.03089	1.53E-05	0.202	0.03089	2.58E-05	0.312

6. CONCLUSIONS

NEOProp2 is not only an upgraded and overall improved version of NEOProp, but it implements many new features which have widened the capabilities of the tool.

The software is now able to perform orbit determination not only of NEOs, but also of any other celestial body, including comets, due to newly implemented perturbation models (e.g. outgassing force and Poynting-Robertson drag). The determination of virtual asteroids has been greatly improved in terms of accuracy and performance. Multiple definitions of LOV are now supported. Meanwhile, the convergence of the orbit determination algorithm has been improved and an outer loop has been added in order to increase the robustness of the Analytical Module.

The Numerical Module also supports the propagation of comets. Moreover, the computational time and the consumption of memory have been reduced. The tool can now assess the impact risks of any object through the computation of important figures as impact probability, Torino and Palermo scale values. Amongst the new features, the backward propagation allows now to use the software not only for risk assessment, but also for other important tasks as object recovery, identification and definition of follow-up observations campaigns.

No big limitation/problems have been identified during the tests, which were also presented in this paper, and the software seems to be flexible (able to tackle objects of various types and with very different orbital elements) and good performing. So far only a system engineering validation and verification could be performed, an extended testing campaign is recommended to really understand potentialities and limitations of the tool.

NEOProp has been thoroughly reviewed and improved and the new NEOProp2 has been shown/proved to be both more robust and faster than its predecessor. The newly developed GUI facilitates the user experience, allowing to easily create and run a new scenario and to compare the results with the other available prediction tools.

7. REFERENCES

1. Zuccarelli, V., Bancelin, D., Weikert, S., Thuillot, W., Hestroffer, D., Valles, C. Y., Koschny, D.,(2013). NEOPROP: a NEO Propagator for Space Situational Awareness, 6th IAASS

2. Pitz, A., Teubert, C., Wie, B., (2011). Earth-Impact Computation of Disrupted Asteroid Fragments using GMAT/STK/CODES, AAS 11-408

3. Binzel, R. P., (2000). The Torino impact hazard scale. Planet. Space Sci. 48, 297–303

4. Chodas, P. W., Yeomans, D. K., (1999). Predicting Close Approaches and Estimating Impact Probabilities for Near-Earth Object, JPL, AAS/AIAA Astrodynamics Specialist Conference

5. Milani, A., (2014). Near Earth Objects: Risk Evaluation and Mitigation Planning, Final SPARC Workshop Space Threats and Critical Infrastructures: Risks and Countermeasures, ASI Rome

GLOBAL COSMIC RISK ASSESSMENT STUDY (COSRAS) BY THE IAASS

Joseph N. Pelton

Executive Committee and Chair Academic Committee, IAASS
joepelton@verizon.net

ABSTRACT

There is currently an effort in the U.S. to create a systematic review of various types of cosmic hazards — natural and manmade — that could create significant harm to vital infrastructure on which the U.S. citizenry now depends for commerce and daily life. This paper proposes that a systematic global assessment of critical infrastructure that is subject to cosmic hazards also be undertaken by the International Association for the Advancement of Space Safety in cooperation with the U.N. Committee on the Peaceful Uses of Outer Space and other potential partners such as the International Academy of Astronautics, COSPAR, and the International Space University.

The purpose of this study would be not only to identify the nature of these various global risks to vital infrastructure by cosmic hazards, but also to identify measures that might be undertaken to lessen these vulnerabilities. The cosmic hazards to be assessed would include coronal mass ejections, solar flares, changes to the earth's protective magnetosphere and Van Allen Belts, asteroid strikes, electromagnetic pulses, orbital debris, as well as other hazards that the study team might identify.

1. INTRODUCTION

There is little general public awareness that we all inhabit a common spaceship that is hurtling through space at over 100,000 kilometer per hour. Further there is little comprehension that life could not exist on planet Earth were it not for a protective but thin atmosphere (proportionately less than the skin of an apple) plus a global magnetosphere that shields us from solar storms. These key protective systems, however, and particularly the geomagnetosphere and the Ozone layer that protect us from solar storms, shift in their protective shielding capabilities over time.

In the United States there are current efforts to enact a "Critical Infrastructure Protection Act" that would assess dangers to U.S. critical infrastructure that might come from solar storms, electromagnetic pulses (EMPs) or other natural or man-made space events that might threaten U.S. electrical grids, computer and telecommunications networks, key satellite networks, etc.

This paper sets forth a process whereby a global critical infrastructure protection study might be undertaken to systematically identify the nature and the possible severity of these hazards and in the second part of the study to characterize preventive or partially protective systems that could be employed to minimize these risks.

2. REVIEW OF THE VARIOUS TYPES OF COSMIC HAZARDS

The most difficult aspect of this ambitious global cosmic risk assessment study (COSRAS) is that there are so many different types of potential risks to be considered and that the scientific and engineering expertise needed for such a study are quire varied. Thus as a first step it might be most useful to identify the various hazards and then form teams suited to consider each of these separately before seeking to integrate the findings into a single unified report. The following sections thus sets forth the various risks that might be considered in a COSRAS global effort.

3. CORONAL MASS EJECTIONS

The most destructive solar events in terms of power and information networks as well as possible damage to satellites are coronal mass ejections. The Carington event of 1859 is the most well-known, but the Montreal event of 1989 and the Halloween solar storms in Scandinavia of 2003 are powerful reminders that not only are these event destructive, but their occurrence are with considerable frequency. The study would provide the latest information about the nature of CMEs, but what might be done to protect power grids, pipeline systems, Supervisory Control and Data Acquisition (SCADA), and vital satellite systems against such solar events. This study would be also give consideration to the earth's weakening geomagnetosphere, the movement of the magnetic poles, and the reduced protective capability of the VanAllen Belts.[1]

4. SOLAR FLARE

The radiation from solar flares, especially during solar max, has the potential to do damage to satellites, electronic systems, and information networks that are vital to transportation systems, global commerce and safety, police, fire, rescue and defense related systems, as well as global commerce in general. The more intense flares can also give rise to genetic mutations to all life forms and also lead to elevated risks of cancer. Space missions such as the NASA Solar Dynamic Observatory (SDO) provides useful information on solar flares. Although solar flares and CMEs are closely related they involve different types of physical events (powerful radiation vs. ionic bombardment). The studies on CME, Solar radiation flares and changes to the geomagenetosphere are closely related and should be carried out in tandem.

5. CHANGING PROTECTIVE GEOMAGNETOSPHERE

Space probes by NASA and ESA in recent years have noted that the Earth's protective geomagnetosphere that helps to prevent the worst damages that can come from solar storms is weakening. ESA's Project Swarm[2] and NASA's Magnetospheric Multi-Scale (MMS) Mission are designed to provide more information about the changes that are occurring as the Earth's magnetic poles shift. These probes are also designed to to solve the mystery of how magnetic fields around Earth connect and disconnect and in the process explosively release energy. This process is known as magnetic reconnection.[3] The fact that the Earth's protective magnetic field has weakened and that the magnetic poles have shifted both downward from the North and upward from the South are of serious concern. The reduced protection from solar radiation and ion particles from the Sun's coronal mass ejections seems to indicate that the world's electrical grids, pipelines, satellites and communications networks are now at increasing risk. Even more importantly these risks seem likely to increase significantly in the decades immediately ahead. Understanding these changes in the Earth's geomagnetosphere is one of the key objectives of this COSRAS effort on a global scale.

6. ASTEROID IMPACT

The NAVSTAR Global Positioning Satellite System include sensors that are designed to detect nuclear explosions, but they can also detect asteroid impact with Earth. This monitoring system has indicated that meteors, bolides and asteroids impact earth with four to ten times greater frequency that was previously thought to be the case. In short there is increasing data that show that asteroid threats are greater than previously thought.

Programs such as NASA's NEOWise Infradred telescope, ground observatory observations and proposed new infrared observatories like NASA's NEOCAM and the B612 Foundation's Sentinel satellite can help identify potentially threatening asteroid impacts with earth. The current NASA guidelines of location asteroids that are 140 meters in diameter as set by the U.S. Congress has not yet been achieved. The fact that 30 meter asteroids traveling at sufficient velocity could be a city-killer and that there are perhaps 500K to a million of these size potentially hazardous asteroids (PHAs) represents a very large risk element. Clearly greater efforts to locate PHAs and dangerous comets and more consideration of planetary defense systems are needed. The UN authorized "NEO WARN" and the "Space Mission Planning Advisory Group" are significant steps forward but much more needs to be done.

7. ORBITAL DEBRIS

The increasing amount of orbital space debris is yet another concern that is being addressed by the Inter-Agency space Debris Committee (IADC) and the UN COPUOS Working Group on the Long Term Sustainability of Outer Space (as well as solar storms)

The projection by former NASA Scientist Don Kessler, who originally warned of the possibility of orbital debris cascading out of control seems to be an increasingly valid space threat concern. His current projection is that with the current environment that there will be a major on-orbit collision every ten years. This projection was made in advance of currently announced plans for new large-scale constellations by One Web, Planet Labs and others such as Space X. Clearly the risk of increasingly orbital debris is one of the key elements of this global COSRAS effort.

8. CONCLUSIONS

The above areas of potential cosmic risks are briefly summarized. The first step is to develop a clear statement of specific areas of cosmic risk to planet Earth and its flora and fauna in coming years. An international committee should be formed, perhaps initially within the IAASS, but also with other international partners that would agree to cooperate in this coordinated multi-year effort. The next step would be to form a coalition of organizations that would undertake this Cosmic Risk Assessment Study (COSRAS) with assigned responsibility for each one of the identified cosmic risks in step one of this effort. A coordinated plan would then need to be developed to undertake detailed studies in each of these areas in terms of not only of the projected level of risk, but also to devise concepts for protective or ameliorative programs that can lessen the level of these risks to infrastructure and to human, animal and vegetative life forms. It is proposed that the Executive Committee of the IAASS add this topic to its agenda for specific discussion as to whether the IAASS in cooperation with other international agencies could agree to formally undertake the COSRAS effort as outlined above.

9. REFERENCES

1. Joseph N. Pelton, Indu Singh and Elena Sitnikova A Global "Pearl Harbor Disaster" Precipitated by Extreme Solar Events: Just How Vulnerable Are Our Worldwide Cyber Networks and Electric Power Grids? Journal of Homeland Security, June, 2015.

2. ESA's Magnetic Field Mission "SWARM", http://esamultimedia.esa.int/multimedia/publications/BR-302/

3. "heliophysics" http://science.nasa.gov/heliophysics/ Accessed on November 1, 2015

PROCEEDINGS OF THE 8TH INTERNATIONAL SPACE SAFETY CONFERENCE

SESSION 32 HUMAN PERFORMANCE - II

HUMAN FACTORS CHECKLIST: THINK HUMAN FACTORS – FOCUS ON THE PEOPLE

Darcy Miller(1), Katrine Stelges(2), Timothy Barth(3), Damon Stambolian(4), Gena Henderson(5), Charles Dischinger(6), Barbara Kanki(7), Ian Kramer(8)

(1) *National Aeronautics and Space Administration, Kennedy Space Center, SA-E2, Kennedy Space Center, FL, USA 32899, Darcy.H.Miller@nasa.gov*
(2) *Aerodyne, Jacobs TOSC-025, Kennedy Space Center, FL, USA 32899, Katrine.S.Stelges-1@nasa.gov*
(3) *National Aeronautics and Space Administration, NESC-2, Kennedy Space Center, FL, USA 32899, Timothy.S.Barth@nasa.gov*
(4) *National Aeronautics and Space Administration, Kennedy Space Center, NE-T1, Kennedy Space Center, FL, USA 32899, Damon.B.Stambolian@nasa.gov*
(5) *National Aeronautics and Space Administration, Kennedy Space Center, NE-EM, Kennedy Space Center, FL, USA 32899, Gena.M.Henderson@nasa.gov*
(6) *National Aeronautics and Space Administration, Marshall Space Flight Center, EV-74, Huntsville, AL, USA 32899, Charles.Dischinger@nasa.gov*
(7) *National Aeronautics and Space Administration, Ames Research Center, Rtd., TH: 262-4, Moffett Field, CA, USA 94035, Barbara.Kanki@nasa.com*
(8) *Jacobs TOSC-025, Kennedy Space Center, FL, USA 32899, Ian.A.Kramer@nasa.gov*

ABSTRACT

A quick-look Human Factors (HF) Checklist condenses industry and NASA Agency standards consisting of thousands of requirements into 14 main categories.

With support from contractor HF and Safety Practitioners, NASA developed a means to share key HF messages with Design, Engineering, Safety, Project Management, and others. It is often difficult to complete timely assessments due to the large volume of HF information. The HF Checklist evolved over time into a simple way to consider the most important concepts. A wide audience can apply the checklist early in design or through planning phases, even before hardware or processes are finalized or implemented. The checklist is a good place to start to supplement formal HF evaluation.

The HF Checklist was based on many Space Shuttle processing experiences and lessons learned. It is now being applied to ground processing of new space vehicles and adjusted for new facilities and systems.

1. INTRODUCTION

If human factors (HF) concepts are not applied, it is often due to the complexity of understanding the large volume of requirements and information available. The sheer quantity of information can make it difficult to know where to begin. Also, with limited resources, it is often difficult to complete timely HF assessments.

To bridge the gap, a simple 14-item HF Checklist was developed to share key HF messages throughout the Kennedy Space Center (KSC) and other NASA Centers. Originators included NASA civil servant and contractor HF Practitioners with a variety of HF training, operations experience, and work experience on different programs at various Centers.

The NASA Constellation Program Ground Operations Project developed a set of HF requirements to cover the key areas of launch processing concerns. These were high-level requirements, such as reach, envelope volume, visual access, damage prevention, lifting, and tool clearances. In 2007, a pathfinder project focused on similar HF issues related to design of ground systems and developed the initial HF checklist as a tool to help engineers identify HF areas of concern. The project applied the checklist to key tasks that would involve multiple personnel or place individuals or teams in complex configurations. [1] A 2014 pathfinder applied an updated checklist to identify and categorize HF aspects with a subset of tasks planned for the Space Launch System (SLS); entailing a multi-purpose crew vehicle atop a core stage rocket and boosters.

The HF Checklist evolved after the pathfinder projects to identify HF risks in new program designs for flight and ground systems. It was revised over time and developed into a flyer for general use within the NASA community. See Fig. 1 and 2.

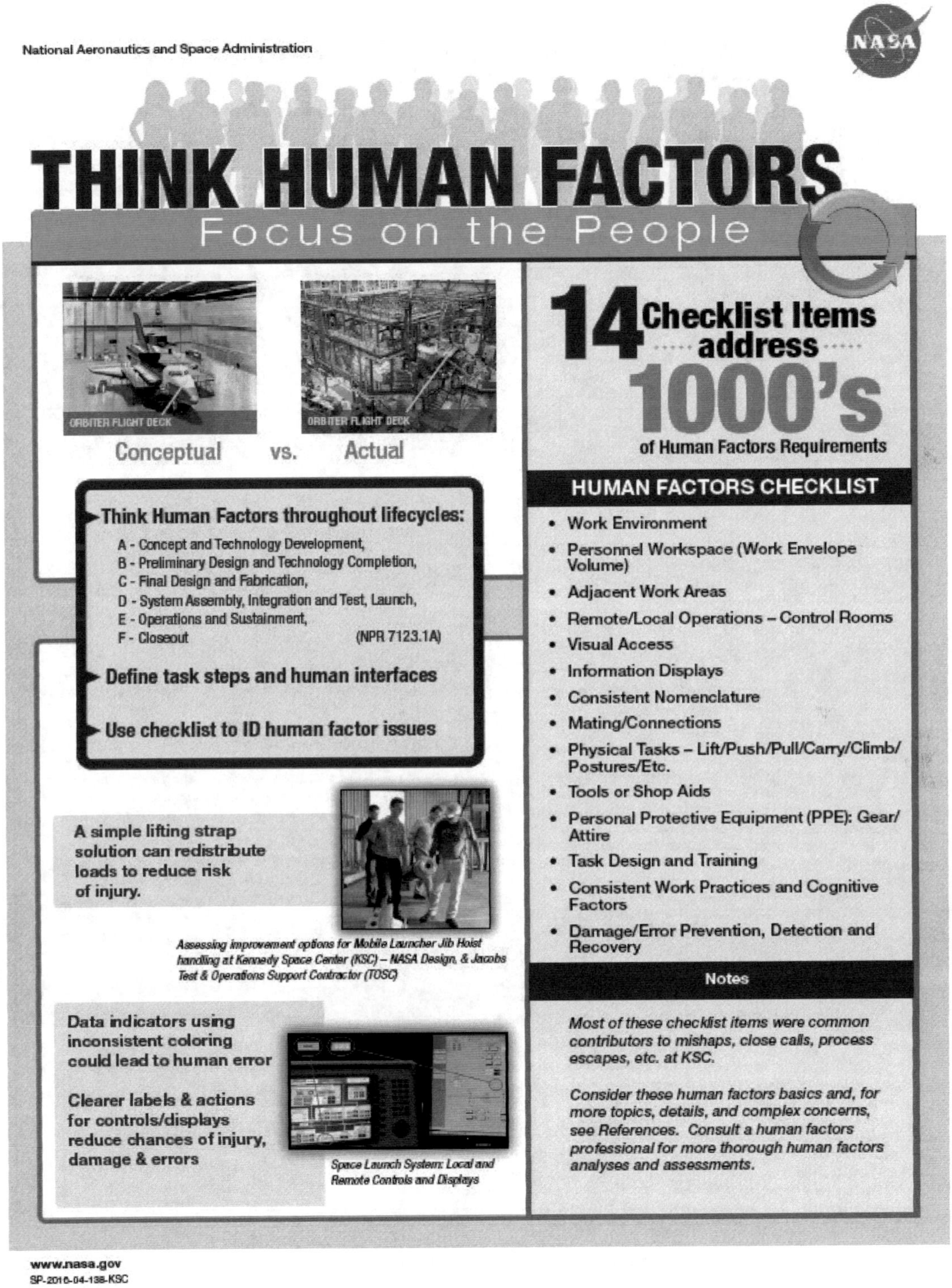

Figure 1: Kennedy Space Center Human Factors Checklist, Side 1

HUMAN FACTORS CHECKLIST

Think about these 14 checklist items

TO IMPROVE YOUR DESIGNS, PLANNING, OPERATIONS AND MORE:

1. **Work Environment** - ACCOUNT FOR WORK LOCATION VARIABLES. Consider: □ Lighting □ Noise □ Temperature □ Vibrations □ Distractions □ Material handling □ Weather □ Wildlife □ Communications □ Travel paths □ Work surfaces □ Elevations □ Portable and fixed items. Mitigate with: □ PPE □ Scheduling □ Training □ Controls □ Barriers.

2. **Personnel Workspace (Work Envelope Volume)** - PROVIDE CLEAR AND ADEQUATE PHYSICAL ACCESS TO PERFORM INTENDED WORK. Consider space to: □ Operate and stage hardware and tools □ Interface with equipment or consoles □ Work adjacent to facility, hardware, or other personnel □ Work individually or in teams.

3. **Adjacent Work Areas** - SEPARATE OR ACCOUNT FOR ACTIVITIES THAT COULD CONFLICT. □ Plan for clear communications □ Provide communication devices □ Prevent conflicting functions (sanding near painting; electrical near plumbing; sensitive work near vibrations, reaching across workstations.) □ Mitigate items crossing work areas (hoses, lines, etc.).

4. **Remote/Local Operations** - Control Rooms - ALIGN DATA AND COMMUNICATIONS AMONG TASK SITES AND CONTROL AREAS. □ Provide console or control room team with data and or camera coverage of task site team □ Supply communication and safety methods for hands-on team.

5. **Visual Access** - PROVIDE DIRECT VISIBILITY TO WORK BEING PERFORMED. Provide: □ Task-specific lighting □ Glare mitigation □ Direct lines of sight □ else, provide displays for viewing □ Avoid visual or physical obstacles □ Avoid positioning items beneath or behind others □ Avoid requiring mirrors to view items.

6. **Information Displays** - PROVIDE CLEAR, LEGIBLE, FUNCTIONAL IDENTIFICATIONS. □ Account for visual variables (impairments, color-blindness, PPE, etc.) □ Employ various indicators (i.e., audio, visual, tactile, movement, color, dimension, size) □ Offer mixture of indicators for safety or time-sensitive systems (i.e., text or alarms accompanying color-coded indicators; categorized maps for inspection coverage, 3D views with 2D drawings).

7. **Consistent Nomenclature** - STANDARDIZE NOMENCLATURE, FORMATS AMONG LOCATIONS, FUNCTIONS, AND COMPONENTS. Maintain consistency in: □ Terminology □ Interface labels □ Signage □ Color-coding □ Types of labels, placards or markings for switches, circuit breakers, panels □ Other human-system interfaces.

8. **Mating/Connections** - PROMOTE STANDARDIZATION, CLEARANCES, CORRECT ALIGNMENTS. □ Provide design, controls, barriers to prevent mis-mating hardware or configurations (i.e., mechanical, electrical, fluid, fasteners, cables, software). Offer clearances for: □ Grasping □ Carrying □ Staging □ Work motions □ Provide checks to match intended configurations, hardware, and documentation.

9. **Physical Tasks** – Lift/Push/Pull/Carry/Climb/Postures/Etc - ACCOUNT FOR PERSONNEL VARIABILITY AND TASK DEMANDS. Consider: □ User populations □ Weight loads □ Workspace sizes □ Postures, □ Travel paths □ Elevations □ PPE □ Other physiology or material handling variables. Provide: □ Handles □ Handholds □ Mobility aids □ Load capacity limits □ PPE □ Padding □ Team communications □ Optimal work heights and reaches. □ Conduct human-system analysis for potential task risks □ Identify areas where loads/grasps are prohibited □ Mitigate multiple physical exertion demands (i.e., twisting while squatting or carrying awkward load while climbing).

10. **Tools or Shop Aids** - ALIGN TOOLS TO THE TASK AND USER POPULATION. □ Limit variety. Consider for standard and any specialized tools/aids: □ Tool or equipment handling □ PPE □ Work environment □ Training □ Other human-system variables.

11. **Personal Protective Equipment (PPE): Gear/Attire** - ACCOUNT FOR GEAR OR ATTIRE THAT MAY ALTER TASK PERFORMANCE. Consider: □ Altered mobility □ Dexterity □ Senses □ Weight □ Size □ Time. Plan for: □ Weather protection □ Clean room attire □ PPE combos □ Material handling □ Travel paths □ Hose management.

12. **Task Design and Training** - CONSIDER USERS AND HUMAN SYSTEM INTEGRATION IN TASK DEVELOPMENT AND EVOLUTIONS. □ Involve workers through concepts, work development, and changes □ Reduce system interface complexities □ Conduct task analysis and human-system-integration reviews □ Conduct on-the job training □ Practice with actual items □ Use common materials or high-fidelity models, simulations, physical mockups □ Document, incorporate lessons learned □ Provide clear, correct, complete work instructions.

13. **Consistent Work Practices and Cognitive Factors** - STANDARDIZE AMONG SIMILAR FUNCTIONS, WORK AREAS. □ Account for psychological, organizational variables □ Use standard conventions □ Inform of irregularities □ Outline work □ Align work with schedules and biological breaks □ Employ situational awareness checks □ Limit multitasking, decision making, complacency □ Provide measurement goals and tolerances □ Avoid mental calculations.

14. **Damage/Error Prevention, Detection and Recovery** - PREVENT HUMAN ERROR AND HUMAN-INDUCED SAFETY OR COLLATERAL DAMAGE OPPORTUNITIES. □ Provide means of detection by inspection or test □ Offer capability to recover □ Define and plan for contingency procedures □ Offer hardware and personnel protection □ Make consequences of errors on safety or system performance clear.

This list is not all inclusive. There may be other areas to consider.

GENERAL REFERENCES

- MIL-STD-1472 Design Criteria Standard Human Engineering
- FAA-HFDS (HF-STD-001) Human Factors Design Standard

Figure 2: Kennedy Space Center Human Factors Checklist, Side 2

2. FOCUS ON THE PEOPLE

All designs and operations can be improved by maintaining a focus on how humans will interface with the entire system or project lifecycle. By defining task steps and human interfaces ahead of time, the HF Checklist can be used to identify HF issues early in the lifecycle, saving time and reducing the cost to make changes.

2.1 System Lifecycles

NASA considers HF throughout system and project lifecycles, including the following phases [2]:

A - Concept and Technology Development
B - Preliminary Design and Technology Completion
C - Final Design and Fabrication
D - System Assembly, Integration and Test, Launch
E - Operations and Sustainment
F - Closeout

2.2 Task Steps and Human Interfaces

For each phase, taking the time to consider the individual task steps and all of the places where people interact with each other or with systems will target areas that may pose HF challenges. As conceptual aspects mature into more detailed renditions, additional structures, features, and human aspects will emerge.

For example, planners conceived that placing a Space Shuttle Orbiter into its horizontal processing hangar would involve only a few floor-based work stands. As processes began, needs arose to add platforms and adjustable components to allow personnel to reach and view each area.

Even with the best planning, events may change in any phase of the lifecycle, resulting in reconsideration of designs, task steps, or human interfaces.

Since predicting future modifications is not always clear in early phases, best practices include designing for flexibility and robustness.

Foreseeing evolutions in designs, maintenance schedules or other components can prevent future human-system-integration risks or delays. For example, initial concepts pointed to leaving Space Shuttle engines installed between flights, but engineering analysis determined that removal was needed more often, changing a task that was expected to occur infrequently into a frequent activity.

2.3 HF Checklist Application

The checklist can aid in addressing HF issues after identifying task steps and human interfaces and prior to finalizing designs or tasks in various applications.

Applications for the checklist can include: Concepts of operations, design reviews, operations planning, task analyses, procedure development, dry runs or prototyping, training, test/checkout, modifications, and communications of HF concerns.

Topics may range from physical operations and maintenance tasks to software and human-computer processes.

2.4 Supplement to HF Expertise

While the checklist is versatile, its intent is not to replace the involvement of HF experts. Many checklist items appear as common contributors to mishaps, close calls, process escapes, or lessons learned. Designers and planners should consider the HF checklist basics and contact HF expertise for more complex concerns.

HF experts recommend that developers and planners refer to the checklist during task analysis activities to involve users and HF expertise in concepts of operations, procedure development, and other task design components. The team can merge subject matter proficiencies to develop improved components and processes for mitigating risks [3].

Figure 3: HF Expertise at Task Analysis Review

3. HF CHECKLIST

The 14 checklist items can be used to address thousands of human factors requirements and best practices. The checklist is not intended to replace the requirements. Nevertheless, it is a good place to start focusing discussions and targeting further analysis. The following HF Checklist item descriptions provide quick overviews and examples. However, they are not all-inclusive.

3.1 Work Environment

Since the work environment can be indoors or outside, ground level or elevated heights, or composed of multiple facets, it becomes one of the first considerations in designing or improving the human-system integration aspects.

Figure 4: Launch Site Work Environment

Account for work location variables, such as lighting, noise, temperature, vibrations, distractions, material handling, weather, wildlife, communications, travel paths, work surfaces, elevations, portable and fixed items. Mitigate with personal protective equipment (PPE), scheduling, training, controls, or barriers.

3.2 Personnel Workspace (Work Envelope Volume)

After addressing the broader aspects of the work environment, the personnel workspace, or envelope, should be the next focus.

Provide clear and adequate physical access to perform intended work including space to operate and stage hardware and tools; interface with equipment or consoles; work adjacent to facility, hardware, or other personnel; and to work individually or in teams. These considerations are also applicable to planning for work in tight quarters to prevent, avoid, or mitigate narrow or limited workspaces.

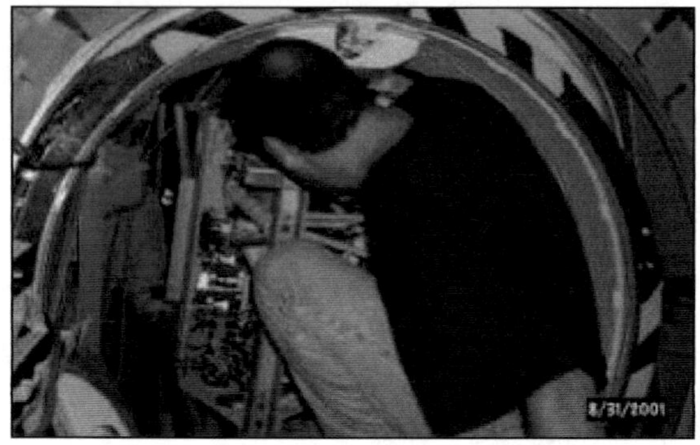

Figure 5: Limited Workspace May Warrant Mobility or Handling Aids

3.3 Adjacent Work Areas

When two or more operations can be performed side-by-side, it is important to look beyond the individual workspaces.

Separate or account for activities that could conflict. Plan for clear communication and provide appropriate communication devices. Prevent conflicting functions (e.g., sanding near painting; electrical near plumbing; sensitive work near vibrations, reaching across workstations). Mitigate items crossing work areas (e.g., hoses, lines).

Figure 6: Communication - Key when Adjacent Work Visibility or Noises are Factors

3.4 Remote/Local Operations – Control Rooms

When operations require teams to be separated due to hazardous or other conditions, the integration between the local hands-on team and the remote operators is essential.

Align data and communications among the task sites and their control areas to promote consistency. Provide console or control room teams with data or camera coverage of task site team. Supply communication and safety methods for hands-on team whether working in different facilities, different rooms in same facility, or in same room or area using verbal or other data commands.

Besides physical tasks or designs, the HF checklist provides benefit to software, programming, and other human-computer interfaces. For example, a visual icon changing to accompany color-coded, categorized text can provide two modalities to aid the decision making process.

Figure 7: Control Room Operations & Communication

3.5 Visual Access

Ensuring visibility to work tasks encompasses several variables.

Provide direct visual access to work tasks. Provide task-specific lighting, glare mitigation, and direct lines of sight. Otherwise, provide displays for viewing remote or hard-to-see operations.

Designs and plans should preclude visual or physical obstacles, positioning items beneath or behind others, or requiring mirrors to view items. Designs or tasks should prevent, avoid, or mitigate limited visibility configurations or conditions.

Together, multiple personnel could provide different vantage points to several Space Shuttle tasks such as lifting valuable flight hardware. At those times, additional communication was key in ensuring information transferred appropriately.

3.6 Information Displays

Viewing various forms of media or indicators can be a key communication aspect.

Provide clear, legible, functional identifications. Account for visual variables such as impairments, color blindness, and PPE. Employ various indicators such as audio, visual, tactile, movement, color, dimension, or size. A mixture of these can further aid safety or time-sensitive systems (e.g., text or alarms accompanying color-coded indicators; categorized maps for inspection coverage, 3D views with 2D drawings).

During Shuttle processing, three similar processing hangars housed the Orbiters between landing and launch operations at KSC. Two facilities were parallel and nearly identical to each other while the third differed by 180 degrees. When Orbiters, work stands, hardware, equipment, and other items were in place, it became challenging to look around and easily determine east, west, north, or south directions as guided on some crane and hoist operations. To improve situational awareness, HF practitioners worked with technical shops to add large directional signage. Located above each Orbiter's payload bay, they clearly identified east, west, as well as forward and aft to note the nose-end and tail-end orientations. To ensure consistent processes, authors updated procedures as well to match the nomenclature of the cost effective information displays. Besides improving operational communications, signage also assisted in potential emergency reporting to better identify incident locations. As noted, information displays can provide multiple benefits. Many HF Checklist components apply.

3.7 Consistent Nomenclature

If confusing or conflicting messages are possible, the outcomes can vary from planned intentions. The goal is for checklist users to ponder and mitigate options to prevent multiple choices or differing results.

Figure 8: Nomenclature and Color-Coding Combined

To avoid this, standardize nomenclature and formats among locations, functions, and components. Ensure consistency in terminology, interface labels and signage, color-coding, and types of labels, placards, or markings for switches, circuit breakers, and panels.

3.8 Mating/Connections

Mating and connector tasks require clear communication and other risk mitigation components. Promote standardization, clearances and correct alignments. Provide design, controls, and barriers to prevent mis-mating hardware or configurations (e.g., mechanical, electrical, fluid, fasteners, cables, software). Offer clearances for grasping, carrying, staging, and work motions, and provide checks to match intended configurations, hardware, and documentation.

Figure 9: Mating – Physical & Visual Access

3.9 Physical Tasks

Physical tasks may include lifting, pushing, pulling, carrying, climbing, postures, and other biomechanical aspects. Challenging or repetitive physical tasks could further compound risks.

Account for personnel variability and task demands. Consider user populations, weight loads, workspace sizes, postures, travel paths, elevations, protective gear, and other physiology or material handling variables. Provide handles, handholds, mobility aids, load capacity limits, PPE, and team communications devices and plans.

Conduct human-system analysis for potential task risks. Identify areas where loads or grasps are prohibited. Mitigate multiple physical exertion demands such as twisting while squatting and carrying awkward loads while climbing.

3.10 Tools or Shop Aids

It seems obvious to choose the right tool or aid for the job at hand, but doing so should encompass several human system integration factors.

Align tools to the task and user population. Limit the variety of tools or aids. For standard and any specialized tools/aids, consider tool or equipment handling, PPE, work environment, training, and other human-system variables, such as frequency, duration, and reach.

With such expansive Space Shuttle surfaces, blends of tools and flexible shop aids brought workers to the hardware (Fig. 10), fitting the job to the person rather than creating individually designed components.

Figure 10: Work Stands with Attachments can Place Personnel/Tools/Equipment at Arm's Reach

3.11 Personal Protective Equipment (PPE) Gear/Attire

Protective gear or attire can add to personnel workspaces and affect task design or planning.

Account for gear or attire that may alter task performance such as altered mobility, dexterity, senses, weight, size, and time. Plan for weather protection, clean room attire, material handling, travel paths, hose management, and PPE combinations, such as communication headsets under helmets.

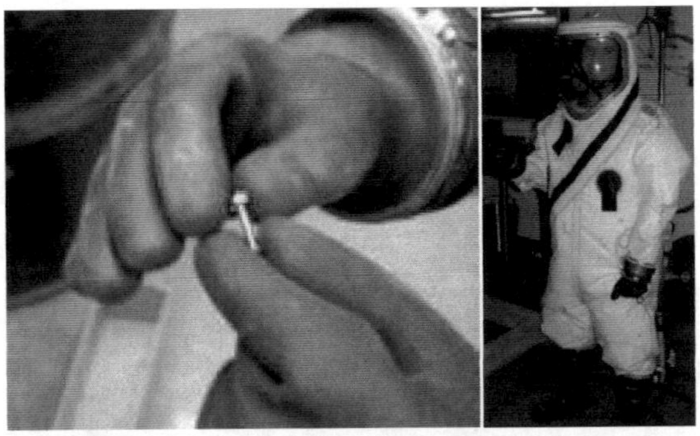

Figure 11: Gloves & Suits Impact Senses/Dexterity

Designers should be encouraged to don PPE to understand first-hand how the protective gear or attire can affect the intended work. Obtaining subject matter expertise from field personnel who have donned or used such PPE can also provide significant lessons learned. Learning the effects of visual perception, mobility, tactile, dexterity, or other challenges in PPE can also be obtained from modelling and simulations if actual practicing with gear or hardware is not possible.

3.12 Task Design and Training

Prior to work execution, diligence to training and task design can impact desired outcomes.

Consider users and human system integration in task development and evolutions. Involve workers through concepts, work development, and changes. Reduce system interface complexities. Conduct task analysis and human-system-integration reviews. Conduct on-the job training. Practice with actual items, or use common materials or high-fidelity models, simulations, physical mock-ups. Document and incorporate lessons learned. Provide clear, correct, and complete work instructions.

Figure 12: Practice with Orion Crew Module

Figure 13: Shuttle Motion Capture Evaluation

Shuttle processing and operations planning for the new Orion crew module employed an example of a task design tool. HF practitioners teamed with computer science entities to develop a means to perform motion capture while assessing real-time human factors and ergonomic data. The tool assisted with a variety of activities ranging from evaluating improvements for the removal of wing leading edge panels to stepping across open flaps or doors on the Shuttle. During design phases of the new crew module, technician teams assisted HF and designers in the validation of collecting dynamic biomechanical, biomedical, fatigue, and collision data. They evaluated did this while performing seat, avionics boxes, suited, and panel handling tasks. Information aided the design, order of operations, staging options, and risk mitigation. Having a quantitative HF tool enhanced the traditional task design and training toolsets [4]. The checklist can assist such tool application or other task design options and training components.

Off-nominal tasks and non-routine situations often compound HF issues. Even with tasks viewed as potentially never or rarely performed, designs and planning phases should consider spending as much or more time designing the procedures and training so that contingency, emergency operations, and troubleshooting aspects will be available allowing focus on the contingency task resolution.

Such practice can occur with acting out actual contingency or emergency tasks such as astronaut rescue at the launch pad (Fig. 14). Other practices or dry runs can start with discussions and use materials and components available.

Figure 14: Astronaut Emergency Egress Simulation

At a minimum, the operators of systems can provide essential information during system designs or execution planning and into procedure development.

The HF Checklist can assist those phases. Utilizing HF practitioners and tools in task designs and training can provide an even broader aspect, especially to more complex tasks.

3.13 Consistent Work Practices and Cognitive Factors

Among systems, personnel, and their environments, consideration to consistency and other cognitive factors must also occur.

Standardize among similar functions and work areas. Account for psychological and organizational variables. Use standard conventions and inform of irregularities. Outline work and break down tasks to align with schedules, biological breaks, and challenges of the steps at hand.

Employ situational awareness checks. Limit multitasking, decision making, and complacency. Provide measurement goals and tolerances. Avoid requiring personnel to execute mental calculations. Instead, provide users with intended goals and acceptable ranges.

Clear labels and actions for controls and displays reduce chances of injury, damage, and errors. On the other hand, data indicators using inconsistent coloring could lead to human error.

3.14 Damage/Error Prevention, Detection, and Recovery

Designing out potential risks or human error opportunities is the most effective approach. Being able to detect any damage during processing and having means for recovery are key to finalizing robust tasks and designs.

Figure 15: Protective Covers Protect Hardware and Personnel

Prevent human error and human-induced safety or collateral damage opportunities. Provide a means of detection by inspection or test while providing the capability to recover. Define and plan for contingency procedures.

Offer hardware and personnel protection. Make consequences of errors on safety or system performance clear.

System experts rely on many lessons learned from Shuttle processing. Through knowledge sharing and HF training, engineers and designers are encouraged to review aerospace and industry lessons learned, gain tribal experiences from legacy experts, as well as conduct activities to increase interfaces with users, practice tasks, and continually evaluate and implement means to mitigate damages and errors.

4. SUMMARY

From initial concepts through designs, testing, implementations, and evolutions, the basic HF Checklist can be a quick tool for highlighting key HF topics while keeping the focus on the people.

Intangible and tangible results may include safer and easier systems to operate and maintain. Cost effective options can also overlay enhancement opportunities.

Summarizing thousands of requirement topics into a manageable set of HF topics makes it easy to use in aerospace activities. Reductions in undetected errors and collateral damage risks can enhance ground crew operations and could ultimately improve flight crew safety [5].

HF Checklist's simplified format also makes it applicable to general industry and outside of the work environment.

As a good place for anyone to start, the checklist provides a beneficial snapshot of many HF topics. Employing HF practitioners in further evaluations provides added advantages to designs and processes.

"Thinking Human Factors and Focusing on the People" throughout lifecycles and evolutions leads to better designs, improved planning and execution, in addition to error reductions.

5. REFERENCES

1. Stambolian, D., Wilson, S., Logsdon, K., Miller, D., Dinally, J., Masse, J., Barth, T. "1-G Human Factors for Optimal Processing and Operability of Constellation Ground Systems" 2009 IEEEACpaper#1286

2. NASA Systems Engineering Processes and Requirements, NPR 7123.1A, (2013), Chapter 5.

3. Barth, T., Kanki, B., Westphal, J. King, J. (2007). Human Factors Overview for Ground System/ Ground System Equipment (GSE) Designers. Kennedy Engineering Academy KEA-11-13-07.

4. Stelges, K., Lawrence, B. Human Engineering Modeling and Performance: Capturing Humans and Opportunities. Advances in Human Modeling and Simulation. CRC Press (2002), 555-564.

5. Kanki, B.G., Barth, T., Miller, D., King, J., Stambolian, D., Westphal, J. (2008). Human Factors Issues in the Design of Ground Systems: A Pathfinder Activity. Proceedings of the 3rd International Association for the Advancement of Space Safety. Rome ITALY: ESA Conference Bureau.

6. NASA Kennedy Space Center, Multimedia Database (2016). Figures noted with photo number references.

7. Jacobs® Host Services Guide (January 2016). Test and Operations Contract (TOSC), Kennedy Space Center, 20, 31.

8. Stelges, K.S. and Kramer, I.A. Jacobs® Test and Operations Contract (TOSC) (2016), Kennedy Space Center. OG0155 Human Factors for Work Authorization Document Authors Training. Figures by hands-on author involvement.

PROCEEDINGS OF THE 8TH INTERNATIONAL SPACE SAFETY CONFERENCE

SESSION 33 SAFETY ON LONG DURATION MISSIONS

DEVELOP GLOBAL SAFETY SYNERGIES FOR LONG-RANGE HUMAN SPACE EXPLORATION, WITH FOCUS ON LAUNCH SYSTEMS

Aline Decadi

He Space Operations BV at European Space Agency, 52 Rue Jacques Hillairet, Paris, France
Email: Aline.Decadi@esa.int

ABSTRACT

The agencies participating in the *International space Exploration Coordination Group* (ISECG) demonstrate a natural synergy for defining and implementing meaningful steps towards advanced human spaceflight. Within the exploration mission themes of the *Global Exploration Roadmap* (GER) [1], crew transportation and operation capabilities are driven by key requirements based on human and robotic mission partnership. They require an acceptable level of safety associated with the risks inherent to live, travel and work in space. This paper pursues the goal to investigate the safety orientations to be targeted by the space agencies, and addresses the main decision factors for implementing global safety rules regarding the space transportation vehicles.

1. INTRODUCTION

Human space exploration will proceed in a step-wise manner, extending proven capabilities to achieve more complex goals while enabling new discoveries with each step. The widespread recognition of the updated *Global Exploration Roadmap* (GER) released in August 2013 [2], provides a common reference, which is then useful for the work of individual space agencies to advance mission concepts and to study associated capabilities.
The GER emphasizes the near-term initiatives in implementing the common international strategy:

1) Full utilisation of the International Space Station (ISS) to prepare for exploration,
2) Continuing efforts to expand on synergies between human and robotics missions,
3) "Proving ground" missions in the lunar vicinity in order to develop capabilities and techniques leading to a sustainable human exploration on the surface of Mars.

ISECG focuses on a continuing definition and elaboration of a common long-term strategy making clear that there is a role for any nation willing to contribute. ISECG also evaluates concepts and opportunities as humans prepare the next steps in a sustainable long-term program of exploration. Human-assisted robotic mission concepts are receiving increased attention by participating agencies within ISECG because they appear to provide opportunities for leveraging the unique capabilities of human exploration to achieve multiple objectives. Work remains to fully quantify and assess this benefit, but several ongoing studies seek to provide data to complete such an assessment. Recent studies in particular, one led by ESA and one led by NASA/JPL [3] are ongoing and shared within ISECG. Results from their analysis will be used in the next iteration of the Global Exploration Roadmap. They provide descriptions of architecture elements through mission campaigns and operational concepts. These studies cover the main categories:

- Prepare human exploration of the lunar surface
- Enable scientific and exploration knowledge gain
- Provide opportunities to advance science mission bringing back benefit on Earth (e.g. Mars sample return scenarios).

Whatever are the different exploration scenarios, the key capabilities related to multi-destination space transportation are [2]:

- Launch vehicle that has the capability to deliver cargo or crew beyond low-Earth orbit. Initial capability evolves with advanced boosters and an upper stage to enable increasingly complex missions with further evolution to support crewed Mars missions (e.g. various launch scenarios including a split launch on Soyuz / Ariane (assembly at the cislunar habitat) or co-manifested launch on Space Launch Vehicle (SLS) Block IB, or Next Generation Space Launch Vehicle (NGSLV))
- Crew (transportation) vehicle capable of delivering a crew to exploration destination and back to Earth (e.g. Multi-purpose Crew Vehicle (MPCV), Orion)
- Cargo logistics vehicles capable of resupplying orbital infrastructures in lunar vicinity with pressurized and unpressurized payloads
- Small Cargo Lander capable of delivering robotics and cargo on the lunar surface to meet lunar exploration objectives (e.g. automated (potentially tele-operated), total wet mass of 10 Tons, multiple missions)
- Crewed Lunar Lander capable of delivering crew and cargo to the lunar surface (may be reusable, and is composed of two or more stages)

- Mars Ascent Vehicle (MAV) capable of transporting crews from the surface to Mars orbit
- Mars Lander (Liquid Oxygen/ Methane Cryogenic Propulsion System) such as a re-usable lander
- Mars orbiter
- Evolvable deep Space Habitat, in-space habitat with relevant subsystems for the purpose of advancing capabilities and systems requiring access to a deep space environment.

These major transportation capabilities for future exploration architectures have different functions to ensure:

- Transport the crew/ cargo (main phases are launch, approach around Mars and landing)
- Propulsion
- Navigation
- Power
- Environmental control and life support
- Communication

These functions are triggered by safety objectives that are developed hereafter.

How does the safety principles apply to the functions related to spaceflight capabilities of the future exploration architectures?

The following chapters present:

- The safety methodology, i.e. objectives and assessment process for mitigation of unacceptable failure conditions (based on Recommended Practices and Lessons Learnt)
- Illustrations of the implementation of the safety assessment on the future exploration transportation capabilities (based on a screening of their main functions)
- The synergies resulting from these Recommended Practices and their use for the elaboration of a global safety standard for space exploration.

2. SAFETY OBJECTIVES

Safety is a major argument on which the agencies commit to secure to crew a safe return to Earth. The objective of Safety is to bring failure probability down to an acceptable level, especially regarding critical functions, which have catastrophic failure conditions. Nevertheless, Safety is confronted to other criteria such as acceptability of budget, schedule and completeness of operational objectives. Safety optimisation is often compared to elements that have different order of magnitude. How combine the residual probability of accident with the overall programme development cost? Several approaches can be addressed in parallel, e.g. with regards to other issues, the use the lessons learnt from previous space flights heritage (e.g. Apollo, ISS, space shuttle), or other hazardous endeavours (e.g. explorations, sporting feat).

The main disadvantaging factors regarding to mission safety when experimenting space exploration to Mars are [6]:

- Mission duration that has an impact on the failure probability of equipment, as well as on crew health (physical and mental)
- Distance associated to confinement during the journey leading to impossibility of receiving help from Earth
- Complexity of the mission, e.g. the number of critical phases/ operations, and number of vehicles to consecutively operate at each mission phase
- Extreme environments (in space and on surface).

The elaboration of safety rules is based on the fundamental principles:

- Redundancy at each level of the mission architecture with regards to degraded cases, with at least one backup lane or unit
- Spare policy, e.g. by splitting the launch dates of cargo and crew vehicles in order to provide additional spare equipment to crew in place to mitigate potential hardware failure cases
- High level of maintainability of systems (e.g. in terms of modularity, ability to repair)
- Simplicity of the mission architecture by limiting the number of modules, phases or critical maneuvers
- Deployment of a maximum of infrastructures on ground (via cargo missions) before sending the crew
- Foreseeing sufficient quantity of supplies in order to wait for the next crew assistance, if required.

This decision-making process for the implementation of the most appropriate safety barriers is based on hazard analysis to be performed in order to propose the best mitigation actions, associated to trade-off and compromise (mainly programmatic such as development uncertainties and cost). Indeed this trade-off makes sense to define the realistic safety needs. Otherwise, results of "safety at all prices" would potentially lead to:

- Redound any element on-board that could lead to such a heavy launch system that the mission would be technically compromised
- Launch dozens of cargo vehicles before sending any crew, such as the mission becomes economically unrealistic.

3. SAFETY ASSESSMENT PROCESS AND UNACCEPTABLE SAFETY-RELATED LOSSES

3.1. Functional mission failure conditions

The functional hazard analysis identifies and classifies the failure conditions associated with the mission level functions. The classification of these failure conditions establishes the safety requirements that any transportation vehicle must validate and verify. The goal is to identify each failure condition along with the rationale for its severity classification. In this frame, any failure condition leading to the loss of life or serious injury is an unacceptable risk, and its severity is classified as catastrophic. The consequence in the mission architecture design is that all failure conditions shall be mitigated down to an acceptable level in order to ensure continued safe launch, journey, landing and return. This analysis will consider single and multiple failures in normal and degraded environments.

3.2 Failure hazard analysis

- Once identified and described, requirements are assigned to these feared event conditions associated with mission level functions. Then, these requirements are cascaded down to the lower level classification, i.e. at spacecraft level, then at system, sub-system, and equipment level. At each level is associated the identification of the method used to verify compliance with the failure condition requirements. The outcome of this analysis is traced down into preliminary design, via fault trees and other analyses used to provide quantitative results.
- Safety requirements intend to limit the effects of the identified function failures. These derived requirements may lead to add design constraints, identify new failure conditions, or recommend crew or systems maintenance actions. By definition [4], derived requirements are additional requirements resulting from design or implementation decisions during the development process. The process is continued and iterative, until the design process is complete.
- The failure hazard analysis is used to complete the list of failure conditions and the corresponding safety requirements. Probabilistic analysis is performed at the sub-system(s) level(s) to show that the failure probability meets the requirement cascaded top-down from the mission level analysis.
- A Common Mode Analysis (CMA) is also conducted in order to substantiate claims of dissymmetry and dissimilarity made in the functional hazard analysis. Item with identical hardware and/or software could be susceptible to generic faults, which could cause malfunctions in multiple items. This analysis aims to consider redundancy via different technologies down to components in order to ensure that same causes cannot lead to the loss of the function. CMA is usually associated to a Zonal Safety Analysis (ZSA). The objective of this analysis is to ensure that the equipment installation meets the safety requirements: the effects of failures of one equipment are analyzed in terms of propagation on the systems and structure of the redounded equipment falling within the same physical sphere of influence. This way, the redounded lanes/ units are located at different places.
- Throughout these steps, additional requirements may need to be identified in order to achieve the mission level budgeted probability, leading to a redesign at some level of the hierarchy. This safety assessment process guarantees the consolidation of the final set of mission requirements.
- This set of mission requirements is ensuring the requested level of mission safety, and is considered as the mission safety baseline for any programmatic trade-off throughout the development.

4. SAFETY ASSESSMENT OF FUTURE EXPLORATION LAUNCH CAPABILITIES

The methodology related to safety assessment provides guidance in implementing fail-safe design, with an emphasis on redundancy and monitoring, including eliminating common mode failures and hazards (causes of unsafe control) completely.

How does the safety process apply for the functions related to spaceflight capabilities of future exploration architectures? This chapter provides several illustrations on the main critical functions that have potential catastrophic failure conditions.

4.1 Transportation

Launch System

The first obstacle is to launch a total mass between 300 and 500 Tons before starting the journey to Mars. The idea is to perform several launches (between 3 and 5) followed by operations of rendez-vous around the Earth, around Mars or on the ground on Mars.

Several options are under study between the Space agencies and their industry. The development of a heavy 100-tons class launcher (same class than Saturn 5) with a reusable deep space transportation system is the one detailed in the GER. NASA is developing the Space Launch

System (SLS) and the Orion crewed spacecraft [5], with the objective to demonstrate logistics reduction in operation, and prepare for human missions into cislunar space. In terms of proving ground, a series of Exploration Missions (EMs), including integrated test of SLS and Orion, are anticipated in 2018. The development is incremental: initial capability evolves with advanced boosters and an upper stage to enable increasingly complex missions with further evolution to support crewed Mars missions. It includes also the development of a Cryogenic Propulsion Stage (CPS): an in-space propulsion capability utilizing cryogenic hydrogen and oxygen as propellants. It aims to provide additional performance for missions to the lunar vicinity, lunar surface or Mars. Mission durations will require long-duration storage of cryogenic propellants.

In all the scenarios under study today, safety aspects are included at each step of the design by developing the Launch System by increments, performing several unmanned launches to test and qualify reliable systems (propulsion, power, communication, life support, entry/ descent & landing systems). Moreover, the Launch Abort System, positioned on a tower atop the crew module Orion, shall be able to activate within milliseconds to propel the crew module to safety away from the launcher in the event of an emergency during launch or climb to orbit.

In-Space Propulsion Technologies

In-space propulsion performs the functions of primary propulsion, reaction control, station keeping, precision pointing, and orbital maneuvering. The main engines used in space provide the primary propulsive force for orbit transfer, planetary trajectories and extra planetary landing and ascent. The reaction control and orbital maneuvering systems provide the propulsive force for orbit maintenance, position control, station keeping, and spacecraft attitude control [13]. So the propulsion system onboard is vital to the safe operation of the spacecraft.

Failure conditions of a propulsion system are manifold (e.g. thrust generation system, turbo-pump, ignition, propellant system storage, line rupture) [14]. Many of these failures will have adverse consequences on safety, cost, life and/or schedule. That's why their failure prevention is necessary, up to the point where the effort of failure prevention no longer weighs up against the consequences of failure. So what could be the reliability rate to be targeted? A single engine constitutes the Apollo lunar module. The failure of this engine would have led to lose the crew and the mission. Nevertheless, only extensive testing has been used as proving ground aiming to reach the targeted reliability level. What could be the transcription of the safety Recommended Practices into space exploration today: redounded propulsion system/ engine?

Several technological developments are under going. Solutions exist or are under development such as:

- Liquid Oxygen/ Methane Cryogenic Propulsion System (Mars Lander)
- Advanced In-Space Cryogenic Propellant Storage & Liquid Acquisition
- Electric Propulsion & Power Processing
- Nuclear Thermal Propulsion (NTP) Engine

There are many space propulsion systems that are being developed. The principles of operation vary considerably between systems. Each system has its own particular types of failure modes.

Guidance, Navigation, and Control (GNC)

Navigation refers to the determination, at a given time, of the vehicle's location, velocity and attitude. The GNC function requires for fundamental sensing, prediction, and controls capabilities. This function is even more critical to enable reconfigurable shapes and multifunctional capabilities such as real-time trajectory tailoring, energy management and propulsion [12].

GNC technologies offer the capabilities to determine anytime the exact position of the space vehicle. They are obviously critical in the way that their functioning offers the possibility to have a safe journey in space and return. The loss of this function may have catastrophic consequences and is considered unacceptable. Due to such complexity, dedicated safety design rules apply such as Failure Detection, Isolation and Recovery (FDIR) using redundancy at each level of the system architecture, reliability computations (i.e. compute probability of success over the required lifetime, based on hardware units MTBF (Mean Time Between Failure)) [15].

Power

The function power is a common mode of failure conditions. It supplies all the other functions, and without energy, an acceptable of safety of the overall space transportation vehicle cannot be guaranteed. It is even worse in degraded case as the recovery procedure cannot be launched as requiring additional energy. The failure modes of the power function can be associated to the power source, the energy storage, the power distribution or the power regulation and control.

Power systems play an integral role in all in-space propulsion systems for both human and robotic missions. In some cases, power is only required for basic functions of instrumentation and controls for which technology advances are not required. But in other cases, the propulsion energy of the technology is derived from electrical power generation, management and distribution; in this case, power system technology advances are critical for the advancement of the propulsion system (e.g., high power solar or nuclear electric propulsion) [13].

Arrival in Mars orbit

The approach of Mars is the step corresponding to inject the spacecraft to orbit. The major hazard during this phase is related to the required accuracy of the spacecraft trajectory. In case of failure of propulsion or navigation systems, two catastrophic consequences can occur:

- It can only touch the high layers of the martian atmosphere: the spacecraft is not low enough and bounce away from the atmosphere following an hyperbolic trajectory that takes it away from Mars. Without any additional propellant on-board for managing this event, the crew is lost.
- It can enter too quickly in the low layers of the atmosphere: the spacecraft is too low; it enters too strongly into the atmosphere. The insufficient atmospheric braking may lead to possible depletion, or even complete burn of the thermal shield, G-force in excess for the crew and/ or eventually high-speed crash on the ground.

Several options may be possible to reach the safety objectives such as aero-capture, aero-braking (such as for Mars Reconnaissance Orbiter in 2006) or retro-propulsion (using propellant) [6]. The technical choice depends on the mission architecture. Any evolution - for example in terms of mass, cost or technological maturity - leads to re-assess the safety constraints that are cascaded in the design: each mission scenario brings to consider a different critical element of the system, so new safety criteria to be included in the trade-off.

Entry, Descent & Landing (EDL)

This last step before landing lasts seven minutes during which several hazards are combined at the same time [16]:

- The spacecraft shall resist to a huge deceleration. It acquires a supersonic speed and the resistance of the air leads to a violent braking. During this phase, the entry trajectory has to be adjusted as much as possible.
- The spacecraft shall resist to a particular thermal warming that can exceed the resistance of certain materials. During this phase, the corridor of entry is key.
- The spacecraft shall resist the sequential of entry at very low altitude (around 10 Km) that starts with a speed around Mach 2, then continue with opening of parachutes, eject the thermal shield, perform engine ignition for braking until a quasi null speed before landing.
- The spacecraft shall resist to the gas ejected from the nozzles while landing. They may dig a crater below the spacecraft leading to its fall.

Any failure during this sequence may lead to catastrophic event and required to develop EDL systems that are tolerant to double failure modes. An additional constraint is that, in case of problem during the descent, it is almost impossible to abort and reacquire altitude to come back into orbit (such a manoeuver would add more complexity and propellant). For this reason, the EDL phase is one of the rare phases where any failure shall be found unacceptable.

The manned mission is associated to a heavy spacecraft to land. And because of the mass, the braking is less efficient. That's why the technologies and procedures used successfully for landing a small robot, is not transposable as is to a spacecraft.

Solutions are under study [6] such as inflatable thermal shield that can reach high diameters (20-30 meters), airbrake in form of petals. Depending on the residual risk, landing two smaller spacecrafts instead of a single one can be an alternative solution.

Regarding the risk of damage and fall-out of the spacecraft while landing, solutions seem implementable, such as a cable that helps the module landing progressively (such as for Curiosity), an inclination of jets outwards. The EDL system is identified as a critical technology for the Mars exploration mission: precision of landing and hazard avoidance are crucial regarding this phase.

4.2 Environmental Control and Life Support

For manned mission, a life support system is constituted by all technical means ensuring the crew survival in space, i.e. a controlled/ recycled environment and supplies (food, water, fresh air); management of waste recycling. So the reliability of this system is crucial for any launch and space transportation vehicle, as well as for the space habitat.

In particular, human presence in deeper space implies the elaboration of life support systems for food, water, air and waste management in "closed loop" in order to limit and eventually remove any re-supply need.

For the three mission phases - journey to mars, mission on Mars then return back to Earth - major functions aim to monitor the internal crew environment and to ensure the crew survival in the transportation vehicle or habitat. These main functions are [6]:

- Atmosphere monitoring: a sub-system maintains the atmospheric gas within the cabin (the composition is a mix of nitrogen and oxygen under a pressure of 101,3 kPa coming from ISS heritage), with continuous monitoring of parameters e.g. total pressure in the cabin, concentration in CO_2, filtering and elimination of polluting substances (ammoniac, acetone, methane). It includes also fire detection, as well as storage of gas needed for ensuring the life support function into pressurized tanks (e.g. made of cryogenic fluid with continuous thermal monitoring).

- Temperature regulation: a sub-system maintains a confortable temperature (typically between 18 to 27 degC) and an adequate level of humidity.
- Water management: a sewerage system retrieves and recycles wastewater, in order to be used for consumption of drinking water, preparation of food and hygiene.
- Waste recycling: a dedicated sub-system retrieves all kind of waste, such as organic waste, food scraps and the salt water coming from the other sub-systems.

A crisis situation can occur anytime during the mission. This requires the operational maintenance of systems of detection, warning systems and automatic software actions to warn the crew and answer automatically to the incident.

- One typical hazard due to the presence of oxygen of air is the fire within the cabin. The Fire Detection and Suppression System are constituted - in addition of automatic alarm systems - by detection sensors, extinguishers, and portable breathing equipment in case of contamination or smoke in the atmosphere.
- Another typical feared event to be considered is the untimely depressurization of the cabin. The life support systems shall be able to continuously monitor the pressure, taken into account influent factors such as resupply in gas, ambient temperature. It has to detect slow drifts as well as brutal leaks. Breathing masks shall be easily accessible and the closure of internal bulkhead shall be performed in a very short time.

The functional analysis of the life support system and associated mission failure conditions lead to a set of requirements for the Environmental Control and Life Support System to be satisfied [6]:

- Capable of continued operation during the whole mission
- Extensively self-maintainable
- Designed such that maintenance operations are quick and simple
- Comply with the mission payload requirements
- Use only proven technologies that have been demonstrated in laboratory and qualified in a representative environment
- Implement a redundant monitoring system - independent of any failure cases on the main on-board computer
- Easily controlled by the crew and also from Earth (subject to communication delay).

The development of innovative technologies is under maturation for use on the International Space Station, crewed transit vehicles, and surface habitats. They will also evolve pending the mission phase with integration of specificities related to transportation vehicle, space habitat and space suit for Extra Vehicular Activity (EVA).

4.3 Communication

It is essential to communicate with the crew during all the mission phases and be able to send orders that may have safety implications, such as alarm and rescue. With regards to capabilities related to space transportation, two aspects are considered:

- Communication system between Earth and the space transportation vehicle (corresponding to mission phases launch, journey into space and return)
- Communication system between the crew on-ground Mars and the space habitat (corresponding to mission phase is the life/ work on Mars).

The instantaneous communication between Earth and crew is not possible: due to the distance to Earth, and depending on the relative position of Earth and Mars, the travel time of the radio signal at the speed of light varies from 4 to 24 minutes [9]. This is a huge constraint that contributes to the isolation of the crew when a crisis situation occurs. That's why the communication function needs to be continuously relayed by several independent systems. The objective is to ensure that one communication system is always available and working (Fail-Safe requirement)

Three technologies exist for communication with Earth [6]:

- Parabolic antenna that consists in concentrating the emission and reception of radio waves in a specific direction. This system amplifies the emitted/received signal, and enables to exchange data in high speed. For efficiency of the communication system, dedicated frequencies shall be allocated to exchanges with the crew. This way, even if the signal coming from mars is low, there are no interferences with other terrestrial signals. That gives performance to this system. The drawback of this option is that the parabolic antennas have to be oriented with accuracy. It requires a high reliable software system capable to determine the direction of Earth in function of the crew localization, hour and date. These programs are developed as software critical as mandatory for safety. In case of failure of this high-speed system, or of the parabolic antenna, another option exists.
- Omnidirectional antenna that consists in emitting the signal in all directions simultaneously. This system

is less efficient because most of the signal energy is lost. It enables to exchange data only in low speed. Even if this system does not enable to send large data such as video and pictures, it still gives the opportunity to exchange text messages that can be crucial to alert or give recommendations in case of emergency.

- Through the communication system of a space vehicle in orbit around Mars. It can come from a return vehicle, an in space habitat, a telecom satellite or a Mars orbiter dedicated to this function. There is a dual interest with this option: 1) the martial orbit is close to the habitat so the transmission of the signal is efficient, 2) the vehicle into orbit can pursue the Earth with reduced blind periods (including during the martian night when the antennas are out of range). In addition, in this configuration, it is possible to choose an appropriate orbit to ensure that the vehicle stays visible from the crew on-ground, so capable to exchange data with them (even if there are also delays of communication to consider).

The multiplication of independent communication systems is the way to ensure that the function is always fulfilled throughout ach step of the mission. In particular, we can mention the multi-purpose spacecraft in the orbit of Mars that continuously provide data to Earth.

5. GLOBAL SAFETY SYNERGIES

The previous chapters provide various illustrations of the implementation of the safety assessment process on the critical functions that may have catastrophic failure conditions. It is based on safety rules coming from lessons learnt and recommendations from the past (e.g. Apollo, ISS, space shuttle, exploration missions) regarding different aspects such as design, manufacturing, testing, instrumentation, review and control. Most of them are used commonly within the global safety community as best practices. They are particularly valuable for large and complex programs that require a certain codification of the lessons learnt.

These safety rules need to be standardized in order to structure the analysis and avoid engineers to put value judgments. Do we have a common global safety standard in the space exploration domain?

Current International Space Safety Standards are based on several ISO (International Organization for Standardization). They are not suitable mainly because they are developed under the legal authority of specialized national or international agencies (e.g. FAA, FDA, ICAO, IAEA, for aviation, nuclear, pharmaceuticals) and not by national industrial standardization bodies [8]. The need has been identified for the institutional stakeholders of space-faring countries to jointly establish safety consensus standards to become recommended references for national regulations.

Regarding the civil airborne, the standardized process is the *Aerospace Recommended Practices* ARP 4761 [4]. It is the guideline and set of methods for conducting safety assessment process used worldwide. At the same time, it is the world's most severe civil aviation standard. It is enriched by decades of maturation and improvement, as well as millions of flight hours each year. As a rough order of magnitude, the cumulated number of hours flown each year is 60 millions (for average flight duration of 2 hours) [11].

This guideline ARP 4761 is used with the ARP 4754A [10]: this process is devoted to determine Development Assurance Level (DAL), which defines the rigor of complex hardware and software development and verification activities. The level of DAL is defined depending on the architecture and the high-level safety objectives that are cascaded to lower systems and sub-systems.
Space transportation systems and earth aerospace systems have different specificities intrinsic to their respective mission architecture, for example for what concern:

- Quantitative objectives of reliability (failures rates): Space transportation capabilities beyond LEO are constituted of critical systems in operation during tens of thousands of hours in a space environment.
- Maintainability procedures in operation: there are no possibilities during the duration of the space travel to make a landfall in order to repair or replace faulty or worn-out components.
- Flight proving qualifying the overall critical systems at the acceptable level of safety: only few space systems will fly often enough to historically validate the associated calculation models regarding the catastrophic failure conditions.

Nevertheless, space exploration can benefit from the civil airborne processes in order to reach an international consensus on space safety standards. Global synergies exist in terms of implementation of safety rules and lessons learnt, and the standardization of these approaches (space exploration and civil airborne) seems key to reflect the common safety status of the knowledge and the accumulated experiences. The global process is elaborated from the shared safety objective that no single failure will result in a catastrophic failure condition.

Such an endeavor, in the Space Exploration area, requires transnational collaborations and active commitments based on these identified global safety synergies.

The agencies participating in the *International space Exploration Coordination Group* (ISECG) have a leading role in order to promulgate the implementation of global safety management capabilities allowing the effective identification of international standards adequate for mitigating safety deficiencies and resolving safety concerns. This international group ISECG can also facilitate the implementation of appropriate data collection and data sharing initiatives.

Moreover, the development of Space Exploration Recommended Practices enables to establish the conditions and circumstances under which to deal with safety deviations through established enforcement procedures. This particular aspect brings the need for standardization in order to avoid the acceptation of deviance that has to be carefully monitored.

This is a continuous and iterative process where safety indicators and findings are to be re-assessed depending on mission architecture evolutions throughout the overall development.

6. CONCLUSION

Human presence beyond LEO enables new opportunities for reaching space exploration goals. This applies not only to the presence of crew but also to space-based infrastructure such as habitation, communication and servicing systems.

The architecture of the Mars exploration mission pursues two objectives: minimise the mass to send beyond Earth orbit, and ensure the safety during each step of the mission (launch, journey, approach on Mars orbit, entry, decent and landing, and return). The challenge is to bring the failure probability down to an acceptable level mainly for critical functions that have catastrophic failure conditions, but without adding complexity to the overall design in a way that the design is compromise. The reliability of the overall systems for a two-years duration is also a huge challenge. Introduce the adequate redundancies is crucial to avoid catastrophic failure conditions. At the same time, the systems have to be repairable by the crew. The *Global Exploration Roadmap* gives an overview of key technologies to be developed within the next decade [2].

It is interesting to see if safety considerations may bring new design constraints. For example: what about using medium class launchers in order to assemble the spacecraft by piece of 20 tons instead and constitute progressively propellant storage?
Regarding the synergies for space exploration:

- How do we exchange between agencies about lessons learnt?
- How do e share the results of our studies?
- How do we share workload and global safety standards?
- Do we agree on these global safety standards between agencies?

The next steps of the Space Exploration Safety standardization will be based on an international interest and collaboration. A recommendation can me made on the fact that ISECG can be considered as the Space Exploration coordination platform for global safety synergies.

7. REFERENCES

1. Bernhard Hufenbach, Kathleen C. Laurini, Naoki Satoh, Christian Lange, Roland Martinez, Juergen Hill, Francois Spiero (2014). The exploration mission themes of the global exploration roadmap, IAC-14.B3,1,10,x22313 .

2. International Space Exploration Coordination Group. The Global Exploration Roadmap (GER), August 2013, www.globalspaceexploration.org

3. Kathleen C. Laurini, Bernhard Hufenbach, Juergen Hill, Alain Ouellet (2015). The global exploration roadmap and expanding human/robotic exploration mission collaboration opportunities, IAC-15.A3.1.1.

4. SAE (1996). Guidelines and methods for conducting the safety assessment process on civil airborne systems and equipment, ARP 4761.

5. http://www.nasa.gov/sites/default/files/atoms/files/journey-to-mars-next-steps-20151008_508.pdf

6. Jean-Francois pellerin, Richard Heidmann, Alain Souchier, Planète Mars (2012). Embarquement pour Mars.

7. Paul Stephen Dempsey (2016). National Laws Governing Commercial Space Activities: Legislation, Regulation, & Enforcement

8. http://www.hq.nasa.gov/office/codeq/trismac/apr08/day3/Sgobba_ESA.pdf

9. http://blogs.esa.int/mex/2012/08/05/time-delay-between-mars-and-earth

10. SAE (2011). Guidelines for development of Civil Aircraft and Systems, ARP 4754A.

11. http://www.icao.int/safety/state%20of%20global%20aviation%20safety/icao_sgas_book_fr_sept2013_final_web.pdf

12. http://www.nasa.gov/centers/ames/research/technology-onepagers/entrysystemtech.html

13. http://www.nasa.gov/pdf/501329main_TA02-ID_rev3-NRC-wTASR.pdf

14. http://www.lr.tudelft.nl/en/organisation/departments/space-engineering/space-systems-engineering/expertise-areas/space-propulsion/design-of-elements/failures/failure-modes/

15. http://www.et.tudresden.de/ifa/fileadmin/user_upload/www_files/Aktuelles/2014-01-27_ESA_Guidance__Navigation__and_Control_Systems.pdf

16. http://mars.nasa.gov/multimedia/interactives/edlcuriosity/index-2.html

DISCLAIMER

This article was prepared or accomplished by Aline Decadi in his personal capacity. The opinions expressed in this article are the author's own and do not reflect the view of the European Space Agency.

IDENTIFICATION OF HAZARDS ASSOCIATED WITH A ONE-WAY HUMAN MISSION TO MARS

Joao Lousada(1), Aureliano Rivolta(1), Matteo Emanuelli(1), Ali Nasseri(1)

(1) Space Generation Advisory Council,
c/o European Space Policy Institute
Schwarzenbergplatz 6, 1030 Vienna, AUSTRIA
Email: joao.lousada@spacegeneration.org

ABSTRACT

Recently, there has been an increasing amount of discussion regarding the concept of a One-way Human Mars mission. The lack of a return journey could lead to lower mass at launch and reduced initial costs. However, it could also lead to higher risk for the crew due to the increased complexity of the mission.

The work presented in this paper focuses on the identification of the hazards associated with this type of mission and analysis of their severity.

A one way mission to Mars is broken down and studies in three phases: the vicinity of Earth, Earth to Mars transfer, and the Mars surface. Specific hazards are identified for each mission phase and a detailed severity analysis is performed for each of them. Finally, an overview of the achieved results, and the overall risks associated with such a mission are provided.

1. INTRODUCTION

Never before has the idea of a manned mission to the red planet been as tangible as it is today. The various Mars manned missions proposed in recent years, and the continuous advancements in human spaceflight systems (particularly in the USA) have resurrected human Mars exploration and brought it into the forefront of public debates. For decades, many of organizations in the space sector have dreamt of landing a man on Mars. Recent interest and support of the topic seems to bode well for the realization of this goal. However, with this increased enthusiasm and excitement, it is easy to disregard the challenges that such a feat entails. Even though more than 50 years have passed since mankind first ventured into space, landing a man on Mars presents challenges which the space sector has never faced before.

In this paper, we analyse the hazards associated with a manned mission to Mars, with specific focus on "one-way" human missions (OHM). The idea of a one-way mission (also known as "Mars to stay") was first proposed in 1996[1] and has, since then, led to several subsequent proposals.

Despite slight variations in the content of these proposals, they are all based on a common argument: that a one-way mission to Mars (i.e. not involving a return of the crew to Earth) requires less mass at launch which translates to lower initial costs. This advantage could help mankind reach Mars earlier. A one-way mission does, however, require a different approach to the design of the habitation modules and presents additional risks.

There are as many arguments advocating a one-way mission as there are arguments against it. This paper will not discuss the merits or drawbacks of the different proposed Mars missions, but will rather focus on technical safety analysis of a generally defined one-way mission to Mars. The paper is divided into three sections, each representing a specific mission phase that is expected as part of a one-way mission to Mars. These phases are:

- Launch and Earth vicinity operations;
- Transfer to Mars and landing;
- Mars habitation and operations

This categorization allows for a structured approach to hazard identification and mitigation.

2. EARTH VICINITY

Launch is one of the most critical phases of any space mission, as it encompasses some of the mission's greatest risks. In the past, failed launches have brought disastrous consequences to the mission and even caused fatalities. While today's launch vehicles have a better safety record than some previously used ones, none of them is currently is compatible with a Mars mission.

As such, any manned mission to Mars must begin with development of a new rocket with higher payload capability and performance. NASA is currently working on a launcher that might eventually be able to do it –the Space Launch System (SLS) [2].

However, this is a launch vehicle more complex than any launch system developed before. This increased complexity adds to the already perilous operations that rocket launch implies.

Another major risk during launch is related to the possible use of a radioisotope thermoelectric generator (RTG) or nuclear reactor as power source for the Mars mission. Many current proposals are based on such devices, the proper handling of which would pose an added difficulty with inherent hazards during the launch phase [3].

During the spacecraft launch, a series of incidents can lead to the loss of the mission and even of the crew. There is the possibility, for example, of inadvertent ignition caused by induced currents or electrostatic discharges or overheating of the propellants due to a fire or accumulated energy from micro vibrations [4]. Note also that, the new materials introduced in the design of the SLS, such as the insulation material of its solid rocket motor[5] add to the inherent uncertainties of the launch vehicle, both at sub-system level and at system level. Unexpected interactions between the different sub-systems, materials or equipment could lead to undesired behavior of system and an eventual catastrophic event.

Once in Earth orbit, most of the Earth vicinity hazards have been overcome. There is, however, still the challenge of putting the spacecraft in a transfer orbit to Mars. Some proposed missions to Mars involve the separate launch and assembly of modules in Earth orbit, before transfer to Mars. This adds another level of complexity as it would require rendezvous and additional operations in Earth orbit. One way to circumvent this could be to send the supplies and/or the habitat module before the crew on a separate transfer orbit. Some preliminary tests could even be run to make sure all the equipment is working properly before sending the crewed module.

Finally, we need to consider the orbital maneuvers that would set the spacecraft on its way to Mars. Even this presents its own hazards, as was demonstrated by the Phobos-Grunt mission[6].

3. MARS TRANSFER

Leaving the Earth's vicinity, particularly, the magnetosphere, entails hazards that have not been faced before, except partially in the case of the Apollo missions.

The general health risks that space presents, such as the loss of bone and muscle mass or balance and orientation issues, have long been studied. However, only recently have we started looking deeper into the long term health effects of spaceflight, with studies such as Valeri Polyakov's 438 days in orbit[7] or the "One Year Crew" mission from NASA[8]. Furthermore, these studies have only been undertaken in low Earth orbit, leaving many unknowns with respect to deep space flight. Other than changes to the body due to fluid shift, one of the biggest hazards involved in journey between Earth and Mars is associated with the exposure to solar radiation. Fig. 1 provides an estimation of the accumulated dose of radiation in low Earth orbit.

This form of radiation can damage the DNA in human cells, leading to mutations and higher risk of cancer. Studies have demonstrated, the potential for cataracts, cancer, cardiovascular diseases, musculoskeletal issues and damage to the central nervous system[9,10,11]. A recent study examined the potential impact of space radiation on neurodegeneration, in particular, the biological processes in the brain that contribute to the development of Alzheimer's disease[12,13]. Exposure to radiation and high-energy particles were found to be associated with cognitive problems.

Low levels of radiation generally do not pose a significant health risk to astronauts; however the effects of radiation are cumulative, so long term exposure increases the lifetime risk to astronauts. Crews of future long duration interplanetary missions will have to travel through the Van Allen radiation belts, and will be exposed to more high energy radiation exposing them to possibly serious health effects.

And radiation hazards are not limited to physiological effects, since the high radiation can also lead to fatal errors on the on-board electronics.

Lastly, we still need to consider the landing on Martian surface. Throught the history of Mars exploration, many have been the missions that failed on this critical mission phase. Not only do we need to consider the direct risk of the spacecraft by a catastrophic even during landing, but it is also important to account for the expected uncertainties in the landing location. Current missions have landing elipses with areas in the order of 500 km2 (e.g. Mars Science Laboratory) [15].

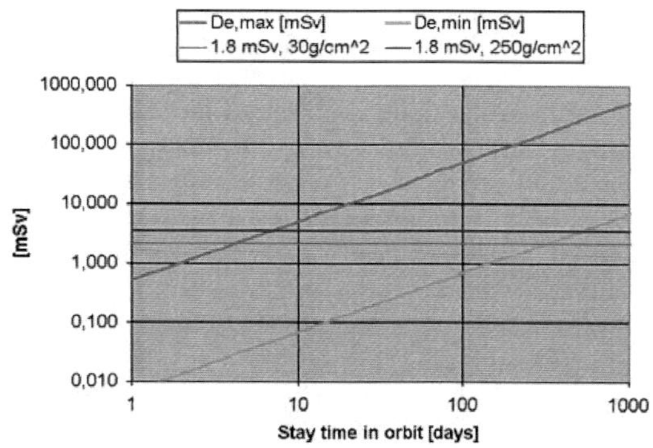

Figure. 1: Dose Equiv. as a function of time in orbit[12].

Table 1: Comparison between the environment on Mars and Earth [16].

Environmental parameters	Earth	Mars
Pressure (atm)	1	0.01
Oxygen (%atm)	21.	0.15
Gravity (g)	1	0.38
Radiation	Low	Medium
Temperature (°C)	-50 to 45	-140 to 20
Weather	Rain, Seasons	Storms, Seasons
Dust	Low	High
Mircrometeorite danger	Low	Low

4. MARS

On a one-way human mission to Mars, the habitat and the surface operations represent the most critical aspect of the mission, not only in terms of time spent, but in terms of possible risks. Mars presents a very hostile environment and to be able living there for an extended time requires very complex and robust systems.

Tab. 1 summarizes some characteristics of the Martian environment when compared with that of Earth[17]. The specific risks of a prolonged stay on Mars can be separated into three distinct types: health risks; habitat specific hazards and hazards associated with extra-vehicular actives (EVAs).

4.1 Human health hazards

The astronauts will be experiencing many stressful conditions: radiation, microgravity, varying g-forces, anxiety, psychosocial stressors, changes in health status, loss of body mass, sudden unexpected illness, physical injuries etc. Health issues can generally be divided into short-term and long-term problems.

Short-term health issues include problems which may appear during the mission but can disappear after stabilization on the Martian surface. But there is a wide range of these effects which are not yet fully understood. A good example is blood circulation and cardiovascular damage. On earth, fluids in the body circulate, working against gravity to prevent blood from pooling in the extremities and bringing blood to the brain. Changes in gravity impact blood circulation by triggering a fluid shift and increasing intracranial pressure. This could lead to cardiovascular and neurological issues, especially on a deep space flight to Mars and in the lower gravity conditions on Mars [17]. Other known short-term health issues include sleep deprivation, fatigue, psychiatric disorders, compromised immunity, and loss of muscle and bone mass.

Another problem in human organism, in the frame of the mission, is the renal stone formation. About 1 in 10 people develop a renal stone during their lifetime on Earth. But that risk is far greater for astronauts because of bone demineralization. That releases salts such as calcium phosphate into the blood. Concentrating in the kidneys, these salts may form stones that range in size from microscopic to as big as a walnut. Kidney stones may cause excruciating pain. But it's also possible for the kidney to stop working if a stone blocks the tube that drains the organ. In addition, many astronauts don't drink the 2 liters (0.5 gal) of water needed to stay completely hydrated and pass the stones in their urine. Even scientists have been working on an ultrasound machine about the size of a desktop computer to detect kidney stones and push them out of the kidney with short bursts of sound waves, this issue is addressing to telemedicine.

Long-term effects are more serious. In many cases they bring irreversible changes to human health. There is no clear solution to reduce the effect of these hazards. For example, recent research implies that gravity helps cells create patterns. In microgravity, the microtubules in developing cells might not organize the way they would on Earth, even after the astronauts return causing different cellular organization and it is unknown how this will affect the Mars crew over the long term. Additionally, some astronauts have developed serious, permanent vision problems from their time in space. The longer the mission - the greater the chance of change. Doctors at the University of Texas performed brain scans on 27 astronauts who'd been in space for over a month. In over 25 percent of them, the back of one or both eyes had flattened. This shortened the affected eyeball, making the person more farsighted. This may be another effect of the rise of fluid that occurs in a person's body in a weightless, or microgravity, environment. In this case, the pressure seems to build inside the skull from the increased flow of cerebrospinal fluid into the head. The fluid can't expand the bone, so it flattens the eyeballs instead. The problem caused by intracranial hypertension also could be solved by on board artificial gravity[18]. Weightlessness during the transfer orbit triggers the human body to excrete calcium and phosphorus (in urine and feces), resulting in rapid bone loss. For example, density in such bones as the pelvis and legs decreases by approximately 1 to 2 percent a month on average, which presents concerns for astronauts during long term missions, because after a stay of more than half a year, the loss is not fully reversible[17]. Like osteoporosis on Earth, bone loss in space can lead to fractures, weakness and painful urinary stones. Exercises can help slow the loss, but it takes long term consistent training upon landing on Mars to repair it. Moreover, weak bones could be health hazard during landing when the astronaut body is under high loads. Artificial gravity in the transfer vehicle could help to mitigate this problem, but this solution is technologically complex.

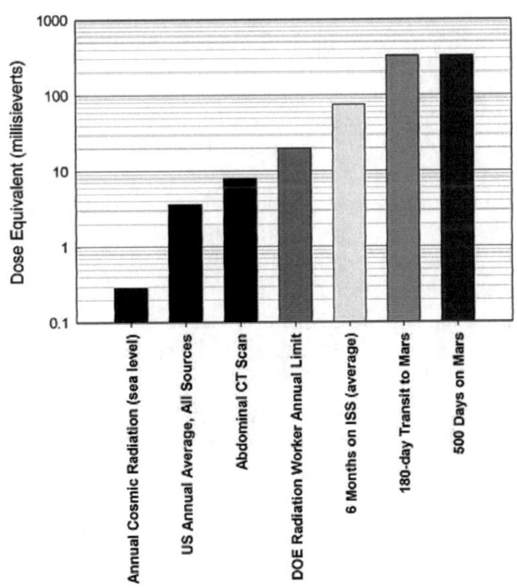

Figure. 2: Comparison of Radiation Doses - includes the amount detected on the trip from Earth to Mars by the RAD on the MSL (2011 - 2013) [15].

Furthermore, radiation exposure on the surface represents an additional critical hazard. Due to the lack of magnetosphere, Mars leaves the astronauts with minimal protection (by its atmosphere) against the radiation coming from the Sun. A more detailed analysis of the effects of long term exposure to radiation has already been presented in section 3 so this section will not go as much into detail. It is however relevant to note, on Fig. 2, the accumulated radiation dose within only 500 days of exposure. The radiation shielding of the outer skin of the habitat would depend strongly on the kind of material and the mass per area of the skin, a long stay on Mars seems to imply accumulated doses above what is currently allowed by most space agencies for their astronauts[14].

4.2 Mars Habitat

It is important to consider the life support system needed to sustain the crew on the Martian surface. The main elements needed for daily crew activities are oxygen and water. The risk of lack of oxygen in the OHM mission is significant due to the low Technology Readiness Level (TRL) of in-situ resource utilization (ISRU) technologies.

It is clear that the only way to sustain human life on Mars for a long time is to use a Closed-Loop Life Support Systems (CLLSS) [15]. The International Space Station (ISS) experience shows that the TRL for oxygen generation is high. Current oxygen generation technologies require water, which can be harvested on Mars from the natural environment. Since ISRU technologies have not been flight tested on Mars, this introduces a high level of uncertainty into any mission that relies on them.

While the risk of lack of oxygen is substantial without resources obtained through ISRU technology, there is also a high risk of fire outbreak if the oxygen content of the habitat's environment exceeds 30% molar fraction, which NASA has defined as its safety threshold. This threshold may be reached due to growing of food crops within the habitat, all of which will produce excess oxygen. Currently, no technology exists to decrease the molar fraction of oxygen within a closed-loop environment. As such, mitigating the risk of fire outbreak depends largely on the development of this system.

Water sources are not the only resource that need to be resupplied. Particularly during the first stages of the mission, the crew would need additional resources, replacement parts, different filters, batteries and construction material. There is then the substantial risk of delays or failure of resupply operations, due to the lack of current knowledge and experience with launching and landing payloads intended for human habitation on Mars. Mars resupply missions are significantly more risky than ISS resupply missions, as a missed launch window for a Mars resupply entails a delay of up to 26 months. In order to minimize the risk of crew death due to lack of supplies, extra resources must be stored in the habitat to account for the possibility of resupply mission delays or failures. It is infeasible to rely solely on resupply missions for crucial resources such as food and spare parts; hence, ISRU and in-habitat manufacturing must be conducted in order to sustain an OHM [16].

According to NASA standards, which define the parameters used for construction of habitats such as the ISS, total cabin pressure must be at least 20.7 kPa (3psi) in order to provide a safe environment for human inhabitants. Previous analyses have demonstrated that crop production within a closed-loop Mars habitat using existing technologies will likely result in either oxygen levels that are above the acceptable threshold for fire safety, or in depressurization due to attempts to expel excess oxygen[16]. Effects of depressurization on the human brain include hypoxia, which may lead to impairment of mental faculties and even unconsciousness. Since current Pressure Control Assemblies (PCAs) cannot selectively remove or inject oxygen into the habitat's atmosphere, the only conceivable way to mitigate pressurization concerns would be to refrain from growing crops inside the habitat, and instead have a separate greenhouse module (further adding to the mission's complexity) or obtain all food supplies from resupply missions. Taking this in consideration, the risk of insufficient food supply is rising. This risk depends on the OHM crew's ability to grow crops and otherwise produce food on Mars. As water is a crucial requirement in any food production, greenhouses are highly dependent on the habitat possessing a sufficient supply of water. Thus, the risk of insufficient food supply can be mitigated by testing and validating ISRU technologies that produce water using Mars-based resources and by recycling crew waste. There is another risk related to greenhouses. As they have not yet been tested in the Mars environment, there is substantial risk involved with deploying a Mars greenhouse and other regenerative systems on which an OHM crew will rely for survival. To mitigate this risk, NASA has previously suggested that inflatable greenhouses be deployed on precursor unmanned Mars lander missions, in order to confirm that such greenhouses are able to withstand the Mars environment. Failure of a greenhouse to withstand the environment during a manned mission could result in depressurization sickness among crew members, or death of crew members by starvation. Differences between the Earth and Martian environments must be taken into account when designing greenhouses and regenerative systems, and when selecting which crops to grow within them.

It is also important to consider the risk of general mass-energy misbalance in the habitat:

- Misbalance can stay unmarked until a serious failure;
- Misbalance can lead to unexpected results
- Failure in one system can result in failure in other systems and cascade effects;
- The crew has to find the issue and act without help or support of mission control centre;
- Misbalance can directly influence the crew, decreasing performance at addressing the risk.

Taking into account human factors, man-machine interaction and systems reliability, this risk can be reduced only with an integrative approach including preliminary studies on system survivability, running tests and researching operator's behaviors in critical situations[19].

4.3 Extravehicular Activities (EVA) hazards

EVA capabilities are defined by many different reports[15,20,21] as a critical element for future human space exploration programs. On Mars, where the gravity is respectively 3/8 of the Earth's, exploration missions would require continuous effort for a long period of time. Current mission proposal require at least three 8-hour EVAs per week covering a range beyond 10 km[21,22]. Given the Mars terrain's relief (Mont Olympus is culminating at 25 km high, 4 times Mount Everest), EVAs might require intense and complex tasks like rock climbing, running, or excavating[24,25]. Planetary surface dust and volatiles can be a particularly abrasive element (as seen with the Lunar regolith during the Apollo missions) that may lead to injuries, damage to the external layers of the spacesuit, and can lock the mechanical parts of the suit (leading to reduced mobility). Exposure to these volatiles can thus impair mobility, cause injuries, damage equipment, and reduce visibility. Dust exposure may also lead to respiratory, or ocular damage. Overall, insufficient data on the nature of the dust and volatiles limits a risk assessment[26]. The severity of the risk is ranked relatively low. The EVA risks can be reduced by protecting spacesuit joints and convolutes from the volatiles, using abrasion resistant fabric for the external layer, and developing a clearance system for the helmet of the spacesuit. The health risks can be reduced through operational procedures such as post-EVA clean up in airlocks, and isolation of contaminated suits from respirable environments.

Mars exploration would likely have an intense schedule that can quickly lead to exhaustion. Moreover, the astronauts will likely still experience loss of muscle mass, reduced physical performance and alterations of the cardiovascular system[24]. These risks are associated with the impairment in mobility of pressurized suits[26,27,28,29].

Improved design for planetary exploration is required to mitigate these risks. In order to do so, a better understanding of the human-suit interactions, and a knowledge of the functional tasks required in planetary exploration are required[30]. Reducing the pressurization of the suit is also primordial to reduce the impairment in mobility. A different paradigm of mechanical counter pressure suits can be studied[31]. Additionally, depending on the space suit design, a liquid cooling garment may be used to reduce the risk of over-heating or helmet fogging due to sweat. Injuries such as fractures or falls are very likely to happen in a planetary exploration scenarios. Injuries can be divided into injuries related to the environment and injuries inside the spacesuit. With the current bone mass loss, the risk of fracture on the Mars surface increases greatly[25]. Gas pressurized spacesuits may cause injuries to the astronaut inside the suit[32,33]. In current designs, the most severe injury observed is a shoulder injury requiring shoulder chirurgical operation. The most frequent injuries are fingernail delamination, abrasions and contusions. The intense EVA schedule reinforces the risk of EVA and compromise the long-term EVA capabilities. On more problem caused by the spacesuit is the risks of decompression sickness. However, these risks are partially controlled[26]. Better suit design is required to mitigate these risks. Better anthropometric fit to the astronaut should be envisaged as well as the use of high tech foams and pads to protect sensible regions of the human body from the suit contact[34]. Using different breathable gases can mitigate the risk of decompression sickness, as well as developing a procedure of adaptation[24].
Another risk is that of hypobaric hypoxia, a condition

where the low supply of oxygen to the body affects its ability to transfer oxygen from the lungs to the bloodstream, and can lead to loss of cognition. The risk is controlled by portable life support systems, regulating the partial pressure of oxygen environment inside the spacesuit.

Breathable air leaks can lead to suffocation, pressure gas leaks can lead to depressurization of the suit and hazards due to the exposure to the vacuum of space. Liquid leaks inside the suit can also be hazardous for astronauts because of the potential toxicity of the liquids or for deteriorating in-suit electrical systems. Mitigating these risks can be achieved by improving the design of the component responsible for the transport of air and liquids. Backup components for these critical elements should be considered.

Constant communication between different members of an EVA team as well as the members in the base is required. The communications between Earth and Mars will experience delays between 8 to 48 minutes, depending on the two planets location along their orbit. The loss of communication can lead to confusion, geographic disorientation, and even loss of life. The risk is partially controlled[26]. The communication systems should be redundant and powerful. Different types of communication should be allowed: long-range communication, emergency mission control emergency, in between astronaut's communication. Clear procedures should be implemented in case of communication loss during EVA and a sign language can be envisaged to replace the voice interface in case of communication loss.

The collaboration between rovers, robots and humans is a critical element of efficient planetary exploration. Poor interactions can lead to inefficiency, but also geographic disorientation, and loss of astronauts, robots or rovers. It can also lead to accidents. However, analysis seems to indicate that the risk can be controlled[27]. Designing the interfaces between these elements taking into account human factors such as situational awareness, cognitive workload, and trust in automation is primordial to allow efficient human-computer interactions. Enhanced reality devices should be implemented into spacesuit helmets.

The risk of fire is one of the most critical risks during EVA, but it has been well controlled. In an environment naturally rich in oxygen, the risk of fire increases. The risk of fire also plays into the trade-off between pressurization and partial pressure of oxygen inside the suit involving the risk of decompression sickness and the risk of mobility [35]. Even though the severity of the risk is high, it has been mitigated by using specific material inside the suit, and controlling oxygen partial pressure.

Another extreme situation is power outage during an EVA which can lead to loss of communication, loss of life support systems and most probably death.

5. DISCUSSION

Tab. 2 was formulated considering all the mission phases analysed in this paper. Each hazard was identified, assessed individually and then included in the overall risk of the phase.

From the Earth vicinity mission phase analysis, no major hazards or risks were detected. It is, however, relevant to point out that separating the mission into different launches and transfers for crewed modules vs. supplies/habitat modules seems to offer a safer approach. Moreover, sending the supplies/habitat modules before the crewed spacecraft offers the possibility to remotely test them before sending the crew.

Table 2: Overview of hazards identified for a one-way mission to Mars

Mission Phase	Earth Vicinity		Earth-Mars Journey		Mars					
	Earth Vicinity risks	Hazard level	Earth-Mars Journey Risks	Hazard level	Human medical risks	Hazard level	Habitat risks	Hazard level	Extravehicular activities risks	Hazard level
	Improper handling of the nuclear material	Controllable	Loss of bone and muscle mass	Not enough data	Intracranial hypertension/ Vision alterations	High (>1%)	Lack of oxygen and water	High (>1%)	Exposure to dust and volatiles	Not enough data
	Inadvertent ignition of launch system	Controllable	Balance and orientation issues	Controllable	Blood circulation and cardiovascular damage	Not enough data	Depressurizatio n	Controllable	Air or liquid leak inside spacesuit	Not enough data
	Overheating of the propellants during launch	Controllable	Radiation effects	Not enough data	Renal stone formation	High (>1%)	Lack of supplies	Controllable	Injuries related to environment	Controllable
	Excess accumulated energy from micro vibrations during launch	Controllable	Radiation triggered fatal errors on the on-board electronics.	Controllable	Radiation exposure	High (>1%)	Insufficient food supply	Controllable	Injuries inside the suit	Controllable
	Unpredicted interactions of launch sub-systemi	Controllable	Catastrophic event during descent to Mars	Controllable	Compromised immune system	High (>1%)	Mass-energy misbalance	Controllable	Decompression sickness	Controllable
	Failure of rendezvous and/or in-orbit assembly	Controllable	Uncertainty in landing area	Not enough data	Bone loss	High (>1%)	Risk of fire outbreak	Controllable	Fire	Controllable
	Failure to inject spacecraft into trans-martian trajectory	Controllable			Muscle loss	Controllable	Problems with greenhouses	Controllable	Communication loss	Controllable
					Cellular organization	Not enough data			Poor human-system interaction	Controllable
					Adverse cognitive or behavioural conditions & psychiatric disorders	Not enough data			Power outage	Controllable
					Sleep deprivation, circadian desynchronization	Controllable			Hypobaric hypoxia	Controllable
					Fatigue	Controllable			Exhaustion and over-exercise	Controllable
Mission Phase	**Earth Vicinity**		**Earth-Mars Journey**		**Mars**					
Overal assessment	**Controllable**		**Generally controllable** but need deeper research in some areas		**High risk** of loss of crew or loss of mission as well as need for deeper research in some areas		**High risk** of loss of crew or loss of mission as well as need for deeper research in some areas		**Generally controllable** but need deeper research in some areas	

Throughout the transfer to Mars, the biggest hazard was found to be direct exposure to solar radiation. While further studies are needed, it seems the accumulated dose can be kept to acceptable levels.

Mars landing presents a series of risks, be it from incidents leading to a catastrophic event, or poor landing precision leading to a large separation between the crewed module and the supplies/habitation modules; thus leaving the crew in dire situation.

The long stay on Mars, which is one of the main characteristics of such mission architecture, implies its own set of medical hazards. Medical hazards from long exposure to the environment include intracranial hypertension, vision alterations, renal stone formation, radiation exposure, compromised immune system and bone loss, among others. All these mentioned conditions were determined to have a high risk (higher than 1%) of loss of crew or loss of mission. Other medical conditions such as cellular organization or blood circulation and cardiovascular damage can a risk. However, not enough data exists today to make this assessment.

Extravehicular activities (EVAs) will be absolutely necessary for such a mission (for maintenance, for repair, etc.). There is still more studied needed to understand aspects such as the effects that the exposure to dust and volatiles on the suits and crew. However, generally speaking, EVAs appear not to present a high risk to the mission or crew.

The primary concerns associated with a long-term Mars habitation involve environmental and life support system (ECLS) technologies and resupply missions. The risk of lack of oxygen in the OHM mission is significant due to the low technology readiness level (TRL) of in-situ resource utilization (ISRU) technologies.

Furthermore, the long exposure of equipment to the Martian environment requires the use of an either extremely robust or extremely redundant system. A way to circumvent these issues would be to perform frequent re-supply missions. However, it has not been demonstrated that we have the capability to perform these reliably. There we have to consider this to be a high risk situation.

6. CONCLUSION

An overall assessment of the hazards involved in a one-way mission to Mars were provided. Several hazards were identified not to have enough studies performed and not enough data available. This does not allow for a proper safety assessment and conclusion regarding one-way missions to Mars. That are several technologies that need to be further developed and medical conditions that need to be further studied before this mission architecture can be implemented. Future work should focus on these developments and studies.

7. REFERENCES

1. George W. Herbert (1996). "One Way to Mars" – The Case for Mars VI Workshop
2. William Harwood, (2014). "NASA commits to $7 billion mega rocket, 2018 debut" – CBS News
3. Joao Lousada et al., (2014). "Report on National Research on Space Debris, Safety of Space Objects with Nuclear Power Sources on Board and Problems of their Collision with Space Debris" – United Nations
4. George P. Sutton, (2010). "Rocket Propulsion Elements" – John Wiley & Sons, Inc.
5. NASA, (2012). "NASA sub-scale solid-rocket motor tests material for Space Launch System." – ScienceDaily
6. Louis D. Friedman, (2012). "Phobos-Grunt Failure Report Release" – The Planetary Society
7. H.G. Stratmann, (2015). "Using Medicine in Science Fiction" - Springer
8. Guy Gugliotta, (2015). Scott Kelly's Year in Space – Air & Space Magazine
9. Palinkas L., (2001). "Psychosocial issues in long-term space flight: overview" – Gravitational and space biology bulletin. 14(2), pp. 25-33.
10. AstroChix, (2012). "Risk of Human Space Flight to Mars" [Online] http://www.astrochix.com/2012/12/19/the-risk-of-human-space-flight-to-mars/
11. NASA, (2015). "About space Radiation", [Online] http://spaceradiation.usra.edu/about/
12. University of Rochester Medical Center, (2012). "Space travel may be harmful to the brain, study suggests" [Online] http://www.sciencedaily.com/releases/2012/12/121231180632.htm.

13. Universities Space Research Association Division of Space Life Sciences (2015). "The Health Risks of Extraterrestial Enviromnents", 2015 [Online], http://three.usra.edu/#section=main.

14. Emily Lakdawalla, (2012). "Curiosity's shrinking landing elipse" [Online] http://www.planetary.org/blogs/emily-lakdawalla/2012/20120611-curiosity-landing-ellipse.html

15. Waldie, M. A. J., (2005). "Mechanical Counter Pressure Space Suits: Advantages, Limitations and Concepts for Martian Exploration" – The Mars Society

16. Ramsey H., (2014). "Astronaut health risks that threaten deep space missions", [Online] http://listverse.com/2014/12/29/10-astronaut-health-risks-that-threaten-deep-space-missions/

17. Race to Mars, (2015). "Known effects of longterm space flights on the human body", [Online] http://www.racetomars.ca/mars/article_effects.jsp

18. Kurmazenko E., Kamaletdinova G., et al., (2013). "Hardware and software Complex System Operation Crew Service for the Interaction Analysis in the Crew-Environment- Integrated Life-Support System Functional System" –29th International Symposium on Space Technology and Science, Nagoya-Aichi, Japan, 2013.

19. ISECG, (2013). "The Global Exploration Roadmap" – International Space Exploration Coordination Group.

20. NRC, (2014). "Pathways to Exploration – Rationales and Approaches for a U.S. Program of Human Space Exploration" – National Research Council of the National Academies

21. Gernhardt, M., et al., (2009). "Risk of Compromised EVA Performance and Crew Health Due to Inadequate EVA Suit Systems" – Houston, TX, Human Research Program

22. Walz, C. and M. Gernhardt, (2008). "Extravehicular Activity - Challenges in Planetary Exploration" – 3rd Space Exploration Conference & Exhibit

23. Buckey, J., C., (2006). "Space Physiology" – New-York, NY, Oxford University Press

24. HRP, 2005). "Bioastronautics Roadmap: A Risk Reduction Strategy for Human Space Exploration" – NASA. SP-2004-6113.

25. Reinhardt, A., (1989). "Results and Applications of a Space Suit Range-of-Motion Study" – NASA Technical Memorandum 102204

26. England, S., et al., (2010). "Functional Mobility Testing: Quantification of Functionally Utilized Mobility among Unsuited and Suited Subjects" – TP-2010- 216122, NASA

27. Aitchison, L. T., (2012). "A Comparison of Methods for Assessing Space Suit Joint Ranges of Motions" – 42nd International Conference on Environmental Systems. A. I. o. A. a. Astronautics

28. Bertrand, P. J., et al., (2014). "Feasibility of Spacesuit Kinematics Characterization and Human-Suit Interactions" – International Conference of Environmental Systems. AIAA. Tuscon, AZ

29. Anderson, A., et al., (2014). "In-Suit Sensor System for Characterizing Human-Space Suit Interaction" – International Conference of Environmental Systems. AIAA. Tucson, AZ

30. Newman, D. J., et al., (2007). "Revolutionary Design for Astronaut Exploration – Beyond the Bio-Suit System" – 4th Symposium on New Frontiers and Future Concepts. Space Technology and Applications International Forum.

31. Anderson, A., (2014). "Understanding Human-Space Suit Interaction to Prevent Injury During Extravehicular Activity" – Aeronautics and Astronautics. Cambridge, MA, Massachusetts Institute of Technology. Ph.D.

32. Strauss, S., (2004). "Extravehicular Mobility Unit Training Suit Symptom Study Report" – Houston, TX, NASA Johnson Space Center.

33. Carr C.E. and N. D.J., (2008). "Characterization of a lower-body exoskeleton for simulation of space-suited locomotion." – Acta Astronautica

34. Scheuring, R. A., et al., (2008). "Risk Assessment of Physiological Effects of Atmospheric Composition and Pressure in Constellation Vehicles." – Acta Astronautica 63(727-739)

PROCEEDINGS OF THE 8TH INTERNATIONAL SPACE SAFETY CONFERENCE

PLENARY CLOSING SESSION

HYPERSONIC, SPACE TRANSIT, AND SPACE ACCESS FLIGHT TEST

James Vasil Souders (1), **Timothy R. Jorris** (2)

(1) *Air Force Safety Center, Kirtland AFB, NM, Email: james.souders@us.af.mil*
(2) *Lockheed Martin Aeronautics Company, Palmdale, CA, Email: timothy.r.jorris@lmco.com*
412 TW-PA-15264: Approved for public release; distribution is unlimited.

ABSTRACT

Hypersonic flight test applies to suborbital, space transit, and space access vehicles. Recent programs are being initiated by the US Air Force, Defense Advanced Research Programs Agency (DARPA), NASA, and commercial companies. The USAF, DARPA, and Industry are pursuing technologies to meet the objectives of hypersonic vehicles. DARPA is also looking for new ways to satisfy the goal of Operationally Responsive Space; a means of inserting a payload to orbit in hours instead of months. Beyond rapid response to space, affordability is another technology motivator; therefore, DARPA has yet another program to pursue that objective by exploring reusability. Many of the hypersonic test vehicles require an airborne launch platform, e.g. the B 52 for the X-51A Scramjet Demonstrator. However, finding a launch platform that has the performance and availability can be challenging. NASA is working towards a towed-glider concept aimed to develop a high capacity, easy to maintain, affordable launch platform. NASA and commercial customers are looking at novel solutions to transit astronauts to and from the International Space Station. And, the X-37B Orbital Test Vehicle reenters the atmosphere for a horizontal landing and must pass through the hypersonic flight regime. Recent and upcoming technologies, and hypersonic flight test challenges are discussed.

1. INTRODUCTION

The advancement of hypersonic technologies is designed to support the Conventional Prompt Global Strike [1] and the High-Speed Strike Weapon (HSSW) [2] programs being managed by the Air Force Research Lab (AFRL) and Defense Advanced Research Projects Agency (DARPA) [3]. While these advancements present challenges in vehicle design in aerodynamics, flight controls, structural loads, heating and propulsion, they also present a challenge to collecting the data generated by these high speed vehicles during flight.

The premise of this paper is to reacquaint the academic and flight test community with past data collection technologies and introduce them to future potential methods for collecting data on high speed, unmanned test vehicles. Past projects' data collection systems and future potential data collection systems are reviewed.

2. PREVIOUS PROGRAM DATA COLLECTION SYSTEMS

Conventional or legacy air vehicle data collections systems are well known and in use on numerous test aircraft programs such as F-35, F-22, B-2, and C-17. Sub- and Supersonic air vehicle instrumentation systems are typically able to transmit/receiver data at establish constrained bandwidth. Transmission/reception max range limits are typically on the order of hundreds to thousands of miles which can be extended through telemetry relay antennas. Past test programs that pushed the boundaries of air vehicle data collection technologies include AFRL X-51 Scramjet Engine Demonstrator and DARPA Hypersonic Technology Vehicle (HTV-2). TM data transmission/reception for these programs was very difficult and costly due to 1) the high speed at which the vehicles were traveling and 2) the extended ranges over which the vehicles flew over water.

2.1 X-51A Scramjet Engine Demonstrator

The X-51A was launched from a B-52 over the Point Mugu Sea Range. TM data were transmitted/received over an air vehicle powered and glide flight range of approximately 350 nm from the point of release from the B-52. The X-51A program utilized three different TM instrumentation platforms to collect data from the X-51A. Data was collected on the B-52 launch platform (Figure 1 below) both prior to and after X-51A release. Data were also collected by a ground stations at Vandenberg AFB and on San Nicholas Island within the Pt Mugu Sea Range. Data were also collected by a specially instrumented Navy P-3 aircraft stationed approximately 260 nm from the initial launch location at an altitude of approximately 20,000 ft. The transmit/receive range of the P-3 instrumentation suite was approximately 200 nm.

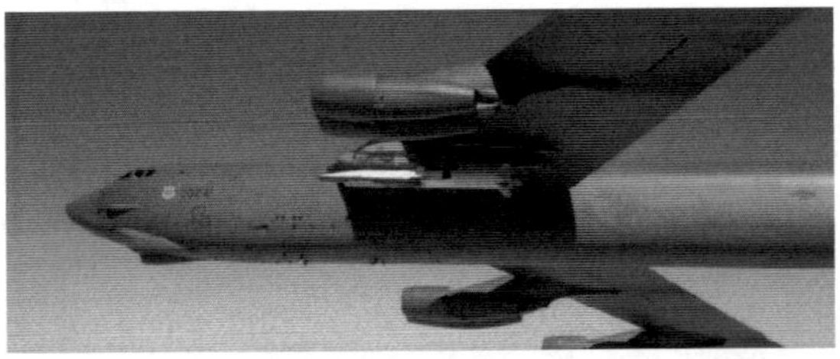

Figure 1. X-51A mounted under B-52 at Edwards AFB

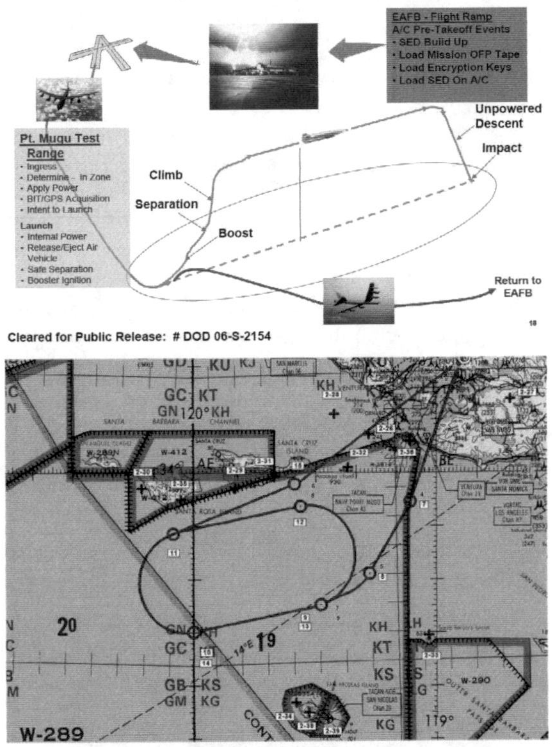

X-51A Launch profile over restricted water range

Valuable data were collected and relayed on all four powered flights by these TM assets. Flight one on 26 May 2010, had 200 seconds of scramjet burn accelerating to Mach 5 [4]. For the second flight on 13 June 2011, the scramjet lit successfully but failed to transition to full power [5]. For the third flight on 14 Aug 2012, the X-51A safely separated from the B-52 and the rocket booster fired as planned. However after 16 seconds, a fault was identified with one of the cruiser control fins. Once the X-51A separated from the rocket booster, approximately 15 seconds later, the cruiser was not able to maintain control due to the faulty control fin and was lost [6]. The fourth and final flight was a record-setting 210 seconds of scramjet combustion reaching a hypersonic speed of 5.1 Mach number [7]. On powered flight #4 the X-51A transmitted TM continuously from its release location from the B-52 to its eventual splash-down point in the Pacific Ocean.

2.2 Boost Glide Vehicles

A boost glide vehicle, for this paper, refers to an unpowered hypersonic vehicle that is launched from a ground based booster and then glides after separation. The Air Force flew the Hypersonic Technology Vehicle 2 (HTV-2) and the Army flew the Advanced Hypersonic Weapon (AHW). The two programs have similar challenges except their mission distances are significantly different. HTV-2 was launched from Vandenberg AFB with a planned flight to Kwajalein Atoll [8] approximately 3000 nm away, whereas AHW only flew from Kauai in Hawaii to Kwajalein Atoll [9], as seen in Figure 2.

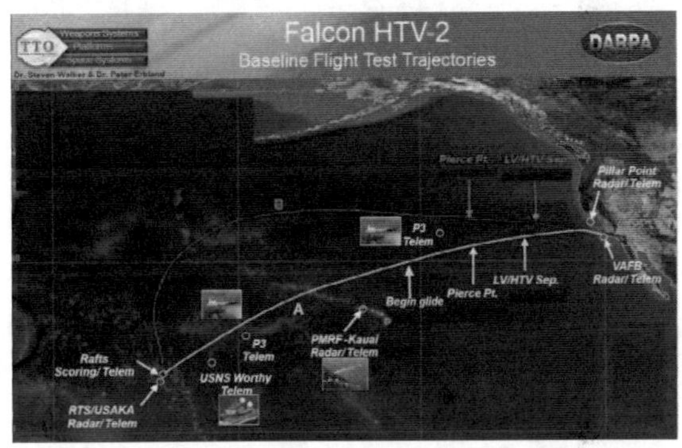

Figure 2. HTV-2 (VAFB), AHW (PMRF) Launch Points

2.3 Hypersonic Technology Vehicle (HTV-2)

The depiction in Figure 3 shows the entire HTV-2 flight profile [10]. The vehicle was launched atop a Minotaur IV Lite from Vandenberg AFB, CA [11]. The first flight on 22 Apr 2010 collected 9 minutes of unique flight data but didn't complete its full mission. "The Engineering Review Board (ERB) concluded that the anomaly resulted from flight control authority limitations to operate at the angle of attack the vehicle was programmed to fly for the speed and altitude of the flight. Detailed analysis conducted by the ERB revealed that the most probable cause of the HTV-2 flight anomaly was higher-than-predicted yaw, which coupled into roll thus exceeding the available control capability at the time of the anomaly". [12]

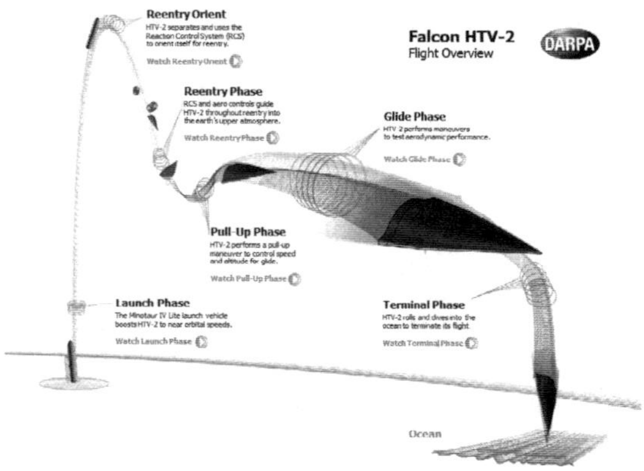

Figure 3. DARPA Falcon HTV-2 Flight Profile

The second flight on 11 August 2011 successfully demonstrated stable aerodynamically-controlled flight at speeds up to Mach 20 for nearly three minutes; however this flight also ended prematurely. "The ERB concluded that the 'most probable cause of the HTV-2 Flight 2 premature flight termination was unexpected aeroshell degradation, creating multiple upsets of increasing severity that ultimately activated the Flight Safety System.'" [13]

The data collected for each of these flights was instrumental in being able to determine the causes of each vehicle failure during flight and will be extremely valuable in guiding the design of future hypersonic flight vehicles. Collecting the data to support each of these flights was again, a challenge due to both the speed of the vehicle and the distance traveled during its flight. In order to continuously collect that TM data throughout the vehicles flight trajectory it was necessary to assemble a "string of pearls" instrumentation assets over the course of the planned trajectory. This string of pearls consisted of land, sea and air based instrumentation platforms.

Specifically, the land assets, like the X-51A resources, consisted of Vandenberg and Pt Mugu Sea Range ground stations. The sea assets consisted of specially instrumented ships deployed to sea for a period of several days to weeks to allow for travel time to the pre-planned location and to accommodate any flight delays. The air instrumentation assets again included specially equipped P-3 aircraft. As one would expect, the cost of operating all of these assets for the period of time required was in the millions of dollars for just one flight.

2.4 Army's Advanced Hypersonic Weapon (AHW)

The flight of the AHW lasted less than 30 minutes on 17 November 2011. "The AHW is a first-of-its-kind glide vehicle, designed to fly within the earth's atmosphere at hypersonic speed and long range, was launched from the Pacific Missile Range Facility, Kauai, Hawaii, to the Reagan Test Site, U.S. Army Kwajalein Atoll, Marshall Islands." [9] Figure 2 shows the relationship between Vandenberg AFB, Kauai, and Kwajalein.

3. SPACE ACCESS, HYPERSONIC VEHICLES

3.1 DARPA Airborne Launch Assist Space Access (ALASA)

The goal of a project funded by DARPA called the Airborne Launch Assist Space Access (ALASA) program, started in late 2011, is to develop an air launch system that can place a 45-kilogram (100-pound) payload into orbit for less than $1 million per mission [14]. Achieving that cost goal led DARPA to focus on what the real cost drivers of launch operations are. "If I'm going to hit a million-dollar-per-flight target, I've got to do things in such a fundamentally different way that I change the cost equation for all the stuff in a space launch that has nothing to do with the rocket," said Mitchell Burnside Clapp, the DARPA ALASA program manager [14]. The Boeing ALASA concept intends to use F-15 aircraft that require no modifications. The rocket, for example, will use the same communications protocols as weapons systems typically mounted on that aircraft. "ALASA is not a weapon, but it does talk the same language as weapons speak," Clapp said. "There are no software changes needed for the F-15. That is a huge advantage." [14]

Airboune Launch Assist Space Access (ALASA)

Since the ALASA payload will be deployed in low earth orbit (LEO) the payload launch platform instrumentation system will require a range equal to its eventual payload deployment altitude and cross range from the instrumentation relay and endpoint reception systems on the ground.

3.2 AFRL/DARPA Reusable Booster System

The Reusable Booster System (RBS) Pathfinder was "a promising approach to meet the Air Force's future space-lift needs. The RBS consists of an autonomous, reusable, rocket-powered first stage with an expendable upper stage stack. The reusable first stage launches vertically and carries the expendable stack to the staging point. From the staging point, the reusable first stage returns directly to the launch base, landing aircraft-style on a runway." [15] Figure 4 shows a graphic of the boost back concept.

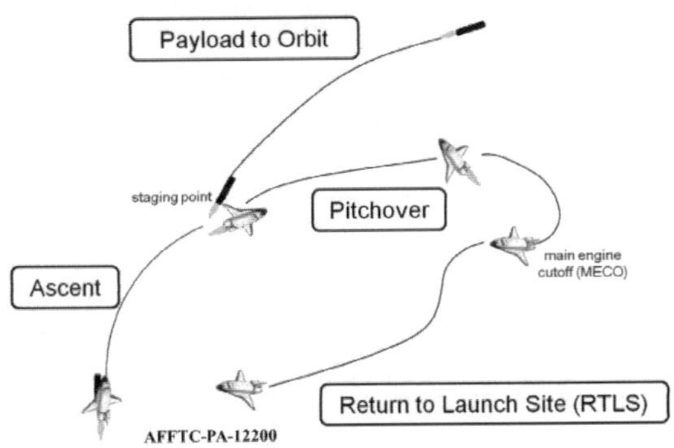

Figure 4. RBS Conceptual Profile

A potential test flight profile of the reusable first stage is shown below in Figure 9. Telemetry data collection and relay assets will have to not only be capable of transmitting/receiving data over a long range horizontally but also vertically. This presents new challenges because air assets used on previous programs such as the P-3 has a service ceiling altitude of only 23,000 ft.

Figure 5. RBS System Example Trajectory

3.3 Experimental Spaceplane (XS-1)

The Experimental Spaceplane (XS-1) [16], solicited by DARPA, is aimed at developing a reusable first stage to deploy a 3k – 5k lb payload to orbit for under $5M per launch. Take-off and landing can be horizontal, vertical or a combination of both. An artist concept is shown in Figure 6. Again, the instrumentation transmit/receive assets would have to be capable to receiving data over a very long range at low to high altitudes.

Figure 6. Artist Concept of XS-1

3.4 High Speed Strike Weapon

AFRL and DARPA have teamed up to manage two different hypersonic propulsion technology development programs; Tactical Boost Glide (TBG) and Hypersonic Air-breathing Weapon Concept (HAWC). Both programs are designed to develop necessary technologies to achieve hypersonic or near-hypersonic speed in a cruise missile sized platform. AFRL is in the early stages of developing the required technologies to support these end products.

3.5 Tactical Boost Glide (TBG)

Boost glide vehicles achieve their high speed by being boosted by a high thrust rocket to high altitude and gliding back down to a lower altitude. TBG will have an intended range that is "tactical" and less than HTV-2. The Army's Advanced Hypersonic Weapon (AHW) is a boost glide vehicle and flew 3,700 km on its first test flight [32]. Both HTV-2 and the AHW required a long line of aerial and ship-based TM collection and relay platforms to transmit the data back to the control room. This type of TM support is very expensive. Despite the TBG reduced flight range, testing in this flight regime will still require range assets that are able to support delivering TM data across long ranges and at high altitudes. Many present day and future programs are not funded to a level that would support repeating the "string of pearls" TM collection and relay system that supported HTV-2. In order to avoid this high cost new TM collection and relay systems have to be developed to provide the same level of performance at a much cheaper cost.

3.6 Hypersonic Air-Breathing Weapon Concept

Due to the long distance of the anticipated HAWC test flight profiles HAWC testing will also require range assets that are able to deliver TM data across long ranges while avoiding the cost of traditional "string of pearls" TM data collection and relay systems.

4. PROPOSED TM DATA COLLECTION

4.1 Aircraft Platforms

A number of UAVs have already served as sensor data and TM relay platforms supporting a specific program or purpose. However, they have typically been single individual systems not a group of systems working together supporting one flight vehicle or program. In order to adequately support a hypersonic or space transit flight test program these UAVs would have to be able to stay aloft for potentially days at a time to support launch delays. There would have to be sufficient numbers of them to support the long distances traveled by hypersonic test vehicles. The instrumentation systems onboard those UAVs would have to have sufficient power to reach boost glide test vehicles at very high altitudes.

Another area of potential use for UAV TM platforms is air vehicle imaging. Obtaining infrared or electro-optical images of a hypersonic or space transit vehicle during flight can greatly enhance anomaly investigation and resolution. Vehicle heating can be observed and used to determine structure materials deterioration or failures during flight. The imagery can also be compared with existing models and updated so as to influence future designs. Atmospheric sensing is another capability that when paired with a UAV platform can contribute to instrumentation.

Potential UAV platforms that could support the above-described functions are Global Hawk, Predator, and Lockheed Martin's Integrated Sensor in Structure Blimp. One program supportability issue would include the cost of sustaining these systems when not actually supporting test execution. These costs would include the proficiency training required of the UAV operators and maintaining and upgrading the UAV and ground station hardware and software.

4.2 Space-based Platforms

There are already in place several satellite based options to collecting and relaying TM data. One is the NASA Tracking and Data Relay System (TDRS). The initial setup cost can range anywhere from \$250K - \$1M [32]. However, once the operation configuration is set the system can support data collection/relay at a cost of only \$130 per minute. Other potential satellite networks that could possibly support the hypersonic and space transit flight test TM requirements are InmarSat [33], Iridium [34], Orbcomm [35], ViaSat [36] and Tactical Targeting Network Technology [37]. So while TDRS and similar networks of satellites are continually being updated with faster and larger bandwidth capability, existing and future capability must be assessed to determine if they will meet future hypersonic and space transit vehicle test requirements.

5. OTHER FLIGHT TEST INSTRUMENTAION

The Office of the Secretary of Defense (OSD) High Speed Systems Test (HSST) program is developing atmospheric sensing technology that could replace weather balloons and chase aircraft; particularly since a hypersonic chase aircraft is currently not available. The High Altitude LIDAR Atmospheric Sensing (HALAS) system is a ground based, portable LIDAR system using an ultra-violet laser and scanner system to measure atmospheric conditions (density, temperature, pressure, wind speed/direction, O_2 content) along a vehicle's flight path at altitudes up to 80 km (250k feet). This will result in significantly more accurate flight test data and vehicle performance estimates with lower uncertainties and better spatial and temporal resolution. The system also has the capability to provide near real-time and highly accurate measurements at lower altitudes to support range safety operations.

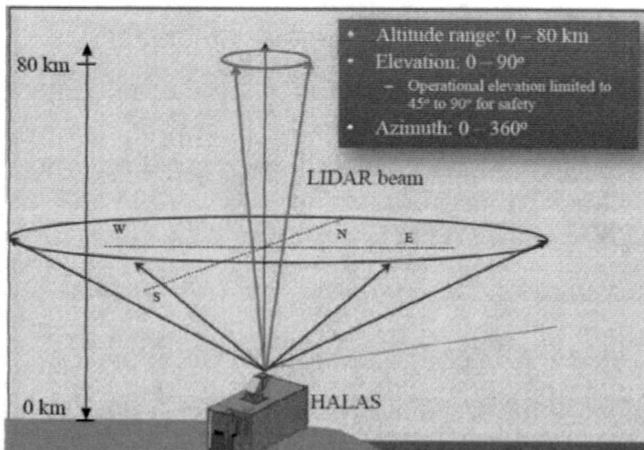

Figure 7. High Accuracy Atmospheric Sensing

Based on predicted performance, the HALAS uncertainty is significantly better than traditional Rawinsonde balloon data. This comparison only considers the atmosphere in the near vicinity of the balloon, it does not take into account uncertainty improvements gained by multiple balloons or modelling. However, such balloon corrections traditionally do not result in an order of magnitude improvement, as demonstrated with the HALAS.

Altitude (kft)	Balloon Uncertainty	HALAS Uncertainty
20	0.47%	0.023%
30	0.49%	0.021%
40	0.63%	0.022%
50	0.71%	0.027%
60	1.01%	0.036%
70	1.30%	0.050%
80	1.87%	0.069%
90	2.86%	0.092%
100	4.57%	0.120%

Figure 8. HALAS to Balloon Comparison

6. CONCLUSIONS

There are multiple hypersonic and space transit vehicle technology programs working to satisfy the Department of Defense and NASA objectives. Each is addressing the challenges inherent in the hypersonic arena. One of these challenges is the collection and relay of vehicle TM data for real-time and post flight data analysis. Recent programs, such as the X-51A and HTV-2 achieved these objectives at a high cost which cannot be overcome by future test programs. It is the role of the flight test community to explore and identify those TM data collection/relay technologies that can deliver the required performance to support numerous different types of test vehicles at a reasonable cost.

7. REFERENCES

1. A. F. Woolf, "Conventional Prompt Global Strike and Long-Range Ballistic Missiles: Background and Issues," 6 July 2012. http://fpc.state.gov/documents/organization/167962.pdf.

2. AFRL-Eglin Research Site, "High Speed Strike Weapon (HSSW) Demonstration Program Industry Day," 30 May 2012. https://www.fbo.gov/index?s=opportunity&mode=form&id=129942a86380453c656ded23b1a6a9e7&tab=core&_cview=0.

3. DARPA, "Hypersonics-The New Stealth," 6 July 2012. http://www.darpa.mil/NewsEvents/Releases/2012/07/06.aspx.

4. Wright-Patterson Air Force Base , "X-51 Waverider makes historic hypersonic flight," 26 May 2010. http://www.wpafb.af.mil/news/story.asp?id=123206524.

5. Air Force Materiel Command, "Second hypersonic flight ends prematurely, brings new flight test data," 15 Jun 2011. afmc.af.mil/news/story.asp?id=123260080.

6. Inside AF.mil, "X-51A flight ends prematurely," 15 August 2012. afmc.af.mil/news/story.asp?id=123314235.

7. D. Mayer, "X-51A Waverider Achieves History in Final flight," 88th Air Base Wing Public Affairs, 3 May 2013. http://www.wpafb.af.mil/news/story.asp?id=123346970.

8. DARPA, "Environmental Assessment for Hypersonic Technology Vehicle 2 Flight Tests," April 2009. http://www.dtic.mil/cgi-bin/GetTRDoc?AD=ADA544343.

9. U.S. Army, "Army successfully launches Advanced Hypersonic Weapon demonstrator," 23 November 2011. http://www.army.mil/article/69855/Army_successfully_launches_Advanced_Hypersonic_Weapon___/.

10. T. R. Jorris, "Hypersonic Vehicles," in AIAA Speaker's Dinner, Lancaster, CA, 2012.

11. DARPA, "Falcon HTV-2 Launch Tests Hypersonic Vehicle Flight Capabilities," 23 April 2010. http://www.darpa.mil/WorkArea/DownloadAsset.aspx?id=2147484133.

12. DARPA Public Affairs, "DARPA Concludes Review of Falcon HTV-2 Flight Anomaly," 16 November 2010. http://www.darpa.mil/WorkArea/DownloadAsset.aspx?id=2147484134.

13. DARPA, "Engineering Review Board Concludes Review of HTV-2 Second Test Flight," 20 April 2012. http://www.darpa.mil/NewsEvents/Releases/2012/04/20.aspx.

14. J. Foust, "Air launch, big and small," The Space Review, 30 June 2014. http://www.thespacereview.com/article/2543/1.

15. Air Force Materiel Command, "Reusable Booster System (RBS) Pathfinder," Reusable Booster System (RBS) Pathfinder, 05 Apr 2010. https://www.fbo.gov/index?s=opportunity&mode=form&id=6ef1c3c8f3ef6732bf86e3eae914f5c6&tab=core&_cview=1.

16. DARPA, "Experimental Spaceplane Shooting for "Aircraft-Like" Operations in Orbit," 17 September 2013. http://www.darpa.mil/NewsEvents/Releases/2013/09/17.aspx.

17. Inside AF.mil, "X-37B Orbital Test Vehicle," 3 March 2011. http://www.af.mil/information/ factsheets/factsheet.asp?id=16639.

18. C. Thomas, "NASA's Commercial Crew Program Progressing for Future of U.S. Human Spaceflight," 3 August 2012. http://www.nasa.gov/exploration/commercial/crew/ccicap-announcement.html.

19. B. L. Reed, "Flight Test Challenges & Risks for the X-51A Scramjet Engine Demonstration Program," in SETP Dayton Symposium, Wright-Patterson AFB, OH, 2012.

20. R. W. Carr, T. R. Jorris and Jacob J. Downey, "Trajectory Analysis Program for Determining Range Safety Considerations using Multi-Phase Pseudospectral Optimization," in 28th Aerodynamic Measurement Technology, Ground Testing, and Flight Testing Conference, New Orleans, Louisiana, 25-28 June 2012.

21. B. M. Hellman, "Comparison of Return to Launch Site Options for a Reusable Booster Stage," Georgia Institute of Technology, August 2005.

22. NASA, "Boeing Tests Parachute System for CST-100 Spacecraft," 4 May 2012. http://www.nasa.gov/exploration/commercial/crew/boeingdroptest.html.

23. Space Exploration Technology Corp., "Dragon Overview," May 2012. http://www.spacex.com/dragon.php.

24. Systems and Sierra Nevada Corporation's Space Systems, "Sierra Nevada Corporation's Dream Chaser Space System Passes Preliminary Design Review," 6 June 2012. http://sncspace.com/press_more_info.php?id=335.

25. NASA, "NASA Range Safety Annual Report," 2011. http://kscsma.ksc.nasa.gov/Range_Safety/Annual_Report/2011/PDFDocuments/V_Emerging_Technology.pdf.

26. U.S. Air Force, "Air Force space vehicle comes in for a landing," 18 June 2012. http://www.af.mil/news/story.asp?id=123306243.

27. T. R. Jorris, "Common Aero Vehicle Autonomous Reentry Trajectory Optimization Satisfying Waypoint and No-Fly Zone Constraints," in Air Force Institute of Technology, Wright-Patterson AFB, OH, 2007.

28. R. W. Carr, T. R. Jorris and A. V. Rao, "Propagation of Uncertainty in Optimal Trajectory Generation," in 11th Annual AIAA Southern California Aerospace Systems and Technology Conference, Santa Ana, CA, 3 May 2014.

29. S. N. Corporation, "Statement from Sierra Nevada Corporation Regarding Dream Chaser Free-Flight Test," Sierra Nevada News & Press Releases, 26 October 2013. http://www.sncorp.com/AboutUs/NewsDetails/569.

30. DARPA, "Broad Agency Announcement, Tactical Boost Glide," DARPA-BAA-14-24, 25 March 2014.

31. J. J. Downey, R. W. Carr, K. I. Bull and T. R. Jorris, "Design of Experiments Using Response Surface Methods to Opimized Simulated Reentry Trajectories," in 28th Aerodynamic Measurement Technology,Ground Testing, and Flight Testing Conference, New Orleans, Louisiana, 2012.

32. NASA Tracking and Data Relay Satellites (TDRS) nasa.gov/content/tracking-and-data-relay-satellite-tdrs/ and personal comm with TDRS program personnel.

33. Inmarsat - http://www.inmarsat.com/global-xpress-us-government/

34. Iridium Satellite network - https://gigaom.com/2012/08/27/how-iridium-took-a-chance-on-spacex-and-won/

35. ORBCOMM - http://www.orbcomm.com/networks/satellite

36. ViaSat - https://www.viasat.com/products/high-capacity-satellites

37. Rockwell Collins Tactical Targeting Network Technology (TTNT) - http://rockwellcollins.com/~/media/Files/Unsecure/Products/Product%20Brochures/Communcation%20and%20Networks/Networks/Tactical%20Targeting%20Network%20Technology/TTNT%20brochure.aspx

International Association for the Advancement of Space Safety (IAASS)
Kapteynstraat 1
2201BB Noordwijk
The Netherlands

ISBN 978-1-5108-7530-2

CURRAN ASSOCIATES INC.
proceedings.com